Lecture Notes in Computer Scienc

Vol. 270: E. Börger (Ed.), Computation Theory and Logic. IX, 442 pages. 1987.

Vol. 271: D. Snyers, A. Thayse, From Logic Design to Logic Programming. IV, 125 pages. 1987.

Vol. 272: P. Treleaven, M. Vanneschi (Eds.), Future Parallel Computers. Proceedings, 1986. V, 492 pages. 1987.

Vol. 273: J.S. Royer, A Connotational Theory of Program Structure. V, 186 pages. 1987.

Vol. 274: G. Kahn (Ed.), Functional Programming Languages and Computer Architecture. Proceedings. VI, 470 pages. 1987.

Vol. 275: A.N. Habermann, U. Montanari (Eds.), System Development and Ada. Proceedings, 1986. V, 305 pages. 1987.

Vol. 276: J. Bézivin, J.-M. Hullot, P. Cointe, H. Lieberman (Eds.), ECOOP '87. European Conference on Object-Oriented Programming. Proceedings. VI, 273 pages. 1987.

Vol. 277: B. Benninghofen, S. Kemmerich, M.M. Richter, Systems of Reductions. X, 265 pages. 1987.

Vol. 278: L. Budach, R.G. Bukharajev, O.B. Lupanov (Eds.), Fundamentals of Computation Theory. Proceedings, 1987. XIV, 505 pages. 1987.

Vol. 279: J.H. Fasel, R.M. Keller (Eds.), Graph Reduction. Proceedings, 1986. XVI, 450 pages. 1987.

Vol. 280: M. Venturini Zilli (Ed.), Mathematical Models for the Semantics of Parallelism. Proceedings, 1986. V, 231 pages. 1987.

Vol. 281: A. Kelemenová, J. Kelemen (Eds.), Trends, Techniques, and Problems in Theoretical Computer Science. Proceedings, 1986. VI, 213 pages. 1987.

Vol. 282: P. Gorny, M.J. Tauber (Eds.), Visualization in Programming. Proceedings, 1986. VII, 210 pages. 1987.

Vol. 283: D.H. Pitt, A. Poigné, D.E. Rydeheard (Eds.), Category Theory and Computer Science. Proceedings, 1987. V, 300 pages. 1987.

Vol. 284: A. Kündig, R.E. Bührer, J. Dähler (Eds.), Embedded Systems. Proceedings, 1986. V, 207 pages. 1987.

Vol. 285: C. Delgado Kloos, Semantics of Digital Circuits. IX, 124 pages. 1987.

Vol. 286: B. Bouchon, R.R. Yager (Eds.), Uncertainty in Knowledge-Based Systems. Proceedings, 1986. VII, 405 pages. 1987.

Vol. 287: K.V. Nori (Ed.), Foundations of Software Technology and Theoretical Computer Science. Proceedings, 1987. IX, 540 pages. 1987.

Vol. 288: A. Blikle, MetaSoft Primer. XIII, 140 pages. 1987.

Vol. 289: H.K. Nichols, D. Simpson (Eds.), ESEC '87. 1st European Software Engineering Conference. Proceedings, 1987. XII, 404 pages. 1987.

Vol. 290: T.X. Bui, Co-oP A Group Decision Support System for Cooperative Multiple Criteria Group Decision Making. XIII, 250 pages. 1987.

Vol. 291: H. Ehrig, M. Nagl, G. Rozenberg, A. Rosenfeld (Eds.), Graph-Grammars and Their Application to Computer Science. VIII, 609 pages. 1987.

Vol. 292: The Munich Project CIP. Volume II: The Program Transformation System CIP-S. By the CIP System Group. VIII, 522 pages. 1987.

Vol. 293: C. Pomerance (Ed.), Advances in Cryptology – CRYPTO '87. Proceedings. X, 463 pages. 1988.

Vol. 294: R. Cori, M. Wirsing (Eds.), STACS 88. Proceedings, 1988. IX, 404 pages. 1988.

Vol. 295: R. Dierstein, D. Müller-Wichards, H.-M. Wacker (Eds.), Parallel Computing in Science and Engineering. Proceedings, 1987. V, 185 pages. 1988.

Vol. 29 ... oceedings, 1987. ...

Vol. 29 ... onopoulos (Eds.), ... 1988.

Vol. 298: ... (Eds.), Mathematical Foundations of Programming Language Semantics. Proceedings, 1987. VIII, 637 pages. 1988.

Vol. 299: M. Dauchet, M. Nivat (Eds.), CAAP '88. Proceedings, 1988. VI, 304 pages. 1988.

Vol. 300: H. Ganzinger (Ed.), ESOP '88. Proceedings, 1988. VI, 381 pages. 1988.

Vol. 301: J. Kittler (Ed.), Pattern Recognition. Proceedings, 1988. VII, 668 pages. 1988.

Vol. 302: D.M. Yellin, Attribute Grammar Inversion and Source-to-source Translation. VIII, 176 pages. 1988.

Vol. 303: J.W. Schmidt, S. Ceri, M. Missikoff (Eds.), Advances in Database Technology – EDBT '88. X, 620 pages. 1988.

Vol. 304: W.L. Price, D. Chaum (Eds.), Advances in Cryptology – EUROCRYPT '87. Proceedings, 1987. VII, 314 pages. 1988.

Vol. 305: J. Biskup, J. Demetrovics, J. Paredaens, B. Thalheim (Eds.), MFDBS 87. Proceedings, 1987. V, 247 pages. 1988.

Vol. 306: M. Boscarol, L. Carlucci Aiello, G. Levi (Eds.), Foundations of Logic and Functional Programming. Proceedings, 1986. V, 218 pages. 1988.

Vol. 307: Th. Beth, M. Clausen (Eds.), Applicable Algebra, Error-Correcting Codes, Combinatorics and Computer Algebra. Proceedings, 1986. VI, 215 pages. 1988.

Vol. 308: S. Kaplan, J.-P. Jouannaud (Eds.), Conditional Term Rewriting Systems. Proceedings, 1987. VI, 278 pages. 1988.

Vol. 309: J. Nehmer (Ed.), Experiences with Distributed Systems. Proceedings, 1987. VI, 292 pages. 1988.

Vol. 310: E. Lusk, R. Overbeek (Eds.), 9th International Conference on Automated Deduction. Proceedings, 1988. X, 775 pages. 1988.

Vol. 311: G. Cohen, P. Godlewski (Eds.), Coding Theory and Applications 1986. Proceedings, 1986. XIV, 196 pages. 1988.

Vol. 312: J. van Leeuwen (Ed.), Distributed Algorithms 1987. Proceedings, 1987. VII, 430 pages. 1988.

Vol. 313: B. Bouchon, L. Saitta, R.R. Yager (Eds.), Uncertainty and Intelligent Systems. IPMU '88. Proceedings, 1988. VIII, 408 pages. 1988.

Vol. 314: H. Göttler, H.J. Schneider (Eds.), Graph-Theoretic Concepts in Computer Science. Proceedings, 1987. VI, 254 pages. 1988.

Vol. 315: K. Furukawa, H. Tanaka, T. Fujisaki (Eds.), Logic Programming '87. Proceedings, 1987. VI, 327 pages. 1988.

Vol. 316: C. Choffrut (Ed.), Automata Networks. Proceedings, 1986. VII, 125 pages. 1988.

Vol. 317: T. Lepistö, A. Salomaa (Eds.), Automata, Languages and Programming. Proceedings, 1988. XI, 741 pages. 1988.

Vol. 318: R. Karlsson, A. Lingas (Eds.), SWAT 88. Proceedings, 1988. VI, 262 pages. 1988.

Vol. 319: J.H. Reif (Ed.), VLSI Algorithms and Architectures – AWOC 88. Proceedings, 1988. X, 476 pages. 1988.

Vol. 320: A. Blaser (Ed.), Natural Language at the Computer. Proceedings, 1988. III, 176 pages. 1988.

Vol. 321: J. Zwiers, Compositionality, Concurrency and Partial Correctness. VI, 272 pages. 1989.

Vol. 322: S. Gjessing, K. Nygaard (Eds.), ECOOP '88. European Conference on Object-Oriented Programming. Proceedings, 1988. VI, 410 pages. 1988.

Vol. 323: P. Deransart, M. Jourdan, B. Lorho, Attribute Grammars. IX, 232 pages. 1988.

Lecture Notes in Computer Science

Edited by G. Goos and J. Hartmanis

355

N. Dershowitz (Ed.)

Rewriting Techniques and Applications

3rd International Conference, RTA-89
Chapel Hill, North Carolina, USA
April 3–5, 1989
Proceedings

Springer-Verlag
Berlin Heidelberg New York London Paris Tokyo

Editor

Nachum Dershowitz
Department of Computer Science, University of Illinois
1304 W. Springfield Ave., Urbana, IL 61801-2987, USA

CR Subject Classification (1987): D.3, F.3.2, F.4, I.1, I.2.2–3

ISBN 3-540-51081-8 Springer-Verlag Berlin Heidelberg New York
ISBN 0-387-51081-8 Springer-Verlag New York Berlin Heidelberg

Printed in Germany

Printing and binding: Druckhaus Beltz, Hemsbach/Bergstr.
2145/3140-543210 – Printed on acid-free paper

PREFACE

This volume contains the proceedings of the **Third International Conference on Rewriting Techniques and Applications** (RTA-89). The conference was held April 3-5, 1989, in Chapel Hill, North Carolina, U.S.A. Professor Garrett Birkhoff of Harvard University delivered an invited lecture.

The two previous conferences were held in France:

- Dijon, May 1985 (*Lecture Notes in Computer Science* **202**, Springer-Verlag);
- Bordeaux, May 1987 (*Lecture Notes in Computer Science* **256**, Springer-Verlag).

For the third conference, papers were solicited in the following (or related) areas:

Term rewriting systems	Symbolic and algebraic computation
Conditional rewriting	Equational programming languages
Graph rewriting and grammars	Completion procedures
Algebraic semantics	Rewrite-based theorem proving
Equational reasoning	Unification and matching algorithms
Lambda and combinatory calculi	Term-based architectures

The 34 regular papers in this volume were selected from 84 submissions by a ten-member committee, consisting of the following individuals:

Bruno Courcelle (Bordeaux)	Deepak Kapur (Albany)
Nachum Dershowitz (Urbana)	Claude Kirchner (Nancy)
Jean Gallier (Philadelphia)	Jan Willem Klop (Amsterdam)
Jieh Hsiang (Stony Brook)	Dallas Lankford (Ruston)
Jean-Pierre Jouannaud (Orsay)	Mark Stickel (Menlo Park)

Each paper was graded by two to six people (committee members and outside referees).

Also included in this volume are short descriptions of a dozen of the implemented equational reasoning systems demonstrated at the meeting.

Local arrangements were organized by:

David A. Plaisted
Department of Computer Science
University of North Carolina at Chapel Hill
Chapel-Hill, NC 27599
U.S.A.

The meeting was in cooperation with the Association for Computing Machinery and received support from the University of Illinois at Urbana-Champaign and the University of North Carolina at Chapel Hill.

Nachum Dershowitz
Program Chairperson

The committee acknowledges the help of the following referees:

Leo Bachmair
J. C. M. Baeten
M. Bezem
Maria Paola Bonacina
Alexandre Boudet
Hans-Jürgen Burckert
Robert Cori
Hervé Devie
François Fages
Isabelle Gnaedig
Miki Hermann
Bharat Jayaraman
Simon Kaplan
Stéphane Kaplan
Hélène Kirchner
Pierre Lescanne
D. Luggiez
A. Megrelis
Yves Metivier
Aart Middeldorp
David Musser
Paliath Narendran
Tobias Nipkow
Friedrich Otto
Uday Reddy
Jean-Luc Rémy
Michael Rusinowitch
Manfred Schmidt-Schauß
Sabine Stifter
Jonathan Stillman
F.-J. de Vries
R. C. de Vrijer
Hantao Zhang

CONTENTS

Characterizations of Unification Type Zero

Franz Baader
IMMD 1, Universität Erlangen-Nürnberg
Martensstraße 3, D-8520 Erlangen (West Germany)

Abstract

In the literature several methods have hitherto been used to show that an equational theory has unification type zero. These methods depend on conditions which are candidates for alternative characterizations of unification type zero. In this paper we consider the logical connection between these conditions on the abstract level of partially ordered sets. Not all of them are really equivalent to type zero.
The conditions may be regarded as tools which can be used to determine the unification type of given theories. They are also helpful in understanding what makes a theory to be of type zero.

1. Introduction.

Let E be an equational theory and $=_E$ be the equality of terms, induced by E. A substitution θ is called an *E-unifier* of the terms s, t iff $s\theta =_E t\theta$. The set of all E-unifiers of s, t is denoted by $U_E(s,t)$. We are mostly interested in complete sets of E-unifiers, i.e. sets of E-unifiers from which $U_E(s,t)$ may be generated by instantiation. More formally, we extend $=_E$ to $U_E(s,t)$ and define a quasi-ordering $\leq_E$ on $U_E(s,t)$ by

$\sigma =_E \theta$ iff $x\sigma =_E x\theta$ for all variables x occurring in s or t.

$\sigma \leq_E \theta$ iff there exists a substitution λ such that $\sigma =_E \theta \circ \lambda$.

In this case σ is called an E-instance of θ. As usual the quasi-ordering $\leq_E$ induces an equivalence relation $\equiv_E$ on $U_E(s,t)$, namely $\sigma \equiv_E \theta$ iff $\sigma \leq_E \theta$ and $\theta \leq_E \sigma$. The $\equiv_E$-class of

Term Rewriting and Universal Algebra in Historical Perspective
—Invited Lecture—

Garrett Birkhoff
Harvard University
Department of Mathematics
Science Center 325
One Oxford Street
Cambridge, MA 02138
USA

Abstract

"Term rewriting" is obviously concerned with the logic of algebra. Closely related to symbolic logic and mathematical linguistics, it derives much of its stimulus from the idea of exploiting the enormous power of modern computers.

My talk will fall into two parts. The first part will review term rewriting from the general standpoint of "universal algebra" in historical perspective. The perspective suggests as especially interesting and/or challenging.

an E-unifier σ of s, t shall be denoted by $[\sigma]$.

A *complete set* $cU_E(s,t)$ *of E-unifiers* of s, t is defined as

(1) $cU_E(s,t) \subseteq U_E(s,t)$,

(2) For all $\theta \in U_E(s,t)$ there exists $\sigma \in cU_E(s,t)$ such that $\theta \leq_E \sigma$.

A *minimal complete set* $\mu U_E(s,t)$ is a complete set of E-unifiers of s, t satisfying the minimality condition

(3) For all $\sigma, \theta \in \mu U_E(s,t)$ $\sigma \leq_E \theta$ implies $\sigma = \theta$.

This notion was introduced by Plotkin (1972). In the same paper he conjectured that there exist an equational theory E and terms s, t such that $\mu U_E(s,t)$ does not exist. We say that such a theory has *unification type zero*. The first example of a type zero theory is due to Fages and Huet (1983). They constructed a theory E and terms s, t such that $U_E(s,t)$ contains a strictly increasing chain with respect to $\leq_E$ which is a complete set of E-unifiers of s, t. Under these circumstances, a complete set of E-unifiers can not satisfy the minimality condition. Schmidt-Schauß (1986) and the present author (1986) showed that the theory of idempotent semigroups is of unification type zero and in Baader (1987) I have proved that almost all varieties of idempotent semigroups are defined by type zero theories. In these cases, complete sets of unifiers are not simply chains. Thus other methods had to be developed to obtain the results. In this paper we consider some of these methods and their logical connection. We shall thus obtain alternative characterizations of unification type zero and sufficient conditions (resp. necessary conditions) for a theory to be type zero. This provides tools which may be used to determine the unification types of given theories. It also helps to understand what it means that a theory is of type zero.

2. Minimal complete sets and maximal elements.

For terms s, t the quasi-ordering $\leq_E$ on $U_E(s,t)$ induces a partial ordering (po) $\leq$ on $U := \{ [\sigma]; \sigma \in U_E(s,t) \}$, namely $[\sigma] \leq [\tau]$ iff $\sigma \leq_E \tau$. The existence of a set $\mu U_E(s,t)$ can be described with the help of the set of all $\leq$-maximal elements in U. But first we extent the notion of complete and minimal complete sets to subsets of partially ordered sets (see also Büttner (1986), Schmidt-Schauß (1986a) and Bürckert (1987)). Parts of our proofs can be done on this abstract level.

Definition 2.1. Let $\leq$ be a po on a set U and let W be a subset of U.

(1) We call W a *complete subset* of U iff for any $u \in U$ there exists some $w \in W$ such that $u \leq w$ holds.

(2) W is a *minimal complete subset* of U iff no proper subset of W is complete.

(3) U has *type zero* (w.r.t. $\leq$) iff there does not exist a minimal complete subset of U.

It is easy to see that a complete subset W of U is minimal iff for any $v, w \in W$ $w \leq v$ implies $w = v$. If $U = \{ [\sigma]; \sigma \in U_E(s,t) \}$ and $\leq$ is the partial ordering induced by $\leq_E$ then U has type zero iff there does not exist a minimal complete set of E-unifiers of s, t. The next lemma describes the connection between minimal complete sets and maximal elements in partially ordered sets.

Lemma 2.2 Let $\leq$ be a po on a set U and let M be the set of all $\leq$-maximal elements in U.

(1) If W is a minimal complete subset of U then W = M. Consequently, if U does not have type zero then M is complete.

(2) If M is complete then M is a minimal complete subset of U and hence U does not have type zero.

Proof. (1) Let W be a minimal complete subset of U and let w be an element of W. Then w is $\leq$-maximal. Otherwise there would be an element v of U such that $w < v$. Since W is complete there is $w' \in W$ with the property $v \leq w'$. But then $w < w'$ is a contradiction to the minimality of W. On the other hand, let m be an element of M. Since W is complete there exists $w \in W$ with $m \leq w$. But m is $\leq$-maximal which yields $m = w \in W$.

(2) Let M be complete. Since different $\leq$-maximal elements are not $\leq$-comparable, M is also minimal. ❑

As an immediate consequence we have

Proposition 2.3 Let s, t be terms, $\leq$ be the partial ordering on $U = \{ [\sigma]; \sigma \in U_E(s,t) \}$ induced by $\leq_E$ and M be the set of $\leq$-maximal elements of U.

(1) A minimal complete set of E-unifiers of s, t exists iff M is complete.

(2) Let $\mu U_E(s,t)$ be a minimal complete set of E-unifiers of s, t. Then $M = \{ [\sigma]; \sigma \in \mu U_E(s,t) \}$. Conversely, any set of representatives of M is a minimal complete set of E-unifiers of s, t. ❑

Obviously, (2) of the proposition yields

Corollary 2.4. If a minimal complete set of E-unifiers of s, t exists, it is unique up to $\equiv_E$-equivalence. ❑

3. Characterizations.

Let $\le$ be a po on a set U. We consider the following properties for U:

(P1) U has type zero w.r.t $\le$.

(P2) There is $u_1 \in U$ such that for all $u \in U$ with $u_1 \le u$ there exists some $\hat{u} \in U$ such that $u < \hat{u}$.

(P3) There is an increasing chain $u_1 \le u_2 \le u_3 \le \ldots$ in U without upper bounds in U which has the property: for all $u \in U$ with $u_n \le u$ there is $\hat{u} \in U$ such that $u \le \hat{u}$ and $u_{n+1} \le \hat{u}$.

(P4) There is a properly increasing chain $u_1 < u_2 < u_3 < \ldots$ in U such that $\{ u_1, u_2, u_3, \ldots \}$ is complete.

(P5) There is an increasing chain in U without upper bounds in U.

Theorem 3.1 The following implications hold:

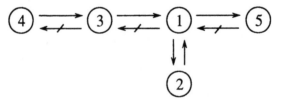

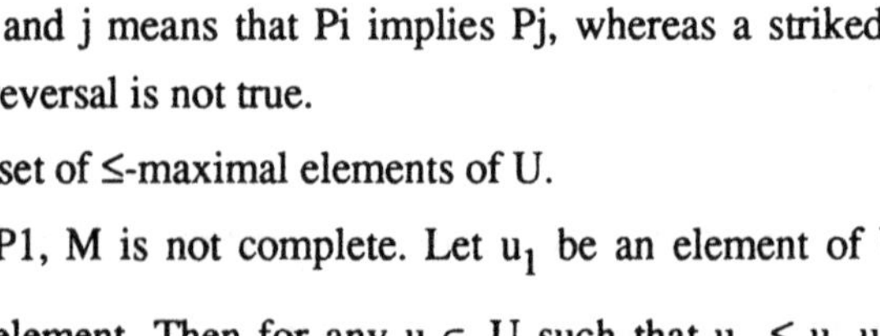

An arrow between i and j means that Pi implies Pj, whereas a striked out arrow between j and i means that the reversal is not true.

Proof. Let M be the set of $\le$-maximal elements of U.

"1→2": Because of P1, M is not complete. Let u_1 be an element of U which does not lie below a $\le$-maximal element. Then for any $u \in U$ such that $u_1 \le u$, u is not maximal.Thus there exists $\hat{u} \in U$ such that $u < \hat{u}$.

"2→1": Assume P2. If M is complete then there is $u \in M$ such that $u_1 \le u$. Because P2 holds there is $\hat{u} \in U$ such that $u < \hat{u}$. Thus u was not maximal.

"1→5": Assume that P5 does not hold. Then any increasing $\le$-chain in U has an upper bound in U. Thus, by Zorn's Lemma, any element of U lies below a maximal element of U. But that means that M is complete.

"5-↛1": We consider $U := \{ a_i; i \geq 0 \} \cup \{ b_i; i \geq 0 \}$. The po $\leq$ on U is defined by

$$a_i \leq a_j \text{ iff } i \leq j,\ b_i \leq b_j \text{ iff } i = j,\ a_i \leq b_j \text{ iff } i \leq j \text{ and } b_j \text{ is never below an } a_i.$$

Then $a_1 < a_2 < a_3 < \ldots$ is a properly increasing chain in U without upper bounds in U. But $\{ b_1, b_2, b_3, \ldots \}$ is a minimal complete subset of U.

"3→1": Let M be complete and let $u_1 \leq u_2 \leq u_3 \leq \ldots$ be a chain with the properties claimed in P3. Since M is complete and the chain does not have an upper bound in U, there is an $n \geq 1$ and an element u of M such that $u_n \leq u$ and $u_{n+1} \not\leq u$. Now P3 yields an $\hat{u} \in U$ such that $u \leq \hat{u}$ and $u_{n+1} \leq \hat{u}$. Since u is maximal we have $u = \hat{u}$. This is a contradiction to $u_{n+1} \not\leq u$.

"1-↛3": Let U be the set of all words over the alphabet $\{ 0, 1 \}$, i.e. $U = \{ 0, 1 \}^*$. The po $\leq$ is defined to be the prefix ordering on $\{ 0, 1 \}^*$, i.e.:

$$u \leq v \text{ iff there exists } w \in \{ 0, 1 \}^* \text{ such that } v = uw.$$

U does not contain maximal elements and thus the set of all maximal elements is not complete. Assume that U contains a chain $u_1 \leq u_2 \leq u_3 \leq \ldots$ with the properties claimed in P3. Since this chain does not have an upper bound in U there exists an $n \geq 1$ such that $u_n < u_{n+1}$. But then $u_{n+1} = u_n a v$ for some $a \in \{ 0, 1 \}$, $v \in \{ 0, 1 \}^*$. Now we consider $u := u_n b$, where $b \in \{ 0, 1 \} \setminus \{ a \}$. Obviously $u_n \leq u$, but there does not exist an element $\hat{u}$ such that $u \leq \hat{u}$ and $u_{n+1} \leq \hat{u}$.

"4→3": Let $u_1 < u_2 < u_3 < \ldots$ be a properly increasing chain in U such that $W := \{ u_1, u_2, u_3, \ldots \}$ is complete. Assume that $b \in U$ is an upper bound of this chain. Since W is complete there is an $n \geq 1$ such that $b \leq u_n$. But then $u_n = u_{n+1} = \ldots$, which is a contradiction. Thus the chain does not have an upper bound in U. Let $u_n \leq u$ for some $u \in U$. Completeness of W yields an $m \geq 1$ such that $u \leq u_m$. Since the chain is properly increasing we know that $n \leq m$. Now $u \leq u_{m+1}$ and $u_{n+1} \leq u_{m+1}$. Thus we can take $\hat{u} = u_{m+1}$.

"3-↛4": Let U be $\{ 0 \}^+ \cup \{ 1 \}^+$ and let $\leq$ be the prefix ordering on U (see 1-↛3). Obviously, $0 < 00 < 000 < \ldots$ is a properly increasing chain in U satisfying the properties claimed in P3. But no chain in U is complete. ❑

Let s, t be terms and let $\leq$ be the partial ordering on $U = \{ [\sigma]; \sigma \in U_E(s,t) \}$ induced by $\leq_E$. If one wants to show that a minimal complete set of E-unifiers of s, t does not exist it suffices to proof that U has one of the properties P4, P3 or P2. The properties P4 and P3 are sufficient but not necessary, whereas P2 is necessary and sufficient. Note that P5 is necessary but not sufficient. P2 was used in Baader (1986) to show that the theory of idempotent semigroups has unification type zero and property P3 was used in Baader (1987) to prove the same for almost all varieties of idempotent semigroups. Fages-Huet (1983) constructed a theory E and terms s, t such that $U = \{ [\sigma]; \sigma \in U_E(s,t) \}$ has property P4. In some papers property P5 is erroneously claimed to be equivalent to P1.
In the remainder of this paper we shall show that there are really partially orderd sets (U, $\leq$) induced by unification problems (i.e. $U = \{ [\sigma]; \sigma \in U_E(s,t) \}$ for some terms s, t and $\leq$ is the partial ordering on U induced by $\leq_E$) such that U has property P3 but not P4 (resp. P1 and not P3, P5 and not P1). This turns out to be much harder than proving the implications of Theorem 3.1. In the following we shall also say "$U_E(s,t)$ satisfies Pi" if U has property Pi.

4. Examples.

We shall only consider theories with several unary function symbols. The fact that theories of this kind are very useful for the construction of examples in unification theory was first mentioned by M. Schmidt-Schauß (see Bürckert-Herold-Schmidt-Schauß (1987)). He called these theories monadic.

Definition 4.1 Let Ω be a signature and V be a denumerable set of variables. We denote the set of Ω-terms with variables in V by $T(\Omega,V)$.
An equational theory $E \subseteq T(\Omega,V)\times T(\Omega,V)$ is *regular* iff $(s,t) \in E$ implies that s and t contain the same variables. E is called *monadic* if, in addition, Ω only consists of unary function symbols.

Ω can also be considered as alphabet. Every term $t = f_1(f_2(...f_n(x)...))$ may be written as $w_t(x_t)$ for $w_t = f_1f_2...f_n \in \Omega^*$ and $x_t = x \in V$. For the congruence $\sim_E$ on Ω^* generated by $\{ (w_s,w_t); (s,t) \in E \}$ we have $s =_E t$ iff $w_s \sim_E w_t$ and $x_s = x_t$.

Example 4.2 A unification problem satisfying P5 but not P1.

Let E be the set $\{\ f_1g_1(x) = g_2f_1(x),\ f_2g_1(x) = g_2f_2(x),\ f_1k(x) = f_2k(x),\ g_1khl(x) = kh(x)\ \}$. It can be shown (see Bürckert-Herold-Schmidt-Schauß (1987)) that E does not have unification type zero. The E-unification problem for $f_1(x)$, $f_2(x)$ has the following properties:

(1) For $\theta_n := \{\ x \mapsto g_1{}^n k(z)\ \}$ the set $\{\ \theta_n;\ n \geq 0\ \}$ is a minimal complete set of E-unifiers of $f_1(x)$, $f_2(x)$.

(2) For $\sigma_n := \{\ x \mapsto g_1{}^n kh(z)\ \}$ the chain $\sigma_1 \leq_E \sigma_2 \leq_E \sigma_2 \leq_E \ldots$ is a properly increasing $\leq_E$-chain without an upper bound in $U_E(f_1(x),f_2(x))$.

A similar example can be found in Baader (1988).

Example 4.3 A unification problem satisfying P1 but not P3.

Let E be the set $\{\ f_1g_0(x) = h_0f_1(x),\ f_1g_1(x) = h_1f_1(x),\ f_2g_0(x) = h_0f_2(x),\ f_2g_1(x) = h_1f_2(x),$

$f_1k(x) = f_2k(x),\ g_0kl(x) = k(x),\ g_1kl(x) = k(x)\ \}$.

The Knuth-Bendix completion procedure (with a Knuth-Bendix ordering where $g_0 < g_1 < h_0 < h_1 < f_2 < f_1 < k < l$ and the function symbols have weight 1) yields the following canonical term rewriting system for E:

$R = \{\ f_1g_0(x) \rightarrow h_0f_1(x),\ f_1g_1(x) \rightarrow h_1f_1(x),$	Commuting f_i and g_j, whereby
$f_2g_0(x) \rightarrow h_0f_2(x),\ f_2g_1(x) \rightarrow h_1f_2(x),$	g_j becomes h_j.
$f_1k(x) \rightarrow f_2k(x),$	When touching k, f_1 is changed to f_2.
$g_0kl(x) \rightarrow k(x),\ g_1kl(x) \rightarrow k(x),$	In g_ikl, g_i and l can be erased.
$h_0f_2kl(x) \rightarrow f_2k(x),\ h_1f_2kl(x) \rightarrow f_2k(x)\ \}$	These two rules come out of the completion of the other rules.

We consider the E-unification problem for $f_1(x)$, $f_2(x)$.

Proposition 4.3.1 If θ is an E-unifier of $f_1(x)$, $f_2(x)$ then $x\theta =_E wk(u)$, where w is a word in $\{\ g_0, g_1\ \}^*$ and u is an arbitrary term.

Proof. With out loss of generality we may assume that $x\theta$ is R-irreducible. In the following we shall denote $\rightarrow_R$ simply by $\rightarrow$. Since $f_1(x\theta) =_E f_2(x\theta)$ there exists a term t such that $f_1(x\theta) \xrightarrow{*} t \xleftarrow{*} f_2(x\theta)$. Because $f_1(x\theta) \neq f_2(x\theta)$ at least one of the derivations has a length > 0. Since $x\theta$ is R-irreducible the first step of this derivation is at occurrence ε.

Thus there are two possible cases: either $x\theta$ has g_i as first symbol or it starts with k.

In the second case $x\theta$ is of the desired form. Otherwise $x\theta = g_i(s)$ for some term s. Then we have $f_1g_i(s) \to h_if_1(s) \overset{*}{\Rightarrow} t \overset{*}{\Leftarrow} h_if_2(s) \leftarrow f_2g_i(s)$. If it could be deduced that $t = h_i(t_1)$ und $f_1(s) \overset{*}{\Rightarrow} t_1 \overset{*}{\Leftarrow} f_2(s)$ the proposition would be proved by induction on term size. Thus it suffices to prove the following assertion:

Let $f_jg_i(s) \to h_if_j(s) \overset{*}{\Rightarrow} t$ and let $g_i(s)$ be R-irreducible. Then $t = h_i(t_1)$ and $f_j(s) \overset{*}{\Rightarrow} t_1$.

The assertion is trivially satisfied if $h_if_j(s) = t$ or if in the derivation $h_if_j(s) \overset{*}{\Rightarrow} t$ no rule is applied at occurrence ε. Otherwise the derivation can be decomposed into $h_if_j(s) \overset{m}{\rightarrow} h_i(t_2) \overset{+}{\Rightarrow} t$, where $f_j(s) \overset{m}{\rightarrow} t_2$ and the first step in $h_i(t_2) \overset{+}{\Rightarrow} t$ is at occurrence ε. Obviously, this step is done with the rule $h_if_2kl(x) \to f_2k(x)$.

For $m = 0$ it follows $s = kl(s_1)$ and $j = 2$. But then $g_i(s) = g_ikl(s_1)$ is not R-irreducible. Hence $m > 0$, i.e. $f_j(s) \overset{+}{\Rightarrow} t_2$. Since s is a subterm of the R-irreducible term $g_i(s)$ the first derivation step must be at occurrence ε. Hence there remain two cases:

Case 1: $s = g_{i'}(s')$.

Then $f_j(g_{i'}(s')) \to h_{i'}(f_j(s')) \overset{*}{\Rightarrow} t_2$. Induction yields $t_2 = h_{i'}(t_2')$ and $f_j(s') \overset{*}{\Rightarrow} t_2'$. Hence the rule $h_if_2kl(x) \to f_2k(x)$ can not be applied at occurrence ε of the term $h_i(t_2) = h_ih_{i'}(t_2')$.

Case 2: $s = k(s')$ und $j = 1$.

Then $h_if_jk(s') \to h_if_2k(s') \overset{*}{\Rightarrow} h_i(t_2) \overset{+}{\Rightarrow} t$. Since s' is R-irreducible $t_2 = f_2k(s')$. The term $h_i(t_2)$ can only be reduced if $s' = l(s'')$. But then $g_i(s) = g_ikl(s'')$ is not R-irreducible. ❑

For $w \in \{ g_0, g_1 \}^*$ we define substitutions θ_w by $x\theta_w := wk(x)$. Proposition 4.3.1 and the fact that all θ_w are E-unifiers of $f_1(x), f_2(x)$ imply:

Corollary 4.3.2

The set $T = \{ \theta_w; w \in \{ g_0, g_1 \}^* \}$ is a complete set of E-unifiers of $f_1(x), f_2(x)$. ❑

We shall now determine how the elements of T are ordered by $\geq_E$.

Lemma 4.3.3 Let $w \in \{ g_0, g_1 \}^*$ and $a \in \{ g_0, g_1 \}$. Then we have $\theta_w \leq_E \theta_{wa}$.

Proof. By definition $x\theta_w = wk(x)$ and $x\theta_{wa} = wak(x)$. The substitution $\lambda = \{ x \mapsto l(x) \}$ yields $x\theta_{wa}\lambda = wakl(x) =_E wk(x) = x\theta_w$. ❑

Thus for all $w, v \in \{ g_0, g_1 \}^*$ $\theta_w \leq_E \theta_{wv}$. The next lemma can be used to show that elements of T are only comparable in this way.

Lemma 4.3.4 Let $w, v \in \{ g_0, g_1 \}^*$ and let t_1, t_2 be terms such that $wk(t_1) \xrightarrow{*} vk(t_2)$. Then v is a prefix of w.

Proof by noetherian induction w.r.t. $\rightarrow_R$.

If $wk(t_1)$ is R-irreducible $wk(t_1) = vk(t_2)$, i.e. $w = v$ and $t_1 = t_2$.

If the first step of $wk(t_1) \xrightarrow{*} vk(t_2)$ is inside t_1 we have $wk(t_1) \rightarrow wk(t_1') \xrightarrow{*} vk(t_2)$. Thus the assertion follows by induction. Otherwise $w = w_1g_i$, $t_1 = l(t_1')$ and $wk(t_1) = w_1g_ikl(t_1') \rightarrow w_1k(t_1') \xrightarrow{*} vk(t_2)$. By induction v is a prefix of w_1 and thus of $w = w_1g_i$. ❑

By Proposition 4.3.1, for any $\theta, \tau \in U_E(f_1(x),f_2(x))$ there exist words $u, v \in \{ g_0, g_1 \}^*$ and terms t_1, t_2 such that $x\theta =_E uk(t_1)$, $x\tau =_E vk(t_2)$, where $uk(t_1)$, $vk(t_2)$ are R-irreducible.

Lemma 4.3.5 If $\theta \leq_E \tau$ then u is a prefix of v.

Proof. Let $\theta =_E \tau \circ \lambda$, i.e. $x\theta =_E x\tau\lambda$. Since $uk(t_1)$ is R-irreducible we have $vk(t_2\lambda) \xrightarrow{*} uk(t_1)$. Thus Lemma 4.3.4 applies. ❑

Particularly for $\theta_u, \theta_v \in T$, Lemma 4.3.3 and 4.3.5 yield: $\theta_u \leq_E \theta_v$ iff u is prefix of v.

We can now show that $U_E(f_1(x),f_2(x))$ satisfies P1.

Proposition 4.3.6

There does not exist a minimal complete set of E-unifiers of $f_1(x)$, $f_2(x)$.

Proof. Let M be a complete set of E-unifiers of $f_1(x)$, $f_2(x)$ and let μ be an element of M. Since T is complete there is an element θ_w of T such that $\mu \leq_E \theta_w$. But then $\theta_w \leq_E \theta_{wg_0}$ and $\theta_{wg_0} \not\leq_E \theta_w$. Since M is complete there exists $\mu' \in M$ with the property $\theta_{wg_0} \leq_E \mu'$. Now $\mu \leq_E \mu'$ and $\mu' \not\leq_E \mu$. Hence M is not minimal. ❑

Proposition 4.3.7 $U_E(f_1(x),f_2(x))$ does not satisfy P3.

Proof. Assume that $U_E(f_1(x),f_2(x))$ contains a chain $\theta_1 \leq_E \theta_2 \leq_E \theta_3$ ... satisfying the properties claimed in P3. Since this chain does not have an upper bound in $U_E(f_1(x),f_2(x))$ and since T is complete there exist $n \geq 1$ and $w \in \{ g_0, g_1 \}^*$ such that $\theta_n \leq_E \theta_w$ and $\theta_{n+1} \not\leq_E \theta_w$. We apply P3 for $\theta_n \leq_E \theta_{wg_0}$ and for $\theta_n \leq_E \theta_{wg_1}$. This yields unifiers τ_0 and τ_1 such that $\theta_{wg_i} \leq_E \tau_i$ und $\theta_{n+1} \leq_E \tau_i$ $(i = 0, 1)$.

Since T is complete there exist θ_u, θ_v with the property $\tau_0 \leq_E \theta_u$ and $\tau_1 \leq_E \theta_v$. Lemma 4.3.5 implies that wg_0 is a prefix of u and wg_1 is a prefix of v, i.e. $u = wg_0u'$ and $v = wg_1v'$ for $u', v' \in \{ g_0, g_1 \}^*$. By Proposition 4.3.1 there exist $s \in \{ g_0, g_1 \}^*$ and a term t such that $x\theta_{n+1} = sk(t)$. Lemma 3.4.5 and $\theta_{n+1} \leq_E \theta_u$ as well as $\theta_{n+1} \leq_E \theta_v$ imply that s is prefix of $u = wg_0u'$ and of $v = wg_1v'$. Hence s is prefix of w. Thus $\theta_{n+1} \leq_E \theta_s \leq_E \theta_w$, which is a contradiction. ❑

Example 4.4 A unification problem satisfying P3 but not P4.

Let E be the set $\{ \; f_1g_1g_1(x) = h_1f_1g_1(x), f_1g_2g_2(x) = h_2f_1g_2(x),$

$$f_2g_1g_1(x) = h_1f_2g_1(x), f_2g_2g_2(x) = h_2f_2g_2(x),$$

$$f_1g_1k(x) = f_2g_1k(x), f_1g_2k(x) = f_2g_2k(x),$$

$$g_1g_1kl(x) = g_1k(x), g_2g_2kl(x) = g_2k(x) \quad \}.$$

The Knuth-Bendix completion procedure (with a Knuth-Bendix ordering where $h_0 < h_1 < g_0 < g_1 < f_2 < f_1 < k < l$ and the function symbols have weight 1) yields the following canonical term rewriting system for E:

$R = \{ f_1g_1g_1(x) \rightarrow h_1f_1g_1(x), f_1g_2g_2(x) \rightarrow h_2f_1g_2(x),$

$$f_2g_1g_1(x) \rightarrow h_1f_2g_1(x), f_2g_2g_2(x) \rightarrow h_2f_2g_2(x),$$

$$f_1g_1k(x) \rightarrow f_2g_1k(x), f_1g_2k(x) \rightarrow f_2g_2k(x),$$

$$g_1g_1kl(x) \rightarrow g_1k(x), g_2g_2kl(x) \rightarrow g_2k(x)$$

$$h_1f_2g_1kl(x) \rightarrow f_2g_1k(x), h_2f_2g_2kl(x) \rightarrow f_2g_2k(x) \}.$$

We consider the E-unification problem for $f_1(x), f_2(x)$.

Proposition 4.4.1 Let θ be an E-unifier of $f_1(x)$, $f_2(x)$. Then $x\theta =_E wk(u)$, where $w \in \{g_1\}^+ \cup \{g_2\}^+$ and t is a term.

Proof. With out loss of generality we may assume that $x\theta$ is R-irreducible. As before we shall abbreviate $\rightarrow_R$ by $\rightarrow$. Since $f_1(x\theta) =_E f_2(x\theta)$ there exists a term t such that $f_1(x\theta) \xrightarrow{*} t \xleftarrow{*} f_2(x\theta)$. At least one of the derivations has length > 0. Again the first derivation step is at occurrence ε. The top symbol of $x\theta$ may be g_1 or g_2. Since g_1 and g_2 behave symmetrically w.r.t. R, it suffices to show:

Let $g_1(s)$ be R-irreducible. Then $f_1g_1(s) \xrightarrow{*} t \xleftarrow{*} f_2g_1(s)$ implies $s = g_1{}^nk(s_1)$ for some $n \geq 0$ and some term s_1.

If $s = k(s')$ then s is of the desired form. Otherwise, $s = g_1(s')$ and we have $f_1g_1g_1(s') \rightarrow h_1f_1g_1(s') \xrightarrow{*} t \xleftarrow{*} h_1f_2g_1(s') \leftarrow f_2g_1g_1(s')$. As in the proof of Proposition 4.3.1 it follows that $f_1g_1(s') \xrightarrow{*} t' \xleftarrow{*} f_2g_1(s')$, where $t = h_1(t')$. Thus by induction on the size of terms we get $s' = g_1{}^nk(s_1)$ and hence $s = g_1{}^{n+1}k(s_1)$. ❑

For $n \geq 1$ we define substitutions σ_n and θ_n by $x\sigma_n := g_1{}^nk(x)$ and $x\theta_n := g_2{}^nk(x)$.

Proposition 4.4.1 and the fact that the substitutions σ_n and θ_n are E-unifier of $f_1(x)$, $f_2(x)$ imply:

Corollary 4.4.2 The set $T = \{ \theta_n;\ n \geq 1 \} \cup \{ \sigma_n;\ n \geq 1 \}$ is a complete set of E-unifiers of $f_1(x)$, $f_2(x)$. ❑

Evidently, $\sigma_n \leq_E \sigma_{n+1}$ and $\theta_n \leq_E \theta_{n+1}$. On the other hand, σ_n and θ_m are incomparable w.r.t. $\leq_E$. This is an immediate consequence of the next lemma.

Lemma 4.4.3 Let $n, m \geq 1$ and let t_1, t_2 be terms. Then $g_1{}^nk(t_1) \neq_E g_2{}^mk(t_2)$.

Proof. From $g_1{}^nk(t_1) =_E g_2{}^mk(t_2)$ we could deduce that there exists a term t such that $g_1{}^nk(t_1) \xrightarrow{*} t \xleftarrow{*} g_2{}^mk(t_2)$. But any term which can be derived from $g_1{}^nk(t_1)$ is of the form $g_1{}^ik(t_1')$ for some i, $1 \leq i \leq n$. Analogously any term which can be derived from $g_2{}^mk(t_2)$ is of the form $g_2{}^jk(t_2')$ for some j, $1 \leq j \leq n$. Thus we have a contradiction. ❑

Now we can show that $U_E(f_1(x),f_2(x))$ does not satisfy P4.

Proposition 4.4.4 Let M be a complete set of E-unifiers of $f_1(x)$, $f_2(x)$. Then M contains at least two elements which are not comparable w.r.t. $\leq_E$.

Proof. Since M is complete there exist σ, $\theta \in$ M such that $\sigma_1 \leq_E \sigma$ and $\theta_1 \leq_E \theta$, i.e. there exist substitutions λ_1, λ_2 satisfying $g_1k(x) =_E x\sigma\lambda_1$ and $g_2k(x) =_E x\theta\lambda_2$. Proposition 4.4.1 and Lemma 4.4.3 imply that $x\sigma = g_1{}^m k(t_1)$ and $x\theta = g_2{}^n k(t_2)$ for some m, n $\geq$ 1. Hence, by Lemma 4.4.3, σ and θ are incomparable w.r.t. $\leq_E$. ❑

On the other side, we have

Proposition 4.4.5 $U_E(f_1(x),f_2(x))$ satisfies P3.

Proof. We consider the chain $\sigma_1 \leq_E \sigma_2 \leq_E \sigma_3 \leq_E \ldots$.

(1) This chain does not have an upper bound in $U_E(f_1(x),f_2(x))$.

Assume that $\sigma \in U_E(f_1(x),f_2(x))$ is an upper bound of $\sigma_1 \leq_E \sigma_2 \leq_E \sigma_3 \leq_E \ldots$. Proposition 4.4.1 and Lemma 4.4.3 imply that $x\sigma = g_1{}^m k(t)$ for some $m \geq 1$ and some term t. Since $\sigma_{m+1} \leq_E \sigma$ there is a substitution λ such that $g_1{}^{m+1}k(x) = x\sigma_{m+1} =_E x\sigma\lambda =_E g_1{}^m k(t\lambda)$. Now $g_1{}^m k(t\lambda) \not\overset{*}{\rightarrow} g_1{}^{m+1}k(x)$, since $g_1{}^{m+1}k(x)$ is R-irreducible. This contradicts the fact that any term which can be derived from $g_1{}^m k(t\lambda)$ is of the form $g_1{}^i k(t')$ for some i, $1 \leq i \leq$ m and some term t'.

(2) Let $\sigma_n \leq_E \sigma$. Then $x\sigma =_E g_1{}^m k(t)$ for some $m \geq 1$ and some term t. We define r := max{ m, n+1 }. Then $\sigma_{n+1} \leq_E \sigma_r$ and $\sigma \leq_E \sigma_m \leq_E \sigma_r$. ❑

5. Conclusion

We have extended the notion of complete and minimal complete sets to arbitrary subsets of partially ordered sets. Thus the characterizations (resp. necessary or sufficient conditions) for type zero could be discussed on a more abstract level. Only for the counter examples of Section 4 real unification problems had to be considered.

In Section 4 we have shown that the partial orderings used in $5 \not\to 1$, $1 \not\to 3$ and $3 \not\to 4$ of the proof of Theorem 3.1 can be realized as complete sets of E-unifiers. Thus the following more general question arises: Which kind of partial ordering is representable in this way ?

6. References

Baader, F. (1986). The Theory of Idempotent Semigroups is of Unification Type Zero. *J. Automated Reasoning* **2.**

Baader, F. (1987). Unification in Varieties of Idempotent Semigroups. *Semigroup Forum* **36.**

Baader, F. (1988). A Note on Unification Type Zero. *Information Processing Letters* **27.**

Bürckert, H.J. (1987). Matching – A Special Case of Unification? SEKI Technical Report, Universität Kaiserslautern.

Bürckert, H.J., Herold, A., Schmidt-Schauß, M. (1987). On Equational Theories, Unification and Decidability. Proceedings of the RTA '87 Bordeaux (France). *Springer Lec. Notes Comp. Sci.* **256.**

Büttner, W. (1986). Unification in the Data Structure Multiset. *J. Automated Reasoning* **2.**

Fages, F., Huet, G. (1983). Complete Sets of Unifiers and Matchers in Equational Theories. Proceedings of the CAAP '83 Pisa (Italy). *Springer Lec. Notes Comp. Sci.* **159.**

Fages, F., Huet, G. (1986). Complete Sets of Unifiers and Matchers in Equational Theories. *Theor. Comp. Sci.* **43.**

Plotkin, G. (1972). Building in Equational Theories. *Machine Intelligence* **7.**

Schmidt-Schauß, M. (1986). Unification under Associativity and Idempotence is of Type Nullary. *J. Automated Reasoning* **2.**

Schmidt-Schauß, M. (1986a). Unifikation Properties of Idempotent Semigroups. SEKI Technical Report, Universität Kaiserslautern.

Siekmann, J. (1988). Unification Theory. To appear in *J. Symbolic Computation.*

PROOF NORMALIZATION FOR RESOLUTION AND PARAMODULATION

Leo Bachmair
Department of Computer Science
State University of New York at Stony Brook
Stony Brook, New York 11794, U.S.A.

Abstract. We prove the refutation completeness of restricted versions of resolution and paramodulation for first-order predicate logic with equality. Furthermore, we show that these inference rules can be combined with various deletion and simplification rules, such as rewriting, without compromising refutation completeness. The techniques employed in the completeness proofs are based on proof normalization and proof orderings.

1 Introduction

The design of efficient methods for equational reasoning is a major goal in automated theorem proving. Rewriting is one of the most effective techniques that has been suggested for dealing with equations. In this paper, we discuss the application of rewriting techniques to resolution and paramodulation. In particular, we are interested in how various deletion and simplification rules used by theorem provers, such as simplification by rewriting (sometimes called demodulation), relate to the deductive inference rules of resolution and paramodulation. We have found the concept of proof normalization to be useful in this context.

We assume that F is a given set of function symbols, V a set of variables, and G a set of predicate symbols (including a distinguished binary symbol $=$ denoting equality). *Terms* are expressions built from function symbols and variables. Variable-free terms are said to be *ground*. We assume that F contains at least one constant. Thus, the set of all ground terms (also called the *Herbrand base*) is non-empty.

An *atomic formula* (or *atom*) is an expression $p(t_1, \ldots, t_n)$, where p is a predicate symbol and $t_1, \ldots, t_n$ are terms. If $t_1, \ldots, t_n$ are ground, we speak of a *ground atom*. If p is the equality symbol, the atomic formula is called an *equation* and written in infix notation $t_1 = t_2$. The letters A, B, and C are used to denote atomic formulas; Γ and Δ, to denote sets of atomic formulas.

We write $A[s]$ to indicate that the term s occurs as a subexpression in the atom A and (ambiguously) denote by $A[t]$ the result of replacing a particular occurrence of s by t. The notation $u[s]$ is used in a similar way. By $t\sigma$ we denote the result of applying the substitution σ to the term t, and call $t\sigma$ an *instance* of t. If $t\sigma$ is ground, it is said to be a *ground instance* of t. Similarly, we shall speak of instances and ground instances of atoms and sets of atoms. For example, if Γ is $\{p(x,b), p(a,y), q(x,y)\}$, then $\{p(a,b), q(a,b)\}$ is a ground instance of Γ.

A *sequent* is a pair of sets of atomic formulas, written $\Gamma \rightarrow \Delta$. The set Γ is called the *antecedent* of the sequent $\Gamma \rightarrow \Delta$; the set Δ, the *succedent*. By a *ground sequent* we mean a sequent in which all atoms are ground. For simplicity, we usually write Γ_1, Γ_2 instead of $\Gamma_1 \cup \Gamma_2$; Γ, A or A, Γ instead of $\Gamma \cup \{A\}$; and $\Gamma - A$ instead of $\Gamma - \{A\}$. When listing atoms explicitly, we write $A_1, \ldots, A_m \rightarrow$

$B_1, \ldots, B_n$ instead of $\{A_1, \ldots, A_m\} \rightarrow \{B_1, \ldots, B_n\}$. A sequent $A_1, \ldots, A_m \rightarrow B_1, \ldots, B_n$ is meant to represent an implication $A_1 \wedge \cdots \wedge A_m \supset B_1 \vee \cdots \vee B_m$.

The set of all ground atoms is also known as the *Herbrand universe*. By an *interpretation* we mean a subset of the Herbrand universe. An interpretation I is said to *satisfy* a ground sequent $\Gamma \rightarrow \Delta$ if and only if either $\Gamma \not\subseteq I$ or else $\Delta \cap I \neq \emptyset$. An interpretation is said to satisfy a (possibly non-ground) sequent $\Gamma \rightarrow \Delta$ if and only if it satisfies all ground instances $\Gamma\sigma \rightarrow \Delta\sigma$. We say that a set of sequents N is *unsatisfiable* if and only if there is no interpretation I satisfying all sequents of N. For instance, every set containing the *inconsistent sequent* $\rightarrow$ (i.e., the sequent $\Gamma \rightarrow \Delta$ with $\Gamma = \Delta = \emptyset$) is unsatisfiable.

A set of sequents N is said to be *equality unsatisfiable* if $N \cup EQ$ is unsatisfiable, where EQ is the set of all equality axioms

$$\begin{array}{rcl} & \rightarrow & x = x \\ x = y & \rightarrow & y = x \\ x = y,\ y = z & \rightarrow & x = z \\ x = y & \rightarrow & f(\ldots x \ldots) = f(\ldots y \ldots) \\ x = y,\ p(\ldots x \ldots) & \rightarrow & p(\ldots y \ldots) \end{array}$$

with f ranging over all function symbols and p over all predicate symbols.

Various inference systems have been proposed for first-order predicate logic with equality. Most automated theorem provers are based on resolution and paramodulation (plus factoring).

$$\textit{Resolution:} \quad \frac{\Gamma_1 \rightarrow \Delta_1, A \quad B, \Gamma_2 \rightarrow \Delta_2}{\Gamma_1\sigma, \Gamma_2\sigma \rightarrow \Delta_1\sigma, \Delta_2\sigma}$$

where σ is a most general unifier of A and B. The sequent $\Gamma_1\sigma, \Gamma_2\sigma \rightarrow \Delta_1\sigma, \Delta_2\sigma$ is called a *resolvent* of $\Gamma_1 \rightarrow \Delta_1, A$ and $B, \Gamma_2 \rightarrow \Delta_2$. We call $A\sigma$ the *principal formula* of the resolution inference.

$$\textit{Factoring:} \quad \frac{\Gamma_1, A, B \rightarrow \Delta_1}{\Gamma_1, A\sigma \rightarrow \Delta_1} \qquad \frac{\Gamma_1 \rightarrow \Delta_1, A, B}{\Gamma_1 \rightarrow \Delta_1, A\sigma}$$

where σ is a most general unifier of A and B. The sequent $\Gamma_1, A\sigma \rightarrow \Delta_1$ (resp. $\Gamma_1 \rightarrow \Delta_1, A\sigma$) is said to be an *immediate factor* of $\Gamma_1, A, B \rightarrow \Delta_1$ (resp. $\Gamma_1 \rightarrow \Delta_1, A, B$.) A sequent is said to be a *factor* of S if it is either an immediate factor of S or a factor of a factor of S.

$$\textit{Paramodulation:} \quad \frac{\Gamma_1 \rightarrow \Delta_1, s \doteq t \quad A[u], \Gamma_2 \rightarrow \Delta_2}{A[t\sigma], \Gamma_1\sigma, \Gamma_2\sigma \rightarrow \Delta_1\sigma, \Delta_2\sigma} \qquad \frac{\Gamma_1 \rightarrow \Delta_1, s \doteq t \quad \Gamma_2 \rightarrow \Delta_2, A[u]}{\Gamma_1\sigma, \Gamma_2\sigma \rightarrow \Delta_1\sigma, \Delta_2\sigma, A[t\sigma]}$$

where σ is a most general unifier of s and u and $s \doteq t$ ambiguously denotes $s = t$ or $t = s$. The sequent $A[t\sigma], \Gamma_1\sigma, \Gamma_2\sigma \rightarrow \Delta_1\sigma, \Delta_2\sigma$ is said to be a *paramodulant* of $\Gamma_1 \rightarrow \Delta_1, s \doteq t$ and $A[u], \Gamma_2 \rightarrow \Delta_2$. The sequent $\Gamma_1\sigma, \Gamma_2\sigma \rightarrow \Delta_1\sigma, \Delta_2\sigma, A[t\sigma]$ is a paramodulant of $\Gamma_1 \rightarrow \Delta_1, s \doteq t$ and $\Gamma_2 \rightarrow \Delta_2, A[u]$. We call $A\sigma[s\sigma]$ the principal formula and $s \doteq t$ the *side formula* of a paramodulation inference.

We implicitly assume that in each inference rule

$$\frac{S_1 \quad \cdots \quad S_n}{S}$$

the sequents $S_1, \ldots, S_n$, and S do not share any variables. (If necessary, variables are renamed.) The sequents $S_1, \ldots, S_n$ are called the *premises* of the inference rule; the sequent S is called the *conclusion*. Resolution and factoring were introduced by Robinson (1965); paramodulation, by Robinson and Wos (1969).

By a (formal) *proof* we mean a tree, the nodes of which are labelled by sequents in such a way that the sequent labelling a non-leaf node can be obtained from the sequents labelling its two successor nodes by application of a given inference rule (in our case, resolution, factoring, or paramodulation). For simplicity, we usually identify nodes with their labels. We shall mostly be concerned with *ground proofs*, i.e., proofs in which all sequents are ground.

Let P be a proof. We say that P is a *proof of S* (and write $P : S$) to indicate that the root of P is labelled by S. We say that P is a *proof from N* if and only if each leaf of P is labeled by some sequent S of N. We also say that S can be *derived* (or *deduced*) from N if there is a proof P of S from N. By a *refutation* we mean a proof of the inconsistent sequent.

For example,

$$\dfrac{\dfrac{A \to B, C \quad B \to C, D}{A \to C, D} \qquad C \to}{A \to D}$$

is a proof of $A \to D$ from sequents $A \to B, C$ and $B \to C, D$ and $C \to$.

We write $P[Q]$ to indicate that Q is a subtree of P. Resolution, factoring, and paramodulation are sound inference rules:

Theorem 1 (Soundness) *Any resolvent or paramodulant of S_1 and S_2 logically follows from S_1 and S_2. Any factor of S logically follows from S.*

However, not every logical consequence of a set of sequents N can be deduced from N by resolution, factoring, and paramodulation. The inference rules are only refutationally complete. That is, the inconsistent sequent can be derived from any set of sequents N that is equality unsatisfiable.

Let R be the set consisting of $\to x = x$ and the "functional-reflexive axioms" $\to f(x_1, \ldots, x_n) = f(x_1, \ldots, x_n)$, where f ranges over all function symbols.

Theorem 2 *If a set of sequents N is equality unsatisfiable, then there is a refutation (by resolution, factoring, and paramodulation) from $N \cup R$.*

The theorem is due to Robinson and Wos (1969); for a proof see Wos and Robinson (1973).

Brand (1975) and also Peterson (1983) have shown that the set R can be replaced by the singleton $\{\to x = x\}$. Paramodulation can also be combined with demodulation (Wos, et al. 1967), a technique based on using equations in a directed way, as *rewrite rules*, to replace subterms in an expression until a simplest expression is obtained. If the set of equations used for simplification forms a canonical rewrite system, so that every term has a unique normal form, then simplification can be shown to be correct, in the sense that its combination with resolution, factoring, and paramodulation is refutationally complete. Results along this line have been proved by Slagle (1974), but under the assumption that all functional-reflexive axioms be included. Knuth and Bendix (1970) designed a procedure, called completion, that attempts to construct a canonical rewrite system for a given set of equations. Lankford (1975) has shown that (a modified version of) the Knuth-Bendix completion procedure can be combined with resolution and paramodulation to yield a refutationally complete proof method. But his completeness proof only applies to theories in which the equality predicate occurs just in sequents $s = t \to$ or $\to s = t$. Brown (1975) studied rewrite techniques in the context of Horn clauses.

Peterson (1983) proved refutation completeness of a similar inference system in which occurrences of the equality predicate are not limited in any way, but with a weak form of simplification by rewriting only. Usually a well-founded ordering is used to decide in which way an equation is to be used as a directed rewrite rule. The ordering guarantees that the rewriting process terminates.

Peterson's completeness proof is based on assumptions that exclude most general-purpose orderings used for rewriting. (For a survey on orderings for rewriting see Dershowitz 1987). Hsiang and Rusinowitch (1986, 1988) introduced transfinite semantic trees to show that more general orderings can be dealt with. Rusinowitch (1988) further refines the transfinite tree technique. (Zhang and Kapur (1988) also describe a theorem prover that employs very powerful simplification techniques, but do not give any completeness proof.)

In this paper, we formalize simplification by rewriting and general deletion rules and describe corresponding techniques for proving refutation completeness. Our approach is based on proof theoretic techniques—*proof normalization* and *proof orderings*—rather than on (transfinite) semantic trees.

2 Reduction Orderings

By an (strict partial) *ordering* we mean an irreflexive and transitive binary relation. An ordering $\succ$ is *well-founded* if and only if there is no endless sequence $E_1 \succ E_2 \succ E_3 \succ \cdots$. An ordering $\succ$ on terms and atoms is said to be *invariant under context application* if and only if $s \succ t$ implies $u[s] \succ u[t]$ and $A[s] \succ A[t]$, for all terms s, t, and u and all atoms $A[s]$. It is *invariant under instantiation* if $s \succ t$ implies $s\sigma \succ t\sigma$ and $A \succ B$ implies $A\sigma \succ B\sigma$, for all terms s and t, atoms A and B, and substitutions σ. An ordering satisfying both properties is simply called *invariant*.

An invariant, well-founded ordering is called a *reduction ordering*. We say that an ordering $\succ$ is *complete* if it is total on ground expressions (i.e., whenever two ground expressions E and E' are distinct, then either $E \succ E'$ or $E' \succ E$). We will consider reduction orderings that can be extended to complete reduction orderings. (All general-purpose orderings used in rewriting can be extended to complete orderings.) The following additional conditions are imposed on the orderings $\succ$ we use:

(i) $A[s] \succ (s = t)$, whenever $s \succ t$ and $A[s]$ is not an equation;

(ii) $(u[s] = v) \succ (s = t)$, whenever $s \succ t$ and $s \neq u[s]$;

(iii) $(s = t) \succ (s = u)$, whenever $t \succ u$.

These conditions guarantee that the side formula is simpler than the principal formula in all paramodulation steps in which we are interested. Complete orderings that satisfy these conditions include, for instance, the *predicate-first orderings* described by Hsiang and Rusinowitch (1988). We assume that the reader is familiar with *multiset orderings* (Dershowitz and Manna 1979) and *recursive path orderings* (Dershowitz 1982).

Reduction orderings can be used to define restricted versions of resolution and paramodulation, for which the principal formula is maximal. (An atomic formula A is said to be *maximal* with respect to an ordeering $\succ$ in a set of atoms Γ if and only if there is no atom B in Γ with $B \succ A$. An atom is said to be maximal in a sequent if and only if it is maximal in both antecedent and succedent.) We shall also use reduction orderings to prove the refutation completeness of these restricted inference rules.

3 Proof Normalization

In this section we consider ground proofs only. (Note that factoring is not needed on the ground level.) Let P be a (ground) refutation. Since the root of P is labelled by the inconsistent sequent, every atom that occurs in any sequent in P is eliminated by serving as the principal or side formula of some inference step. This suggests that atoms be eliminated in some systematic order. For instance, one may specify an ordering on atoms and apply only inferences in which the principal formula is maximal. A variety of orderings have been proposed for that purpose (e.g., Joyner 1976, Peterson 1983, Hsiang and Rusinowitch 1986).

Let $>$ be a complete reduction ordering. A *resolution* inference is said to be in *normal form* if and only if its principal formula is maximal in both premises. For example, if $D > C > B > A$, then

$$\frac{A \to B, C \quad B \to C, D}{A \to C, D}$$

is not in normal form as the principal formula B is not maximal in $A \to B, C$. On the other hand, the inference

$$\frac{A \to B, C \quad C \to}{A \to B}$$

is in normal form.

A *paramodulation* inference, with principal formula $A[s]$ and side formula $s = t$, is said to be in *normal form* if and only if (i) the principal formula is maximal in both premises and (ii) $s > t$. For example, given that $B > A > s > t$, the inference

$$\frac{A \to s = t \quad \to B[s]}{A \to B[t]}$$

is in normal form, while

$$\frac{A \to s = t \quad \to B[t]}{A \to B[s]}$$

is not.

A *proof* is said to be in *normal form* (with respect $>$) if and only if all its inference steps are in normal form.

We shall prove that whenever a refutation P from N is not in normal form, then there is another refutation P' from N that is simpler than P with respect to some well-founded ordering $>_{\mathcal{P}}$. To define the ordering $>_{\mathcal{P}}$, we first introduce new symbols $\langle A \rangle_1$, $\langle A \rangle_0$, and $\langle S \rangle$, for all (ground) atoms A and (ground) sequents S. The *complexity* $\langle P \rangle$ of a proof P is defined as follows.

(a) If P is a proof

$$\frac{P_1 : \Gamma_1 \to \Delta_1, A \quad P_2 : A, \Gamma_2 \to \Delta_2}{\Gamma \to \Delta}$$

then $\langle P \rangle = \langle A \rangle_1(\langle P_1 \rangle, \langle P_2 \rangle)$.

(b) If P is a proof

$$\frac{P_1 : \Gamma_1 \to \Delta_1, s \doteq t \quad P_2 : A[s], \Gamma_2 \to \Delta_2}{A[t], \Gamma \to \Delta}$$

or

$$\frac{P_1 : \Gamma_1 \to \Delta_1, s \doteq t \quad P_2 : \Gamma_2 \to \Delta_2, A[s]}{\Gamma \to \Delta, A[t]}$$

then $\langle P \rangle = \langle A[s] \rangle_0(\langle P_1 \rangle, \langle P_2 \rangle)$.

(c) If P is a proof consisting of a single sequent S, then $\langle P \rangle = \langle S \rangle$.

Intuitively, $\langle A \rangle_1$ is the "weight" of a resolution inference with principal formula A; $\langle A \rangle_0$ the "weight" of a paramodulation inference with principal formula A; and $\langle S \rangle$ the "weight" of a leaf labelled with sequent S. Inference steps are compared relative to this "weight".

Let δ be a mapping from sequents to some set D and $>^d$ be a well-founded ordering on D. We identify $\langle A \rangle_0$ with the triple $(\bot, A, \bot)$; $\langle A \rangle_1$ with the triple $(\bot, A, \top)$; $\langle \to t = t \rangle$ with the triple

$(\perp, \perp, t = t)$; and $\langle S \rangle$ with the triple $(\delta(S), \perp, \perp)$, if S is not of the form $\rightarrow t = t$. Such triples are compared lexicographically, using the ordering $>^d$ in the first component and the reduction ordering $>$ in the second and third component. (It is understood that in either ordering $\top$ is the largest element and $\perp$ the smallest.) The resulting well-founded ordering is denoted by $>^c$. Using the corresponding recursive path ordering $>^c_{rpo}$, we define: $P >_{\mathcal{P}} Q$ *if and only if* $\langle P \rangle >^c_{rpo} \langle Q \rangle$.

The mapping δ and corresponding ordering $>^d$ are irrelevant for proof normalization, but are essential in formalizing deletion rules. They will be discussed in detail in a later section. Note that $Q >_{\mathcal{P}} Q'$ implies $P[Q] >_{\mathcal{P}} P[Q']$, for all proofs $P[Q]$, Q, and Q'.

The following properties of the ordering $>^c$ are essential for proof normalization:

(i) $\langle A \rangle_1 >^c \langle A \rangle_0 >^c \langle \rightarrow t = t \rangle$ and

(ii) $\langle A \rangle_i >^c \langle B \rangle_j$, whenever $A > B$ and $0 \leq i, j \leq 1$.

The following lemma will be used repeatedly.

Lemma 1 *Let $P : \Gamma, \Gamma' \rightarrow \Delta, \Delta'$ and $P' : \Gamma \rightarrow \Delta$ be ground proofs with $P >_{\mathcal{P}} P'$. Then for every ground proof $Q[P]$ there is a ground proof Q', such that $Q[P] >_{\mathcal{P}} Q'$.*

To illustrate the kind of proof transformations we consider, we show that any resolution inference in which the principal formula is an equation $s = t$, where s and t are distinct, can be eliminated from (ground) refutations.

Let P be a proof

$$\frac{P_1 : \Gamma_1 \rightarrow \Delta_1, s = t \quad P_2 : s = t, \Gamma_2 \rightarrow \Delta_2}{\Gamma_1, \Gamma_2 \rightarrow \Delta_1, \Delta_2}$$

where $s > t$, $\langle P_1 \rangle = X_1$, and $\langle P_2 \rangle = X_2$. The complexity of P is $\langle s = t \rangle_1(X_1, X_2)$. The proof P',

$$\frac{\rightarrow t = t \qquad \dfrac{P_1 : \Gamma_1 \rightarrow \Delta_1, s = t \quad P_2 : s = t, \Gamma_2 \rightarrow \Delta_2}{t = t, \Gamma_1 - t = t, \Gamma_2 - t = t \rightarrow \Delta_1, \Delta_2}}{\Gamma_1 - t = t, \Gamma_2 - t = t \rightarrow \Delta_1, \Delta_2}$$

has complexity $\langle t = t \rangle_1(\langle \rightarrow t = t \rangle, \langle s = t \rangle_0(X_1, X_2))$. Since $(s = t) > (t = t)$, we have $\langle s = t \rangle_1 >^c \langle t = t \rangle_1$. Moreover, $\langle s = t \rangle_1 >^c \langle s = t \rangle_0$ and $\langle s = t \rangle_1 >^c \langle \rightarrow t = t \rangle$. Thus we may infer that $\langle P \rangle >^c_{rpo} \langle P' \rangle$ and therefore $P >_{\mathcal{P}} P'$. By Lemma 1, any refutation $Q[P]$ containing P as a subproof can be transformed to a simpler refutation.

Theorem 3 (Proof Normalization) *Let $>$ be a complete reduction ordering. For every ground refutation Q from N, there exists a ground refutation Q' from N, such that $Q \geq_{\mathcal{P}} Q'$ and Q' is in normal form with respect to $>$.*

Proof. Since the ordering $>_{\mathcal{P}}$ is well-founded, it suffices show that whenever a ground refutation Q is not in normal form, then there is a ground refutation Q' with $Q >_{\mathcal{P}} Q'$.

If a refutation Q is not in normal form, it must contain a subproof of the form

$$\frac{\dfrac{P_1 : S_1 \quad P_2 : S_2}{S''} \qquad P_3 : S_3}{S}$$

in which the second inference (with conclusion S) is in normal form, but the first inference (with conclusion S'') is not. (A refutation that is not in normal form must contain at least one such subproof, as any inference with the inconsistent sequent as conclusion is in normal form.) We shall prove that for each such proof $P : S$, where S is a sequent $\Gamma, \Gamma' \rightarrow \Delta, \Delta'$, there is a proof $P' : S'$, such that $P >_{\mathcal{P}} P'$ and S' is $\Gamma' \to \Delta'$. Using Lemma 1, we can then infer that there is a proof Q' with $Q >_{\mathcal{P}} Q'$.

Since each of the two inference steps in P may be by resolution or paramodulation, we distinguish four different cases of proof transformations. We describe representative proof transformations for all four cases. For simplicity, we shall write S_1 instead of $P_1 : S_1$ and assume that $\langle P_1 \rangle = X_1$, etc.

Resolution and resolution. Let P be a proof

$$\dfrac{\dfrac{\Gamma_1 \to \Delta_1, A \quad A, \Gamma_2 \to \Delta_2, B}{\Gamma_1, \Gamma_2 \to \Delta_1 - B, \Delta_2, B} \qquad B, \Gamma_3 \to \Delta_3}{\Gamma_1, \Gamma_2, \Gamma_3 \to \Delta_1 - B, \Delta_2, \Delta_3}$$

where A occurs neither in Δ_1 nor in Γ_2, B occurs neither in Δ_2 nor in Γ_3, and $B > A$. The complexity of this proof is $\langle B \rangle_1(\langle A \rangle_1(X_1, X_2), X_3)$.

If B is not in Δ_1 (and therefore $\Delta_1 - B = \Delta_1$), then the two inference steps can be "commuted". More precisely, let P' be the proof

$$\dfrac{\Gamma_1 \to \Delta_1 - B, A \qquad \dfrac{A, \Gamma_2 \to \Delta_2, B \quad B, \Gamma_3 \to \Delta_3}{A, \Gamma_2, \Gamma_3 - A \to \Delta_2, \Delta_3}}{\Gamma_1, \Gamma_2, \Gamma_3 - A \to \Delta_1 - B, \Delta_2, \Delta_3}$$

with complexity $\langle A \rangle_1(X_1, \langle B \rangle_1(X_2, X_3))$. Since $B > A$ implies $\langle B \rangle_1 >^c \langle A \rangle_1$, we may infer that the proof P' is simpler than the original proof P.

If the atom B is in Δ_1, it has to be eliminated by an additional resolution step. Let P' be the proof

$$\dfrac{\dfrac{\Gamma_1 \to \Delta_1 - B, A, B \quad B, \Gamma_3 \to \Delta_3}{\Gamma_1, \Gamma_3 \to \Delta_1 - B, \Delta_3 - A, A} \qquad \dfrac{A, \Gamma_2 \to \Delta_2, B \quad B, \Gamma_3 \to \Delta_3}{A, \Gamma_2, \Gamma_3 - A \to \Delta_2, \Delta_3}}{\Gamma_1, \Gamma_2, \Gamma_3 \to \Delta_1 - B, \Delta_2, \Delta_3}$$

with complexity $\langle A \rangle_1(\langle B \rangle_1(X_1, X_3), \langle B \rangle_1(X_2, X_3))$. Again, the proof P' is simpler than P.

Resolution and paramodulation. Let P be the proof

$$\dfrac{\dfrac{\Gamma_1 \to \Delta_1, A \quad A, B[s], \Gamma_2 \to \Delta_2}{\Gamma_1 - B[s], \Gamma_2, B[s] \to \Delta_1, \Delta_2} \qquad \Gamma_3 \to \Delta_3, s = t}{\Gamma_1 - B[s], \Gamma_2, \Gamma_3, B[t] \to \Delta_1, \Delta_2, \Delta_3}$$

where A is not in Δ_1 or Γ_2, $B[s]$ is not in Γ_2, $s = t$ is not in Δ_3, $s > t$, and $B[s] > A$. The complexity of this proof is $\langle B[s] \rangle_0(\langle A \rangle_1(X_1, X_2), X_3)$.

If $B[s]$ is not in Γ_1, let P' be the proof

$$\dfrac{\Gamma_1 - B[s] \to \Delta_1, A \qquad \dfrac{A, B[s], \Gamma_2 \to \Delta_2 \quad \Gamma_3 \to \Delta_3, s = t}{A, \Gamma_2, \Gamma_3 - A, B[t] \to \Delta_2, \Delta_3}}{\Gamma_1 - B[s], \Gamma_2, \Gamma_3 - A, B[t] \to \Delta_1, \Delta_2, \Delta_3}$$

the complexity of which is $\langle A \rangle_1(X_1, \langle B[s] \rangle_0(X_2, X_3))$.

If B is in Γ_1, let P' be the proof

$$\dfrac{\dfrac{\Gamma_3 \to \Delta_3, s = t \quad B[s], \Gamma_1 - B[s] \to \Delta_1, A}{\Gamma_1 - B[s], \Gamma_3, B[t] \to \Delta_1, \Delta_3} \qquad \dfrac{A, B[s], \Gamma_2 \to \Delta_2 \quad \Gamma_3 \to \Delta_3, s = t}{A, \Gamma_2, \Gamma_3 - A, B[t] \to \Delta_2, \Delta_3}}{\Gamma_1 - B[s], \Gamma_2, \Gamma_3, B[t] \to \Delta_1, \Delta_2, \Delta_3}$$

the complexity of which is $\langle A \rangle_1(\langle B[s] \rangle_0(X_1, X_2), \langle B[s] \rangle_0(X_1, X_3))$.

Since $B[s] > A$ implies $\langle B[s]\rangle_0 >^c \langle A\rangle_1$, the proof P' is simpler than P in either case.

Paramodulation and resolution. Suppose that P is a proof

$$\frac{\dfrac{B,\Gamma_1 \to \Delta_1, s=t \quad A[s],\Gamma_2 \to \Delta_2}{B,\Gamma_1,\Gamma_2 - B, A[t] \to \Delta_1,\Delta_2} \qquad \Gamma_3 \to \Delta_3, B}{\Gamma_1,\Gamma_2 - B,\Gamma_3, A[t] \to \Delta_1,\Delta_2,\Delta_3}$$

where B is not in Γ_1 or Δ_3, $s = t$ is not in Δ_1, $A[s]$ is not in Γ_2, $B > A[s]$, and $B > A[t]$. The complexity of this proof is $\langle B\rangle_1(\langle A[s]\rangle_0(X_1, X_2), X_3)$.

If B is in Γ_2, let P' be the proof

$$\frac{\dfrac{B,\Gamma_1 \to \Delta_1, s=t \quad \Gamma_3 \to \Delta_3, B}{\Gamma_1,\Gamma_3 \to \Delta_1,\Delta_3, s=t} \qquad \dfrac{\Gamma_3 \to \Delta_3, B \quad B, A[s],\Gamma_2 - B \to \Delta_2}{A[s],\Gamma_2 - B,\Gamma_3 - A[s] \to \Delta_2,\Delta_3}}{\Gamma_1,\Gamma_2 - B,\Gamma_3, A[t] \to \Delta_1,\Delta_2,\Delta_3}$$

the complexity of which is $\langle A[s]\rangle_0(\langle B\rangle_1(X_1, X_3), \langle B\rangle_1(X_3, X_2))$.

Since $B > A[s]$ implies $\langle B\rangle_1 > \langle A[s]\rangle_0$, the proof P' is less complex than P. The necessary transformation is even simpler if A is not in Δ_1.

Paramodulation and paramodulation. First suppose that P is a proof

$$\frac{\dfrac{\Gamma_1 \to \Delta_1, s=t \quad A[s],\Gamma_2 \to \Delta_2, B[u]}{\Gamma_1,\Gamma_2, A[t] \to \Delta_1 - B[u],\Delta_2, B[u]} \qquad \Gamma_3 \to \Delta_3, u=v}{\Gamma_1,\Gamma_2,\Gamma_3, A[t] \to \Delta_1 - B[u],\Delta_2,\Delta_3, B[v]}$$

where $B[u]$ is not in Δ_2, $s = t$ is not in Δ_1, $A[s]$ is not in Γ_2, $u = v$ is not in Δ_3, $u > v$, $B[u] > A[s]$, and $B[u] > A[t]$. The complexity of this proof is $\langle B[u]\rangle_0(\langle A[s]\rangle_0(X_1, X_2), X_3)$.

If $B[u]$ is not in Δ_1, let P' be the proof

$$\frac{\Gamma_1 \to \Delta_1 - B[u], s=t \qquad \dfrac{A[s],\Gamma_2 \to \Delta_2, B[u] \quad \Gamma_3 \to \Delta_3, u=v}{A[s],\Gamma_2,\Gamma_3 - A[s] \to \Delta_2,\Delta_3, B[v]}}{\Gamma_1,\Gamma_2,\Gamma_3 - A[s], A[t] \to \Delta_1 - B[u],\Delta_2,\Delta_3, B[v]}$$

the complexity of which is $\langle A[s]\rangle_0(X_1, \langle B[u]\rangle_0(X_2, X_3))$.

If $B[u]$ is in Δ_1, the transformation is sligthly more complicated. In either case, the proof P' is less complex than P.

Next, let P be the proof

$$\frac{\dfrac{\Gamma_1 \to \Delta_1, s=t \quad \Gamma_2 \to \Delta_2, A[s,u]}{\Gamma_1,\Gamma_2 \to \Delta_1 - A[t,u],\Delta_2 - A[t,u], A[t,u]} \qquad u=v,\Gamma_3 \to \Delta_3}{\Gamma_1,\Gamma_2,\Gamma_3 \to \Delta_1 - A[t,u],\Delta_2 - A[t,u],\Delta_3, A[t,v]}$$

where $s = t$ is not in Δ_1, $A[s,u]$ is not in Δ_2, $u = v$ is not in Γ_3, $t > s$, and $u > v$. The complexity of this proof is $\langle A[t,u]\rangle_0(\langle A[s,u]\rangle_0(X_1, X_2), X_3)$.

We have to distinguish various cases, depending on whether or not $A[t,u]$ is in Δ_1 and/or Δ_2. The most complicated transformation, shown below, is necessary if $A[t,u]$ is both in Δ_1 and Δ_2.

Let P_1 be the proof

$$\frac{\Gamma_1 \to \Delta_1 - A[t,u], s=t, A[t,u] \quad u=v,\Gamma_3 \to \Delta_3}{\Gamma_1,\Gamma_3 \to \Delta_1 - A[t,u],\Delta_3, s=t, A[t,v]}$$

and P_2 be

$$\frac{\Gamma_2 \to \Delta_2 - A[t], A[s,u], A[t,u] \quad u=v,\Gamma_3 \to \Delta_3}{\Gamma_2,\Gamma_3 \to \Delta_2 - A[t],\Delta_3, A[s,u], A[t,v]}$$

The complexity of these proofs is $\langle A[t,u]\rangle_0(X_1, X_3)$ and $\langle A[t,u]\rangle_0(X_2, X_3)$, respectively.

Let P' be

$$\dfrac{P_1 \quad \dfrac{P_2 \quad u = v, \Gamma_3 \to \Delta_3}{\Gamma_2, \Gamma_3 \to \Delta_2 - A[t,u], \Delta_3, A[s,v], A[t,v]}}{\Gamma_1, \Gamma_2, \Gamma_3 \to \Delta_1 - A[t,u], \Delta_2 - A[t,u], \Delta_3, A[t,v]}$$

The complexity of this proof is $\langle A[s,v]\rangle_0(\langle A[t,u]\rangle_0(X_2, X_3), \langle A[t]\rangle_0(\langle A[t]\rangle_1(X_1, X_3), X_3))$. Hence P' is simpler than P.

In the preceding case we assumed that both paramodulation steps applied to the same atom A, but did not "overlap". The following transformation covers an "overlap" cases. The other cases are similar.

Let P be the proof

$$\dfrac{\dfrac{\Gamma_1 \to \Delta_1, s = t[u] \quad \Gamma_2 \to \Delta_2, A[s]}{\Gamma_1, \Gamma_2 \to \Delta_1 - A[t[u]], \Delta_2 - A[t[u]], A[t[u]]} \quad u = v, \Gamma_3 \to \Delta_3}{\Gamma_1, \Gamma_2, \Gamma_3 \to \Delta_1 - A[t[u]], \Delta_2 - A[t[u]], \Delta_3, A[t[v]]}$$

where $s = t[u]$ is not in Δ_1, $A[s]$ is not in Δ_2, $u = v$ is not in Γ_3, $t > s$, and $u > v$. The complexity of this proof is $\langle A[t,u]\rangle_0(\langle A[s,u]\rangle_0(X_1, X_2), X_3)$.

Let us assume, this time, that $A[t[u]]$ is in Δ_1 but not in Δ_2. Let P_1 be the proof

$$\dfrac{\Gamma_1 \to \Delta_1 - A[t[u]], s = t[u], A[t[u]] \quad u = v, \Gamma_3 \to \Delta_3}{\Gamma_1, \Gamma_3 \to \Delta_1 - A[t[u]], \Delta_3, s = t[u], A[t[v]]}$$

the complexity of which is $\langle A[t[u]]\rangle_0(X_1, X_3)$.

Let P' be

$$\dfrac{\dfrac{P_1 \quad u = v, \Gamma_3 \to \Delta_3}{\Gamma_1, \Gamma_3 \to \Delta_1 - A[t[u]], \Delta_3, s = t[v], A[t[v]]} \quad \Gamma_2 \to \Delta_2, A[s]}{\Gamma_1, \Gamma_2, \Gamma_3 \to \Delta_1 - A[t,u], \Delta_2 - A[t,u], \Delta_3, A[t[v]]}$$

The complexity of this proof is $\langle A[s[v]]\rangle_0(\langle A[s = t[u]]\rangle_0(\langle A[t[u]]\rangle_0(X_1, X_3), X_2))$. We may again infer that P' is simpler than P. $\square$

4 First-Order Inference Rules

The normalization techniques outlined in the preceding section apply to *ground* proofs. Ground inferences in normal form can be regarded as ground instances of restricted versions of resolution and paramodulation.

Let $\succ$ be a reduction ordering that can be extended to a complete ordering.

Ordered resolution: $$\dfrac{\Gamma_1 \to \Delta_1, A \quad B, \Gamma_2 \to \Delta_2}{\Gamma_1\sigma, \Gamma_2\sigma \to \Delta_1\sigma, \Delta_2\sigma}$$

where σ is a most general unifier of A and B and $A\sigma$ is maximal with respect to $\succ$ in $\Gamma_1\sigma$, $\Gamma_2\sigma$, $\Delta_1\sigma$, and $\Delta_2\sigma$.

Oriented paramodulation:

$$\dfrac{\Gamma_1 \to \Delta_1, s \doteq t \quad A[u], \Gamma_2 \to \Delta_2}{A[t\sigma], \Gamma_1\sigma, \Gamma_2\sigma \to \Delta_1\sigma, \Delta_2\sigma}$$

$$\dfrac{\Gamma_1 \to \Delta_1, s \doteq t \quad \Gamma_2 \to \Delta_2, A[u]}{\Gamma_1\sigma, \Gamma_2\sigma \to \Delta_1\sigma, \Delta_2\sigma, A[t\sigma]}$$

where σ is a most general unifier of s and u, $A[s\sigma]$ is maximal with respect to $\succ$ in $\Gamma_1\sigma$, $\Gamma_2\sigma$, $\Delta_1\sigma$, and $\Delta_2\sigma$, and $t\sigma \not\succeq s\sigma$.

Ordered resolution and different versions of oriented paramodulation have been described by Hsiang and Rusinowitch (1986, 1988). Let us also remark that factoring can be restricted in a similar way to "ordered factoring".

Lemma 2 *Let $\succ$ be a reduction ordering that can be extended to a complete ordering $>$ and let P be a ground proof*

$$\frac{\Gamma_1\sigma_1 \rightarrow \Delta_1\sigma_1 \quad \Gamma_2\sigma_2 \rightarrow \Delta_2\sigma_2}{\Gamma \rightarrow \Delta}$$

that is in normal form with respect to $>$. Then there is a proof P'

$$\frac{\Gamma'_1 \rightarrow \Delta'_1 \quad \Gamma'_2 \rightarrow \Delta'_2}{\Gamma' \rightarrow \Delta'}$$

by ordered resolution and oriented paramodulation, such that $\Gamma'_1 \rightarrow \Delta'_1$ and $\Gamma'_2 \rightarrow \Delta'_2$ are factors of $\Gamma_1 \rightarrow \Delta_1$ and $\Gamma_2 \rightarrow \Delta_2$, respectively, and $\Gamma \rightarrow \Delta$ is a ground instance of $\Gamma' \rightarrow \Delta'$.

This lemma corresponds to standard "lifting" lemmata (e.g., Chang and Lee 1973) and indicates the correspondence between ordered resolution, oriented paramodulation, and factoring and ground normal form proofs. (Observe that an atomic formula A has to be maximal in a sequent S, whenever some ground instance $A\sigma$ is maximal in $S\sigma$.) We emphasize that the lemma is only valid if *paramodulation into variables* is allowed.

For example, given sequents $\rightarrow s = t$ and $\rightarrow p(x,x)$ where $s > t$, the ground proof

$$\frac{\rightarrow s = t \quad \rightarrow p(s,s)}{\rightarrow p(s,t)}$$

is in normal form. The corresponding oriented paramodulation step

$$\frac{\rightarrow s = t \quad \rightarrow p(x,x)}{\rightarrow p(s,t)}$$

requires that the variable x be instantiated by s, so that one occurrence of s can be replaced by t.

Paramodulation into variables would be difficult to control in an automated theorem prover. We shall use proof transformation techniques to prove that it is unnecessary (as are the functional-reflexive axioms). Before we outline the proof we discuss deletion and simplification rules.

5 Deletion and Simplification

A theorem prover deduces new sequents by applying given inference rules to a given set of sequents N. We write $N \vdash N'$ to indicate that $N = \{S_1, \ldots, S_k\}$ and $N' = \{S_1, \ldots, S_k, S_{k+1}\}$, where S_{k+1} is obtained from sequents in N by application of a deductive inference. The deductive inference rules we consider are ordered resolution, factoring, and oriented paramodulation, all assumed to be defined in terms of a reduction ordering $\succ$ that can be extended to a complete ordering $>$.

Some sequents may not be needed in constructing a refutation; hence can be deleted. We shall also write $N \vdash N'$ to indicate that such a deletion has taken place, that is, $N = \{S_1, \ldots, S_k\}$ and $N' = \{S_1, \ldots, S_{k-1}\}$.

Proper subsumption is an example of such a deletion rule. A sequent $\Gamma \rightarrow \Delta$ is said to *subsume* a sequent $\Gamma' \rightarrow \Delta'$ if and only if $|\Gamma| + |\Delta| \leq |\Gamma'| + |\Delta'|$, $\Gamma' = \Gamma\sigma, \Gamma''$, and $\Delta' = \Delta\sigma, \Delta''$, for some substitution σ and sets Γ'' and Δ''. We say that S *properly subsumes* S' (and write $S' \rhd S$) if and only if S subsumes S' but not vice versa.

Subsumption: $N \cup \{S, S'\} \vdash N \cup \{S\}$

if S properly subsumes S'.

Using orderings on proofs, we can formulate a general deletion rule:

Deletion: $N \cup \{S\} \vdash N$

if for every ground sequent $S\sigma$ there exist a set N' of ground instances of sequents in N and a proof $P : S'$ from N', such that $S\sigma >_{\mathcal{P}} P$ and $S\sigma$ can be written as $\Gamma, \Gamma' \rightarrow \Delta, \Delta'$ with S' being $\Gamma' \rightarrow \Delta'$.

We shall prove that subsumption is a deletion rule in this sense. Simplification by rewriting (also called demodulation or reduction) can be viewed as paramodulation combined, in a certain way, with deletion.

Reduction:

$$N \cup \{\rightarrow s = t\} \cup \{\Gamma, A[s\sigma] \rightarrow \Delta\} \vdash N \cup \{\rightarrow s = t\} \cup \{\Gamma, A[t\sigma] \rightarrow \Delta\}$$

$$N \cup \{\rightarrow s = t\} \cup \{\Gamma \rightarrow A[s\sigma], \Delta\} \vdash N \cup \{\rightarrow s = t\} \cup \{\Gamma \rightarrow A[t\sigma], \Delta\}$$

if $s\sigma \succ t\sigma$ and $s\sigma = t\sigma$ is not maximal in $\Gamma, A[s\sigma] \rightarrow \Delta$.

Reduction consists of a paramodulation step followed by a deletion step:

$$\begin{aligned} & N \cup \{\rightarrow s = t\} \cup \{\Gamma, A[s\sigma] \rightarrow \Delta\} \\ & \quad \vdash \ N \cup \{\rightarrow s = t\} \cup \{\Gamma, A[s\sigma] \rightarrow \Delta\} \cup \{\Gamma, A[t\sigma] \rightarrow \Delta\} \\ & \quad \vdash \ N \cup \{\rightarrow s = t\} \cup \{\Gamma, A[t\sigma] \rightarrow \Delta\}. \end{aligned}$$

Deletion depends on the mapping δ and the corresponding ordering $>^d$ in terms of which the ordering $>_{\mathcal{P}}$ is defined. We shall specify a specific mapping and ordering so that both subsumption and the deletion part of reduction are deletion rules in the above sense. For that purpose, we will have to prove, among other things, the existence of ground proofs P, such that $S\sigma >_{\mathcal{P}} P$, where $S\sigma$ and all leaves of P are ground instances of given sequents. We shall therefore assume that with each leaf S of a ground proof P a sequent S' is associated, such that S is a ground instance of S'.

Let S' be the sequent $\Gamma \rightarrow \Delta$ and S be a ground instance $S'\sigma$. We define $\delta(S)$ to be the triple $(\Gamma\sigma \uplus \Delta\sigma, |\Gamma| + |\Delta|, S)$, where $X \uplus Y$ denotes multiset union. (For example, $\{A, B\} \uplus \{B, C\} = \{A, B, B, C\}$.) The ordering $>^d$ used to compare such triples is defined to be the lexicographic combination of the multiset extension of the complete reduction ordering $>$, the usual greater-than ordering on the natural numbers, and the proper subsumption ordering $\rhd$.

Now consider a subsumption step $N \cup \{S, S'\} \vdash N \cup \{S\}$, where S properly subsumes S'. Let us assume that S is $\Gamma \rightarrow \Delta$ and S' is $\Gamma\sigma, \Gamma' \rightarrow \Delta\sigma, \Delta'$. Let $S'\tau$ be a ground instance of S', M' be the multiset of all sequents in the antecedent and succedent of $S'\tau$, and P be the proof $S'\tau$. Then $\langle P \rangle = \langle S'\tau \rangle = ((M', k, S'), \bot, \bot)$, where $k = |\Gamma\sigma| + |\Gamma'| + |\Delta\sigma| + |\Delta'|$. Moreover, let P' the proof consisting of a single sequent $S\sigma\tau$, where $\sigma\tau$ denotes the composition of the two substitutions σ and τ, and let M the multiset of all sequents in the antecedent and succedent of $S\sigma\tau$. Then $\langle P' \rangle = \langle S\sigma\tau \rangle = ((M, l, S), \bot, \bot)$, where $l = |\Gamma| + |\Delta|$. Since M is a sub-multiset of M', $k \geq l$, and S properly subsumes S', we may conclude that $P >_{\mathcal{P}} P'$, which shows that subsumption is a deletion rule.

Next let us consider the deletion part of reduction. Let S be the sequent $A[s\sigma], \Gamma \rightarrow \Delta$; S' the sequent $A[t\sigma], \Gamma \rightarrow \Delta$; and S'' the sequent $\rightarrow s = t$. We assume that $s\sigma \succ t\sigma$ and that $s\sigma = t\sigma$ is not maximal in S. If P is a ground proof consisting of one node $S\tau$, define P' to be the ground proof

$$\frac{\rightarrow s\sigma\tau = t\sigma\tau \qquad A\tau[t\sigma\tau], \Gamma\tau \rightarrow \Delta\tau}{A\tau[s\sigma\tau], \Gamma\tau \rightarrow \Delta\tau}$$

Then $\langle P\rangle = \langle S\tau\rangle = ((A\tau[(s\sigma)\tau] \uplus \Gamma\tau \uplus \Delta\tau, k, S), \bot, \bot)$ and $\langle P'\rangle = \langle A\tau[(t\sigma)\tau]\rangle_0(\langle\rightarrow s\sigma\tau = t\sigma\tau\rangle, \langle S'\tau\rangle)$. By definition of $>^c$, we have $\langle S\tau\rangle >^c \langle A\tau[(t\sigma)\tau]\rangle_0$. Since $s\sigma = t\sigma$ is not maximal in S, we also have $\langle S\tau\rangle >^c \langle\rightarrow s\sigma\tau = t\sigma\tau\rangle$. Finally, since S' is obtained from S by replacing $A[s\sigma]$ by $A[t\sigma]$ and $s\sigma > t\sigma$, we have $\langle S\tau\rangle >^c \langle S'\tau\rangle$. In sum, $P >_{\mathcal{P}} P'$, which proves that the deletion part of reduction is an instance of the general deletion rule.

Rusinowitch (1988) has also studied subsumption and reduction in the context of resolution and paramodulation. More powerful reduction rules, but for purely equational theories only, have been formulated by Bachmair, Dershowitz, and Hsiang (1986) and Bachmair (1987). Ganzinger (1988) has introduced deletion and reduction rules for conditional rewriting. His techniques for simplification by conditional rewriting can be adapted to the above inference rules in the form of a "conditional reduction" rule that also combines deductive inferences with deletion.

6 Refutation Completeness

Let $N_0 \vdash N_1 \vdash \cdots$ be a sequence of sets of sequents in which every set N_{i+1} is obtained from N_i by a deductive inference (resolution, factoring, or paramodulation) or by deletion. We say that such a sequence is *fair* if and only if all sequents that can be derived (in one step) from $\bigcup_i \bigcap_{j\geq i} N_j$ by ordered resolution, factoring, or oriented paramodulation are contained in $\bigcup_i N_i$.

In other words, fairness simply requires that deductive inference rules are applied to all "persisting" sequents. Evidently, fair sequences do exist.

Theorem 4 (Herbrand) *A finite set of sequents N is unsatisfiable if and only if some finite set N' of ground instances of sequents in N is unsatisfiable.*

Theorem 5 (Refutation completeness) *Let ordered resolution, oriented paramodulation, ordered factoring, and deletion be defined with respect to a reduction ordering $\succ$ that can be extended to a complete ordering. If $N_0 \vdash N_1 \vdash \cdots$ is a fair sequence, where N_0 is equality unsatisfiable and contains the sequent $\rightarrow x = x$, then some set N_k contains the inconsistent sequent.*

Sketch of proof. Let $>$ be a complete reduction ordering containing $\succ$. We shall prove that for any fair sequence $N_0 \vdash N_1 \vdash \cdots$ there is a corresponding sequence of ground refutations $P_0, P_1, \ldots$ such that

(i) P_i is a proof from N_i', where N_i' is a set of ground instances of N_i,

(ii) $P_i \geq_{\mathcal{P}} P_{i+1}$, and

(iii) if P_i consists of more than one node, then $P_i >_{\mathcal{P}} P_{i+j}$, for some j.

Since the ordering $>_{\mathcal{P}}$ is well-founded, properties (i), (ii), and (iii) imply that there is some refutation P_k consisting of a single node. Consequently, the set N_k must contain the inconsistent sequent.

Since N_0 is equality unsatisfiable there is a suitable proof P_0, by Herbrand's Theorem. Given any refutation P_i from N_i', we first construct a refutation Q_{i+1}.

If N_{i+1} is obtained from N_i by deleting some sequent S, then Q_{i+1} is obtained from P_i by replacing each leaf labelled with an an instance $S\sigma$ of S by a suitable proof Q of $S\sigma$. Deletion has been defined in such a way that there is a proof Q simpler than the single node proof $S\sigma$. If no leaf of P_i is labelled by a sequent $S\sigma$ then Q_{i+1} is P_i. We have $P_i \geq_{\mathcal{P}} Q_{i+1}$.

If N_{i+1} is obtained from N_i by adding a resolvent or paramodulant, then Q_{i+1} is obtained from P_i by replacing every subproof

$$\frac{S_1 \quad S_2}{S\sigma}$$

for which $\langle S_1\rangle >^c \langle S\rangle$ or $\langle S_2\rangle >^c \langle S\rangle$, by a proof consisting of a single node $S\sigma$. Again, we have $P_i \geq_{\mathcal{P}} Q_{i+1}$. (Most ground inferences in normal form satisfy the requirement that $\langle S_1\rangle >^c \langle S\rangle$ or

$\langle S_2 \rangle >^c \langle S \rangle$. The requirement may be violated if S_1 or S_2 are tautologies; that is, sequents of the form $\Gamma, A \to A, \Delta$. Such tautologies are dealt with separately.)

If N_{i+1} is obtained from N_i by adding a factor $S\sigma$ of S, then Q_{i+1} differs from P_i only in that instead of S the sequent $S\sigma$ is associated with each leaf $(S\sigma)\tau$. Again, we have $P_i \geq_{\mathcal{P}} Q_{i+1}$.

The proof P_{i+1} is essentially a normalized version of Q_{i+1}, except that certain tautologies and inferences requiring paramodulation into variables are eliminated. We sketch the necessary proof transformations for the latter.

Suppose $S[x, x, x]$ is a sequent $\Gamma[x] \to \Delta, A[x, x]$ and P is a ground proof

$$\frac{\Gamma' \to \Delta', s = t \quad \Gamma[u[s]] \to \Delta, A[u[s], u[s]]}{\Gamma', \Gamma[u[s]] \to \Delta', \Delta, A[u[s], u[t]]}$$

in normal form, where $A[u[s], u[s]]$ does not appear $\Gamma' \to \Delta', s = t$. The proof P corresponds to instantiating the variable x by a term $u[s]$ and then replacing one occurrence of $u[s]$ by $u[t]$. Lifting it would require paramodulation into the variable x. Instead of lifting the proof, we can instantiate the variable x by $u[t]$ and then replace all occurrences of $u[t]$, except one, by $u[s]$.

Thus let P' be the proof

$$\frac{\Gamma' \to \Delta', s = t \qquad \dfrac{\Gamma' \to \Delta', s = t \quad \Gamma[u[t]] \to \Delta, A[u[t], u[t]]}{\Gamma', \Gamma[u[s]] \to \Delta', \Delta, A[u[t], u[t]]}}{\Gamma', \Gamma[u[s]] \to \Delta'', \Delta, A[u[s], u[t]]}$$

where $\Delta'' = \Delta' - \{A[u[t], u[t]]\}$.

In general, if S is a sequent containing n occurrences of the variable x, then P' consists of $n - 1$ paramodulation steps. The proof P' is not in normal form, but is simpler than P because $\langle \Gamma[u[s]] \to \Delta, A[u[s], u[s]] \rangle >^c \langle \Gamma[u[t]] \to \Delta, A[u[t], u[t]] \rangle$ and $\langle \Gamma[u[s]] \to \Delta, A[u[s], u[s]] \rangle >^c \langle \Gamma' \to \Delta', s = t \rangle$. By Lemma 1, whenever Q_{i+1} contains a problematic subproof P, there is a proof Q'_{i+1} with $Q_{i+1} >_{\mathcal{P}} Q'_{i+1}$. Repeated application of such proof transformations, followed by proof normalization, until all problematic subproofs are eliminated will eventually result in the desired proof P_{i+1}.

The sequence of proofs thus constructed satisfies (i) and (ii). If property (iii) is not satisfied, then there is a proof P_k consisting of more than one node, such that $P_k = P_{k+1} = P_{k+2} = \cdots$. It can be shown that P_k must contain a subproof

$$\frac{S_1\sigma \quad S_2\sigma}{S\sigma}$$

where S_1 and S_2 are contained in $\bigcup_i \bigcap_{j \geq i} N_j$ and S can be obtained from S_1 and S_2 by ordered resolution or oriented paramodulation (but not into variables). Fairness guarantees that the sequent S is deduced, which would imply that $P_k >_{\mathcal{P}} P_{k+j}$, for some j, contradicting $P_k = P_{k+j}$. We conclude that the sequence of proofs satisfies (iii), which completes the proof. □

Acknowledgements. Part of this research was done during a visit at the University of Nancy, France. I have also had the benefit of many stimulating discussions with Jieh Hsiang, Jean-Pierre Jouannaud, and Michael Rusinowitch.

References

[1] BACHMAIR, L. 1987. Proof methods for equational theories. Ph.D. diss., University of Illinois, Urbana-Champaign.

[2] Bachmair, L., Dershowitz, N., and Hsiang, J. 1986. Orderings for equational proofs. In *Proc. Symp. Logic in Computer Science*, Boston, Massachusetts, 346–357.

[3] Brand, D. 1975. Proving theorems with the modification method. *SIAM J. Comput.* **4**:412–430.

[4] Brown, T. 1975. A structured design-method for specialized proof procedures. Ph.D. diss., California Institute of Technology, Pasadena.

[5] Chang, C., and Lee, R. C. 1973. *Symbolic logic and mechanical theorem proving.* New York, Academic Press.

[6] Dershowitz, N. 1982. Orderings for term-rewriting systems. *Theor. Comput. Sci.* **17**:279–301.

[7] Dershowitz, N. 1987. Termination of rewriting. *J. Symbolic Computation* **3**:69–116.

[8] Dershowitz, N., and Manna, Z. 1979. Proving termination with multiset orderings. *Commun. ACM* **22**:465–476.

[9] ganzinger, H. 1988. A completion procedure for conditional equations. To appear in *J. Symbolic Computation.*

[10] Hsiang, J., and Rusinowitch, M. 1986. A new method for establishing refutational completeness in theorem proving. In *Proc. 8th Int. Conf. on Automated Deduction*, ed. J. H. Siekmann, Lect. Notes in Comput. Sci., vol. 230, Berlin, Springer-Verlag, 141–152.

[11] Hsiang, J., and Rusinowitch, M. 1988. Proving refutational completeness of theorem proving strategies, Part I: The transfinite semantic tree method. Submitted for publication.

[12] Joyner, W. 1976. Resolution strategies as decision procedures. *J. ACM* **23**:398–417.

[13] Knuth, D., and Bendix, P. 1970. Simple word problems in universal algebras. In *Computational Problems in Abstract Algebra*, ed. J. Leech, Oxford, Pergamon Press, 263–297.

[14] Lankford, D. 1975. Canonical inference. Tech. Rep. ATP-32, Dept. of Mathematics and Computer Science, University of Texas, Austin.

[15] Peterson, G. 1983. A technique for establishing completeness results in theorem proving with equality. *SIAM J. Comput.* **12**:82–100.

[16] Robinson, G., and Wos, L. T. 1969. Paramodulation and theorem proving in first order theories with equality. In *Machine Intelligence 4*, ed. B. Meltzer and D. Michie, New York, American Elsevier, 133–150.

[17] Robinson, J. A. 1965. A machine-oriented logic based on the resolution principle. *J. ACM* **12**:23–41.

[18] Rusinowitch, M. 1988. Theorem proving with resolution and superposition: An extension of the Knuth and Bendix procedure as a complete set of inference rules.

[19] Slagle, J. R. 1974. Automated theorem proving for theories with simplifiers, commutativity, and associativity. *J. ACM* **21**:622–642.

[20] Wos, L. T., and Robinson, G. 1973. Maximal models and refutation completeness: Semidecision procedures in automatic theorem proving. In *Word Problems*, ed. W.W. Boone et al., Amsterdam, North-Holland, 609–639.

[21] Wos, L. T., Robinson, G. A., Carson, D. F., and Shalla, L. 1967. The concept of demodulation in theorem proving. *J. ACM* **14**:698–709.

[22] Zhang, H., and Kapur, D. 1988. First-order theorem proving using conditional rewrite rules. In *Proc. 9th Conf. Automated Deduction*, Lect. Notes in Comput. Sci. Berlin, Springer-Verlag, 1–20.

Complete Sets of Reductions Modulo Associativity, Commutativity and Identity

Timothy B. Baird

Department of Mathematics and Computer Science
Harding University, Searcy, Arkansas 72143

Gerald E Peterson

McDonnell Douglas Corporation, W400/105/2/206
P. O. Box 516, St. Louis, Missouri 63166

Ralph W. Wilkerson

Department of Computer Science, University of Missouri—Rolla,
Rolla, Missouri 65401

Abstract: We describe the theory and implementation of a process which finds complete sets of reductions modulo equational theories which contain one or more associative and commutative operators with identity (ACI theories). We emphasize those features which distinguish this process from the similar one which works modulo associativity and commutativity. A primary difference is that for some rules in ACI complete sets, restrictions are required on the substitutions allowed when the rules are applied. Without these restrictions, termination cannot be guaranteed. We exhibit six examples of ACI complete sets that were generated by an implementation.

1. Introduction

In 1970 Knuth and Bendix published a remarkable paper [KB70] in which a process for finding "complete sets of reductions" was described and such a set for groups was exhibited. A complete set of reductions lets you solve the word problem for an equational theory simply by reducing each side of a given equation to normal form using the reductions in the set and then comparing the two sides to see if they are identical. The Knuth-Bendix approach requires each reduction in the complete set to be oriented with the most complex, in some sense, on the left. For this reason they were unable to handle commutativity since neither side could be construed more complex than the other. In 1981 Peterson and Stickel [PS81] (see also [LB77]) showed that the problem with commutativity could be handled by incorporating it along with associativity into the unification mechanism rather than having them appear explicitly as rules. They were able to generate complete sets of reductions for Abelian groups, rings, and other structures. The purpose of the present research is to take the idea of building-in one step further by building the identity law as well as associativity and commutativity into the unification mechanism.

We felt that building in the identity would accomplish at least the following four things.

1. We knew that Jouannaud and Kirchner [JK86] and Bachmair and Dershowitz [BD87] had simplified, generalized, and created a better theoretical framework for complete sets of reductions when some of the equations were built into the unification mechanism, i.e. *modulo* the set E of equations. However, these theories do not cover the ACI case because they are restricted to sets of reductions which are E terminating and we will show herein that E termination in the ACI case is not possible without restrictions. Thus we wanted to consider the ACI case separately in order to find out how to modify the theory when dealing with equational theories for which E termination does not hold.
2. We felt that the process of building inferences into the reasoning mechanism rather than explicitly representing them was useful because larger inferencing steps could then be taken. In fact, we felt it would be useful to build in as much as possible. For example, if all of Abelian group theory is to be built in, then building in the identity is the logical next step.
3. We knew that the number of unifiers in a complete set for ACI unification is always much smaller than the number of unifiers in an AC complete set. We reasoned that ACI unification would, therefore, probably be faster that AC unification and this, along with the larger step size, would speed up the whole process.
4. We knew that if ACI complete sets could be found, they almost surely would be smaller than their AC counterparts, giving a more compact representation of the "computational essence" of a theory.

The extent to which we achieved these goals is considered in the conclusion.

The following four paragraphs describe interesting phenomena which arose during the course of the research.

The main problem was to find a way to ensure E termination. This was solved by restricting the substitutions that are allowed when a reduction is applied. For example, in the ACI complete set for Abelian groups, the reduction $-(x+y) \rightarrow (-x)+(-y)$ may not be applied if either x or y is 0.

Another problem arose in the generation process. We could not make the process work efficiently without adding another inference rule which we call *identity substitution*. It is simply this: in any equation, any variable may be replaced with the identity in order to obtain another valid equation.

When finding AC complete sets it was found necessary to add "extensions" ([PS81] [LB77]) of reductions in order to complete a set. This somehow compensated for the absence of the associativity axiom. For ACI complete sets, extensions are still necessary; however, unlike the AC case, the reduction which is extended need not also appear in the complete set because it can be obtained from the extension by identity substitution. For example, $x+(-x) \rightarrow 0$ can be obtained from $u+x+(-x) \rightarrow u$ by substituting 0 for u.

Finally, when computing complete sets of ACI unifiers, we noticed that often one unifier would be very similar to another and seemed to carry the same information. For example, if $x + y + (-y)$ is ACI unified with itself, i.e. with $u + v + (-v)$ after renaming variables, two of the unifiers in the complete set are:

$v \leftarrow z1 + z3$ $u \leftarrow -(-(z1 + z3) + z1 + z2) + z2 + z4$ $y \leftarrow -(z1 + z3) + z1 + z2$ $x \leftarrow z3 + z4$	and	$v \leftarrow -(z1 + z2) + z1 + z3$ $u \leftarrow z2 + z4$ $y \leftarrow z1 + z2$ $x \leftarrow -(-(z1 + z2) + z1 + z3) + z3 + z4$

There appears to be some symmetry that can possibly be exploited. In fact, the original unification problem is not changed if y is interchanged with v and x is interchanged with u. If these interchanges are made in one of these unifiers, the unifiers then become identical except for the names of the z variables. Thus, they do indeed carry the same information. The computer program which we constructed incorporates some symmetry checking in order to improve efficiency. For a thorough discussion of the exploitation of symmetry in the completion process, see the dissertation by Mayfield [Ma88].

A more detailed discussion of much of what we present here may be found in [Ba88]. Furthermore, a chapter of [Ba88] is devoted to the important topic of termination which is only lightly treated here.

2. Testing a Set of Reductions for Completeness

The Knuth-Bendix algorithm and its derivatives are of the "generate and test" variety. That is, first a set of reductions is generated and then it is tested for completeness. If the test fails, then the set is modified and tested again, etc. Thus the algorithm can be split into the part which generates sets of reductions and the part which tests them. In this section we develop a test for completeness; in the next we will consider the generation process.

The test will be based on an abstract Church-Rosser theorem due to Jouannaud and Kirchner [JK86]. The notation in this section is similar to that in [JK86] and the interested reader should refer to that paper for more details.

Let S be a set. Let $=^1_E$ be a symmetric relation on S and let $=_E$ be its reflexive, transitive closure. Let R, also written $\rightarrow_R$, be a relation on S, and let R^+ (or $\xrightarrow{+}_R$) and R^* (or $\xrightarrow{*}_R$) be its transitive and reflexive transitive closure, respectively. Let R/E (or $\rightarrow_{R/E}$) be the relation $=_E \circ \rightarrow_R \circ =_E$. *We assume that R/E is well-founded.* Let $t{\downarrow}_R$ be a normal form obtained from t using the well-founded relation R.

Let $=^1_{R \cup E}$ be the relation $=^1_E \cup \rightarrow_R \cup {}_R\!\leftarrow$. Let $=_{R \cup E}$ be the reflexive transitive closure of $=^1_{R \cup E}$. Let R^E (or $\rightarrow_{R^E}$) be a relation which satisfies $R \subseteq R^E \subseteq R/E$.

Definition 1. To say that a relation R^E is E *complete* means $s =_{R \cup E} t$ if and only if $s{\downarrow}_{R^E} =_E t{\downarrow}_{R^E}$. That is, when R^E is E complete, two terms are equal in the $R \cup E$ sense if and only if their R^E normal forms are equal in the E sense.

Definition 2. R^E is *locally coherent modulo E* if whenever $t =^1_E s$ and $t \rightarrow_{R^E} t_1$, it follows that there is a term s_1 such that $s \rightarrow_{R^E} s_1$ and t_1 and s_1 have a common R/E successor.

Definition 3. R^E is *locally confluent modulo E* if whenever $t \rightarrow_{R^E} t_1$ and $t \rightarrow_{R^E} t_2$, it follows that t_1 and t_2 have a common R/E successor.

E Church-Rosser Theorem. *The following two statements are equivalent:*

(1) *R^E is E complete.*

(2) *R^E is locally coherent and locally confluent modulo E.*

See [JK86] for the proof.

We apply this theorem to term rewriting systems in order to obtain the test for completeness. In doing so, R is a set of reductions and E is a set of equations. We are particularly interested in the case in which E is ACI. The set S is the set of terms. The relation $=^1_E$ is a single application of an equality in E and the relation $\rightarrow_R$ is a single application of one of the rewrite rules in R.

The above theorem requires the relation R/E to be well-founded. But if we stick with the ordinary definitions as in the previous paragraph and if E is ACI, then it is easy to find examples in which R/E is not well-founded. One example is given by the rule $-(x+y) \rightarrow (-x)+(-y)$ which is true in Abelian groups. If this rule is in R and if E=ACI, then the following is a non-terminating sequence of R/E reductions:

$$\begin{aligned} -0 &=_{ACI} -(0+0) \\ &\rightarrow_R (-0)+(-0) \\ &=_{ACI} -(0+0)+(-0) \\ &\rightarrow_R (-0)+(-0)+(-0) \\ &\dots \end{aligned}$$

In order to overcome this problem, we found it necessary to restrict the applicability of the reductions in R. For example, for $-(x+y) \to (-x)+(-y)$, if we insist that neither x nor y take on an ACI equivalent of 0, then R/E *is* well-founded. These restrictions are made manifest in the theory by associating with each element $\lambda \to \rho$ of R a set $\Theta(\lambda \to \rho)$ of forbidden substitutions.

For any set of substitutions Θ, let $\iota_E\Theta$ be the set of all E instances of substitutions in Θ, where an E instance of θ is a substitution which is E equal to an instance of θ. Then $\theta \notin \iota_E\Theta$ means that θ is not an E instance of a substitution in Θ. For a term t, $dom(t)$ is the set of all positions of subterms of t, ε is the element of $dom(t)$ at which t occurs, $sdom(t)$ is the subset of $dom(t)$ at which non variables occur, if $m \in dom(t)$, then t/m is the subterm occurring at m, and $t[m \leftarrow s]$ is the term obtained from t if the subterm at m is replaced with s.

We now state the definition of $\to_R$ which embodies the restrictions. We say that $t \to_R s$, for terms s and t, if there is an element $\lambda \to \rho$ of R, a position m in $dom(t)$, and a substitution θ such that

$$\theta \notin \iota_E\Theta(\lambda \to \rho),$$
$$t/m = \lambda\theta, \quad \text{and}$$
$$s = t[m \leftarrow \rho\theta].$$

The restriction $\theta \notin \iota_E\Theta(\lambda \to \rho)$ in this definition is the only difference between it and the usual definition of reduction.

Let $t \to_{R^E} s$ mean that there exist $\lambda \to \rho \in R$, $m \in dom(t)$, and θ such that

$$\theta \notin \iota_E\Theta(\lambda \to \rho),$$
$$t/m =_E \lambda\theta, \quad \text{and}$$
$$s = t[m \leftarrow \rho\theta].$$

The reason for using R^E instead of R/E is that when reducing a term t using R^E, it is necessary only to examine each of the subterms of t; it is not necessary to access E equivalents of those subterms. Note that $R \subseteq R^E \subseteq R/E$.

We are now ready to state the characterizations of local coherence and local confluence which form the basis of our test for completeness. We regret that space does not permit inclusion of their proofs. We assume from here on that there exists a finite and complete E unification algorithm. When E is ACI such an algorithm exists and has been described by Stickel [St81].

Local Coherence Theorem. *The following two statements are equivalent.*

(1) *R^E is locally coherent modulo E.*

(2) *Whenever $\ell = r \in E$ (or $r = \ell \in E$), $\lambda \to \rho \in R$, $m \in sdom(\ell)$ but $m \neq \varepsilon$, and σ is an element of a complete set of E unifiers of ℓ/m and λ such that $\sigma \notin \iota_E\Theta(\lambda \to \rho)$*

it follows that there is a term u such that $r\sigma \to_{R^E} u$ and $\ell\sigma[m \leftarrow \rho\sigma]$ and u have a common R/E successor.

We will say that $\lambda \to \rho$ *coheres* with $\ell = r$ when (2) above holds for all appropriate values of m and σ.

Based on the Local Coherence Theorem we can implement the following test for local coherence: First find all $\ell = r$, $\lambda \to \rho$, m, and σ, which satisfy conditions (2) of the theorem. Then test each $r\sigma$ for R^E reducibility. If some $r\sigma$ is not reducible then coherence fails. If $r\sigma$ is reducible using some $\lambda' \to \rho'$, σ', and m', then the *coherence critical pair*

$$< \ell\sigma[m \leftarrow \rho\sigma],\ r\sigma[m' \leftarrow \rho'\sigma'] >$$

can be reduced to normal form via R^E. If both sides of the normal form are identical, then coherence succeeds, otherwise coherence fails. (It can be shown that here and in the following test for confluence, reduction to normal form via R^E is a valid way to proceed even though the theorem specifies R/E reduction.)

When the set E is an ACI equational theory, it can be shown that all reductions will cohere with identity and commutative laws. Thus, the procedure for assuring coherence for an ACI equational theory simplifies to the following: for each reduction $\lambda \to \rho \in R$ together with the associative law for each ACI operator in λ perform the coherence test described above.

Local Confluence Theorem. *If R^E is locally coherent modulo E, then the following two statements are equivalent.*

(1) *R^E is locally confluent modulo E.*

(2) *Whenever $\lambda_1 \to \rho_1 \in R$, $\lambda_2 \to \rho_2 \in R$, $m \in sdom(\lambda_1)$, and σ is an element of a complete set of E unifiers of λ_1/m and λ_2 such that $\sigma \notin \iota_E\Theta(\lambda_1 \to \rho_1)$ and $\sigma \notin \iota_E\Theta(\lambda_2 \to \rho_2)$, it follows that $\lambda_1\sigma[m \leftarrow \rho_2\sigma]$ and $\rho_1\sigma$ have a common R/E successor.*

The test for local confluence then proceeds as follows: First test for local coherence. If successful, then generate all of the *confluence critical pairs*

$$< \lambda_1\sigma[m \leftarrow \rho_2\sigma],\ \rho_1\sigma >$$

for which the hypotheses of (2) are satisfied. Both sides of the critical pair are then reduced to normal forms via R^E and compared. If the normal forms are identical, the pair *conflates*, and the test for confluence succeeds, otherwise it fails.

By the E Church-Rosser Theorem, the above test for confluence is also a test for completeness.

3. The Generation Process

The essence of the generation phase of the program is to decide what to do if the completeness test fails. If completeness fails, then some coherence or confluence critical pair failed to conflate. The usual remedy is to take the fully reduced form of the pair, orient it if possible, and add it as a new reduction. Then, after inter-reducing all the reductions, try the completeness test again. This is normally what we do, but there are three unusual aspects we would like to discuss.

Firstly, when a coherence critical pair fails to conflate you have the option of adding an extension rather than the new critical pair. This is discussed in Section 4 below.

Secondly, when a critical pair fails to conflate and its fully reduced form cannot be oriented, what do we do? The original Knuth-Bendix algorithm simply halted with failure in this case. We proceed in two ways. Firstly, as in [FG84], we "shelve" such pairs and continue to process the other pairs. Some pair which is generated later may be able to reduce a shelved pair. If all shelved pairs are eventually taken off the shelf, then we need not halt. Secondly, we try substituting the identity for variables in shelved pairs. Often this will result in a new critical pair which can not only be oriented, but can also reduce the shelved pair from which it came. For example, in the Abelian group problem, the critical pair $< -(x+y)+y+z,\ -(x+w)+w+z >$ was generated. It could be neither further reduced nor oriented. However, if w is replaced by 0, we obtain the pair $< -(x+y)+y+z, (-x)+z >$ which is both orientable and can be used to conflate the pair from which it came.

Thirdly, when we orient a critical pair to form a reduction, it may be necessary to add restrictions in order to ensure termination. All of our restrictions are logical combinations of statements which restrict some variable from being assigned an identity. These restrictions are calculated by assigning the identity to variables and finding out whether or not the resulting pair is still orientable. For example, consider the pair $< -(x+y),\ (-x)+(-y) >$. We would like to orient this into the rule $-(x+y) \rightarrow (-x)+(-y)$. If x is 0, however, the pair becomes in an ACI system, $< -y,\ (-0)+(-y) >$ which is oriented the wrong way! Thus in forming the rule from the pair, we add the restriction that x is not allowed to be 0. When adding the restriction, we must make sure that we do not change the underlying equational theory. That is, if we split a critical pair into a restricted portion and the part that remains, then the part that remains must be able to be reduced away. For example, we may split $< -(x+y),\ (-x)+(-y) >$ into $< -y,\ (-0)+(-y) >$, $< -x,\ (-x)+(-0) >$ and the pair $< -(x+y),\ (-x)+(-y) >$ in which x and y are restricted to be non zero. In the context of Abelian groups, the first two of these will reduce using $-0 \rightarrow 0$ to pairs in which both sides are identical and then will be discarded.

4. Results of an Implementation

The ACI completion procedure has been implemented in a computer program. This program is written in Common LISP and has been run successfully under a number of different hardware and software configurations including a DEC Microvax II, a XEROX 1109, a Symbolics, and an IBM PC-RT. Those interested may obtain a copy of the program source code in machine readable form by contacting Ralph Wilkerson. In this section we discuss the implementation issues of term symmetry and the use of extensions. Then we present some of the complete sets of reductions which have been generated by the program.

Dealing with Term Symmetry

One observation that we have made from running the completion procedure and examining the critical pairs produced is that many of the critical pairs which are generated are redundant. Mayfield [Ma88] has a detailed discussion of term symmetry and its effect on both unifiers and critical pairs. For example, the pairs

$$< x + (-z) + y,\ y + z + x + (-z) + (-z) >$$
$$< u + v + (-w),\ v + (-w) + (-w) + u + w >$$

are symmetric modulo associativity and commutativity. Mayfield shows that when two pairs are symmetric, only one of them needs to be processed by the completion algorithm. The other may be discarded before attempting conflation with no effect on the completion process. We have seen that as many as three-fourths of the critical pairs generated by ACI unification may be discarded due to symmetry. Mayfield has developed a test which determines whether or not two terms are symmetric. The cost of using the test is usually much smaller than attempting to conflate the redundant critical pairs.

Using Extensions

Although we have mentioned previously that we do not need the concept of extended reductions in order to perform ACI completion, we may still want to consider using them because they often add to the efficiency of the completion process. If a coherence critical pair which comes from a reduction and an associative law will not conflate, then the reduction can be extended with conflation assured for both the original reduction and its extension. Adding the extension is usually more efficient than further processing of the coherence critical pairs. Because of this gain in efficiency, we have retained the use of extensions in our completion program.

When a reduction needs to be extended we simply add the needed extension variables to the original reduction. For example, if $\lambda \to \rho$ needs to be extended for both ACI operators $+$ and $*$, we replace it with the extended reduction $u*(v+\lambda) \to u*(v+\rho)$.

Clearly this one reduction provides the functionality of the three reductions $\lambda \rightarrow \rho$, $v + \lambda \rightarrow v + \rho$, and $u * \lambda \rightarrow u * \rho$ since u and v may each take on an identity and collapse away. Such an extended reduction will cohere with the associative laws for both ACI operators $+$ and $*$. Unfortunately, it is possible that this extended reduction will be simplified by one of the other reductions and the simplified form will not have the ability to provide the coherence that the original possessed. For example, consider the situation in which the reduction $x * 0 + x * y \rightarrow x * y$ is being checked for coherence with multiplicative associativity and the set R contains the distributive law $x * (y + z) \rightarrow x * y + x * z$. In order to satisfy coherence, the reduction is extended to become $v * (x * 0 + x * y) \rightarrow v * x * y$. However, the distributive law simplifies this extension and it becomes $v * x * 0 + v * x * y \rightarrow v * x * y$ and this new reduction does *not* satisfy the coherence property with the associative law for $*$. For this reason, other researchers have enforced protection schemes which protect certain extended reductions from simplification and/or deletion during the completion process.

We have chosen to avoid treating extensions differently than other reductions and to avoid any sort of protection schemes by simply ignoring this potential problem during the completion process. This creates the possibility that our program may halt with what should be a complete set of reductions when, in fact, it is not. This situation is easy to detect, however, by running a coherence check on the final set of reductions. If all reductions satisfy the coherence property, then the reduction set is actually complete. If they do not satisfy coherence, then it is not. In our experience the program has never found a potentially complete set of reductions which was not complete. This holds in spite of the fact that our program has encountered the situation described above where the distributive law potentially destroyed the coherence property. As long as the troublesome reduction does not make it into the final complete set this will never cause a problem.

Results

We now present examples which show the results of running the ACI completion program for five different algebraic structures. A sixth is shown in the next section in the right half of Table 2. All of the examples were run on an IBM PC-RT with ten megabytes of main memory, using LUCID COMMON LISP under the AIX operating system.

In Example 3, $\oplus$ is the operator for exclusive or, $*$ is *and*, 1 is *true* and 0 is *false*. In Example 4, H is a homomorphism from a group with operators $(+,\ -,\ 0)$ to one with operators $(\oplus,\ \ominus,\ 0')$ and in Example 5, H is a homomorphism from a ring with operators $(+,\ *,\ -,\ 0,\ 1)$ to one with operators $(\oplus,\ \otimes,\ \ominus,\ 0'\ 1')$.

1. $u + (-v) + v \to u$
2. $-(-u) \to u$
3. If $u \neq 0$ and $v \neq 0$ then $-(u+v) \to (-u) + (-v)$

Example 1. ACI complete set for a commutative group

1. $u + (-v) + v \to u$
2. $-(-u) \to u$
3. If $u \neq 0$ and $v \neq 0$ then $-(u+v) \to (-u) + (-v)$
4. If $w \neq 1$ and $u \neq 0$ and $v \neq 0$ then $w * (u+v) \to w * u + w * v$
5. If $u \neq 1$ then $u * 0 \to 0$
6. If $u \neq 1$ then $(-v) * u \to -(v * u)$

Example 2. ACI complete set for a commutative ring with unit

1. If $u \neq 0$ then $v \oplus u \oplus u \to v$
2. If $u \neq 1$ then $v * u * u \to v * u$
3. If $w \neq 1$ and $u \neq 0$ and $v \neq 0$ then $w * (u \oplus v) \to (w * u) \oplus (w * v)$
4. If $u \neq 1$ then $u * 0 \to 0$

Example 3. ACI complete set for a Boolean algebra

1. $u + (-v) + v \to u$
2. $-(-u) \to u$
3. If $u \neq 0$ and $v \neq 0$ then $-(u+v) \to (-u) + (-v)$
4. $u \oplus v \oplus (\ominus v) \to u$
5. $\ominus(\ominus u) \to u$
6. If $u \neq 0'$ and $v \neq 0'$ then $\ominus(v \oplus u) \to (\ominus v) \oplus (\ominus u)$
7. If $u \neq 0$ and $v \neq 0$ then $H(u+v) \to H(u) \oplus H(v)$
8. $H(0) \to 0'$
9. $H(-u) \to \ominus H(u)$

Example 4. ACI complete set for a group homomorphism

1. $u + (-v) + v \to u$
2. $-(-u) \to u$
3. If $u \neq 0$ and $v \neq 0$ then $-(u+v) \to (-u) + (-v)$
4. If $w \neq 1$ and $u \neq 0$ and $v \neq 0$ then $w * (u+v) \to w * u + w * v$
5. If $u \neq 1$ then $u * 0 \to 0$
6. If $u \neq 1$ then $(-v) * u \to -(v * u)$
7. $u \oplus (\ominus v) \oplus v \to u$
8. $\ominus(\ominus u) \to u$
9. If $u \neq 0'$ and $v \neq 0'$ then $\ominus(u \oplus v) \to (\ominus u) \oplus (\ominus v)$
10. If $w \neq 1'$ and $u \neq 0'$ and $v \neq 0'$ then $w \otimes (u \oplus v) \to (w \otimes u) \oplus (w \otimes v)$
11. If $u \neq 1'$ then $u \otimes 0' \to 0'$
12. If $u \neq 1'$ then $(\ominus v) \otimes u \to \ominus(v \otimes u)$
13. If $u \neq 1$ and $v \neq 1$ then $H(u * v) \to H(u) \otimes H(v)$
14. $u \otimes H(1) \otimes H(v) \to u \otimes H(v)$
15. If $u \neq 0$ and $v \neq 0$ then $H(u+v) \to H(u) \oplus H(v)$
16. $H(0) \to 0'$
17. $H(-u) \to \ominus H(u)$

Example 5. ACI complete set for a ring homomorphism

In order to illustrate some features of the program we will now present an overview of how the ACI complete set of Example 1 was calculated. The program was given the additive inverse axiom $x + (-x) = 0$ as a critical pair equation along with the declaration that $+$ is an ACI operator with identity 0. It generated the complete set as shown in Figure 1.

R1: $u + v + (-v) \to u$	from input equation
R2: If $u \neq 0$ then $v + (-(w+u)) + u \to v + (-w)$	from R1 with itself
R3: $(-0) \to 0$	from R1 simplified
R4: $-(-u) \to u$	from R1 with itself
R1 deleted	
R5: $u + (-((-v) + w)) + (-(v + x)) \to u + (-(x + w))$	from R2 with itself
R6: $u + (-((-(v + w)) + x)) + (-v) \to u + (-x) + w$	from R2 with itself
R7: $-((-u) + v) \to (-v) + u$	from R2 with itself
R8: If $u \neq 0$ and $v \neq 0$ then $-(u+v) \to (-u) + (-v)$	from R5 simplified
R5, R6 and R7 deleted	
R9: $u + (-v) + (-w) + v \to u + (-w)$	from R2 simplified
R2 deleted	
R10: $u + (-v) + v \to u$	from R3 with R9
R3 and R9 deleted	

Figure 1. Generating the ACI complete set for an Abelian group

This computation illustrates the following:

- Rule R1 was generated as an extended form of the additive inverse law in order to satisfy coherence.
- Rules R2 and R8 have conditions in order to maintain termination of the rewriting relation.

- Rule R1 was removed from the set of reductions early in the completion process, but then reappeared as R10 near the end.
- Rule R3 was obtained from R1 and R8 was obtained from R5 by setting u and v to 0. This illustrates the use of the identity substitution rule which is not a feature of AC completion, but is essential in ACI completion.
- The inferences are not easy for a human to reproduce because of the complex nature of ACI unification.

5. ACI Completion versus AC Completion

We performed a side-by-side comparison of ACI completion and AC completion for the same algebraic structures. To generate the data for these comparisons we ran the E completion procedure on the same algebraic structures for which we generated ACI complete reductions sets in Examples 1 through 6. For each problem we changed the appropriate operators from ACI to AC and added the necessary equations to handle the identity properties as rewrite rules.

Size of Complete Sets

A comparison of the number of reductions in the final complete sets is given in Table 1.

Example Problem	Reductions in Complete Set AC	ACI
1	6	3
2	10	6
3	9	5
4	15	9
5	26	17
6	11	3

Table 1. Comparison of the number of reductions in complete sets

The ACI complete sets contain fewer reductions than the AC complete sets in every case. On average the number of reductions in an ACI complete set is 53% of the number in the corresponding AC complete set. Some gain in this regard was expected because the identity laws are being removed and incorporated into the unification mechanism. However, it was not expected that the size of the complete set would be almost halved on average. In some cases, the reduction in the number of rules in the complete set can be quite dramatic. One example is the complete sets of reductions for distributive lattices with identities. A side-by-side comparison of the ACI complete set and the AC complete set in this case (Example 6) is shown in Table 2.

AC	ACI
1. $x * (y + z) \rightarrow x * y + x * z$	1. If $x \neq 1$ and $y \neq 0$ and $z \neq 0$, then $x * (y + z) \rightarrow x * y + x * z$
2. $x * y * y \rightarrow x * y$	2. If $y \neq 1$, then $x * y * y \rightarrow x * y$
3. $x + y + (y * z) \rightarrow x + y$	3. If ($y \neq 1$ or $z \neq 0$) and ($y \neq 0$ or $z \neq 1$), then $x + y + (y * z) \rightarrow x + y$
4. $x + (x * y) \rightarrow x$	
5. $x * 1 \rightarrow x$	
6. $1 + x \rightarrow 1$	
7. $x + x \rightarrow x$	
8. $x + y + y \rightarrow x + y$	
9. $0 * x \rightarrow 0$	
10. $x * x \rightarrow x$	
11. $x + 0 \rightarrow x$	

Table 2. AC and ACI complete sets for distributive lattices

The decrease in the number of reductions in a complete set modulo ACI as compared to the complete set modulo AC comes in two ways:

1. Removal of identity laws from the AC complete set and incorporating them into the unification mechanism. Rules 5 and 11 in Table 2 are examples.
2. Obtaining AC rules from ACI rules by substituting an identity for a variable in the ACI rule. Rules 4, 6, 7, 8, 9, and 10 in Table 2 are examples. As a special case of this, in the AC case, if a rule needs to be extended to maintain coherence, then both it and its extension need to be retained in the complete set. In the ACI case, if a rule needs to be extended, then only the extension needs to be retained because the original can be obtained from the extension by substituting the identity for the extension variable. Rules 4, 7 and 10 in Table 2 are examples of this special case.

A complete set of reductions wraps the mathematical essence of an equational theory up into a neat little package such that equational conjectures can simply be calculated true or false. It would appear, therefore, that they have some mathematical interest quite apart from their utility in the process of automating proofs. In this regard, complete sets of ACI reductions have the advantage of compactness of representation when compared to complete sets of AC reductions.

Step Size of Deductions

It is interesting to compare the number of steps for a derivation of an ACI complete set to the number of steps for a derivation of an AC complete set for the same algebraic structure. Table 3 compares the number of inferences and related steps for our example problems.

Example Problem	Inferences Made		Matches Attempted		Reductions Applied		Reductions Added	
	AC	ACI	AC	ACI	AC	ACI	AC	ACI
1	28	14	6588	4284	231	138	8	10
2	243	31	256256	71368	1668	1735	46	17
3	39	13	9401	2517	143	41	15	9
4	105	36	27852	14310	214	77	23	18
5	555	148	443983	309378	866	1193	62	44
6	68	8	71998	20377	892	1110	17	6

Table 3. Comparison of number of inferences

The uniformly smaller number of inferences required for ACI completion still took us to essentially the same point as the corresponding AC completion. This clearly indicates that more distance was covered by each step, on the average. The consistently lower numbers seen in Table 3 reflect that the branching factor of the search space is indeed smaller when more is built into a single step. This is one result which we had hoped to see. Unfortunately, this does not automatically translate into a gain in efficiency.

Efficiency

Even though an ACI completion run takes fewer logical steps than its AC counterpart, the run times shown in Table 4 indicate that this does not, at least for our implementation, automatically give an increase in efficiency.

Example Problem	Total Time		Terminal Form		Critical Pair		Unification	
	AC	ACI	AC	ACI	AC	ACI	AC	ACI
1	53.9	56.2	25.9	38.8	17.6	8.5	12.6	4.0
2	1528	3044	1084	2017	116	737	82	316
3	51.7	40.1	23.6	27.8	6.4	2.2	2.4	1.3
4	141	144	79	67	30	43	26	19
5	2382	2872	1596	2114	116	226	78	92
6	414	1627	293	1142	68	435	35	245

Table 4. Comparison of run times in seconds

For most of the problems the AC completion procedure runs in less time. This is especially true of the larger problems, where we had hoped that the exact opposite would happen. Examining the time spent in computing terminal forms we see that our ACI completion procedure generally spent more time performing fewer attempted matches. We attribute this to the fact that ACI matching is slower than AC matching. We also see that, for the larger problems, the ACI completion procedure spent considerably more time in the formation of critical pairs. Furthermore, only part of that difference can be attributed to ACI unification, as reflected in the last column of Table 4.

Table 5 gives some insight into what is going on in the formation of critical pairs.

Example Problem	Critical Pairs AC	Critical Pairs ACI	Asymmetric Pairs AC	Asymmetric Pairs ACI	Percent Asymmetric AC	Percent Asymmetric ACI
1	132	63	93	38	70	60
2	625	1381	518	224	83	16
3	79	12	59	11	75	92
4	96	202	74	38	77	19
5	410	837	309	246	75	29
6	385	891	275	214	71	24

Table 5. Comparison of critical pairs

It seems that the AC completion process generates far fewer symmetric, and therefore redundant, critical pairs. For the larger problems we often see that 75% of the critical pairs generated by AC completion are kept after the symmetry test. For ACI completion, however, the program often spends a great deal of time generating a very large number of critical pairs, only to expend more time eliminating them via the symmetry test. For example, Table 5 shows that for the commutative ring problem of Example 2, 1381 critical pairs were computed yet only 224, or 16% survived the symmetry test. The other 84% were redundant. This suggests that we may make a real improvement in this portion of the run time if we are able to directly generate only the asymmetric critical pairs. This issue is addressed in [Ma88].

6. Conclusion

The extent to which we achieved the goals stated in the introduction is considered first. Then we discuss some unforeseen accomplishments.

1. We definitely extended the theory of [JK86] to cover the ACI case; primarily through the adding of restrictions to rules. We postulate that restrictions will always be necessary when dealing with an E which has infinite congruence classes.
2. We demonstrated that an ACI process takes larger inference steps than its AC counterpart. We are closer to building in all of Abelian group theory.
3. Our implementation of the ACI process was not faster than that of the AC process when generating complete sets for the same algebraic structures. Unfortunately, therefore, the goal of improving efficiency was not achieved. We believe, however, that further exploitation of symmetry in the unification algorithm could make this goal achievable.
4. ACI complete sets definitely provide a more compact representation of a given algebraic theory than their AC counterparts.

When we started this research we thought it would probably be fairly simple to move the identity from the set of rules to the unification mechanism. After all, an ACI unification algorithm was known and it was not much different from AC unification, the method for finding complete sets modulo AC was well understood, and the identity is a very simple rule. Thus, we were surprised at the rich new features which greeted us. Namely, we needed restrictions on substitutions and a new inference rule. Furthermore, we found that symmetry played a significant role. Learning about these new ideas, which we had not anticipated, was every bit as beneficial as reaching the goals we set out to achieve.

References

[BD87] L. Bachmair and N. Dershowitz, "Completion for rewriting modulo a congruence," *Rewriting Techniques and Applications, Lecture Notes in Computer Science* **256**, Springer-Verlag (1987), pp. 192–203.

[Ba88] T. Baird, "Complete sets of reductions modulo a class of equational theories which generate infinite congruence classes," Ph.D. Dissertation, University of Missouri—Rolla, Rolla, MO, (1988).

[FG84] R. Forgaard and J. V. Guttag, "A term rewriting system generator with failure-resistant Knuth-Bendix," Technical Report, MIT Laboratory for Computer Science, Massachussets Institute of Technology, Cambridge, MA (1984).

[JK86] J.-P. Jouannaud and H. Kirchner, "Completion of a set of rules modulo a set of equations," *SIAM Journal of Computing*, **15** (1986), pp. 1155–1194.

[KB70] D. Knuth and P. Bendix, "Simple word problems in universal algebras," *Computational Problems in Abstract Algebras*, J. Leech, ed., Pergamon Press, Oxford, England, (1970), pp. 263–297.

[LB77] D. Lankford and A. Ballantyne, "Decision procedures for simple equational theories with commutative-associative axioms: complete sets of commutative-associative reductions," Memo ATP-39, Dept. of Mathematics and Computer Science, University of Texas, Austin, Texas (1977).

[Ma88] B. Mayfield, "The role of term symmetry in equational unification and completion procedures," Ph.D. Dissertation, University of Missouri—Rolla, Rolla, MO, 1988.

[PS81] G. Peterson and M. Stickel, "Complete sets of reductions for some equational theories," *J. ACM*, **28** (1981), pp. 233–264.

[St81] M. Stickel, "A unification algorithm for associative-commutative functions," *J. ACM*, **28** (1981), pp. 423–434.

Completion-Time Optimization of Rewrite-Time Goal Solving*

Hubert Bertling
Harald Ganzinger
Fachbereich Informatik, Universität Dortmund
D-4600 Dortmund 50, W. Germany
uucp, bitnet: hg@unido

Completion can be seen as a process that transforms proofs in an initially given equational theory to rewrite proofs in final rewrite rules. Rewrite proofs are the normal forms of proofs under these proof transformations. The purpose of this paper is to provide a framework in which one may further restrict the normal forms of proofs which completion is required to construct, thereby further decreasing the complexity of reduction and goal solving in completed systems.

1 Introduction

Completion computes final sets of (conditional) equations such that any equational consequence can be proved by a *rewrite proof*. Being able to restrict attention to rewrite proofs does very much decrease the search space for proofs. It may, under certain additional requirements, even lead to an efficient term evaluator and decision procedure for the equational theory. Conceptually, completion can be seen as a process that transforms any proof in the initial system into a rewrite proof of the final system. Rewrite proofs can be regarded as the *normal forms* of proofs under these proof transformations. This idea is due to Bachmair, Dershowitz, and Hsiang [BDH86], [Bac87], and has been carried over to the conditional case by the second author [Gan87], [Gan88a].

The purpose of this paper is to provide a framework in which one may *further restrict the normal forms* of proofs which completion is required to construct, thereby further *decreasing the search space* for proofs in completed systems.

This paper is concerned with *conditional equations* and one of its basic assumptions is that one should not put any syntactic restrictions, e.g. about *extra variables* in the condition or in the right side, on conditional equations. A concept of logic programming with conditional equations in which such extra variables are not admitted would be far too restrictive in practice. The use of extra variables in conditions of Prolog-clauses — often called "local variables" in this context — is essential for Prolog-programming. Also, many of the examples in [DP85], [Fri85], and [GM84], where logic programs consist of conditional equations, exploit this facility of extra variables. Furthermore, extra variables can be generated during transformation of equations, e.g. flattening [BGM87], upon compilation of conditional equations into executable code. If, however, a conditional equation has extra variables in the condition, *goal solving* is needed for checking the applicability of this equation and, hence, for computing the normal forms of terms. Goal solving in conditional equational theories can be very complex. Not even narrowing is a complete goal solving method for canonical systems of conditional rewrite rules (with extra variables in conditions) [GM88]. Even when narrowing is complete, the rewrite relation in a canonical system can be undecidable [Kap84].

Our idea to improve the efficiency of rewriting and goal solving in the conditional case is to make completion transform the system in a way such that not all solutions for the condition of an equation need to be found at rewrite-time. Then, goal solving can possibly be implemented by *incomplete but more efficient* methods. The incompleteness of the goal solving methods which we want to use at rewrite-time will have to be "completed" at completion-time.

Completion in the sense of [KR88] and [Gan87] already restricts the normal forms of proofs to rewrite proofs in which clauses are applied under *irreducible and reductive substitutions* only. This additional property of the normal forms of proofs can immediately be exploited for goal solving. We will show that *restricted*

*This work is partially supported by the ESPRIT-project PROSPECTRA, ref#390.

paramodulation is a *complete* goal solving method in completed systems. Restricted paramodulation will become a restricted kind of narrowing in cases in which the consequent of any final equation can be oriented. Hence, counter examples to the completeness of narrowing like the one presented in [GM88] cannot occur as part of completed systems.

Apart from the standard restriction to reductive substitutions, further restrictions of the rewrite-time application of equations can be useful, if they induce further prunings of paramodulation paths.

Application restrictions will formally be given as sets of *relevant substitutions*. At rewrite-time only relevant substitution instances of equations need to be considered. To apply an equation with an irrelevant substitution is not forbidden but may make reduction and goal solving less efficient. We will prove that restricted paramodulation where superposition is restricted to *potentially relevant* unifiers is a complete goal solving method in complete systems of application-restricted equations. A complete system is one which is canonical and *logical*, where a system is logical if equational consequences can be proved by using relevant substitution instances of equations only. Concrete examples of application restrictions which we will discuss are nonoperational equations, oriented conditions, ignorant conditions, and combinations thereof.

Nonoperational equations do not have relevant substitutions at all. Such equations are often created during completion as superposition instances. To be able to ignore these completely may make completion terminate and improve reduction speed at rewrite-time.

For *oriented conditions*, oriented goal solving is the appropriate method. Solving oriented goals has, for example, been advocated in [DS88] as a method for pruning some redundant paths in general goal solving. Another important application of oriented goals are conditional equations which result from transliterating let-expressions with patterns in functional languages. A familiar example is the definition of quicksort as

$$split(x,l)\doteq(l_1,l_2) \Rightarrow sort(cons(x,l))\doteq append(sort(l_1),cons(x,sort(l_2))),$$

where the intended operational meaning is to normalize $split(x,l)$ and then match the result with (l_1,l_2) to obtain the complete binding of the variables of the replacement

$$append(sort(l_1),cons(x,sort(l_2))).$$

In the presence of equalities on the constructor terms, as may be the case in a MIRANDA-program, it is not at all trivial to guarantee that this operational semantics is logically complete.

The quicksort-definition is an example of an equation with oriented conditions which we will call *quasi-reductive* below. Quasi-reductive equations are a generalization of reductive equations, cf. [Kap84], [JW86], [Kap87]. As for reductive equations, the applicability of a quasi-reductive equation is decidable by a backtrack-free process and no termination proofs are needed at rewrite-time.

An *ignorant* condition is one for which some subset of equations can be partially ignored when searching for solutions. Paramodulation need not compute particular superpositions with these.

The work in this paper is based on earlier results by the second author on the subject of completion of conditional equations [Gan87], [Gan88a] and on the results presented in a companion paper [BG89] on unfailing completion in the presence of application restrictions, cf. section 3 for a short summary of the results of the latter paper. Completion of Horn clauses as studied by Kounalis and Rusinowitch [KR88] appears as a particular case in our framework. It is the case in which the relevant substitutions of any equation are exactly the reductive ones. Our work employs proof orderings as pioneered in [BDH86], [Bac87]. One of the main practical advantages of proof orderings over semantic trees as used in [KR88] seems to be the increased power for simplification and elimination of equations during completion. As a consequence, completion terminates much more often and becomes a practical tool for the compilation of conditional equations.

2 Application-Restricted Equations

2.1 Basic Concepts

In what follows we assume a reduction ordering $>$ to be given on the set of terms with variables over the given signature Σ. $>_{st}$ will denote the transitive closure of $> \cup\ st$, with st the strict subterm ordering. $>_{st}$ is noetherian and stable under substitutions.

As motivated for in the introduction, we do not want to put restrictions on the syntactic form of conditional equations. In particular, conditions and right sides may have extra variables. Equations need not be reductive. To compensate for this permissiveness, the application of equations will be restricted.

In the following, the term "equation" is used for conditional and unconditional equations likewise. Our notation is $C \Rightarrow s \doteq t$ for conditional equations with a condition C. s and t are terms over Σ. Conditions are possibly empty conjunctions of (unconditional) Σ-equations. If C is empty we simply write $s \doteq t$. E will always denote a set of equations. Equations can be used both directions for rewriting. If $e = (C \Rightarrow s \doteq t)$ is an equation, its inverse e^- is obtained by reversing the consequent of e, i.e. $e^- = (C \Rightarrow t \doteq s)$. $E^=$ is the set of equations and their inverses, together with the reflexivity of $\doteq$ as an auxiliary equation:

$$E^= = E \cup \{e^- \mid e \in E\} \cup \{(x \doteq x) \doteq true\}$$

We assume that the operators $\doteq$ and $true$ do not occur in Σ. Moreover, $true$ is supposed to be smaller than any other term in the reduction order.

An equation $u_1 \doteq v_1 \wedge \ldots \wedge u_n \doteq v_n \Rightarrow u \doteq v$ is called *reductive*, if $u > v$, $u > u_i$, and $u > v_i$, for $1 \leq i \leq n$. A substitution σ is called *reductive* for $e = (C \Rightarrow s \doteq t)$, if $e\sigma$ is reductive *and if s is not a variable*. By $O(\phi)$, ϕ a term or a formula, we denote the set of nonvariable occurrences in ϕ.

Definition 2.1 *A system of* **application-restricted equations** *is a pair, denoted E^R, consisting of a set E of conditional equations and of* **application restrictions** *R. R associates with any equation $e = (C \Rightarrow s \doteq t) \in E^=$ a set of substitutions $R(e)$, called the* **relevant substitutions** *for e, such that the following is satisfied:*

1. *$R((x \doteq x) \doteq true)$ is the set of all substitutions.*
2. *If $\sigma \in R(e)$, then σ is reductive for e.*

The complements $\overline{R(e)}$ and $\overline{R(e^-)}$ of $R(e)$ and $R(e^-)$, i.e. the set of substitutions not in $R(e)$ and $R(e^-)$, respectively, are called the **irrelevant substitutions** *of e and e^-.*

Application restrictions specify the substitutions $\overline{R(e)}$ and $\overline{R(e^-)}$ for which an equation need *not* be applied from left to right or from right to left, respectively, at rewrite-time. The intention is not to strictly forbid the use of irrelevant substitutions. Rather, we envisage to use incomplete but more efficient goal solving methods for testing the applicability of an equation. This will then still lead to complete mechanisms for reduction and goal solving as long as the incompleteness of the goal solving method is limited to irrelevant substitutions.

Rewriting with application-restricted equations is just classical recursive rewriting [Kap84] in which only relevant substitution instances of equations are used:

Definition 2.2 *$N[u\sigma] \rightarrow_{E^R} N[v\sigma]$, iff $e = (u_1 \doteq v_1 \wedge \ldots \wedge u_n \doteq v_n \Rightarrow u \doteq v) \in E^=$, $\sigma \in R(e)$, and σ is a $\rightarrow_{E^R}$-***solution** *of any of the conditions, i.e. $(u_i\sigma \doteq v_i\sigma) \rightarrow^*_{E^R} true$, for $1 \leq i \leq n$. (The latter is equivalent to the existence of terms w_i such that $u_i\sigma \rightarrow^*_{E^R} w_i \leftarrow^*_{E^R} v_i\sigma$, also written as $u_i\sigma \downarrow_{E^R} v_i\sigma$.) By $t \downarrow_{E^R}$ we denote the (any) result of E^R-rewriting t into an irreducible term.*

To achieve completeness of restricted paramodulation, cf. below, the following additional requirement must be met in application-restricted equations:

Definition 2.3 *A system of application-restricted equations is called* **closed under reduction**, *if $\sigma \in R(C \Rightarrow s \doteq t)$ implies $\sigma' \in R(C \Rightarrow s \doteq t)$, whenever $s\sigma = s\sigma'$ and $\sigma \rightarrow^*_{E^R} \sigma'$, that is, $x\sigma \rightarrow^*_{E^R} x\sigma'$, for every variable x.*

If the condition or the right side of an equation e has extra variables, $e\sigma$ can be different from $e\sigma'$, even if $s\sigma = s\sigma'$. Requiring closure under reduction avoids some of the problems which come with extra variables. Note that we do not allow rewriting with equations in which the left side is just a variable. The closure property is trivially satisfied, if only irreducible substitutions are relevant. From now on we will assume that application-restricted equations are closed under reduction.

Rewriting with application-restricted equations is terminating. This follows from the requirement that relevant substitution instances be reductive. Hence, Newman's lemma can be applied to infer confluence from local confluence. For reductive equations, the critical pairs lemma is valid [JW86]. Moreover, narrowing is a complete goal solving method for confluent systems of reductive equations [Kap84]. Below we will see that these properties carry over to our more general case. If we also had admitted nonreductive substitutions to be relevant, these important prerequisites for practical completion, rewriting and goal solving would have been lost immediately. This is shown by counter examples in [DOS88] (critical pairs lemma) and [GM88] (narrowing), respectively.

Clearly, $\equiv_E = \overset{*}{\leftrightarrow}_E$, however, $\equiv_E$ and $\overset{*}{\leftrightarrow}_{E^R}$ need *not* be equal. We will require completion to complete a system of application-restricted equations in this sense, too.

Definition 2.4 *E^R is called* **complete,** *iff $\rightarrow_{E^R}$ is confluent and* **logical,** *that is, $\equiv_E = \overset{*}{\leftrightarrow}_{E^R}$.*

2.2 Examples of Application Restrictions

The most severe application restriction for an equation e is represented by $R(e) = R(e^-) = \emptyset$. This amounts to the wish of being able to completely ignore e for rewriting and goal solving. We will call an equation *nonoperational* in this case. In the examples, we will annotate nonoperational equations by a "-". We call an equation *operational*, if it has at least one relevant substitution.

The other extreme in our framework is to require all reductive substitution instances of an equation e and its inverse e^- to be relevant. We will call such an equation *fully operational*. This will be the default case in the examples. If a fully operational equation e is reductive, $R(e)$ contains all substitutions, and $R(e^-)$ is empty.

An interesting case in between these extremes is represented by the notion of an *oriented condition*. For an equation $e = (C \Rightarrow s \doteq t)$ to have an oriented condition $u \doteq v \in C$ means that $R(e)$ and $R(e^-)$ contain all reductive substitutions σ for e and e^-, respectively, such that $v\sigma$ is $\rightarrow_{E^R}$-irreducible. We will subsequently use the notation $u \stackrel{\rightarrow}{=} v$ for oriented conditions. At rewrite-time it is sufficient to perform oriented goal solving in the sense of [DS88] for an oriented condition. No paramodulation superposition on v is required. The concept of oriented conditions for equations is a generalization of the notion of ground-normal systems of Bergstra and Klop [BK82].

We call a condition equation $u \doteq v$ of an equation e *e'-ignorant*, $e' \in E$, and write $u \doteq_{-e'} v$, if e' needs not be used for the first step of condition solving. More exactly, $R(e)$ and $R(e^-)$ contain all reductive substitutions σ for e and e^-, respectively, such that $x\sigma$ is irreducible, for any $x \in var(u \doteq v)$ and, additionally, $(u \doteq v)\sigma$ is e'-irreducible, i.e. cannot be $\rightarrow_{E^R}$-rewritten using e'.

These last two kinds of application restrictions can be useful in order to restrict the search space of narrowing, cf. examples below.

Note that all these examples are application restrictions which are closed under reduction, as required. Clearly, restrictions can be combined by intersection without loosing this closure property.

As an example for the combination of two restrictions consider the restriction of goal solving for a condition $u \doteq v$ of an equation e to mere unification of u and v. We can model this restriction by replacing the condition $u \doteq v$ by two oriented ones $x \stackrel{\rightarrow}{=} u$ and $x \stackrel{\rightarrow}{=} v$, with a new variable x. We will write $u \equiv v$ and call the condition *syntactic* in this case.

For examples of concrete systems the reader is referred to section 4.

2.3 Restricted Paramodulation

Paramodulation [RW69] is an inference rule for computing solutions of goals. For a complete system of application-restricted equations, paramodulation can be greatly restricted. *Restricted paramodulation* is paramodulation without superpositions at variables and without using the functional reflexive axioms. In our framework, an additional reduction of the search space will be achieved by immediately rejecting superpositions with equations for which the unifying substitution is not potentially relevant according to the given application restrictions.

To prove the completeness of restricted paramodulation by the usual lifting of rewriting or resolution derivations to paramodulation derivations, cf. [Hus85], [Höl87], [BGM87], [Pad88] among others, the ability to restrict attention to reduced substitutions is of crucial importance and not trivially satisfied in the case of conditional rewriting [Höl87], [GM88]. The problems arise from extra variables. Fortunately we can show in the next lemma that the restriction to reduced substitutions does not affect completeness in our case.

To that end, let in the following E^{R_r} be defined by

$$R_r(e) = \{\sigma \in R(e) \mid \sigma(x) = \sigma(x) \downarrow_{E^R}\},$$

for any $e \in E^=$, i.e. in E^{R_r} application of equations is further restricted to irreducible substitutions.

Lemma 2.5 *Let E^R be given such that $\rightarrow_{E^R}$ is confluent. Then,*

1. $\rightarrow_{E^R} \subset (\rightarrow^*_{E^{R_r}} \circ \leftarrow^*_{E^{R_r}})$
2. $\rightarrow_{E^{R_r}}$ *is confluent.*

Hence, the $\rightarrow_{E^R}$- and $\rightarrow_{E^{R_r}}$-normal forms of terms are identical.

Proof: By noetherian induction over $>_{st}$.

1. Suppose, $N = N[s\sigma] \rightarrow_{E^R} N[t\sigma]$ using $C \Rightarrow s\doteq t \in E^=$ and $\sigma \in R(C \Rightarrow s\doteq t)$. Let σ' be obtained by reduction of σ, i.e. $x\sigma' = x\sigma \downarrow_{E^R}$, for any variable x. If $u\doteq v \in C$, $u\sigma \downarrow_{E^R} v\sigma$, hence $u\sigma' \leftarrow^*_{E^R} u\sigma \downarrow_{E^R} v\sigma \rightarrow^*_{E^R} v\sigma'$, with all terms on this chain of equation applications smaller than N wrt. $>_{st}$. Using the induction hypothesis for 1 and 2, we obtain $u\sigma' \downarrow_{E^{R_r}} v\sigma'$. Now, two cases for $s\sigma'$ have to be distinguished.
   (a) $s\sigma \neq s\sigma'$:
   As s cannot be a variable, cf. 2.1 and the definition of reductive substitutions, $s\sigma > s\sigma'$ and any step of rewriting from $s\sigma$ to $s\sigma'$ has a redex smaller than N wrt. $>_{st}$. Consequently, $s\sigma \downarrow_{E^{R_r}} s\sigma'$, by induction hypothesis for 1 and 2. Moreover, $s\sigma' \downarrow_{E^R} t\sigma$, from the confluence of $\rightarrow_{E^R}$, and, therefore, $s\sigma' \downarrow_{E^{R_r}} t\sigma$ by induction hypothesis for 1 and 2. Again applying the induction hypothesis to $s\sigma'$, we infer $s\sigma \downarrow_{E^{R_r}} t\sigma$, as required.
   (b) Otherwise, $\sigma' \in R(C \Rightarrow s\doteq t)$, from the closedness of relevant substitutions under reduction, cf. 2.3. As σ' is reduced, $\sigma' \in R_r(C \Rightarrow s\doteq t)$. Now, $s\sigma = s\sigma' \rightarrow^*_{E^{R_r}} t\sigma' \downarrow_{E^{R_r}} t\sigma$ where the convergence of $t\sigma'$ and $t\sigma$ follows from the induction hypothesis for 1 and 2.
2. This part is immediate from $\rightarrow_{E^{R_r}} \subset \rightarrow_{E^R}$, from 1, and from the confluence of $\rightarrow_{E^R}$.

□

Lemma 2.5 is already incorporated in the following definition:

Definition 2.6 *A substitution σ is called* **potentially reductive** *for e, iff there exists a substitution σ' such that $\sigma\sigma'$ is reductive for e. σ is called* **potentially relevant** *for an equation e, iff there exists a substitution σ' such that $\sigma\sigma'$ is a $\rightarrow_{E^R}$-solution of the conditions of e, and is relevant, i.e. $\sigma\sigma' \in R_r(e)$.*

In particular, potentially relevant substitutions must be irreducible.

Definition 2.7 *Given a system E^R of application-restricted equations,* **restricted paramodulation** *is the derivation relation between of goals which is defined by two inference rules given below. Goals are, as usual, sets (conjunctions) of literals where a literal is an unconditional equation or "true". Note that the reflexivity of $\doteq$ is among the equations in $E^=$.*

1. *Let G be a goal.*
$$G \cup \{u\doteq v\} \mapsto_{[(e,\sigma)]} G\sigma \cup C\sigma \cup \{(u\doteq v)[o \leftarrow t]\sigma\},$$
if $e = (C \Rightarrow s\doteq t) \in E^=$ with new variables, σ the most general unifier of s and $(u\doteq v)/o$, where o is a nonvariable occurrence in $u\doteq v$, such that, moreover, σ is potentially relevant[1] *for e.*
2. *If $G \mapsto_{[(e_1,\sigma_1),\ldots,(e_n,\sigma_n)]} G'$ and $G' \mapsto_{[(e,\sigma)]} G''$, then*
$$G \mapsto_{[(e_1,\sigma_1\sigma),\ldots,(e_n,\sigma_n\sigma),(e,\sigma)]} G'',$$
if $\sigma_i\sigma$ is potentially relevant for e_i, for $1 \leq i \leq n$.

σ is called an **answer substitution** *to G, if $G \mapsto_{[(e,\sigma),\ldots]} \{true\}$.*

[1]It is assumed that if e results from e' by renaming variables, $R(e)$ is $R(e')$, composed with the renaming substitution.

In practice, to keep the search space of paramodulation small, good criteria are required to decide when a substitution cannot be potentially relevant. Of course, nothing can go wrong, if this property is undecidable and an irrelevant answer substitution is computed. The tests for potential relevance should on one hand be efficient to compute and on the other hand approximate the property of potential relevance as precisely as possible. The particular examples of application restrictions which we have described above come with simple criteria of guaranteed irrelevance.

Proposition 2.8 *1. If σ is reducible, σ is not potentially relevant for any equation.*

2. If e is nonoperational, no potentially relevant substitution exists for e and e^-.

3. A substitution σ is not potentially relevant for an equation $C \Rightarrow s \dot{=} t$, if $s\sigma$ is not a maximal term in $(C \Rightarrow s \dot{=} t)\sigma$ with respect to the reduction ordering.

4. If $u \stackrel{\rightarrow}{=} v$ is an oriented condition in e, and $v\sigma$ is $\rightarrow_{E^R}$-reducible, then σ is not potentially relevant for e.

5. If $u \equiv v$ is a syntactic condition in e, and $v\sigma$ or $u\sigma$ is $\rightarrow_{E^R}$-reducible, then σ is not potentially relevant for e.

6. If $u \dot{=}_{-e'} v$ is an e'-ignorant condition in e, and e' can be applied in $(u \dot{=} v)\sigma$ at a nonvariable position in $u \dot{=} v$, then σ is not potentially relevant for e.

For paramodulation, 1 implies that superposition can be restricted to basic occurrences as in basic narrowing, cf. [Hul80]. Nonoperational equations need not be considered for superposition at all, cf. 2. If the unifying substitution obtained from superposing the left side s of the consequent of an equation $C \Rightarrow s \dot{=} t$ creates a substitution instance of $C \Rightarrow s \dot{=} t$ which is not reductive, the paramodulation step can immediately be cut according to 3. In particular we never need to replace a term by a greater term in the reduction ordering. Hence, if all equations are reductive and fully operational, restricted paramodulation reduces to conditional narrowing. New goals which are created during paramodulation from an oriented condition of an equation need not be superposed on their right sides, as follows from 4. Likewise, syntactic goals are only subject to resolution with $(x \dot{=} x) \dot{=} true$, cf. 5. Finally, goals $u \dot{=}_{-e'} v$ need not be superposed by e'.

Restricted paramodulation is a complete goal solving method for complete systems of application-restricted equations.

Theorem 2.9 *Let E^R be a complete system of application-restricted equations. Let G be a goal and σ be an $\rightarrow_{E^R}$-irreducible substitution. $u\sigma \equiv_E v\sigma$, for every $u \dot{=} v \in G$, iff there exists an answer substitution σ' to G obtained from restricted paramodulation such that $\sigma'\sigma''|_{var(G)} = \sigma|_{var(G)}$, for some other substitution σ''.*

Proof: By the usual lifting of rewriting derivations to paramodulation derivations, cf. [Hus85] among others. Lemma 2.5 is of crucial importance for being able to ignore the functional reflexive axioms and superposition into variable occurrences. □

In particular, restricted paramodulation is a complete method for checking the applicability of an equation when computing rewrite derivations according to $\rightarrow_{E^R}$. Note that 2.9 implies the completeness of conditional narrowing in canonical systems of reductive rewrite rules as stated in [Kap84]. To see why the counter example in [GM88] does not apply here, suppose E to be given as

$$a \dot{=} b$$
$$a \dot{=} c$$
$$f(b,x) \dot{=} f(x,c) \Rightarrow b \dot{=} c$$

with the assumption that $a > b > c$. As there exists no reductive substitution for

$$f(b,x) \dot{=} f(x,c) \Rightarrow b \dot{=} c,$$

$\rightarrow_{E^R}$ cannot be confluent, if the first two equations are operational.

3 Completion of Application-Restricted Equations

3.1 Completion Inference Rules

This section is a short summary of the main results of a companion paper [BG89] on completion of systems of application-restricted equations. There are two main new aspects of this method compared to the semantic tree-based approach in [KR88]. One is our more general framework of how a rewrite relation is generated from a system of syntactically unrestricted conditional equations. In [KR88], all reductive substitutions of an equation are considered relevant for rewriting. Besides this, results of the latter paper about rewriting in completed systems assume that final equations do not have extra variables.[2] Another advantage of our proof ordering-based method is that we have much more power to reduce and eliminate conditional equations during completion. This, in turn, leads to a substantially better termination behaviour in practice. The main difference between [Gan87] and [BG89] is that the latter paper uses a slightly modified proof algebra and proof ordering to allow for unfailing completion. Moreover, the latter paper proves the connections between application restrictions and fairness of completion which will be used below.

In this paper we do not have sufficient space to explain the underlying concept of proofs and proof orderings in any detail. We can only briefly describe the basic inferences CC of conditional completion, without indicating in which way they allow to transform proofs into the required normal forms.

Application Restrictions

The completion inference rules are to a large extent independent of particular application restrictions. They, however, consider their general requirements, reductivity of relevant substitutions and closedness of relevant substitutions under reduction. The latter property is achieved by the construction of proof orderings which make the application of equations under a reducible substitution more complex than an application which uses the reduced substitution instead. In other words, the completion inference rules and the proof ordering already guarantee that in any final system which can be obtained as the limit of a fair completion process the proof of any equational consequence can be given as a rewrite proof in which equations are applied under reductive and irreducible substitutions only. Particular application restrictions beyond reductivity and irreducibility only affect the fairness property of a completion derivation.

Formally one may hence take the view that application restrictions are added only when the final system has been constructed. Then, it only remains to verify the corresponding fairness constraints. In practice, application restrictions are defined at the time when an equation is generated during completion and, perhaps, later changed, if the final system is not as one had expected. Although application restrictions refer to the yet unknown final system, the completion procedure can make good use of them to minimize the number of superpositions to be computed. Knowing the restrictions in advance also helps to improve reduction at completion-time, e.g. in the case of quasi-reductive equations.

Adding a Critical Pair

Definition 3.1 *Let $D \Rightarrow u \doteq v$, $C \Rightarrow s \doteq t \in E^=$. Furthermore, assume that their variables have been renamed such that they do not have any common variables. Assume, moreover, that o is a nonvariable occurrence in s such that u/o and s can be unified with a mgu σ. Then,*

$$(C \wedge D \Rightarrow u[o \leftarrow t] \doteq v)\sigma$$

is a **contextual critical pair** *with superposition term $u\sigma$, provided σ is potentially reductive for both $C \Rightarrow s \doteq t$ and $D \Rightarrow u \doteq v$.*

Critical pairs are computed to replace peaks in $\rightarrow_{E^R}$-rewriting. This definition allows to superpose both sides of the consequent of any two operational equations to form critical pairs. However we have restricted attention to such critical pairs which can possibly appear in a peak of two rewrite steps. If σ is not potentially reductive for one of the overlapping equations, no peak in the final system can be an instance of the overlap.

[2]Note, however, that in our more liberal case $\rightarrow_{E^R}$ can be undecidable. Hence, even the existence of a complete E^R does not imply the decidablity of the equational theory of E, cf. interpreter example in section 4.2.

Adding a Paramodulation Instance
Paramodulation inferences on conditions are computed to achieve $\overset{*}{\leftrightarrow}_{E^R} = \equiv_E$ in the limit.

Definition 3.2 *Let* $e = (D \wedge u \dot{=} v \Rightarrow l \dot{=} r) \in E$, *with* $u \dot{=} v$ *the superposition condition of* e, *and* $e' = (C \Rightarrow s \dot{=} t) \in E^=$. *Furthermore, assume that variables have been renamed to achieve disjointness. Let* o *be a nonvariable occurrence in* $u \dot{=} v$ *such that* $(u \dot{=} v)/o$ *and* s *can be unified with a mgu* σ. *Then,*

$$(C \wedge D \wedge (u \dot{=} v)[o \leftarrow t] \Rightarrow l \dot{=} r)\sigma$$

is called **an instance of** $D \wedge u \dot{=} v \Rightarrow l \dot{=} r$ **by paramodulating** $C \Rightarrow s \dot{=} t$ **on the condition** $u \dot{=} v$, *if* σ *is potentially reductive for* e'.

Simplification by Contextual Reduction
The simplification inference rules allow for the reduction of terms in a conditional equation by using the condition equations as additional hypotheses.

Definition 3.3 *Let* E *be a set of conditional equations and* C *be a set of unconditional equations in which the variables are considered as constants, i.e.* $\equiv_C$ *is assumed to be the closure of* C *under equivalence and contexts, but not under substitutions.* $s \rightarrow_{E,C} t$ *iff* $s \equiv_C K[l\sigma]$, $s \geq K[l\sigma]$, $t = K[r\sigma]$, $e = (D \Rightarrow l \dot{=} r \in E^=)$, σ *reductive for* e, *and* $u\sigma \downarrow_{E,C} v\sigma$, *for any* $u \dot{=} v \in D$. *In this case,* $\downarrow_{E,C}$ *stands for* $u\sigma \rightarrow^*_{E,C} u'$, $v\sigma \rightarrow^*_{E,C} v'$, *and* $u' \equiv_C v'$.

A condition of an equation $C \wedge u \dot{=} v \Rightarrow s \dot{=} t$ may be reduced to $C \wedge w \dot{=} v \Rightarrow s \dot{=} t$, if $u \rightarrow_{E,C} w$. The consequent of $C \Rightarrow s \dot{=} t$ may be reduced to $C \Rightarrow u \dot{=} t$, if $s \rightarrow_{E,C} u$ and if some additional restrictions are satified. These restrictions refer to the size of the redex, the proof that has lead to the construction of $C \Rightarrow s \dot{=} t$, and to the reducing equation. For details see [BG89].

Deleting a Trivial Equation
An equation $C \Rightarrow s \dot{=} t$ may be eliminated if the current set E of equations admits a proof of $C \vdash s \dot{=} t$ which is simpler wrt. the proof ordering than the proof in which $C \Rightarrow s \dot{=} t$ is applied under the identity substitution to the hypotheses C. Again we refer to [Gan87] and [BG89] for details. In particular, proof orderings can be defined in a way such that a simpler proof exists, if $s \equiv_C t$ — variables in C considered as constants. If E contains an equation which properly subsumes $C \Rightarrow s \dot{=} t$, the latter can be eliminated. These two particular cases of simpler proofs, together with simplification by contextual rewriting, already provide quite substantial power for the elimination of equations. In practice, a more general mechanism which enumerates proofs of $C \vdash_E s \dot{=} t$ to a certain bound and compares their complexities against the complexity of the proof that has caused the creation of $C \Rightarrow s \dot{=} t$ is required to improve the termination behaviour of completion, cf. [Gan88a].

3.2 Fair Completion Derivations

A completion procedure is a mechanism which applies the CC inference rules, starting from an initial set of equations E, to produce a sequence $E = E_0, E_1, \ldots$ with a limit system $E_\infty = \cup_j \cap_{k \geq j} E_k$. The final system E_∞ is equipped with application restrictions R. A completion derivation is called *fair*, if E^R_∞ is complete. The way the application restrictions R are defined determines which superpositions (critical pairs, paramodulation superpositions) have to be computed during completion.

Particularly simple fairness results are obtained if a particular pattern in the proof of an instance of a condition may generate an irrelevant substitution for the corresponding conditional equation and if any irrelevant substitution can be generated that way. It is the first step of a condition rewrite proof which we are particulary interested in. For that purpose, let first(P) denote the pair (o, e), if the first step in the sequence $P = s \rightarrow_{E^R} \ldots \rightarrow_{E^R} t$ of rewritings applies the equation e at the occurrence o in s. We call (o, e) the *start pattern* of P.

Definition 3.4 *Let* E^R *and* $e = (u_1 \dot{=} v_1 \wedge \ldots \wedge u_n \dot{=} v_n \Rightarrow u \dot{=} v) \in E$ *be given. A set*

$$\alpha_{E^R}(e) \subset O(u_1 \dot{=} v_1 \wedge \ldots \wedge u_n \dot{=} v_n) \times E^=$$

is called an **approximation** *of* $\overline{R(e)}$ *and* $\overline{R(e^-)}$ *(by irrelevant condition proof patterns), iff for any* E^R*-irreducible* $\sigma \in \overline{R(e)} \cup \overline{R(e^-)}$ *for which* $(u_i \dot{=} v_i)\sigma \rightarrow^*_{E^R}$ *true,* $1 \leq i \leq n$, *there exists* $1 \leq j \leq n$ *and a particular rewrite proof* $P = (u_j \dot{=} v_j)\sigma \rightarrow_{E^R} \ldots \rightarrow_{E^R}$ *true such that* first(P) $\in \alpha_{E^R}(e)$.

The basic observation behind the following theorem is that E_∞^R is logical, if, for a given approximation $\alpha_{E_\infty^R}$ of the irrelevant substitutions of E_∞^R, all paramodulation superpositions which are specified by $\alpha_{E_\infty^R}$ have been computed during completion.

Theorem 3.5 ([BG89]) *Let $E_0, E_1, \ldots$ be a CC-derivation. and let R define the application restrictions for the final system E_∞. Moreover, assume that for each $e \in E_\infty$ we have $\alpha_{E_\infty^R}(e)$, an approximation of its application restrictions by illegal condition proof patterns. The derivation is fair, if following holds true:*

1. *If e is an unconditional equation in E_∞, either e or e^- must be reductive and fully operational according to R.*

2. *$\cup_k E_k$ contains every critical pair between final equations in E_∞ such that the unifying substitution of the overlap is potentially relevant for both equations.*

3. *If $e \in E_\infty$, $e' \in E_\infty^=$ and $(o, e') \in \alpha_{E_\infty^R}(e)$, then $\cup_k E_k$ contains all paramodulation instances of e which are obtained by superposing e' at o in e.*

Note that because of the first fairness constraint completion may fail, if an application-restricted equation without conditions cannot be reduced or eliminated during completion. Completion becomes unfailing and, hence, refutationally complete for the equational theory, if the given reduction order is total on ground terms and if any final unconditional equation is fully operational, cf. [BG89].

3.3 Don't Repeat at Rewrite-Time What You Have Done at Completion-Time

Condition proof patterns which approximate application restrictions can be ignored by restricted paramodulation at rewrite- and goal solving-time:

Theorem 3.6 ([BG89]) *Let 1–3 of 3.5 be satisfied. In E_∞^R, restricted paramodulation can be further restricted. If $u \doteq v \in C\sigma$ is a subgoal which is generated by a paramodulation inference using $e = (C \Rightarrow s \doteq t) \in E_\infty$, the first paramodulation step in $u \doteq v$ need not consider any superposition corresponding to a $(o, e') \in \alpha_{E_\infty^R}(e)$.*

Note that this fact can lead to substantial further pruning of paramodulation paths, in particular if the approximation of E_∞^R is very crude.

Let us now describe irrelevant condition proof patterns which approximate the particular examples of application restrictions which we have described above.

1. Suppose e is a nonoperational equation in E^R. In this case, both $\overline{R(e)}$ and $\overline{R(e^-)}$ contain all substitutions. Then, whatever condition $u \doteq v$ of e is selected, any proof of an $(u \doteq v)\sigma$, σ irreducible, will generate an irrelevant substitution. Vice versa, if σ is an irreducible solution of $u \doteq v$, the associated rewrite proof must start with applying a relevant substitution instance of an equation in $E^=$ somewhere at a nonvariable occurrence of $u \doteq v$. Hence, an approximation of $\overline{R(e)}$ and $\overline{R(e^-)}$ is given by the set of nonvariable occurrences o in $u \doteq v$, together with those equations $e' \in E^=$ which superpose at o with an mgu σ which is potentially relevant for e'. (Remember that nonoperational equations e' do not have relevant substitutions σ and, therefore, need not be superposed on the selected condition of e.) Hence, if an equation e is declared as nonoperational during completion, a condition for completion-time superposition should be selected, too. Alternatively, one may, as in [KR88], adopt the strategy of superposing on maximal condition literals. In many cases the strategy of selecting a particular condition for paramodulation will lead to less superpositions.

2. Let e be a fully operational equation in E^R. If e and e^- are not reductive their application is restricted to reductive substitutions. In this case, we can again select an arbitrary condition $u \doteq v$ of e and consider the nonvariable occurrences o in $u \doteq v$, together with those equations $e' \in E^=$ which superpose at o with an mgu σ which is potentially relevant for e' and not reductive for both e and e^-. This set of proof patterns, denoted $\mathrm{NRED}_e(\mathrm{u} \doteq \mathrm{v})$, approximates the irrelevant nonreductive substitutions of e and e^-. $\mathrm{NRED}_e(\mathrm{u} \doteq \mathrm{v})$ need only contain pairs (o, e') such that o is inside the larger of the two terms, in case one of the condition terms $v\sigma$ and $u\sigma$ is smaller than the other in the reduction order. $\mathrm{NRED}_e(\mathrm{u} \doteq \mathrm{v}) = \emptyset$, if e or e^- is reductive.

3. If e contains an oriented condition $u \stackrel{\rightarrow}{=} v$, the set of occurrences o in v and equations $e' \in E^{=}$ which superpose at o with an mgu σ which is potentially relevant for e', is an approximation of this application restriction, provided e or e^- is quasi-reductive, cf. 4.1 below. Otherwise, to make the nonreductive applications irrelevant, the patterns in $\mathrm{NRED}_e(s \doteq t)$, for some arbitrarily selected condition $s \doteq t$, have to be added.

4. If e contains an e'-ignorant condition $u \doteq_{-e'} v$, an approximation of this application restriction is, as expected, $O(u \doteq v) \times \{e'\}$, together with some $\mathrm{NRED}_e(s \doteq t)$, in case both e and e^- are not reductive.

Combinations of restrictions can be approximated by the union of their approximations.

4 Applications and Examples

4.1 Quasi-Reductive Equations

One application of oriented conditions is to model the operational meaning of let-expressions with patterns in functional languages such as MIRANDA or ML. The oriented condition in

$$split(x,l) \stackrel{\rightarrow}{=} (l_1,l_2) \Rightarrow sort(cons(x,l)) \doteq append(sort(l_1), cons(x, sort(l_2)))$$

means operationally that application of the equation (from left to right) can be restricted to substitutions which can be found by first matching $sort(cons(x,l))$ with the current subterm and, then, rewriting $split(x,l)$ to a normal form which matches (l_1,l_2). After performing the match, the replacement will be uniquely determined. Another application of oriented conditions is the handling of non-sort-decreasing rules in order-sorted specifications. The equation

$$i * i \stackrel{\rightarrow}{=} i_{nat \subset int}(n) \Rightarrow square(i) \doteq n,$$

where $i_{nat \subset int}$ is an injection from nat to int, represents rewrite-time type checking during computation of squares and is generated from superposing the non-sort-decreasing definition of $square$

$$i_{nat \subset int}(square(i)) \doteq i * i$$

on the injectivity axiom for $i_{nat \subset int}$. In the latter applications, there may be more that one way of going from a subsort to a supersort. This leads to equalities between injection terms, making oriented goal solving incomplete, if not appropriately completed. For more details cf. [Gan88b]. The latter example is also interesting as alternative axiomatizations without extra variables in the condition would have to involve (partial) retracts from integers to natural numbers, thereby introducing junk into the initial algebra.

For both examples, the matching substitutions can be computed fully deterministically. In addition, reductivity of the substitution is guaranteed. E.g., if $i * i$ is smaller in the reduction order than $square(i)$, which we can assume here, its normal form $i_{nat \subset int}(n)$ and, hence, the replacement n will also always be smaller than $square(i)$. No reductivity tests are required at rewrite-time. Hence, rewriting with equations of this kind is similarly efficient as rewriting with reductive equations as stated in [Kap87]. Equations of this form will be called *quasi-reductive* below. The definition of quicksort above is another example of a quasi-reductive equation.

To simplify the formal treatment, let us assume that equations have oriented conditions only. (If an equation has an unoriented condition $u \doteq v$, we can replace the latter by the two oriented conditions $u \stackrel{\rightarrow}{=} x$ and $v \stackrel{\rightarrow}{=} x$, where x is a new variable.)

We call a conditional equation $u_1 \doteq v_1 \wedge \ldots \wedge u_n \doteq v_n \Rightarrow s \doteq t$ (with oriented conditions) *deterministic*, if, after appropriately changing the orientation of the literals and the order of the condition equations, the following holds true:

$$var(u_i) \subset var(s) \cup \bigcup_{1 \le j < i} (var(u_j) \cup var(v_j)),$$

and

$$var(t) \subset var(s) \cup \bigcup_{j=1}^{n} (var(u_j) \cup var(v_j)).$$

To arrive at a concept for avoiding reductivity proofs at rewrite-time, let us now assume the existence of some enrichment $\Sigma' \supseteq \Sigma$ of the signature such that the given reduction ordering on $T_\Sigma(X)$ can be extended to a reduction ordering on $T_{\Sigma'}(X)$.

Definition 4.1 *A deterministic equation* $u_1 \doteq v_1 \wedge \ldots \wedge u_n \doteq v_n \Rightarrow s \doteq t$, $n > 0$, *is called* **quasi-reductive**, *if there exists a sequence* $h_i(\xi)$ *of terms in* $T_{\Sigma'}(X)$, $\xi \in X$, *such that* $s > h_1(u_1)$, $h_i(v_i) \geq h_{i+1}(u_{i+1})$, $1 \leq i < n$, *and* $h_n(v_n) \geq t$. *An unconditional equation* $s \doteq t$ *is quasi-reductive, if* $s > t$.

The equation

$$i * i \doteq i_{natCint}(n) \Rightarrow square(i) \doteq n$$

becomes quasi-reductive under a recursive path ordering, if $square > *$ in precedence. Choosing, $h_1(\xi) = \xi$, the inequalities $square(i) > i * i$ and $i_{natCint}(n) \geq n$ are obvious. Also quasi-reductive is

$$split(x, l) \doteq (l_1, l_2) \Rightarrow sort(cons(x, l)) \doteq append(sort(l_1), cons(x, sort(l_2))),$$

with $h_1(\xi) = f(\xi, x)$, where f is a new auxiliary function symbol. The termination proofs can be given by an appropriately chosen polynomial interpretation. The interpretation has to be chosen such that $sort(cons(x, l)) > f(split(x, l), x)$ and $f((l_1, l_2), x) \geq append(sort(l_1), cons(x, sort(l_2)))$.[3] Quasi-reductivity is a proper generalization of reductivity:

Proposition 4.2 *If the equation* $u_1 \doteq u_{n+1} \wedge \ldots \wedge u_n \doteq u_{2n} \Rightarrow s \doteq t$ *is reductive, then the equation*

$$u_1 \doteq x_1 \wedge u_{n+1} \doteq x_1 \wedge \ldots \wedge u_n \doteq x_n \wedge u_{2n} \doteq x_n \Rightarrow s \doteq t,$$

is quasi-reductive, if the x_i *are new, pairwise distinct variables.*

Proof: Choose $h_i(\xi) = M(\xi, u_{i+1}, \ldots, u_{2n}, t)$ with M being the multiset operator. □

One other interesting class of quasi-reductive equations is obtained through the process of flattening reductive equations in the sense of [BGM87], cf. section 4.3.

Proposition 4.3 *Let* E *be finite and* $u_1 \doteq v_1 \wedge \ldots \wedge u_n \doteq v_n \Rightarrow s \doteq t \in E$ *a quasi-reductive equation.*

1. *If* σ *is a substitution such that* $u_i\sigma \rightarrow^*_{E^R} v_i\sigma$, $1 \leq i \leq n$, *then,* $s\sigma > t\sigma$.
2. *If* $N' \rightarrow_{E^R} N''$ *is decidable for all terms* N' *such that* $N >_{st} N'$, *then the applicability of the equation* $u_1 \doteq v_1 \wedge \ldots \wedge u_n \doteq v_n \Rightarrow s \doteq t$ *in* N *is decidable.*

Proof: The proof of 1. follows immediately from the assumptions and the fact that $u_i\sigma \geq v_i\sigma$.

For the proof of 2. we note that because the given equation is deterministic, any σ which solves the condition is obtained by rewriting the $u_i\sigma_i$ and matching the rewrites against $v_i\sigma_i$, where σ_i is that part of σ which has been obtained after having matched s against a subterm of N and after having solved the condition equations up to index $i-1$. Because of termination of $\rightarrow_{E^R}$ there are only finitely many irreducible r with $u_i\sigma \rightarrow^*_{E^R} r$ that need to be matched against $v_i\sigma_i$, and, hence, finitely many σ_{i+1} which need to be considered for the next condition. The quasi-reductivity implies that $N \geq_{st} s\sigma = s\sigma_i >_{st} u_i\sigma_i$. Hence any of these r can be effectively computed, as we have assumed the decidability of $\rightarrow_{E^R}$ in terms smaller than N. □

Corollary 4.4 *Let* E^R *be a set of application-restricted equations in which any equation is either nonoperational or quasi-reductive and fully operational. Then,* $\rightarrow_{E^R}$ *is decidable.*

For confluent $\rightarrow_{E^R}$, the normal forms of the instances of the left sides of an oriented condition is uniquely determined. Computing the substitution is, then, completely backtrack-free. Moreover, no termination proofs are required at rewrite-time. The following system is complete and satisfies the requirements of 4.4:

[3] Note that if one defines quicksort by an unconditional equation such as

$$sort(cons(x, l)) \doteq append(sort(first(split'(cons(x, l)))), cons(x, sort(second(split'(cons(x, l))))))$$

there is no way of using simplification orderings for a termination proof because $sort(cons(x, l))$ is embedded in the right side of the equation. (Replacing the unary $split'$ by the binary $split$ does not help.) This problem is essentially the same for any other divide-and-conquer algorithm and can be solved by the use of quasi-reductive equations.

Example 4.5

```
1   0=<n     = true
2   n=<n     = true
3   s(n)=<0 = false
4   s(n)=<s(m) = n=<m
5   n=<m = true => n=<s(m) =true.
6   append([],l) = l
7   append([e|l1],l2) = [e|append(l1,l2)]
8   sort([]) = []
9   split(x,[]) = ([],[])
10  x=<y = false and split(x,l) =→ (l1,l2) => split(x,[y|l]) = ([y|l1],l2)
11  x=<y = true  and split(x,l) =→ (l1,l2) => split(x,[y|l]) = (l1,[y|l2])
12  split(x,l) =→ (l1,l2) => sort([x|l]) = append(sort(l1),[x|sort(l2)])
13- x=<y = false => y=<x = true
14- x=<y = true and y=<z = true => x=<z = true
15- (l1,l2) = (l5,l4) and split(x1,l3) = (l1,l2) =>
            append(sort(l1),[x1|sort(l2)]) = append(sort(l5),[x1|sort(l4)])
16- (l1,l2) = (l5,l4) and x1=<y1 = false and split(x1,l3) = (l1,l2) =>
            ([y1|l1],l2) = ([y1|l5],l4)
17- (l1,l2) = (l5,l4) and x1=<y1 = true  and split(x1,l3) = (l1,l2) =>
            (l1,[y1|l2]) = (l5,[y1|l4])
18-  s(y)=<z = true and x1=<y = true => x1=<z = true
19-  s(y)=<z = true and x=<y = true => s(x)=<z = true
```

This is a complete specification of quicksort over =< on natural numbers. The last seven equations are nonoperational, where 15–17 are some of the contextually reduced, less trivial critical pairs which are generated according to the fairness theorem 3.5. 18 is generated from superposing on the first condition of 14. 19 is generated from superposing on the first condition of 18. Being able to ignore equations 13 and 14 at rewrite-time is very helpful to improve efficiency. Superposition on the right side of the oriented conditions in equations 10–12 is not possible as there are no equations for the pairing operator (_, _).

The concept of quasi-reductive equations has been implemented in the CEC-system [BGS88]. The last example, the example in 4.3, and many others have been successfully run on the system.

4.2 An Interpreter for While-Programs

In this section we describe a little interpreter for while-programs to demonstrate that one can also apply completion to nonterminating mechanisms in the framework of application-restricted equations.

Example 4.6

```
...
final :  state  -> stateAfterNSteps
run   :  program * state * nat -> stateAfterNSteps
run   :  program * state -> state
...
run(p,s,n)=final(s')  =>  run(p,s) = s'
run(p,s,n)=final(s')  =>  run(p;q,s,n+m) = run(q,s',m)
val(b,s)=true and run(p,s,n)=final(s')
                 =>  run(while b do p,s,n+m) = run(while b do p,s',m)
val(b,s)=false =>  run(while b do p,s,n) = final(s)
run(x:=e,s,n+1) = final(assign(s,x,val(e,s)))
...
```

In the example, the first equation requires to guess an upper bound for the complexity of program execution. If such an upper bound exists, reductive rewriting will compute the correct final state. All equations are assumed to be fully operational, i.e. all reductive substitutions are considered relevant. Dershowitz and Plaisted in [DP85] have emphasized this way of implementing partiality and infinite objects in their notion of rewrite programs.

4.3 Pruning Paths of Narrowing Derivations

The following example is a complete system of conditional equations which has been obtained by flattening a canonical system of unconditional equations for defining addition of integers. Flattening has been proposed in [BGM87] as a technique for implementing basic narrowing through resolution. The purpose of this example is to illustrate that application restrictions can make basic restricted paramodulation behave like resolution and that further application restrictions may further prune some paths in narrowing.

Example 4.7

```
1   s(p(x)) = x.
2   p(s(x)) = x.
3   x + 0 = x.
4   x + y ≟₋₅ u and s(u) ≟ v => x + s(y) = v.
5   x + y ≟₋₄ u and p(u) ≟ v => x + p(y) = v.
```

It is obvious that equations 4 and 5 are quasi-reductive, as is any conditional equation which is obtained from flattening a reductive conditional equation. Basic restricted paramodulation, which is restricted basic narrowing in this case, will produce a subset of the inferences which resolution would compute from this set of clauses (with the addition of the reflexivity axiom for $\doteq$). To solve the subgoals $x + y \doteq_{-5} u$ and $x + y \doteq_{-4} u$ of equation 4 and 5 respectively, the equations 5 and 4, respectively, need not be tried for resolution, although they will match in many cases. This follows from the completeness of the system wrt. the indicated application restrictions.

5 References

[Bac87] Bachmair, L.: Proof methods for equational theories. PhD-Thesis, U. of Illinois, Urbana Champaign, 1987.

[BDH86] Bachmair, L., Dershowitz, N. and Hsiang, J.: Proof orderings for equational proofs. Proc. LICS 86, 346-357.

[BG89] Bertling, H. and Ganzinger, H.: Completion of Application-Restricted Equations. Report 289, FB Informatik, U. Dortmund, 1989.

[BGM87] Bosco, P.G., Giovannetti, E. and Moiso, C.: Refined Strategies for Semantic Unification. TAPSOFT 87, Springer LNCS 250, 276–290, 1987.

[BGS88] Bertling, H., Ganzinger, H. and Schäfers, R.: CEC: A system for conditional equational completion. User Manual Version 1.4, PROSPECTRA-Report M.1.3-R-7.0, U. Dortmund, 1988.

[BK82] Bergstra, J. and Klop, J.W.: Conditional Rewrite Rules: Confluence and Termination. Report IW198/82, Mathematisch Centrum, Amsterdam, 1982.

[DOS88] Dershowitz, N., Okada, M. and Sivakumar, G.: Confluence of conditional rewrite systems. Proc. 1st Int'l Workshop on Conditional Term Rewriting, Orsay, 1987, Springer LNCS 308, 1988, 31–44.

[DP85] Dershowitz, N. and Plaisted, D.A.: Logic Programming cum Applicative Programming. Proc. IEEE Symp. on Logic Programming, Boston 1985, 54–67, 1985.

[DS88] Dershowitz, N. and Sivakumar, G.: Solving Goals in Equational Languages. Proc. 1st Int'l Workshop on Conditional Term Rewriting, Orsay, 1987, Springer LNCS 308, 1988, 45–55.

[Fri85] Fribourg, L.: SLOG: A Logic Programming Language Interpreter Based on Clausal Superposition. Proc. IEEE Symp. on Logic Programming, Boston 1985, 172–184, 1985.

[Gan87] Ganzinger, H.: A Completion procedure for conditional equations. Report 234, U. Dortmund, 1987, also in: Proc. 1st Int'l Workshop on Conditional Term Rewriting, Orsay, 1987, Springer LNCS 308, 1988, 62–83 (revised version to appear in J. Symb. Computation).

[Gan88a] Ganzinger, H.: Completion with History-Dependent Complexities for Generated Equations. In Sannella, Tarlecki (eds.): Recent Trends in Data Type Specifications. Springer LNCS 332, 1988, 73–91.

[Gan88b] Ganzinger, H.: Order-Sorted Completion: The Many-Sorted Way. Report 274, FB Informatik, Univ. Dortmund, 1988. Extended abstract to appear in Proc. TAPSOFT(CAAP) '89, Barcelona.

[GM88] Giovannetti, E. and Moiso, C.: A Completeness Result for E-Unification Algorithms Based on Narrowing. Proc. Workshop on Foundations of Logic and Functional Programming, Springer LNCS 306, 157–167, 1988.

[GM84] Goguen, J.A. and Meseguer, J.: Eqlog: Equality, Types, and Generic Modules for Logic Programming. J. Logic Programming, 1,2, 179–210, 1984.

[Höl87] Hölldobler, St.: From Paramodulation to Narrowing. Univ. der Bundeswehr, München, 1987.

[Hul80] Hullot, J.M.: Canonical Forms and Unification. Proc. CADE 1980, 318–334, 1980.

[Hus85] Hussmann, H.: Unification in Conditional-Equational Theories. Proc. EUROCAL, Springer LNCS 204, 543–553, 1985.

[JW86] Jouannaud, J.P. and Waldmann, B.: Reductive conditional term rewriting systems. Proc. 3rd TC2 Working Conference on the Formal Description of Programming Concepts, Ebberup, Denmark, Aug. 1986, North-Holland.

[Kap84] Kaplan, St.: Fair conditional term rewrite systems: unification, termination and confluence. Report 194, U. de Paris-Sud, Centre d'Orsay, Nov. 1984.

[Kap87] Kaplan, St.: A compiler for conditional term rewriting. Proc. RTA 1987, LNCS 256, 1987, 25-41.

[KR88] Kounalis, E. and Rusinowitch, M.: On word problems in Horn theories. Proc. 1st Int'l Workshop on Conditional Term Rewriting, Orsay, 1987, Springer LNCS 308, 1988, 144–160.

[Pad88] Padawitz, P.: Computing in Horn Clause Theories. EATCS Monographs in Computer Science, 16, 1988.

[RW69] Robinson, G.A. and Wos, L.: Paramodulation and Theorem Proving in First-Order Theories with Equality. In: Meltzer, Mitchie (Eds): Machine Intelligence 4, 1969.

Computing Ground Reducibility and Inductively Complete Positions*

Reinhard Bündgen
Wilhelm-Schickard-Institut
Universität Tübingen
D-7400 Tübingen
Federal Republic of Germany
⟨iabu001@DTUZDV5A.BITNET⟩

Wolfgang Küchlin
Computer and Information Sciences
The Ohio State University
Columbus, OH 43210-1277
U.S.A.
⟨Kuechlin@cis.ohio-state.EDU⟩

Abstract: We provide the extended ground-reducibility test which is essential for induction with term-rewriting systems based on [Küc89]: Given a term, determine at which sets of positions it is ground-reducible by which subsets of rules. The core of our method is a new parallel *cover* algorithm based on recursive decomposition. From this we obtain a separation algorithm which determines constructors and defined function symbols in a term-algebra presented by a rewrite system. We then reduce our main problem of extended ground-reducibility to *separation* and *cover*. Furthermore, using the knowledge of algebra separation, we refine the bounds of [JK86] for the size of ground reduction test-sets. Both our cover algorithm and our extended ground-reducibility test are engineered to be adaptive to actual problem structure, i. e., to allow for lower than the worst case bounds for test-sets on well conditioned problems, including well conditioned subproblems of difficult cases.

1 Introduction

Our overall interest lies in inductionless induction. In particular, we show how to implement the induction theory of [Küc86, Küc89], which also covers the achievements of [Fri86]. This theory is based on an extended ground-reducibility test which is now capable of exhibiting induction positions in the sense of [BM79]. For lack of space, we refer the reader to [JK86, Küc89, KM87] for an introduction and discussion of inductionless induction. (An approach which does not use the ground-reducibility test is presented in [KNZ86]; we do not know whether it can be refined to exhibit induction positions.)

As a starting basis, we use the algorithms of [JK86]. The main technical difference between the induction theory of [JK86] and [Fri86, Küc86, Küc89] is the replacement of

*A substantial part of this work was done while the first author was supported by a Fulbright scholarship and both authors were at the Department for Computer and Information Sciences, University of Delaware, Newark, Delaware.

confluence by ***ground confluence***. In the approach of [Küc86, Küc89], ground confluence is detected by the ***ground subconnectedness criterion***, a straightforward specialization of the general subconnectedness criterion (see e. g. [Küc85, BD88]) to the ground case. Algorithmically, it reduces to the ***extended ground reducibility test*** which computes all sets $P_i \subseteq O(t)$ of positions of a term t, together with the rules $\mathcal{R}_i \subseteq \mathcal{R}$ applicable at these positions, such that t is ground reducible at P_i by $\mathcal{R}_i$. Thus even *sets* of inductively complete positions are exhibited.

The test which we present in this paper is an extension of, and improvement over, the JK-test in [JK86]. It always works if $\mathcal{R}$ is left-linear. If we have left-linearity only in rules whose left-hand sides consist of constructors and variables only, our method still always detects ground-reducibility, but may fail to detect some cases of extended ground-reducibility. It is implemented[1] (see [Bün87a, Bün87b]) in the ALDES/SAC-2 IC-system as part of the induction prover outlined in [Küc89].

A trivial approach to test ground reducibility of a term t is to try to reduce all possible ground instances of t. This is in general impossible as the set of ground terms is always infinite in algebras where proofs by induction are of interest. Instead, [JK86] exhibits a finite ***test set*** S of substitutions with the property: any term t is ground reducible iff all instances $t\sigma$ of t with $\sigma \in S$ are reducible. The JK-test works with test set $GR_{\mathcal{R}}(d)$ of top portions (up to depth d) of irreducible ground terms. The depth $d(\mathcal{R})$ is the depth of the deepest left-hand side in $\mathcal{R}$ and is guaranteed to be sufficiently large for any t. Since $GR_{\mathcal{R}}(d)$ grows exponentially with d, it can become prohibitively large even for those t which should be easy to test.

Example 1 *An algebra as simple as* $\mathcal{P}_+ = \{\langle +(0,b) \rightarrow b\rangle, \langle +(S(a),b) \rightarrow S(+(a,b))\rangle\}$ *already has* 33,490 *ground terms of depth* 4, *and* 1,133,870,930 *of depth* 5. *Observe* $d(\mathcal{P}_+) = 1$. *Now enrich* $\mathcal{P}_+$ *to* $\mathcal{P}'_+$, *say with a complex definition for some* f, *such that* $d(\mathcal{P}'_+) = 5$. *As a consequence, the JK test-set will grow by some* 9 *orders of magnitude,* even for testing "old" terms which are already ground-reducible by $\mathcal{P}_+$. *With our refined bounds, a test-set of depth* 0 *will be sufficient for* $\mathcal{P}_+$, *and also for* $\mathcal{P}'_+$ *if* f *is not a constructor. If* f *is a (non-free) constructor, only terms with a ground instance that can be reduced with a new rule will possibly need a deeper test-set.* □

In Section 3 we refine the test set bounds of [JK86] in order to overcome pathological behaviour such as that exhibited in Example 1. If the separation $\mathcal{F} = \mathcal{C} \uplus \mathcal{D}$ of function symbols into constructors (not necessarily free) and defined symbols is known, then $d = d(\mathcal{R}')$, where $\mathcal{R}' \subseteq \mathcal{R}$ is the set of rules with left hand sides in $T(\mathcal{C}, \mathcal{X})$. Since $GR_{\mathcal{R}}(d)$ is exponential in d, the savings over $d = d(\mathcal{R})$ from [JK86] are substantial in practice; if $\mathcal{C}$ is free, then $d = 0$ and we automatically have an efficient solution for this easy case. Furthermore, symbols from $\mathcal{D}$ need not be considered in constructing $GR_{\mathcal{R}}(d)$.

In Section 4, we develop the core of our method: a new parallel *cover* algorithm based on recursive problem decomposition and dynamic programming. The algorithm reduces the cover problem for term sequences (see [HH82, Thi84, Kou85]) to the cover problem for term sets. Frequently, several of the subproblems generated at any level of recursion are equivalent, so that substantial savings occur. Furthermore, the test sets needed for the

[1] With the exception that a different cover algorithm was used.

solution of "good" subproblems may be substantially smaller than that for the "worst" subproblem.

In Section 5, we solve the extended ground reducibility problem by reducing it to the cover test as follows: whether t is ground-reducible at a position p depends only on the rules that superpose with t at p (see [Thi84]). This leads to the problem whether $\vec{x}\mu$ covers all possible irreducible ground instances of the variables $\vec{x}$ of t. Thus the necessary test set depends only on the depth of $\vec{x}\mu$, i. e. the depth of the terms substituted for t's variables by the superposition, which for deep t may be substantially smaller than $d(\mathcal{R}')$. The savings from our adaptive cover algorithm carry over to this problem. The reducing rule-set $\mathcal{R}_i$ is simply collected from the superposing rules.

In Section 6 we use the cover test for an algorithm that determines the algebra separation $\mathcal{F} = \mathcal{C} \uplus \mathcal{D}$. The idea is to assume inductively that some $\mathcal{C}' \subseteq \mathcal{C}$ is known, and to check definedness of all $f \in \mathcal{F} \setminus \mathcal{C}'$ with the cover test. Separation itself needs a test set of depth $d(\mathcal{R})$, since the cover algorithm must test the function in whose definition this depth occurs. However, other, shallower, function definitions only need smaller test sets. In our implementation we use this algorithm as a one-time preprocessing step. All subsequent ground-reducibility tests then benefit from the improved bounds of Section 3.

2 Notation

Our notation is essentially the standard one for term manipulation (see e. g. [Gal86]) and term rewriting systems (see e. g. [HO80]). Nevertheless, we will briefly repeat some of the terminology concerning *ground terms* with which we will be mainly concerned since we are interested in initial models. We shall consistently use the prefix *ground-* to refer to general properties specialized to ground terms, such as *ground-confluence* etc.

$O(t)$ is the set of all positions (occurrences) in the term t and $O'(t)$ is the set of all non-variable positions of t. The symbol at position o of term t is denoted by $t(o)$, and $t|_o$ denotes the subterm of t with root $t(o)$.

A *substitution* $\sigma : \mathcal{X} \rightarrow T(\mathcal{F}, \mathcal{X})$ is a *ground substitution* if $\forall t \in im(\sigma) : t \in T(\mathcal{F})$, where $im(\sigma)$ is the image (range) of σ. Any $t\sigma \in T(\mathcal{F})$ is called a *ground instance* of t. We will often denote ground substitutions by γ.

Let $\mathcal{R}$ be a *term-rewriting system.* A term t is in $\mathcal{R}$-*normal form* iff there is no term $t' \in T(\mathcal{F}, \mathcal{X})$ such that $t \rightarrow_{\mathcal{R}} t'$; otherwise t is *reducible.* If a term t in normal form is ground, we say t is in *ground normal form.* A substitution σ is in normal form if all terms in $im(\sigma)$ are in normal form.

We have the usual definitions of confluence, critical pairs, etc. A term rewriting system is *ground confluent* if it is confluent on all reduction sequences consisting of ground terms (notice that ground terms can only be reduced to ground terms). In the same sense, we speak of ground confluent critical pairs. A term rewriting system $\mathcal{R}$ is *(ground) convergent* if it is (ground) confluent and terminating, and it is called *(ground) canonical* if it is (ground) convergent and interreduced. A *rule* $\langle l \rightarrow r \rangle$ *is left-linear* if l is linear, and $\mathcal{R}$ *is left-linear* if all its rules are left-linear.

In contrast to convergence, ground convergence is no overly restrictive condition: it simply requires that our recursive equations define functions in such a way that every recursive evaluation terminates and every function call yields a unique result.

Note: In the sequel, unless explicit conditions for $\mathcal{R}$ are stated, we shall always assume that $\mathcal{R}$ is *ground convergent!*

3 Ground Reduction Test Sets

In our approach to induction, the initial algebra is (implicitely) given as a set S of irreducible ground terms. Induction only makes sense if S is infinite; so ground reduction test sets (GRTS's) are used as finite (non-ground) representatives of S. The computability of such sets has been shown in [Pla85] and [KNZ87], but the algorithms are far too complex to be practical. For left-linear term rewriting systems, more efficient approaches have been proposed in [JK86] and [KNZ86].

In this section we shall refine some results from [JK86] concerning test sets to the case where the separation $\mathcal{F} = \mathcal{C} \uplus \mathcal{D}$ of an algebra into constructors and defined functions is known. Our observation is essentially that the depth of a ground reduction test set does not depend on any left-hand sides which contain defined function symbols. In many cases this observation decreases the size of a test set considerably.

While in [HH82] and in [JK86] this classification is used to define, respectively optimize, the inductive completion algorithm, we will use it in section 5 to improve the ground reducibility test itself. In Section 6 we present an inductive separation algorithm which we use in our theorem prover as a preprocessing step to obtain the constructors and defined functions of an algebra.

First we introduce some definitions and results from [JK86].

A *ground reduction test set* is a finite set S of substitutions with the property: any term t is ground reducible iff all instances $t\sigma$ of t with $\sigma \in S$ are reducible. To compute such a test set, we need the definition of the function *top* from [NW82].

Definition 1 *The* top of a term t at depth d, $top(t,d)$, *is a term such that*

$$top(t,d) = t \text{ if } depth(t) \leq d$$

$$top(f(t_1,\ldots,t_n),0) = f(x_1,\ldots,x_n) \text{ for distinct new variables } x_1,\ldots,x_n$$

$$top(f(t_1,\ldots,t_n),d) = f(top(t_1,d-1),\ldots,top(t_n,d-1)) \text{ otherwise.}$$

Example 2 *With* $t = +(+(S(0),x)),S(y))$, $top(t,4) = t$, $top(t,2) = +(+(S(z),x),S(y))$, *and* $top(t,1) = +(+(u,v),S(w))$, *where* $u,v,w,x,y,z \in \mathcal{X}$. □

Now let $S(\mathcal{R}) = \{top(t,d) | t \in T(\mathcal{F})$ is in normal form, and d is the depth of the deepest left-hand side of a rule in $\mathcal{R}\}$. In the sequel, we use the term *test set* for any finite set that is meant to be a description of an infinite set. Thus, $S(\mathcal{R})$ is also a test set, a set of terms. It will become clear from the context which test set we are speaking about. The following theorem shows that we can build from $S(\mathcal{R})$ a test set of substitutions S_t for a term t as mentioned above:

$$S_t = \{\sigma |\, dom(\sigma) = Var(t) \text{ and } \forall x \in Var(t) : \sigma(x) \in S(\mathcal{R})\}$$

It is obvious that $|S_t|$ grows exponentially with $|Var(t)|$.

Theorem 1 (Th. 9 in [JK86]) *A term t is ground-reducible by a left-linear term rewriting system $\mathcal{R}$ iff all instances obtained by substituting terms of $S(\mathcal{R})$ into its variables are reducible.* □

The next theorem shows that the set $S(\mathcal{R})$ can be computed effectively (but not efficiently, according to Example 1).

Theorem 2 (Th. 10 in [JK86]) *Let $S_i = \{top(t,d) \mid t \in T(\mathcal{F}), t$ irreducible, $depth(t) < i\}$, where $\mathcal{R}$ is left-linear and the depth of the deepest left-hand side of a rule in $\mathcal{R}$ is d. Then $S(\mathcal{R}) = S_k$ as soon as $S_k = S_{k+1}$ for some integer k.* □

As a first step to improve their algorithms, Jouannaud and Kounalis propose to exploit the knowledge of a classification of the elements of $\mathcal{F}$:

Definition 2 *An n-ary function $f \in \mathcal{F}$ is a* defined function *iff $f(x_1, x_2, \ldots, x_n)$ is ground reducible by $\mathcal{R}$, otherwise f is a* constructor. *Let C be the set of constructors of the algebra. If for all t consisting of constructors and variable symbols, t is irreducible, the constructors in C are called* free *and the algebra is said to be* freely generated by C.

Example 3 *For $\mathcal{R} = \mathcal{P}_+$ we have $C = \{0, S\}$ and $\mathcal{D} = \{+\}$.* □

From now on, $C \subseteq \mathcal{F}$ and $\mathcal{D} \subset \mathcal{F}$ will respectively denote the set of constructors and defined functions of an algebra.

Definition 3 *A* d-topset $GR_{\mathcal{R}}(d)$ of an algebra presented by $\mathcal{R}$ *is the set of terms $t = top(t', d)$ where t' is a ground term irreducible by $\mathcal{R}$. We simply write $GR(d)$ if $\mathcal{R}$ is clear from the context.*

Note that $S(\mathcal{R}) = GR_{\mathcal{R}}(d)$ where d is the depth of the deepest left-hand side of a rule in $\mathcal{R}$.

Example 4 $GR_{\mathcal{P}_+}(1) = \{0, S(0), S(S(x))\}$. □

Definition 4 *A set S is called a* ground reduction test set (GRTS) for an algebra $T(\mathcal{F}, \mathcal{X})$ and term rewriting system $\mathcal{R}$, *if any term $t \in T(\mathcal{F}, \mathcal{X})$ is ground reducible by $\mathcal{R}$ iff all its instances obtained by substituting terms of S to its variables are reducible by $\mathcal{R}$.*

With such a set S every term $t \in T(\mathcal{F}, \mathcal{X})$ can be tested for ground reducibility by trying to reduce all possible instances of t obtained by substituting terms of S for the variables of t, instead of examining all possible ground instances.

The following two lemmas are specializations of Theorems 1 and 2 which are applicable to test whether a term in $T(C, \mathcal{X})$ is ground reducible. Thus we can say that $S(\mathcal{R})$ is a GRTS for $T(\mathcal{F}, \mathcal{X})$ whereas the $GR_{\mathcal{R}}(d)$ of Lemma 3 is a GRTS for $T(C, \mathcal{X})$. Note that any term containing a defined function symbol is ground reducible by definition, so testing terms in $T(C, \mathcal{X})$ is all we need. The proofs of these lemmas are analogous to those in [JK86] and, like all other missing proofs, can be found in [Bün87b] or [Bün87a].

Lemma 3 *Let $\mathcal{R}$ be a left-linear term rewriting system and let d be the depth of the deepest left-hand side of all rules $\langle l \rightarrow r\rangle \in \mathcal{R}$ where l is in $T(C, \mathcal{X})$, if such a rule exists, and 0 otherwise. Then a term $t \in T(C, \mathcal{X})$ is ground reducible iff all instances obtained by substituting terms of $GR_{\mathcal{R}}(d)$ into its variables are reducible by $\mathcal{R}$.* □

Lemma 4 *Let $\mathcal{R}$ be a term rewriting system where all rules $\langle l \rightarrow r\rangle \in \mathcal{R}$ with $l \in T(C, \mathcal{X})$ are left-linear, and let d be greater than or equal to the depth of the deepest left-hand side of any such rule in $\mathcal{R}$. Then $GR_{\mathcal{R}}(d) = \{top(t, d) | t \in T(C)$ and t is in normal form with respect to $\mathcal{R}\}$ can be computed as follows: Let $GR(d)_i = \{top(t, d) | t \in T(C)$ is in normal form and $depth(t) < i\}$. Then $GR_{\mathcal{R}}(d) = GR(d)_k$ if $GR(d)_k = GR(d)_{k+1}$.* □

A program based on Theorem 2 of Jouannaud and Kounalis creates all possible ground terms up to a certain depth and then tests whether they are in normal form. Using Lemma 4 instead, the number of ground terms to be created can be reduced substantially because only the constants and operators in C must be considered to create ground terms, and the depth needed as an upper bound for the terms to be created is generally lower. Note that for a freely generated algebra, like for example $\mathcal{P}_+$, the GRTS needed is the 0-topset, which is particularly small. Hence we have re-gained the usual efficiency of [HH82] for this important case.

4 A Parallel Cover Algorithm

The problem of *complete definition*, or *sufficient completeness*, asks whether f is defined on all argument (i. e. ground term) sequences of a set A [HH82, Thi84, Kou85]. Following Definition 2, if f is given by a rewrite system $\mathcal{R}$, this happens if $\forall(t_1, \ldots, t_n) \in A$: $f(t_1, \ldots, t_n)$ is reducible.

Therefore, as observed in [JK86], we are really facing a special case of ground-reducibility which is also called the *cover problem*. It is important in its own right because it tells us whether we "forgot something" when defining a new function. It is also at the heart of algebra separation [JK86], and it is crucial in the approach of [ZKK88] to translate the induction principle of Boyer and Moore ([BM79]) into an induction principle for an equational environment. Furthermore, we will reduce the entire extended ground reducibility problem to *cover* in Section 5. The main result here is a new *parallel* cover algorithm.

Example 5 *Let*
$$\mathcal{R} = \left\{ \begin{array}{llll} f(& 0, & y, & z) \rightarrow \cdots \\ f(& S0, & 0, & z) \rightarrow \cdots \\ f(& S0, & Sy, & z) \rightarrow \cdots \\ f(& SSx, & y, & 0) \rightarrow \cdots \\ f(& SSx, & y, & Sz) \rightarrow \cdots \end{array} \right\}$$

No matter what the right-hand sides are, f is completely defined on natural numbers, i. e., every $f(t_1, t_2, t_3)$ with t_i either 0 or $S^k(0)$ is reducible, if

$$\left\{ \begin{array}{llll} (& 0, & y, & z) \\ (& S0, & 0, & z) \\ (& S0, & Sy, & z) \\ (& SSx, & y, & 0) \\ (& SSx, & y, & Sz) \end{array} \right\} \text{ covers all irreducible ground instances of } \{(x, y, z)\}.$$

□

Definition 5 *Given a term-rewriting system $\mathcal{R}$ and n-ary function symbol f, the* matrix of f in $\mathcal{R}$ *is the set of term sequences* $T^* = \{(t_1,\ldots,t_n) \mid \exists\langle l \to r\rangle \in \mathcal{R} : l = f(t_1,\ldots,t_n)\}$.

So we have on the one hand a set of term sequences and on the other hand a corresponding sequence of variables whose irreducible ground instances must be covered.

Definition 6 *Let $T = (t_1,\ldots,t_n)$ be a term sequence. Then T is* linear *if t_i is linear for all i and $Var(t_i) \cap Var(t_j) = \emptyset$ for $i \neq j$, $i,j \in \{1,\ldots,n\}$.*

Definition 7 *The* universe $UNIV_{\mathcal{R}}((t_1,\ldots,t_n))$ of ground normal forms represented by a linear term sequence $(t_1,\ldots,t_n)$ *is the set of all sequences $(\bar{t}_1,\ldots,\bar{t}_n)$ of terms $\bar{t}_i$ in $\mathcal{R}$-ground normal form, where $\bar{t}_i = t_i\gamma$ $\forall i \in \{1,\ldots,n\}$ and some ground substitution γ.*

Note that $UNIV_{\mathcal{R}}((t_1,\ldots,t_n))$ is empty if at least one t_i is ground reducible by $\mathcal{R}$. As usual, we omit the subscript if we do not have to point out a particular $\mathcal{R}$.

Example 6 $UNIV_{\mathcal{P}_+}((Sx, SSy)) = \{(t_1,t_2) \mid t_1 = S^j0, t_2 = S^k0, j \geq 1, k \geq 2\}$. □

Definition 8 $T = (t_1,\ldots,t_n)$ sequence-matches *(short* s-matches*)* $S = (s_1,\ldots,s_n)$, *iff there is a substitution σ such that $t_i\sigma = s_i$ for all $i \in \{1,\ldots,n\}$.*

Example 7 $(Sx, +(0,x))$ *s-matches* $(S0, +(0,0))$. □

Definition 9 *A set of term sequences $T^* = \{T_1,\ldots,T_k\}$* covers *a set of term sequences $S = \{S_1,\ldots,S_l\}$ if the length of all sequences is the same and $\forall S_j \exists T_i : T_i$ s-matches S_j. Hence the empty set is covered by any T^*, but it covers only itself.*

Example 8 *Let $\mathcal{R} = \mathcal{P}_+$. Then $T^* = \{(Sx,y),(x,0)\}$ covers $\{(S0,0),(Sx,Sy),(0,0)\}$ but T^* does not cover $\{(x,y)\}$.* □

Our cover algorithm below will recursively reduce the cover problem to the special case where all term sequences are of length 1. We shall call these *singleton sequences* and generally identify them with their terms. Let us note some obvious properties of singleton sequences.

Lemma 5 *Let T, S be singleton term sequences. If T matches S then $\{T\}$ covers $UNIV(S)$.*

PROOF: If T matches S then it also matches all instances of S. □

Lemma 6 *Let T, S be singleton term sequences and let T be linear. Let S be ground, or else let the depth of the shallowest variable in S be greater than the depth of T. If T does not match S, then either $\{T\}$ does not cover $UNIV(S)$, or $UNIV(S)$ is empty.*

PROOF: Assume the universe of S is not empty. T does not s-match S. Since T is linear, there must exist a non-variable occurrence o in T such that $T(o) \neq S(o)$ for some $i \in \{1,\ldots,n\}$. If S is ground, then either S is reducible and $UNIV(S)$ is empty, or $UNIV(S) = \{S\}$ in which case it is not covered. Otherwise, since the depth of the shallowest variable in S is deeper than the depth of T, for any substitutions σ, γ, $T\sigma(o) \neq S\gamma(o)$. Hence, T does not match any element of the ground universe represented by S, so either $UNIV(S) = \emptyset$ or $\{T\}$ does not cover $UNIV(S)$. □

Hence the universe of a singleton ground sequence S is covered by $T^* = \{T_1, \ldots, T_k\}$ iff either there exists a term sequence in T^* which s-matches S, or S is reducible. If S is not ground and Lemmas 5 and 6 are not applicable, we shall increase the depth of the representation of $UNIV(S)$ until Lemma 6 applies.

Definition 10 *A* σ-successor of a term t *is an instance* $t' = t\sigma$ *of* t, *where* $\sigma(x)$ *is either a constant or* $\sigma(x) = f(x_1, \ldots, x_n)$ *for any variable* $x \in Var(t)$ *where* f *is an n-ary operator symbol in* $\mathcal{F}$ *and* $x_i, \ldots, x_n$ *are distinct new variables.*

Note that the σ-successor of a linear term is again linear. The σ-successor of a (singleton) term sequence is defined accordingly.

Example 9 *In* $\mathcal{P}_+$, $(+(+(v, w), 0))$ *is a* σ*-successor of* $(+(y, z))$. □

Lemma 7 *Let* $S, T_1, \ldots, T_k$ *be singleton term sequences where* S *is linear and not ground. Then* $T^* = \{T_1, \ldots, T_k\}$ *covers the universe represented by* S *iff*

1. *there exists a term sequence in* T^* *which s-matches* S

OR

2. *for each* S' *of the* σ*-successors of* S, T^* *covers the universe represented by* S'.

PROOF:

"$\Leftarrow$" 1. Follows directly from Lemma 5.

2. Assume there exists an $\hat{S} \in UNIV(S)$ and no $T_i \in T^*$ s-matches $\hat{S}$. $\hat{S} = S\gamma$ where γ is a ground substitution. Thus, for every $x \in Var(S)$, $\gamma(x)$ is a constant or $\gamma(x) = f(s_1, \ldots, s_n)$, where f is an n-ary operator symbol and s_i is a ground term for $i \in \{1, \ldots, n\}$. Therefore, S is also a ground instance of the corresponding successor S' of S. Since T^* covers the ground universe of S', and it is not empty, there must be a $T_i \in T^*$ matching $S'\gamma$.

"$\Rightarrow$" We have to show that if T^* covers $UNIV(S)$ and no term sequence in T^* s-matches S, then for each term sequence S' in the set of σ-successors of S, T^* covers $UNIV(S')$. We prove this again indirectly, assuming that for some element $\hat{S}$ of the universe represented by a successor S' of S there exists no T_i in T^* which matches $\hat{S}$. Then $\hat{S} = S'\gamma$ and $S' = S\sigma$ for some substitutions γ and σ. Thus, $\hat{S} = S\sigma\gamma$ and $\hat{S}$ is an uncovered element of $UNIV(S)$, which contradicts the fact that T^* covers that universe. □

So if the representation S of $UNIV(S)$ is not deep enough for Lemma 6 to apply, we can replace S by the set $\{S'\}$ of its σ-successors; each S' then represents a segment of $UNIV(S)$. Two things can then happen: if the depth of the shallowest variable in a successor S' is still less than the maximal depth of a sequence in T^*, one of the $T_i \in T^*$

may match S' (although it did not yet match S); or this depth bound is exceeded by one S' without being matched, whence its universe is not covered. Lemma 6 yields an upper bound for the number of successor generations to be constructed for S, because the depth of the shallowest variable occurrence in a sequence of the i-th generation is greater or equal to i plus the depth of the shallowest variable occurrence in the initial S.

We still need a procedure to determine if the ground universe represented by a term is empty, i. e. if the term is ground reducible. This test is trivial if the algebra is freely generated by a known set of constructors. But in any case we know the depth of the current representation S' of a segment of $UNIV(S)$. Now the following lemma lets us decide whether S' is ground reducible.

Lemma 8 *Let $GR_{\mathcal{R}}(d)$ be a d-topset and t a linear term with $depth(t) \leq d$. Then t is ground reducible by $\mathcal{R}$ iff t does not match any term in $GR_{\mathcal{R}}(d)$.*

PROOF:

"$\Rightarrow$" Assume t matches a term in $GR_{\mathcal{R}}(d)$. Then by definition of $GR_{\mathcal{R}}(d)$ there exists an irreducible ground instance of t, which contradicts its ground reducibility.

"$\Leftarrow$" Assume t is not ground reducible. Then there exists a ground substitution γ such that $t\gamma$ is irreducible. Thus $top(t\gamma, d) \in GR_{\mathcal{R}}(d)$. If $top(t\gamma, d) = t\gamma$, then $t\gamma \in GR_{\mathcal{R}}(d)$, a contradiction. If $top(t\gamma, d) \neq t\gamma$, then, by definition of top, $depth(t\gamma) \geq d+1$. Now let σ be a substitution such that $t\sigma = top(t\gamma, d)$. σ is well defined because t is linear. Thus $t\sigma \in GR_{\mathcal{R}}(d)$, a contradiction. □

So far we have treated the case of singleton term sequences; so we can decide whether a function of one argument, given by a left-linear $\mathcal{R}$, is defined. The following theorem lets us reduce the case of n-ary functions to the unary case. It is the main result of this section.

Theorem 9 (Parallel Cover Algorithm) *Assume $UNIV_{\mathcal{R}}((x_1, \ldots, x_n)) \neq \emptyset$. Let A be a set of linear term sequences of length $n \geq 1$. Then:*

$$A = \left\{ \begin{array}{c} (a_{1_1}, \ldots, a_{1_n}) \\ \vdots \\ (a_{m_1}, \ldots, a_{m_n}) \end{array} \right\} \text{ covers } UNIV_{\mathcal{R}}((x_1, \ldots, x_n))$$

iff

1. $\{(a_{1_1}), \ldots, (a_{m_1})\}$ *covers* $UNIV_{\mathcal{R}}((x_1))$

AND

2. $\forall t \in GR_{\mathcal{R}}(d)$, *where* $d \geq \max\{depth(a_{i_1})\}$:
   $\{(a_{i_2}, \ldots, a_{i_n}) \mid a_{i_1}$ *matches* $t\}$ *covers* $UNIV_{\mathcal{R}}((x_2, \ldots, x_n))$.

PROOF: "$\Rightarrow$":

1. Suppose $\exists t_1 \in UNIV((x_1))$ such that no a_{i_1} matches t_1. Then, because $UNIV((x_1, \ldots, x_n)) \neq \emptyset$, $\exists (t_1, t_2, \ldots, t_n) \in UNIV((x_1, \ldots, x_n))$ such that no $(a_{i_1}, \ldots, a_{i_n})$ s-matches $(t_1, t_2, \ldots, t_n)$, a contradiction.

2. Suppose $\exists t_1 \in GR_{\mathcal{R}}(d)$ such that $\{(a_{i_2}, \ldots, a_{i_n}) \mid a_{i_1}$ matches $t_1\}$ does not cover $UNIV((x_2, \ldots, x_n))$. Hence $\exists \bar{t_2}, \ldots, \bar{t_n}$ which are irreducible ground terms, such that no $(a_{i_2}, \ldots, a_{i_n})$ s-matches $(\bar{t_2}, \ldots, \bar{t_n})$. Now let $\bar{t_1}$ be any irreducible ground instance of t_1 (which exists because $t_1 \in GR_{\mathcal{R}}(d)$). Then $(\bar{t_1}, \bar{t_2}, \ldots, \bar{t_n}) \in UNIV((x_1, \ldots, x_n))$ is not covered by A, because otherwise $\exists (a_{k_1}, \ldots, a_{k_n})$: $(a_{k_1}, \ldots, a_{k_n})$ s-matches $(\bar{t_1}, \bar{t_2}, \ldots, \bar{t_n})$. Now:

   - if $\bar{t_1} = t_1$, then a_{k_1} matches t_1 and $(a_{k_2}, \ldots, a_{k_n})$ s-matches $(\bar{t_2}, \ldots, \bar{t_n})$, a contradiction.
   - if $\bar{t_1} = t_1\gamma$, $\gamma \neq \emptyset$, then t_1 has variables and, by definition of a set $GR_{\mathcal{R}}(d)$ with $d \geq \max\{depth(a_{i_1})\}$, all variables occur at depth $d+1$. Hence, by linearity, a_{k_1} matches t_1 itself, and again $(a_{k_2}, \ldots, a_{k_n})$ s-matches $(\bar{t_2}, \ldots, \bar{t_n})$, a contradiction as above.

"$\Leftarrow$": Suppose $\exists (\bar{t_1}, \ldots, \bar{t_n}) \in UNIV((x_1, \ldots, x_n))$: no $(a_{i_1}, \ldots, a_{i_n})$ s-matches $(\bar{t_1}, \bar{t_2}, \ldots, \bar{t_n})$. By condition (1.), $I = \{i \mid a_{i_1}$ matches $\bar{t_1}\}$ is not empty. By definition of $GR_{\mathcal{R}}(d)$, $\exists!\, t_1 \in GR_{\mathcal{R}}(d) : \bar{t_1} = t_1\gamma$. Since $d \geq \max\{depth(a_{i_1})\}$ and the a_{i_1} are linear, $\forall i \in I : a_{i_1}$ matches t_1. Then, by condition (2.), $\{(a_{i_2}, \ldots, a_{i_n}) \mid i \in I\}$ covers $UNIV((x_2, \ldots, x_n))$. Hence $\exists k \in I : (a_{k_2}, \ldots, a_{k_n})$ s-matches $(\bar{t_2}, \ldots, \bar{t_n})$ and a_{k_1} matches $\bar{t_1}$, whence by linearity $(a_{k_1}, \ldots, a_{k_n})$ s-matches $(\bar{t_1}, \bar{t_2}, \ldots, \bar{t_n})$, a contradiction. □

Example 10 *In Example 5, f is defined because* $\{0, S0, S0, SSx, SSx\}$ *covers* $UNIV(x)$ *AND* $\{(y,z)\}$ *covers* $UNIV((y,z))$ *AND* $\{(0,z),(Sy,z)\}$ *covers* $UNIV((y,z))$ *AND* $\{(y,0),(y,Sz)\}$ *covers* $UNIV((y,z))$. □

Note that the depth d of the necessary test set depends only on the depth of terms in A. This corresponds to our intuition: testing f for definedness should be independent of any other defined g. The construction of the test set itself of course still depends on C.

Furthermore, the size of the test set may vary from column to column: in Example 10, depth 1 is needed only while processing column 1 which contains the deep terms; depth 0 is sufficient for the residual problems. Thus our bounds are adaptive even to the actual *form* of the problem, and well behaved subproblems are not penalized by ill behaved exceptions.

Instead of a d-topset we can of course use any other complete and minimal test set with the obvious restrictions on the contour depth of its non-ground elements. Such sets are described e. g. in [KNZ86].

Our algorithm exhibits two kinds of parallelism: AND-parallelism, where the overall problem is recursively broken down into smaller problems which can be treated in parallel (but all must be solved); the simple case of one argument coverage forms the induction basis.

We also have OR-parallelism, because the ordering of the arguments is immaterial, and we may start with any arbitrary column to decompose the problem. Thus, on a parallel machine, we may also do all column decompositions in parallel, and the first decomposition to finish yields the solution.

Example 11 *Beginning decomposition on the right-most column in Example 5, we get the following conditions for complete definition of f:* $\{z, z, z, 0, Sz\}$ *covers* $UNIV(z)$ *AND* $\{(0,y), (S0,0), (S0,Sy), (SSx,y)\}$ *covers* $UNIV((x,y))$ *AND* $\{(0,y), (S0,0), (S0,Sy), (SSx,y)\}$ *covers* $UNIV((x,y))$. □

This example shows the potential advantages which we may get from exploiting the *dynamic programming* approach facilitated by the recursive decomposition, especially in case true parallel processing is unavailable: decomposition yields two identical subproblems, so that only one of them must be solved, resulting in considerable savings. As usual, a dynamic programming solution may exhibit equivalent subproblems again on any level of the recursive decomposition.

Example 12 *Continuing on Example 10, we first have* $\{(y,z)\}$ *covers* $UNIV((y,z))$ *as a trivial problem; then both of the remaining problems may be decomposed, on column* 1, *and* 2, *respectively, into the equivalent elementary subproblems:* $\{0, Sy\}$ *covers* $UNIV(y)$ *AND* $\{z, z\}$ *covers* $UNIV(z)$, *and respectively* $\{0, Sz\}$ *covers* $UNIV(z)$ *AND* $\{y, y\}$ *covers* $UNIV(y)$. □

We would like to stress that this example was not contrived for this purpose. (It arose as a counter-example to an earlier formulation of the theorem). Recursive decomposition simply exhibited the symmetry which the designer of the system followed, and symmetries are indeed frequently used by designers to keep the rule-sets manageable.

Example 13 *Ackermann's function, given by*

$$\mathcal{R} = \left\{ \begin{array}{llll} A(& 0, & y) & \rightarrow & Sy \\ A(& Sx, & 0) & \rightarrow & A(x, S0) \\ A(& Sx, & Sy) & \rightarrow & A(x, A(Sx, y)) \end{array} \right\}$$

is completely defined on the natural numbers: $\{(0,y), (Sx,0), (Sx,Sy)\}$ *covers* $UNIV((x,y))$ *because* $\{y, 0, Sy\}$ *covers* $UNIV(y)$ *AND* $\{0, Sx\}$ *covers* $UNIV(x)$ *AND* $\{0, Sx\}$ *covers* $UNIV(x)$. *Again, we have two equivalent subproblems after decomposing on the right-most column.* □

5 The Extended Ground Reducibility Test

We now improve the test for ground reducibility from [JK86], and we extend it to compute all inductively complete sets of positions in a term, as needed in [Küc89].

Definition 11 ([Küc89]) *A term t is* ground reducible by $\mathcal{R}$ at $P \subseteq O'(t)$ using $\mathcal{R}' \subseteq \mathcal{R}$ *if for every ground instance* $t\gamma$ *of* t *there exists a* $p \in O(t\gamma)$ *such that either* $p \in P$ *and* $t\gamma$ *is reducible by* $\mathcal{R}'$ *at* p, *or* $p \notin O'(t)$ *and* $t\gamma$ *is reducible by* $\mathcal{R}$ *at* p[2]. *If* $\mathcal{R}$ *is ground convergent we call* P *an* inductively complete set (of positions) for $\mathcal{R}'$ *and if* $P = \{p\}$ *we also call* p *an* inductively complete position.

[2]Equivalently, for every normalized ground substitution γ, $t\gamma$ must be reducible by $\mathcal{R}'$ at some $p \in P$.

Our approach will be to reduce the extended ground-reducibility problem to the special case solved by the parallel cover algorithm in Section 4: after some preconditioning, the problem will be completely independent of the term and its positions. Our main contribution apart from the solution of the extended problem is again a reduction of search space, i. e. of the number of instances of a term t which must be tested for reducibility. This is again accomplished by making the algorithm adaptive to the given problem. These savings are additional to those of Section 4.

Assume now we are given some t and $\mathcal{R}$. We shall have a closer look at the rules which can possibly reduce ground instances of t, and the positions in t where these rules can possibly apply. We first make an easy observation (see [Thi84]):

Lemma 10 *Let $\langle l \rightarrow r\rangle \in \mathcal{R}$ be a rule and t a term. Then for all substitutions σ in normal form, if $\langle l \rightarrow r\rangle$ reduces $t\sigma$, there exists a non-variable occurrence p in t such that $t|_p$ unifies with l.* □

Now let $t\gamma$ be reducible by $\langle l \rightarrow r\rangle$ at position p and let $\mu = mgu(t|_p, l)$; then all instances of $t\mu$ are reducible by $\langle l \rightarrow r\rangle$ at p. The only remaining question is then, whether the instances of t by normalized ground substitutions are the same as the instances of $\{t\mu \mid \mu = mgu(t|_p, l), \langle l \rightarrow r\rangle \in \mathcal{R}\}$ by normalized ground substitutions. If yes, then t is ground-reducible at p by $\mathcal{R}$. We shall now factor out the common parts of t and $t\mu$ and phrase this as a cover problem. This method is the second main result of this paper.

Theorem 11 (Extended Ground-Reducibility Test) *Let t be a term, $P \subseteq O'(t)$. Let $\mathcal{R}' \subseteq \mathcal{R}$ be left-linear. Let $M = \{mgu(t|_p, l) | \langle l \rightarrow r\rangle \in \mathcal{R}', p \in P\}$. Let $S = (x_1, \ldots, x_n)$ be the variables of t in any sequence, and let $T^* = \{T_1, \ldots, T_m\}$ such that $\forall \mu \in M\, \exists T_i \in T^* : T_i = (x_1\mu, \ldots, x_n\mu)$. Then t is ground reducible at P using $\mathcal{R}'$ iff T^* covers $UNIV_{\mathcal{R}}(S)$.*

PROOF:

"$\Leftarrow$" T^* covers $UNIV(S)$. Suppose γ is a normalized ground substitution. Then $G = (x_1\gamma, \ldots, x_n\gamma) \in UNIV(S)$. Therefore $\exists T \in T^* : T$ s-matches G, and $T = (x_1\mu, \ldots, x_n\mu)$ where $\mu = mgu(t|_p, l)$ for some $p \in P, \langle l \rightarrow r\rangle \in \mathcal{R}'$. So there is some σ: $x_i\mu\sigma = x_i\gamma$. Hence $t\gamma = t\mu\sigma$. But $t\mu|_p = l$, so, since $p \in O'(t)$, $t\mu\sigma|_p = (t\mu|_p)\sigma = l\sigma = t\gamma|_p$, so $t\gamma$ is reducible by $\langle l \rightarrow r\rangle$ at p.

"$\Rightarrow$" Let t be ground-reducible at P using $\mathcal{R}'$. We define the normalized ground substitution $\gamma = \{x_1 \leftarrow \bar{t_1}, \ldots, x_n \leftarrow \bar{t_n}\}$. Since γ is normalized, $(\bar{t_1}, \ldots, \bar{t_n}) \in UNIV(S)$. Since t is ground-reducible at P using $\mathcal{R}'$, $\exists p \in P, \langle l \rightarrow r\rangle \in \mathcal{R}'$: $t\gamma$ is reducible at p by $\langle l \rightarrow r\rangle$. Then, by Lemma 10, $t|_p$ unifies with l, with unifier μ, and $\exists\rho : \gamma = \mu\rho$. Then $T = (x_1\mu, \ldots, x_n\mu) \in T^*$. We also note that T is linear, because l is linear. Therefore T s-matches $(\bar{t_1}, \ldots, \bar{t_n})$, and hence T^* covers $UNIV(S)$. □

This proof relies on a technical lemma which for lack of space we state without proof.

Lemma 12 *Let t and l be two terms with distinct variable sets which are unifiable with mgu μ. Let l be linear and let $Var(t) = \{x_1, \ldots, x_n\}$ where the x_i are distinct. Then the term sequence $M = (x_1\mu, \ldots, x_n\mu)$ is linear.* □

Theorem 11 gives us an algorithm to compute for a term t the pairs $(P, \mathcal{R}')$ where $P \subseteq O'(t)$ and $\mathcal{R}' \subseteq \mathcal{R}$ such that t is ground reducible at P using $\mathcal{R}' \subseteq \mathcal{R}$.

Let t, l be two terms with disjoint variable sets and let $\mu = mgu(t,l)$. Then it follows that $depth(x\mu) \leq depth(t)$ if $x \in Var(t)$ and l is linear. But the universes which have to be covered consist only of sequences of irreducible ground terms, and thus only those most general unifiers must be taken into account for which $x\mu \in T(\mathcal{C}, \mathcal{X})\ \forall x \in Var(t)$. Then the depth of $\{x\mu \mid x \in Var(t)\}$ for any most general unifier of any subterm of t and any left hand side of a rule in $\mathcal{R}$ is at most as deep as the deepest constructor subterm in a left-hand side of $\mathcal{R}$, provided the subterm of t or the left-hand side of the applied rule is linear.

In summary, for the test of ground-reducibility the test set has to be at most as deep as any constructor left-hand side in $\mathcal{R}$. For the extended ground-reducibility test, the bound is the deepest constructor subterm in any left-hand side in $\mathcal{R}$. For a given problem, the actually needed depth is given by the maximum depth of T^* computed by Theorem 11. If p is a shallow position in a deep t, these terms can be expected to be considerably less deep than the left-hand sides of $\mathcal{R}'$ were—the part that was common with t to all of them has been factored out. The cover algorithm of Section 4 will in turn be able to solve many subproblems with an even smaller depth.

Example 14 *Let $\mathcal{R} = \mathcal{I}_\pm$ be the following description of the integers:*

$$
\begin{aligned}
\mathcal{I}_\pm = \{\ & -(0) && \rightarrow\ 0,\\
& -(-(x)) && \rightarrow\ x,\\
& S(-(S(x))) && \rightarrow\ -(x),\\
& +(0,x) && \rightarrow\ x,\\
& +(S(x),y) && \rightarrow\ S(+(x,y)),\\
& +(-(S(x)),y) && \rightarrow\ -(S(+(x,-(y))))\}
\end{aligned}
$$

- *Let $t = +(v, S(w))$. The set of all mgu's between t and rules in $\mathcal{I}_\pm$ (restricted to $Var(t)$) is*

  $$\{\{v \leftarrow 0, w \leftarrow w\}, \{v \leftarrow S(x), w \leftarrow w\}, \{v \leftarrow -(S(x)), w \leftarrow w\}\}.$$

  The problem is: does $\{(0,w),(S(x),w),(-(S(x)),w)\}$ cover $UNIV((x,w))$? The answer is "yes," because $\{w\}$ covers $UNIV(w)$ AND $\{0, S(x), -S(x)\}$ covers $UNIV(x)$. Note that the problem has depth 1 whereas the deepest constructor left-hand side is $S(-(S(x)))$ of depth 2.

- *The set of all mgu's between $t|_2$ and rules in $\mathcal{I}_\pm$ (restricted to $Var(t|_2)$) is*

  $$\{\{w \leftarrow -(S(x))\}\}.$$

  Obviously, $\{0\} \subset UNIV(w)$ cannot be covered by $\{-(S(x))\}$. □

6 Algebra Separation

As shown in the previous sections, we can greatly improve the ground reduction test if we know the partition $\mathcal{F} = \mathcal{C} \uplus \mathcal{D}$ into constructors and the defined functions in an algebra. As mentioned in [JK86], this problem can be solved if a cover algorithm is available.

The general idea for our solution is to assume as induction hypothesis that a certain set C' of operators and constants are constructors, and then to test for every other operator $f \in \mathcal{F} \setminus C'$ if its matrix covers the universe of argument sequences in ground normal form constructible with C'.

Thus, we first have to relativize the notion of "cover" with respect to an arbitrary set of constructors. The following is a straightforward generalization of Definition 7.

Definition 12 *Given a set C of function symbols. The C-*universe *C-$UNIV_{\mathcal{R}}((t_1, \ldots, t_n))$ of ground normal forms represented by a linear term sequence $(t_1, \ldots, t_n)$, $t_i \in T(C, \mathcal{X})$, is the set of all sequences $(\bar{t}_1, \ldots, \bar{t}_n)$ of terms $\bar{t}_i \in T(C)$ in $\mathcal{R}$-ground normal form, where $\bar{t}_i = t_i\gamma$ for all $i \in \{1, \ldots, n\}$ and some substitution γ.*

Now we can speak about a situation where some T^* covers[3] only a subset $T(C') \subseteq T(C)$. Once we know that the set C' is at least a subset of C, the following lemma tells us if our assumption is correct. In the other case at least one new constructor is exhibited.

Lemma 13 *Let $C' \subseteq C$ be a set of constants and operators and $\mathcal{D}'$ a set of operators such that $\mathcal{F} = C' \uplus \mathcal{D}'$. Then the following holds:*

1. *If $d \in \mathcal{D}'$, d n-ary, and the matrix of d does not cover C'-$UNIV((x_1, \ldots, x_n))$, then d is a constructor.*

2. *If for all $d \in \mathcal{D}'$, d n-ary, the matrix of d covers C'-$UNIV((x_1, \ldots, x_n))$, then $\mathcal{D}'$ is a set of defined operators.* □

Since the initial algebra of $\mathcal{R}$ is not empty (otherwise proving by induction makes no sense), and since $\mathcal{R}$ is ground confluent and Noetherian, there must be at least one constant or 0-ary function which is irreducible. Thus, we easily find a set of constructors to begin with. Now, by applying Lemma 13, and incrementally adjusting our constructor set, we eventually partition our algebra into constructors and defined operators. The following theorem describes that strategy.

Theorem 14 (Algebra Separation) *Let $C_0 = \{c \in \mathcal{F} \mid$ the arity of c is 0 and c is irreducible$\}$ be the set of constants and 0-ary function symbols which are irreducible. Let $\mathcal{D}_0 = \mathcal{F} \setminus C_0$ the set of all other function symbols. Let $E_i = \{f \in \mathcal{F} \setminus C_i \mid$ the matrix of f does not cover C_i-$UNIV_{\mathcal{R}}(x_1, \ldots, x_n)\}$. Let $C_{i+1} = C_i \cup E_i$ and let $\mathcal{D}_{i+1} \doteq \mathcal{D}_i \setminus E_i$. Then $C_k = C$ and $\mathcal{D}_k = \mathcal{D}$ as soon as $E_k = \emptyset$.*

PROOF: Since $\mathcal{F}$ is finite, the existence of a k with $E_k = \emptyset$ is ensured. By Lemma 13.1, we have $\mathcal{D}_k \subseteq \mathcal{D}$. As $C_k \uplus \mathcal{D}_k = \mathcal{F}$ and $C \uplus \mathcal{D} = \mathcal{F}$, it suffices to show, by induction on k, that $C_k \subseteq C$. □

According to the cover algorithm, the depth of the test set for some $T(C_i)$ depends on the depth of the matrix of the operator f processed. It does not depend on any other operator.

[3]More generally, we could also have defined ground-reducibility with respect to some $T(C') \subseteq T(C)$. We do not quite need this important concept here, but it now follows immediately from restricted cover.

In our preconditioning step we compute $\mathcal{C}$ and $\mathcal{D}$ using Theorem 14 with our cover algorithm. We also compute the d-topset, where d is the depth of the deepest subterm $s \in T(\mathcal{C}, \mathcal{X})$ of a left-hand side of a rule in $\mathcal{R}$. To find out if $\mathcal{C}$ contains only free constructors we may use the following:

Lemma 15 ([HH82]) *The set $\mathcal{C}$ of constructors is free, if $\forall \langle l \to r \rangle \in \mathcal{R} : l \notin T(\mathcal{C}, \mathcal{X})$.* □

In the case of a freely generated algebra, we do not need a topset to check the emptyness of a universe built by a term sequence. However, topsets are also quite revealing for the understanding of induction. Given a ground-confluent subset $\mathcal{R}' \subseteq \mathcal{R}$, let d be the depth of the deepest subterm l' of any rule $\langle l \to r \rangle \in \mathcal{R}'$ with $l' \in T(\mathcal{C}, \mathcal{X})$. Then the ground terms in the topset $GR_{\mathcal{R}}(\mathcal{C}, d-1)$ comprise all bases to be considered for induction using $\mathcal{R}'$, whereas the non-ground terms determine all cases of the induction steps.

Example 15 *Let* $\mathcal{R} = \mathcal{P}_+ \cup \{ \; \frac{1}{2}(0) \to 0, \; \frac{1}{2}(S(0)) \to 0, \; \frac{1}{2}(S(S(x))) \to S(\frac{1}{2}(x))\}$.

Then $d(\mathcal{P}_+) = 1$ determines an associated $GR(1-1) = GR(0) = \{0, S(x)\}$, and hence for an induction proof using the definition $\mathcal{P}_+$ of $+$ we need only the induction basis 0 and case $S(x)$ in the step. For an induction over the definition $\mathcal{R}'$ of $\frac{1}{2}$ we get $d(\mathcal{R}') = 2$ with $GR(2-1) = GR(1) = \{0, S(0), S(S(x))\}$, yielding the bases 0 and $S(0)$ and case $S(S(x))$ in the step. □

7 Summary: The Whole Test

The entire extended ground reducibility test consists of two parts. The first part is the preconditioning step. There we compute $\mathcal{C}$, $\mathcal{D}$ and the d-topset $GR_{\mathcal{R}}(d)$ where d is the depth of the deepest subterm $l' \in T(\mathcal{C}, \mathcal{X})$ in a left-hand side of a rule $\mathcal{R}$. We also compute whether $\mathcal{C}$ is a set of free constructors by implementing Lemma 15.

In the second part, algorithm $COVER$ then uses $GR_{\mathcal{R}}(d)$, which, as discussed, is in general much smaller than the test-set of [JK86]. Passing $\mathcal{C}$ and freedom information to $COVER$ allows efficient computation of σ-successors, and of the test whether a term-universe is empty, in the important cases where all constructors are indeed free.

The computation of inductively complete sets of positions in a term t is done by implementing Theorem 11. For a detailed description of the implementation, including several smaller refinements such as dynamic computation of $GR_{\mathcal{R}}(d)$, see [Bün87a].

References

[BD88] Leo Bachmair and Nachum Dershowitz. Critical pair criteria for completion. *Journ. Symbolic Computation*, 6(1):1–18, August 1988.

[BM79] Robert S. Boyer and J Strother Moore. *A Computational Logic*. Academic Press, Orlando, Florida, 1979.

[Bün87a] Reinhard Bündgen. Design, implementation, and application of an extended ground-reducibility test. Master's thesis, Computer and Information Sciences, University of Delaware, Newark, DE 19716, 1987.

[Bün87b] Reinhard Bündgen. Design, implementation, and application of an extended ground-reducibility test. Technical Report 88-05, Computer and Information Sciences, University of Delaware, Newark, DE 19716, December 1987.

[Cav85] B. F. Caviness, editor. *Eurocal'85*, volume 204 of *LNCS*. Springer-Verlag, 1985. (European Conference on Computer Algebra, Linz, Austria, April 1985).

[Fri86] Laurent Fribourg. A strong restriction of the inductive completion procedure. In *Proc. 13th ICALP*, volume 226 of *LNCS*, Rennes, France, 1986. Springer-Verlag. (To appear in *J. Symbolic Computation.*).

[Gal86] Jean H. Gallier. *Logic for Computer Science.* Harper & Row, New York, 1986.

[HH82] Gérard Huet and Jean-Marie Hullot. Proofs by induction in equational theories with constructors. *J. Computer and System Sciences*, 25:239–266, 1982.

[HO80] Gérard Huet and Derek C. Oppen. Equations and rewrite rules: A survey. In Ron Book, editor, *Formal Languages: Perspectives and Open Problems*, pages 349–405. Academic Press, 1980.

[JK86] Jean-Pierre Jouannaud and Emmanuel Kounalis. Proofs by induction in equational theories without constructors. Rapport de Recherche 295, Laboratoire de Recherche en Informatique, Université Paris 11, Orsay, France, September 1986. (To appear in *Information and Computation*, 1989.).

[KM87] Deepak Kapur and David R. Musser. Proof by consistency. *Artificial Intelligence*, 31(2):125–157, February 1987.

[KNZ86] Deepak Kapur, Paliath Narendran, and Hantao Zhang. Proof by induction using test sets. In Jörg H. Siekmann, editor, *8th International Conference on Automated Deduction*, volume 230 of *LNCS*, pages 99–117. Springer-Verlag, 1986.

[KNZ87] Deepak Kapur, Paliath Narendran, and Hantao Zhang. On sufficient-completeness and related properties of term rewriting systems. *Acta Informatica*, 24(4):395–415, 1987.

[Kou85] Emmanuel Kounalis. Completeness in data type specifications. In Caviness [Cav85], pages 348–362.

[Küc85] Wolfgang Küchlin. A confluence criterion based on the generalised Knuth-Bendix algorithm. In Caviness [Cav85], pages 390–399.

[Küc86] Wolfgang Küchlin. *Equational Completion by Proof Transformation.* PhD thesis, Swiss Federal Institute of Technology (ETH), CH-8092 Zürich, Switzerland, June 1986.

[Küc89] Wolfgang Küchlin. Inductive completion by ground proof transformation. In H. Aït-Kaci and M. Nivat, editors, *Rewriting Techniques*, volume 2 of *Resolution of Equations in Algebraic Structures*, chapter 7. Academic Press, 1989.

[NW82] Tobias Nipkow and G. Weikum. A decidability result about sufficient completeness of axiomatically specified abstract data types. In *Sixth GI Conference on Theoretical Computer Science*, volume 145 of *LNCS*, pages 257–268, 1982.

[Pla85] David Plaisted. Semantic confluence and completion methods. *Information and Control*, 65:182–215, 1985.

[Thi84] Jean-Jacques Thiel. Stop loosing sleep over incomplete data type specifications. In *Proc. 11th PoPL*, Salt Lake City, Utah, 1984. ACM.

[ZKK88] Hantao Zhang, Deepak Kapur, and Mukkai S. Krishnamoorthy. A mechanizable induction principle for equational specifications. In E. Lusk and R. Overbeek, editors, *9th International Conference on Automated Deduction*, volume 310 of *LNCS*, pages 162–181. Springer-Verlag, 1988.

Inductive proofs by specification transformations

Hubert COMON*

Abstract

We show how to transform equational specifications with relations between constructors (or without constructors) into order-sorted equational specifications where every function symbol is either a free constructor or a completely defined function.

This method allows to reduce the problem of inductive proofs in equational theories to Huet and Hullot's proofs by consistency [HH82]. In particular, it is no longer necessary to use the so-called "inductive reducibility test" which is the most expensive part of the Jouannaud and Kounalis algorithm [JK86].

Introduction

Let F be a set of function symbols together with their profile (for example, $F = \{0 :\rightarrow int2;\ succ : int2 \rightarrow int2\}$) and E be a finite set of equational axioms (for example $E = \{succ(succ(0)) = 0\}$). The problem of inductive proofs in equational theories is to decide whether an equation (whose variables are implicitly universally quantified) is valid in $T(F)/=_E$, the quotient algebra of the terms constructed on F by the congruence generated by E. (For example, $succ(succ(x)) = x$ is an inductive theorem in the specification (F, E) but is not an equational consequence of E).

The "proof by consistency" method [KM87] consists in adding to E the theorem to be proved and trying to deduce a contradiction (inconsistency) using equational reasoning. This method has been widely studied. Let us cite among others [Mus80,Gog80,Lan81,HH82,KM87], [Kuc87,JK86,Fri86,KNZ86,Bac88].

All these works use the Knuth-Bendix completion procedure as a basis for equational deduction: E is assumed to be oriented into a ground convergent term rewriting system $\mathcal{R}$. If the completion procedure constructs for $\mathcal{R} \cup \{s = t\}$ (where $s = t$ is the theorem to be proved) a (ground) convergent term rewriting system without deriving an inconsistency, then $s = t$ is an inductive theorem[1]. If an inconsistency is derived, $s = t$ is not an inductive theorem.

The papers cited above essentially differ in the assumptions they make on F, E and in the way they detect inconsistencies. For example, in Musser's paper [Mus80] E is assumed to contain a complete axiomatization of an equality predicate and an inconsistency is derived simply when the completion procedure generates the equation $true = false$.

In Huet and Hullot's method [HH82], F is assumed to be split into two disjoint sets C (constructors) and D (defined operators) with the following conditions :

*Laboratoire d'Informatique fondamentale et d'Intelligence Artificielle, Institut IMAG, 46 Ave. Félix Viallet, 38031 Grenoble cedex, France. E-mail : comon@lifia.imag.fr

[1]In [Bac88] the requirement for the resulting term rewriting system to be convergent has been weakened.

- every term constructed on C only is irreducible by $\mathcal{R}$
- every term in $T(F) - T(C)$ is reducible by $\mathcal{R}$

Then, an inconsistency is detected when the completion procedure generates an equation $s = t$ between two "constructor terms" (i.e built without any symbol of D).

This method was generalized in the so-called "inductive completion procedure" by Jouannaud and Kounalis [JK86] where the requirement on F to be split into constructors and defined operators is dropped. They show that the key concept for detecting inconsistencies is the "inductive reducibility test". A term is said to be inductively reducible when all its ground instances are reducible. For example, $succ(succ(x))$ is inductively reducible by $succ(succ(0)) \to 0$ (but it is not reducible). Then, an inconsistency is detected when the completion procedure generates an equation $s = t$ where $s > t$ (for a given simplification ordering containing $\to_{\mathcal{R}}$) and s is not inductively reducible.

Recently Bachmair [Bac88], refining the equational consequences to be added during the completion procedure, proved that it is not necessary to orient the equations computed by the inductive completion procedure. In this case, an inconsistency is detected when a non inductively reducible *equation* is derived.

Although inductive reducibility has been shown to be decidable [Pla85,Com88a], Plaisted's algorithm as well as others ([KNZ85,KNZ86,JK86] for example) are very complex. Actually, they are at least twice exponential and (except for [Com88a]) cannot be used in practice.

The aim of this paper is to show that it is possible to reduce the general case handled in [JK86,Bac88] to Huet and Hullot's method by transforming the specification. This allows to avoid the inductive reducibility test since such a test is trivial in Huet and Hullot's algorithm.

Given a Term Rewriting System (TRS for short) $\mathcal{R}$, it is shown in [CR87,Com88b,Com88a] how to compute a conditional grammar for the set NF of the ground terms which are irreducible by $\mathcal{R}$. This construction is performed using equational problems simplification [CL88].

In [Com88b,Com88a] a cleaning algorithm for conditional grammars is given. This provides a method for deciding inductive reducibility in the general case but can also be used for computing an order-sorted specification which is equivalent (in some suitable sense) to the original specification and where F is split into constructors and defined operators. Such a construction can be extended in order to handle order-sorted specifications as well.

Another specification transformation was already proposed in [Tha85] in a very specific case. This paper shows that, whenever there are no overlap between left hand sides of the rules, when the set of function symbols is split into constructors and defined functions, and when the TRS is left linear, then the signature can be enriched with new constructor symbols in order to have the additional property that no rule contains "inner" occurrences of a defined symbol. Such a transformation is similar to ours since we actually add some new constructors. However we don't make the above mentioned assumptions and give very different (stronger) results.

Also, Kapur and Musser [KM86] proposed some specification transformations related to proofs by consistency. However, they do not address the same problem. Roughly, they assume some information about "what should be" the initial algebra (i.e. what should be the constructors) and then complete the set of axioms (in some consistent way) in order to indeed get this initial algebra. At the opposite, we want to preserve the initial algebra (up to isomorphism) and we allow some relations between constructors (or, more generally, we do not assume that a set of constructors has been defined at all). Then we show in this paper how to compute free generators of the initial algebra.

We present the transformation in section 2 and state the basic properties of the resulting specification. Theorem 4 is the main (new) result of the paper. Then we show in section 3 how to perform inductive proofs in the resulting order-sorted algebra.

1 Many-sorted and Order-sorted Algebras

We recall in this section most of the basic definitions on many-sorted and order-sorted algebras. The reader is referred to [GM87,SNGM87] for more details. We also introduce a notion of equivalent specifications.

1.1 Many-sorted and Order-Sorted Signatures

A Many-Sorted Signature (MSS for short) is a pair (S, F) where S is a set of*sorts* names (which will be denoted $\underline{s}, \underline{s}_1, \ldots$) and F is a set of function symbols together with a typing function τ which associates to each $f \in F$ a string in S^+. When $\tau(f) = \underline{s}_1\underline{s}_2 \ldots \underline{s}_n\underline{s}$ we write $f : \underline{s}_1 \times \ldots \times \underline{s}_n \to \underline{s}$ and say that f has *profile* $\underline{s}_1 \times \ldots \times \underline{s}_n \to \underline{s}$.

An Order-Sorted Signature (OSS for short) is a triple $(S, \geq, F)$ where S is a set of sort symbols, $\geq$ is an ordering on S and F is a set of function symbols together with a typing function τ which associates to each $f \in F$ a finite non empty subset of S^+. All words in $T(F)$ have the same length $n+1$ and $|f| = n$ is the *arity* of f.As in the many-sorted case, we say that f has profile $\underline{s}_1 \times \ldots \times \underline{s}_n \to \underline{s}$ when $\underline{s}_1 \ldots \underline{s}_n\underline{s} \in \tau(f)$.

In both cases (many-sorted and order-sorted) X is a set of *variable* symbols. A sort is assigned to each variable and we write $x : \underline{s}$ $\tau(x) = \underline{s}$. We assume that there are infinitely many variables of each sort.

In both cases, if SIG is a signature, $T(SIG, X)$ (sometimes written $T(F, X)$ when there is no ambiguity) is the set of "well formed" terms constructed on SIG and X in the usual way (cf [GM87] for example). When X is empty we write $T(SIG)$ (or $T(F)$) instead of $T(SIG, \emptyset)$.

In the following, we always assume that, for every $\underline{s} \in S$, there is at least one $t \in T(SIG)$ such that t has sort $\underline{s}$.

A signature is *finite* when both S and F are finite. An OSS is *regular* when each term $t \in T(SIG, X)$ has a least sort $LS(t)$. This property can be syntactically characterized for finite signatures ([GM87] for example): $(S, \geq, F)$ is regular iff, for every $w_0, w_1, w_2 \in S^*$ such that $w_0 \leq w_1$ [2] and $w_0 \leq w_2$ and every $f \in F$ such that $f : w_1 \to \underline{s}_1$ and $f : w_2 \to \underline{s}_2$, there exists $w_3 \in S^*$ and $\underline{s}_3 \in S$ such that $w_0 \leq w_3 \leq w_1, w_2$, $\underline{s}_3 \leq \underline{s}_1, \underline{s}_2$ and $w_3 \to \underline{s}_3 \in \tau(f)$.

If, in addition, each connected component of $(S, \geq)$ has a top element, then the regular signature is called *coherent* ([GM87]). *All signatures considered here are assumed to be coherent and finite* (except when the contrary is explicitly stated).

[2] the ordering on S is extended to S^* by comparing the sorts componentwise

1.2 Order-Sorted Algebras

Let $SIG = (S, \geq, F)$ be an OSS. A (SIG-)Order-Sorted Algebra (OSA for short) $\mathcal{A}$ is (as defined in [SNGM87])

- a family $(\mathcal{A}_{\underline{s}})_{\underline{s} \in S}$ of non empty sets such that, when $\underline{s} \leq \underline{s}'$, then $\mathcal{A}_{\underline{s}} \subseteq \mathcal{A}_{\underline{s}'}$. Let $C_{\mathcal{A}} = \bigcup_{\underline{s} \in S} \mathcal{A}_{\underline{s}}$.

- for each function symbol f, a mapping $f_{\mathcal{A}}$ from $D_f^{\mathcal{A}} \subseteq C_{\mathcal{A}}^{|f|}$ into $C_{\mathcal{A}}$ such that, if f has the profile $w \rightarrow \underline{s}$, then $\mathcal{A}_w \subseteq D_f^{\mathcal{A}}$[3]. and $f(\mathcal{A}_w) \subseteq \underline{s}_{\mathcal{A}}$.

Given a MSS (S, F), an (S, F)-Many-Sorted Algebra is simply an $(S, \geq, F)$ OSA where $\geq$ is the trivial ordering on S ($\underline{s} \geq \underline{s}'$ iff $\underline{s} = \underline{s}'$). In this way, OSAs strictly generalizes MSAs[4]. Therefore, when we speak about substitutions, rewrite rules, ... without any more specific mention, one should understand "order-sorted substitutions", "order-sorted rewrite rules",...

Homomorphisms are defined in the usual way. Then, for any OSS SIG, $T(SIG)$ is an initial OSA.

Let $\mathcal{A}$ be a SIG-OSA. An $\mathcal{A}$-assignment σ is a morphism from $T(SIG, X)$ into $\mathcal{A}$ which associates with each $x : \underline{s}$ an element $t \in \mathcal{A}_{\underline{s}}$.

A substitution σ is a $T(SIG, X)$-assignment such that $Dom(\sigma) = \{x \in X, x\sigma \neq x\}$ (called the *domain* of σ) is finite. The set of SIG-substitutions is denoted by Σ_{SIG} (or simply Σ when there is no ambiguity). If X_0 is a finite subset of X, a X_0-grounding substitution σ is a substitution whose domain includes X_0 and such that $\forall x \in X_0, x\sigma \in T(SIG)$. Often, we will omit the X_0 prefix, assuming that X_0 contains the variables occurring in the terms to which σ is applied. The set of all grounding substitutions w.r.t. some understood X_0 is denoted by $\Sigma_{SIG,g}$ (or simply Σ_g when there is no ambiguity).

1.3 Equations and Rewrite Rules

An *equation* is a pair of terms $s, t \in T(SIG, X)$ where $LS(s)$ and $LS(t)$ are in the same connected component of $(S, \geq)$. A model of a finite set of equations (axioms) E is defined as usual. The class of models of E is referred to as the equational theory E. In [GM87] (for example) a complete set of inference rules for equational deduction is given. This means that every equation which is valid in the equational theory can be derived using these rules. This allows to construct the congruence relation $=_E$ over $T(SIG, X)$ defined by a finite set of equations E. Then we have the following result:

Theorem 1 [GM87] *If SIG is a coherent signature and E a set of equations, then $T(SIG)/=_E$ is initial in the category of models of E.*

---

[3] If $\underline{s}_1 \ldots \underline{s}_n = w \in S^+$, $\mathcal{A}_w$ is the cartesian product $\mathcal{A}_{\underline{s}_1} \times \ldots \times \mathcal{A}_{\underline{s}_n}$.

[4] Because, here, MSS do not allow "overloaded" declarations.

An Order-Sorted Specification (OSSpec) is a pair (SIG, E) where SIG is an OSS and E a finite set of equations $s = t$ where $s, t \in T(SIG, X)$. A Many-Sorted Specification (MSSpec) is defined in the same way.

A *rewrite rule* is a couple of terms $s, t \in T(SIG, X)$ such that $Var(t) \subseteq Var(s)$. It is written $s \to t$. A Term Rewriting System (TRS) is a finite set of rewrite rules. A TRS $\mathcal{R}$ is *sort decreasing* [KKM88] if, for every rule $s \to t$ in $\mathcal{R}$ and every substitution σ, $LS(s\sigma) \geq LS(t\sigma)$[5]. In such a case, the reduction relation $\to_{\mathcal{R}}$ associated with a TRS $\mathcal{R}$ is defined as in the many-sorted case. $\to^*_{\mathcal{R}}$ is the reflexive transitive closure of $\to_{\mathcal{R}}$. For every relation $\to$, $\leftrightarrow$ is the symmetric closure of $\to$.

A TRS is *noetherian* if there are no infinite chain $t_1 \to_{\mathcal{R}} \ldots t_n \to_{\mathcal{R}} \ldots$. It is *confluent* (resp. *ground confluent*) if, for all $s, t_1, t_2 \in T(SIG, X)$ (resp. $T(SIG)$), $s \to^*_{\mathcal{R}} t_1$ and $s \to^*_{\mathcal{R}} t_2$ implies the existence of a term u such that $t_1 \to^*_{\mathcal{R}} u$ and $t_2 \to^*_{\mathcal{R}} u$. A TRS $\mathcal{R}$ is *convergent* (resp. *ground convergent*) if it is noetherian and confluent (resp. ground confluent). When a TRS is convergent (resp. ground convergent), for every term $t \in T(SIG, X)$ (resp. $t \in T(SIG)$) there is a unique term $t \downarrow_{\mathcal{R}}$ such that $t \to^*_{\mathcal{R}} t \downarrow_{\mathcal{R}}$ and $t \downarrow_{\mathcal{R}}$ is irreducible by $\mathcal{R}$.

A TRS $\mathcal{R}$ is *canonical* if it is convergent and if for every rule $l \to r$ in $\mathcal{R}$, l and r are irreducible by $\to_{\mathcal{R}-\{l \to r\}}$.

$=_{\mathcal{R}}$ is the congruence on $T(SIG, X)$ generated by the set of axioms obtained by considering the rules in $\mathcal{R}$ as equations. Then $\mathcal{I}(\mathcal{R})$ (or $\mathcal{I}(E)$) is another notation for the initial algebra $T(SIG)/=_{\mathcal{R}}$. $=_{\mathcal{I}(\mathcal{R})}$ is the congruence relation defined on $T(SIG, X)$ by :

$$s =_{\mathcal{I}(\mathcal{R})} t \iff \forall \sigma \in \Sigma_g, s\sigma =_{\mathcal{R}} t\sigma$$

1.4 Equivalent Specifications

Let $SIG = (S, \geq, F)$ and $SIG' = (S', \geq', F')$ be two coherent OSS such that $S \subseteq S'$, $\geq \subseteq \geq'$, $F \subseteq F'$ and. for each $f \in F$, $\tau(f) \subseteq \tau'(f)$. Then $T(SIG', X)$ is (canonically) a SIG-algebra. Let ϕ be the unique (injective) SIG-homomorphism from $T(SIG, X)$ into $T(SIG', X \cup X')$ which is the identity on X. Then, the OSSpec (SIG', E') is said to be *equivalent*[6] to (SIG, E) if

$$\forall s, t \in T(SIG, X), \quad (s =_{\mathcal{I}(E)} t \iff \phi(s) =_{\mathcal{I}(E')} \phi(t))$$

This means that, when we only consider the terms built on SIG, the specifications have the same class of inductive theorems.

Finally, an OSSpec $((S, \geq, F), E)$ is said to be *decomposed* if F can be split into two sets C and D such that:

- $\forall s, t \in T(C), s \neq_E t$
- $\forall s \in T(F) - T(C), \exists t \in T(C), s =_E t$

[5]Note that Rewriting Systems are always sort decreasing in MSA.

[6]This "equivalence" is not symmetric. This is an abbreviation for (SIG, E) is (SIG, E)-equivalent to (SIG', E'), the (SIG, E)-equivalence being indeed symmetric.

2 Transformation of Specifications

In this section we show how to transform a MSSpec, the *source specification* into an equivalent decomposed OSSpec: the *target specification*. However, as the simplification of equational problems can be generalized to finite coherent order-sorted signatures [Com88b], the method given in this section also applies to (finite coherent) OSSpec.

The source specification will be denoted by $(SIG_0, \mathcal{R}_0)$ where $SIG_0 = (S_0, F_0)$ and the target specification by $(SIG_T, \mathcal{R}_T)$ where $SIG_T = (S_T, \geq_T, F_T)$. We assume in the following that $\mathcal{R}_0$ is ground convergent. NF will denote the set of ground terms in $T(F_0)$ that are irreducible by $\mathcal{R}_0$.

2.1 Ground Normal Form Grammars

We don't give here the full algorithm that produces a conditional grammar for NF. Let us only sketch on an example the way it is computed.

Example 1
$F_0 = \{s : int2 \rightarrow int2;\ 0 : \rightarrow int2;\ + : int2 \times int2 \rightarrow int2\}$
$\mathcal{R}_0 = \{\ 0 + x \rightarrow x \quad x + 0 \rightarrow x \quad x + x \rightarrow 0 \quad s(s(0)) \rightarrow 0\ \}$
The first set of derivation rules only states that a term in NF has a root symbol in F_0:

$$NF_{int2} \quad \rightarrow \quad NF_0 \ \mid \ NF_{s(x)} \ \mid \ NF_{x_1+x_2}$$

where NF_t denotes both a non terminal and the language it generates: $NF_t = NF \cap \{t\sigma,\ \sigma \in \Sigma_g\}$.

Now we compute the derivation rules, say, for $NF_{s(x)}$: [7]

$$t \in NF_{s(x)} \quad \text{iff} \quad t = s(u), t \neq s(s(0)) \ \text{ and } \ u \in NF$$

Solving $s(u) \neq s(s(0))$ in $T(F_0)$, using the algorithm described in [CL88], leads to the four disjoint solutions:

1. $\exists x_1, x_2,\ u = x_1 + x_2$
2. $u = 0$
3. $\exists x_1, x_2,\ u = s(x_1 + x_2)$
4. $\exists x_1,\ u = s(s(x_1))$

This can be expressed by the four rules:

$$NF_{s(x)} \quad \rightarrow \quad s(NF_{x_1+x_2}) \ \mid \ s(NF_0) \ \mid \ s(NF_{s(x_1+x_2)}) \ \mid \ s(NF_{s(s(x))})$$

Using again the same method, we compute the derivation rules for the non terminals we introduced. There would here remain to compute the derivation rules for NF_0, $NF_{x_1+x_2}$, $NF_{s(s(x))}$, $NF_{s(x_1+x_2)}$.

---

[7] Informally, t is an irreducible ground instance of $s(x)$ iff

1. its root symbol is s
2. it does not match at the root any left hand side of a rule
3. its proper subterms are irreducible

This characterization of $NF_{s(x)}$ can be generalized to any NF_t (see [CR87,Com88a]).

Theorem 2 *This procedure fully described in [Com88a] does always terminate.*

In our example, we get the additional grammar rules:

$$
\begin{array}{lcllll}
NF_0 & \rightarrow & 0 & & & \\
NF_{x_1+x_2} & \rightarrow & NF_{s(x_1)} + NF_{s(x_2)} & \text{IF} & x_1 \neq x_2 & \mid NF_{x_1+x_2} + NF_{s(x)} \\
 & & \mid\ NF_{x_1+x_2} + NF_{x_3+x_4} & \text{IF} & x_1 \neq x_3 & \mid NF_{s(x)} + NF_{x_1+x_2} \\
 & & \mid\ NF_{x_1+x_2} + NF_{x_3+x_4} & \text{IF} & x_2 \neq x_4 & \\
NF_{s(s(x))} & \rightarrow & s(NF_{s(s(x))}) & & & \mid s(NF_{s(x_1+x_2)}) \\
NF_{s(x_1+x_2)} & \rightarrow & s(NF_{x_1+x_2}) & & &
\end{array}
$$

Then the grammar is "cleaned up" using an algorithm described in [Com88b,Com88a]. This algorithm is similar to the usual cleaning algorithm for context free word grammars; the non terminals from which there is no derivation chain reaching a terminal tree (called *useless* non terminals) are removed.

Theorem 3 [Com88b,Com88a] *There is an algorithm producing a conditional grammar of NF which does not contain any useless non-terminal.*

The grammar produced in this way will be called the *reduced grammar of NF*. In our example, we get the following reduced grammar:

$$
\begin{array}{lcll}
NF_{int2} & \rightarrow & NF_0 & \mid NF_{s(x)} \\
NF_{s(x)} & \rightarrow & s(NF_0) & \\
NF_0 & \rightarrow & 0 &
\end{array}
$$

And, indeed, there are only two terms in NF : 0 and $s(0)$.

Of course, this step (cleaning the grammar) is equivalent to an inductive reducibility test since we proved simultaneously that $x_1 + x_2$, $s(x_1 + x_2)$ and $s(s(x))$ are inductively reducible (the corresponding set of irreducible ground instances are empty). However, this computation has to be done only once, whatever inductive completion is performed afterwards.

We give another simple example which illustrates the transformation. This is a specification of the integers.

Example 2

$$
\begin{array}{l}
F = \{\ s, p : int \rightarrow int;\ \ 0 : \rightarrow int\ \ \ + : int \times int \rightarrow int\ \ \} \\
\mathcal{R} = \{\ \begin{array}{rcl}
s(p(x)) & \rightarrow & x \\
p(s(x)) & \rightarrow & x \\
0 + x & \rightarrow & x \\
s(x) + y & \rightarrow & s(x+y) \\
p(x) + y & \rightarrow & p(x+y)\ \ \}
\end{array}
\end{array}
$$

We get the following reduced grammar for NF:

$$
\begin{array}{lcllll}
NF & \rightarrow & NF_0 & \mid & NF_{s(x)} & \mid NF_{p(x)} \\
NF_0 & \rightarrow & 0 & & & \\
NF_{s(x)} & \rightarrow & s(NF_0) & \mid & s(NF_{s(x)}) & \\
NF_{p(x)} & \rightarrow & p(NF_0) & \mid & p(NF_{p(x)}) &
\end{array}
$$

In both examples, the reduced grammar of NF is a regular tree grammar. *We will assume this property in the following.* Note that such a property is ensured by the left linearity of the original TRS (see e.g. [GB85]). However this condition is not necessary as shown by example 1.

More precisely, we call NF-grammar any pair $\mathcal{G} = (NT, P)$ satisfying

- $NT = \{NF_{\underline{s}} | \underline{s} \in S\} \cup \{NF_t | t \in T_0\}$ where T_0 is a finite subset of linear terms in $T(F_0, X)$[8].
- P is a set of derivation rules $N \to f(N_1, \ldots, N_k)$ or $N \to N'$ such that:
  1. $N, N', N_1, \ldots, N_k \in NT$
  2. $f \in F_0$
  3. $\forall N \in NT, N = \bigcup_{N \to N' \in P} N'$
  4. $\forall NF_t, NF_{t'} \in NT$, t is not a variable and t and t' are not equal up to the renaming of their variables.
  5. For each $N \to f(N_1, \ldots, N_k) \in P$, $N = NF_t$ for some t such that $root(t) = f$.
  6. For each $N \to N' \in P$, $N = NF_{\underline{s}}$ for some $\underline{s} \in S_0$
  7. For each $N \in NT$, $N \neq \emptyset$

The reduced grammar of NF is an NF-grammar.

2.2 Constructing $(S_T, \geq_T)$

Let $\mathcal{G} = (NT, P)$ be an NF-grammar. S_T is the set of non terminals NT. For sake of clarity, we rename the terms in NT: in example 2, $NF_{s(x)}$ is usually denoted by *pos* (for strictly positive integers) and $NF_{p(x)}$ is usually denoted by *neg*.

The ordering $\geq_T$ on S_T is defined by:

- If $t, t' \in T(F_0, X)$, $NF_t \geq_T NF_{t'}$ iff $\exists \sigma \in \Sigma, t' = t\sigma$.
- $NF_{\underline{s}} \geq_T NF_t$ when $LS(t) \leq_0 \underline{s}$

F_0 is split into the sets C_0 and D_0 in the following way:

- C_0 is the set of function symbols f such that there is a rule $NF_{\underline{s}} \to NF_{f(x_1,\ldots,x_n)}$ in P
- $D_0 = F - C_0$

Equivalently, D_0 is the set of symbols $f \in F$ such that $f(x_1, \ldots, x_n)$ is inductively reducible. (This is so because of the properties of $\mathcal{G}$).

Now, let C_T be the set C_0 where every symbol has been primed. For each production rule $N \to f(N_1, \ldots, N_k)$ we associate with $f' \in C_T$ the profile $f' : N_1 \times \ldots \times N_k \to N$. We get now an OSS $(S_T, \geq_T, C_T)$[9] Let us show how it works on our two examples:

---

[8] When NF_t is computed by the algorithm, then t is always a linear term. Thus this is not an additional assumption. (See [Com88b,Com88a]).

[9] Note that condition 7 in the definition of an NF-grammar ensures that, for every $\underline{s} \in S_T$, $T(C_T)_{\underline{s}}$, the set of ground terms of sort $\underline{s}$ is not empty (as we required in the definition of an OSS)

Example 1
We associate with each non terminal a sort in the target specification. In this example, *int2* is associated with NF_{int2}, *pos* with $NF_{s(x)}$ and *zero* with NF_0. Then, each rule of the reduced grammar corresponds either to a subsort declaration or a profile declaration. The rules

$$NF \rightarrow NF_0 \mid NF_{s(x)}$$

give the inclusions $int2 > pos$ and $int2 > zero$.
The rule $NF_{s(x)} \rightarrow s(NF_0)$ gives $s' : zero \rightarrow pos$ and the rule $NF_0 \rightarrow 0$ gives $0' :\rightarrow zero$.

Example 2
This leads to the sort structure $S_T = \{int, pos, neg, zero\}$ with the relations $int > neg$, $int > pos$ and $int > zero$ corresponding to the rules

$$NF \rightarrow NF_0 \mid NF_{s(x)} \mid NF_{p(x)}$$

The other rules correspond to the profile declarations:

$$0' :\rightarrow zero \qquad \begin{array}{ll} s': & zero \rightarrow pos \\ & pos \rightarrow pos \end{array} \quad \begin{array}{ll} p': & zero \rightarrow neg \\ & neg \rightarrow neg \end{array}$$

Proposition 1 *Assume that $\mathcal{R}_0$ is ground convergent, then NF is an $(S_T, \geq_T, C_T)$-algebra*

In general $(S_T, \geq_T, C_T)$ is not coherent. In order to guarantee its coherence, we have to construct $(S_T, \geq_T, C_T)$ using a suitable NF-grammar:

Proposition 2 *Assume that $\mathcal{R}_0$ is ground convergent and that there exists an NF-grammar. Then there exists an NF-grammar $\mathcal{G}$ such that $(S_T, \geq_T, C_T)$ is a coherent OSS.*

Sketch of the proof: the NF-grammar is constructed starting from any NF-grammar $\mathcal{G}_0$ and adding some sorts and some profiles in the following way : for every pair of rules

$$\begin{array}{lcl} NF_{f(t_1,\ldots,t_n)} & \rightarrow & f(NF_{u_1},\ldots,NF_{u_n}) \\ NF_{f(t'_1,\ldots,t'_n)} & \rightarrow & f(NF_{u'_1},\ldots,NF_{u'_n}) \end{array}$$

such that

- for all i, u_i and u'_i are unifiable with a most general common instance $u_i \wedge u'_i$
- $NF_{u_i \wedge u'_i} \in NT_0$
- $\exists i$ s.t. u_i is not an instance of u'_i and $\exists j$ s.t. u'_j is not an instance of u_j

add the grammar rule

$$NF_{f(t_1 \wedge t'_1,\ldots,t_n \wedge t'_n)} \rightarrow f(NF_{u_1 \wedge u'_1},\ldots,NF_{u_n \wedge u'_n})$$

and the non terminal $f(t_1 \wedge t'_1, \ldots, t_n \wedge t'_n)$ (if not already in NT_0).
It is not difficult to see that the resulting set of non terminals and production rules constitute an NF-grammar. (Although some non-terminals may not be reachable). With such an NF-grammar, $(S_T, \geq_T, C_T)$ is coherent.

2.3 Computing the target specification

We take $D_T = F$ and $F_T = D_T \cup C_T$ together with the profile declarations $f : NF_{\underline{s}_1} \times \ldots \times NF_{\underline{s}_n} \rightarrow NF_{\underline{s}}$ if $f \in F$ has the profile $\underline{s}_1 \times \ldots \times \underline{s}_n \rightarrow \underline{s}$.

Then, each term in $T(F_0, X_0)$ can be viewed as a term in $T(F_T, X_T)$. In other words:

Lemma 1 *$T(F_T, X_T)$ is a (free) (S_0, F_0)-algebra.*

Indeed, let $[\![\cdot]\!]$ denote the function defined on $T(F_0, X_0)$ by:

- $[\![x : \underline{s}]\!] = x : NF_{\underline{s}}$
- $[\![f(t_1, \ldots, t_n)]\!] = f([\![t_1]\!], \ldots, [\![t_n]\!])$

$[\![\cdot]\!]$ is an injective (S_0, F_0)-morphism.

Let $\mathcal{R}_1$ be the set of rules

$$f(x_1 : \underline{s}_1, \ldots, x_n : \underline{s}_n) \rightarrow f'(x_1 : \underline{s}_1, \ldots, x_n : \underline{s}_n)$$

for every $f \in C_0$ and every profile $f' : \underline{s}_1 \times \ldots \times \underline{s}_n \rightarrow \underline{s}$. Such a construction is "well formed" since, if $f : \underline{s}'_1 \times \ldots \times \underline{s}'_n \rightarrow \underline{s}'$ in SIG_0, then, for every index i, $\underline{s}_i \geq_T NF_{\underline{s}'_i}$.

Lemma 2 *$\mathcal{R}_1$ is canonical and sort decreasing.*

This is indeed a consequence of proposition 2.

Let $\mathcal{R}_2$ be the set of rewrite rules $[\![t]\!] \downarrow_{\mathcal{R}_1} \rightarrow [\![u]\!] \downarrow_{\mathcal{R}_1}$ for every rule $t \rightarrow u$ in $\mathcal{R}_0$.

A *decreasing renaming* of a term t is a substitution θ which associates to each variable a variable with a lower sort in such a way that there is at least one variable x in t such that $sort(x\theta) < sort(x)$.

$\mathcal{R}_T = \mathcal{R}_1 \cup \mathcal{R}_2 \cup \mathcal{R}_2^*$ where $\mathcal{R}_2^*$ is the set of rules $l\theta \downarrow_{\mathcal{R}_1} \rightarrow r\theta \downarrow_{\mathcal{R}_1}$ for each rule $l \rightarrow r \in \mathcal{R}_2$ and each decreasing renaming of l. (Such a set of rules is finite as S is finite). Of course, rules in $\mathcal{R}_T$ which are instances of some other rule in $\mathcal{R}_T$ can be removed.

Example 1 [10]

$$\begin{array}{rll}
\mathcal{R}_1 = \{ & s(x : zero) \rightarrow s'(x : zero) & \\
 & 0 \rightarrow 0' & \} \\
\mathcal{R}_2 = \{ & s(s'(0')) \rightarrow 0' & \\
 & 0' + x \rightarrow x & \\
 & x + 0' \rightarrow x & \\
 & x + x \rightarrow 0' & \}
\end{array}$$

and $\mathcal{R}_T = \mathcal{R}_1 \cup \mathcal{R}_2$ (every rule in $\mathcal{R}_2^*$ is an instance of a rule in $\mathcal{R}_2$).

[10] When the sort of a variable is not mentioned, it must be understood that it has the greatest sort of its connected component.

Example 2

$$\mathcal{R}_1 = \{ \begin{array}{ll} s(x:zero) \to s'(x:zero) & s(x:pos) \to s'(x:pos) \\ p(x:zero) \to p'(x:zero) & p(x:neg) \to p'(x:neg) \\ 0 \to 0' & \end{array} \}$$

$$\mathcal{R}_2 = \{ \begin{array}{ll} s(p(x)) \to x & p(s(x)) \to x \\ 0' + x \to x & s(x) + y \to s(x+y) \\ p(x) + y \to p(x+y) & \end{array} \}$$

$$\mathcal{R}_2^* = \{ \begin{array}{ll} s(p'(x:zero)) \to x:zero & s(p'(x:neg)) \to x:neg \\ p(s'(x:zero)) \to x:zero & p(s'(x:pos)) \to x:pos \\ s'(x:zero) + y \to s(x+y) & s'(x:pos) + y \to s(x+y) \\ p'(x:zero) + y \to p(x+y) & p'(x:neg) + y \to p(x+y) \end{array} \}$$

2.4 Properties of the target specification

There are basically two mappings linking the source and the target specification. $[\![\cdot]\!]$ has already been mentioned. Now, for $t \in T(SIG_T, X)$ let $\tilde{t}$ be the term in $T(SIG_0, X_0)$ obtained by replacing each variable x by a variable x' whose sort is the greatest sort in the connected component of $sort(x)$ and replacing each primed function symbol by the unprimed one. (Then $\widetilde{[\![t]\!]} = t$).

Our construction using NF-grammars has the following main property:

Lemma 3 *$u \to_{\mathcal{R}_T} v$ iff $u \to_{\mathcal{R}_1} v$ or $\tilde{u} \to_{\mathcal{R}_0} \tilde{v}$.*

Corollary 1 *For every $t \in T(SIG_T)$, $\tilde{t}$ is irreducible by $\mathcal{R}_0$ iff t is irreducible by $\mathcal{R}_2 \cup \mathcal{R}_2^*$.*

Corollary 2 *For every $t, u \in T(SIG_T)$, $\tilde{t} \downarrow_{\mathcal{R}_0} = \tilde{s} \downarrow_{\mathcal{R}_0}$ iff $t \downarrow_{\mathcal{R}_T} = s \downarrow_{\mathcal{R}_T}$*

By construction, the target OSSpec is coherent[11]. The rewrite system $\mathcal{R}_T$ has also the desired properties:

Lemma 4 *If $\mathcal{R}_0$ is ground convergent, then so is $\mathcal{R}_T$.*

Sketch of the proof
When $t \to_{\mathcal{R}_T} u$, either $\tilde{t} \to_{\mathcal{R}_0} \tilde{u}$ or $\tilde{t} = \tilde{u}$. In the latter case $t \to_{\mathcal{R}_1} u$. This proves that $\mathcal{R}_T$ does terminate. The ground confluence proof is more involved; assuming that $s, t_1, t_2 \in T(SIG_T)$ and $s \to_{\mathcal{R}_T} t_1$ and $s \to_{\mathcal{R}_T} t_2$, there are three cases to investigate:

1. $\tilde{s} = \tilde{t_1} = \tilde{t_2}$. Then we use lemma 2
2. $\tilde{s} = \tilde{t_1} \neq \tilde{t_2}$. Then we use the construction of $\mathcal{R}_T$
3. $\tilde{t_1} \neq \tilde{s} \neq \tilde{t_2}$. Then we use the ground confluence of $\mathcal{R}_0$.

Lemma 5 *$\mathcal{R}_T$ is sort-decreasing.*

[11]This is easy to deduce from proposition 2. Note that this proves the existence of the initial algebra, as recalled in section 1.

This can be easily verified. As shown in [KKM88], when a term rewriting system is convergent and sort decreasing, for every equational order-sorted deduction of $s = t$, there is a rewrite proof of $s = t$. This result easily extends to ground convergent TRS:

Lemma 6 *If $\mathcal{R}$ is ground convergent and sort decreasing, then $s =_{I(\mathcal{R})} t$ iff, for every grounding substitution σ $s\sigma\downarrow_{\mathcal{R}} = t\sigma\downarrow_{\mathcal{R}}$.*

Now what we expect for inductive proofs is the equivalence between the two specifications:

Theorem 4 *If $\mathcal{R}_0$ is ground convergent, then $(SIG_T, \mathcal{R}_T)$ is equivalent to $(SIG_0, \mathcal{R}_0)$.*

In other words

$$s =_{I(\mathcal{R}_0)} t \quad \Leftrightarrow \quad [\![s]\!] =_{I(\mathcal{R}_T)} [\![t]\!]$$

Sketch of the proof
When $\sigma \in \Sigma_{SIG_0}$, $[\![\sigma]\!]$ is the substitution whose domain is $Dom(\sigma)$ and such that, for every $x \in Dom(\sigma)$, $x[\![\sigma]\!] = [\![x\sigma]\!]$. In the same way, when $\sigma \in \Sigma_{SIG_T}$, $\tilde{\sigma}$ is defined by $Dom(\tilde{\sigma}) = \{\tilde{x}, x \in Dom(\sigma)\}$ and, for every $x \in Dom(\sigma)$, $\tilde{x}\tilde{\sigma} = \widetilde{x\sigma}$. Now the theorem follows from the equivalences:

- $[\![s]\!] =_{I(\mathcal{R}_T)} [\![t]\!] \quad \Leftrightarrow \quad \forall \sigma \in \Sigma_{SIG_T,g},\ s\tilde{\sigma}\downarrow_{\mathcal{R}_0} = t\tilde{\sigma}\downarrow_{\mathcal{R}_0}$
- $s =_{I(\mathcal{R}_0)} t \quad \Leftrightarrow \quad \forall \sigma \in \Sigma_{SIG_0,g},\ [\![s]\!][\![\sigma]\!]\downarrow_{\mathcal{R}_T} = [\![t]\!][\![\sigma]\!]\downarrow_{\mathcal{R}_T}$

Let us show the second equivalence. $s =_{I(\mathcal{R}_0)} t$ iff for every grounding substitution σ, $s\sigma\downarrow_{\mathcal{R}_0} = t\sigma\downarrow_{\mathcal{R}_0}$. On the other hand, $s\sigma\downarrow_{\mathcal{R}_0} = \widetilde{[\![s\sigma]\!]}\downarrow_{\mathcal{R}_0}$ and, by corollary 2, $\widetilde{[\![s\sigma]\!]}\downarrow_{\mathcal{R}_0} = \widetilde{[\![t\sigma]\!]}\downarrow_{\mathcal{R}_0}$ iff $[\![s\sigma]\!]\downarrow_{\mathcal{R}_T} = [\![t\sigma]\!]\downarrow_{\mathcal{R}_T}$. Now, the equivalence follows from the identity $[\![s\sigma]\!] = [\![s]\!][\![\sigma]\!]$.

For the first equivalence, by lemmas 4 and 6, $[\![s]\!] =_{I(\mathcal{R}_T)} [\![t]\!]$ iff for every grounding substitution σ, $[\![s]\!]\sigma\downarrow_{\mathcal{R}_T} = [\![t]\!]\sigma\downarrow_{\mathcal{R}_T}$. Now, following the identity $[\![s]\!]\sigma\downarrow_{\mathcal{R}_T} = [\![s\tilde{\sigma}]\!]\downarrow_{\mathcal{R}_T}$, we have the equivalences

$$\begin{aligned} [\![t]\!]\sigma\downarrow_{\mathcal{R}_T} = [\![s]\!]\sigma\downarrow_{\mathcal{R}_T} &\Leftrightarrow [\![s\tilde{\sigma}]\!]\downarrow_{\mathcal{R}_T} = [\![t\tilde{\sigma}]\!]\downarrow_{\mathcal{R}_T} \\ &\Leftrightarrow \widetilde{[\![s\tilde{\sigma}]\!]}\downarrow_{\mathcal{R}_0} = \widetilde{[\![t\tilde{\sigma}]\!]}\downarrow_{\mathcal{R}_0} \\ &\Leftrightarrow s\tilde{\sigma}\downarrow_{\mathcal{R}_0} t\tilde{\sigma}\downarrow_{\mathcal{R}_0} \end{aligned}$$

□

It is thus possible to perform inductive proofs in the target algebra instead of the source algebra. As announced, we have also the following property which states that the target algebra is "simpler" than the source one:

Proposition 3 *$(SIG_T, \mathcal{R}_T)$ is a decomposed OSSpec.*

This indeed is a consequence of lemma 3.

3 Inductive Proofs in Order-Sorted Algebras

Now, it remains to show how to perform inductive proofs in decomposed order-sorted algebras. The aim of this paper is not to give results in this field. Therefore, we only show how to perform inductive proofs in our target algebra and sketch a general method.

The only difficulty with OSA is that equational reasoning may lead to ill formed terms (see for example [SNGM87]). Such a problem does not occur when dealing with sort decreasing term rewriting systems. And, by lemma 5, our system is sort-decreasing.

However, if we use the order-sorted completion (as in [KKM88]), an equational consequence $u = v$ where neither $LS(u) \leq LS(v)$ nor $LS(v) \leq LS(u)$ may be derived. In such a case, it is not possible to orient the equation and keep the sort-decreasing property.

Let us assume that the source algebra is a MSSpec. In this case, the target specification has some additional properties which ensure that such a sort problem cannot occur:

Lemma 7 *Assume that the source specification is a MSSpec. Let $s = t$ be an equation in the source specification. Then every equational consequence $u = v$ of $\mathcal{R}_T \cup \{[\![s]\!] = [\![t]\!]\}$ derived by a completion procedure satisfies either $LS(u\sigma) \geq LS(v\sigma)$ or $LS(v\sigma) \geq LS(u\sigma)$ for every substitution σ or $u, v \in T(C_T, X)$.*

This indeed can easily be proved using the fact that a term whose root symbol is in D_T has necessarily a sort which is maximal in its connected component[12].

Now three situations can occur when an equation $u = v$ is derived by the (inductive) completion procedure :

1. $u, v \in T(C_T, X)$ and it is possible to derive a contradiction
2. for every substitution σ, $LS(u\sigma) > LS(v\sigma)$ (or $LS(v\sigma) > LS(u\sigma)$) in which case the equation can be oriented, provided that constructor terms are smaller than non constructor ones for the reduction ordering.
3. for every substitution σ, $LS(u\sigma) = LS(v\sigma)$

Therefore, it is possible to use completion procedures (as in [KKM88]) in this case, without modifying the sort structure.

When the source specification is an OSSpec there are some more difficulties since an equation between two non constructor terms with uncomparable sorts can be derived by a completion procedure.

However, in order to solve complement problems in OSA, it is necessary to transform the sort structure ([Com88b]). In the result of this transformation (which is mainly a tree automaton determinization), two distinct sorts have disjoint carriers in $T(F)$. (But functions symbols remain "overloaded"). If we assume this additional property of the source specification, then lemma 7 holds for order sorted specifications. Therefore, no sort problem can occur during a completion procedure.

Examples 1 and 2

In examples 1 and 2 (previously defined) it is not possible to use directly the results of [HH82,Lan81] since there are some relations between constructors. However, using the target specification (see above), it is possible to use these methods.

[12] Our target signature may be compared with the *compatible signatures* of [SNGM87].

For example, in example 1, the commutativity $x+y=y+x$ is an inductive theorem since there is no (proper) critical overlap between $x+y$ and a rule in $\mathcal{R}_T$. This is sufficient for ensuring $x+y =_{I(\mathcal{R}_T)} y+x$ as shown in [Bac88].

Let us show how it is proved that $s(x)+y=y+0$ is not an inductive theorem:

$$\frac{\dfrac{\dfrac{s(x)+y=y+0}{s(x)=0'}\ \text{overlap with the rule } x+0'\to x}{s'(x:zero)=0'}\ \text{overlap with the rule } s(x:zero)\to s'(x:zero)}{\text{DISPROOF}}\ \text{since } s'(x:zero),0'\in T(C_T,X)$$

4 Concluding remarks

For any term rewriting system it is possible to compute a conditional grammar for NF without useless non-terminals. However, the method presented in this paper requires some more hypothesis. First, it requires the term rewriting system to be ground convergent (may be this hypothesis could be weakened to ground confluence). Secondly, it requires the clean grammar of NF to be a regular one. This means that the language of reducible ground terms is regular. There is no hope to weaken this hypothesis since the set of well formed ground terms in a finite OSS is a regular tree language: profile declarations provide a (bottom-up) finite tree automaton for recognizing it. Therefore, our transformation into a decomposed OSSpec seems to be optimal in some sense: whenever such a transformation exists, then it is computed[13].

A drawback of the method is that there does not exist any simple check for the regularity of the language of reducible ground terms. It is known that left linearity is a sufficient condition (see e.g. [GB85]). However, many examples can be built showing that this condition is not necessary[14]. Of course, it is possible to simply compute the cleaned up grammar and see if it is regular. But, as shown in this paper, this is not an easy computation. An open question is to broaden the left linearity condition in order to have some more general (syntactic) sufficient condition for regularity.

Let us also note that we could not use any regular tree grammar for our transformation since we actually use additional properties of the grammars produced by our algorithm for proving a stronger property of the target specification: the *equivalence* with the source one. Indeed, the isomorphism $T(F)/=_E \sim T(F')/=_{E'}$ does not provide in itself a way for deducing inductive theorems in $T(F,X)$ from inductive theorems in $T(F',X)$.

Anyway, at least in the left linear case (and others, see above), our method proves that it is possible to use Huet and Hullot's algorithm and avoid inductive reducibility checks. This is very useful since the test set of Plaisted's inductive reducibility test is always huge. That is the reason why we think our approach is well suited to the implementation of inductive proofs in equational theories without constructors.

[13] This optimality result will be detailed in a forthcoming paper.

[14] For example $F=\{a,f,h\}$ and the left hand sides are $\{h(f(x,x)), f(f(x_1,x_2),x_3), f(h(x_1),x_2)\}$. The language of reducible ground terms is regular.

Also, it must be noted that inductive proofs in order-sorted decomposed specifications is not harder than in the unsorted case. Indeed, this is the meaning of lemma 7. Therefore, our method is a real improvement over classical ones.

References

[Bac88] L. Bachmair. Proof by consistency in equational theories. In *Proc. 3rd IEEE Symp. Logic in Computer Science, Edinburgh*, July 1988.

[CL88] H. Comon and P. Lescanne. *Equational Problems and Disunification*. Research Report Lifia 82 Imag 727, Univ. Grenoble, May 1988. To appear in J. Symbolic Computation.

[Com88a] H. Comon. An effective method for handling initial algebras. In *Proc. 1st Workshop on Algebraic and Logic Programming, Gaussig*, 1988.

[Com88b] H. Comon. *Unification et Disunification: Théorie et Applications*. Thèse de Doctorat, I.N.P. de Grenoble, France, 1988.

[CR87] H. Comon and J.-L. Remy. *How to Characterize the Language of Ground Normal Forms*. Research Report 676, INRIA, June 1987.

[Fri86] L. Fribourg. A strong restriction of the inductive completion procedure. In *Proc. 13th ICALP, Rennes, LNCS 226*, pages 105–115, Springer-Verlag, 1986.

[GB85] J. H. Gallier and R. V. Book. Reductions in tree replacement systems. *Theoretical Computer Science*, 37:123–150, 1985.

[GM87] J. Goguen and J. Meseguer. *Order-Sorted Algebra I: Partial and Overloaded Operators, Errors and Inheritance*. Draft, Computer Science Lab., SRI International, 1987.

[Gog80] J. A. Goguen. How to prove inductive hypothesis without induction. In *Proc. 5th Conf. on Automated Deduction, LNCS 87*, 1980.

[HH82] G. Huet and J.-M. Hullot. Proofs by induction in equational theories with constructors. *Journal of Computer and System Sciences*, 25(2), 1982.

[JK86] J.-P. Jouannaud and E. Kounalis. Automatic proofs by induction in equational theories without constructors. In *Proc. 1st IEEE Symp. Logic in Computer Science, Cambridge, Mass.*, June 1986.

[KKM88] C. Kirchner, H. Kirchner, and J. Meseguer. Operational semantics of obj-3. In *Proc. 15th Int. Conf on Automata, Languages and Programming, LNCS 317*, Springer-Verlag, July 1988.

[KM86] D. Kapur and R. D. Musser. Inductive reasoning with incomplete specifications. In *Proc. 1st IEEE Symp. Logic in Computer Science, Cambridge, Mass.*, June 1986.

[KM87] D. Kapur and D. Musser. Proof by consistency. *Artificial Intelligence*, 31(2), February 1987.

[KNZ85] D. Kapur, P. Narendran, and H. Zhang. *On Sufficient Completeness and Related Properties of Term Rewriting Systems*. Research Report, General Electric Company, October 1985. Preprint.

[KNZ86] D. Kapur, P. Narendran, and H. Zhang. Proof by induction using test sets. In *Proc. 8th Conf. on Automated Deduction, Oxford, LNCS 230*, Springer-Verlag, July 1986.

[Kuc87] W. Kuchlin. *Inductive Completion by Ground Proofs Transformation*. Research Report, University of Delaware, February 1987.

[Lan81] D. Lankford. *A simple explanation of inductionless induction*. Technical Report MTP-14, Mathematics Department, Louisiana Tech. Univ., 1981.

[Mus80] D. Musser. Proving inductive properties of abstract data types. In *Proc. 7th ACM Symp. Principles of Programming Languages, Las Vegas*, 1980.

[Pla85] D. Plaisted. Semantic confluence tests and completion methods. *Information and Control*, 65:182–215, 1985.

[SNGM87] G. Smolka, W. Nutt, J. Goguen, and J. Meseguer. *Order-Sorted Equational Computation*. SEKI Report SR-87-14, Univ. Kaiserslautern, December 1987.

[Tha85] S. R. Thatte. On the correspondance between two classes of reductions systems. *Information Processing Letters*, 20:83–85, February 1985.

Narrowing And Unification In Functional Programming —An Evaluation Mechanism For Absolute Set Abstraction

John Darlington Yi-ke Guo*
Department of Computing
Imperial College of Science and Technology
180 Queen's Gate, London SW7 2BZ, England

Abstract

The expressive power of logic programming may be achieved within a functional programming framework by extending the functional language with the ability to evaluate absolute set abstractions. By absolute set abstraction, logical variables are introduced into functional languages as first class objects. Their set-valued interpretations are implicitly defined by the constraining equations. Narrowing and unification can be used to solve these constraining equations to produce the satisfying instantiations of the logic variables. In this paper, we study an execution mechanism for evaluating absolute set abstraction in a first order (non-strict) functional programming setting. First, we investigate the semantics of absolute set abstraction. It is shown that absolute set abstraction is no more than a set-valued expression involving the evaluation of function inversion. Functional equality is defined coinciding with the semantics of the continuous and strict equality function in functional programming. This new equality means that the well known techniques for equation solving can be adopted as a proper mechanism for solving the constraining equations which are the key to the evaluation of absolute set abstraction. The main result of this paper lies in the study of a particular narrowing strategy, called lazy pattern driven narrowing, which is proved to be complete and optimal for evaluating absolute set abstraction in the sense that a complete set of minimal solutions of the constraining equations can be generated by a semantic unification procedure based on this narrowing strategy. This indicates that a mechanism for equation solving can be developed within a functional programming context, producing a more expressive language.

1. Introduction

Recently, many efforts have been devoted to integrating functional and logic programming languages. A number of approaches have been proposed. A general survey can be found in [Bellia86]. Among them a stimulating approach is based on the idea that the additional power of logic programming can be achieved within a functional programming framework by introducing some "logic programming facilities" [Darl 85]. In [Darl 85] a new construct, called **absolute set abstraction,** was proposed for this purpose. An absolute set abstraction is of the form {E(x,y) with y | P(x,y)}, where x stands for the bound variables and y for the free variables, and denotes the set of elements formed by evaluating E(x, y) for all y's satisfying certain constraints P(x,y) . For example, evaluating the A.S.A. expression :

$\{\ (l_1,l_2)\ \text{with}\ l_1,l_2\ |\ \text{append}\ (\ l_1,l_2\ )\ =\ [1,2]\ \}$

will give a set of all the pairs of lists which concatenated together yield [1, 2]. That is:

```
{([],[1,2]), ([1],[2]), ([1,2],[]) }
```

So, we can define a function, in terms of A.S.A., to compute all the possible splits of a list:

---

* On leave from Dept. of Computer Science, Tsinghua University, Beijing, PRC.

```
dec split : list alpha -> set (list alpha x list alpha);
--- split ( L ) <= {(l1,l2) with l1,l2 | append ( l1,l2 ) = L }
```

In this case, the function append doesn't simply get applied to its arguments, but is run backwards to generate the appropriate values satisfying the constraining equation. This novel construct introduces the "solve" ability into functional programming languages. In this paper, we investigate the semantics of this construct within the setting of a first order (non-strict) functional programming language. Its declarative semantics is presented in terms of function inversion. Based on this semantics, we present a unification approach to perform the inverted evaluation of expressions via equation solving. To achieve this, the conventional E_unification is modified by defining equality on the data object domain, rather than on the equality theory derived by regarding the functional program as the axioms. This semantic unification procedure is based on narrowing. It is shown that this modified unification procedure, which generates the complete minimal solution set for equation solving in a functional programming context, provides a semi_decidable procedure for evaluating absolute set abstractions.

Using narrowing to solve equations on a non-classical equality is widely investigated in the literature. Fribourg[Frib85], Van Emden and Yakawa[EmYu86] and Levi et.al[Giov86] have proposed complete semantic unification mechanisms to solve equations on the equality, similar to our definition , for integrating functional programming notion into Horn clause logic. The completeness of narrowing, in their case, depends on the reconstruction of the semantic domains to establish some new models for their extended logical system. Our research concentrates on a somewhat different subject: solving equations in a functional programming framework. In this area, Reddy[Reddy85] has suggested using narrowing as the operational semantics of functional languages and outlined a lazy narrowing strategy. The main achievements of our work is two fold: As to languages, we provide a declarative semantics for absolute set abstraction and construct a semantically correct execution mechanism for the construct.This reveals an efficient way to incorporate narrowing and unification into functional programming. As to narrowing, we refine and reconstruct precisely Reddy's lazy narrowing strategy and rigorously prove its completeness and minimality for equation solving.

We present, in section 2, an abstract first order functional language using the terminology of term rewrite systems in order to provide the notations used in this paper. In section 3 we discuss the semantics of absolute set abstraction. In section 4 we first investigate the equality interpretation in functional programming and define an equality called functional equality. We then extend Huet's standardization theorem of reduction to narrowing in the functional programming setting and show that, for solving equations on functional equality, enumerating all standard narrowing derivations generates the complete minimal solution set. A complete and optimum unification procedure, based on a lazy pattern driven narrowing strategy, is proposed in section 5 and shown to be able to enumerate all and only all standard narrowing derivations. Conclusion and discussions about related works are presented in section 6 .

2. A First Order Functional Language

In this section, a first order functional language is abstractly defined using the terminology of term rewriting system so that a functional program can be regarded as a TRS of a certain restricted form. We assume readers are familiar with the basic notions of many-sorted algebras and term rewriting systems.

Definition 2.1 *A signature* $\Sigma=<S, F>$ *consists of a set of sorts S and a set of function symbols F with type* $S^* \rightarrow S$. *We use* F_n *to denotesall n-ary function symbols. The operator set F can be partitioned into two disjoint parts,* $F=F_d+F_c$, *we call the function symbols in* F_c *free constructors and those in* F_d *defined functions. So* F_{c0} *is the set of all constants .Given a set of variable symbols V (disjoint from F), the set* $T_\Sigma(V)$ *is the minimal set of first order terms containing V which satisfies the condition:* $\forall f \in F_n, \forall t_1.....t_n \in T_\Sigma(V) \Rightarrow f(t_1........t_n) \in T_\Sigma(V)$. $\vartheta(t)$ *denotes the set of variables in term t .*

Definition 2.2 *For all* $t \in T_\Sigma(V)$

if $t \in T_{\Sigma C}$ *i.e. t is constructed only by constructors, then t is called a ground value.*
if $t \in T_{\Sigma C}(V)$ *i.e. t is composed of constructors and variables, then t is called a data object.*
if $t \in \{ f(t_1........t_n) \mid t_i \in T_\Sigma(V)$ *and* $f \in F_{d_n} \}$, *then t is called an (application) expression.*
if $t \in \{ f(t_1........t_n) \mid t_i \in T_\Sigma(V)$ *and* $f \in F_{cn} \}$, *then t is an expression in WHNF.*

Definition 2.3 *A term rewriting system over* Σ *is a set of rewrite rules :* $\{ L_i \rightarrow R_i \}_{i \in I}$, *where* $L_i, R_i \in T_\Sigma(V)$ *and* $L_i \notin V, \vartheta(R_i) \subseteq \vartheta(L_i)$. L_i *is called the LHS of the rule i and* R_i *the RHS.*

A functional program, written in a first order constructor-based functional language, is a term rewriting system of a certain restricted form with restrictions imposed on the LHS's of the rewrite rules. Such a term rewriting system , called *a Functional TRS* , is defined as follows:

Definition 2.4 *A TRS is a functional TRS iff it satisfies the following restrictions on the rewriting rules:*

1. Left_linearity: No variable appears more than once in the LHS of any rewrite rule.
2. Nonambiguity: There are no critical pairs (since all LHS's are ununifiable).
3. Following "constructor discipline": For every rule $F(t_1,...,t_n) \rightarrow E$, $F \in F_d$ *and* $t_i \in T_\Sigma(V)$. *That is, F is the only defined function symbol on the LHS.*

A model-theoretic semantics of a functional TRS is defined by an interpretation (D, [| |]), where D is the semantic domain on which the functions are defined and [| |] is a meaning function assigning to every function symbol a continuous function on the semantic domain D. For modeling lazy evaluation D is organized as an algebraic CPO [Giov86]. Following the notion of Herbrand interpretation, D can be derived from the standard Herbrand universe, which is simply the set of all ground data values, by incorprating ground partial values and ground infinite values. Such an extended Herbrand universe is called the complete Herbrand universe (CU). The interpretation of $\bot$ can be regarded as the least defined data structure. So the set $F_c \cup \{\bot\}$ is ordered by the reflexive closure of the relation: $\forall c \in F_c\ \bot \leq c$. Thus, the ordering on CU is defined by following:

$d \leq d'$ iff $\forall u \in D(d)\ d(u) \leq d'(u)$ where D is the occurrence set of a term (see [Sec.4]).

By this construction, D=(CU, ≤) is an algebraic CPO [Giov86]. On this semantic domain,the meaning of a ground term is derived by imposing structural compositionality, i.e.

$$[\,/\,f(e_1,......e_n)\,/]\,] = [/\,f\,/]\,(\,[/\,e_1\,/]\,,......[/\,e_n\,/]\,) \qquad for\ f \in \Sigma_{F_n + C_n}\ and\ e_i \in T_\Sigma.$$

The interpretation of constructors is defined as themselves:

$$[/\,c(e_1,......e_n)\,/]\,] = [/\,c\,/]\,(\,[/\,e_1\,/]\,,......[/\,e_n\,/]\,)$$
$$= c\,(\,[/\,e_1\,/]\,,......[/\,e_n\,/]\,) \qquad for\ c \in \Sigma_c\ and\ e_1,......e_n \in T_\Sigma.$$

The interpretation of non-ground terms involves an environment which is a function ρ: V→ D assigning to variables the values in D. So we have:

$$[/\,v\,/]_\rho = \rho v\ for\ v \in V$$
$$[/\,f(t_1,......t_n)\,/]_\rho = [/\,f\,/]\,(\,[/\,t_1\,/]_\rho\,,......[/\,t_n\,/]_\rho) \quad for\ f \in \Sigma_{F_n + C_n}\ and\ t_i \in T_\Sigma(V).$$

An interpretation satisfies a rule $L_i \to R_i$ if, for every environment ρ, $[|L_i|]_\rho = [|R_i|]_\rho$ under this interpretation, where "=" denotes the "semantic equality" [Reddy85] admitting $\bot = \bot$. An interpretation is called a *model* if it satisfies all the rules of the program. The semantics of the program is defined by *the initial model* of the model class. We will not delve deeper into the semantics, many details can be found in [Reddy85] [Giov86].

3. Absolute Set Abstraction

The syntax of a rule involving absolute set abstraction is defined as follows:

$$f(d_1,....d_n) \to \{Exp\ \textbf{with}\ l_1,.....,l_k \mid e_{11}=e_{12}\,,\ ....,\ e_{m1}=e_{m2}\}$$

where $\vartheta(Exp) \subseteq \Omega$ and $\vartheta(e_{i1}) \cup \vartheta(e_{i2}) \subseteq \Omega$ for $\Omega = \vartheta(d_1) \cup .... \cup \vartheta(d_n) \cup \{l_1,.....,l_k\}$.

In this syntax, $l_1,....l_k$ are the logical variables whose values are implicitly defined by the constraining equations. Exp is called the *head expression* computing the elements of the set. Intuitively, an A.S.A. is a set of values resulting from the evaluation of the head expression for all instantiations of the logical variables satisfying the constraining equations. We formalize this semantics in terms of function inversion.

The semantics of absolute set abstraction can also be defined by the interpretation (D,[| |]):

$$[/\,\{\,Exp\ \textbf{with}\ l_1,.....l_k \mid e_{11}=e_{12},\ e_{21}=e_{22},\ .....,e_{n1}=e_{n2}\}\,/]_\rho$$
$$= \{\,\textbf{map}\,(\,\lambda l_1...l_k\,.[/\,Exp\,/]_\rho\,)\ [/\{l_1,........l_k \mid e_{11}=e_{12},\ e_{21}=e_{22},.....,e_{n1}=e_{n2}\}\,/]_\rho\}$$

where map is a meta function which maps a function over a set of values , i.e.

$map\ f\ \{\} = \{\}$

$map\ f\ \{\ a \cup S\ \} = f(a\ ) \cup map\ (\ f, S\ )$

$\{\ l_1,\ldots..l_k \mid e_{11}=e_{12},\ \ldots,e_{n1}=e_{n2}\ \}$ is called *the abstraction body*. Its semantics is defined as: $[/\{\ l_1,\ldots..l_k \mid e_{11}=e_{12},\ \ldots..e_{n1}=e_{n2}\}\ /]_r = \{\ <t_1,\ldots.t_k> \mid f\ t_1\ldots.t_k = true\ \}$

where f is a boolean valued function defined by an equality function:

$f = \lambda\ l_1\ldots\ldots l_k\ .\ [/\ (\ (\ e_{11},\ e_{21},\ldots..e_{n1}) = (\ e_{12},\ e_{22},\ \ldots\ldots e_{n2})\ )\ /]_\rho$

$= \lambda\ l_1\ldots\ldots l_k.\ eq\ ([/\ (\ e_{11},\ e_{21},\ldots..e_{n1})\ /]_\rho\ ,\ [/\ (\ e_{12},\ e_{22},\ \ldots\ldots e_{n2})\ /]_\rho\ )$

where eq is a continuous equality function defining the interpretation of equality in the functional programming setting (see Sec 4). In the semantic definition, the instantiations of the tuple of logical variables $(l_1,\ldots\ldots l_k)$ are denoted by the set of inverse images of this derived function, based on the equality function, at the value true .

For example, consider the following Hope function, position, defined by A.S.A., to compute the positions where a particular element occurs in a list :

```
dec position: alpha x list alpha → set(num)
---position(a,l)<=
     {length(l₁)+1 with l₁,l₂|append (l₁,append([a],l₂))=l}
```

where append and length are two functions computing the concatenation of two lists and the the length of a list respectively. By the semantics defined above, for a given element a and list l, this A.S.A. has the meaning:

$map\ (\lambda\ l_1,\ l_2.\ length\ (l_1) + 1\ )\ \{\ <t_1\ ,\ t_2> /\ f\ t_1\ t_2 = true\ \}$

$where\ f = \lambda\ x_1,\ x_2.\ eq(append\ (x_1,\ append\ ([\ a]\ ,\ x_2\ ))\ ,\ l\ )$

So, the expression position(b, [b,c,a,b]) should be evaluated by following steps:

$map\ (\lambda\ l_1,\ l_2.\ length\ (l_1)+1)\ \{\ <[\ ]\ ,\ [c,a,b]>,\ <[b,c,a],\ []>\ \} = \{\ 1,4\ \}$

4. Equation Solving In Functional Programming

In this section, we aim to design an efficient mechanism for solving constraining equations which is the key to implementing A.S.A.. We investigate first the interpretation of equality in the functional programming framework in order to define a special equality called *functional equality*. This ensures that the inverse images of a function based on the continuous equality function, eq, at value true can be effectively evaluated by computing the complete data solution set of the equation over this equality. In the second part of the section, we generalize Huet's standardization theorem of reduction in the regular TRS to narrowing in the functional TRS and show that the complete and minimal solution set of an equation on functional equality can be efficiently computed by enumerating all standard narrowing derivations.

4.1 On the Equality in Functional Programming

In traditional equational logic, an equality theory can be defined by a term rewriting system which regards rewrite rules as equational axioms. The equality, called E_equality, denoted as $=_E$, is generated by an axiom set E as the finest congruence containing all equations A=B in E and closed under replacement and instantiation [HuOp80]. For a confluent TRS, R, the completeness of equational deduction via reduction is expressed by the well known Church-Rosser property: $S =_E T$ iff there exists a term P such that $S \to^* P$ and $T \to^* P$ where E is the equality theory defined by R. The handing of an equation on E_equality, say $E_1 = E_2$, can be performed by the procedure, called E_unification, to find all the substitutions, σ, such that $\sigma E_1 =_E \sigma E_2$ [HuOp 80][GaSn87][Mate86].

The notion of equality in functional programming framework has been widely studied in the literature [Reddy85] [Giov86]. To guarantee computability, the equality of two expressions is captured by a continuous equality function, eq, which is based on the identity of the data values denoted by the component expressions. Such data values are maximal (containing no $\perp$) and algebraic (finite) elements of the semantic domain D (i.e. data objects) [Giov86]. We call such an equality *functional equality* . Its semantics are defined by the following fixed interpretation :

$$[/ E_1 = E_2 /]_\rho = eq[/(E_1,E_2)/]_\rho = \begin{cases} True \text{ when } [/E_1/]_\rho \in D' \; [/E_2/]_\rho \in D' \; [/E_1/]_\rho = [/E_2/]_\rho \\ False \text{ when } [/E_1/]_\rho \in D' \; [/E_2/]_\rho \in D' \; [/E_1/]_\rho \neq [/E_2/]_\rho \\ \perp \text{ otherwise} \end{cases}$$

where $D' \subseteq D$ is the set of all maximal finite data value element of D.

Note that $eq(E_1, E_2)$ is strict in both its arguments, we can't say anything about the equality of two expressions before they are evaluated to data values without uncertain information, i.e. their equality can only be decided by their values. This is the basic difference between the classical E_equality and functional equality. We can't check the equality of expressions through equational deduction, since, if we do so, the monotonicity of the equality function may be destroyed [Reddy85]. For this reason, we adopt functional equality for evaluating absolute set abstractions. Following the soundness and completeness of reduction in our semantic setting, we have the following equivalent proof-theoretical definition of functional equality:

Definition 4.1 *For any two expressions E_1 and E_2 in a functional program P, they are said to be functionally equal, denoted as $E_1 =_F E_2$, iff $\exists D_1, D_2 \in T_{\Sigma C}(V)$ such that $E_1 \to^* D_1$ and $E_2 \to^* D_2$ and $D_1 = D_2$ where "=" denotes identity up to variable renaming.*

This definition provides a semi-decidable proof procedure for functional equality. The equation $E_1 = E_2$ can be solved on functional equality by finding all substitutions σ, such that $\sigma E_1 =_F \sigma E_2$.We will show that the inverse images of a function based on the equlity function can be derived by computing the complete set of data solutions of the corresponding equation on functional equality. To prove this, we first organize the data objects as a complete lattice.

Lemma 4.1 *Let Sp be the data object set of sort P which is the quotient set of all data objects of the sort modulo the equivalence relation of variable renaming, then the data object domain of P, D_p, is a complete lattice $< Sp_{\perp}, \leq, \wedge, \vee >$ where $\leq$ is the instantiation order and the joi operation $\vee$ is first order generalization and the meet operation $\wedge$ is first order unification.*

Proof: Note that all the data objects in *Sp* are first order terms. This lemma directly results from Reynold's lattice of atomic formula in predicate logic [Reyno70].

The meaning of a data object can be defined on the semantic domain $<2^{T_{\Sigma}}c, \subseteq, \cap, \cup>$ as a set of ground values by a semantic function ζ, which is an order homomorphism, computing all "well typed" ground instantiations of a data object. This "data object as set " semantics establishes the relationship between the complete solution set of an equation and the inverse images of the related function. For a program P, we define its data object domain *Cp* as the separate sum of all data object domains of the sorts in the program. The following lemma claims the existence of the solution on *Cp* for a solvable equation .

Theorem 4.1 *Given an equation $E_1=E_2$, when it is solvable, there must exist a substitution θ such that $\theta E_1 =_F \theta E_2$ and $\forall x \in D(\theta)\ \theta x \in Cp$ where $D(\theta)$ is the substitution domain.*

Proof: For each $x_i \in D(\rho)$ $\rho x_i \notin Cp$, if $\exists d_i \in Cp$ such that $\rho x_i \rightarrow^* d_i$, the new solution is derived by replacing all ρx_i with d_i. Otherwise, providing $\rho x_j = s \notin Cp$ and there is no finite reduction sequence to reduce s to a data object, x_j is an "irrelevant variable" to the reductions of the two expressions since ρE_1 and ρE_2 are nonstrict in ρx_j. Therefore, x_j can be deleted from $D(\rho)$ without any impact on proving equality.

This proof has shown that all the meaningful solutions for solving equations on functional equality are *data solutions*. i.e solutions on the data object domain. Therefore, what we need to do to solve an equation is find a complete set of data solutions. In the remainder of this paper, solutions will always mean data solutions. The comparison of substitutions on *Cp* is obviously based on syntactical identity so we use =, $\leq$ when comparing substitutions.

Definition 4.2 *Let $\Pi(E_1,E_2)$ be the set of all solutions of an equation $E_1=E_2$, Σ is a complete solution set of the equation iff: 1) $\Sigma \subseteq \Pi(E_1,E_2)$ (Correctness). 2) $\forall \rho \in \Pi(E_1, E_2)$ $\exists \rho' \in \Sigma\ \rho' \leq \rho$ (completeness). A complete solution set Σ is minimal if $\forall \rho_1, \rho_2 \in \Sigma\ \rho_1 \neq \rho_2 \Rightarrow \rho_1, \rho_2$ are incomparable, denoted as $\rho_1 <> \rho_2$ (Minimality).*

Now, we come to the main result of this subsection: the inverted evaluation of expressions can be performed by equation solving on functional equality. The following theorem can be easily proved by the notion of complete solution set and the "data object as subset" semantics.

Theorem 4.2 *Given an equation E: $E_1=E_2$ with $\vartheta(E_1, E_2) = \{ l_1,...l_n\}$. Let Σ be the complete solution set of E on functional equality and ξ be the function computing all "well typed " ground instances of the data objects, then $\cup\{\xi <v_1..v_i..v_n> / v_i = \rho l_i$ and $\rho \in \Sigma\} = \{ t / f\ t = true \}$ where $f = \lambda l_1...l_n.\ eq\ (E_1, E_2)$.*

This theorem indicates what we should do for a semantically correct implementation: design a mechanism for solving equations to generate the complete data solution set.

4.2 Equation Solving By Narrowing

This subsection is devoted to investigating the computation required to solve equations over functional equality. Narrowing is proved to be complete in the sense that a complete solution set of an equation can be computed by a semantic unification algorithm based on it. The well known standardization theorem of reduction derivations in a regular TRS is extended to narrowing derivations in a functional TRS by proving that the narrowing derivation space is subsumed by a set of *standard narrowing derivations* . The main result is that the complete and minimal solution set of an equation over functional equality can be computed by enumerating all standard narrowing derivations. First of all, we define some related notation.

Definition 4.3 *Given a term, we define its occurrence set, D(E), as a finite subset of integer sequences which satisfies : a) $\varepsilon \in D(E)$, b) $u \in D(e_i) \Rightarrow i.u \in D(F(e_1, .., e_n))$ for any $F \in F_{dn}$. D (E) is partially ordered by the prefix ordering : $u \le v$ iff $\exists w\ u.w = v$. In this case, we define v-u = w. We say that u is outer to v (v is inner to u) iff $u \le v$ and $u \ne v$ (denoted as $u < v$). Two incomparable occurrences u,v are denoted by u<>v. D* (E) is the non-variable occurrence set of E .We define E/u as the subexpression of E at occurrence u, E(u) as the symbol at u and E [$u \leftarrow E'$] the expression obtained by replacing E/u by E' in E.*

Definition 4.4 *Given a TRS P and an expression E, we say that E narrows to E' at u by rule $\alpha_k \to \beta_k$, denoted as $E \surd\!\!\to_{[u,k,\rho]} E'$ (or simply $E \surd\!\!\to_\rho E'$) where $u \in D^*(E)$ and $E' = \rho(E[u \leftarrow \beta_k])$, iff E/u is unifiable with α_k and ρ is the most general unifier of E/u and α_k ($\alpha_k \to \beta_k$ is renamed so that $\vartheta(\alpha_k) \cap \vartheta(E) = \varnothing$).*

In [Hull80] narrowing was proved to be complete for E_unification in a canonical TRS. In this proof, confluency is essential for completeness whereas termination only plays a role in finding normalized solutions. In our case, all intended solutions are data solutions which are non-narrowable. So narrowing is complete with respect to solving equations in a functional programming setting even when the program is non-terminating.

Theorem 4.3 *Let P be a functional TRS and E be an equation, $E_1 = E_2$ and S the set of all substitutions such that $\rho \in S$ iff there exists a narrowing derivation : $eq(E_1, E_2) \surd\!\!\to_\sigma^* eq(D_1, D_2)$ where $D_1, D_2 \in Cp$ such that $\exists\theta\ \theta D_1 = \theta D_2$ (θ is m.g.u.) and $\sigma\theta = \rho$, then $\Sigma = S\downarrow_{\vartheta(E)}$ is a complete solution set of E where $\sigma\downarrow_{\vartheta(t)}$ denotes the restriction of σ to the variables in t.*

By this theorem, an equation $E_1 = E_2$, can be solved over functional equality by narrowing component expressions to data objects and then unifying them using standard syntactic unification. No cross checking of unifiability of all the intermediate expressions is necessary during narrowing (c.f. Hullot's algorithm). This makes the narrowing algorithm practical for equation solving. However, generating a complete solution set by enumerating all narrowing derivations is still not practical for implementing a language construct since many redundant computations still exist. We wish to design a narrowing strategy that can prune redundant narrowing derivations without losing completeness. We will move towards such a complete and optimal strategy by generalizing Huet's standardization theorem, [Huet86], on reduction in a regular TRS to narrowing in a functional TRS setting.

Definition 4.5 *Let a: $E_0 \to_{[u,k]} E_1$ be a one step reduction, $v \in D(E_0)$. We define the residual map r as a function of a and v computing a subset of $D(E_1)$ being the residual set of v by a thus:*

$r[E_0 \to_{[u,k]} E_1]\, v \quad = \{ u.w.(v-u.v') \mid \alpha_k(v') = \beta_k(w) = x \in V \text{ and } v \geq u.v' \}$ *if* $v > u$

$\quad = \{v\}$ *if* $u <> v$ *or* $v < u$

$\quad = \{\}$ *otherwise*

When reducing an expression A to B, every occurrence in B has been either created by the reduction step if it is inside the RHS of the rule, or it pertains to the residual set of an occurrence of A . So the function r indicates how B shares the subexpression of A. We can extend this notion to the residual map by a derivation :

Definition 4.6 *Let $U = \{u_0, \ldots, u_n\}$ and A denote the derivation $E_0 \to_U^* E_n$, (A_1, A_2) is a partition of A, that is, $A_1 = E_0 \to_{U'}^* E_i$, $A_2 = E_i \to_{U''}^* E_n$. The residual map function r* is defined as :*

$r^*[E_i \to_u E_{i+1}]\, v = r[E_i \to_u E_{i+1}]\, v$

$r^*[A_1, A_2]\, v = \cup\{ r^*[A_2]\, w \mid w \in r^*[A_1]\, v\}$

Given two "co-initial "derivations A, B, where A: $E \to^* E_n$ and B: $E \to_u E'$ is elementary, we can extend the residual map to projections of the two co-initial derivations by defining the derivation residual map rd(B, A) of B by A as the derivation from E_n and reducing redexes in r*(A, u).This can be extended to two arbitrary co-initial derivations A, B by using the *parallel moves theorem* (see [HuLe79]). Any two co-initial derivations A,B, A≡B, are said to be *equivalent by permutation* , iff for all co-initial derivation C, rd(C, A) = rd(C, B).

Definition 4.7 *Suppose E can be reduced by the rule $\alpha_k \to \beta_k$ at occurrence u . For all $v \in D(E)$ and $u<v$, they can be partitioned between the occurrences internal to α_k, denoted as $C_E(u)$ i.e. $C_E(u) = \{ u.w \in D(E) \mid w \in D^*(\alpha_k) \}$ and the ones which pertain to the substitution part, denoted as $S_E(u) = \{ u.w \in D(E) \mid w \notin D^*(\alpha_k) \}$.*

Definition 4.8 *Let A: $E \to_U^* E_n$ be a reduction derivation of E, and $u \in D(E)$, we say that $u \in \chi(A)$, the external occurrences set of A , iff : 1) $\forall i<n$ there exists no $u_i \in U$ and $u_i < u$ (called A preserves u) or 2) A can be decomposed to A_1, A_2, A_3 such that $\exists v\ v<u$ A_1 preserves u, A_2 is $E' \to_v E''$ where $u \in C_{E'}(v)$, $v \in \chi(A_3)$.*

Intuitively, $u \in \chi(A)$ iff A does not reduce at any step a redex E/v which is outer to u ($v < u$) for which u does not contribute ($u \notin C_E(v)$).

Definition 4.9 *Let A be a reduction derivation $E\text{->}_U{}^*E_n$, R(E) be the set of redex occurrences of E. For any E/w $\in$ R(E), w is an initial redex occurrence pertaining to A, denoted by w $\in$ R(A), iff for some $i \leq n$, $U \cap r^*A_{[i-1]}w \neq \emptyset$ where $A_{[i-1]}$ denotes the first i reduction steps in the derivation and we define the external redex occurrences of A ,$W(A) = \{v \mid v \in R(A) \cap \chi(A)\}$.*

Definition 4.10 *Given a reduction derivation, A, it is said to be outside_in iff either A= [] or $A=A_1A_2$,where A_1 is a one step reduction reducing a redex at the occurrence in W(A) and A_2 is outside_in.*

So we have following procedure for standardizing a given derivation to produce a derivation with outside_in property by reducing at every step a redex external to the rest of the derivation.

Standardization (A : $E \rightarrow_U E_n$))

if A is an empty derivation then A

else A_1A_2 where $A_1 = E \rightarrow_v E_1$ and E/v is the leftmost redex in W(A))

A2 = Standardization (rd (A , A_1))

Fig. 4.1 Standardization Algorithm

Such a procedure, which is always terminating, constructs a new derivation by rearranging the original one to compute redexes in an outside-in manner. If we impose a fixed choice function over W(A) (all redexes in W(A) are disjoint) such a derivation is unique. We call such a constructed derivation, using a "leftmost" choice function, *standard.* We have following standardization theorem for a regular TRS (see[Huet79]).

Theorem 4.4 *Let R be a regular TRS, E_0 be expressions. For every reduction derivation $E_0 \rightarrow^* E_n$, there exists an standard derivation $E_0 \rightarrow E_m'$ such that $E_n = E_m'$, and for every derivation class (modulo permutation equivalence) such a standard derivation is unique.*

Now, we are able to show that this nice standardization theorem for reduction derivation can be extended to narrowing derivation in a functional TRS (which is a particular regular TRS).

Definition 4.11 *Given a narrowing derivation S: $E_0 \surd\!\!\rightarrow_{[u_1,\rho_1]} \ldots E_{n-1} \surd\!\!\rightarrow_{[u_n,\rho_n]} E_n$, Let S_1 and S_2 be a partition of S, that is $S_1 = E_0 \surd\!\!\rightarrow_{U'}{}^* E_i$, $S_2 = E_i \surd\!\!\rightarrow_{U''} E_n$. The residual map of v by the narrowing derivation S is a function r_n^* :*

$r_n^*[E_i \surd\!\!\rightarrow_{[u_0,\rho_0]} E_{i+1}]\, v = r[\rho_0(E_i) \rightarrow_{u_0} E_{i+1}]\, v$

$r_n^*[S_1,S_2]\, v = \cup\{ r_n^*[S_2]w \mid w \in r_n^*[S_1]\, v\}$.

Definition 4.12 *Suppose E/u is narrowable by the rule: $\alpha_k\text{->}\beta_k$,t he narrowing scheme of E/u is a set of occurrences $N_E(u)$, $N_E(u) = \{u.w \mid u.w \in D^*(E) \text{ and } w \in D^*(\alpha_k)\}$.*

Definition 4.13 *Let A: $E \surd\!\!\rightarrow_U{}^* E_n$ be a narrowing derivation of E and $u \in D^*(E)$, $u \in \chi_n(A)$, the external occurrence set of A, iff : 1) $\forall i<n$ there exists no $u_i \in U$ and $u_i<u$ (called A preserves u) or 2) A can be decomposed to A_1,A_2,A_3 such that $\exists v\ v<u$ A_1 preserves u, A_2 is $E' \surd\!\!\rightarrow_v E''$ and $u \in N_E(v)$, $v \in \chi_n(A_3)$.*

By this definition, $u \in \chi_n(A)$ iff A does not narrow at any step a narrowable redex E/v which is outer to u ($v < u$) for which u does not contribute ($u \notin N_E(v)$).

Definition 4.14 *Let A : $E\surd\!\!\rightarrow_U{}^*E_n$ be a narrowing derivation, $R_n(E)$ is the set of narrowable redex occurrences of E. For any $w = <v, k> \in \{< u, i > / u \in R_n(E)$ and E/u is narrowable by rule : $\alpha_i\text{->}\beta_i\}$, we say v is an initial narrowable occurrence pertaining to A, denoted by $v \in R_n(A)$, iff for some $i \leq n$, $E_i \surd\!\!\rightarrow_{[u_i,k,\rho_i]} E_{i+1}$ is in A and $u_i \in r_n{}^*A_{[i-1]}v$. We define the external narrowable occurrence set of E by $W_n(A) = R_n(A) \cap \chi_n(A)$.*

Lemma 4.2 *Consider a functional TRS P. Let A : $E_0\surd\!\!\rightarrow_{[U,K,\sigma]}{}^* E_n$ be a narrowing derivation, A^*: $\sigma E_0 \rightarrow_{[U,K]}{}^* E_n$ be the corresponding reduction derivation, then $W_n(A) \neq \emptyset$ and $W_n(A) = W(A^*)$.*

Proof: First we prove that $W_n(A) \neq \emptyset$ simply by induction on the length of the derivation. We then prove that $\chi_n(A) \subseteq \chi(A^*)$ by induction on the length of the derivation and then show that $\{\chi(A^*) - \chi_n(A)\} \cap R(A^*) = \emptyset$ since $\forall u \in \{\chi(A^*) - \chi_n(A)\}$ and $\sigma E_0/u \in Cp$. Due to the "constructor principle" of the functional TRS, we have $R_n(A) = R(A^*)$ (This may be not true for a regular TRS). Thus, $\chi_n(A) \cap R_n(A) = \chi(A^*) \cap R(A^*)$ i.e. $W_n(A) = W(A^*)$.

This lemma is essential to extend the previous standardization theorem to narrowing derivation in a functional TRS. We first define the notion of standard narrowing derivation.

Definition 4.15 *Given a narrowing derivation, A, it is said to be outside-in iff either A = [] or $A = A_1A_2$ where A_1 is an one step narrowing at an occurrence in $W_n(A)$, and A_2 is outside_in . If A_1 always narrows the redex at the leftmost occurrence in $W_n(A)$, A is called standard.*

Lemma 4.3 *Consider a functional TRS P. Let A: $E_0\surd\!\!\rightarrow_{[U,K,\sigma]}{}^* E_n$ be a standard narrowing derivation, then the reduction derivation : $A^* = \sigma E_0 \rightarrow_{[U,K]}{}^* E_n$ is standard. Conversely, for a standard reduction derivation B: $E_0' \rightarrow_{[U,K]}{}^*E_n'$ where $E_0' = \rho E_0$ and ρ is on Cp , a narrowing derivation B': $E_0\surd\!\!\rightarrow_{[U,K,\eta]}{}^* E_n''$ is standard and for θ, such that $E_n' = \theta E_n''$ and $\rho\downarrow_{\vartheta(E)} = \eta\theta\downarrow_{\vartheta(E)}$.*

Proof: Since $W_n(A) = W(A^*)$ and noting that the leftmost choice function is used for defining standard narrowing derivations, it is straightforward that A* is standard. Since ρ is on *Cp* , by Hullot's "lifting" theorem [Hull80], B' exists. Furthermore, in a similar way to the proof of lemma 4.2, we can prove that $W(B) = W_n(B')$. So B' is a standard narrowing derivation.

The following theorem is the direct consequence of the above lemma and extends the standardization theorem to narrowing derivations in a functional TRS. That is, given a narrowing derivation in a functional TRS, it can be subsumed by a standard derivation which computes narrowable redexes in an outside-in manner.

Theorem 4.5 *Let P be a functional TRS. For each narrowing derivation A: $E_0\surd\!\!\rightarrow_{[U,\sigma]}{}^*E_n$, there exists a standard narrowing derivation B: $E_0\surd\!\!\rightarrow_{[U',\sigma']}{}^*E_n'$ such that $\theta\rho\downarrow_{\vartheta(E)} = \sigma\downarrow_{\vartheta(E)}$ and $E_n = \theta E_n'$.*

Proof: For a derivation A, we have a reduction derivation: $\sigma E_0 \rightarrow_{[U,K]}{}^*E_n$. Following theorem 4.4, there exist a standard reduction derivation $\rightarrow\!\!\rightarrow^*$ such that: $\sigma E \rightarrow\!\!\rightarrow_{[U,K,]}{}^*E_n$.

Note that $\sigma\downarrow_{\vartheta(E)}$ is on Cp, by the theorem above, then we get a standard narrowing derivation: $E\surd\!\rightarrow_{[U',\rho]}{}^*E'_n$ such that $E_n = \theta E'_n$ and $\theta\rho\downarrow_{\vartheta(E)}=\sigma\downarrow_{\vartheta(E)}$.

Following this theorem, every narrowing derivation is subsumed by its corresponding standard derivation. By using theorem 4.5 the following lemma can be simply proved which indicates that the semantic unification based on enumerating standard narrowing derivations computes most general solutions.

Lemma 4.4 *Let P be a functional TRS, E an equation* $E_1=E_2$. *For any narrowing derivation:* $eq(E_1, E_2)\surd\!\rightarrow_\sigma{}^* eq(D_1, D_2)$ *where* $D_1, D_2 \in Cp$ *and are unifiable with m.g.u.* θ,
Then there exists a standard narrowing derivation:
$eq(E_1, E_2)\surd\!\rightarrow_\rho{}^* eq(D_1', D_2')$ *where* $D_1', D_2' \in Cp$ *are unifiable with m.g.u.* β, *then* $\rho\beta\downarrow_{\vartheta(E)} \leq \sigma\theta\downarrow_{\vartheta(E)}$.

It turns out that, if we have a narrowing strategy which enumerates only all standard narrowing derivations, it obviously keeps completeness.

Theorem 4.6 *The complete set of solutions of an equation in a functional TRS can be computed by only enumerating all standard narrowing derivations.*

To prove the minimality of the complete solution set computed by enumerating all standard narrowing derivations, we should indicate that such a complete minimal solution set of an equation over functional equality does exist.

Lemma 4.5 *Let* Σ *be the complete solution set of an equation* $E_1=E_2$, *then there exists a complete minimal solution set* $\Sigma_m \subseteq \Sigma$, *that is,* $\forall \rho_1,\rho_2 \in \Sigma_m\, \rho_1 <> \rho_2$.

Proof: Straightforward since all substitutions are on Cp and Cp is a complete lattice.

Now the minimality is proved based on the uniqueness of standard reduction derivation to every permutation equivalent class of derivations.

Lemma 4.6 *Given a functional program P, a non-ground expression E and a data object D* $\in Cp$, *for any two substitutions* σ_1,σ_2 *on* Cp , *if there are two different standard reduction derivations such that : 1)* $\sigma_1 E \rightarrow_1{}^* D$, *2)* $\sigma_2 E \rightarrow_2{}^* D$, *then* $\sigma_1\downarrow_{\vartheta(E)}$, $\sigma_2\downarrow_{\vartheta(E)}$ *are independent.*

Proof : Assume that σ_1 and σ_2 are not independent. Without losing generality, we suppose that $\sigma_1 \leq \sigma_2$. So we have a substitution η on Cp such that $\eta\,\sigma_1 = \sigma_2$. Since η is on Cp , $\eta\sigma_1 E = \sigma_2 E \rightarrow_1{}^* \eta(D) \in Cp$ is still a standard derivation. Since P is confluent, we have $\sigma_2 E \rightarrow_1{}^* D$. Since $D \in Cp$, it is obvious that $\rightarrow_1{}^* \equiv \rightarrow_2{}^*$. So $\rightarrow_1{}^*$ and $\rightarrow_2{}^*$ are the same by the uniqueness of standard derivation to a permutation equivalent class of derivations. So σ_1 and σ_2 must be independent.

Theorem 4.7 *Let P be a functional program and E be an equation ,* $E_1=E_2$, *then enumerating all standard narrowing derivations generates the complete minimal solution set of E.*

Proof: By theorem 4.6, we have the completeness. We denote the generated complete solution set as Σ. Suppose a, b are two solutions in Σ, a is generated by the standard derivation: $eq(E_1, E_2)\surd\!\rightarrow_\rho{}^* eq(D_1, D_2)$ so $a=\rho\theta\downarrow_{\vartheta(E)}$ (θ is the m.g.u. of D_1,D_2).b is generated by the standard narrowing derivation: $eq(E_1, E_2)\surd \rightarrow_\sigma{}^* eq(D_3,D_4)$ so $b=\sigma\eta\downarrow_{\vartheta(E)}$ (η is the m.g.u. of D_3,D_4). By lemma 4.3, we have two corresponding standard reductions : 1) $\theta\rho eq(E_1, E_2) \rightarrow^* \theta eq(D_1,D_2)=eq(\theta D_1,\theta D_2)\rightarrow true$ and 2) $\eta\sigma eq(E_1,E_2) \rightarrow^* \eta eq(D_3,D_4)=eq(\eta D_3,\eta D_4)\rightarrow true$. Therefore, $a=\theta\rho\downarrow_{\vartheta(E)}$ and $b=\eta\sigma\downarrow_{\vartheta(E)}$ are incomparable (by lemma 4.6). Thus the complete solution set computed by enumerating standard narrowing derivations has the minimality property.

5. Lazy Pattern Driven Narrowing

In this section, we will present a narrowing strategy, called lazy pattern driven narrowing, that enumerates all standard narrowing derivations of an expression. The basic idea of the algorithm lies in the fact that all inner expressions should only be narrowed when they are needed for fitting the pattern of an outer expression. It will be shown that for narrowing an expression to its WHNFs lazy pattern driven narrowing enumerates all and only all standard derivations. Thus, the algorithm for solving equation on functional equality based on the lazy pattern driven narrowing can be shown to generate a complete minimal solution set.

Definition 5.1 *Given an expression E and pattern P, E is fittable with P, iff* $\forall u \in D^*(P)$, $u \in D^*(E)$ *implies* $E(u) = P(u)$.

Because of the left-linearity of the rewriting rules, when narrowing an expression E by $\alpha_k \rightarrow \beta_k$ failure of unification between E and α_k can only occur due to function symbol conflicts. These conflicts may happen when fitting E with α_k or when performing substitution. Imagine that the unification process is composed of two parts, one for testing fittability and the other for substitution. Narrowing at some inner subexpressions can be driven by fitting the pattern of an outer expression. We illustrate this " Pattern Fitting " procedure as following:

Pattern Fitting (S).

if $S = \varnothing$ *then* ***return*** $\varnothing$
    else if $\exists a=a \in S$ *then* ***return*** *Pattern Fitting* ($S - \{a=a\}$)
      elseif $\exists t = s \in S$ *where* $t(\varepsilon) = C_1$ *and* $s(\varepsilon)=C_2$
        then if $C_1 = C_2 \in F_{cn}$
        then ***return*** *Pattern Fitting* ($S[\{t=s\} \leftarrow \{t/i = s/i \mid t(i)\ \&\ s(i) \notin V, i \in [1,n]\}]$)
          else ***return*** $\perp$
      else ***return*** $\{ S \}$.

In the algorithm, S is an ordered set of equations derived from fitting the pattern of an expression. Suppose the LHS of a defining rule is $f(p_1,\ldots.p_n)$, when narrowing an expression $f(e_1,\ldots.e_n)$ the pattern fitting between them can be performed by applying the procedure to an ordered set of equations $\{e_i=p_i \mid e_i\&p_i \notin V, i\in[1,n]\}$.The pattern fitting succeeds if an empty set is returned. So narrowing can be performed at the top level. If the expression can't be fitted with the LHS due to constructor conflicts, then the expression can't be narrowed by this rule. Otherwise an ordered set of equations will be returned which are of the form $Gd_1=Cd_2$ where $G\in F_d$ and $C\in F_c$. This indicates that the unification between $f(e_1,....e_n)$ and $f(p_1,....p_n)$ is unsuccessful because symbol conflict between some defined functors and constructors. To resolve these conflicts, narrowing at some inner occurrences is performed. We outline such a narrowing strategy, called lazy pattern driven narrowing, as follows to reveal the pattern driven properties of the strategy. This algorithm is similar to Martelli's transformation based E-unification algorithm for canonical TRS [MaMr86].

Narrowing (E, ρ)
if $E(\varepsilon)\in F_c$ *then* **return** { <E, ρ> } *else* $E(\varepsilon) = f \in F_{dn}$
 For each defining rule $\alpha_k\to\beta_k \in \{\alpha_i\to\beta_i \mid \alpha_i\to\beta_i \in P \text{ and } \alpha_i(\varepsilon)=f\}$
  CASE Pattern Fitting ($\{E/i = \alpha_k/i \mid E(i)\&\alpha_k(i)\notin V, i\in[1,n]\}$) IN
 1) ∅:   /* E and α_k are fittable */
  if $\theta_k = m.g.u(E, \alpha_k)$ exists then **return** Narrowing ($\theta_k\beta_k$, $\theta_k\rho$)
     else **return** ∅.
 2) ⊥: **return** ∅.   /* E and α_k are never fittable */
 3) an ordered set of equation $U=\{E/u^i = \alpha_k/u^i\}$ where $E(u^i)\in F_{dn}$, $\alpha_k(u^i)\in F_c$:
  choosing E/u in the leftmost equation $E/u^i = \alpha_k/u^i$ in U
 return ∪{ Narrowing(ρ'(E[u←E']), ρ'ρ) | <E', ρ'>∈ Narrowing (E/u ,{})}

We are now at the position to prove that this lazy pattern driven narrowing is the intended narrowing strategy, enumerating all standard narrowing derivations for equation solving.

Lemma 5.1 *Let A be an enumerated derivation* $E_0\sqrt{\to}_{[U',K',\sigma]}{}^*E_n$ *by lazy pattern driven narrowing, where* E_n *is in WHNF, then A is standard* .

Proof : Suppose the first step of narrowing happens at u, following the algorithm it can be shown that $u \in \chi_n(A)$, so $u \in W_n(A)$. Moreover, u is the leftmost occurrence in $W_n(A)$ due to the ordering imposed on the equations derived from pattern fitting. By induction on the length of A, A is shown to be standard.

Lemma 5.2 *Given a standard narrowing derivation A:* $E_0\sqrt{\to}_{[U,K,\sigma]}{}^*E_n$ *where* E_n *is in WHNF, A can be enumerated by the lazy pattern driven narrowing algorithm.*

Proof: Since A is standard, the reduction derivation A': $\sigma E_0\to_{[U,K]}{}^* E_n$ is standard. Note that σE_0 can be reduced by the lazy pattern driven narrowing procedure resulting in a standard reduction derivation B: $\sigma E_0\to_{[U',K']}{}^* E_n$. Since the final step of reduction in A' and B is at the top level, it can be proved that A'≡ B (see lemma 3.18 in [HuLe79]). By uniqueness of standard derivations, A'=B. So A can be enumerated by the algorithm.

By these two lemmas, the following theorem, which concludes that lazy pattern driven narrowing will enumerate all and only all narrowing derivations when narrowing an expression to its WHNF is straightforward.

Theorem 5.1 *In narrowing an expression to its WHNFs, the lazy pattern driven narrowing enumerates all but only all standard derivations.*

Finally, we establish the ideal property of this intended narrowing strategy: lazy pattern driven narrowing enumerates all but only all standard narrowing derivations for equation solving in functional programming context.

Lemma 5.3 *Let A :* $E_0 \surd\!\!\rightarrow_{[U,\sigma]}{}^* E_{n-1} \surd\!\!\rightarrow_{[\varepsilon,\sigma]} E_n$ *be a narrowing derivation with a final narrowing step at the top occurrence, for any B :* $E_n \surd\!\!\rightarrow_{[w,\sigma]}{}^* E_{n+m}$, *AB is standard iff A and B are standard.*

Proof: First we can prove that $\chi_n(AB) = \chi_n(A)$ and $W_n(AB) = W_n(A)$ by induction on the length of A .Then the lemma is straightforward.

Theorem 5.2 *Given a narrowing derivation, A:* $E_0 \surd\!\!\rightarrow_{[U,\sigma]}{}^* D$ *where* $D \in Cp$, *A is a standard derivation iff A is enumerated by successively using the lazy pattern driven narrowing to narrow some outermost expression.*

Proof : Suppose A is enumerated by successively using the lazy pattern driven narrowing. A is standard, by lemma 5.3, since subderivations enumerated by the lazy pattern driven narrowing for narrowing outer expressions are standard and end with a top level narrowing to get the WHNFs. Also from lemma 5.3, if A is a standard derivation, A can be split into some standard derivations by some top level narrowing steps. Every split subderivation is responsible for narrowing an outer expression to its WHNF. So they can be enumerated by the lazy pattern driven narrowing by theorem 5.1.

By this theorem, when solving an equation $E_1=E_2$, the semantic unification can be performed lazily by narrowing E_1 and E_2 to their WHNFs and then checking their unifiablity. So if the check fails due to constructor conflict at some outermost level, no further narrowing will happen. Otherwise, lazy pattern driven narrowing is successively used to narrow some outer expressions for further solving. Such an algorithm enumerates all and only all standard narrowing derivations in solving the equation on functional equality and so computes the complete minimal solution set. The full algorithm is given in [DG88] which gives an algorithm for solving a set of constraining equations based on the lazy pattern driven narrowing approach.

6.Conclusion And Related Works

We have defined the declarative semantics of absolute set abstraction. An efficient execution mechanism for its implementation has been designed and shown to be both optimal and correct. It provides a semi_decidable procedure for solving the constraining equations in A.S.A. that results in a set of complete minimal solutions. We can still claim the overall determinism of a program involving absolute set abstraction from the fact that the complete minimal solution set is unique up to variable renaming.

E_unification and narrowing are general means for equality reasoning in equational logic [GaSn87][Hull80] and can be used for embedding equality into logic languages [EmYu86] and for equational programming [Ders 85][JaSi 86]. Our work concentrates on embedding the "logic programming" features into functional programming by enhancing the functional programming languages with the ability to solve equations. In our case, equality is defined in terms of a continuous equality function and the required semantic unification, which should be modulo this equality, can be designed. Narrowing is still the basic means for such a semantic unification. It seems to us that this research is closely related to recent work on incorporating some unification-complete first-order theory into Horn clause logic programming [JaLaMa86]. Of particular relevance to our work are other works on integrating equality theories coincidental with the functional equality. In his abstract, [You88], You outlined the approach of embedding a special equality theory, capturing the notion of functional equality, into Prolog (This approach is also discussed in [EmYu 86]). Since only data solutions are necessary for this theory the minimal unifiers modulo the theory can be computed by enumerating all outer narrowing derivations which can be viewed as the general form of any standard narrowing derivation. You proved his results by the process of transforming narrowing derivations whereas our work on standard narrowing derivation is quite different from this "transformative" approach. The completeness and minimality of enumerating standard narrowing derivations has been proved by generalizing Huet's standardization theorem. By this "constructive" approach, a precise lazy narrowing strategy has been derived and its correctness proved rigorously.

Absolute set abstraction has been successfully embedded into the functional programming language Hope [Burs80]. The resulting language, called Hope with Unification, has been implemented by Helen Pull at Imperial College [Darl88].

Note that the semantics and the execution mechanism presented here implies that no function involved in the constraining equations can itself be defined by an A.S.A.. This restricts the use of A.S.A. to behave in the way of a "logic query". By releasing this restriction, we come to a new approach towards a functional programming based integrated language by regarding the program as a more general first order theory (e.g. a theory for set inclusion) in order to capture the notion of non-deterministic programming and release the confluent requirement imposed on rewriting. This approach is related to some current research on non-deterministic algebraic specification [Huss88] and general equational languages (e.g. EqL [JaSi 86]). Following this approach, narrowing and unification can be naturally embedded in functional programming languages producing a functional constraint programming language, [DG88-2], of even greater expressive power.

References

[Bellia 86] Bellia,M. and Giorgio, L." The Relation Between Logic and Functional Language: A Survey" The Journal of Logic Programming 3, Oct. 1986.

[Burs 80] Burstall R. et.al. " Hope : an experimental applicative language" Proc. Lisp Conf. Stanford. 1980.

[Darl 86] Darlington, J. Field, A.J. and Pull, H." The unification of Functional and Logic languages" in Logic Programming: Functions, Relations and Equations, ed. Doug Degroot and G. Lindstrom, Prentice-Hall , 1986.

[Darl 88] Darlington, J. et.al. " A Functional Programming Environment Supporting Execution, Partial Execution and Transformation " To appear.

[Ders 85] Dershowitz, N. and Plaisted, D.A. "Equational programming" in Machine Intelligence, D.Michie, J.E. Hayes and J. Richards, eds.,1986.

[DG88] Darlington, J, Yi-ke Guo "Narrowing and unification in Functional Programming" Research Report, FPS, DoC, Imperial College, Oct. 1988.

[DG88-2] Darlington, J, Yi-ke Guo " On Constraint Functional Programming" To appear.

[EmYu86] Van Emden, Maarten. H. and Yukawa, K. "Logic Programming with Equations" Journal of Logic Programming 4, 1986.

[Fay79] Fay, M. J., "First-order Unification in an Equational Theory " in 4th Workshop on Automated Deduction, 1979.

[Frib85] Fribourg, L.,"SLOG: A logic programming language interpreter based on clausal superposition and rewriting" in Proc. 1985 Symposium on Logic Programming, Boston, Mass.

[GaSn87] Gallier, J. H. and Snyder, W. " A general Complete E-unification procedure," in Proc. of RTA87, LNCS 256,1987

[Giov86] E. Giovannetti, G. Levi, C. Moiso and C.Palamidessi, Kernel LEAF: An experimental logic plus functional language-its syntax, semantics and computational model, ESPRIT Project 415, Second year report 1986.

[Hull80] Hullot, J. M., " Canonical Forms and Unification ," in Proc. 5th conference on Automated Deduction, LNCS 80, 1980.

[HuLe79] Huet, G. Levy, L.L. "Computations in Nonambiguous Linear Rewriting Systems" Research Report. INRIA. 1979.

[HuOp80] Huet,G. and Oppen , D.C., " Equations and Rewrite Rules : A Survey " TR -CSL-111 , SRI, Jan, 1980.

[Huss88] Hussmann, H, " Nichtdeterministische Algebraische Spezifikationen (in German), Ph.D. thesis, Univ. of Passau, 1988.

[JaLaMa86] Jaffar, J., Lassez, J. and Maher, M. " Constraint Logic Programming" in Proc. of. 14th ACM symp. POPL. 1987.

[JaSi 86] Jayaraman, B. and Silberman,F., "Equations, Sets and Reduction Semantics for Functional and Logic Programming" in Proc. of. ACM Conf. on LISP and Functional Programming, Aug.1986.

[Mate 86] Martelli, A. et,al "An algorithm for unification in equational theories", in Proc.1986 International Symposium on Logic Programming, Salt Lake City, Utah, 1986.

[Reddy85] Reddy, U.S. " Narrowing as the Operational Semantics of Functional Languages" Proc. of the 1985 Symp. on Logic Programming, July , 1985.

[Reyno 70] Reynolds, J.C. " Transformational Systems and the Algebraic Structure of Atomic Formulas." Machine Intelligence 5. ed. Michie, D. Edinburgh Univ. Press, 1970.

[You 88] You, Jia-Huai "Unification Modulo an Equality Theory for Embedding Functional Programming into Prolog" Extended Abstract, Dept.of Computing Science, Univ.of Alberta, Canada, May,1988.

SIMULATION OF TURING MACHINES BY A LEFT-LINEAR REWRITE RULE ¥

Max Dauchet
LIFL (URA 369-CNRS), Université de Lille-Flandres-Artois.
UFR IEEA, 59655 VILLENEUVE D 'ASCQ Cedex FRANCE.
e mail: dauchet@frcitl71.bitnet

ABSTRACT: We prove in this paper that for every Turing machine there exists a left-linear, variable preserving and non-overlapping rewrite rule that simulates its behaviour. The main corollary is the undecidability of the termination for such a rule. If we suppose that the left-hand side can be unified with an only subterm of the right-hand side, then termination is decidable.

INTRODUCTION.

The first part (the main one) of this paper proves that a left-linear, complete and non-overlapping rewrite rule R_M can be associated with every Turing machine M simulates that Turing machine M. More precisely, let us call *move history* a (multi-rooted) attributed tree $\Omega(t_1,\ldots,t_k)$ defined as follows: an instantaneous description (ID) of M is attributed to the root of every t_i. An inherited ID is attributed to every node, the ID of the Ith son is deduced of the ID of its father by application of the M-instruction I. So, a move history can be seen as a set of computations, starting from a finite set of ID. A branch describes a computation, it can be finite (finite computation) or infinite (infinite computation). Our main theorem is the identification of the set of move stories with the set of R_M derivations.

Intuitively, this result means
that an one rewrite rule system is as powerfull as a Turing machine! Furthermore, this rule is of a very special type (left-linear, variable preserving and non-overlapping).

As a corollary, we prove that termination is undecidable even in the one (left-linear, variable preserving and non-overlapping) rule case. This result improves a previous result of Dershowitz (1985,1987), who proved

---

¥ Supported in part by the "GRECO de Programmation" and the PRC "Mathématiques et Informatique". Part of the ESPRIT Basic Research Actions, BRA "Algebraic and Syntactic Methods in Computer Science" and BRA "Computing by Graph Transformations".

the undecidability of the termination of two rule systems. Huet pointed out that using operators K and S of combinatory logic lead also to this result. But the interest of Dershowitz's systems remains that they work "almost at the top of terms". Jouannaud (1987) pointed out the one rule case and we recently got a first proof of the undecidability in this case (Dauchet, 1988).

In the present paper, we improve this result on several points: the proof is simpler, and its real meaning is given, namely the simulation of Turing machines; overall, we obtain undecidability for left-linear and variable preserving rules. Only six variables are necessary. So, several questions arise with respect to one rule systems: is termination decidable in the linear case? Is termination decidable in the semi-Thue case (when functions are monadic) ? Is termination decidable if one restricts the unique left-hand side to be unifiable with only one subterm of the right-hand side (this case is called the deterministic case)?

The two first problems remain open. Jouannaud and H. Kirchner (1984) proved the termination of a rule if the left- hand side does not unify with any non variable term embedded in the right- hand side. We improve this result and we prove the decidability of the third problem by reducing it to the "Prolog-like rewriting" (i. e. one rewrites only at the top (= the root) of the trees). Indeed, several communications prove that problems related to that case are also decidable (Devienne Ph.&Lebègue P.1986, Dauchet M., Devienne Ph.&Lebègue P.1988, and Devienne Ph. 1988).

Part I: The simulation theorem:

The goal of this part is to explain and justify the following figure. {IDi} denotes an instantaneous description of a Turing machine M; a branch can be identified to a M-Turing computation and to "deep-only derivation" of a rewrite rule R_M that we will define.

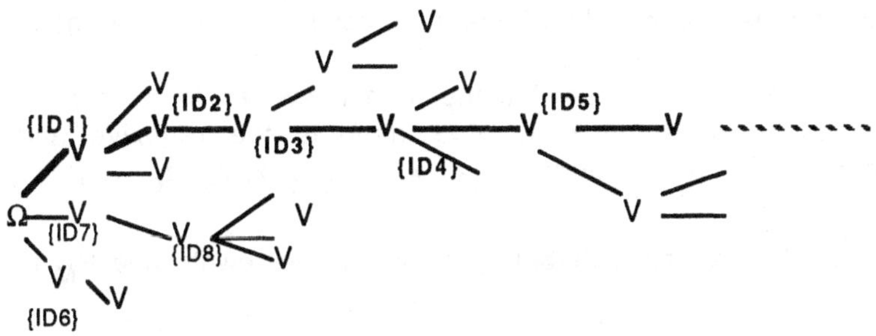

Turing machine: notations. Let M be a *Turing machine*, with a single tape. Let $Q = \{q_1, q_2, \dots q_n\}$ be the set of *states*. States are also denoted q, q', q", and so on. Letters of the *tape alphabet* $\Sigma = \{a_1, a_2, \dots, a_p\}$ are also denoted by first small latin letters a,b,c,...., b_1..,c_i,... Without lack of generality, we suppose that the non-blank portion of the tape is always enclosed between special symbols $\#_L$ and $\#_R$. An *instantaneous description* (ID) of M is written

$$(c_m...c_1\#_L, q, a, d_1...d_n\#_R)$$

which means that M is in state q, the head scans the symbol a, with non-blank left portion of tape $\#_L c_1...c_m$ (from the left end to the symbol preceding the read head) and right-portion $d_1...d_n\#_R$ (from the symbol following the read head). An empty non-blank portion is denoted by NIL. (q, a, b, q', L) denotes the *left-moving M-instruction* . It means "if M is in state q reading the symbol a, then replace on the tape a by the symbol b, move left and go into state q' ". Right-moving instructions are defined the same way. To increase the non-blank portion of a tape, there are *special-left moving instructions* of the form (q, $\#_L$, $\#_L$b, q', L). They mean "if M is in state q reading the symbol $\#_L$, then insert the symbol b, move left to position the head on $\#_L$ again, and go into state q' ". Special-right moving instructions are defined the same way. Obviously, moving instructions such as (q, $\#_L$, b, q', L) are forbidden. Without lack of generality, we do not considere "standing" instructions.

ID1 |-- ID2 denotes a *computation step.* It means that M moves from ID1 to ID2 by application of an instruction. If that instruction is J, we sometimes write ID1 |-J- ID2. |-*- is the reflexive and transitive closure of |--; if S is a sequence of instructions, ID1|-S-ID2 denotes the corresponding *computation.*

Rewrite rules: notations. Let us recall that a rule is *left-linear* (respectively *right-linear*) if no variable occurs more than once on the left-hand side (resp. right-hand side) of the rule; a rule is *linear* if it is both left- and right- linear. A rule is *variable -preserving* if every variable which occurs on the left-hand side of the rule occurs on the right-hand side.

We write t ==> u to indicate that the term t rewrites to the term u by a single application of some rule. =*=> denotes the reflexive and transitive closure of ==>. For any term, x, x', x", x_h, x_t, z, z_s, z_c, y, y', ... denote variables.

Let us define the ranked alphabet $\mathbb{A}$, on which the rewrite rule R_M is defined:

- Each state symbol and tape symbol of M will be a constant of $\mathbb{A}$.
- Each name I,...,J,...., K of instruction is an unary operator.
- A, Q, V, T, *, NIL are new operators. We define their rank by

  rank(A) = 1 + cardinality of the tape alphabet Σ .

  rank(Q) = 1 + cardinality of the set of states.

rank(V) = cardinality of the set of instructions.
rank(T) = 4; rank(*) = 2; rank(NIL) = 0.

The rule R_M associated with a Turing machine M

LEFT --> RIGHT will be an abbreviation of R_M. To explicit the six variables, we write
LEFT($x_h, x_t, z_s, z_c, y_h, y_t$) --> RIGHT($x_h, x_t, z_s, z_c, y_h, y_t$).

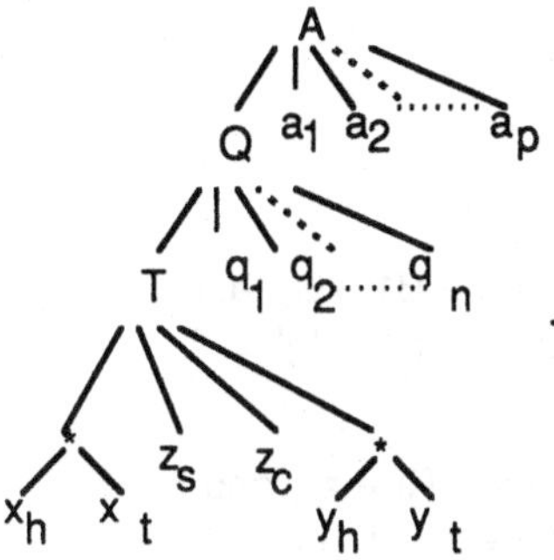

List of subterms $t_I \ldots t_J \ldots t_K$
associated with the set of M-instructions I,..., J,..., K

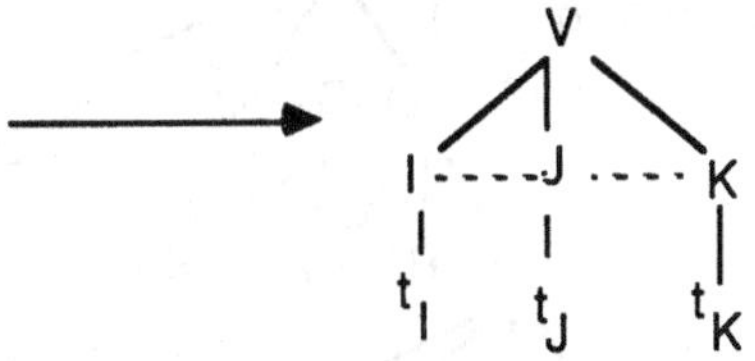

With subterms $t_I, \ldots, t_J, \ldots, t_K$ defined as follows

If J is a right-moving instruction (q_i, a_j, b, q_k, R)

then t_J =

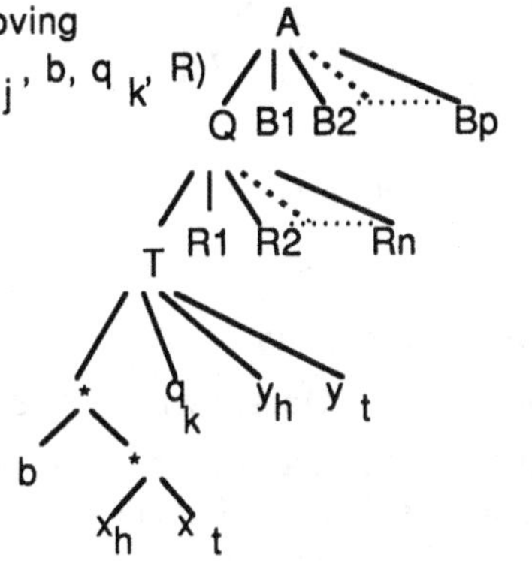

Where Bj = z_c and Bj' = $a_{j'}$ if j' ≠ j

Where Ri = z_s and Ri' = $q_{i'}$ if i' ≠ i

If J is a left-moving instruction (q_i, a_j, b, q_k, L)

then t_J =

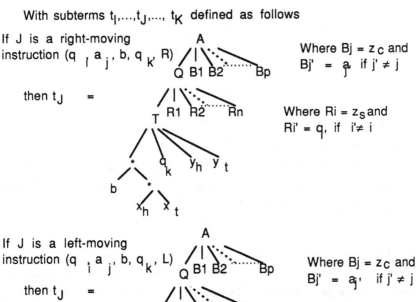

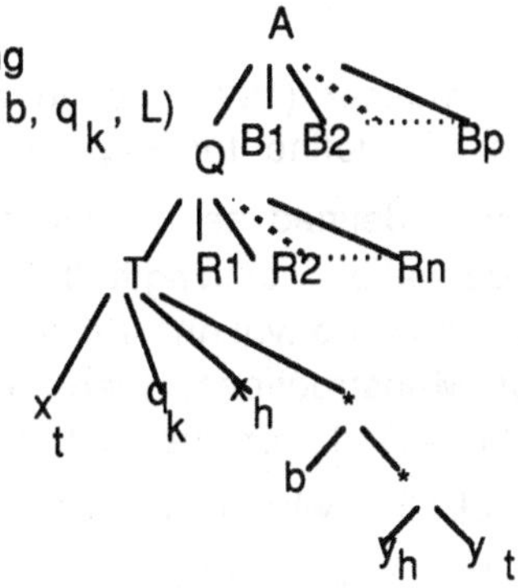

Where Bj = z_c and Bj' = $a_{j'}$ if j' ≠ j

Where Ri = z_s and Ri' = $q_{i'}$ if i' ≠ i

If J is a special right-moving instruction (q_i, #R, b#R, q_k, R)

then t_J =

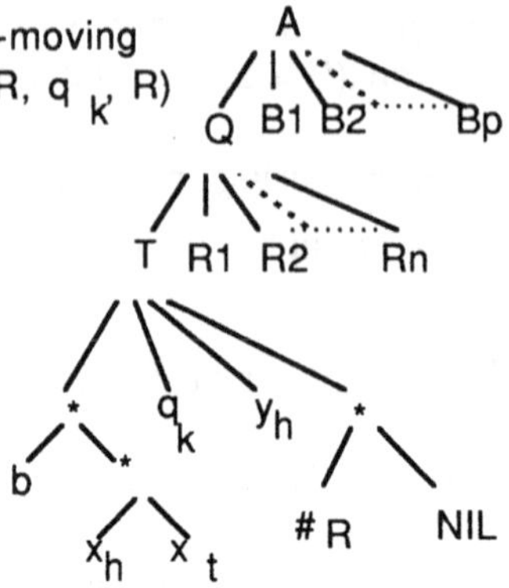

Where Bp = z_c and Bj' = q_j, if j' ≠ p

Where Ri = z_s and Ri' = q_i if i'≠ i

If J is a special left-moving instruction (q_i, #L, #Lb, q_k, L)

then t_J =

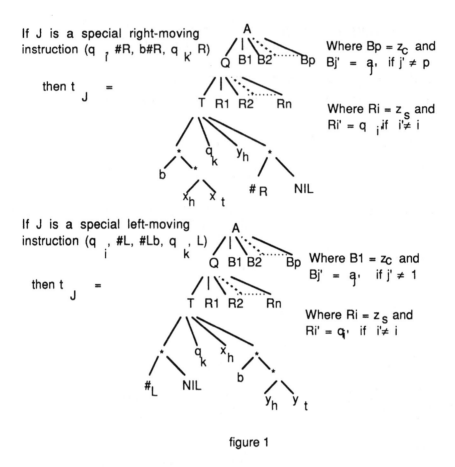

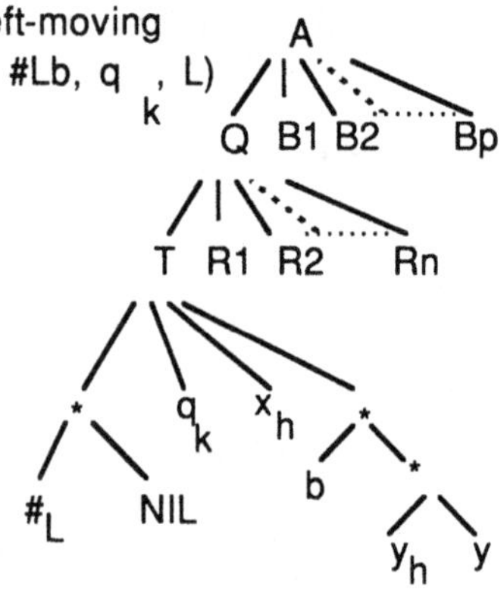

Where B1 = z_c and Bj' = q_j if j' ≠ 1

Where Ri = z_s and Ri' = q_i if i'≠ i

figure 1

First properties of R_M.

(R1) R_M is linear and variable preserving.

(R2) R_M is *non-overlapping* (Dershowitz, 1987, def. 23), i.e. "non-ambiguous", i.e. "without critical pairs", i.e. the left-hand side does not overlap a non-variable proper subterm of itself.

(R3) If LEFT *overlaps* RIGHT, then LEFT is unified with some t_J .

Definition: Lankford & Musser (1978) define the notion of forward closure. By the same way, let us define for R_M the notion of *deep-only derivation* ; it may be inductively defined as follows: every derivation t==>u of length 1 is an ε-deep-only derivation of length 0 (ε denotes the empty word). Let t=S=>u be a S-deep-only derivation (of length l), where S is a finite sequence of names of M-instructions; then t=S=>u==>v is a S;J-deep-only derivation (of length l+1) if, for the last application of R_M (i.e. the last rewrite), t_J matches LEFT , where t_J occurs in the occurence of RIGHT corresponding to the last rewrite in t=S=>u. Let us remark that our definition induces a total order on the rewrites of the derivation.

<u>Definition:</u> Let t==>u be a rewrite with t=t'.LEFT.∂ and u=t'.RIGHT.∂ (that we also denote t'.∂(RIGHT)). The substitution ∂ is called the *assignment of the rewrite* (indeed, it is the "inside assignment" and t' is the "outside assignment", but we only use the inside assignment). In a derivation, we will consider assignments corresponding to its different rewrites.

<u>Convention:</u> A portion of tape C = $c_m \ldots c_1 \#_L$ of the Turing machine M *will be identified to a list* <C> = *(c_m, *(... *(c_1, *($\#_L$, NIL))...)). Such a list is said of *type list* (abbreviation of tape-list type).

<u>Definition:</u> a rewrite assignment is said of *type ID* iff

∂ (x_h), ∂ (z_c) and ∂ (y_h) are tape symbols ; ∂ (x_t) and ∂ (y_t) are of type list; ∂ (z_s) is a state.

Intuitively, a assignment ∂ of type ID can be identified with the ID (∂ (x_h) ∂(x_t) , ∂ (z_s), ∂ (z_c) , ∂ (y_h) ∂ (y_t)), where ∂ (x_h) denotes the first symbol of the left assignment of the IC, and so on.

<u>Definition:</u> If t is a term, the finite set of its *M-components* associated with t is defined as follows (see figure 2).

Let LEFT. ∂ be a subterm of t such that ∂ (x_h), ∂ (z_c) and ∂ (y_h) are tape symbols and ∂ (z_s) is a state. To every such occurence of LEFT in t, we associate a M-component t'= LEFT. ∂ ', with:

∂'(x_t) is "the bigger term of type list obtained by replacing in ∂ (x_t) a subterm with $\#_L$"; ∂ '(y_t) is "the bigger term of type list obtained by replacing in ∂ (y_t) a subterm with $\#_R$"; ∂' (w) = ∂ (w) for w=x_h, z_s, z_c and y_h .

So, ∂' is of type ID. The substitution ∂ ' of some M-component LEFT. ∂ ' is called *the ID of this component.*

Unformaly, we obtain components of t by cutting t at the top of every LEFT and selecting the "ID-part" of the corresponding inside assignment.

Simulation lemma (technical version):

Let [S] be a finite sequence of M-Instructions.

(S1) If M(C, q, a, D) |-S- M(C', q', a', D')

then there is some S-deep-only derivation, the first assignment of which is

(head<C>, tail<C>,q, a, head<D>, tail<D>)

and the last assignment of which is

(head<C'>, tail<C'>,q', a', head<D'>, tail<D'>)

(S2) Every S-deep-only derivation on a term t can be identified to a S-deep-only derivation on some M-component t' of t, and then

a/ All the contextes of this derivation are of type ID

b/ if the first assignment is
(head<C>, tail<C>,q, a, head<D>, tail<D>)
and the last (head<C'>, tail<C'>,q', a', head<D'>, tail<D'>),
then M(C, q, a, D) |-S- M(C', q', a', D'). ◊

Proof of (S1): by induction on the length of [S]. Let us suppose that (S1) is true for [S] and let J be the right moving instruction (q', a', b, q", R).
We obviously get M(C, q, a, D) |-S;J- M(bC', q", head(D'), tail(D')).

It is easy to check that the substitution

$\partial\,(x_h,x_t,z_s,z_c,y_h,y_t)$ = (b, <C'>, q", head<D'>, head(tail<D'>), tail(tail<D'>)

matchs LEFT with the S-occurrence of t_J in the S-deep-only derivation considered by induction. Eventually, we have buit a S;J-deep-only derivation and the substitution S is by definition the last assignment of this derivation; so this S;J-deep-only derivation satisfies the induction. We leave to the reader the proof in the three other cases (other types of instructions).

Proof of (S2): by induction on the length of [S]. Let us suppose that (S2) is true for [S] and let us consider a S;J-deep-only derivation on a term t. Let t' be the M-component of t in which the first LEFT of the deep-only occurs. By construction of M-components, the first assignment is of type ID. Let us suppose that, for the corresponding J-deep-only derivation, the first assignment was
(head<C>, tail<C>,q, a, head<D>, tail<D>), the last one
(head<C'>, tail<C'>,q', a', head<D'>, tail<D'>) and that
M(C, q, a, D) |-S- M(C', q', a', D'). Then, if a S;J-deep-only derivation is possible on t, it is also possible on t', because matching in t works only on "well typed" subterms (i.e. components of type ID). Let us suppose that, for the last step of the derivation, LEFT is matched with a t_J corresponding to a right moving instruction. Then it is easy to check that the derivation is possible iff

S $(x_h,x_t,z_s,z_c,y_h,y_t)$ = (b, <C'>, q", head<D'>, head(tail<D'>), tail(tail<D'>)

matchs LEFT with the S-occurrence of t_J

and J denotes the instruction (q', a', b, q", R). *It is importante to remark that the "old state* q' " *of the instruction is forced to be equal to the state of the current* ID *by matching of the sons of Q in* LEFT *and* t_J. *The same remark holds for the current scanned symbol a'.*

Then, it is easy to check that
M(C, q, a, D) |-S;J- M(bC', q", head(D'), tail(D')).
We leave to the reader the proof in the three other cases. ◊

Let us identify an ID (C, q, a, D) of the Turing machine M with the assignment (head<C>, tail<C>,q, a, head<D>, tail<D>) of a deep-only derivation. Then, we can formulate by a new way the simulation lemma:

Simulation lemma (second version):

ID |-S- ID' *iff*

there is a S-deep-only derivation on a term t, the first assignment (respectively the last) of which is ID *(resp.* ID'*) in the corresponding* M*-component of t.* ◊

Definition: Let t be a M-component and t=*=>u be a derivation applying R_M. It is easy to check (by induction on the length of the derivation) that u can be decomposed into u'.u", where each node of u' is labeled either by V or by an instruction name, and every composant of the substitution u" is rooted by the instruction name just preceeding a node labeled by A. The sequence of rewrites spelt along a branch defines a deep-only derivation; let us attribute to each node V the assignment of the corresponding rewrite (i.e. the assignment of the LEFT which rewrites this occurrence of V); then, we can define the *derivation history* of t=*=>u as the tree u', to each node of which is attributed the assignment defined above (it is like an inherited attribut). We denote DS(t=*=>u) this "attributed" tree .

Now, let t be any term; let Ω be a new ranked letter and let $t_1,\ldots,t_k$ be the components of t. Then $\Omega(t_1,\ldots,t_k)$ is called *canonical decomposition* of t , denoted Ω(t), and the *derivation history* of t=*=>u, also denoted DS(t=*=>u) is defined by

Ω(DS(t_1=*=>u_1),...., DS(t_i=*=>u_i),...., DS(t_k=*=>u_k),), where t_i=*=>u_i is the derivation corresponding to the component t_i . (Ω(t) is only a part from the order of the sons of Ω). It is important to remark that the definition is sound because of (R1), (R2) and the part (S2) of the simulation lemma.

Definition: A *computation history* is an $\Omega(t_1,\ldots,t_k)$, where Ω has been defined above and the t_i are elementary computation histories defined below:

For every ID of the Turing machine M, the ranked letter V attributed by ID is an elementary computation history .

Let t_i(....,{ID1}V,.....) be an elementary computation history , where ID1 is attributed to a leaf V (in a tree, we denote attributes between {}).

If ID1 |-J-ID2 , then t_i(....,{ID1}V(....,J({ID2}V),.....) is an elementary computation history . (J({ID2}V) is the J-th son of V, according the implicite order on the instruction set used along this paper).

Simulation theorem:

An attributed tree is a R_M-derivation history iff it is a M-computation history .

Such attributed trees may be called "*move histories* "!

Proof: The sequence of rewrites spelt along a branch of an R_M-derivation history defines a deep-only derivation and the attributes of this branch are the corresponding contexts; the sequence of instructions spelt along a branch of an M-computation history defines a computation and the attributes of this branch are the corresponding ID; eventually, we derive only on canonical decomposition. Then, the proof is an obvious consequence of the simulation lemma.

Corollary: termination of rewriting is undecidable in the one rule case, even if the rule is left-linear, variable preserving and non-overlapping.

Proof: The uniform halting problem is undecidable for Turing machines.

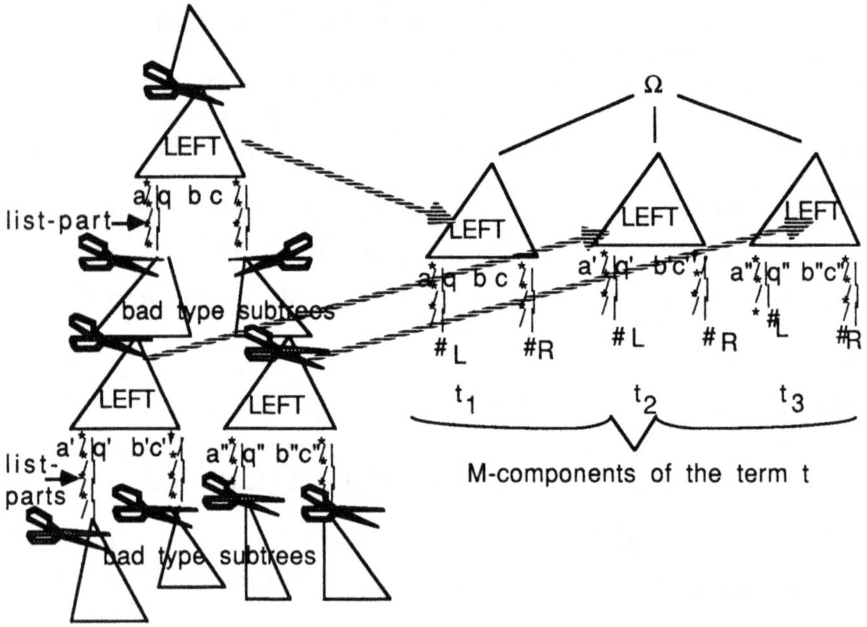

Part II: The deterministic case.

Definition: Let R: l--> r be a rewrite rule. We will say that R is *deterministic* iff it satisfies the two following properties:

(R'1) R is *non-overlapping* (Dershowitz, 1987, def. 23), i.e. "non-ambiguous", i.e. "without critical pairs", i.e. the left-hand side l does not overlap any non-variable proper subterm of itself.

(R'2) There is at the most a non-variable subterm r' of r unifiable with l (it is said that l *overlaps* r) or, excusively, at most a non-variable subterm l' of l unifiable with r (it is said that r overlaps l) (as usually, variables of different occurrences of rules are considered disjoint).

The following rewrite rules are deterministic.

Example 1: some like-program schemes...

F(s(x)) --> if zero(x) then 1 else F(x)

Example 2: some Prolog-like programs...

P(+(x,y),z)) --> P(x,+(y,z))

Counter-example 1: A(x,A(by,z)) --> A(ax, A(x,y))

there are two different overlaps between the left-hand side and the right-hand side.

Counter-example 2: R_M is non-overlapping but non-deterministic!

Chain lemma: *Let* l--> r *be a deterministic rule. If there is only one occurrence of* l *in a term t then, for every derivation t=*=>u, there is at the most one occurrence of* l *in u and the derivation is a deep-only derivation.*

Proof: By induction on the derivation.

Let us suppose the lemma proved for t=*=>u, and let a rewrite u==>v. For some tree w and some substitution w', we get u= w.l.w' and v =w.r.w'. Let us suppose that there are in v two occurrences l' and l" of the left-hand side l; by definition of the deterministic rewrite systems, there is no overlapping between r and l' or between r and l"; then l' or l" occurs in w or w', and then occurs in u. That is impossible, because l occurs also in u and is different from l' and l".

We will reduce the termination problem for a deterministic rule to the case of *top-derivations*, i.e. *"Prolog-like derivations"*. More precisely, a derivation t=*=>u is a top-derivation if, for each step u.l.v ==>u.r.v, u is reduced to a variable (i.e. we can write l.v ==>r.v).

If l overlaps r (and then r does not overlap l) in a deterministic rule R (R = l-->r) , let us define as follows the *top-rule* R^t associated to R: R^t = l--> r', where r' is the non-variable subterm of r unifiable with l .

Reduction lemma: *A deterministic rule* l-->r *is non terminating iff* l *overlaps* r *and its top-rule is not top-terminating* .

Sketch of the proof:

Let l --> r be a deterministic rule. If l does not overlap r, the rule terminates (Jouannaud & H. Kirchner (1984). If there is an infinite top-derivation of the top-rule associated to R, it is easy to built by induction an infinite derivation of R. So, the only delicate point is to prove that if R does not terminate, then there is a top-derivation of R^t which does not terminate. Let us point out the main steps of that proof.

Notations: l can be unified with the subterm r' of r. Let us decompose r into r°.r', where r' is subtituted to a distinguish variable X of r°.

Let t_0 ==> t_1 ==>....==>t_i ==> t_{i+1} ==>.... be an infinite derivation starting from a **minimal** term t_0 (i.e. there is no infinite derivation starting from a proper subterm of t_0). Let us exhibit in this infinite derivation its "top-deep-only sub-derivation" defined by induction as follows: $t_i=u_i.l.v_i$ ==> $u_i.r.v_i=t_{i+1}$ is a "top-deep-only rewrite" iff no future rewritten occurrence of l will be rooted strictly above on the branch of u_i leading to the root of l in the ith rewrite.

First fact: u_0 is reduced to a variable (i.e. t_0=l.v_0).
Proof: t_0 is minimal. ◊

Now, for every integer n, let us associate to the rule l-->r its *n-deep-only rule* defined (if possible), as follows: l-->r is the 0-deep-only rule and r' is its *handle* ; if l_n-->r_n is the n-deep-only rule and the occurrence s of r' in r_n its handle, then, l_{n+1}-->r_{n+1} is the critical pair associated to r_n-->l_n and l--> r, where l matches the handle s; the handle is the occurrence of r' rewritten by the last rewrite.

Second fact: For every i, there is some n_i such that t_0 = $l_{ni}.w_0$ and t_i = $r_{ni}.w_i$. n_i top-deep-only rewrites are used, they are a deep-only derivation and every rewrite which is not a top-deep-only rewrite rewrites a derived term of w_0 .

Sketch of proof: Use the fact that every subterm of the l_n and the r_n are obtained by unifications based on l and r, and then that they match subterms of l or r different from r'. As l-->r is deterministic, if l overlaps r_{ni}, it overlaps the handle of the top-deep-only rewrite. ◊◊

Third fact: If l is linear, we can deduce from the infinite derivation an infinite top-deep-only derivation.

Sketch of proof: check that $l_{ni}.w_0$ =*=> $r_{ni}.w_0$, using only top-deep-only rewrite, because other rewrites do not overlap. If the top deep-only derivation was not infinite, there would be some infinite derivation starting from w_0, but t_0 is minimal. ◊◊◊

Fourth fact: Then we deduce of this derivation an infinite top-derivation of l-->r'

Proof: obvious. ◊◊◊◊

If l is not linear, some derivations can be usefull to get equalities of subterms. Nevertheless, the result can be extended to the general case:

Fifth fact: In the general case we can deduce from the infinite derivation an infinite top-deep-only derivation.

Sketch of proof: Let us suppress as in the third fact all

"useless" derivations. Let us only considere subterms of w_0 which are never "forbidden" i.e. such as at least one copy is substituted to a variable which occurs in the right-hand side (even if the rule is non variable preserving). Then, each necessary rewrite on a term derived from w_0 is necessary to get an equality; the right copy is then the rewritten term and this is possible only finitely many times; because t_0 is minimal. ◊◊◊◊◊

Corollary:

Termination is decidable for a deterministic rule.

Proof: The reduction lemma reduces the problem to the problem of a single "like-Prolog" rule. We solve this problem in (Devienne Ph.&Lebègue P.1986, Dauchet M., Devienne Ph.&Lebègue P.1988, and Devienne Ph. 1988)

Aknowledgements: I would like to thank Pierre Lescanne for helping to improve the presentation of the paper.

REFERENCES

Book, R.V. (1987). Thue Systems as Rewriting Systems. *J. Symbolic computation*, **3** p. 39-68.

Dauchet,M., Devienne Ph.&Lebègue P. (1988). Décidabilité de la terminaison d'une règle de réécriture en tête, *Journées AFCET-GROPLAN*, Bigre+Globule **59**, p. 231-237.

Dauchet, M. (1988). Termination of rewriting is undecidable in the one-rule case. MFCS 1988, Carlsbad. *Springer Lec. notes Comp. Sci.* **324,** p. 262-268.

Dershowitz, N. (1987). Termination. *J. Symbolic computation*, **3** p.69-116.

Devienne Ph. & Lebègue P. (1986). Weighted graphs, a tool for logic programming, CAAP 86, Nice, *Springer Lec. notes Comp. Sci.* **214,** p.100-111.

Devienne Ph. (1988). Weighted Graphs, a tool for expressing the Behaviour of Recursive Rules in Logic Programming, in proceeding of *FGCS' 88,* Tokyo. 397-404. Extended paper to appear in TCS, special issue.

Huet, G. Personal communication (1988).

Huet, G. & Lankfork D.S. (1978). On the uniform halting problem for term rewriting systems. *Rapport Laboria* 283, INRIA.

Huet, G. & Oppen D. C. (1980). Equations and rewrite rules: A survey, in R. V. Book, ed., New York: Academic Press. *Formal Language Theory: Perspectives and Open Problems*, pp. 349-405.

Jouannaud, J. P. (1987). Editorial of *J. Symbolic computation*, **3**, p.2-3.

Jouannaud, J. P. & Kirchner, H. Construction d'un plus petit ordre de simplification. *RAIRO Informatique Théorique/ Theorical Informatics*, **18-3**, p. 191-207.

Lankford, D.S. & Musser, D.R. (1978). A finite termination criterion. Unpublished draft, Information Sciences Institut, University of South California.

Higher-order Unification with Dependent Function Types*

Conal M. Elliott

Department of Computer Science
Carnegie Mellon University
Pittsburgh, Pennsylvania 15213-3890

Internet: conal@cs.cmu.edu

Abstract

Roughly fifteen years ago, Huet developed a complete semidecision algorithm for unification in the simply typed λ-calculus ($\lambda_{\rightarrow}$). In spite of the undecidability of this problem, his algorithm is quite usable in practice. Since then, many important applications have come about in such areas as theorem proving, type inference, program transformation, and machine learning.

Another development is the discovery that by enriching $\lambda_{\rightarrow}$ to include *dependent function types*, the resulting calculus (λ_{Π}) forms the basis of a very elegant and expressive Logical Framework, encompassing the syntax, rules, and proofs for a wide class of logics.

This paper presents an algorithm in the spirit of Huet's, for unification in λ_{Π}. This algorithm gives us the best of both worlds: the automation previously possible in $\lambda_{\rightarrow}$, and the greatly enriched expressive power of λ_{Π}. It can be used to considerable advantage in many of the current applications of Huet's algorithm, and has important new applications as well. These include automated and semi-automated theorem proving in encoded logics, and automatic type inference in a variety of encoded languages.

1. Introduction

In the past few years, higher-order unification ("HOU$_{\rightarrow}$", *i.e.*, unification in the simply typed λ calculus, "$\lambda_{\rightarrow}$") has found a rapidly increasing number of applications. In spite of the undecidability of the problem [7], experience has shown that Huet's semidecision algorithm [13] is quite usable in practice. The first applications were to theorem proving in higher-order logic, using resolution [12] and later matings [1]. Another was to use $\lambda_{\rightarrow}$ to encode the syntax of programming languages, and in particular the scopes of variable bindings. One can then express program transformation rules without many of the previously required complicated side conditions, and use HOU$_{\rightarrow}$ to apply them [14,22,8,4]. A related area of application is to encode the syntax and inference rules of various logics and use HOU$_{\rightarrow}$ to apply inference rules (in either direction) [19,5]. Quite recently, it has been shown that HOU$_{\rightarrow}$ is exactly what is required for

*This research was supported in part by the Office of Naval Research under contract N00014-84-K-0415, and in part by NSF Grant CCR-8620191.

solving the problem of partial type inference in the ω-order polymorphic λ-calculus [21]. $HOU_{\rightarrow}$ has even been applied to machine learning to extract justified generalizations from training examples [3]. In applications such as these, $HOU_{\rightarrow}$ is usually coupled with some other search mechanism. To meet the needs of such applications, Nadathur and Miller developed a programming language "λProlog", which is an extension of traditional Prolog both in using a richer logic (including explicit quantification and implication), and in using $\lambda_{\rightarrow}$ terms instead of the usual first-order terms (and thus requiring $HOU_{\rightarrow}$) [18,15].

A parallel development is the Edinburgh Logical Framework (LF) [10]. Here, $\lambda_{\rightarrow}$ is enriched with *dependent function types.* The resulting calculus, "λ_{Π}", together with the *judgments as types* principle forms the basis of a very elegant and expressive system encompassing the syntax, rules, and proofs for a wide class of *object-logics.*[1]

This paper presents an algorithm for HOU_{Π}, *i.e.*, unification in λ_{Π}, that was inspired by Huet's algorithm for $HOU_{\rightarrow}$. Many of the applications of $HOU_{\rightarrow}$ listed above would benefit from the addition of dependent types. For instance, the encoding of a typed object-language in $\lambda_{\rightarrow}$ usually cannot directly capture its typing rules. To account for the object-language's type system generally requires extra nontrivial mechanisms for specifying object-language typing rules and checking or inferring object-language types. In λ_{Π}, one can represent object-language typing rules directly in the language's signature. For some object-languages, HOU_{Π} can then do automatic object-language type checking and inference. For object-languages with considerably more complicated type systems, we can still do some of the object-type checking or inference automatically, and return the remaining work as a theorem proving task. It is important to note, though, that the necessary theorem proving can be handled in the logical framework provided by λ_{Π}, (and can therefore be automated or semi-automated).

As outlined in [10], LF is a "first step towards a general theory of interactive proof checking and proof construction." HOU_{Π} can be of significant assistance in two ways. First, it allows us to go beyond purely interactive theorem proving, to do automated (and semi-automated) theorem proving in LF encoded logics. These theorem provers could be expressed, *e.g.*, in a λProlog based on λ_{Π}, or in Elf, a language for logic definition and verified meta-programming [20]. (Implementations of both of these languages are currently under development.) Experience with typed programming languages has shown automatic type inference to be of considerable practical value [9,2,28]. In λ_{Π}, a new problem arises: It is also often possible to infer some of the subterms of a term, in the sense that no other subterms would result in a well-typed term. It is precisely this λ_{Π} *term inference* that provides for object-language type inference. As we will show, HOU_{Π} turns the λ_{Π} type checking algorithm into a type checking and term inference algorithm.[2] By a simple extension of HOU_{Π} to handle type variables, which we have implemented, one can do automatic type inference as well. (See [4, Chapter 6] for a similar extension to $HOU_{\rightarrow}$.)

The rest of this paper is structured as follows: Section 2 covers some preliminaries, including the language and typing rules of λ_{Π}, the notion of *approximate well-typedness* and some concepts involved in unification. Section 3 discusses the difficulties in adapting Huet's algorithm to λ_{Π}. Section 4 develops our algorithm for HOU_{Π} based on a collection of transformations on unification problems. Section 5 contains a proof that the "solved form" unification problems

[1]We refer to an encoded logic as an *object-logic* to stress the difference between the language being represented, and the calculus in which it is represented. The latter is often referred to as the *meta-logic*. When the encoded language is not necessarily a logic, we will often use the more general terms *object-language* and *meta-language*.

[2]Term inference does not subsume theorem proving in object-logics because it may leave free variables in the resulting terms.

constructed by our algorithm are always unifiable. Section 6 presents an algorithm for λ_Π type checking and term inference. Finally, Section 7 discusses related work, and Section 8 summarizes the paper and points to future work.

2. Preliminaries

2.1. The Language and its Typing Rules

The calculus in which we do unification is essentially the one used by LF [10]. However, to simplify the exposition, and because it does not add any expressive power to the calculus, we omit the LF type λ, which explicitly forms functions from terms to types.[3] Letting the meta-variables M and N range over terms, A and B (and sometimes α, σ, and τ) over "types" (including type families), and K over kinds, the language is as follows:

$$M ::= \mathsf{c} \mid v \mid M\,N \mid \lambda v{:}\,A.\,M$$

$$A ::= \mathsf{c} \mid \Pi v{:}\,A.\,B \mid A\,M$$

$$K ::= \mathsf{Type} \mid \Pi v{:}\,A.\,K$$

We will often use the abbreviation "$A \to B$" for "$\Pi v{:}\,A.\,B$" when v is not free in B, and similarly for kinds.

The typing rules for λ_Π may be found in [10]. Below we reproduce just enough to understand the rest of this paper. A typing judgment involves a *signature* Σ assigning types and kinds to constants, a *context* Γ assigning types to variables. We omit well-formedness considerations for signatures, contexts, and kinds. Constants and variables are looked up in Σ and Γ respectively. The other rules are, for the kind of a type,[4]

$$\frac{\Gamma \vdash_\Sigma A \in \mathsf{Type} \qquad \Gamma\,,v{:}\,A \vdash_\Sigma B \in \mathsf{Type}}{\Gamma \vdash_\Sigma \Pi v{:}\,A.\,B \in \mathsf{Type}}$$

$$\frac{\Gamma \vdash_\Sigma A \in \Pi v{:}\,B.\,K \qquad \Gamma \vdash_\Sigma M \in B}{\Gamma \vdash_\Sigma A\,M \in [\,M/v\,]K}$$

$$\frac{\Gamma \vdash_\Sigma A \in K \qquad K =_{\beta\eta} K' \qquad \Gamma \vdash_\Sigma K' \text{ kind}}{\Gamma \vdash_\Sigma A \in K'}$$

and, for the type of a term,

$$\frac{\Gamma \vdash_\Sigma A \in \mathsf{Type} \qquad \Gamma\,,v{:}\,A \vdash_\Sigma M \in B}{\Gamma \vdash_\Sigma \lambda v{:}\,A.\,M \in \Pi x{:}\,A.\,B}$$

$$\frac{\Gamma \vdash_\Sigma M \in \Pi v{:}\,A.\,B \qquad \Gamma \vdash_\Sigma N \in A}{\Gamma \vdash_\Sigma M\,N \in [\,N/v\,]B}$$

$$\frac{\Gamma \vdash_\Sigma M \in A \qquad A =_{\beta\eta} B \qquad \Gamma \vdash_\Sigma B \in \mathsf{Type}}{\Gamma \vdash_\Sigma M \in B}$$

---

[3]These λ's never appear in normal form terms. Of course, we do allow constants of functional kind.

[4]We write "$\Gamma\,, v{:}\,A$ for the result of extending the context Γ by the assignment of the type A to the variable v and, for simplicity, we will assume that v is not already assigned a type in Γ. (This can always be enforced by α-conversion.) Also, in these rules, the $=_{\beta\eta}$ relation refers to convertability between types or kinds using the β or η conversion rules on the terms contained in these types or kinds. We take α-conversion for granted.

2.2. Approximate Typing and Normalization

Huet's algorithm depends on normalizability and the Church-Rosser property, and both may fail with ill-typed terms.[5] However, as we will see, in HOU_Π we are forced to deal with ill-typed terms. A key factor in the design of our algorithm is how we deal with ill-typedness, and to that end, we define a notion of *approximate well-typedness,* and state some of its properties.

We begin by defining two approximation functions, one that maps a term M into a simply typed term $\overline{M}$, and another that maps a type A into a simple type $\overline{A}$:[6]

Definition 1 *The* approximation *of a term or type is given by*

$$\begin{aligned} \overline{c} &= c \\ \overline{v} &= v \\ \overline{MN} &= \overline{M}\,\overline{N} \\ \overline{\lambda x{:}A.\,M} &= \lambda x{:}\overline{A}.\,\overline{M} \end{aligned}$$

and

$$\begin{aligned} \overline{c} &= c \\ \overline{\Pi x{:}A.\,B} &= \overline{A}\rightarrow\overline{B} \\ \overline{AM} &= \overline{A} \end{aligned}$$

Approximation is extended to kinds by setting $\overline{K} = \mathsf{Type}$ *for all* K*, and to contexts and signatures in the obvious way.*

Definition 2 *A term* M has the approximate type A in Γ *iff* $\overline{\Gamma} \vdash_{\overline{\Sigma}} \overline{M} \in A$*. (Note then that* A *is a simple type.) In this case, we say that* M *is* approximately well-typed *in* Γ.

Even approximately well-typed terms do not have the strong normalization or Church-Rosser properties, but they do have weaker normalization and uniqueness properties, which will suffice for our algorithm.

Definition 3 *A term is in* long $\beta\eta$ head normal form *(*LHNF *for short) if it is*

$$\lambda x_1{:}\sigma_1.\ \cdots \lambda x_k{:}\sigma_k.\ a\,M_1 \cdots M_m$$

for some $k \geq 0$ *and* $m \geq 0$*, where* a *is a variable or constant, and* $a\,M_1 \cdots M_m$ *is not of approximate function type. Given a term in LHNF as above, its* head *is* a*, and its* heading *is* $\lambda x_1.\ \cdots \lambda x_k.\ a$*. (Note that if the term is approximately well-typed, the heading determines* m*, but not* $\sigma_1, \ldots, \sigma_k$ *or* $M_1, \ldots, M_m$*.)*

We can now state our normalization and uniqueness properties:

Theorem 4 *Every approximately well-typed term has a LHNF. Furthermore, every LHNF of an approximately well-typed term has the same heading (modulo α-conversion).*

---

[5] For non-normalizability, consider $(\lambda x{:}\sigma.\ x\,x)\,(\lambda x{:}\sigma.\ x\,x)$. This is ill-typed for any choice of σ, and has no normal form. For loss of Church-Rosser, consider $(\lambda x{:}\sigma.\,(\lambda y{:}\tau.\,y)\,x)$, which β-reduces to $(\lambda x{:}\sigma.\,x)$, but η-reduces to $(\lambda y{:}\tau.\,y)$.

[6] This translation is similar to the translation in [10] of λ_Π terms into untyped terms.

Proof sketch: The properties follow from the normalizability and Church-Rosser properties of well-typed terms in $\lambda_{\rightarrow}$ [11,23], since given an approximately well-typed term M, any reduction sequence on $\overline{M}$ (which is well-typed in $\lambda_{\rightarrow}$) can be paralleled on M, and noting that we can follow normalization by enough η-expansions to convert to long form (where the body $a\, M_1 \cdots M_m$ is not of approximate function type). □

Because of this uniqueness property, we will apply the words "head" and "heading" to an arbitrary approximately well-typed term M to mean the head or heading of any LHNF of M.

This normal form allows us to make an important distinction (as in HOU$_{\rightarrow}$) among approximately well-typed terms:

Definition 5 *Given an approximately well-typed term M whose heading is $\lambda x_1. \cdots \lambda x_n. a$, we will call M* rigid *if a is either a constant or one of the x_i, and* flexible *otherwise.*

The value of this distinction is that applying a substitution cannot change the heading of a rigid approximately well-typed term.

2.3. Unification

Our formulation of higher-order unification is a generalization of the usual formulation that is well suited for exposition of the algorithm. First we need to define well-typed and approximately well-typed substitutions:

Definition 6 *The set $\Theta_{\Gamma}^{\Gamma'}$ of* well-typed substitutions from Γ to Γ' *is the set of those substitutions θ such that for every $v\colon A$ in Γ, we have $\Gamma' \vdash_{\Sigma} \theta v \in \theta A$. The set Θ_{Γ} of* well-typed substitutions over Γ *is the union over all well-formed contexts Γ' of $\Theta_{\Gamma}^{\Gamma'}$. Similarly the* approximately well-typed substitutions *from Γ to Γ' are those θ such that $\overline{\Gamma'} \vdash_{\overline{\Sigma}} \overline{\theta v} \in \overline{\theta A}$.*

For our purposes, a unification problem is with respect to a *unification context* Γ, and is made up of an initial substitution θ_0 and a disagreement set[7] D whose free variables are in Γ. The intent is to compose the unifiers of D with θ_0. More precisely,

Definition 7 *A* unification problem *is a triple $\langle \Gamma, \theta_0, D\rangle$ consisting of a context Γ, a substitution $\theta \in \Theta_{\Gamma_0}^{\Gamma}$ for some context Γ_0, and a set D of pairs of terms whose free variables are typed by Γ.*

Definition 8 *The* set of unifiers *of a unification problem $\langle \Gamma, \theta_0, D\rangle$ is*[8]

$$\mathcal{U}(\langle \Gamma, \theta_0, D\rangle) \;\triangleq\; \{\, \theta \circ \theta_0 \mid \theta \in \Theta_{\Gamma} \wedge \forall \langle M, M'\rangle \in D.\, \theta M =_{\beta\eta} \theta M' \,\}$$

Note that when θ_0 is an identity substitution and D is a singleton set, we have the usual problem of simply unifying two terms.

This specification serves as an organizational tool. What we really want to implement is something like the most general unifiers computed in first-order unification. However, in the higher-order case, (a) there are not unique most general unifiers, (b) even producing a "complete set of unifiers", the set of whose instances forms the set of all solutions, becomes at a certain

[7] We adopt the standard term ***disagreement pair*** for a pair of terms to be unified, and ***disagreement set*** for a set of disagreement pairs to be simultaneously unified.

[8] We use functional order composition, *i.e.*, $(\theta \circ \theta_0)M = \theta(\theta_0 M)$.

point too undirected to be useful, and (c) it is not possible to eliminate redundant solutions [12]. Huet's idea of *pre-unification* [12] (implicit in [13]) solved these difficulties. For us, each pre-unifier of a unification problem $\langle \Gamma, \theta_0, D \rangle$ is a new unification problem $\langle \Gamma', \theta', D' \rangle$, such that each solution of $\langle \Gamma', \theta', D' \rangle$ is a solution of $\langle \Gamma, \theta_0, D \rangle$, and $\langle \Gamma', \theta', D' \rangle$ is in *solved form, i.e.*, D' contains only pairs of flexible terms. Huet showed (constructively) that in $\lambda_{\rightarrow}$ such D' are always unifiable [13, Section 3.3.3], and hence the existence of a pre-unifier implies the existence of a unifier. This is not always the case in λ_{Π}, but by maintaining a certain invariant on unification problems, we will show that the pre-unifiers constructed by our algorithm do always lead to unifiers.

3. Problems with Adapting Huet's Algorithm

Huet's algorithm relies on two important invariants that we are forced to abandon in HOU_{Π}. This section motivates our algorithm by explaining why we cannot maintain these invariants, and discussing how we handle the resulting difficulties.

In $\mathrm{HOU}_{\rightarrow}$, one can can require all disagreement pairs to be "homogenous", *i.e.*, relating terms of the same type. In λ_{Π}, since substitution affects types, two terms might be unifiable even if they have different types. For example, the disagreement pair $\langle \lambda x{:}\sigma.\, x, \lambda y{:}\tau.\, y \rangle$ is unified by the unifiers of the types σ and τ. Also, even if we do not start out with heterogenous disagreement pairs, they arise naturally in the presence of dependent types. Consider, for instance, a disagreement pair $\langle \mathsf{q}\, M\, N, \mathsf{q}\, M'\, N' \rangle$ of well-typed terms in the signature

$$\langle\ \mathsf{a}{:}\mathsf{Type}\ ,\ \mathsf{b}{:}\mathsf{a}{\rightarrow}\mathsf{Type}\ ,\ \mathsf{c}{:}\mathsf{Type}\ ,\ \mathsf{q}{:}\Pi x{:}\mathsf{a}.\, (\mathsf{b}\, x){\rightarrow}\mathsf{c}\ \rangle$$

Note that both terms have type c. As in the SIMPL phase of Huet's algorithm, we can replace this disagreement pair by the unification-equivalent[9] set of pairs $\{\ \langle M, M' \rangle, \langle N, N' \rangle\ \}$. Note, however, that N has type $(\mathsf{b}\, M)$, while N' has type $(\mathsf{b}\, M')$.

The second invariant we have to abandon is that disagreement pairs contain only well-typed terms. To see why, consider a disagreement set $D' \cup \{\langle v, \mathsf{c} \rangle\}$, where $v{:}\,\sigma$ in Γ and $\mathsf{c}{:}\,\tau$ in Σ. In Huet's algorithm, assuming the situation were allowed to arise, the treatment of this disagreement set would involve applying the substitution $[\,\mathsf{c}/v\,]$ to D' and continuing to work on the result. However, if σ and τ are different types, and v occurs (freely) in D' this substitution will construct ill-typed terms.[10] A similar phenomenon can happen in $\lambda_{\rightarrow}$ but is carefully avoided by Huet's algorithm. In the MATCH phase, if the flexible head has type $\alpha_1{\rightarrow}\cdots{\rightarrow}\alpha_m{\rightarrow}\alpha_0$, for a base type α_0, then among the m possible "projection substitutions", only the well-typed ones are tried. Because substitution does not affect types in $\lambda_{\rightarrow}$, this is just a matter of comparing type constants. When substitutions instantiate types though (as in λ_{Π}), we need to unify types, not just compare them.

These considerations might suggest that we can regain our invariants if, at certain points in the algorithm, we do type unification before continuing to unify terms. The fatal flaw in this idea is that we are doing *pre-unification*, not full unification, and after pre-unifying types, there may still be some remaining flexible-flexible pairs. Thus, some heterogeneity or ill-typedness may still be present.

[9] We use *unification-equivalent* to mean having the same set of unifiers.

[10] On the other hand, if v does not occur in D', for instance if $D' = \{\ \}$, we will have forgotten that we must unify σ and τ, and so the result would be wrong.

Our solution is to perform just enough of the type unification to insure that the substitution we are about to apply is approximately well-typed and cannot therefore destroy head normalizability. We do this by defining a partial function R_Γ that converts a pair of types, into a (unification-equivalent) set of disagreement pairs. If the function fails (is undefined) then the types are nonunifiable. R_Γ is defined by the following cases, and is undefined if no case applies. For brevity, we will write "$\lambda\Gamma.\, M$" for a context $\Gamma = [\, x_1{:}\alpha_1, \ldots, x_n{:}\alpha_n \,]$ to mean $\lambda x_1{:}\alpha_1. \cdots \lambda x_n{:}\alpha_n.\, M$.

$$
\begin{aligned}
\mathsf{R}_\Gamma(\mathsf{c},\mathsf{c}) &= \{\,\} \\
\mathsf{R}_\Gamma(A\,M, A'\,M') &= \mathsf{R}_\Gamma(A,A') \cup \{\langle \lambda\Gamma.\, M, \lambda\Gamma'.\, M'\rangle\} \\
\mathsf{R}_\Gamma(\Pi v{:}A.\, B, \Pi v'{:}A'.\, B') &= \mathsf{R}_\Gamma(A,A') \cup \mathsf{R}_{(\Gamma, v:A)}(B, [\, v/v' \,]B')
\end{aligned}
$$

We will use R as follows: Before performing a substitution of a term of type τ to a variable of type σ, in a unification context Γ, we compute $\mathsf{R}_{[]}(\sigma,\tau)$. If this is undefined, we know that the substitution cannot lead to a (well-typed) unifier. If it yields a disagreement set $\hat{D}$, we perform the substitution (which we now know to be approximately well-typed), and add $\hat{D}$ to the resulting disagreement set, to account for any ill-typedness introduced by the substitution.

In order to prove unifiability of the disagreement sets in the pre-unifiers produced by our algorithm, we will need to keep track of which disagreement pairs account for which others. This relationship and the required approximate well-typedness conditions are embodied in the following invariant, on which our algorithm depends and which it maintains while transforming disagreement sets.

Definition 9 *A unification problem $\langle \Gamma, \theta_0, D\rangle$ is* acceptable, *which we will write as "$\mathcal{A}(Q)$", iff the following conditions hold:*

1. *Each of the disagreement pairs in D relates terms of the same approximate type (which are therefore approximately well-typed).*
2. *There is a strict partial order*[11] *"$\sqsubset$" on D such that for any disagreement pair $P \in D$, every unifier of $\{\, P' \in D \mid P' \sqsubset P \,\}$ instantiates P to a pair of terms of the same type (which are therefore well-typed).*
3. *Any well-typed unifier of D, when composed with θ_0, yields a well-typed substitution.*

It is important to note that the algorithm only maintains the *existence* of these strict partial orders, but never actually constructs them.

4. The Dependent Pre-unification Algorithm

This section presents an abstract algorithm for HOU_Π, based on collection of transformations from unification problems to sets of unification problems, which preserve sets of unifiers and maintain our invariant of acceptability. The goal of the transformations is to eventually construct unification problems in solved form.

We define the property required of our transformations as follows:

Definition 10 *For a unification problem Q and a set of unification problems $\mathcal{Q}$, we say that "$Q \lhd \mathcal{Q}$" iff when Q is acceptable, (a) so are all of the members of $\mathcal{Q}$, (b) the set of unifiers of*

[11] *i.e.*, a transitive, antisymmetric, *non*reflexive relation

Q is the union of the sets of unifiers of the members of $\mathcal{Q}$, and (c) the members of $\mathcal{Q}$ have no unifiers in common. More formally, $Q \lhd \mathcal{Q}$ iff $\mathcal{A}(Q)$ implies the following three conditions

$$\forall Q' \in \mathcal{Q}.\, \mathcal{A}(Q')$$

$$\mathcal{U}(Q) = \bigcup_{Q' \in \mathcal{Q}} \mathcal{U}(Q')$$

$$\forall Q', Q'' \in \mathcal{Q}.\, Q' \neq Q'' \Rightarrow \mathcal{U}(Q') \cap \mathcal{U}(Q'') = \{\ \}$$

Our algorithm is based on three transformations. Others may be added as optimizations, but these three suffice for completeness.[12] The transformations deal with unification problems containing a rigid-rigid, rigid-flexible, or flexible-rigid pairs. When no transformation applies, we have a unification problem in solved form. Collectively, these three transformations form a subrelation $\lhd_{\mathcal{U}}$ of $\lhd$.

4.1. The Transformations

For brevity, in all cases we assume for our unification problem Q that

$$\begin{aligned} Q &= \langle \Gamma, \theta_0, D \rangle \\ D &= \{\, \langle M, M' \rangle \,\} \cup D' \end{aligned}$$

Without loss of generality, we can assume that M and M' are in LHNF (if not, convert them), so let

$$\begin{aligned} M &= \lambda x_1{:}\alpha_1.\, \cdots \lambda x_k{:}\alpha_k.\, a\, M_1 \cdots M_m \\ M' &= \lambda x_1'{:}\alpha_1'.\, \cdots \lambda x_k'{:}\alpha_k'.\, a'\, M_1' \cdots M_{m'}' \end{aligned}$$

(The invariant that M and M' have the same approximate type insures that they start with the same number k of λ's.)

The rigid-rigid transformation. Assume that M and M' are rigid, *i.e.*, a is a constant or some x_j, and a' is a constant or some $x_{j'}'$. If a and a' are the same modulo their binding contexts[13] then approximate well-typedness insures that $m = m'$, and we have

$$Q \ \lhd_{\mathcal{U}} \ \{\, \langle \Gamma, \theta_0, D' \cup \{\, \langle \hat{M}_i, \hat{M}_i' \rangle \mid 1 \leq i \leq m \,\} \rangle \,\}$$

where $\hat{M}_i = \lambda x_1{:}\alpha_1.\, \cdots \lambda x_k{:}\alpha_k.\, M_i$ and $\hat{M}_i' = \lambda x_1'{:}\alpha_1'.\, \cdots \lambda x_k'{:}\alpha_k'.\, M_i'$ for $1 \leq i \leq m$. Otherwise,

$$Q \ \lhd_{\mathcal{U}} \ \{\ \}$$

This transformation corresponds to one step of Huet's SIMPL phase. It preserves the set of unifiers because substitution does not affect the heading of M or M'. Thus rigid terms with distinct heads are nonunifiable, and rigid terms with the same head are unified exactly by the unifiers of their corresponding arguments. As for the invariant, (1) comes from M and M' being approximately well-typed, and (3) follows because the initial substitution is the same and the new disagreement set has the same unifiers as D. For (2), a new strict partial order can be

---

[12] Of particular value is the variable-term case, using a *rigid path* check [13]. Several others for $\mathrm{HOU}_{\rightarrow}$ are found in [16].

[13] *i.e.*, they are the same constant, or $a = x_j$ and $a' = x_j'$ for some j

derived from an old one by replacing the old disagreement pair by the new ones, and adding $\langle \hat{M}_i, \hat{M}'_i \rangle \sqsubset \langle \hat{M}_j, \hat{M}'_j \rangle$ for $1 \leq i < j \leq m$. The reason for adding these is that each of the M_i can appear in the types of later M_j, and similarly for the M'_i.

As an example, consider the earlier case of $\{ \langle \mathsf{q}\, M\, N, \mathsf{q}\, M'\, N' \rangle \}$. The rigid-rigid transformation replaces this pair by $\{ \langle M, M' \rangle, \langle N, N' \rangle \}$. The disagreement pair $\langle M, M' \rangle$ accounts for the difference between the type $(\mathsf{b}\, M)$ of N and the type $(\mathsf{b}\, M')$ of N'. As another example, consider the disagreement set $\{\langle \lambda x{:}\sigma.\ x, \lambda y{:}\tau.\ y \rangle\} \cup D'$. The rigid-rigid transformation yields simply D', which is correct, because the invariant insures that the difference between σ and τ is already accounted for in D'.

The flexible-rigid transformation. Assume that M is flexible and M' is rigid. In this case, we will form a set of substitutions, each of which partially instantiates the flexible head a in such a way that completely determines its head but leaves its arguments completely undetermined. These are the imitation and projection substitutions generated by Huet's MATCH phase.

These substitutions are motivated by consideration of the LHNF of θa for any unifier θ of M and M'. Let the type of a be $\Pi y_1{:}\tau_1. \cdots \Pi y_m{:}\tau_m. \tau_0$, where τ_0 is not a Π type. Then for any well-typed substitution θ, any LHNF of θa has the form

$$\lambda y_1{:}\hat{\tau}_1. \cdots \lambda y_m{:}\hat{\tau}_m. b\, N_1 \cdots N_n$$

for some variable or constant b. Since the head of θM has to be the rigid head a' of M', the only possibilities for b are (1) a', if a' is a constant, or (2) some y_i, for $1 \leq i \leq m$. For each such b, we can capture this restriction on θa by equivalently saying that θ has the form $\hat{\theta} \circ [\, N_b/a\,]$, for some $\hat{\theta}$, where

$$N_b \quad = \quad \lambda y_1{:}\tau_1. \cdots \lambda y_m{:}\tau_m. b\,(v_{b1}\, y_1 \cdots y_m) \cdots (v_{bl_b}\, y_1 \cdots y_m)$$

Here $v_{b1}, \ldots, v_{bl_b}$ are distinct variables not in Γ or among the y_j.[14] (See [4, Section 3.2.2] for details.) Letting B be the set of possible b's described above, one can then show that[15]

$$\mathcal{U}(Q) \quad = \quad \bigcup_{b \in B} \mathcal{U}([\, N_b/a\,]Q)$$

where by an application "$\theta Q'$" of a substitution $\theta \in \Theta_{\Gamma'}^{\Gamma''}$ to a unification problem $Q' = \langle \Gamma', \theta'_0, D' \rangle$, we mean the unification problem

$$\langle \Gamma'', \theta \circ \theta'_0, \{ \langle \theta M, \theta M' \rangle \mid \langle M, M' \rangle \in D' \} \rangle$$

Now, to reestablish the invariant, we must add disagreement pairs to account for any difference between the types of a and N_b. Let σ_a and σ_{N_b} be the types in Γ of a and N_b, and, for each $b \in B$ such that $\mathsf{R}_{[]}(\sigma_a, \sigma_{N_b})$ is defined, let

$$Q_b \quad = \quad ([\, N_b/a\,]\langle \Gamma, \theta_0, D \rangle) \cup \mathsf{R}_{[]}(\sigma_a, \sigma_{N_b})$$

[14] In order to account for the v_{b_i}, we assume a nonstandard definition of substitutions and their composition that causes temporary variables to be eliminated appropriately [4, Section 2.6].

[15] The key step is that $\mathcal{U}(\langle \Gamma, \theta_0, D \rangle)$ can be re-expressed as

$$\bigcup_{b \in B} \{ (\hat{\theta} \circ [\, N_b/a\,]) \circ \theta_0 \mid \hat{\theta} \in \Theta_{\Gamma_b} \wedge \forall \langle M, M' \rangle \in D.\, (\hat{\theta} \circ [\, N_b/a\,])M =_{\beta\eta} (\hat{\theta} \circ [\, N_b/a\,])M' \}$$

where Γ_b is the result of removing a and adding $v_{b1}, \ldots, v_{bl_b}$ to Γ. Then by using associativity of substitution composition, and contracting the definition of $\mathcal{U}$, the result follows.

where by the union of a unification problem $\langle \Gamma', \theta'_0, D' \rangle$ with a disagreement set D'', we mean $\langle \Gamma', \theta'_0, D' \cup D'' \rangle$. Then we have

$$Q \ \triangleleft_{\mathcal{U}} \ \{ Q_b \mid b \in B \text{ and } \mathsf{R}_{[]}(\sigma_a, \sigma_{N_b}) \text{ is defined} \}$$

Note that this addition of $\mathsf{R}_{[]}(\sigma_a, \sigma_{N_b})$ is the only place where our set of transformations differs from Huet's. For simply typed terms, these $\mathsf{R}_{[]}(\sigma_a, \sigma_{N_b})$ (when defined) are always empty, so our algorithm does no more work than Huet's.

It is important to note that the generated unification problems have no solutions in common. This is because, for each b, any unifier of Q_b is an instance of $[\, N_b/a \,] \circ \theta_0$, and since each N_b has a different rigid head (namely b), no two of the $[\, N_b/a \,] \circ \theta_0$ can have any instances in common. This property is what guarantees minimality (see Theorem 13).

A new strict partial order can be derived from an old one by replacing each disagreement pair by its newly instantiated version, and by adding $P \sqsubset P'$ for each $P \in \mathsf{R}_{[]}(\sigma_a, \sigma_{N_b})$ and $P' \in [\, N_b/a \,]D$, since we added $\mathsf{R}_{[]}(\sigma_a, \sigma_{N_b})$ to account for any introduced ill-typedness.

The rigid-flexible transformation. The rigid-flexible case can be handled simply by reflecting it into the flexible-rigid case: If M is rigid and M' is flexible then

$$Q \ \triangleleft_{\mathcal{U}} \ \{ \langle \Gamma, \theta_0, \{ \langle M', M \rangle \} \cup D' \rangle \}$$

4.2. The Algorithm

Now that we have presented the three transformations, together defining the relation $\triangleleft_{\mathcal{U}}$, we will describe a search process that operates on a set of unification problems and enumerates a set of pre-unifiers. Informally, the process goes as follows: If there are no unification problems left, stop. Otherwise, choose a unification problem to work on next. If it is in solved form, add it the the set of solutions. Otherwise, apply one of the transformations, in some way, to replace the unification problem by a finite set of new unification problems. Then continue.

Note that two kinds of choices are made in this process. First, there is the choice of which unification problem to work on next, and second, there is the choice of which transformation to apply and how to apply it. It turns out that the second kind of choice may be made completely arbitrarily, but, in order to have completeness, the first kind must be done in a fair way.[16] Huet formulated this difference by constructing "matching trees", in which the nodes are disagreement sets and the edges are substitutions, and then showed that all matching trees are complete. His pre-unifiers are constructed by composing substitutions along edges that form a path from the original disagreement set to one in solved form. In our formulation, these composed substitutions are part of the unification problem.

Definition 11 *For a relation ρ between unification problems and sets of unification problems, and a unification problem Q, a ρ* search tree *from Q is a tree T of unification problems such that*

- *The root of T is Q.*
- *For every node Q' in T, the set of children of Q' in T is either empty if Q' is in solved form, or is some $\mathcal{Q}$ satisfying $Q' \ \rho \ \mathcal{Q}$ if Q' is not in solved form.*

[16] In implementation terms, this means that we can use *e.g.*, breadth-first search or depth-first search with iterative deepening, but not simple depth-first search.

Definition 12 *For a relation ρ between unification problems and sets of unification problems, we define the relation ρ^{**} as follows: $Q \rho^{**} \mathcal{Q}$ iff there is some ρ search tree from Q whose set of solved nodes is $\mathcal{Q}$.*

We can then show the following

Theorem 13 *Let Q be a unification problem such that $\mathcal{A}(Q)$, and let $\mathcal{Q}$ be any set of unification problems such that $Q \lhd_{\mathcal{U}}^{**} \mathcal{Q}$. Then*

1. *$\mathcal{A}(Q')$ for each $Q' \in \mathcal{Q}$.*
2. *Every $Q' \in \mathcal{Q}$ is a pre-unifier of Q, i.e., it is in solved form and $\mathcal{U}(Q') \subseteq \mathcal{U}(Q)$.*
3. *$\mathcal{Q}$ is minimal, i.e., for any two distinct members Q', Q'' of $\mathcal{Q}$, $\mathcal{U}(Q') \cap \mathcal{U}(Q'') = \{\ \}$.*
4. *$\mathcal{Q}$ is complete, i.e., for any unifier θ of Q, there is a $Q' \in \mathcal{Q}$ such that $\theta \in \mathcal{U}(Q')$.*

Proof sketch:

1. Each transformation maintains the invariant for each constructed unification problem.
2. The transformations do not introduce new unifiers, and $\mathcal{Q}$ contains only solved form unification problems.
3. As noted in the discussion of the flexible-rigid transformation, when the search for pre-unifiers branches, the new unification problems have no unifiers in common.
4. Because our invariant insures head normalizability, the completeness proof goes much as in [13], and has two main parts: (a) For a given unification problem Q, there can be only finitely many successive applications of the rigid-rigid and rigid-flexible transformations. (b) For any unification problem $\langle \Gamma, \theta_0, D \rangle$ to which the flexible-rigid transformation applies, and any unifier θ of D, there is flexible-rigid-successor $\langle \Gamma', \theta_0', D' \rangle$ of $\langle \Gamma, \theta_0, D \rangle$ and a unifier θ' of D' such that θ' has strictly lower complexity than θ, where complexity is defined in terms of sizes of the long $\beta\eta$ normal forms involved. Another consideration, not required in $\mathrm{HOU}_{\rightarrow}$, is that R_{Γ} always terminates.

□

Given a pair of terms M and M' to unify, we can satisfy the invariant initially in either of two ways. The first is to simply check that M and M' are well-typed and have the same type. (This is possible because type checking is decidable [10].) This method is simple but does not allow for terms that will become well-typed or disagreement pairs that will become homogeneous after substitution.[17] The second method is much more flexible. Instead of type-checking the terms, we perform only approximate type-checking, and at the same time, construct a disagreement set whose unifiers (if any) instantiate the terms to well-typed terms of the same type. This process is defined in Section 6.

[17] Ill-typedness and heterogeneity can still arise during unification though.

5. Unifiability of Flexible-flexible Disagreement Sets

The value of pre-unification in $\lambda_{\rightarrow}$ is that solved disagreement sets (ones containing only flexible-flexible pairs) are always unifiable, and so pre-unifiability implies unifiability [13]. This is not true in general for λ_Π, but it is true of solved sets satisfying our invariant. By making vital use the strict partial order in the definition of $\mathcal{A}$, we can generalise Huet's constructive proof of this fact to λ_Π. For the simply typed subset of λ_Π, the substitution that we use specializes to Huet's.

Definition 14 *For a context* Γ, *the* canonical unifier θ_Γ^C *over* Γ *is the substitution assigning to each variable* $v{:}\Pi x_1{:}\sigma_1.\cdots\Pi x_m{:}\sigma_m.\,\mathsf{c}\,Q_1\cdots Q_n$ *in* Γ, *the term* $\lambda x_1{:}\sigma_1.\cdots\lambda x_m{:}\sigma_m.\,h_\mathsf{C}\,Q_1\cdots Q_n$, *where* h_C *is a variable of type* $\Pi y_1{:}\tau_1.\cdots\Pi y_n{:}\tau_n.\,\mathsf{c}\,y_1\cdots y_n$.

Theorem 15 *If* Q *is a acceptable unification problem in solved form with unification context* Γ, *then* $\theta_\Gamma^C \in \mathcal{U}(Q)$.

Proof: Let $\sqsubset$ be the strict partial order imposed on disagreement sets by our invariant. Since disagreement sets are always finite, $\sqsubset$ is a well founded ordering, and thus we will give an inductive argument. Let $\langle M, M'\rangle$ be an arbitrary member or our disagreement set for

$$\begin{aligned} M &= \lambda z_1{:}\alpha_1.\cdots\lambda z_k{:}\alpha_k.\,v\,M_1\cdots M_m \\ M' &= \lambda z'_1{:}\alpha'_1.\cdots\lambda z'_k{:}\alpha'_k.\,v'\,M'_1\cdots M'_{m'} \end{aligned}$$

where v and v' are variables with types

$$\begin{aligned} v &: \Pi x_1{:}\sigma_1.\cdots\Pi x_m{:}\sigma_m.\,\mathsf{c}\,Q_1\cdots Q_n \\ v' &: \Pi x'_1{:}\sigma'_1.\cdots\Pi x'_{m'}{:}\sigma'_{m'}.\,\mathsf{c}\,Q'_1\cdots Q'_n \end{aligned}$$

and

$$\mathsf{c} : \Pi y_1{:}\tau_1.\cdots\Pi y_n{:}\tau_n.\,\mathsf{Type}$$

The reason that the types of both v and v' must involve the same type constant c, is that our invariant insures that M and M' have the same approximate type. Now, for $1 \le j \le n$, let

$$\begin{aligned} N_j &= \lambda z_1{:}\alpha_1.\cdots\lambda z_k{:}\alpha_k.\,[\,(\theta_\Gamma^C M_1)/x_1,\ldots,(\theta_\Gamma^C M_m)/x_m\,]Q_j \\ N'_j &= \lambda z'_1{:}\alpha'_1.\cdots\lambda z'_k{:}\alpha'_k.\,[\,(\theta_\Gamma^C M'_1)/x_1,\ldots,(\theta_\Gamma^C M'_m)/x_m\,]Q'_j \end{aligned}$$

Then, for some choice of $\hat{\alpha}_1,\ldots,\hat{\alpha}_k$ and $\hat{\alpha}_1,\ldots,\hat{\alpha}_k$, we have

$$\begin{aligned} \theta_\Gamma^C M &= \lambda z_1{:}\hat{\alpha}_1.\cdots\lambda z_k{:}\hat{\alpha}_k.\,h_\mathsf{C}\,(N_1\,z_1\cdots z_k)\cdots(N_n\,z_1\cdots z_k) \\ \theta_\Gamma^C M' &= \lambda z'_1{:}\hat{\alpha}'_1.\cdots\lambda z'_k{:}\hat{\alpha}'_k.\,h_\mathsf{C}\,(N'_1\,z'_1\cdots z'_k)\cdots(N'_n\,z'_1\cdots z'_k) \end{aligned}$$

By induction, assume that θ_Γ^C unifies all disagreement pairs below $\langle M, M'\rangle$ in the ordering. Thus, by our invariant, $\theta_\Gamma^C M$ and $\theta_\Gamma^C M'$ are well-typed terms of the same type, so

$$\begin{aligned} &\Pi z_1{:}\hat{\alpha}_1.\cdots\Pi z_k{:}\hat{\alpha}_k.\,\mathsf{c}\,(N_1\,z_1\cdots z_k)\cdots(N_n\,z_1\cdots z_k) \\ =\ &\Pi z'_1{:}\hat{\alpha}'_1.\cdots\Pi z'_k{:}\hat{\alpha}'_k.\,\mathsf{c}\,(N'_1\,z'_1\cdots z'_k)\cdots(N'_n\,z'_1\cdots z'_k) \end{aligned}$$

It then follows that $\theta_\Gamma^C M =_{\beta\eta} \theta_\Gamma^C M'$. □

6. Automatic Term Inference

It is well known that first-order unification provides for type inference in $\lambda_{\rightarrow}$ with type variables and in similar languages [17]. Recently, it has been shown that $\mathrm{HOU}_{\rightarrow}$ is the key ingredient for the corresponding problem in the ω-order polymorphic λ-calculus [21]. In λ_Π there is a new problem of interest, namely *term inference*, which requires HOU_Π. This problem has two important applications. One is making our unification algorithm more widely applicable, by initially establishing the required invariant, as mentioned at the end of Section 4, and made precise below. The other is to provide automatic object-language type inference. This section gives a very simple algorithm for λ_Π term inference, using HOU_Π.

We will construct the term inference algorithm using two partial functions. The first one, Mi_Γ, for a given context Γ and signature Σ (the latter of which we will leave implicit) takes a term M and, if defined, yields a pair consisting of a type A and a disagreement set D. $\mathsf{Mi}_\Gamma(M)$ is undefined when M is not even approximately well-typed. Otherwise, for every unifier θ of D, it is the case that $\Gamma \vdash_\Sigma \theta M \in \theta A$. The second partial function Ai_Γ, takes a type A and, if defined, yields a pair consisting of a kind K and a disagreement set D. If $\mathsf{Ai}_\Gamma(A)$ is undefined then A has no well-kinded instance. Otherwise, for every unifier θ of D, we have $\Gamma \vdash_\Sigma \theta A \in \theta K$. The structure of these definitions is determined by the typing rules in Section 2, and uses the partial function R defined in Section 3.

$$
\begin{aligned}
\mathsf{Mi}_\Gamma(v) &= \langle A, \{\ \}\rangle \text{ where } v{:}A \text{ in } \Gamma\\
\mathsf{Mi}_\Gamma(\mathrm{c}) &= \langle A, \{\ \}\rangle \text{ where } \mathrm{c}{:}A \text{ in } \Sigma\\
\mathsf{Mi}_\Gamma(M\,N) &= \langle [\,N/v\,]B, D \cup D' \cup \mathsf{R}_{[]}(A, A')\rangle\\
&\quad \text{where } \begin{cases} \mathsf{Mi}_\Gamma(M) = \langle(\Pi v{:}A.\ B), D\rangle^{18}\\ \mathsf{Mi}_\Gamma(N) = \langle A', D'\rangle\end{cases}\\
\mathsf{Mi}_\Gamma(\lambda v{:}A.\ M) &= \langle \Pi v{:}A.\ B, D \cup D'\rangle\\
&\quad \text{where } \begin{cases} \mathsf{Ai}_\Gamma(A) = \langle \mathsf{Type}, D\rangle\\ \mathsf{Mi}_{\Gamma, v:A}(M) = \langle B, D'\rangle\end{cases}
\end{aligned}
$$

$$
\begin{aligned}
\mathsf{Ai}_\Gamma(\mathrm{c}) &= \langle K, \{\ \}\rangle \text{ where } \mathrm{c}{:}K \text{ in } \Sigma\\
\mathsf{Ai}_\Gamma(\Pi v{:}A.\ B) &= \langle \mathsf{Type}, D \cup D'\rangle\\
&\quad \text{where } \begin{cases} \mathsf{Ai}_\Gamma(A) = \langle \mathsf{Type}, D\rangle\\ \mathsf{Ai}_{\Gamma, v:A}(B) = \langle \mathsf{Type}, D'\rangle\end{cases}\\
\mathsf{Ai}_\Gamma(A\,M) &= \langle [\,M/v\,]K, D \cup D' \cup \mathsf{R}_{[]}(B, B')\rangle\\
&\quad \text{where } \begin{cases} \mathsf{Ai}_\Gamma(A) = \langle(\Pi v{:}B.\ K), D\rangle\\ \mathsf{Mi}_\Gamma(M) = \langle B', D'\rangle\end{cases}
\end{aligned}
$$

Given a term M in a context Γ we do type-checking/term inference as follows. If $\mathsf{Mi}_\Gamma(M)$ is undefined, then M is not approximately well-typed, and hence it has no well-typed instance, so we indicate failure. Otherwise, let $\langle A, D\rangle = \mathsf{Mi}_\Gamma(M)$, and let $\mathcal{Q}$ be such that $\langle \Gamma, \theta_\Gamma^{id}, D\rangle \lhd_{\mathcal{U}}^{**} \mathcal{Q}$, where θ_Γ^{id} is the identity substitution over Γ. If $\mathcal{Q}$ is empty, then M has no well-typed instance.

[18] The intended interpretation is that if the type part of $\mathsf{Mi}_\Gamma(M)$ is not a Π type, then $\mathsf{Mi}_\Gamma(M\,N)$ is undefined.

Otherwise, for each $\langle \Gamma', \theta', D' \rangle \in \mathcal{Q}$, we return the instantiated term $\theta' M$ together with the "constraint" D'.[19]

If on the other hand we have two terms M and M' to be unified in a context Γ, we can proceed as follows: If $\mathsf{Mi}_\Gamma(M)$ or $\mathsf{Mi}_\Gamma(M')$ is undefined, then M or M' is not approximately well-typed, so we indicate typing error. Otherwise, let $\langle A, D \rangle = \mathsf{Mi}_\Gamma(M)$ and $\langle A', D' \rangle = \mathsf{Mi}_\Gamma(M')$. Then if $\mathsf{R}_{[]}(A, A')$ is undefined, we indicate typing mismatch. Otherwise, let $D'' = \mathsf{R}_{[]}(A, A')$. Then apply the pre-unification algorithm to the unification problem $\langle \Gamma, \theta_\Gamma^{id}, \{\langle M, M' \rangle \cup D \cup D' \cup D''\} \rangle$.

The reason that λ_Π term inference often gives object-language type inference is that we can use λ_Π terms to encode object-language types and then construct object-language terms using constants whose (dependent) types record the object-language's typing system. One example of this is in the encoding of higher-order logic, given in [10]. Another is a simple first-order typed expression language [4, Section 7.3.3].

7. Related Work

Our algorithm is clearly influenced by the ideas underlying Huet's. A related transformational approach is Snyder and Gallier's for $\mathrm{HOU}_\rightarrow$ and equational unification [27,26]. One minor difference is that, rather than carrying along a substitution as part of their unification problems, they represent these substitutions as a "solved" part of their disagreement sets. A more important difference is that their transformations map a unification problem to a single unification problem, rather than a set of unification problems. However, doing so prevents an important distinction between the two kinds of "nondeterminism" in the algorithm, namely between the multiplicity of pre-unifiers in a complete set, and the multiplicity of ways in which transformation rules can be chosen and applied (resulting in different complete sets of preunifiers). We do not see how one could construct *minimal* sets of pre-unifiers using their approach.

Recently, Pym [24] reported an independently developed algorithm for HOU_Π.

8. Conclusions and Further Work

In this paper, we have presented an algorithm for HOU_Π, *i.e.*, higher-order (pre-)unification in a typed λ-calculus with dependent function types. This algorithm makes possible many valuable extensions to current applications of $\mathrm{HOU}_\rightarrow$, as well as mechanized theorem proving in object-logics encoded as in the Edinburgh Logical Framework (LF). We also presented a particularly useful application of HOU_Π to perform λ_Π *term inference*. This algorithm makes HOU_Π more widely applicable and allows for automatic type inference in a variety of object-languages.

Our algorithm has good efficiency properties. For simply typed examples, it does the same work as Huet's algorithm. Thus the additional power of the algorithm is only payed for where it is used.

A critical property of pre-unification in $\lambda_\rightarrow$ is that pre-unifiability is a sufficient condition for unifiability. Although this is not generally true in λ_Π under the relaxed typing conditions that we are forced to allow, we showed that the pre-unifiers constructed by our algorithm do indeed lead to unifiers.

[19] Depending on the application, if $\mathcal{Q}$ has more than one element, and/or if D' is nonempty for some $\langle \Gamma', \theta', D' \rangle \in \mathcal{Q}$, it may be appropriate to request a user to provide a more constrained term.

We have implemented a prototype version of an extension of our HOU_Π algorithm, which also handles type variables. Although the treatment of type variables is incomplete, it is quite useful in practice. We also plan to add treatment of the dependent version of Cartesian product types (often called "strong sum" or "Σ" types).[20] This implementation will form the basis of (a) a generalization of the programming language λProlog [18] to λ_Π, to serve as a convenient implementation language for applications of HOU_Π, and (b) the new language Elf for logic definition and verified meta-programming [20].

An area for future work is to develop a complete treatment of type variables, and if this succeeds, explicit polymorphism as in the second- or ω-order polymorphic λ-calculus [6,25].

9. Acknowledgment

I am very grateful to Frank Pfenning for originally suggesting the problem in λ_Π, and for several very helpful discussions yielding many useful ideas, in particular the idea of approximate well-typedness.

References

[1] Peter B. Andrews, Dale Miller, Eve Cohen, and Frank Pfenning. Automating higher-order logic. *Contemporary Mathematics*, 29:169–192, August 1984.

[2] R. M. Burstall, D. B. MacQueen, and D. T. Sanella. *HOPE: an Experimental Applicative Language.* Technical Report CSR-62-80, Department of Computer Science, University of Edinburgh, Edinburgh, U.K., 1981.

[3] Michael R. Donat and Lincoln A. Wallen. Learning and applying generalised solutions using higher order resolution. In Ewing Lusk and Ross Overbeek, editors, *9th International Conference on Automated Deduction, Argonne, Illinois*, pages 41–60, Springer-Verlag LNCS 310, Berlin, May 1988.

[4] Conal Elliott. *Some Extensions and Applications of Higher-order Unification: A Thesis Proposal.* Ergo Report 88–061, Carnegie Mellon University, Pittsburgh, June 1988. Thesis to appear June 1989.

[5] Amy Felty and Dale A. Miller. Specifying theorem provers in a higher-order logic programming language. In Ewing Lusk and Ross Overbeek, editors, *9th International Conference on Automated Deduction, Argonne, Illinois*, pages 61–80, Springer-Verlag LNCS 310, Berlin, May 1988.

[6] Jean-Yves Girard. Une extension de l'interpretation de Gödel a l'analyse, et son application a l'elimination des coupures dans l'analyse et la theorie des types. In J. E. Fenstad, editor, *Proceedings of the Second Scandinavian Logic Symposium*, pages 63–92, North-Holland Publishing Co., Amsterdam, London, 1971.

[7] Warren D. Goldfarb. The undecidability of the second-order unification problem. *Theoretical Computer Science*, 13:225–230, 1981.

[8] John Hannan and Dale Miller. Uses of higher-order unification for implementing program transformers. In Robert A. Kowalski and Kenneth A. Bowen, editors, *Logic Programming: Proceedings of the Fifth International Conference and Symposium, Volume 2*, pages 942–959, MIT Press, Cambridge, Massachusetts, August 1988.

[9] Robert Harper. *Standard ML.* Technical Report ECS-LFCS-86-2, Laboratory for the Foundations of Computer Science, Edinburgh University, March 1986.

[20] Preliminary work on this appears in [4].

[10] Robert Harper, Furio Honsell, and Gordon Plotkin. A framework for defining logics. In *Symposium on Logic in Computer Science*, pages 194–204, IEEE, June 1987.

[11] J. Roger Hindley and Jonathan P. Seldin. *Introduction to Combinators and λ-calculus*. Cambridge University Press, 1986.

[12] Gérard Huet. *Résolution d'équations dans des langages d'ordre* $1, 2, \ldots, \omega$. PhD thesis, Université Paris VII, September 1976.

[13] Gérard Huet. A unification algorithm for typed λ-calculus. *Theoretical Computer Science*, 1:27–57, 1975.

[14] Gérard Huet and Bernard Lang. Proving and applying program transformations expressed with second-order patterns. *Acta Informatica*, 11:31–55, 1978.

[15] Dale Miller, Gopalan Nadathur, Frank Pfenning, and Andre Scedrov. Uniform proofs as a foundation for logic programming. *Journal of Pure and Applied Logic*, 1988. Submitted.

[16] Dale A. Miller. Unification under mixed prefixes. 1987. Unpublished manuscript.

[17] Robin Milner. A theory of type polymorphism in programming. *Journal of Computer and System Sciences*, 17:348–375, August 1978.

[18] Gopalan Nadathur and Dale Miller. An overview of λProlog. In Robert A. Kowalski and Kenneth A. Bowen, editors, *Logic Programming: Proceedings of the Fifth International Conference and Symposium, Volume 1*, pages 810–827, MIT Press, Cambridge, Massachusetts, August 1988.

[19] Lawrence C. Paulson. *The Representation of Logics in Higher-Order Logic*. Technical Report 113, University of Cambridge, Cambridge, England, August 1987.

[20] Frank Pfenning. *Elf: A Language for Logic Definition and Verified Meta-Programming*. Ergo Report 88–067, Carnegie Mellon University, Pittsburgh, Pennsylvania, October 1988.

[21] Frank Pfenning. Partial polymorphic type inference and higher-order unification. In *Proceedings of the 1988 ACM Conference on Lisp and Functional Programming*, ACM Press, July 1988.

[22] Frank Pfenning and Conal Elliott. Higher-order abstract syntax. In *Proceedings of the SIGPLAN '88 Symposium on Language Design and Implementation*, pages 199–208, ACM Press, June 1988. Available as Ergo Report 88–036.

[23] Garrel Pottinger. Proof of the normalization and Church-Rosser theorems for the typed λ-calculus. *Notre Dame Journal of Formal Logic*, 19(3):445–451, July 1978.

[24] David Pym. A unification algorithm for the logical framework. November 1988. Laboratory for Foundations of Computer Science, University of Edinburgh. To appear as LFCS report.

[25] John Reynolds. Towards a theory of type structure. In *Proc. Colloque sur la Programmation*, pages 408–425, Springer-Verlag LNCS 19, New York, 1974.

[26] Wayne Snyder. *Complete Sets of Transformations for General Unification*. PhD thesis, University of Pennsylvania, 1988.

[27] Wayne Snyder and Jean H. Gallier. Higher-order unification revisited: complete sets of transformations. *Journal of Symbolic Computation*, 1988. To appear in the special issue on unification.

[28] David A. Turner. Miranda: a non-strict functional lanugage with polymorphic types. In *Functional Programming Languages and Computer Architecture*, Springer-Verlag, Berlin, September 1985.

An Overview of LP, The Larch Prover

Stephen J. Garland and John V. Guttag *
Massachusetts Institute of Technology †

LP is a theorem prover, based on equational term-rewriting, for a fragment of first-order logic. It has been used to analyze formal specifications written in Larch [8], to reason about algorithms involving concurrency [6], and to establish the correctness of hardware designs [6, 18]. The intended applications of LP motivate several departures in features and design from other term-rewriting programs and, in particular, from its (by now distant) ancestor REVE [14].

- LP permits users to define theories by means of equations, induction schemes, and other (nonequational) rules of deduction.
- LP supports a variety of proof methods beyond those found in conventional equational term-rewriting. In our applications, conventional techniques such as normalization, completion [12, 17], and proof by consistency [15, 10, 11] are often inapplicable, awkward, or computationally too expensive. Other techniques, such as proofs by cases or traditional induction, often lead to better results.
- LP provides facilities that allow users to establish subgoals and lemmas during the proof of a theorem.
- LP is designed to work efficiently with large sets of large equations, without having to find convergent sets of rewrite rules. REVE was used primarily to investigate relatively small axiomatizations of algebraic varieties. LP has been used to verify circuit designs described by several hundred equations. Even if such sets of equations could be completed, the computation would be prohibitively expensive.

1 Style of use

To provide a context for the technical details of LP, we discuss first the style of use that LP is intended to support.

---

*Research supported in part by the Advanced Research Projects Agency of the Department of Defense, monitored by the Office of Naval Research under contract N00014-83-K-0125, by the National Science Foundation under grant CCR-8706652, and by NYNEX.

†Authors' address: MIT Laboratory for Computer Science, 545 Technology Square, Cambridge, MA 02139. E-mail: garland@lcs.mit.edu, guttag@lcs.mit.edu

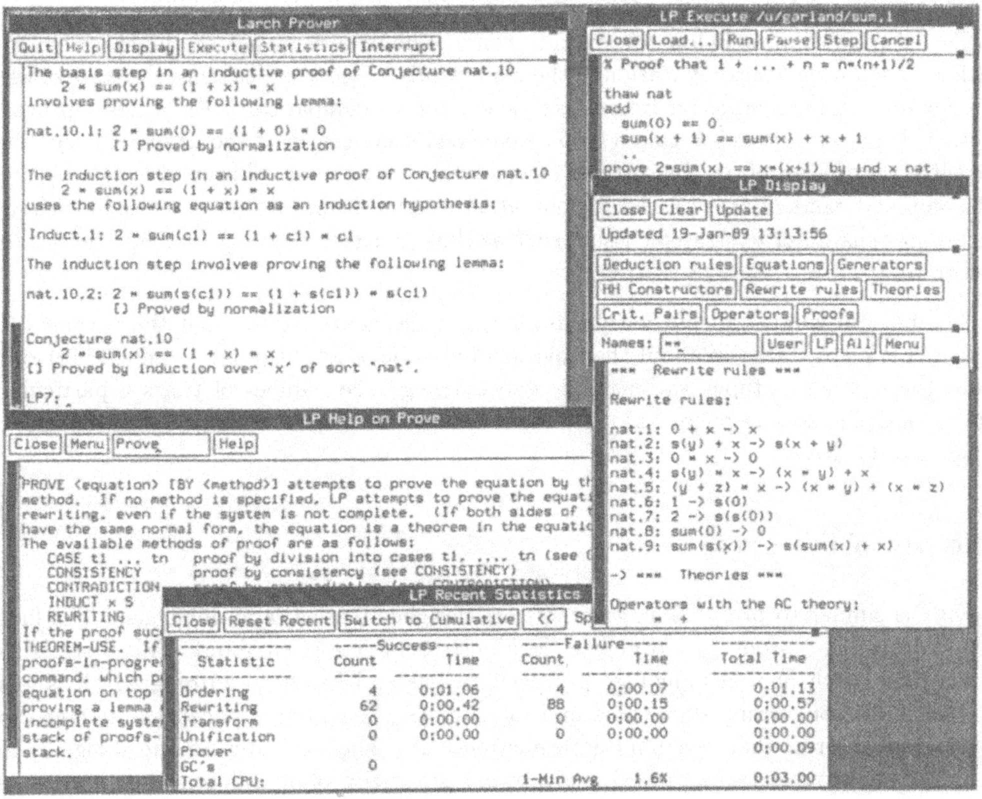

Figure 1: X Window interface to LP

User interface

LP has two user interfaces, an environment-independent "dumb-terminal" interface and an X Window interface. Figure 1 contains a snapshot of an LP session with the X Window interface. The windows can be sized, moved, iconified, selected, etc., in the usual way. Buttons at the top of the windows can be clicked with a mouse to initiate actions in the windows or to create other windows. Those windows in which users can enter text support emacs-style editing.

The upper left window is the main interaction window. Every LP session involves exactly one main interaction window, which records all commands that change the state of the prover. The amount of other information appearing in this window depends upon the tracing level selected by the user. Everything appearing in this window can be recorded automatically in a file.

The upper right window contains a sequence of LP commands to be executed in the main window. Commands can be loaded into this execute window from a file, or they can be entered directly and edited. The mouse can be used to select a command at which to begin execution. As LP executes commands in the window, it highlights the current command. LP also contains a log facility for recording sessions in ".lp" files that can be replayed in the execute window; commands in .lp files can invoke other .lp files.

The display window below the execute window holds a snapshot of the state of LP. Without disrupting the computation in the main window, users can capture the state of LP in a display window, either for immediate viewing or for comparison with states captured in other display windows. Within a display window, users can view the axioms in use, the current state of a proof, or the settings of various LP parameters. Parameters, set by the **set** command, control aspects of LP's operation ranging from incidental features such as the prompt character to strategic issues such as the procedure used to order equations into rewrite rules.

The bottom window displays asynchronously-updated statistics about the current LP session. LP keeps statistics about the time spent in various activities (e.g., unification) and about the number of times various actions occur (e.g., the number of times a particular rule is used for various purposes). The final window in Figure 1 supports an extensive online help facility.

The proof life cycle

Proving is similar to programming: proofs are designed, coded, and debugged. The first step in designing a proof is to formalize the objects being reasoned about. This is straightforward for Larch Shared Language [8] specifications and has been automated. It is less automatic for concurrent algorithms and for circuits. For large applications, it helps to identify subtheories that resemble data abstractions. The next step in the design is to formalize conjectures to be proved, e.g., formal properties of specifications or invariants to be maintained by concurrent algorithms. The last step in the design is to outline a structure for the proof, including key lemmas.

Designs for proofs translate into sequences of LP commands much as program designs translate into code in a programming language. Details of this translation are discussed later in this paper. LP permits users to preserve some of the modularity of designs when coding a proof. The **set name** command provides control over the names of user-entered and LP-generated objects. Names make it easy to refer to sets of objects. For example, the command **critical-pairs** *lemma.1* **with** *case* causes LP to compute all critical pairs between the rewrite rule named *lemma.1* and all hypotheses in a nested proof by cases.

Once part of a proof has been coded, LP is used to debug it. Proofs of interesting conjectures hardly ever go through the first time. Sometimes the conjecture is wrong. Sometimes the formalization is incorrect or incomplete. Sometimes the proof strategy is flawed or not detailed enough. When an attempted proof does fail, we use a variety of LP facilities (e.g., case analysis) to try to understand the problem. Because most proof attempts do fail, LP is designed to fail relatively quickly and to provide useful information when it does. It is not designed to find difficult proofs automatically. Unlike [2], it does not perform heuristic searches for a proof. Unlike [16], it does not allow users to describe complicated search strategies. Strategic decisions such as to try induction must appear as explicit LP commands. On the other hand, LP is more than a "proof checker" since it does not require proofs to be described in minute detail. In many respects, LP is described best as a "proof debugger."

One ramification of our emphasis on debugging proofs is that we try to use axiomatizations that simplify terms rather than merely normalize them. This is often leads to incomplete axiomatizations, thus increasing the need for the kinds of auxiliary proof mechanisms described in Section 3.

When debugging proofs, we frequently reformulate axioms and conjectures. When verifying a circuit, for example, we may discover that some important property does not follow from the description of the circuit (after all, discovering such things is the whole point of the process). When we change an axiomatization, we must recheck not only the conjecture whose proof uncovered a problem, but also any conjecture proved with the old axiomatization. LP has facilities that support such regression testing.

Particularly useful is a facility for having LP annotate proofs in .lp files with [ ]'s ("boxes") to indicate that some phase of the proof (e.g., the basis of an induction) has been completed. In box-checking mode, as LP executes an annotated (and edited) proof, it halts execution and prints an error message if a box appears at an inappropriate place or if it completes a proof and does not find a box on the next line of the .lp file. The check for omitted boxes is useful when changes in an axiomatization cause some step of a proof to go through with less help than expected. Without the check, LP might, for example, apply a tactic intended for the basis step in an induction to the proof of the induction step.

Figure 2 displays a .lp file for a trivial proof. Commands on the first four lines open a file in which to record all user input and LP output, retrieve a theory frozen during an earlier LP session, set the root of names for new objects derived by LP, and start a proof by induction. The box on the fifth line indicates that the user expects the basis case of the induction to go through without further interaction. The sixth line starts a subsidiary induction, and the boxes on the remaining lines indicate that the user expects LP to complete the remaining steps in the proof without further interaction.

set script *example*
thaw *boolnat*
set name *lemma*
prove $(x < y) \Rightarrow (x < (y + z))$ **by induction** y *nat*
    [ ] % basis step
    resume by induction x *nat*
        [ ] % basis step
        [ ] % induction step
    [ ] % induction over x for induction step
[ ] % induction over y

Figure 2: Sample .lp file containing proof commands

2 Defining theories

The basis for proofs in LP is a logical system or *theory*. This section contains an overview of the components of an LP theory, which consists of equations, rewrite rules, operator theories, inconsistencies, induction schemes, and deduction rules. Section 3 describes how these components interact with various proof techniques. Section 4 illustrates how both are used in a complete proof.

Equations

Equations are either entered directly by the user or produced by LP as consequences of the theory. The following sample LP commands enter some axioms about sets.

set name *set*
add
$not(in(x,\ empty))$
$in(x, single(y)) == x = y$
$in(x,\ y{+}{+}z) == in(x,y) \mid in(x,z)$ % ++ is union
$insert(x,\ y) == single(x){+}{+}y$
..

LP uses the logical symbol == for equality in an equation. Equations may also be entered using the connective → instead of == to constrain the way in which they may be ordered into rewrite rules. The symbol == is implicit in axioms such as $not(in(x, empty))$, which are shorthands for equations with right-hand side *true*. The symbol = is an operator.

Rewrite rules

LP orders equations into rewrite rules, as described below. LP keeps rewrite rules normalized with respect to one another (with some exceptions, as discussed in Section 3), and it uses these rules to normalize terms appearing in equations, conjectures, and deduction rules.

Despite the obvious advantages of convergent rewriting systems, we usually work with nonconvergent systems. Some interesting theories—in particular, all undecidable ones—cannot be described by convergent systems. Some can be described by convergent systems, but finding or using one is impractical. For example, we do not axiomatize booleans using exclusive-or as in [9, 21]. While this axiomatization provides a decision procedure for boolean identities, terminal forms often take a long time to compute; furthermore, terminal forms of nonidentities can be large and hard to read. Instead, we use incomplete axiomatizations that permit relatively quick computations of (usually) smaller normal forms.

Nonconvergence implies incompleteness. LP may reduce terms to different normal forms even though they are in the rewriting relation. Furthermore, the behavior of LP can be nonmonotonic, i.e., two terms that reduce to the same normal form with respect to a rewriting system may reduce to different normal forms when a new rule is added. LP's auxiliary proof mechanisms, described in Section 3, help compensate for these problems.

Operator theories

Assertions about operators are entered by commands such as **operator ac** +, which asserts that + is associative and commutative. Logically, these assertions are merely abbreviations for equations. Operationally, LP uses them in equational term-rewriting to perform all matching, unification, and completion modulo one of three operator theories: the empty theory, the commutative theory, or the associative-commutative theory. Assertions about different operators are combined using the algorithm described in [20].

Equational term-rewriting is a powerful facility. Not only does it increase the number of theories that LP can reason about, but it also reduces the number of axioms required to describe various theories, the number of reductions necessary to derive identities, and the need for certain kinds of user interaction, e.g., case analysis. Its main drawback is that equational matching can be much slower than conventional matching.

Inconsistencies

LP uses inconsistencies in proofs by cases, consistency, and contradiction. LP treats the equations *true* == *false* and $x == t$, where t is a term not involving the variable x, as inconsistent. Users can specify further inconsistencies by commands such as

add-inconsistency $single(x) == single(y)$

If an inconsistency arises during a proof by cases, the current case is deemed impossible (and therefore succeeds). If it arises during a proof by contradiction, that proof succeeds. If it arises during a proof by consistency, that proof fails.

Induction schemes

LP uses user-supplied induction schemes to generate subgoals to be proved for the basis and induction steps in proofs by traditional induction. The following sample LP command specifies that finite sets of natural numbers are generated by the empty set, singleton sets, and union:

add-generators $empty: \rightarrow set,\ single: nat \rightarrow set,\ ++ : set\ set \rightarrow set$

Deduction rules

LP uses deduction rules to deduce new equations from existing equations and rewrite rules. Deduction rules in LP generalize cancellation laws of the form $z + x = z + y \Rightarrow x = y$ used by Stickel [19] in reasoning about rings. The LP command

add-deduction-rule when $z + x == z + y$ **yield** $x == y$

specifies the cancellation law for addition. More powerful rules, such as

add-deduction-rule when forall $z : in(z, x) == in(z, y)$ **yield** $x == y$

for set extensionality, allow explicit universal quantification of variables occurring in the hypothesis of the implication. Logically, these deduction rules are equivalent to universal existential axioms such as $\forall x \forall y [\forall z (z \in x \Leftrightarrow z \in y) \Rightarrow x = y]$. Some deduction rules, such as the built-in &-splitting law

when $x \,\&\, y == true$ **yield** $x == true,\ y == true$

or the cancellation law for addition, serve solely to improve performance or to reduce user interaction. Others, such as set extensionality, serve to increase the strength of theories that can be formulated in LP; they enable LP to deduce equations such as $x == x ++ x$ from equations such as $in(y, x) == in(y, x ++ x)$. LP automatically applies deduction rules to equations and rewrite rules whenever they are normalized. The extended example in Section 4 illustrates the logical power of deduction rules, as well as the benefits of applying them automatically to case and induction hypotheses in a proof.

Built-in operators and theories

LP treats several operators (e.g., =) as having prescribed operator theories (e.g., commutativity), rewrite rules (e.g., $x = x \rightarrow true$), and deduction rules (e.g., **when** $x = y == true$ **yield** $x == y$). Built-in theories serve primarily to increase efficiency and to encourage uniform use of certain operators. Their use can be controlled by the **set** command.

An important special case is the built-in operator *if*. The built-in *if*-distribution metarules

$$g(\ldots, if(x, y, z), \ldots) \rightarrow if(x, g(\ldots, y, \ldots), g(\ldots, z, \ldots))$$

apply to all operators g other than *if*, saving users the trouble of entering separate distribution rules for each operator. Although these rules are equivalent to rewrite rules, the *if*-simplification metarule

$$if(t_1, t_2[t_1], t_3[t_1]) \rightarrow if(t_1, t_2[true], t_3[false])$$

(in which t_1 occurs as a subterm of t_2 or t_3) allows LP to perform reductions that cannot be expressed by rewrite rules. These rules are similar to ones in Affirm [7].

Ordering equations into rewrite rules

Some of LP's inference mechanisms work directly with equations. Most, however, require that equations be oriented into rewrite rules. It is usually essential that the rewriting relation be uniformly terminating. LP's ordering mechanisms can partitioned into three classes:

- Mechanisms that give the user complete control over how equations are oriented, but provide no guarantees about termination.
- Registered orderings, which yield simplification orderings [3, 13] and guarantee termination of sets of rules that do not contain commutative or associative-commutative operators. These orderings are derived directly from those in REVE [4] and are particularly easy to use. When LP cannot order an equation using the current registry, it suggests minimal extensions to that registry; users need only select one of these suggestions.
- A polynomial ordering [1], which guarantees termination in all cases but requires considerable assistance from the user.

Most users rely on LP's simplification orderings to order all equations, even when they contain commutative or associative-commutative operators.[1] To our knowledge, no user working on an application has encountered difficulties because one of LP's simplification orderings produced a nonterminating set of rules.

[1]LP will not attempt proofs by consistency unless the rules are guaranteed to terminate. Other proof methods do not require termination.

3 Proof techniques

LP provides mechanisms for proving theorems using both forward and backward inference. Forward inferences produce consequences from a theory; backward inferences produce lemmas that can be proved to establish a conjecture. LP applies some forward inferences automatically each time new information is added to a theory; users invoke others explicitly. Backward inferences are always invoked explicitly by the user.

Forward inference

LP provides four primary methods of forward inference: reduction to normal form, critical pairs and completion, application and generation of deduction rules, and instantiation.

Reduction to normal form

We usually think of reduction to normal form as a method of backward inference used to simplify a conjecture. If the conjecture reduces to an identity, it is a theorem.

Reduction to normal form is also a method of forward inference. It can be used to derive new consequences from a theory, e.g., upon entering a new case in a proof by cases. LP automatically normalizes rewrite rules, equations, and hypotheses of deduction rules whenever it adds a new rewrite rule to the theory. If an equation or rewrite rule reduces to an identity, it is discarded. If the hypothesis of a deduction rule reduces to an identity, the deduction rule is replaced by the equations in its conclusions.

Users can "immunize" equations, rewrite rules, and deduction rules to protect them from being reduced, both to enhance the performance of LP and to aid in proofs. When sets of rules are combined, the combined set is sometimes a conservative extension of one or both of the original sets; in this case, immunization improves performance (sometimes considerably) by preventing LP from trying to reduce rules that the user believes to be irreducible. Immunization can also assist in proofs by preserving rules that would otherwise normalize to identities, thereby making them available for critical pair computations.

Critical pairs and completion

The computation of critical pairs and the Knuth-Bendix completion procedure [17] are used to derive consequences from nonconfluent rewriting systems. For reasons cited before, we rarely try to complete our theories.

The LP command **critical-pairs** $rules_1$ **with** $rules_2$ finds new critical pairs between the rewrite rules named by $rules_1$ and those named by $rules_2$. Experience has shown that it pays to have LP compute critical pairs between smaller rules first. LP disposes of any nontrivial critical pairs (i.e., nonidentities) according to the current setting of **theorem-use**. If **theorem-use is order-equations-into-rules**, LP orders the critical pairs as they are generated and attempts to prove the current conjecture by normalization. If that succeeds, the critical pair computation is terminated.

Application and generation of deduction rules

LP uses deduction rules to perform automatic forward inferences whenever equations and rewrite rules are normalized. Deduction rules can also be invoked explicitly by the command **apply** *deduction-rules* **to** *equations*, e.g., to apply newly added deduction rules to existing equations.

Operationally, LP applies a deduction rule to an equation if there is a substitution that matches the rule's hypothesis to the equation and that maps each variable bound in the hypothesis to a variable that does not occur in any other term in the range of the substitution. The result of applying a deduction rule is the set of equations obtained by instantiating each of its conclusions by the substitution(s) that matched its hypothesis (strengthened by substituting fresh variables, if necessary, for variables that occur in a conclusion, but do not occur unbound in the hypothesis). For example, the deduction rule for set extensionality applies to the equation $in(x, f(y)) == in(x, g(y))$, yielding $f(y) == g(y)$, but does not apply to $in(x, f(x)) == in(x, g(x))$.

Another method of forward inference is provided by the command **set deduction-generation on**, which causes LP to automatically add a deduction rule of the form **when** $t_1 == \mathit{true}$ **yield** $t_2 == \mathit{true}$ whenever it adds a rewrite rule of the form $not(t_1)|t_2 \rightarrow \mathit{true}$ in which t_1 contains a free variable. This forward inference provides an alternative to critical pairing for producing consequences of implications. For example, when LP adds the equation $f(x) \Rightarrow g(x) == \mathit{true}$, deduction generation automatically produces the deduction rule **when** $f(x) == \mathit{true}$ **yield** $g(x) == \mathit{true}$; if the user later adds the equation $f(a) == \mathit{true}$, this deduction rule yields the equation $g(a) == \mathit{true}$. Without deduction generation, LP would have merely added and ordered the two equations. Critical pairing could then have been used to generate the equation $\mathit{true} \Rightarrow g(a) == \mathit{true}$, which would normalize to $g(a) == \mathit{true}$.

Instantiation

LP provides explicit instantiation as another alternative to the computation of critical pairs. The command **instantiate** *variable* **by** *term* [, *variable* **by** *term*]* **in** *names* substitutes (simultaneously) the specified terms for variables in the named equations, rewrite rules, and deduction rules. It then normalizes the results and disposes of the resulting equations and deduction rules, if they differ and are not identities, according to the current **theorem-use**. When instantiating a rewrite rule, LP does not use that rule in normalizing the result until it has been normalized with respect to the rest of the theory.

Backward inference

LP provides four primary methods of backward inference (in addition to reduction to normal form): proof by cases, induction, consistency, and contradiction. In each method, LP generates subgoals, i.e., lemmas to be proved (sometimes assuming additional axioms).

Proof by cases

Conjectures can often be simplified by dividing a proof into cases. When a conjecture reduces to an identity in all cases, it becomes a theorem. Proofs by cases have two primary uses. When the conjecture is a theorem, a proof by cases can be used to circumvent a lack of confluence in the rewrite rules. When the conjecture is not a theorem, a proof by cases can be used to simplify the conjecture to make it easier to understand where the proof has failed.

The command **prove** e **by cases** $t_1 \ldots t_n$ directs LP to prove an equation e by division into cases $t_1, \ldots, t_n$ (or into two cases, t_1 and $not(t_1)$, if $n = 1$). First, LP attempts to prove $t_1 | \ldots | t_n == true$. If that succeeds, then, for each i from 1 to n, LP substitutes new constants for the variables of t_i in both t_i and e to form t_i' and e', adds the case hypothesis $t_i' == true$ to the equations (with a name beginning with *Case*), and attempts to prove e'. In each case of a proof by cases, LP first adds the case hypothesis without using it to reduce the other rewriting rules. Only if this action fails to reduce e' to an identity does LP use the case hypothesis to reduce the other rules. If an inconsistency results from adding a case hypothesis, LP accepts e' as vacuously true in that case.

Proofs by cases are often used to simplify implications, terms involving *if*, and terms involving repeated boolean subexpressions. For example, given two axioms $a \Rightarrow b \rightarrow true$ and $b \Rightarrow c \rightarrow true$, the command **prove** $a \Rightarrow c$ **by case** a first adds $a \rightarrow true$ as a case hypothesis; this new rule helps normalize the first axiom to $b \rightarrow true$, which in turn helps normalize the second to $c \rightarrow true$, which helps normalize the conjecture to *true*. When LP adds $not(a) \rightarrow true$ as the hypothesis of the other case, a built-in deduction rule yields $a \rightarrow false$, which helps normalize the conjecture to *true*, thereby completing the proof of the conjecture.

Proof by induction

LP supports two methods of proof by induction, traditional and inductionless induction. For proofs by traditional induction, users supply induction schemes that list generators for sorts over which they want to induct. This information does not affect the rewriting theory, but does define the inductive theory as described in [5].

The command **prove** e **by induction** x S directs LP to prove the equation e by traditional induction on the variable x using an induction scheme named S. LP first generates lemmas to prove as the basis of the induction by substituting each basis generator of S for x in e (basis generators are those with no arguments of the sort being inducted over; new variables are used as their arguments in the lemmas for the basis case). When LP has proved these lemmas, it generates equations (with names beginning with *Induct*) to serve as the induction hypothesis by substituting new constants for x in e. LP then generates lemmas to prove in the induction step by substituting terms involving these new constants, and headed by the nonbasis generators of S, for x in e.

LP also allows multilevel induction, as described in [5].

Proof by consistency (inductionless induction)

The command **prove** *e* **by consistency** directs LP to prove an equation *e* by consistency or "inductionless induction" [15, 10, 11]. For such proofs to be sound, the theory must satisfy the Huet-Hullot principle of definition. A proof by consistency proceeds by adding *e* to the theory and running the completion procedure. If that procedure terminates without generating an inconsistency, *e* is an inductive consequence of the theory; if it terminates with an inconsistency, *e* is not; if it does not terminate, *e* may or may not be an inductive consequence.

LP's implementation of proof by consistency has proved useful largely for reasoning about algebraic varieties. For reasons discussed in [5], it has not been widely used in other applications.

Proof by contradiction

The command **prove** *e* **by contradiction** directs LP to prove an equation *e* by contradiction. It does this by substituting new constants for the variables in *e* to form an equation e', adding the equation e' == *false* (with a name beginning with *Contra*), and attempting to find an inconsistency, e.g., a proof of *true* == *false*. If this succeeds, then *e* is a theorem.

4 Extended example

The following transcript shows how LP can be used to prove a simple theorem. The transcript was produced with a high tracing level (to show the effects of applying deduction rules and computing critical pairs) and then condensed manually. User input is underlined.

LP1: thaw finiteSets

LP2: display equal set exten

Rewrite rules:

```
equal.1:  x = x -> true
set.1:    in(x, empty) -> false
set.2:    in(x, single(y)) -> x = y
set.3:    in(x, y ++ z) -> in(x, y) | in(x, z)
set.4:    insert(x, y) -> single(x) ++ y
```

Deduction rules:

```
equal.2:  when x = y == true yield x == y
exten.1:  when forall x: in(x, y) == in(x, z) yield y == z
```

Operators with the AC theory: & | ++
Operators with the commutative theory: =

Generators for sort 'finiteSet':

```
Basis:       empty: -> finiteSet
             single: nat -> finiteSet
Inductive:   ++: finiteSet finiteSet -> finiteSet
```

LP3: set name thm

LP4: prove w ++ w == w

Conjecture thm.1, w == w ++ w, is not provable using the current partially completed system.

Proof of Conjecture thm.1 suspended.

LP5: instantiate y by w, z by w ++ w in exten

Instantiated deduction rule exten.1 to deduction rule exten.1.1
    when forall x: in(x, w) == in(x, w ++ w) yield w == w ++ w
which was normalized to equation exten.1.1.1, w == w ++ w, which was ordered into rewrite rule exten.1.1.1, w ++ w -> w

Conjecture thm.1, w == w ++ w
[] Proved by rewriting.

LP6: prove in(x, y) => insert(x, y) = y by induction y finiteSet

The basis step in an inductive proof of Conjecture thm.2 involves proving 2 lemmas:
    thm.2.1: in(x, empty) => (insert(x, empty) = empty) == true
             [] Proved by normalization
    thm.2.2: in(x, single(vi2)) => (insert(x, single(vi2)) = single(vi2)) == true

Proof of Lemma thm.2.2 suspended.

LP7: resume by case vi2 = x

Case.1.1, c1 = c2 == true, involves proving Lemma thm.2.2.1
    in(c2, single(c1)) => (insert(c2, single(c1)) = single(c1)) == true

Equation Case.1.1, c1 = c2 == true, was transformed by deduction rule equal.2 into equation Case.1.1.1, c1 == c2, which was ordered into rewrite rule Case 1.1.1, c1 -> c2

Lemma thm.2.2.1 in the proof by cases of Lemma thm.2.2
[] Proved by rewriting (with unreduced rules).

Case.1.2, not(c1 = c2) -> true, involves proving Lemma thm.2.2.1
    in(c2, single(c1)) => (insert(c2, single(c1)) = single(c1)) == true
[] Proved by rewriting (with unreduced rules).

Lemma thm.2.2 for the basis step in the proof of Conjecture thm.2
[] Proved by cases (vi2 = x) | not(vi2 = x)

The induction step in an inductive proof of Conjecture thm.2 uses the following equations as induction hypotheses:
    Induct.1: in(x, c1) => (insert(x, c1) = c1) == true
    Induct.2: in(x, c2) => (insert(x, c2) = c2) == true

The induction step involves proving Lemma thm.2.3
    in(x, c1 ++ c2) => (insert(x, c1 ++ c2) = (c1 ++ c2)) == true
which reduces to the equation
    ((c1 ++ c2) = (c1 ++ c2 ++ single(x))) | not(in(x, c1) | in(x, c2)) == true

Proof of Lemma thm.2.3 suspended.

LP8: resume by cases in(x, c1) in(x, c2) not(in(x, c1 ++ c2))

Case.2.1, in(c3, c1) −> true, involves proving Lemma thm.2.3.1
    in(c3, c1 ++ c2) => (insert(c3, c1 ++ c2) = (c1 ++ c2)) == true
which reduces to the equation (c1 ++ c2) = (c1 ++ c2 ++ single(c3)) == true

Proof of Lemma thm.2.3.1 suspended.

LP9: critical-pairs case with induct

A critical pair between rule Case.2.1 and rule Induct.1 is
    thm.3: ((c1 ++ single(c3)) = c1) | not(true) == true
which normalized to equation thm.3, (c1 ++ single(c3)) = c1 == true, which was transformed by deduction rule equal.2 into equation thm.3.1, c1 ++ single(c3) == c1, which was ordered into rewrite rule thm.3.1, c1 ++ single(c3) −> c1

Lemma thm.2.3.1 in the proof by cases of Lemma thm.2.3
[ ] Proved by rewriting.

Critical-pair computation abandoned because a theorem has been proved.

Case.2.2, in(c3, c2) −> true, involves proving Lemma thm.2.3.1
    in(c3, c1 ++ c2) => (insert(c3, c1 ++ c2) = (c1 ++ c2)) == true
which reduces to the equation (c1 ++ c2) = (c1 ++ c2 ++ single(c3)) == true

Proof of Lemma thm.2.3.1 suspended.

LP10: critical-pairs case with induct

... Case.2.2 concludes in the same fashion as Case.2.1 ...

Case.2.3, not(in(c3, c1 ++ c2)) −> true, involves proving Lemma thm.2.3.1
    in(c3, c1 ++ c2) => (insert(c3, c1 ++ c2) = (c1 ++ c2)) == true
[ ] Proved by rewriting (with unreduced rules).

Lemma thm.1.3 for the induction step in the proof of Conjecture thm.1
[ ] Proved by cases in(x, c1) | in(x, c2) | not(in(x, c1 ++ c2))

Conjecture thm.1: in(x, y) => (insert(x, y) = y) −> true
[ ] Proved by induction over 'y' of sort 'finiteSet'.

5 Current development

LP is written in CLU and runs on DEC VAXes as well as Sun and HP workstations. It has been in relatively heavy use at several sites for the past year and a half. During that time LP has changed dramatically, primarily in response to the needs of its users. These needs have been for more logical power, improved performance, and more user amenities. Thus far, we have been able to add logical power without hurting performance. In fact, some increases in logical power, e.g., deduction rules, have led to significant performance improvements. User amenities, e.g., checking for [ ]'s, may increase slightly the machine time required to run a successful proof, but reduce greatly the time (both human and machine) required to debug a proof.

LP continues to change in response to user needs. Front-ends are being developed, e.g., to assist in checking Larch specifications. Front-ends assist users with formalizations and syntactic details; they also lessen the need for explicit sort declarations within LP. Features for proving and using first-order formulas are also being developed, as are extensions of the notions of critical pairs and completion to encompass deduction rules. Our primary concern is to preserve the basic style and efficiency of proofs in LP.

Although LP is much faster than its ancestor REVE, performance continues to be an issue. Each increase in speed tempts LP's users to try larger examples. Frequently, these examples suggest other desirable user amenities or further opportunities for improvements in performance. The sample proof in Section 4 took five seconds; more ambitious proofs, such as the transparency of the DEC Firefly cache [18], take several hours. A major goal for LP is to reduce as much as possible the costs of developing and executing such ambitious proofs.

6 Acknowledgements

By using LP in their own work, James Horning, Andres Modet, James Saxe, and Jørgen Staunstrup have helped greatly with the design and evaluation of LP.

References

[1] Ben Cherifa, A. and Lescanne, P. "An Actual Implementation of a Procedure that Mechanically Proves Termination of Rewriting Systems Based on Inequalities between Polynomial Interpretations," *Proc. 8th Int. Conf. on Automated Deduction*, Oxford, England, **LNCS 230**, July 1986, 42–51.

[2] Boyer, R. S. and Moore, J. S. **A Computational Logic**, Academic Press, 1979.

[3] Dershowitz, N. "Orderings for Term-Rewriting Systems," *Theoretical Computer Science* 17:3 (March 1982), 279–301.

[4] Detlefs, D. and Forgaard, R. "A Procedure for Automatically Proving the Termination of a Set of Rewrite Rules," *Proc. 1st Int. Conf. on Rewriting Techniques and Applications*, Dijon, France, **LNCS 202**, May 1985, 255–270.

[5] Garland, S. J. and Guttag, J. V. "Inductive Methods for Reasoning about Abstract Data Types," *Proc. 15th ACM Conf. on Principles of Programming Languages*, San Diego, California, January 1988, 219–228.

[6] Garland, S. J., Guttag, J. V. and Staunstrup, J. "Verification of VLSI Circuits Using LP," *Proc. Int. Workshop on Design for Behavioral Verification*, July, 1988.

[7] Guttag, J.V., Horowitz, E. and Musser, D. R. "Abstract Data Types and Software Validation," *CACM* 21:12 (December 1978), 1048–1064.

[8] Guttag, J. V. and Horning, J. J. "Report on the Larch Shared Language" and "A Larch Shared Language Handbook," *Sci. Computer Programming* 6:2 (March 1986), 103–157.

[9] Hsiang, J. and Dershowitz, N. "Rewrite Methods for Clausal and Nonclausal Theorem Proving," *Proc. 10th EATCS Int. Colloquium on Automata, Languages, and Programming*, Barcelona, Spain, **LNCS 154**, July 1983, 331–346.

[10] Huet, G. and Hullot, J. M. "Proofs by Induction in Equational Theories with Constructors," *J. Computer and System Sciences* 25:2 (October 1982), 239–266.

[11] Kapur, D. and Musser, D. R. "Proof by Consistency," *Artificial Intelligence* 31 (1987), 125–157.

[12] Knuth, D. E. and Bendix, P. B. "Simple Word Problems in Universal Algebras," in *Computational Problems in Abstract Algebra*, J. Leech (ed.), Pergamon Press, Oxford, England, 1969, 263–297.

[13] Lescanne, P. "Uniform Termination of Term Rewriting Systems: Recursive Decomposition Ordering with Status," *Proc. 6th Colloquium on Trees in Algebra and Programming*, Bordeaux, France, Cambridge University Press, March 1984.

[14] Lescanne, P. "REVE: A Rewrite Rule Laboratory," *Proc. 8th Int. Conference on Automated Deduction*, Oxford, England, **LNCS 230**, July 1986, 695–696.

[15] Musser, D. R. "On Proving Inductive Properties of Abstract Data Types," *Proc. 7th ACM Conference on Principles of Programming Languages*, Las Vegas, Nevada, January 1980, 154–162.

[16] Paulson, L. C. "The Foundation of a Generic Theorem Prover," Technical Report No. 130, University of Cambridge Computer Laboratory, March 1988.

[17] Peterson, G. L. and Stickel, M. E. "Complete Sets of Reductions for Some Equational Theories," *JACM* 28:2 (Apr. 1981), 233–264.

[18] Saxe, J. Private communication.

[19] Stickel, M. E. "A Case Study of Theorem Proving by the Knuth-Bendix Method: Discovering that $x^3 = x$ Implies Ring Commutativity," *Proc. 7th Int. Conf. on Automated Deduction*, Napa, California, **LNCS 170**, May 1984, 248–258.

[20] Yelick, K. A. "Unification in Combinations of Collapse-Free Regular Theories," *J. Symbolic Computation* 3 (1987), 153–181.

[21] Zhegalkin, I.I. "On a technique of evaluation of propositions in symbolic logic," *Matematisheskii Sbornik* 34:1 (1927), 9–27.

GRAPH GRAMMARS, A NEW PARADIGM FOR IMPLEMENTING VISUAL LANGUAGES

Herbert Göttler

Lehrstuhl für Programmiersprachen
Universität Erlangen-Nürnberg
Martensstr. 3
D-8520 Erlangen
F. R. Germany

Abstract: This paper is a report on an ongoing work which started in 1981 and is aiming at a general method which would help to considerably reduce the time necessary to develop a syntax-directed editor for any given diagram technique. The main idea behind the approach is to represent diagrams by (formal) graphs whose nodes are enriched with attributes. Then, any manipulation of a diagram (typically the insertion of an arrow, a box, text, coloring, etc.) can be expressed in terms of the manipulation of its underlying attributed representation graph. The formal description of the manipulation is done by programmed attributed graph grammars.

Keywords: CR classification system (1987): D.1 Programming Techniques, D.2.1 Requirements/Specification, D.2.2 Tools and Techniques, D.2.6 Programming Environments, E.1 Data Structures, F.4.2 Grammars and other Rewriting Systems;
additional: Graph Grammars, Syntax-directed Editors, Diagram Languages.

1 INTRODUCTION / MOTIVATION

This contribution makes propaganda for two issues: for *visual languages* and for *graph grammars*. It might not be necessary to beat a big drum for the first topic because it seems to get increasing attention lately. However, much more advertising has to be done for the second and, hopefully, the rest of the paper will proof that it deserves it. If a software engineer accepts the message which is supposed to be conveyed by this report on an ongoing work at the University of Erlangen-Nürnberg he/she can get a considerable amount of help for his/her work, especially since the proposed ideas can be transferred to other areas of application.

Software development should be used as a synonym to *model development*. When software engineers are assigned the task to implement a *system* ("... a collection of objects which have a certain influence on each other ...") they have to communicate with their customers to find out what the problem is. At first they should develop a *model as an abstraction* of the system which is to be implemented. Then, in terms of the model, the relevant facts can be described and the requirements specification can be settled.

At this crucial and for the success of the project very sensitive phase of the software life-cycle where especially on the customers' side people might be involved who are not trained in formal, mathematical notations the question arises what kind of sytax to use to state the facts. A piece of software almost never is a stand-alone product. It must be seen within the context of, say, a department whose people have to work with the program. So the software developer has to scrutinize what is going on in this particular department.

"A picture is worth a thousand words" is an old saying which should be considered by the software developer. It expresses that an issue under discussion becomes clear or is at least better understood if a picture, a *diagram*, related to the subject is drawn. Now, if the task is to automize some office procedures then, with the help of the members of the department, a diagrammatical paper model of the department should be developed which could aid as a reference in case of questions. Within and by the terms of this first diagram model the objects and functions to program can be identified. The rest of the work a software developer has to accomplish is the *transformation of the model* into an *executable model*, the (final) program. Yes, in this sense, the machine code has to be understood as the final model of the system which was to be implemented!

The morphology of the models of the system under discussion will change considerably, perhaps from a diagrammatic syntax like SADT ("Structured Analysis and Design Technique", see [11]) or Petri-nets, via a module specification in the IPSEN-style (see [1] or [12]) to the source code of a high level programming language and finally to the bit strings of the machine code. In the course of reading this paper the dogmatic sounding point of view will become clear that, at any time, these models *can be modelled* (i.e. represented) as (mathematical) *graphs*.

Thus, graphs are a very general vehicle to *model concepts*, they allow to *represent ideas*. In this sense, the (real) world of human thinking consists of objects - modelled by the nodes of a graph - and the relations between the objects become the edges.

However, a graph by itself is a static object. If the objects of the real world are subject to a change according to a prescribed procedure this change can be modelled, too, in terms of graphs (in general, see [10], "*conceptual modelling by graphs*"). This is the place where graph grammars come in: they are a means to describe how the structure of a graph can be changed.

The ideas mentioned so far in a rather vague manner are exemplified by an application to the field of developing syntax directed editors for diagram languages.

2 VISUAL LANGUAGES

In software engineering, a new notion became more and more important within the last years: "*Visual programming*". Its meaning is not very precise but it stands for any effort to use graphic in the process of software development. Why is it useful to use *visual languages* or *diagram languages* as we want to call them in the sequel? Skimming through the proceedings of the CHI-conferences ("Human Factors in Computing Systems") one can find many strong pro-arguments for the use of diagrams: While text has an intrinsic linear order (although it uses a second dimension on a sheet of paper or a third if written in a book) diagrams allow some kind of random-access to the information. Text looks uniform, monotonous; there are just a few means (underlining, different fonts, etc.) to make important information conspicuous. A diagram language is more flexible; some kind of preprocessing of the information is possible: important information can be highlighted by adequate shape, color, spatial arrangement, etc. And there are even more arguments favoring the use of diagrams: Human memory seems to be of a graphical nature; it is easier to remember things which were seen than to remember things which were heard ('acoustic text'). The part of the human brain which is the 'hardware' for visual thinking is also responsible for creative abilities. The use of diagrams seems to stimulate creativity.

However, the use of diagrams can be of disadvantage sometimes, it can be misleading, deceiving. Diagrams tend to make the observer believe to 'see the facts'. But, strictly speaking, this is not the fault of graphic! Texts are subject to misinterpretations, too. There is no evidence that it might be impossible to work with diagrams as precisely as with a mathematical text-formula for example. What has to be done is the provision of an exact syntactical and semantical description of the diagram language.

Many methods especially in the field of commercial computer applications used for program construction are heavily based on diagrams. Working with diagrams is a very time consuming affair if no tool is available. An ordinary graphical system for doing the drawing does not suffice since the results of such systems are amorphous collections of pixels to be lightened. There is the same lack of structure as in an ordinary source program - which is just a string - before it is processed by the syntactical and semantical phase of a compiler. Developing tools for diagram languages is as laborious as compiler writing and therefore it is worth the while to design a system which aids the fast production of the tools in a manner similar to compiler generators which help to develop a (perhaps less efficient) compiler for a specific language more quickly than the traditional way of writing it by hand.

There is an abundance of diagram languages and the question is how they can be treated in a uniform manner. Fig. 1 shows a way how the *concrete syntax* can be represented by an *abstract* one. What is abstract about the right side of fig. 1? There are just the diagrams which are well-known as the circle/arrow-representation of (mathematical) graphs. The adjective "abstract" serves just for the purpose to stress that a unification has been taken place in the sense that the graphical objects of the diagrams on the left-hand side are now the labels of the nodes of the graphs on the right-hand side. The key idea is *to represent* (to model) each graphical object by a node. What is considered to be a graphical object is somewhat arbitrary: A box could be a single object or it might be considered to consist of four lines. An extreme point of view is to take each pixel to be an object. But let's proceed step by step: The diagram on the left side in fig. 1a is the diagrammatical representation of a graph. Therefore it could be considered to be its own representation. Whether this is correct or not will turn out soon. In fig. 1b the two squares are conspicuous. At a first glance the right side might be a useful representation. But in fact it is not. Much better is fig. 1c where the arrow is also modelled explicitly. Fig. 1d shows a more complicated diagram (SADT-example). The arrows on the right side which bear no label in order not to overload the representing diagram can be considered to express the relation "connected with". One fact should be stressed: Of course, it is not possible to reconstruct the diagrams on the left side just out of the representation graphs. Additional information - say, for the length of the sides of the boxes - is necessary in form of attributes which are attached to the nodes representing the graphical objects. Such an information is relevant only for the layout; for a further processing of the information, in most cases, the representation graph suffices. The problem how to edit diagrams has now turned into the problem how to construct *attributed representation graphs.* Actually, these are two problems, first, how to construct the graph, discussed in section 3 and 5 and, secondly, how to handle the attributes, discussed in section 4.

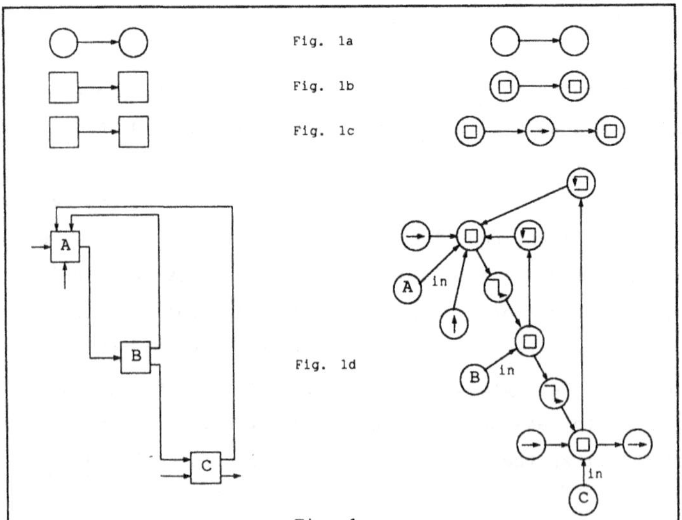

Fig. 1

3 GRAPH GRAMMARS

How can graphs which are subject to certain rules be automatically generated? For this purpose the theory of graph grammars has been developed within the last two decades. One of its main features is the design of prescriptions which state *how a graph can be derived from another graph by applying a production rule.* There are two major approaches to graph rewriting, an algebraic and a set theoretic one, see [10]. The one shown in this paper is a modification of [9]. It is, how else could it be, described in a diagram language. The general idea is demonstrated by an example. Definitions can be found in [3] and more details in [6].

We assume finite labelled directed graphs to be well-known mathematical concepts and confine us to the description of how those graphs can be 'manipulated'. So we get to the inherent problem of 'graph-rewriting'. No problem causes to say: "Substitute in the string 'abc' the character 'b' by 'de'!", which results in 'adec'. If one wishes to substitute the node 1 in fig. 2 by the graph of fig. 3, there is the problem to describe explicitly how the two graphs are to be connected (cf. fig. 4).

A *graph production* consists of three parts: (1) A graph which is to be substituted (e.g. node 1 in fig. 2), (2) a graph which has to be inserted (e.g. the graph of fig. 3) and (3) some prescription how the 'new' graph is to be attached to the rest of the old graph. We use the sign Y as a connector (or separator), cf. fig. 5, to make the three mentioned parts of a graph production visible. Basically, any graph (connected or not) can be used as a graph production. Its effect is depending on the way how the nodes are assigned to the three parts of Y. There is only one restriction: The placement of the nodes may not be such that there is an edge between a node of the left and right side.

We now *apply* the graph production r of fig. 5 to node 1 in fig. 2 as follows:

(1) If you find, in the given graph g, the left side of r, thus a node labelled with a, construct a new graph g', resulting from g, by taking out this particular node (together with the adjacent edges) first.

(2) Insert into the partially constructed g' the right side of r.

(3) Consider in r the edges cut by the right hand of Y independently from each other (In our example they are labelled with k, m and n, respectively.) Then r requires the following actions to take place:

    (3k) Go back to g (fig. 2) and look for a node labelled with d which is connected with the node 1 to be taken out in the following manner: Start at 1 and go in the reverse direction of an (incoming) edge labelled with i to a node labelled with c. From this node advance, if possible, in the direction of a j-labelled edge to a node labelled with d. Thus we get via node

2 to node 4. Now draw an edge labelled with k from 4 to the f-labelled node in the partially constructed graph g'.

(3m) Go back to g and look for a node labelled with b which is connected with node 1 via a j-labelled edge starting in 1. This yields node 3. To this node we draw an edge labelled with m from the f-labelled node in the intermediate construction for g'.

(3n) Analogously to (3m) we get node 3. As fig. 5 says, we have to draw to node 3 from the e-labelled node 6 in the partially constructed g' an edge labelled with n.

This completes the construction of g'. The result is shown in fig. 6 and is denoted by *appl(r,g,1)* (the *application of r to g in 1*).

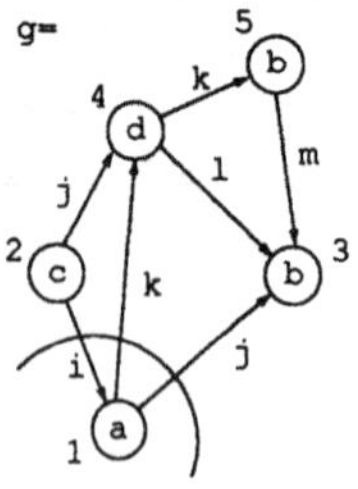

Fig. 2

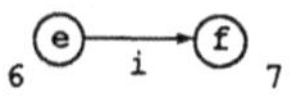

Fig. 3

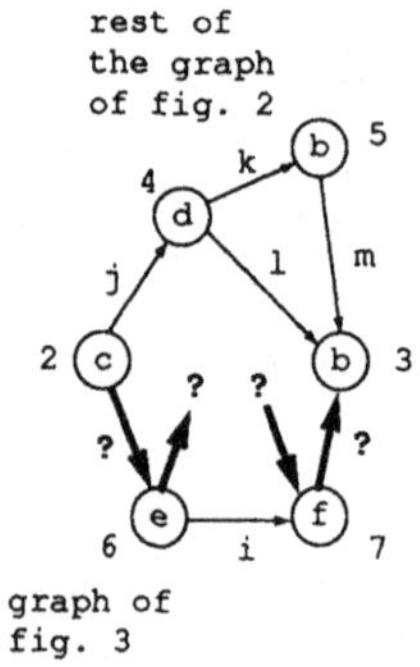

Fig. 4

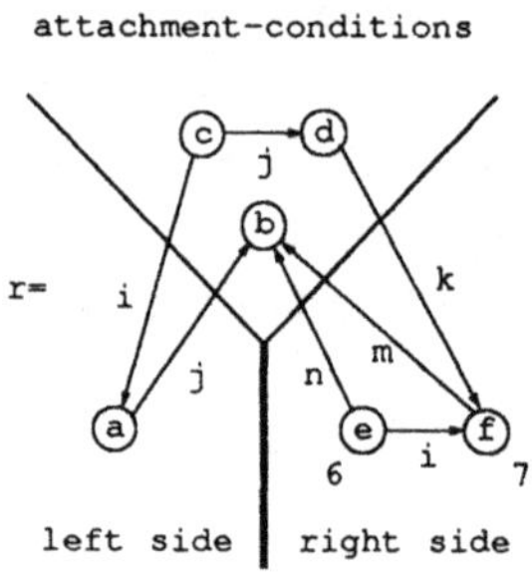

Fig. 5

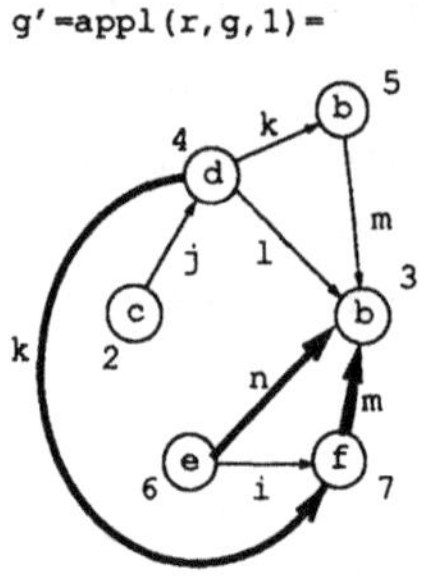

Fig. 6

One should notice that the substitution of a partial graph of g which is isomorphic to the left side of r will always take place if there is one. If there are more occurancies of the left side of r in g then one is chosen nondeterministically. But the connecting edges between the rest of g and the inserted right side of r are drawn iff the conditions exemplified in step (3), (3k), (3m) and (3n) hold, otherwise they are not drawn. If no isomorphic partial graph of the left side of r could be found in graph g, of course, nothing is changed in g. This is a *trivial application.*

Implementing the substitution mechanism exactly the way as it was described above is not wise. It is inherently inefficient. This can be demonstrated in fig. 7 where only the relevant labels are shown. Let's assume, only a second edge between node 1 and node 2 of graph g is to be constructed. The graph production p would accomplish this. However, according to the definitions above the nodes have to be taken out of the graph and then are resubstituted together with the additional edge. The upper part of p guarantees an identical embedding. It was the first time in [7] when the problem was discussed how this unnecessary work can be avoided. The main idea is to identify in some way the parts which remain 'constant'.

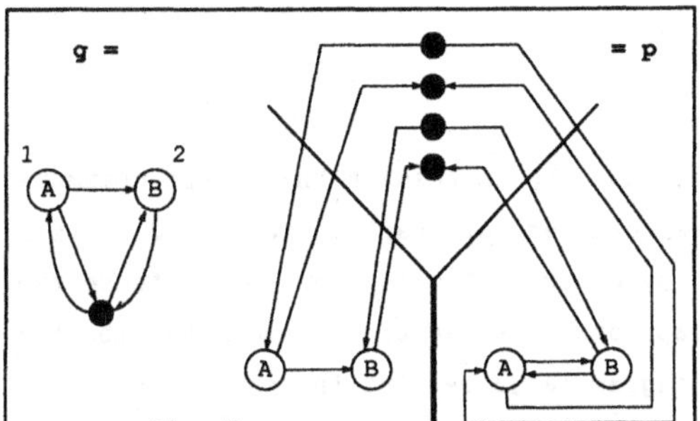

Fig. 7

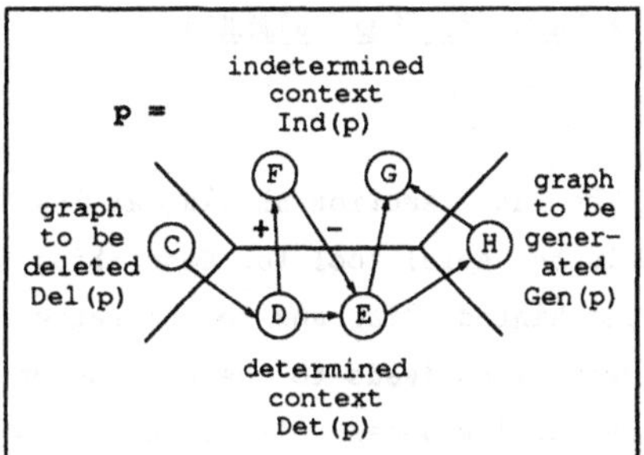

Fig. 8

Fig. 8, the X-notation of a graph production (called *graph operation* in the sequel), takes its rise from this consideration. The lower part of the X denotes the subgraph which would be in common to the left and to the right side in the Y-notation. When a X-shaped graph operation p is applied to a graph g, a partial graph is searched which is isomorphic to the graph consisting of Del(p)∪Det(p). Then in g the part isomorphic to Del(p) is deleted and Gen(p) is added. Similarly to the Y-notation the edges connecting the restgraph to the inserted one is constructed. So the left side of the Y is corresponding to Del(p)∪Det(p) and the right side to Gen(p)∪Det(p).

Det(p), being in the intersection, has the desired advantage having not to be restored. However, sometimes this can turn into the disadvantage if edges of g which should be erased are not, and, similarly, edges which one wants to be generated are not, either. The practical applications show that this ability of the Y-notation should not be given up and a synthesis of the expressive power of the two notations (the 'X-efficiency' and the 'Y-structural-power') should be strived for. Thus the edges between Det(p) and Ind(p) can get a label, one is a "-" for "to be removed" and the other one is a "+" for "to be added". This is expressed by fig. 8, too, where, for the sake of simplicity, edge-labels have been omitted.

Of course, there are more possibilities to describe the issue how to wisely combine the effects of the Y- and X-notation. One can think of using the same "+"- and "-"-labels also in Det(p) to express that these edges are to be added or removed, respectively. Then, no nodes have to be erased if just an edge is to be added or removed. Truely, to a certain extent it is irrelevant which kind of mechanism to use. The Y-notation is sufficient, theoretically. But one point has to be stressed! As it was mentioned in section 1, the purpose of the whole project was to *prove the usefulness of the graph grammars approach for practical purposes.* A main goal was that *the implementation of the graph grammars must agree with the theory.* So, no 'real programmer' could say theory is useless if it comes to a practical problem. The improvement in performance between an implementation of the "efficient" X-notation and a comparable implementation of the Y-notation were orders of magnitude. Clearly it just boils down to avoiding unnecessary substitutions.

4 ATTRIBUTED GRAPH GRAMMARS

The last section should at least have given the impression that graph grammars could be useful tool for describing structural changes of graphs. Section 2 of this paper hinted diagrams to be representable by graphs. The combination of these two observations leads to the conclusion that graph grammars can be used to describe the structural changes of the elements of a diagram language. But the end of section 2 suggests this to be only half the story. It is not possible to reconstruct a diagram out of its representation graph in general. Additional information is necessary which has to be associated with the nodes representing the graphical objects of the diagram. A similar problem was solved in formal language theory when the symbols of the alphabet of a grammar were associated with *attributes to express semantics. Now the nodes of the graph grammars are attributed in a similar way.*

A very simple but instructive example are "Fibonacci-Diagrams", see fig. 10a. The elements of this language are boxes where the length of one of the sides grow like Fibonacci-numbers, defined by $f_1=0$, $f_2=1$, and $f_{n+2}=f_{n+1}+f_n$. In addition, the boxes are connected by a line of unit length. It is nonsense to try to encode in some way the size of the boxes in structure by adding nodes representing the size like it is done in fig. 10b.

Fig. 11a shows the relevant graph operation where the node labeled with a triangle should be considered as the pen which does the drawing. As it is stated, the pen has to be taken out and must be restored. It would be more efficient, of course, to place this particular node into the lower part of the X. The nodes of the graph operations in fig. 11a are labelled and, by this, express a *type* which characterizes what it is used for: A type is associated with a set of attributes, which again are associated with an evaluation rule for each. Fig. 11 demonstrates this. Two types, one for the graphical object box and one for the graphical object line, can be found. The necessary attributes for box are the (constant) height and the cartesian x- and y-coordinates of the left lower corner. The types of the nodes in the graph operation are added in fig. 11 as comments. The next important part of such a specification is the BODY-part where the formulas for the evaluation are stated. The assignment "3.breadth <- 4.breadth + 5.breadth" is due to Fibonacci's formula.

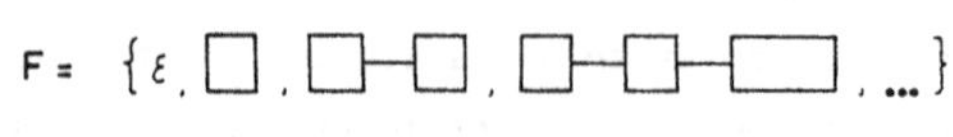

Fig. 10a

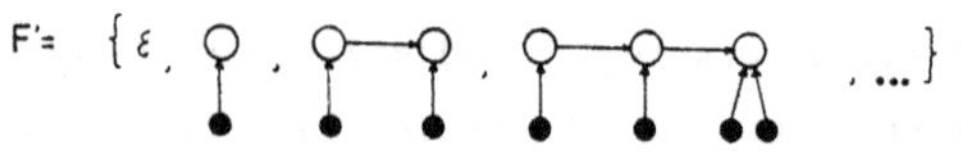

Fig. 10b

The set of *attributed graph operations*, the *NODETYPE*-part, and the *BODY*-part form the *attributed graph grammar*. Now, the application of an attributed graph operation has two effects, a structural change of the representation graph, and the construction of a program for the evaluation of the attributes requiring a simple substitution mechanism. The following one is just an example of what could be used. If the graph operation znK (for drawing an additional box) is applied to the axiom graph in fig. 11b then its name, znK, together with an index designing the number of already applied graph operations, say "(1)" because it was the first application, forms a prefix, thus "znK(1)". The other prefix is the name of the graph. These measures must be taken to prevent naming clashes.

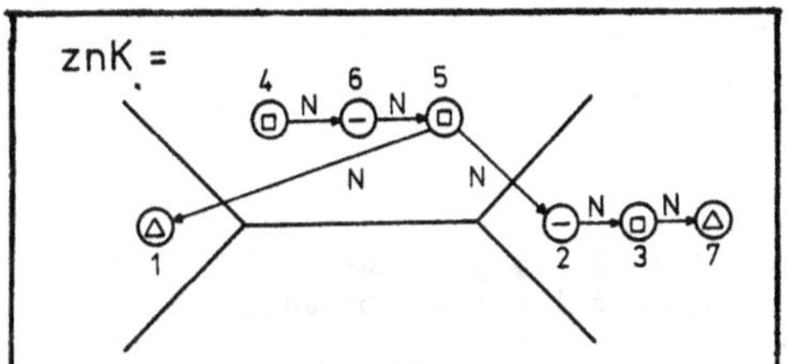

Fig. 11a

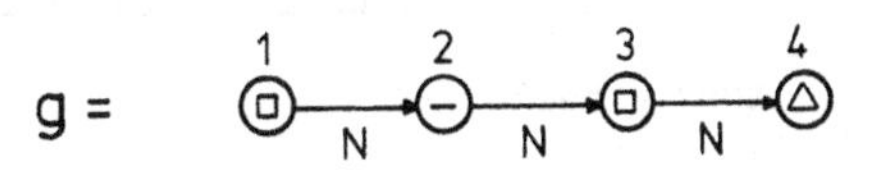

Fig. 11b

```
PROJECT fibonacci;

NODETYPE  box =  heighth <- 1    ;     /* constant for all boxes */
                 breadth <- #    ;     /* default value "#" */
                 x_coord_llc     ;
                 y_coord_llc     ;

NODETYPE  line = x_coord_sp      ;     /* defined by its starting point */
                 y_coord_sp      ;

NODETYPE  pen                    ;     /* gets no attributes */


GRAPHOP   znK;

/*      NODES    1, 7  :  pen ;      4 , 5 , 3  :  box ;      2 , 6  :  line ;    */

        BODY = begin
                    3.breadth     <- 4.breadth + 5.breadth              ;
                    3.x_coord_llc <- 5.x_coord_llc + 5.breadth + 1      ;
                    3.y_coord_llc <- 5                                  ;
                    2.x_coord_sp  <- 5.x_coord_sp + 5.breadth           ;
                    2.y_coord_sp  <- 5.5                                ;
               end ;

ENDOFPROJECT ;
```

Fig. 11

The result of applying znK the first time to graph g of fig. 11b can be found in fig. 12 together with the modified BODY of znK. As stated above the application of an attributed graph operation changes the structure of the representation graph and generates prescriptions out of the BODY-parts how the values of the attributes are to be calculated. The concatination of these BODY-parts forms the attribute evaluation program of the *attribute evaluation language AEL.*

```
GRAPHOP   znK(1);          /* modified by doing some renaming */

/*      NODES    g.4, znK(1).7         : pen  ;
                 g.1, g.3, znK(1).3    : box  ;
                 znK(1).2, g.2         : line ;

                 Since for the (first) application of znK the following
                 correspondences are established
                        znK(1).1  to  g.4
                        znK(1).5  to  g.3
                        znK(1).6  to  g.2
                        znK(1).4  to  g.1           */

        BODY = begin

                   znK(1).3.breadth      <- g.1.breadth + g.3.breadth           ;
                   znK(1).3.x_coord_llc  <- g.3.x_coord_llc + g.3.breadth + 1   ;
                   znK(1).3.y_coord_llc  <- 5                                   ;
                   znK(1).2.x_coord_sp   <- g.3.x_coord_sp + g.3.breadth        ;
                   znK(1).2.y_coord_sp   <- 5.5                                 ;

               end ;
```

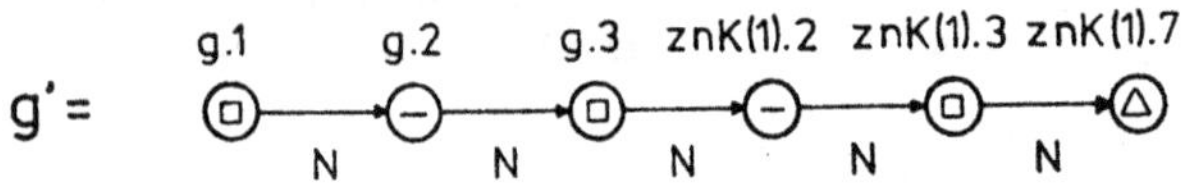

Fig. 12

AEL has an PASCAL-like style as it was shown. There exists an AEL-compiler which transforms the AEL-programs into LISP-programs which are then executed by the LISP-interpreter. The AEL-programs are subject to a permanent change. Each application of a graph production adds BODY-parts to them. This is the reason while an interpreted language like LISP is best suited for the target language of the translation of AEL-programs. The application of an attributed graph operation ago to a graph g is just a LISP-operation of the kind "(ago g)".

5 PROGRAMMED ATTRIBUTED GRAPH GRAMMARS

Sometimes it can be difficult to model a user function of a diagram technique by a single graph production. This is the case when the user function causes a very complicated structural change of the diagram. Then it might be a good advice to break the intended graph operation into several parts. This 'trick' has also a counterpart in formal language theory: programmed grammars. To get the wished effect, finally, the simpler graph operations are applied in a prescribed manner. There is no good diagram technique yet which allows to denote the combinations, the *programs*, of attributed graph operations. However there are three types of *control structures*: *sapp*, the sequential application of two ore more graph operations; *capp*, the conditional application of the graph operations listed after capp, meaning that the first graph operation with a nontrivial application (see section 3) is used; *wapp*, the application of an attributed graph operation as long as it is nontrivially (see section 3) applicable. The three control structures can arbitrarily be nested. The whole system which allows to use attributed graph operations, AEL and the three control structures will be called *PAGG-system* (standing for "programmed attributed graph grammar").

6 GRAPH TECHNOLOGY

The idea to develop syntax directed diagram editors by means of PAGGs can be transferred to other areas of application. [1] and [12] for instance use attributed graph grammars to specify the implementation of IPSEN. In this project all the data structures involved are modelled by attributed graphs. The change of a program, for example during the time of editing, results in the changes of the representation graph. There is a simple, general principle when using attributed graph grammars to solve implementation problems. First, choose an appropriate data structure and then describe the operations necessary by means of attributed graph operations. These ideas are not new. They were first mentioned in [13]. The efficient graph operations together with the AEL-programs go a step further as the work of [1] and [12]. Here, the programmed attributed graph operations are *not just* used as a method to specify the solution which has to be programmed finally in some executable language - in the IPSEN-project it was MODULA 2 - but can be used as an *executable specification*.

The PAGG-system wants to serve as a frame-work in which other software engineering principles can be incorporated. One for example, is "stepwise refinement". When designing a syntax directed editor the decision on the exact placement of the graphical objects can be postponed for quite a long time. The placement could qualitatively be expressed first by relations of the type "neighbouring to". Then later in the course of developing the whole system this neighbouring-relation will be refined into formulas. Another principle, "modularity" is enforced, too, by PAGG due to a rule "one user function has to correspond to an attributed graph operation or to a sapp/wapp/capp-combination". There are even more software engineering principles provided for in PAGG, like "abstract datatypes", "assertion mechanism" and so forth.

7 PAGGs AS A TOOL

The question could arise whether these ideas are practically used, or, put in other terms, how does an environment for developing syntax directed editors look like? The first interpreter for attributed 'Y-shaped' graph grammars (attention: "programmed" is ommitted!) was finished at Erlangen in 1983 and is described in [8]. Its performance was unacceptable. Based on this experience a new implementation (now for 'X-shaped' programmed attributed graph grammars) was attempted and finished in 1986. The system was written in Franz Lisp and several diagram editors were implemented. The basic ideas for dealing with the system is elucidated in fig. 13.

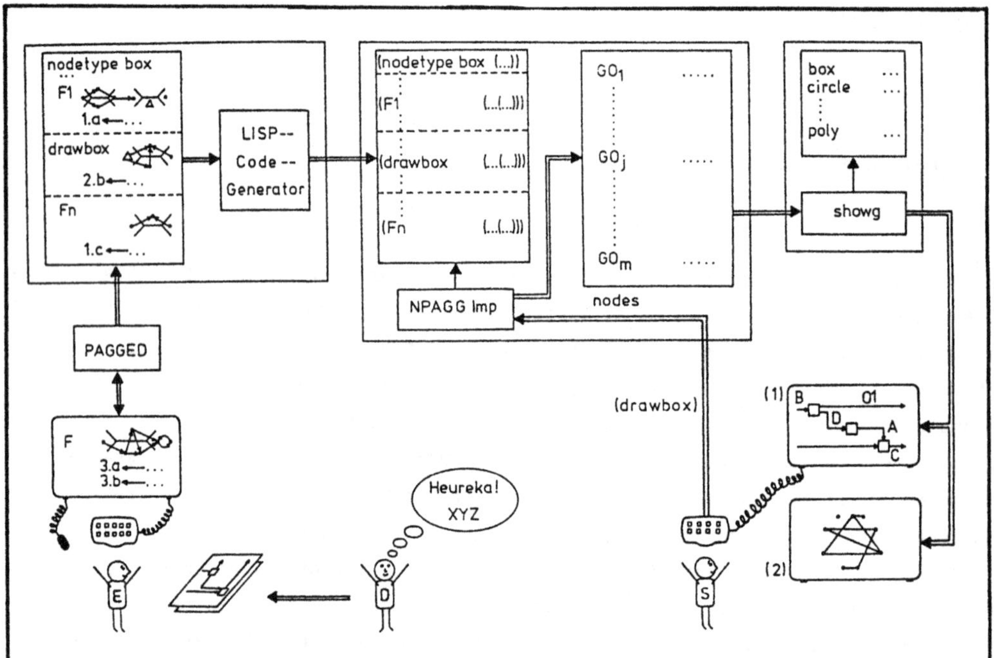

Fig. 13

Assume, the designer D of a new diagram based software engineering method XYZ wants to convince the colleagues of the usefulness of XYZ. Therefore he/she needs a good tool for it which shows a satisfying performance no matter how stupidly the final user acts. D has heard of a system which allows to quickly produce a graphical user interface. (Of course, it's the one described in here.) So D consults E, an expert for programmed attributed graph grammars. D explains to E the features of XYZ. By pondering XYZ-examples (first on a sheet of paper) expert E works out the programmed attributed graph grammar.

To ease the effort designing the graph grammar E uses a graphical display with a mouse served by PAGGED. The system PAGGED ("Programmed Attributed Graph Grammar EDitor", a program which is written in C) interacts with E. The functions of two menues can be used to fill nodes and edges into the X-connector. An ordinary UNIX vi-editor is used for typing in the formulas for the attributes. There are also options to read productions from a library, or to write them to it, or to make a hardcopy of them, etc.

For each of the n intended user functions F1,...,Fn in Fig. 13 either a single attributed graph production is developed, like for Fn, or a program involving a series sapp of attributed graph operations, like for F1, is designed. (Likewise in the case of capp and wapp). Of course, the formulas for the placement of the graphical objects (like "1.a <- ...") must be thought about, too. To have a meaningful example for a user function assume there exists "drawbox". The result of E's activity is the input to a program called "Lisp-Code-Generator" (a program written in Franz Lisp) which transforms the X-diagrams of the graph into (interpretable) Lisp code. At this point a graphical interface for XYZ is established!

Now the software engineer S can use XYZ. S sits down at the terminal and commands "(drawbox)" for example. (The parentheses are due to Lisp.) The program NPAGGImp (which is written in Franz Lisp; the acronym stands for "New Programmed Attributed Graph Grammars Implementation") accepts the command and looks up the XYZ-specific table of user functions. NPAGGImp interprets this entry, applies it to the (intermediate) representation graph of the diagram to be developed, generates the assignments for the attributes, and evaluates them. This activity finally results in a file "nodes" which keeps the graphical objects in form of their evaluated attributes. A program "showg" takes these values and displays them at the graphical terminal (1). For frequently used graphical objects like "box" there are descriptors in a library. The process for developing a graphical user interface for XYZ is not always straightforward. Errors are made and it is not always easy to cast aesthetical considerations into mathematical formulas. So there are also intermediate steps of the constructions which is indicated in (2). It is only important to E to have such an option.

Meanwhile, there exists also a syntax directed editor for graph operations based on PAGG; so, some kind of bootstrapping has been made.

8 PRESENT FIELDS OF APPLICATIONS

The PAGG-tool has been used to derive several diagram editors, for resistor nets,

SADT (including its refinement mechanism and the so-called ICOM-code), HIPO, Petri-nets, Fibonacci-diagrams (of course), EADT (A method to describe the state of a distributed algorithm on microprocessor networks. The algorithm analyzes the EEG of little children with hearing deficiencies.) and for several diagram editors in the field of process automation. Presently, the PAGG-method and the PAGG-tool for prototyping are used to develop a CAD system for lighting installations.

9 OTHER APPROACHES

This described project is not the only one, of course, which stresses the importance to model concepts by graphs. The IPSEN-project has already be mentioned. ***Knowledge representation***, a field of AI, is heavily based on graph models. For the purpose of producing syntax directed diagram editors other approaches can be found in literature. There are ***table driven systems*** like the ones of [15] or [17]. A system which uses a grammar system (but not graph grammars), too, has been developed by [16]. [14] is based on the work described in this paper. However, it is restricted to ***precedence graph grammars*** but allows attributed edges. To say a word to the limitations of PAGG: It surely is not useful for implementing window systems or for using it as an Icon-editor. Annotations to the PAGG-system can be found in [4] and [5].

10 SUMMARY

Two recommendations have been given. One was concerning the use of diagram languages for software development. It was pointed out that diagrams are the right means for modelling a system to be implemented together with the customers not trained in formal mathematical thinking. But also for the experienced software engineer working with diagrams can be of considerable value. The other recommendation was on the use of graphs as a general data structure, whose expressive power, if combined with graph grammars, is going far beyond the abilities of the well-known trees, queues, stacks, etc. The generality can only be achived by adding the feature "attribution" to the graphs and the graph operations.

Using attributed graphs brings an advantage which must highly be estimated: It was sufficiently elucidated that they are a means to represent any diagram language. If one consideres the different morphologies of the models which a piece of software has to go through, from the phase of requirements specification to the phase of implementation yielding the source code, then it should be clear that the transformation between the models via their representation graphs can also be described by means of programmed attributed graph grammars.

11 LITERATURE

[1] Engels,G.: "Graphen als zentrale Datenstrukturen in einer Software-Entwicklungsumgebung", PhD-thesis, Techn, Rept. Universität Osnabrück, 1986.

[2] Göttler,H.: "Zweistufige Graphmanipulationssysteme für die Semantik von Programmiersprachen", PhD-thesis, Techn. Rep. Vol. 10 Nr. 12, IMMD University of Erlangen-Nürnberg, 1977.

[3] Göttler,H.: "Semantical Description by Two-Level Graph Grammars for Quasihierarchical Graphs", Proc. WG'78 'Graphtheoretic Concepts in Computer Science, Hanser Verlag, München, 1978.

[4] Göttler,H.: "Attributed Graph Grammars for Graphics", 2nd Intern. Workshop on Graph Grammars and their Application to Computer Science 1982, Osnabrück 1982 FRG, Lect. Notes in Computer Science Nr. 153, H.Ehrig - M.Nagl - G.Rozenberg (Eds.), Springer Verlag, New York, 1982.

[5] Göttler,H.: "Graph Grammars and Diagram Editing", 3rd Intern. Workshop on Graph Grammars and their Application to Computer Science, Warrenton, VA. USA, Lect. Notes in Computer Science Nr. 291, H.Ehrig - M.Nagl - G.Rozenberg (Eds.), Springer Verlag, Heidelberg, 1987.

[6] Göttler,H.: "Graphgrammatiken in der Softwaretechnik", Informatik Fachberichte, Nr. 178, Springer Verlag, Heidelberg, 1988.

[7] Grabska,E.: "Pattern Synthesis by Means of Graph Theory", PhD-thesis, Uniwersitet Jagiellonski (Instytut Informatyki), Krakow, 1982.

[8] Heindel,A.: "Implementierung attributierter Graphgrammatiken", Master Thesis, IMMD University Erlangen-Nürnberg, 1983.

[9] Nagl,M.: "Formale Sprachen von markierten Graphen", PhD-thesis, Techn. Rpt. Vol. 7, Nr. 4, IMMD University Erlangen-Nürnberg, 1974.

[10] Nagl,M.: "Set theoretic approaches to graph-grammars", 3rd Intern. Workshop on Graph Grammars and their Application to Computer Science, Warrenton, VA. USA, Lect. Notes in Computer Science Nr. 291, H.Ehrig - M.Nagl - G.Rozenberg (Eds.), Springer Verlag, Heidelberg, 1987.

[11] Ross,D.T.: "Structured Analysis (SA): A Language for Communicating Ideas", IEEE Transactions on Software Engineering, Vol. SE-3, No. 1, 1977.

[12] Schäfer,W.: "Eine integrierte Softwareentwicklungsumgebung: Konzepte, Entwurf und Implementierung", PhD-thesis, VDI-Verlag, Reihe 10: Informatik/-Kommunikationstechnik, Düsseldorf, 1986.

[13] Schneider,H.J.: "Syntax-directed Description of Incremental Compilers", 4. GI-Jahrestagung, Springer LNCS Bd. 26, Heidelberg, 1974.

[14] Schütte,A.: "Spezifikation und Generierung von Übersetzern für Graphsprachen durch attributierte Graphgrammatiken", Express Edition Verlag, Berlin, 1987.

[15] Sommerville,I. - Welland,R.- Beer,S.: "Describing Software Design Methodologies", The Computer Journal, Vol. 30, No. 2, 1987.

[16] Szwillius, G.: "GEGS - A System for Generating Graphical Editors", Proc. INTERACT'87 (Human-Computer Interaction), Bullinger,H.-J. - Shackel,B. (Eds.), Elsevier Science Publishers B. V. (North Holland), Amsterdam, 1987.

[17] Tichy,W.F. & Newbery,F.J.: "Knowledge-Based Editors for Directed Graphs", Proc. ESEC'87 (1st European Software Engineering Conference), Nichols,H.K. - Simpson,D. (Eds.), Lecture Notes in Computer Science Nr. 289, Springer Verlag, Heidelberg, 1987.

[18] Jones,C.V.: "An Introduction to Graph-Based Modeling Systems", Proc. TIMS/ORSA-Meeting, Denver, 1988.

Termination proofs and the length of derivations (Preliminary version)

Dieter Hofbauer
Technische Universität Berlin

Clemens Lautemann *
Universität Bremen

Abstract

The derivation height of a term t, relative to a set R of rewrite rules, $dh_R(t)$, is the length of a longest derivation from t. We investigate in which way certain termination proof methods impose bounds on dh_R. In particular we show that, if termination of R can be proved by *polynomial interpretation* then dh_R is bounded from above by a doubly exponential function[1], whereas termination proofs by *Knuth–Bendix ordering* are possible even for systems where dh_R cannot be bounded by any primitive recursive functions. For both methods, conditions are given which guarantee a singly exponential upper bound on dh_R. Moreover, all upper bounds are tight.

1 Introduction

The complexity of a term rewriting system R can be measured in different ways. In this paper, we consider the ***derivation height***, expressed in the function Dh_R, where $Dh_R(n)$ is the length of a longest derivation which starts from a term of size n. Thus Dh_R describes the worst possible behaviour of R.

Obviously, Dh_R is partial recursive, and is total if and only if R is terminating. The rate of growth of Dh_R then quantifies termination of R in much the same way as the rate of convergence of a series (q_n) quantifies its convergence.

As termination is one of the basic concepts in term rewriting theory, crucial e.g. in Knuth–Bendix completion procedures or for inductive proofs, a lot of techniques have been developed to support or even automatically generate termination proofs (see [De87] for a survey).

Proving termination with one of these specific techniques in general proves more than just the absence of infinite derivation sequences. It turns out that such a proof in many cases implies an upper bound on Dh_R, so that the rate of growth of Dh_R can be used for measuring the power of termination proof methods.

A similar point of view is taken in [DO88], where Dershowitz and Okada investigate the order type of certain simplification orderings. However, the order type of such an ordering as such does not place a bound on its power for termation proofs, as it does not say anything about the intersection of the simplification ordering with the rewrite relation induced by a finite term rewriting system.

*in the early stages of this research supported by ESPRIT project PROSPECTRA, ref. #390

[1]this result was independently obtained by Oliver Geupel

In this paper we have a closer look at two widely used methods for proving termination: *polynomial interpretation*, introduced by Lankford in [Ln75], and the *Knuth–Bendix ordering* of [KB70]. For both of them some work has been done to fully automatise such proofs, so that they are particularly useful for practical applications (cf. [Ln79], [CL87], [Ma87]).

Throughout, we will consider rewrite systems defined by *finite* sets of rules over *finite* signatures. For a signature Σ, we denote by Σ_n its subset of n–ary operation symbols. The size of a term t is denoted by $|t|$.
For two functions $f, g : \mathbb{N} \longrightarrow \mathbb{N}$, we write $f = O(g)$, or, equivalently, $g = \Omega(f)$, if there is a constant $c > 0$ such that, for all large enough n, $f(n) \leq c \cdot g(n)$.

For a set R of rules over Σ, we define the function $dh_R : T_\Sigma \longrightarrow \mathbb{N}$, by
$dh_R(t) := \max\{m \in \mathbb{N} \,/\, \text{ there is a term } t', \text{ such that } t \longrightarrow^m t'\}$.
The function $Dh_R : \mathbb{N} \longrightarrow \mathbb{N}$ then maximises dh_R over terms of the same size:
$Dh_R(n) := \max\{dh_R(t) \,/\, |t| = n\}$.
For further notation, and for general background, see, e.g., [HO80], or [De87].

We conclude this section by giving a few examples for rewrite systems which will be discussed in detail in later sections. In order to get a flavour of the kind of problem we are dealing with, the reader is invited to try and find derivations of maximal length for these examples.

Example 1
$\Sigma_2 = \{+\}$, $\Sigma_1 = \{s,\, d,\, q\}$, $\Sigma_0 = \{0\}$.
R consists of the following rules:

(1)	$x + 0$	$\rightarrow$	x
(2)	$x + sy$	$\rightarrow$	$s(x + y)$
(3)	$d0$	$\rightarrow$	0
(4)	dsx	$\rightarrow$	s^2dx
(5)	$q0$	$\rightarrow$	0
(6)	qsx	$\rightarrow$	$qx + sdx$

Thus, R denotes an arithmetical system with successor (s), double (d), and square (q) operations.

Example 2
$\Sigma_2 = \{+\}$, $\Sigma_1 = \{s\}$, $\Sigma_0 = \{0\}$.
R consists of the following two rules:

(1)	$sx + (y + z)$	$\rightarrow$	$x + (ssy + z)$
(2)	$sx + (y + (z + w))$	$\rightarrow$	$x + (z + (y + w))$

□

Example 3
Here we consider the string rewriting system over $\{a,\, b\}$, given by the two rules

(1)	a	$\rightarrow$	b
(2)	abb	$\rightarrow$	baa

□

2 Polynomial interpretation

The fact that a termination proof by polynomial interpretation implies a bound on the length of derivations has been well-known for some time; the order of magnitude of this bound, however, has remained unclear. Different claims were made (polynomial [HO80], exponential [Le86], super exponential [De87]), but no proofs provided. In [Le86], Lescanne showed by example that it must be at least exponential. In this section, we will show that, for general polynomials, the correct bound is doubly exponential[2], whereas it is of the form $2^{O(n)}$ in case of a linear interpretation.

2.1 General polynomial interpretations

Let $\tau : T_\Sigma \longrightarrow \mathbb{N}$ be a homomorphism obtained by mapping each n-ary function symbol f onto a monotonic n-place polynomial f_τ with integer coefficients. τ is called a *polynomial interpretation*; if every f_τ is either a constant, or is linear in every argument, we speak of a *linear interpretation*. A set R of rules over Σ is *reducing under* τ, if for every rule $l \to r \in R$, and for every ground substitution σ, $\tau(l\sigma) > \tau(r\sigma)$.

Fact
If R is reducing for a polynomial interpretation τ then R is terminating. □

Termination of example 1 can be shown by polynomial interpretation.

Example 1 (continued)
Consider the following interpretation of the function symbols.
$0_\tau = 2$, $s_\tau(a) = 1 + a$, $a +_\tau b = a + 2b$, $d_\tau(a) = 3a$, $q_\tau(a) = a^3$.
All rules are reducing under τ: for, e.g., rule 6, we obtain
$\tau(qsx) = (1 + \tau(x))^3 > \tau(x)^3 + 2(1 + 3\tau(x)) = \tau(qx + sdx)$. □

2.1.1 Proposition

There are Σ, R, and a polynomial interpretation τ, such that R is reducing under τ, and $Dh_R(n) = 2^{2^{\Omega(n)}}$.

Proof:

We show that, in the system of example 1, for the infinitely many terms of the form $t = q^{n+1}(s^2(0))$, $dh_R(t) \geq 2^{2^{c|t|}}$, for some $c > 0$. Clearly, t is reducible to $q(s^{2^{2^n}}(0))$, and from this term, rule 6 alone allows a derivation of length 2^{2^n}. It follows that
$dh_R(t) > 2^{2^n} = 2^{2^{|t|-4}} \geq 2^{2^{c|t|}}$, for $c \leq \frac{1}{4}$. □

Essentially, example 1 is worst possible.

2.1.2 Proposition

Let $\tau : T_\Sigma \longrightarrow \mathbb{N}$ be a polynomial interpretation, and let R be a set of rules over Σ. If R is reducing under τ then $Dh_R(n) = 2^{2^{O(n)}}$.

Proof:

Since $0 \leq dh_R(t) \leq \tau(t)$, it suffices to show that $\tau(t) \leq 2^{2^{c|t|}}$, for all $t \in T_\Sigma$, and some constant c.

Choose $c \geq k \cdot \log d$, where k, d are constants such that for every $f \in \Sigma$,
$f_\tau(a_1, \ldots, a_m) \leq d \cdot \prod_{i=1}^m a_i^k$.
We now proceed by induction on $|t|$.
Clearly, if t is a constant then $\tau(t) \leq d \leq 2^c < 2^{2^{c|t|}}$.
If $t = f(t_1, \ldots, t_m)$ then
$\tau(t) = f_\tau(\tau(t_1), \ldots, \tau(t_m)) \leq d \cdot \prod_{i=1}^m \tau(t_i)^k \leq d \cdot \prod_{i=1}^m \left(2^{2^{c|t_i|}}\right)^k$ (by induction hypothesis)
$= d \cdot 2^{k\Sigma_{i=1}^m 2^{c|t_i|}} \leq 2^{c\Sigma_{i=1}^m 2^{c|t_i|}} \leq 2^{2^{c(1+\Sigma_{i=1}^m |t_i|)}} = 2^{2^{c|t|}}$. □

---

[2] This result will also be published in [La88], and it was independently obtained by Oliver Geupel, [Ge88].

2.2 Linear interpretations

For a linear interpretation τ, the upper bound of proposition 2.1.2 can be improved.

2.2.1 Proposition
If, under the same assumptions as in proposition 2.1.2, τ is a linear interpretation then $Dh_R(n) = 2^{O(n)}$. □

The proof is very similar to that of proposition 2.1.2, and is omitted.

Again, an example shows that this bound is best possible.

Example 4
$\Sigma_0 = \{0\}$, $\Sigma_1 = \{s, d\}$, $R = \{d0 \to 0, dsx \to s^2dx\}$, i.e., R consists of rules 3 and 4 of example 1. Restriction of the interpretation used there to T_Σ gives a linear interpretation under which R is reducing. □

2.2.2 Proposition
There are Σ, R, and a linear interpretation τ such that R is reducing under τ, and $Dh_R(n) = 2^{\Omega(n)}$.

Proof:

Clearly, $t = d^n(s(0))$ is reducible to $s^{2^n}(0)$. A single reduction step either eliminates one d, or increases the number of s' by one, so $2^n + n - 1$ steps are necessary, i.e., $dh_R(t) \geq 2^n \geq 2^{c|t|}$, for $c \leq \frac{1}{4}$. □

3 Knuth–Bendix ordering

The Knuth–Bendix ordering (KBO) introduced in [KB70], is a simplification ordering widely used for proving termination of rewriting systems (cf., e.g., [De87], [Ma87]). In this section, we investigate how a termination proof for R via KBO can affect the rate of growth of Dh_R. It turns out that, in general, no primitive recursive upper bound on Dh_R can be derived, but for a number of syntactically restricted systems exponential bounds are obtained.

Definition

Let Σ be a signature, $>$ an ordering of Σ, and $w : \Sigma \longrightarrow \mathbb{N}$ a weight function. We call the triple $[\Sigma, >, w]$ an *extended signature*, and say that it has the *KBO property*, if the following hold:
- $w(c) > 0$, for every constant c, and
- if $w(f) = 0$ for some *unary* f then $f > g$ for all $g \neq f$.

A rule $l \to r$ over Σ is *reducing under KBO for* $[\Sigma, >, w]$, if, for every ground substitution σ, $l\sigma >_{\text{KBO}} r\sigma$, where $>_{\text{KBO}}$ is recursively defined as follows:

$t >_{\text{KBO}} s \iff$ (1) $w(t) > w(s)$, or
(2) $w(t) = w(s)$, and
(2.1) $t = f(t_1, \ldots, t_n)$, $s = g(s_1, \ldots, s_m)$, and $f > g$, or
(2.2) $t = f(t_1, \ldots, t_n)$, $s = f(s_1, \ldots, s_n)$, and
$t_k >_{\text{KBO}} s_k$, for some $k \leq n$, and $t_i = s_i$, for $i = 1, \ldots, k-1$. □

$>_{\text{KBO}}$ is a simplification ordering and thus well-founded. It is total, if the ordering $>$ is; since every partial order can be extended to a total one, and since KBO is incremental in the sense that such an extension of $>$ implies an extension of the corresponding $>_{\text{KBO}}$ we can assume $>$ to be total without any loss of generality.

Fact
R is terminating if there are $>$, w such that R is reducing under KBO for$[\Sigma, >, w]$. □

In that case we say that R *has a termination proof via KBO.*

3.1 A lower bound

Example 2 (continued)
Define $>$ by $s > + > 0$, and w by $w(s) = w(+) = 0$, $w(0) = 1$. Then $[\{+, s, 0\}, >, w]$ has the KBO property, and R can easily be seen to be reducing.

We want to show that very long derivations are possible over R. To this end, let us consider terms of the form

$$s^{i_0}(0) + (s^{i_1}(0) + (\ldots + (s^{i_r}(0) + 0)\ldots))$$

(note that R preserves this form).
For notational convenience we will identify such a term with the list

$$[i_0, i_1, \ldots, i_r].$$

The effect of applying a rule to such a term can then be described as follows:
(1) $[\ldots, n+1, m, \ldots] \to [\ldots, n, m+2, \ldots]$
(2) $[\ldots, n+1, m, k, \ldots] \to [\ldots, n, k, m, \ldots]$

A strategy for building very large terms (and hence very long derivations) can be based on the following three sequences.

Starting with $[n+1, 0, \ldots, 0]$, repeated application of rule (1) yields
$[n+1, 0, \ldots, 0] \to^* [1, 2n, 0, \ldots, 0] \to^* [1, 0, \ldots, 0, 2^r n]$.

Given a term of the form $[1, 0, \ldots, 0, m]$ we can apply rule (1) r times:
$[1, 0, \ldots, 0, m] \to [0, 2, 0, \ldots, 0, m] \to [0, 1, 2, 0, \ldots, 0, m] \to^* [0, 1, 1, \ldots, 1, 0, m+1]$[3].

Now, we make use of rule (2), and move $m+1$ to the front of the list.
$[0, 1, 1, \ldots, 1, 1, 0, m+1] \to [0, 1, 1, \ldots, 1, 0, m+1, 0] \to^* [0, 0, m+1, 0, \ldots, 0]$.
Throughout, the length of the list remains unchanged. □

3.1.1 Theorem

There is a set R of rules over $\{+, s, 0\}$ such that R has a termination proof via KBO, but Dh_R cannot be bounded from above by any primitive recursive function.

Proof:

Consider example 2. Let
$f(x, y) := \max\{z \,/\, [y+1, 0, \ldots, 0] \to^* [0, 0, \ldots, z]$, where the length of the list is $2x+2\}$. Application of the strategy described above gives the following recurrence for f:
$f(0, y) \geq 2y + 2 > y + 1$
$f(x+1, 0) \geq f(x, 1)$, since $[1, 0, 0, \ldots] \to^* [0, 0, 2, \ldots]$

[3]Really, the last two entries should be $1, m+2$, but $0, m+1$ will simplify the exposition, without leading to longer derivations.

$f(x+1,y+1) \geq f(x,f(x+1,y))$,
since $[y+2,0,0,\ldots] \rightarrow^* [1,0,\ldots,f(x+1,y)] \rightarrow^* [0,0,f(x+1,y)+1,0,\ldots,0]$.

Thus, f grows faster than the Ackermann function A. For every constant c, if $t_{c,n}$ is the term $[n+1,0,\ldots,0]$ (list length $2c+2$), we have $|t_{c,n}| \leq 2n$, for almost all n, and consequently, for every upper bound B on Dh_R, $B(2n) \geq Dh_R(2n) \geq f(c,n) \geq A(c,n)$, so B cannot be primitive recursive (cf., e.g., [He78]). □

3.2 Syntactical restrictions: the signature

As example 2 shows, the rate of growth of Dh_R can be gigantic, even if termination of R has been proved by KBO. We now show that such behaviour cannot occur over a signature which does not contain either a unary function symbol with weight 0, or a n–ary function symbol for some $n \geq 2$; in fact, either restriction will imply an exponential upper bound on Dh_R.

Before proceeding, we will simplify our task by a construction which, essentially, shows that the signature of example 2 can be considered a canonical signature for our considerations.

3.2.1 Theorem
Let $[\Sigma,<,w]$ be the extended signature of example 2. Let $[\Sigma',>',w']$ be an extended signature with the KBO property, and let R' be a set of rules over Σ'. Then there is a generalised homomorphism $h: T_{\Sigma'}(X) \longrightarrow T_\Sigma(X)$, such that the following hold:

1. R' is reducing under KBO for $[\Sigma',>',w']$ if and only if
   $R := \{h(l) \rightarrow h(r) \,/\, l \rightarrow r \in R'\}$ is reducing under KBO for $[\Sigma,>,w]$.
2. $Dh_{R'}(n) \leq Dh_R(c{\cdot}n)$, for some constant c.
3. If Σ' contains no unary function symbol f with $w'(f)=0$ then $h(T_{\Sigma'}(X)) \subseteq T_{\{+,0\}}(X)$.

The proof of the theorem is in two steps.

Lemma 1
There is an extended signature $[\Sigma,>,w]$ with the KBO property such that

(i) Σ contains only one constant, c, and

(ii) $w(c)=1, w(f)=0$, for all $f \neq c$, and

(iii) the conclusion of theorem 3.2.1 holds.

Lemma 2
Theorem 3.2.1 holds if

(i) Σ' contains only one constant, c, and

(ii) $w'(c)=1, w'(f)=0$, for all $f \neq c$.

Proof of Lemma 1

We only give the construction, verification of (iii) is straightforward (though tedious in places). Without loss of generality, we assume that
$w'(g) > 0 \Longrightarrow w'(g) \geq 2$.
$[\Sigma,>,w]$ is now defined as follows.

$\Sigma := \bigcup_{n\geq 0} \Sigma_n \uplus \{F\}$, where F is a new binary function symbol; and
$\Sigma_0 := \{c\}$,
$\Sigma_{n+1} := \{f_d \,/\, d \in \Sigma'_0,\ w(d) = n+1\} \cup \{f_g \,/\, g \in \Sigma'_{n+1},\ w'(g) = 0\} \cup \{f_g \,/\, g \in \Sigma'_n,\ w'(g) > 0\}$

$f_g > f_r \iff g >' r$, and $F < f_g$, for all $g,\ r \in \Sigma'$.

$w(c) = 1, w(F) = 0$, and $w(f_g) = 0$, for all $g \in \Sigma$.

Finally, h is defined as
$h(x) := x$, for $x \in X$,
$h(d) := f_d(c,\ldots,c)$, for $d \in \Sigma'_0$, and
$h(g(t_1,\ldots,t_n)) := f_g(h(t_1),\ldots,h(t_n),w_g)$,
where w_g is the (unique) term
$F(c, F(c,\ldots,F(c,c)\ldots))$ of size $2w'(g) - 1$. □

Proof of Lemma 2

Again, we only give the construction. First, we change w' to a weight function w'' as follows. Let $k \in \mathbb{N}$ be big enough that the number of binary trees with $k-2$ leaves is greater than $|\ \Sigma'\ |$. Define
$w''(c) := k$, and for $g \in \Sigma'_{n+1}$,
$w''(g) := k{\cdot}n$.
Then, for every $t \in T_{\Sigma'}$, $w''(t) = k{\cdot}(2w'(t) - 1)$, and, consequently,
$w'(t) > w'(t') \iff w''(t) > w''(t')$.
For every $i \leq |\Sigma'|$, and $n, m \geq 2$ define terms p_i, a_n, w_m as follows:

- p_i is the i-th largest $\{+,0\}$–term (with respect to $>_{\mathrm{KBO}}$) with exactly $k-2$ occurences of 0;
- a_n is the "partial term" $(\ldots((\circ + \circ) + \circ) + \ldots + \circ)$ with n empty positions (indicated by $\circ$),
- w_m is the $\{+,0\}$–term $0 + (0 + (\ldots(0+0)\ldots))$ with exactly m occurences of 0.

h is now defined as follows (let $\Sigma'_1 = \{f\}$).
$h(c) := p_i + w_2$, if c is the i-th largest symbol under $>'$,
$h(f(t)) := s(h(t))$,
$h(g(t_1,\ldots,t_n)) := p_i + (a_n(h(t_1),\ldots,h(t_n))) + w_m)$, if g is the i-th largest symbol under $>'$, and $m = w''(g) - (k-2)$. □

Using theorem 3.2.1 we can now show exponential upper bounds on Dh_R, over restricted signatures.

3.2.2 Proposition
Let $[\Sigma, >, w]$ be an extended signature with the KBO property, and let R be a set of rules which is reducing under KBO for $[\Sigma, >, w]$.
Then $Dh_R(n) = 2^{0(n)}$, if one of the following holds:

1. $w(f) > 0$, for every unary function symbol f.
2. $\Sigma_n = \emptyset$, for all $n \geq 2$.

Proof

1. The construction of theorem 3.2.1 yields a set $\tilde{R}$ of rules over $\{+,0\}$, which is reducing under KBO for $[\{+,0\}, + > 0, w(t) = 0, w(0) = 1]$, and a constant c for which $Dh_R(n) \leq Dh_{\tilde{R}}(cn)$.

It therefore suffices to show that for every such $\tilde{R}, Dh_{\tilde{R}}(n) = 2^{0(n)}$.
If $t \longrightarrow^*_{\tilde{R}} t'$, then $t \geq_{\mathrm{KBO}} t'$. This, in turn, implies $w(t) \geq w(t')$, and therefore, $dh_{\tilde{R}}(t) \leq |\{t' / w(t') \leq w(t)\}|$. The number of terms t' with $w(t') = k$ is equal to the number of binary trees with k leaves, i.e., the k–th Catalan number,

$$C(k) = \frac{1}{k}\binom{2k-2}{k-1} < 2^{2k-2},$$

hence

$$dh_{\tilde{R}}(t) < \sum_{k=1}^{w(t)} 2^{2k-2} < 2^{2w(t)-1} = 2^{|t|}$$

$\Longrightarrow Dh_{\tilde{R}}(n) < 2^n$.

2. Let $\Sigma = \{a_1, \ldots, a_m\}$, where $a_1 < \ldots < a_m$, and let $w(a_m) = 0$ (otherwise 1 applies). Set $d := \max\{|r| / l \to r \in R\}$, and choose $c > m \cdot d$.
Every term $t = a_{i_1}(a_{i_2}(\ldots(a_{i_n})\ldots))$ can be identified with the Σ–string $a_{i_1} \cdots a_{i_n}$. Define a weight function $B : T_\Sigma \longrightarrow \mathbb{N}$ by

$$B(t) := \sum_{j=1}^{n} i_j \cdot c^{w(a_{i_j} \ldots a_{i_n})}.$$

We claim that R is reducing under B.
If $t \longrightarrow_{(l \to r)} t'$, then either
(i) $t = uvy$, and $t' = uv'y$, where $l \to r$ is $vx \to v'x$ or $v \to v'$, or
(ii) $t = uvy$, and $t' = uv'$, where $l \to r$ is $vx \to v'$, and $|y| \geq 1$.
In either case, it can easily be seen that $B(t) > B(t')$. Consider (i).
If $w(v) > w(v')$, then
$B(v) \geq c^{w(v)} > m \cdot d \cdot c^{w(v')} \geq B(v')$, and therefore

$$\begin{aligned} B(t) &= B(u) \cdot c^{w(vy)} + B(v) \cdot c^{w(y)} + B(y) \\ &> B(u) \cdot c^{w(v'y)} + B(v') \cdot c^{w(y)} + B(y) = B(t'). \end{aligned}$$

(ii) is similar.
It follows that $dh_R(t) \leq B(t) \leq |t| \cdot m \cdot c^{w(t)} \leq 2^{e|t|}$, for some constant e, and hence $Dh_R(n) = 2^{O(n)}$. □

We conclude this part by showing that none of the above upper bounds can be improved.

Example 3 (continued)
Identify the string $a^{i_1} b^{j_1} \ldots a^{i_n} b^{j_n}$ with the term $a^{i_1}(b^{j_1}(\ldots a^{i_n}(b^{j_n}(0))\ldots))$, and the string rewriting rule $l \to r$ with the term rewriting rule $l(x) \to r(x)$.
This systems satisfies both restrictions of proposition 3.2.2, and all rules are reducing under KBO for $[\{0, a, b\}, a > b > 0, w(0) = w(b) = w(a) = 1]$. □

3.2.3 Proposition
$Dh_R(n) = 2^{\Omega(n)}$, where R is the system of example 3.

Proof:
Clearly, a^n can be reduced to ba^{n-1}, and this reduction can be done in the following way.
$a^n \longrightarrow^* aba^{n-2} \longrightarrow^* abba^{n-3} \longrightarrow ba^{n-1}$.
This gives the following recurrence for $L(n)$, the length of the longest derivation of ba^{n-1} from a^n:
$L(n) \geq L(n-1) + L(n-2) + 1$.

Hence $L(n)$ is greater than the n-th Fibonacci number, and therefore,
$Dh_R(n) \geq L(n) = 2^{\Omega(n)}$. □

3.3 Syntactical restrictions: the rewrite rules

In the last section, we considered the case where an extended signature contains no unary function symbol of weight 0 (proposition 3.2.2). But even if such a symbol is present we get an exponential bound on the derivation height if the rewrite rules satisfy a certain condition.

3.3.1 Proposition
$Dh_R(n) = 2^{O(n)}$, if no rule of R has more occurences of the (only) unary function symbol of weight 0 on its right-hand side than on its left-hand side.

Proof:

First note that the construction to standardise signatures in the proof of theorem 3.2.1 preserves the above restriction. Therefore, w.l.o.g., we can assume the underlying extended signature to be $[\{+, s, 0\},\ s > + > 0,\ w(s) = w(+) = 0,\ w(0) = 1]$.

Every term t over this signature with $w(t) = m$, and r s-symbols can be thought of as composed from a binary tree on m leaves with r s-symbols distributed over its $2m - 1$ nodes. There are $C(m) < 2^{2m-1} \leq 2^{|t|}$ binary trees with m leaves (cf. proposition 3.2.2), and for each there are
$\binom{2m-2+r}{r} < 2^{|t|}$
ways of distributing the r s-symbols over its $2m - 1$ nodes. Since no rule can increase the size of the underlying binary tree, or the number of s-symbols, we conclude that
$dh_R(t) \leq \sum_{i=0}^{|t|} 2^{2i-1} < 2^{|t|}$, i.e., $Dh_R(n) = 2^{O(n)}$. □

Restricting the number of rules in R or the number of variables in rules of R does not yield such an upper bound on the derivation height in general. As example 2 shows, two rules are enough to produce derivations of rather impressive length. Even the one rule system

$$s(x) + (y + (z + (u + v))) \rightarrow x + (s(s(y)) + (u + (z + v)))$$

has no primitive recursively bounded derivation height (proof omitted).

On the other hand rules with no more than 3 variables are enough to simulate the system of example 2. Just replace its second rule by the two rules
$s(x) + (y + z) \rightarrow x + (z + y)$
$(x + y) + z \rightarrow x + (z + y)$.

The case of rewrite systems where each rule contains at most two variables remains an open problem.

Obviously, it would be desirable to have criteria which guarantee polynomial upper bounds on Dh_R. As yet, no such restrictions are known, although there are examples, where polynomial bounds can be proved.

Example 5
Consider the famous rewrite system for group theory (cf. [KB70, p.279]). This system G consists of 10 rules, the crucial ones being (using our standard signature)
(g1) $(x + y) + z \rightarrow x + (y + z)$
(g2) $s(x + y) \rightarrow s(y) + s(x)$

It can be shown that $Dh_G(n) = O(n^3)$, but (so far) we can't prove this with direct reference to a KBO–termination proof for G (which does exist). Instead we use mappings of terms to naturals, defined as follows: Let
$f_0(t)$ be the number of +–symbols in t ,
$f_1(t)$ the sum of all $f_0(t')$ over all occurences of t' as a left son of a +–symbol in t ,and
$f_2(t)$ the sum of all $f_0(t')$ over all occurences of t' as a son of a s–symbol in t.

Thus for ground terms s, t we get $f_0(t) = f_0(s)$, $f_1(t) < f_1(s)$, $f_2(t) = f_2(s)$ if t results from applying (g1) once to s, and
$f_0(t) = f_0(s)$, $f_1(t) < f_1(s) + f_0(s)$, $f_2(t) < f_2(s)$ if t results from applying (g2) once to s.

That is, f_1 can only be increased, by an amount of at most f_0, while decreasing f_2, and f_0 and f_2 are never increased. (The other 8 rules of G never increase any of f_0, f_1, f_2, as is easily seen.)

For a term t of size n we have $f_0(t) = O(n)$, $f_1(t) = O(n^2)$, $f_2(t) = O(n^2)$, and, therefore, f_1 can be increased at most $O(n^2)$ times by at most $O(n)$, and can thus be decreased, altogether, at most $O(n^3)$ times, which implies $Dh_G = O(n^3)$.

This (rather sketchy) proof reveals the best possible upper bound. Consider a term of size $3n + 1$ of the form
$s^n(...(((0 + 0) + 0) + 0)... + 0)$ (with n +–symbols). An innermost reduction of such a term, using the rules (g1) and (g2) only yields a derivation of length $n^2 \cdot \frac{n-1}{2}$. □

4 Conclusion

This paper discusses the complexity of term rewriting systems in terms of their *derivation height*, Dh_R. In particular, we investigated the interaction between termination proofs and derivation height, finding that proofs by polynomial interpretation are possible only for systems with a doubly exponential upper bound on Dh_R, whereas termination via KBO is more powerful, allowing Dh_R to grow faster than any primitive recursive function.

In addition to providing insight into the power of termination proof methods, our approach also leads to a number of conditions under which a termination proof via KBO (resp. polynomial interpretation) implies a small upper bound on Dh_R. Results of this type can be of use for implementations.

This is but a first step towards a quantitative termination theory for term rewriting systems. We feel that this approach should be pursued further, and believe that it will yield further insight into the nature of term rewriting.

We end with a list of suggestions as to further research along the lines of this paper.

- Find similar results for other termination proof methods, e.g., recursive path ordering, or for classes of proof methods, e.g., simplification orderings.
- Find conditions under which a termination proof by KBO (polynomial interpretation, RPO, etc.) implies a *polynomial* upper bound on Dh_R.
- Consider other complexity measures instead of dh_R. Particularly interesting would be the length of a shortest terminating derivation (in accordance with the usual complexity measure for nondeterministic models).

References

[CL87] Ahlem Ben Cherifa and Pierre Lescanne, *Termination of Rewriting Systems by Polynomial Interpretations and its Implementation.* Sci. of Comp. Prog. 9, pp. 137–159.

[De87] Nachum Dershowitz, *Termination of rewriting.* J.Symbolic Computation 3, pp. 69–116.

[DO88] Nachum Dershowitz and Mitsuhiro Okada, *Proof-theoretic techniques for term rewriting theory.* Proc. 3rd Ann. Symp. on Logic in Computer Science, pp. 104–111.

[Ge88] Oliver Geupel, *Terminationsbeweise bei Termersetzungssystemen.* Diplomarbeit, Sektion Mathematik, TU Dresden.

[He78] Hans Hermes *Aufzählbarkeit, Entscheidbarkeit, Berechenbarkeit.* 3rd ed., Springer.

[HO80] Gérard Huet and Derek Oppen, *Equations and rewrite rules: a survey.* In *Formal languages, perspectives and open problems*, ed. Ronald Book, Academic Press.

[KB70] Donald E. Knuth and Peter B. Bendix, *Simple Word Problems in Universal Algebras.* In: J. Leech, Ed., *Computational Problems in Abstract Algebra*, Oxford, Pergamon Press, pp. 263–297.

[Ln75] Dallas Lankford, *Canonical algebraic simplification in computational logic.* Report ATP-25, University of Texas.

[Ln79] Dallas Lankford, *On proving term rewriting systems are Noetherian.* Report MTP-3, Louisiana Tech University.

[La88] Clemens Lautemann, *A note on polynomial interpretation.* EATCS Bulletin 36, to appear.

[Le86] Pierre Lescanne, *Divergence of the Knuth–Bendix completion procedure and termination orderings.* EATCS Bulletin 30, pp. 80–83.

[Ma87] Ursula Martin, *How to choose the weights in the Knuth Bendix ordering.* Proc. of the Second Int. Conf. on Rewriting Techniques and Applications, LNCS 256, pp. 42–53.

ABSTRACT REWRITING

WITH CONCRETE OPERATORS

Stéphane Kaplan [1,2,3], Christine Choppy [3]

Now, what is intuition, except that gnostic reproduction of the drastic act by which the infinite complexity is offered as indivisible simplicity?
V. Jankélévitch, *Le Je-ne-sais-quoi et le Presque-Rien*

1. INTRODUCTION

Formal specifications are widely considered as crucial in early stages of software development. Their essential purpose is to state precise and non-ambiguous definitions of the problem under consideration, allowing to check later realizations against them. The interest of formal specifications is still greater when they can be used *effectively* at various stages in the software life cycle, such as prototyping, program construction, testing, and software re-use [CG 86]. Among formalisms available, algebraic specifications allow to write modular specifications – provided adequate specification building primitives (such as in [Gau 85], [Bid 88]) are given. Therefore, complex specifications (that correspond to complex and/or large systems) may be decomposed into smaller units (called specification modules) that are more understable and manageable. Tools and integrated environments (cf. [BCV 85], [GHW 85], [FGJM 85], etc.) for the development of algebraic specifications are now available, providing among others *symbolic evaluation* tools; they allow to execute partial experiments on a given specification, while it is still at an abstract, non-implemented stage.

Experience with large specifications shows that it is desirable to have a strategy with freedom of development: when dealing with a complex problem, it may be uneasy to use a strictly top-down approach, since it may not be clear yet how to analyze the problem as a whole. Moreover, this may lead to make

[1] Computer Science Department, Hebrew University, Givat Ram, Jerusalem (Israel)
[2] Computer Science Department, Bar-Ilan University, Ramat-Gan (Israel)
[3] L.R.I., Bât. 490, Université de Paris-Sud, Orsay (France)

assumptions about subsystems that are checked only in later stages, and errors are discovered too late. On the contrary, using a bottom-up approach leads to start the development process with too elementary specifications, i.e. with a granularity that entails an unnatural design process, and with increased probability of having to modify early decisions. It is then important to develop specifications starting "somewhere in the middle", at points where the degree of complexity is apprehensible without having to deal with too low-level details. Moreover, such "middle-out" approach promotes the re-use of specifications ([CG 86], [GM 88]). In this article, we concentrate on how to mix modules at an early development stage (there are still abstract, i.e. non-implemented), with modules that have already received a concrete implementation.

In the following, we consider executable specifications, in which positive conditional axioms may be viewed as *term rewriting systems*. Prototypes are thus easily obtained for such specifications; and checking the system behaviour becomes possible using a symbolic evaluator. Now, once a module has been checked, it may be interesting to provide an efficient code implementation for it, so as to speed up further evaluations. Operations of this module are then called *built-in* operators. If such an implemented module is used by another non-implemented specification module, evaluation of a term has to be *mixed*: via direct code evaluation for the built-in operators, and via rewriting for the other operators. Another advantage of this approach is that it allows to provide treatment for data structures with complex equations between constructors (s.a. sets, arrays, circular lists), that still remain out of the scope of rewriting modulo equations (cf. [Jouannaud 83], [JK 86], [DJ 89]). Mixed evaluation raises various issues [Cho 87] among which type-checking problems, parsing and displaying mixed expressions, etc. Also, mixing code and rewriting evaluations may, in some cases, block the evaluation. This situation occurs when both implemented and non-implemented operation symbols appear in the left-hand side of a rewriting rule, as illustrated in the following example.

Example 1 :

Assume a specification module NAT_{basic}, featuring the operations 0, s, +, and $*$, has been specified, and implemented in a standard fashion (for instance, $s\ s\ s\ s\ 0$ yields 4). Now, a new specification module NAT_{fact} is constructed above NAT_{basic}, enriching it with the operation symbol *fact* and the rules: $\{fact(0) \rightarrow s(0);\ fact(s(x)) \rightarrow s(x) * fact(x)\}$. Evaluation of $fact(4)$ cannot be performed as it is, since no rule matches. There is here a need for an explicit inverse to the constructor s, to discover that $4 = s(3)$. Evaluation may then proceed: $fact(4) \rightarrow 4 * fact(3) \rightarrow 4 * 3 * fact(2) \rightarrow ... \rightarrow 12$.

In this article, we define a formal framework for mixed evaluation, where such an evaluation process may be performed using inverse operations for the constructors. Section 2 defines specifications, and implementations, with built-in operators. Section 3 introduces a notion of rewriting in a concrete model; we then define built-in pattern-matching (via the inverses of the implementation of the constructors), and the notion of mixed rewriting. We state the correctness of mixed rewriting w.r.t. the intended specification, and show that it yields an optimally abstract implementation (w.r.t. a family of fixed built-in operators). We assume the reader has general knowledge about the fields of term rewriting systems and algebraic specifications; we refer respectively to [HO 80], [Klop 87], [JK 88] and to [ADJ 78], [EM 85], [Wirsing 89] for basic references. Our notations closely follow the ones that are standard in these fields. Along the paper, we denote vectors (of variables, expressions) by overlined symbols, s.a. $\overline{x}$, $\overline{\omega}$. In order to improved readability, our definitions have been presented for one-sorted specifications; extension to the multi-sorted case are straightforward. Also, we mainly put emphasis here on the case of a single module being partially implemented via built-in operators and via rewriting. Extension to the case of several modules is easy; one can then freely combine parts of concrete and abstract modules between them. These two extensions are used liberally in the paper (cf. e.g. example 2).

2. BUILT-IN SPECIFICATIONS AND IMPLEMENTATIONS

We systematically adopt a hierarchical approach of the specification and of the implementation processes. To this effect, the family Σ of operators is divided into a set C of *constructors* and a set of *derived operators*. Each constructor receives a built-in implementation. The derived operators are further divided into the set D_{bi} of operators that have a built-in implementation, and into the set D_{rew} that will be interpreted via mixed rewriting.

2.1. Built-in specifications

Formally, a *built-in specification* is a specification $<C \cup D_{bi} \cup D_{rew}\ ,\ E_C \cup E_{bi} \cup R>$, where:

- C is a set of operators called *constructors*. E_C is a set of *equations between constructors*, i.e. positive conditional equations between terms of $T_C(X)$.
- D_{bi} is a set of *built-in derived operators*. E_{bi} is a set of equations defining the operators of D_{bi}, i.e. positive conditional equations between terms of $T_{C \cup D_{bi}}(X)$. The specification $<C \cup D_{bi}\ ,\ E_C \cup E_{bi}>$ must be hierarchically consistent and sufficiently complete with respect to $<C\ ,\ E_C>$. This means respectively that $\forall t,t' \in T_C,\ t \equiv_{E_C \cup E_{bi}} t'$ *iff* $t \equiv_{E_C} t'$ and that $\forall t \in T_{C \cup D_{bi}}$, $\exists t' \in T_C$ such that $t \equiv_{E_C \cup Eder} t'$.
- D_{rew} is a second set of derived operators. R is a term rewriting system defining D_{rew}. Its rules are constrained to the form $P \Rightarrow d(\overline{\omega}) \rightarrow \tau$, where $\overline{\omega}$ is a vector of terms of $T_C(X)$, τ is in $T_{C \cup D_{bi} \cup D_{rew}}(X)$, and P is a conjunction of equalities between terms of $T_{C \cup D_{bi} \cup D_{rew}}(X)$.

Example 1 (continued): back to the example presented in section 1, we have $C = \{0, s\}$, $D_{bi} = \{+, *\}$, $D_{rew} = \{fact\}$, $E_C = \emptyset$, $E_{bi} = \{x+0=x;\ x+s(y)=s(x+y);\ x*0=0;\ x*s(y)=x+(x*y)\}$, $R = \{fact(0) \rightarrow s(0);\ fact(s(x)) \rightarrow s(x)*fact(x)\}$.

Example 2 : we consider here a specification of sets of natural numbers (which are supposed to have been specified previously. We do not concentrate in this article on how a new specification is built on top of another – as mentioned in the introduction). Then: $C = \{\varnothing:\rightarrow set,\ ins:nat\times set\rightarrow set\}$, $D_{bi}=\varnothing$, $D_{rew}=\{Inf:set\rightarrow nat\}$ (intended to compute the smallest element of a set), $E_C = \{ins(x,ins(x,S))=ins(x,S);\ ins(x,ins(y,S))=ins(y,ins(x,S))\}$, $E_{bi}=\varnothing$ and $R = \{Inf(\varnothing)\rightarrow 0;\ Inf(ins(x,\varnothing))\rightarrow x;\ (x\leq y)=T \Rightarrow Inf(ins(x,ins(y,S))) \rightarrow Inf(ins(x,S))\}$. It should be noted that the specification is not sufficiently complete in the usual sense [no rule allows to compute $Inf(ins(x,ins(y,S)))$ when $(x\leq y)=F$]. However, with the built-in implementation of it that is considered hereafter (section 3), a weaker form of completeness is satisfied. We do not know whether this example may be treated using a classical approach such as rewriting modulo equations (which necessitates, in particular, unification and pattern-matching algorithms for E_C. Obtaining such algorithms for an *ad hoc* E_C is still a challenge).

2.2. Built-in implementations

In a built-in implementation, the operators of C and D_{bi} receive built-in interpretations, that must satisfy the equations of $E_C \cup E_{bi}$. We also suppose that the user provides *inverse functions* for the built-in interpretation of the constructors of C, as explained below. On the other hand, the operators of D_{rew} are purposedly left uninterpreted: their implementation will be by means of mixed rewriting. Formally, given a built-in specification as above, an associated *built-in implementation* consists of:

- a semi-computable domain A
- for every constructor c in C, a semi-computable function $c^A : A^{ar(c)} \rightarrow A$. It is assumed that A and the c^A's satisfy the axioms of E_C
- for every $c \in C$, an *inverse* function to c^A, i.e. a function $c^{-A} : A\times N \rightarrow A^{ar(c)} \cup \{\perp\}$. c^{-A} is so that, for a given $y\in A$, the sequence $(c^{-A}(y,n))_{n\geq 0}$ gives all the solutions $\bar{x}$ to the equation $c^A(\bar{x})=y$ in A. Moreover, if $c^{-A}(y,n)=\perp$, then all the solutions should have been obtained for $(c^{-A}(y,n'))_{n'<n}$
- for any d in D_{bi}, a semi-computable function $d^A : A^{ar(d)} \rightarrow A$. It is assumed that A, the c^A's $(c\in C)$ and the d^A's $(d\in D_{bi})$ satisfy the axioms of E_C and E_{bi}

Note: inversion of the constructors may yield an infinite number of solutions. This is why we chose to let $c^{-A}(y,n)$ stand for the nth solution to the equation $c^{-A}(\bar{x})=y$. If the solutions happen to be in finite number, then $c^{-A}(\bar{x})$ is identically equal to $\perp$, for n large enough. Also remark that in the above definition, $ar(c)$ – the arity of the constructor c – is actually a *word* on the sort names (so that, for instance, $A^{ar(ins)}$ actually denotes $A^{nat}\times A^{set}$).

Example 1 (continued) : A is taken as being the classical domain N of natural numbers. '0', 's', '+', '$*$' receive their usual interpretation. The inverses for '0' and 's' are defined by:

$$0^{-N}(x,0) = \textbf{if } x \neq 0_N \textbf{ then } \bot \textbf{ else } () \qquad \text{[the empty list of argument] and}$$
$$s^{-N}(x,0) = \textbf{if } x = 0_N \textbf{ then } \bot \textbf{ else } x - 1_N$$

[$0^{-N}(x,n)$ and $s^{-N}(x,n)$ are equal to $\bot$ for $n \geq 1$].

Example 2 (continued) : consider the above specification of sets. We take for A the classical sets of natural numbers. The constructor '$\varnothing$' is implemented by the empty set, and '*ins*' is implemented by $ins_A(x,S) = \{x\} \cup S$. The inverse function ins^{-A} is defined as follows:

$$ins^{-A}(\{e_1,\ldots,e_m\}, n) = \textbf{if } n \in [1..m] \textbf{ then } (\{e_1,\ldots,e_{n-1},e_{n+1},\ldots,e_m\}, e_n)$$
$$\textbf{elseif } n \in [m+1..2m] \textbf{ then } (\{e_1,\ldots,e_m\}, e_{n-m})$$
$$\textbf{else } \bot$$

3. REWRITING WITH BUILT-IN OPERATORS

3.1. Rewriting in a concrete model

In this paragraph (only), we simply suppose that a signature Σ is given, together with a term rewriting system R on Σ. Let A be a Σ-algebra. $T_\Sigma(A)$ stands for the Σ-algebra generated by A. Its elements may be seen as a partially evaluated expressions, where abstract entities (operators of Σ) and built-ins (elements of A) are mixed. There exists a unique morphism from $T_\Sigma(A)$ into A. Its application to an expression t of $T_\Sigma(A)$ is denoted by t^A.

Given a Σ-term rewriting system R, we define the following *built-in rewrite* relation on $T_\Sigma(A)$:

$t \rightarrow_{R|A} t'$ *iff* there exists a rule $r: \lambda \rightarrow \rho$ in R, a context K in $T_\Sigma[X]$ and a substitution $\sigma: X \rightarrow T_\Sigma(A)$ such that

$$t^A = (K[\lambda\sigma])^A \text{ and } t'^A = (K[\rho\sigma])^A$$

Notes: $\rightarrow_{R|A}$ being a binary predicate, classical notions s.a. noetherianity, confluence and normal forms are defined as usual. Actually, $\rightarrow_{R|A}$ is noetherian (resp. confluent) iff its restriction to A is. Notice also that if A happens to satisfy a set of equations E, then $\rightarrow_{R|A}$ may be viewed as implementing rewriting *modulo E !*

3.2. Pattern-matching with built-in constructors

We consider a built-in specification and an associated built-in implementation A, as in section 2. In order to compute the application of an operator d of D_{rew} to an expression, one needs to perform pattern-matching in A. For instance, to apply the rule $fact(s(x)) \rightarrow s(x) * fact(x)$ to the expression $fact(4)$, one needs to detect that $4 = s^N(3)$. So far, one knows how to compute the solutions $\bar{x}$ in A to the equation $c^A(\bar{x}) = y$, when c is a *constructor* of C; this is precisely the role of the inverse function c^{-A}. We now extend this to equations of the form $t^A(\bar{x}) = y$, where t is any context of $T_C(X)$, i.e. made of constructors only. To this effect, define the function $t^{-A}: A \times N \rightarrow A^{ar(t)} \cup \{\bot\}$ in the following bottom-up fashion (where as before $t^{-A}(y,n)$ stands for the nth solution to $t^A(\bar{x}) = y$):

```
function t^{-A}(y,n)
  if t = c[X̄] for some c∈C then return( c^{-A}(y,n) )
  else
    -- t may be written under the form c(ω_1,...,ω_m), with each ω_i in T_C(X).
    (p,q) := the n^th couple of natural numbers according to diagonal enumeration ;
    -- Compute the q^th solution to c^A(ξ_1,...,ξ_m) = y
    (ξ_1,...,ξ_m) := c^{-A}( y , q ) ;
    (n_1,...,n_m) := the p^th m-tuple of natural numbers ;
    -- Compute for each i the n_i^th solution to ω_i(x̄_i) = ξ_i
    for i = 1 to m do x̄_i := ω_i^{-A}( ξ_i , n_i ) ;
    -- This provides m partial assignements to the variables of t. Return their "union"
    -- (and '⊥' if for some i,j, x̄_i and x̄_j assign different values to a variable of t)
    return( ∪_{i=1}^m x̄_i )
```

Note: the role of p, q and of the n_i's is purely technical, allowing to obtain precisely the nth solution to the equation. We freely use Cantor's diagonal enumeration for tuples of natural numbers.

Theorem

$(t^{-A}(y,n))_{n\geq 0}$ is the set of solutions in A to the equation $t^A(\overline{x}) = y$.

The proof is by structural induction on t, and will appear in the full paper. Note that if the set of solutions to $t^A(\overline{x}) = y$ is finite, then $t^{-A}(y,n)$ is always equal to $\perp$ for n large enough.

3.3. Evaluation via mixed rewriting

Given a built-in implementation A, we now indicate how to evaluate expressions in $T_{C\cup D_{bi}\cup D_{rew}}(A)$. A function $Val_A : T_{C\cup D_{bi}\cup D_{rew}}(A) \rightarrow A\cup\{Error\}$ is defined as follows:

```
function Val_A(t)
  if t∈A then return(t)
  elseif t=f(t_1,...,t_m) with f∈C∪D_bi then return( f^A(Val_A(t_1),...,Val_A(t_m)) )
  elseif t=d(t_1,...,t_m) with d∈D_rew then
    for i=1 to m do t_i := Val_A(t_i) ;
    if there exists a rule P ⇒ d(ω_1,...,ω_m) → τ and natural numbers n_1,...,n_m, s.t.
      x̄ = ∪_{i=1}^m ω_i^{-A}(t_i,n_i) ≠ ⊥ and P[x̄] holds
    then return( Val_A(τ(x̄)) )
    else return 'Error'
```

Note that the reduction strategy is call-by-value. This follows naturally the structure of the rules: a rule applies at a given occurrence if the children under this occurrence match the ω_i, which are made of constructors only. Hence the choice to reduce the children first.

We say that the rewrite definitions of the operators of D_{rew} are *built-in complete w.r.t. A* if for any $d \in D_{rew}$, for any $t_1,\ldots,t_n$ in A, there exists a rule $P \Rightarrow d(\omega_1,\ldots,\omega_m) \rightarrow \tau$ that rewrites $d(t_1,\ldots,t_n)$; this means that there exist natural numbers $(n_i)_{1 \le i \le m}$ such that $\overline{x} = \cup_{i=1}^{m} \omega_i^{-A}(t_i,n_i) \neq \bot$ and such that $P[\overline{x}]$ is true.

Example 2 (continued) : the derived operator definitions are actually complete w.r.t. to our implementation of sets. When S has zero or one element, *Inf* (S) may be reduced by the first or by the second rule of R. If S has two elements or more, we can always assume that $S = ins_A(x, ins_A(y,S'))$, with $x \le y$. The third rule then applies to *Inf* (S). For example, *Inf*$(\{5,4,3\})$ may be rewritten by the third rule for one of the substitutions $\sigma_1 = [x \leftarrow 3, y \leftarrow 4, S' \leftarrow \{5\}]$, $\sigma_2 = [x \leftarrow 3, y \leftarrow 5, S' \leftarrow \{4\}]$, $\sigma_3 = [x \leftarrow 4, y \leftarrow 5, S' \leftarrow \{3\}]$. This example shows that built-in completeness – as defined above – is indeed relative to a given implementation A. If natural numbers had been implemented by built-ins *not* totally ordered via '≤', we could not have assumed that any set S with two elements or more is expressible as $S = ins_A(x, ins_A(y,S'))$, with $x \le y$. In such a case, built-in completeness would not be satisfied.

We now have the following correctness results:

Theorem

Let A be a built-in implementation. Suppose that $\rightarrow_{R|A}$ is confluent and noetherian, and that the rewrite definitions of D_{rew} are complete w.r.t. A. Then:

(1) For any t in $T_\Sigma(A)$, all the possible executions of $Val_A(t)$ terminate, eventually yielding the normal form of t for $\rightarrow_{R|A}$

(2) A may be viewed as a $<C \cup D_{bi} \cup D_{rew}, E_C \cup E_{bi} \cup R>$-model, interpreting any $d \in D_{rew}$ via $\lambda \overline{x}\,.\, Val_A(d(\overline{x}))$

(3) Let F stand for the forgetful functor from the $C \cup D_{bi} \cup D_{rew}$-algebras into the $C \cup D_{bi}$-algebras. Let C_A be the class of all the $<C \cup D_{bi} \cup D_{rew}, E_C \cup E_{bi} \cup R>$-models M such that $F(M)$ is isomorphic to A. Then A is the initial model of C_A.

Under the above hypothesis, **(1)** states the total correctness of the function Val_A. **(2)** means that the implementation of the operators of D_{rew} via $\rightarrow_{R|A}$-rewriting is correct w.r.t. R. And **(3)** means that among all the models for which the interpretation of C and D_{rew} is frozen into the built-in implementation A, A is initial (and may thus be considered as optimal, from the abstraction point of view).

4. CONCLUSION

In this article, we have presented a formal basis for mixing fully abstract specifications and concrete ones.

The overall approach thus allows to work at the same time with specifications at different stages of development, and to run– still– prototyping experiments with them. In this framework, evaluation is performed in a mixed fashion: the abstract parts are computed via rewriting, whereas the concrete ones are dealt with via direct code execution.

The ideas presented here are being currently implemented in Lisp, within the ASSPEGIQUE integrated environment for the development of algebraic specifications [BCV 85]. Special attention is devoted to the actual implementation of the built-in pattern-matching process. We have adopted a demand-driven strategy, such that the $(n+1)^{\text{th}}$ solution of an equation $c^A(\overline{x}) = y$ is computed only if the nth solution has lead to failure (non-satisfaction of the premises of the current rule). Our first experiments are particularly encouraging; in particular, we are able to run, in a still abstract fashion, tests with specifications that could not be dealt with so far (e.g. with the sets of natural numbers specification presented in example 2, that has been implemented exactly as presented in the article). Implementation issues will receive further attention in the next future, together with other questions: possibility of *compiling* (cf. [Kap 87]) mixed rewriting, means of defining inverses in a clean and systematic fashion, etc.

Acknowledgments :

This work is partially supported by ESPRIT project No. 432, and by CNRS PRC Programmation avancée et outils de l'intelligence artificielle.

REFERENCES

[ADJ 78] J.A. Goguen, J.W. Thatcher, E.G. Wagner, *An Initial Algebra Approach to the Specification, Correctness, and Implementation of Abstract Data Types,* Current Trends in Programming Methodology, Vol. 4, Ed. Yeh R., Prentice-Hall, pp. 80-149 (1978)

[Bid 88] M. Bidoit, *The stratified loose approach: a generalization of the initial and loose semantics,* in Recent Trends in Data Type Specification, LNCS 332, Springer Verlag (1988)

[BCV 85] M. Bidoit , C. Choppy, F. Voisin, *The ASSPEGIQUE specification environment, Motivations and design,* Proc. of the 3rd Workshop on Theory and Applications of Abstract data types, Bremen, Nov 1984, Recent Trends in Data Type Specification (H.-J. Kreowski ed.), Informatik-Fachberichte 116, Springer Verlag, Berlin-Heidelberg, pp. 54-72 (1985)

[CG 86] C. Choppy, M.-C. Gaudel, *Impact des spécifications formelles sur le développement de logiciel,* Recueil des conférences de la Convention Informatique, Tome A, Paris, 19 pp 335-339 (1986)

[Cho 87] C. Choppy, *Formal specification, prototyping and integration testing,* Proc. of the 1st European Software Engineering Conference, Strasbourg, pp. 185-192 (1987)

[DJ 89] N. Dershowitz, J.-P. Jouannaud, *Rewriting systems,* Handbook of Theoretical Computer Science, A. Meyer, M. Nivat, M. Peterson, D. Perrin eds., (to appear 1989)

[EM 85] H. Ehrig, B. Mahr, *Fundamentals of algebraic specifications. I : Equations and initial semantics,* EATCS monographs on Theoretical Computer Science, Springer Verlag (1985)

[FGJM 85] K. Futatsugi, J.A. Goguen, J.-P. Jouannaud, J. Meseguer, *Principles of OBJ2*, Proc. 12th ACM Symp. on Principle of Programming Languages, New Orleans, pp. 52-66 (1985)

[Gau 85] M.-C. Gaudel, *Towards structured algebraic specifications*, Proc. ESPRIT Technical Week, Bruxelles, Springer Verlag (1985)

[GM 88] M.-C. Gaudel, T. Moineau, *A theory of software reusability*, Proc. of the ESOP'88 Conf., LNCS 300, Springer Verlag (1988)

[GHW 85] J.V. Guttag, J.J. Horning, J.W. Wing, *The Larch family of specification languages*, IEEE Software, **2**,4 (1985)

[HO 80] G. Huet, D.C. Oppen, *Equations and rewrite rules : a survey*, Formal languages : Perspective and open problems, R. Book Ed., Academic Press (1980)

[Jouannaud 83] J.-P. Jouannaud, *Confluent and coherent sets of reductions with equations*, Proc. of the 8th ICALP Conference, L.N.C.S. 53, pp. 269-283 (1983)

[JK 86] J.-P. Jouannaud, H. Kirchner, *Completion of a set of rules modulo a set of equations*, SIAM J. on Computing **15**, pp. 1155-1194 (1986)

[Kap 87] S. Kaplan, *A compiler for conditional rewriting systems*, Proc. 2nd Conf. on Rewriting techniques and applications, Bordeaux, L.N.C.S. 256, Springer Verlag, pp. 25-41 (1987)

[Klop 87] J.W. Klop, *Term rewriting systems : a tutorial*, Bulletin of the EATCS, **32**, pp. 143-183 (1987)

[Wirsing 89] M. Wirsing, *Algebraic specifications*, Handbook of Theoretical Computer Science, A. Meyer, M. Nivat, M. Peterson, D. Perrin eds., (to appear 1989)

On How To Move Mountains 'Associatively and Commutatively'

Mike Lai*
Department of Computer Science
Royal Holloway and Bedford New College
University of London

Abstract

In this paper we give another characterization of a set of rules which defines a Church-Rosser reduction on the term algebra specified by some associative and commutative equations. This characterization requires fewer conditions to be satisfied than those previously given in the literature do. As a result, when the required conditions are satisfied, the word problem in the term algebra defined by the set of rules and the set of associative and commutative equations can be solved by successive applications of rewriting to the elements in question.

In addition, what makes this approach different from the others is that notions such as AC-compatibility or coherence modulo $\mathcal{AC}$ of reductions induced by sets of rules, which are essential in [Pe-St] or [Jo-Ki] respectively, are not required here. Consequently, a proof of correctness of the completion algorithm (given in [Lai 2]) for constructing a desired set of rules based on this approach can be compared directly with that of Huet in [Hu 2]. In fact, it turns out that all we have to do is to replace terms in [Hu 2] by AC-equivalence classes of terms. The main reason is that all the complications due to AC-compatibility or coherence modulo $\mathcal{AC}$ simply are not present here.

Finally, we shall discuss how to minimize the unnecessary computation of some critical pairs during the completion.

1 Introduction

In the early part of 70's, Knuth and Bendix [Kn-Be] developed a characterization of the set of rewrite rules which would solve the word problem in the quotient Δ of a free algebra Γ defined by the given set of equations. Intuitively, rules are just equations with some extra conditions being satisfied; and so, if an element in the free algebra Γ can be rewritten into another by successive applications of the set of rules, then they must represent the same element in the quotient algebra Δ defined by the set of rules. However, there is no way to make an equation such as $f(x,y) = f(y,x)$ (commutative equation) into a rule as some element can be rewritten for ever. So, certainly, Knuth and Bendix's characterization has its limitations.

*Supported by the U.K. Science and Engineering Research Council under grant GR/E83634

Since then a number of authors such as Lankford and Ballantyne [La-Ba 8], Huet [Hu 1], Peterson and Stickel [Pe-St], Pederson [Pe], Jouannand and Kirchner [Jo-Ki], and Bachmair [Ba] have offered their own variations and improvements. In particular, they overcame some of the limitations of the original characterization of Knuth and Bendix. For instance, instead of computing within the free algebra Γ, it was proposed that computations should take place among the elements of the quotient algebra Δ' defined by a subset of the given set of equations and this subset of equations would not be regarded as rules. In that case, obviously, the original quotient algebra Δ is isomorphic to the quotient of Δ' defined by the rest of the equations. However, a special unification algorithm is required in this case (unification is one of the basic tools in most characterizations of a set of rules, including that of Knuth and Bendix).

Let us now introduce some basic ideas by using the well-known example, a free abelian group. In the presence of the associative and commutative equations, $x + (y + z) = (x + y) + z$ and $x + y = y + x$ respectively, a set of rules R which are essentially the following is given in [La-Ba 8], [Pe-St], [Jo-Ki] and [Ba] to solve the word problem in the free abelian group Δ.

$$\begin{array}{rcll} x+0 & \rightarrow & x & (1) \\ -0 & \rightarrow & 0 & (2) \\ -(-x) & \rightarrow & x & (3) \\ -(x+y) & \rightarrow & (-x)+(-y) & (4) \\ x+(-x) & \rightarrow & 0 & (5) \end{array}$$

In fact, the Church-Rosser property of R is one of the conjectures originally stated in [La-Ba 4] and then proved later in [La-Ba 8]. Most importantly, a complete proof of the correctness of an algorithm to support the construction of R is found in [Jo-Ki] and [Ba]. Here the underlying algebra Δ' is the quotient of Γ by the associative and commutative equations. The rewriting or the reduction $[\mathcal{R}]_{AC}$ on Δ' defined by R is the following: let S and T be two distinct elements of Δ', we say S is $[\mathcal{R}]_{AC}$-rewritten to T by a rule r in R if there exist a representative $s \in S$ and a representative $t \in T$ such that s is rewritten to t in Γ (in the usual sense of rewriting in a free term algebra) by r. For example, let S and T be the equivalence classes of $(x + y) + (-y)$ and $0 + x$ respectively, then S is $[\mathcal{R}]_{AC}$-rewritten to T since $s = x + (y + (-y))$, which is a representative of S, is rewritten to $t = x + 0$, which is a representative of T, in Γ by rule (5) of R. Consequently, the word problem in Δ, which is isomorphic to the quotient of Δ' by R, can be solved by the rewriting $[\mathcal{R}]_{AC}$.

Although the flexibility of representations of elements in Δ' allows such a powerful rewriting $[\mathcal{R}]_{AC}$, none of the above approaches give a direct proof of the Church-Rosser property of $[\mathcal{R}]_{AC}$. Instead, the ideas of extensions of rules and coherence are introduced. Firstly, a set of rules R^e (called the extension of R), which contains R, is given more or less as follows:

$$\begin{array}{rcll} x+0 & \rightarrow & x & (1) \\ (x+0)+y & \rightarrow & x+y & (1^e) \\ -0 & \rightarrow & 0 & (2) \\ -(-x) & \rightarrow & x & (3) \\ -(x+y) & \rightarrow & (-x)+(-y) & (4) \\ x+(-x) & \rightarrow & 0 & (5) \\ x+(-x)+y & \rightarrow & 0+y & (5^e) \end{array}$$

where rule (1^e) and rule (5^e) are called the extension of rule (1) and rule (5) respectively (note that, with the notion of left linear rule, $(x + 0) + y \rightarrow x + y$ is replaced by $0 + x \rightarrow x$ in [Jo-Ki] and [Ba]). It is clear that the rewritings $[\mathcal{R}]_{AC}$ and $[\mathcal{R}^e]_{AC}$ on Δ' defined by R and R^e are identical although, for instance, if an element S is $[\mathcal{R}^e]_{AC}$-rewritten to another element T by rule (5), it does not imply that S can also be $[\mathcal{R}^e]_{AC}$-rewritten to T by rule (5^e) in Δ'.

Secondly, another rewriting $\mathcal{R}^e.\mathcal{AC}$ on Γ induced by R^e is defined as follows: let s, t be terms in Γ, s is $\mathcal{R}^e.\mathcal{AC}$-rewritten to t if s has a subterm s' such that an AC-equivalent of s', say s'', can be rewritten to t' by a rule of R^e in Γ, and t is the result of replacing s' by t' in s. In addition, the notion of coherence modulo $\mathcal{AC}$ (where AC-compatibility is simply a special case) of $[\mathcal{R}^e]_{\mathcal{AC}}$ is given as follows: let t be a term in Γ and $[t]_{\mathcal{AC}}$ be its AC-equivalence class in Δ'. If $[t]_{\mathcal{AC}}$ is $[\mathcal{R}^e]_{\mathcal{AC}}$-rewritten to an element S by a rule in R^e, then there exist terms u and v such that t is $\mathcal{R}^e.\mathcal{AC}$-rewritten to u by a rule in R^e; and either (a) $S = [v]_{\mathcal{AC}} = [u]_{\mathcal{AC}}$, or (b) $S = [v]_{\mathcal{AC}}$ and $[u]_{\mathcal{AC}}$ can be $[\mathcal{R}^e]_{\mathcal{AC}}$-rewritten to $[v]_{\mathcal{AC}}$ by rules in R^e, or (c) $[v]_{\mathcal{AC}} = [u]_{\mathcal{AC}}$ and S can be $[\mathcal{R}^e]_{\mathcal{AC}}$-rewritten to $[v]_{\mathcal{AC}}$ by rules in R^e or (d) $[u]_{\mathcal{AC}}$ and S can be $[\mathcal{R}^e]_{\mathcal{AC}}$-rewritten to $[v]_{\mathcal{AC}}$ respectively by rules in R^e where $[u]_{\mathcal{AC}}$ and $[v]_{\mathcal{AC}}$ are the AC-equivalence classes of u and v respectively in Δ'. For example, let $t = (x + y) + (-y)$, and $S = [x + 0]_{\mathcal{AC}}$ be the result of the $[\mathcal{R}^e]_{\mathcal{AC}}$-rewriting on $[t]_{\mathcal{AC}}$ by rule (5). Then t is $\mathcal{R}^e.\mathcal{AC}$-rewritten by rule ($5^e$) to $u = 0 + x$, which is AC-equivalent to $x + 0$, since t is AC-equivalent to the left-hand side of rule (5^e).

As a consequence, this example demonstrate also that coherence modulo $\mathcal{AC}$ of $[\mathcal{R}]_{\mathcal{AC}}$ is not satisfied. However, it is showed in [La-Ba 8], [Pe-St], [Jo-Ki] and [Ba] that $[\mathcal{R}^e]_{\mathcal{AC}}$ has the property of coherence modulo $\mathcal{AC}$, from which the Church-Rosser property of $[\mathcal{R}^e]_{\mathcal{AC}}$ and hence that of $[\mathcal{R}]_{\mathcal{AC}}$ follow. Nevertheless, one would ask, for instance, whether the extension of rule (5) is really necessary. As indicated in the previous example, since $[t]_{\mathcal{AC}}$ is already $[\mathcal{R}^e]_{\mathcal{AC}}$-rewritten to S by rule (5), there is not much point to have $[t]_{\mathcal{AC}}$ also $[\mathcal{R}^e]_{\mathcal{AC}}$-rewritten to S by rule (5^e). Furthermore, during the procedure of the completion algorithm based on this similar approach, it is the Church-Rosser property of $[\mathcal{R}^e]_{\mathcal{AC}}$ that the algorithm is checking, rather than that of $[\mathcal{R}]_{\mathcal{AC}}$. Particularly, special care is needed for the extensions of rules in the one described in [Jo-Ki] or [Ba]; especially, if we require the algorithm to deliver a set of rules which is as "simple" as possible in terms of the $[\mathcal{R}]_{\mathcal{AC}}$-reducibility. Moreover, the Church-Rosser property of the rewriting $[\mathcal{R}]_{\mathcal{AC}}$, induced by a set of rules R, in Δ' does not really require the property of coherence modulo $\mathcal{AC}$ of $[\mathcal{R}]_{\mathcal{AC}}$ itself as clearly demonstrated in the case of free abelian group above. Consequently, this becomes the main motivation of this paper.

We show that the well-foundedness of $[\mathcal{R}]_{\mathcal{AC}}$ in Δ' and the $[\mathcal{R}]_{\mathcal{AC}}$-reducibility of "critical peaks" (with their corresponding critical pairs) between all pairs of rules in R are enough to guarantee the Church-Rosser property of $[\mathcal{R}]_{\mathcal{AC}}$. A peak with apex $[t]$ is roughly a triple of elements $([s], [t], [u])$ such that $[t]$ is $[\mathcal{R}]_{\mathcal{AC}}$-reduced to a pair of elements $[s]$ and $[u]$ by a pair of rules r and ℓ respectively; and if $([s], [t], [u])$ is a "critical peak", then $[s] = [u]$ is a critical pair. Three main ideas of this paper are the following:

1. instead of the complicated noetherian induction which is a combination of subterm and subsumption orderings modulo $\mathcal{AC}$, a simple induction on "the derived lengths of terms, with certain tree structure, containing a given proper subterm" is employed. The author feels that this method may be useful also in situations where infinite equivalence classes are allowed.
2. besides unifying a pair of terms, we discover that there is another job that unification in the quotient algebra Δ' defined by a set of associative and commutative equations can do for us. Namely, it can tell us what the "disguises" of an element in Δ' "roughly" look like.
3. we do not need to consider the rewriting by the set of rules directly but a similarly defined rewriting by a set of super rules. Intuitively, the super rule of a rule r is a subset of the many "disguises" of r in Δ'. Actually, if we replace the left- and right-hand sides of a rule by their respective $\mathcal{AC}$-equivalence classes, we can view R as a set of rules in Δ'. In that case, the super rule of a rule r are just images of r under some substitutions. It is shown that a peak involving r can be factored through a peak involving one of r's "disguises" in its super rule. The amount of work (unification of the subterms of its left-hand side with the left-hand sides of other rules) r does to construct "critical peaks" involving r is shared out by its "disguises". It can be seen

from the definition of "critical peaks" (Section 4) that the work each "disguise" of r does is lesser than that r itself does using the traditional definition of critical pair. Moreover, it gives an insight on how to eliminate the "useless" critical peaks.

In fact, it turns out that a rewriting in Δ' by a rule r is the same as a rewriting in Δ' by an element of the super rule of r. In that case, exactly as the original characterization of Knuth and Bendix, the one that we are developing here also depends on the well-foundedness of rewriting and the reducibility of peaks by applications of rewriting in Δ'. Locally confluent is a just special case of peak reducing which has a slight advantage when we prove the confluence property of a rewriting. In addition, we give a definition of critical peak in Δ' which depends on a pair of super rules; but in contrast to the similar definitions of critical peak or critical pair elsewhere, we have a total of four types of critical peak. They are especially designed in such a way that the number of suitable candidates for proving our results is as small as possible. For example, if the left-hand side of a rule has the form $t = (((x + x) + x) + x)$, then peaks involving proper subterms of t will never be considered as critical peaks. Moreover, we shall sketch the proof of the Peak-Reduction Lemma which states that if every critical peak in Δ' is reducible by applications of rewriting, then so is any other peak in Δ'. Consequently, we also have a characterization of a set of rules R which would solve the word problem in the required quotient algebra, namely the quotient of Δ' by R. As a result, there is an effective way to check whether a given set of rules R does indeed meet this characterization which is by checking the well-foundedness of the rewriting defined by the set of rules and the reducibility of critical peaks.

As a second motivation, a phenomenon of critical peak is pointed out in [Ka-Mu-Na], namely, if the apex $[t]$ of a "critical peak" $([s],[t],[u])$ is also $[\mathcal{R}]_{\mathcal{AC}}$-reduced, in a certain way, to an element different from $[s]$ and $[u]$, then the computation of $([s],[t],[u])$ is not required. As a result, unlike other algorithms, there is a checking of the $[\mathcal{R}]_{\mathcal{AC}}$-reducibility of the unified terms after the process of the unification of the subterms of one left-hand side with the other left-hand side of a pair of rules. This is shown to be worth doing. However, it can be seen in [Lai 2] that the $[\mathcal{R}]_{\mathcal{AC}}$-reduction of $[t]$ to an element different from $[s]$ and $[u]$ often arises from the $[\mathcal{R}]_{\mathcal{AC}}$-reducibility, in a certain way, of a "disguise" of one of the rules which $[\mathcal{R}]_{\mathcal{AC}}$-reduces $[t]$ to $[s]$ and $[u]$. In other words, if a "disguise" of a rule is discovered to be $[\mathcal{R}]_{\mathcal{AC}}$-reducible at the ith stage of the algorithm then the unification and hence the computation of the critical peaks involving this "disguise" is not required from the ith stage onward. In a way, checkings of the $[\mathcal{R}]_{\mathcal{AC}}$-reducibility of all the "disguises" of a rule r can almost replace the checkings of the $[\mathcal{R}]_{\mathcal{AC}}$-reducibility of the unified terms of the subterms of the left-hand side of r with the left-hand sides of other rules from the ith stage onward. A complete proof of the correctness of the associative and commutative completion algorithm in co-operate with this strategy is given in [Lai 2].

The rest of this paper is organized as follows. In Section 2, we define the concept of relations. In Section 3, we give the necessary terminologies about term algebras and, also, state the relevant lemmas or theorems. In particular, if a term s contains a proper subterm u in a certain way, then, in the tree structure of s, the path from the top to the point where u occurs will be crucial to our main proof in Section 4. Furthermore, the two main roles of unification will be discussed. In Section 4, we formally define the super rule of a rule; and also the rewriting by a set of rules and the one by a set of super rules of rules. It will then be shown that the two rewritings are identical in the quotient algebra defined by a given set of associative and commutative equations. In addition, the formal definition of critical peaks are stated and we outline how they can used to prove our main result, the Peak-Reduction Lemma. The application to the free abelian group is given as an example in Section 5. Finally, we shall discuss the effectiveness of the completion algorithm based on this approach.

2 Relation

Let S be a set, a *relation* $\Re$ on S is a subset of $S \times S$. We write $x \overset{\Re}{\mapsto} y$ to mean $(x,y) \in \Re$; and also $x \xrightarrow{\Re} y$ to mean $x \overset{\Re}{\mapsto} \cdots\cdots \overset{\Re}{\mapsto} y$.

A relation $\mathcal{E}$ is called an *equivalence relation* if it is symmetric, reflexive and transitive; and we write $x \overset{\mathcal{E}}{\sim} y$ to mean $(x,y) \in \mathcal{E}$. In this case we say that x is *$\mathcal{E}$-equal* or *$\mathcal{E}$-equivalent* to y. Now if we write $[x]_{\mathcal{E}}$ to be the set of elements *$\mathcal{E}$-equal* to x in S, then the *quotient* of S by $\mathcal{E}$, denoted by $S/\mathcal{E}$, is defined to be the set $\{[x]_{\mathcal{E}} \mid x \in S\}$.

On the other hand, a relation $\mathcal{R}$ is called a *reduction* if it is anti-reflexive and we say that x is *$\mathcal{R}$-reduced* to y if $x \xrightarrow{\mathcal{R}} y$ or simply x is *$\mathcal{R}$-reducible* if there exists an element y such that $x \xrightarrow{\mathcal{R}} y$. Otherwise x is said to be *$\mathcal{R}$-irreducible* or *$\mathcal{R}$-reduced.* In addition, a reduction $\mathcal{R}$ is *well-founded* if there is no infinite sequence $(x_1, x_2, \cdots\cdots)$ of elements in S such that $x_i \overset{\mathcal{R}}{\mapsto} x_{i+1}$ for every $i \geq 1$, in which case y is said to be a *$\mathcal{R}$-normal form* of x if $x \xrightarrow{\mathcal{R}} y$ and y is *$\mathcal{R}$-reduced.* Also, by convention, any $\mathcal{R}$-reduced element is the $\mathcal{R}$-normal form of itself. Finally, a pair of elements (x,y) are *$\mathcal{R}$-related* and written as $x \xleftrightarrow{\mathcal{R}} y$ if there is a finite sequence of elements $(x_1, \cdots\cdots, x_n)$ such that $x = x_1$, $x_n = y$; and if $x_i \neq x_{i+1}$, then either $x_i \overset{\mathcal{R}}{\mapsto} x_{i+1}$ or $x_{i+1} \overset{\mathcal{R}}{\mapsto} x_i$ for every i, $1 \leq i < n$ with $n \geq 1$. Let us call such a finite sequence an *$\mathcal{R}$-mountain range.* Moreover, an ordered subsequence $(x_i, \cdots, x_j, \cdots, x_k)$ with $1 \leq i < j < k \leq n$ is called an *$\mathcal{R}$-mountain* if

(a) either $i = 1$ or $x_{i-1} \overset{\mathcal{R}}{\mapsto} x_i$

(b) $x_j \overset{\mathcal{R}}{\mapsto} x_{j-1} \overset{\mathcal{R}}{\mapsto} \cdots \overset{\mathcal{R}}{\mapsto} x_{i+1} \overset{\mathcal{R}}{\mapsto} x_i$

(c) $x_j \overset{\mathcal{R}}{\mapsto} x_{j+1} \overset{\mathcal{R}}{\mapsto} \cdots \overset{\mathcal{R}}{\mapsto} x_{k-1} \overset{\mathcal{R}}{\mapsto} x_k$

(d) either $k = n$ or $x_{k+1} \overset{\mathcal{R}}{\mapsto} x_k$.

In this case, we call x_j the *apex* of the mountain $(x_i, \cdots, x_j, \cdots, x_k)$.

Suppose that $\mathcal{R}$ is a reduction on S, an *$\mathcal{R}$-peak* is an ordered triple (u, w, v) of elements in S such that $w \overset{\mathcal{R}}{\mapsto} u$ and $w \overset{\mathcal{R}}{\mapsto} v$. So every $\mathcal{R}$-mountain has a unique $\mathcal{R}$-peak, but there need not be an $\mathcal{R}$-mountain in an $\mathcal{R}$-mountain range! On the other hand, an *$\mathcal{R}$-valley* is a triple (x, z, y) of elements in S satisfying any one of the following:

(a) $x = z = y$

(b) $x = z$ and $y \xrightarrow{\mathcal{R}} z$

(c) $y = z$ and $x \xrightarrow{\mathcal{R}} z$

(d) $x \xrightarrow{\mathcal{R}} z$ and $y \xrightarrow{\mathcal{R}} z$.

Now, suppose that $\mathcal{R}'$ and $\mathcal{R}''$ are also reductions on S, not necessarily different from $\mathcal{R}$. We say an $\mathcal{R}$-peak (u, w, v) is *$\mathcal{R}'$-reducible* if there exists an $\mathcal{R}'$-mountain range $(x_1, \cdots, x_n)$ such that $x_1 = u$, $x_n = v$ and w can be $\mathcal{R}''$-reduced[1] to the apex of every mountain in this mountain range. Finally, we say a reduction $\mathcal{R}$ *allows peak reduction using* $\mathcal{R}''$ if every $\mathcal{R}$-peak is $\mathcal{R}$-reducible. However, we just say $\mathcal{R}$ allows peak reduction if $\mathcal{R}'' = \mathcal{R}$.

A reduction $\mathcal{R}$ on S is *Church-Rosser* if, for any $\mathcal{R}$-related pair $x \xleftrightarrow{\mathcal{R}} y$ of elements in S, there exists an element z such that the triple (x, z, y) forms a $\mathcal{R}$-valley. On the other hand, $\mathcal{R}$ is *confluent* if whenever an ordered triple (u, w, v) of elements such that $w \xrightarrow{\mathcal{R}} u$ and $w \xrightarrow{\mathcal{R}} v$ ("a mountain range with only one mountain") is given, then there always exists an element z such that the triple (u, z, v) forms an $\mathcal{R}$-valley.

Theorem 2.1 *A reduction $\mathcal{R}$ on S is Church-Rosser if and only if it is confluent.* □

---

[1] rightly pointed out by one of the referees, this reduction $\mathcal{R}''$ can be replaced by any ordering on S.

Lemma 2.1 *Given a well-founded reduction $\mathcal{R}''$, a reduction $\mathcal{R}$ on S is confluent if and only if it allows peak-reduction using $\mathcal{R}''$.* □

Let us now suppose that an equivalence relation $\mathcal{E}$ and a reduction $\mathcal{R}$ on S are given. Then $\mathcal{R}$ induces a relation $[\mathcal{R}]_\mathcal{E}$ on $S/\mathcal{E}$ vis : $[x]_\mathcal{E} \overset{[\mathcal{R}]_\mathcal{E}}{\mapsto} [y]_\mathcal{E}$ if $\exists x' \in [x]_\mathcal{E}$ and $\exists y' \in [y]_\mathcal{E}$ such that $x' \overset{\mathcal{R}}{\mapsto} y'$. Clearly, $[\mathcal{R}]_\mathcal{E}$ is well-defined. Further, it is anti-reflexive and therefore is a reduction if $\mathcal{E}$ and $\mathcal{R}$ intersect trivially. So Theorem 2.1 and Lemma 2.1 can be translated to

Theorem 2.2 *Suppose that $\mathcal{E}$ and $\mathcal{R}$ are an equivalence relation and a reduction on S respectively. The induced reduction $[\mathcal{R}]_\mathcal{E}$ on $S/\mathcal{E}$ is Church-Rosser if and only if it is confluent.* □

Lemma 2.2 *Suppose that $\mathcal{E}$ and $\mathcal{R}$ are an equivalence relation and a reduction on S respectively. The reduction $[\mathcal{R}]_\mathcal{E}$ on $S/\mathcal{E}$ is confluent if it is well-founded and it allows peak-reduction.* □

3 Term Algebra and One of its Quotients

Given a set $\mathcal{V}$ of variables and a graded set $\mathcal{F}$ of function symbols, $\Gamma(\mathcal{F},\mathcal{V})$ denotes the *free algebra* over $\mathcal{F}$. Variables have arity 0 by convention. Elements of $\Gamma(\mathcal{F},\mathcal{V})$, called *terms*, are viewed as *labelled trees* in the following way: a term t is a partial application of the free monoid on the natural numbers $\mathcal{M}(\aleph)$ into $\mathcal{F}\bigcup\mathcal{V}$ with domain $\mathcal{D}(t)$ such that the empty word $* \in \mathcal{D}(t)$; and $\nu{\cdot}i \in \mathcal{D}(t)$ if and only if $\nu \in \mathcal{D}(t)$ and $i \in [1, \text{arity}(t(\nu))]$. In that case, $\mathcal{D}(t)$ is the set of *occurrences* of t, $\mathcal{V}(t) = \{\nu \in \mathcal{D}(t) \mid t(\nu) \in \mathcal{V}\}$ is the set of *variable occurrences* of t, and $\mathcal{G}(t) = \mathcal{D}(t) \setminus \mathcal{V}(t)$ is the set of *ground occurrences* of t. Hence, we say a function $f \in \mathcal{F}$ or a variable $x \in \mathcal{V}$ *occurs* in a term t if $f = t(\nu)$ or $x = t(\nu)$ where $\nu \in \mathcal{F}(t)$ or $\nu \in \mathcal{V}(t)$ respectively. Further, for every term t, we write $\mathcal{V}_t$ for the set of variables $\{x \in \mathcal{V} \mid t^{-1}(x) \in \mathcal{V}(t)\}$ in t. Now, since any term t has a tree structure, if ν and η are two occurrences in $\mathcal{D}(t)$, then either

(a) $\nu \geq \eta$, that is, $\nu \cdot \mu = \eta$ for some $\mu \in \mathcal{M}(\aleph)$

(b) $\nu \leq \eta$, that is, $\eta \cdot \mu = \nu$ for some $\mu \in \mathcal{M}(\aleph)$

(c) ν and η are unrelated, written as $\nu \perp \eta$, that is, neither (a) nor (b) holds.

Further, in either (a) or (b), we write $[\nu,\eta]_t$ to be the unique reduced path in t with initial occurrence ν and terminal occurrence η. Let $\nu \in \mathcal{D}(t)$, we denote by t/ν the *subterm* of t at ν; and we write $t[\nu \leftarrow t']$ to mean the term obtained from t by replacing t/ν by t'. Notice that if $s = t/\nu$ is a subterm of t and η is an occurrence of t such that $\nu \geq \eta$, then $s(\nu\backslash\eta) = t(\eta)$ and $t/\eta = s/(\nu\backslash\eta)$ where $\nu{\cdot}(\nu\backslash\eta) = \eta$.

Definition 3.1 A *homomorphism* $\alpha : \Gamma(\mathcal{F},\mathcal{V}) \to \Gamma(\mathcal{F},\mathcal{U})$ of *term algebras over* $\mathcal{F}$ is nothing more than a $\mathcal{F}$-map. A homomorphism ϕ of $\Gamma(\mathcal{F},\mathcal{V})$ onto itself is a called a *substitution* if it fixes all but finitely many variables in $\mathcal{V}$, and it is called a *variable-renaming* if it permutes a finite subset of $\mathcal{V}$ and fixes the rest of variables. So given a substitution ϕ and a term $t \in \Gamma(\mathcal{F},\mathcal{V})$, if $\mathcal{V}(t) = \{\nu_1, \cdots\cdots, \nu_n\}$ and $x_i = t(\nu_i)$, $1 \leq i \leq n$, then $\phi(t) = t[\nu_1 \leftarrow \phi(x_1)]\cdot\cdots\cdot[\nu_n \leftarrow \phi(x_n)]$. Moreover, if s and t are terms in $\Gamma(\mathcal{F},\mathcal{V})$, we say s is an *instance* of t by ϕ if $s = \phi(t)$. On the other hand, s and t are *unifiable* by ϕ if $\phi(s) = \phi(t)$. Now, it is not too difficult to show the following lemma from the definitions. Further, we shall write Γ instead of $\Gamma(\mathcal{F},\mathcal{V})$ whenever the meaning is clear.

Lemma 3.1 *Let s and t be terms in Γ, $\nu \in \mathcal{D}(s)$ and ϕ a substitution on Γ. Then $\nu \in \mathcal{D}(\phi(s))$, $\phi(s/\nu) = \phi(s)/\nu$, and $\phi(s[\nu \leftarrow t]) = (\phi(s))[\nu \leftarrow \phi(t)]$.* □

Definition 3.2 A *relation* $\Re$ *on a term algebra* $\Gamma(\mathcal{F},\mathcal{V})$ is a relation on the set $\Gamma(\mathcal{F},\mathcal{V})$ such that if $(t,t') \in \Re$, then so does $(\, s[\nu \leftarrow \phi(t)]\,,\, s[\nu \leftarrow \phi(t')]\,)$ where $\nu \in \mathcal{D}(s)$, $s \in \Gamma(\mathcal{F},\mathcal{V})$ and ϕ is a substitution. So everything we have said about relations on sets in the last section can be extended to relations on term algebras.

Moreover, let $\mathcal{E}$ be an equivalence relation on $\Gamma(\mathcal{F},\mathcal{V})$ and $\alpha : \Gamma(\mathcal{F},\mathcal{V}) \to \Gamma(\mathcal{F},\mathcal{U})$ be a homomorphism of term algebras. Then we call α an $\mathcal{E}$-homomorphism if $t \overset{\mathcal{E}}{\sim} t'$ implies $\alpha(t) = \alpha(t')$, for all pairs (t,t') in $\Gamma(\mathcal{F},\mathcal{U})$. Clearly, in that case, α induces homomorphism $[\alpha]_{\mathcal{E}} : \Gamma(\mathcal{F},\mathcal{V})/\mathcal{E} \to \Gamma(\mathcal{F},\mathcal{U})/\mathcal{E}$.

Definition 3.3 An *equation in* Γ is an element of $\Gamma\times\Gamma$. The equation (s,t) is written as $s = t$. A *equational theory* is a set of equations. In particular, we are interested in a set of *associative* and *commutative* (AC-) *equations* of the form $f_{AC}(x,y) = f_{AC}(y,x)$ and $f_{AC}(x, f_{AC}(y,z)) = f_{AC}(f_{AC}(x,y),z)$ where $f_{AC} \in \mathcal{F}_{AC}$ which is a subset of $\mathcal{F}$ and is called the set of *AC-functions*. Now, let AC be the smallest equivalence relation on Γ, written as $\overset{AC}{\sim}$, generated by the set of AC-equations as follows: a pair of terms s,t are AC-equivalent and so $s \overset{AC}{\sim} t$ if there exist an occurrence ν in $\mathcal{D}(s)$ and an AC-equation $(c = d)$ such that $s/\nu = \phi(c)$ and $t = s[\nu \leftarrow \phi(d)]$ for some substitution ϕ. Now from the definitions, we have

Lemma 3.2 *Suppose that s and t are terms in Γ and ϕ is a substitution. If $s \overset{AC}{\sim} t$, then $\phi(s) \overset{AC}{\sim} \phi(t)$.* □

Lemma 3.3 *Suppose that s is a term in Γ and $\{\nu_1, \cdots, \nu_n\}$ is a pairwise unrelated subset of $\mathcal{D}(s)$. If $s/\nu_i \overset{AC}{\sim} t_i$ for $1 \le i \le n$, then $s \overset{AC}{\sim} s[\nu_1 \leftarrow t_1]\cdot\cdots\cdot[\nu_n \leftarrow t_n]$.* □

Notice that because the set of occurrences are unrelated, it does not matter which order we take to perform the replacement of s/ν_i by t_i.

Definition 3.4 Suppose that s and t are terms in Γ. Then s is an *AC-instance* of t if there is a substitution ϕ such that $s \overset{AC}{\sim} \phi(t)$; in other words, there exists $s' \in [s]_{AC}$ such that s' is an instance of t by ϕ. Furthermore, s and t are *AC-unifiable* if there is a substitution ψ such that $\psi(s) \overset{AC}{\sim} \psi(t)$ and we call ψ an *AC-unifier* of s and t. On the other hand, given substitutions θ and ϕ, we say that they are *AC-equal*, written as $\theta \overset{AC}{\sim} \phi$, if $\theta(x) \overset{AC}{\sim} \phi(x)$, for all $x \in \mathcal{V}$.

Now, suppose that s and t are terms. A *complete set of AC-unifiers of s and t* is a set $\Psi(s,t)$, whose elements are said to be *most general*, of substitutions such that

(a) if $\psi \in \Psi(s,t)$, then ψ is a AC-unifier of s and t

(b) if θ is a AC-unifier of s and t, then there exist $\psi \in \Psi(s,t)$ and a substitution ϕ such that $\theta \overset{AC}{\sim} \phi\cdot\psi$.

Actually, we had implicitly assumed that $\mathcal{V}_s \cap \mathcal{V}_t = \emptyset$. Nevertheless, this can always be achieved by renaming the variables occurring in s and t. Also note that, by (b), $\Psi(s,t)$ is unique up to AC-equivalence. The following theorem proved by Fages, Huet, Peterson and Stickel is well-known.

Theorem 3.1 *Given any pair of terms (s,t), the complete set of AC-unifiers, $\Psi(s,t)$, is always a finite set. Moreover, there is an algorithm to find $\Psi(s,t)$.* □

Remark 3.1 Let t be a term and Φ be the set of all substitutions up to AC-equivalence. Suppose that we are given a term u, can one determine that u is a subterm of an element in the set $\{\ s \in \Gamma \mid s \overset{AC}{\sim} \phi(t)\ ,\ \phi \in \Phi\ \}$? Note that if $\mathcal{D}(t) = \{*\}$, then it is obvious. However, in general, this set is not the same as $\bigcup_{\phi \in \Phi} \phi([t]_{AC})$. Nevertheless, we can still make the following observations: suppose that $\mathcal{D}(t) \supset \{*\}$. Then $t(*)$ must be a function with non-zero arity. Let $f_* = t(*)$. There are two cases; (a) $f_* \notin \mathcal{F}_{AC}$ and (b) $f_* \in \mathcal{F}_{AC}$.

Assume (a), if $s \overset{AC}{\sim} \phi(t)$, then $s(*) = f_* \notin \mathcal{F}_{AC}$. Hence $s = f_*(t_1, \cdots, t_n)$ where $t_i \overset{AC}{\sim} \phi(t/i)$ for all $i \in [1, n]$ and $n = \text{arity}(f_*)$. Therefore, a proper subterm of s must be a subterm (not necessarily proper) of an element belonging to $[\phi(t/i)]_{AC}$ for some $i \in [1, n]$.

For (b), if $s \overset{AC}{\sim} \phi(t)$, then $s(*) = f_* \in \mathcal{F}_{AC}$; in particular, its arity is two. So $s = \phi'(f_*(x, y))$ for some substitution ϕ' and some variables $x, y \notin \mathcal{V}_t$. Therefore, $f_*(x, y)$ and t are $\mathcal{AC}$-unifiable by $\phi'' = \phi{\cdot}\phi'$. Hence there exists a most general $\mathcal{AC}$-unifier $\psi \in \Psi(f_*(x, y), t)$ such that $\phi'' \overset{AC}{\sim} \sigma{\cdot}\psi$ for some substitution σ. As a result, $s = f_*(t_1, t_2)$ where $t_1 \overset{AC}{\sim} \sigma(\psi(x))$ and $t_2 \overset{AC}{\sim} \sigma(\psi(y))$. Therefore, a proper subterm of s must be a subterm (not necessarily proper) of an element belonging to $[\sigma(\psi(x))]_{AC}$ or $[\sigma(\psi(y))]_{AC}$. However, the pair $\psi(x)$ and $\psi(y)$ are determined by t; and there are only finitely many candidates by Theorem 3.1.

From now on, if t is a term in Γ, we shall write $[t]$ for $[t]_{AC}$.

Definition 3.5 A term $t \in \Gamma$ is said to have *the leftmost tree-structure containing s at ν with respect to $\mathcal{F}_{AC}$* if t contains a series, called a *derived series of length n*, of proper subterms

$$t = \rho_0 > \rho_1 > \cdots\cdots > \rho_n \geq s ;$$

and $\mathcal{G}(t)$ contains two strictly descending sequences of occurrences

$$\mu_0 > \mu_1 > \cdots\cdots > \mu_n \text{ and}$$

$$\eta_0 > \eta_1 > \cdots\cdots > \eta_m ,$$

where the second sequence may be empty (in that case, we shall assign m to be -1 by convention), such that

1. $\forall i \in [0, n]$, $\rho_i = t/\mu_i$, and so let $\rho_i(*) = t(\mu_i) = f_i \in \mathcal{F}$
2. by considering $[*, \mu_n]_t$ as a set of occurrences, $\{\eta_0, \cdots, \eta_m\}$ is the maximal subset of $[*, \mu_n]_t$ subject to the condition that all of its images under t, $\{t(\eta_0), \cdots, t(\eta_m)\}$, belong to $\mathcal{F}_{AC}$
3. $\forall i \in [1, n]$, $\mu_i = \eta_{i-1}{\cdot}1$; and $\eta_{i-1} \leq \mu_{i-1}$ (*i.e.* $\mu_i < \eta_{i-1} \leq \mu_{i-1}$)
4. for $i = 0$,
   (a) if $f_0 \in \mathcal{F}_{AC}$, then $\mu_0 = \eta_0 = *$
   (b) if $f_0 \notin \mathcal{F}_{AC}$, then $\mu_0 = *$; and $\eta_0 < *$ or $m = -1$
5. $\forall i \in [1, n-1]$,
   (a) if $f_i \in \mathcal{F}_{AC}$, then $\eta_i = \mu_i$; and $f_i \neq t(\eta_{i-1})$.
   (b) if $f_i \notin \mathcal{F}_{AC}$, then $\eta_i < \mu_i$
6. if $\rho_n > s$, then $f_n \neq t(\eta_{n-1})$
7. n is maximal with respect to the above properties.

Proposition 3.1 (The Existence of the Leftmost Tree Structure of a Term Containing a Proper Subterm)

Let u in Γ be a term containing a proper subterm s at ν. Then there is a term u' in $[u]$ such that u' has the leftmost tree structure containing s at some non-trivial occurrence ν' w.r.t. $\mathcal{F}_{AC}$. In other words, we can transform u into u' by AC-equality steps. In particular, these AC-equality steps take place in $u[\nu \leftarrow x]$ where $x \notin \mathcal{V}_u$. □

We remark that one can also establish the uniqueness, up to $\mathcal{AC}$-equivalence, of the leftmost tree structure of a term containing a proper subterm.

Example 3.1 *The following is an example of a term t in Γ having the leftmost tree structure of derived length n containing a proper subterm s at ν w.r.t. $\mathcal{F}_{AC}$ where $\{+, \times, f_{AC}, f'_{AC}\} \subseteq \mathcal{F}_{AC}$, $g, h \in \mathcal{F}$, $h \neq f'_{AC}$ and $\{h_1, \cdots, h_k, g_1, \cdots, g_\ell\} \subseteq \mathcal{F} \setminus \mathcal{F}_{AC}$ with $0 < k$ and $0 < \ell$.*

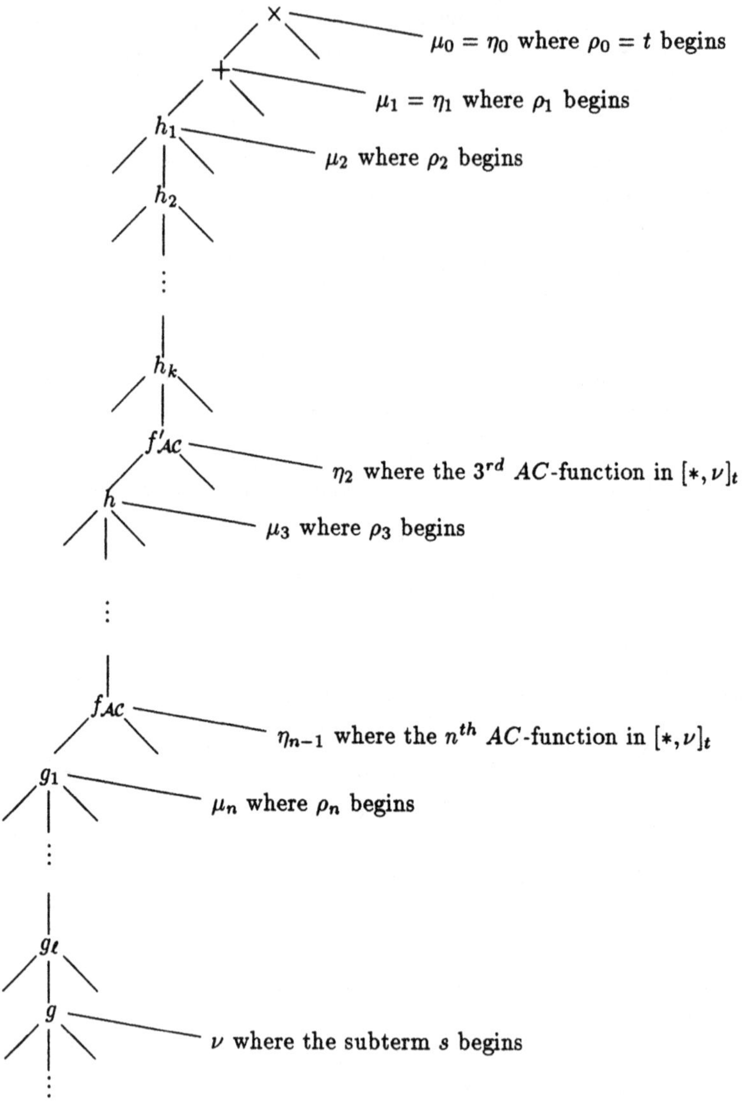

4 Rewriting in the Term Algebra and its Quotient by AC-Equations

Let Γ be a term algebra. We often encounter a set of equations R in Γ of the form that if $a = b$ belongs to R, then (1) a and b are distinct terms in Γ; (2) every variable which occurs in b also occurs in a; and (3) the equation $b = a$ does not belong to R. We call an equation satisfying (1) and (2) a *rule*, and R is called a *set of rules*. Further, if $r : a = b$ is a rule in R, then we shall denote it by $r : a \rightarrow b$ instead; and a is called the *left-hand side of* r while b is called the *right-hand side of* r.

Now, given a set of rules R in Γ, a relation $\mathcal{R}$ on Γ can be generated by R as follows: suppose s and t are terms in Γ. Then $s \overset{\mathcal{R}}{\mapsto} t$ if and only if there exist an occurrence ν in $\mathcal{D}(s)$ and a rule $r : a \rightarrow b$ in R such that $s/\nu = \phi(a)$ and $t = s[\nu \leftarrow \phi(b)]$ for some substitution ϕ. In this case, we say s *is rewritten to* t *at* ν *by* r. Note that this is the same as saying that s is reduced to t at ν by r. Moreover, the reduction $\mathcal{R}$ induces a relation $[\mathcal{R}]_{\mathcal{AC}}$ on $\Gamma/\mathcal{AC}$ as follows: $[s] \overset{[\mathcal{R}]_{\mathcal{AC}}}{\mapsto} [t]$ if $\exists s' \in [s]$ and $\exists t' \in [t]$ such that $s' \overset{\mathcal{R}}{\mapsto} t'$; and similarly, we say $[s]$ *is rewritten to* $[t]$ *by* r.

Recall that a relation $\mathcal{R}$ is a reduction if it is anti-reflexive. Clearly this is the case if $\mathcal{R}$ is well-founded. Further, $[\mathcal{R}]_{\mathcal{AC}}$ is a reduction on $\Gamma/\mathcal{AC}$ if $\mathcal{R}$ is a reduction and $\mathcal{R} \cap \mathcal{AC} = \emptyset$. So, from now on, R is always a set of rules which generates a reduction $\mathcal{R}$, which intersects $\mathcal{AC}$ trivially, on Γ. Furthermore, the induced relation $[\mathcal{R}]_{\mathcal{AC}}$ on $\Gamma/\mathcal{AC}$ is well-founded. Note that the well-foundedness of $\mathcal{R}$ is always guaranteed by the well-foundedness of $[\mathcal{R}]_{\mathcal{AC}}$ in the assumption.

Suppose that $r : a \rightarrow b$ is a rule in R. Let κ be an occurrence in $\mathcal{G}(a)$ such that the function $a(\kappa) = f_\kappa$ belongs to $\mathcal{F}_{\mathcal{AC}}$. We shall call κ an *exit* ("to paradise") of a if

(a) $\kappa = *$ or

(b) $\kappa \neq *$ and $f_\kappa \neq a(\kappa')$ where (κ, κ'), the path consisting of just these two occurrences, is the initial segment of the unique path $[\kappa, *]_a$.

Now, if κ is an exit of a, clearly, we can unify a/κ with $f_\kappa(v, w)$, where $v \neq w \in \mathcal{V} \setminus \mathcal{V}_a$, to obtain a finite non-empty set of *most general unified terms* $a_{\psi\kappa} = f_\kappa(\psi_\kappa(v), \psi_\kappa(w))$ where ψ_κ is a most general $\mathcal{AC}$-unifier in the complete set $\Psi_\kappa(a/\kappa, f_\kappa(v, w))$.

Definition 4.1 Let $r : a \rightarrow b$ be a rule in R, and let $\mathcal{K}(a)$ be the set of all exits of a. We define *the super rule of* r to be the set of equations

$$[r : a \rightarrow b]_{\mathcal{AC}} = \{ \psi_\kappa(a)[\kappa \leftarrow a_{\psi\kappa}] = \psi_\kappa(b) \mid \psi_\kappa \in \Psi_\kappa(a/\kappa, f_\kappa(v, w)), \kappa \in \mathcal{K}(a) \}$$

and, by convention, this set will consist of just the equation $a = b$ if $\mathcal{K}(a) = \emptyset$. Notice that $\psi_\kappa(a)$ is well-defined because we can always assume that if $x \in \mathcal{V}(a) \setminus \mathcal{V}_{a/\kappa}$, then $\psi_\kappa(x) = x$. Further, since $\kappa \in \mathcal{D}(a)$, $\psi_\kappa(a)[\kappa \leftarrow a_{\psi\kappa}] = \psi_\kappa(a[\kappa \leftarrow f_\kappa(v, w)])$. On the other hand, we should be aware that some variable y may appear at both occurrences ν and ν' of a where $\nu < \kappa$ and $\nu' \perp \kappa$ respectively. In such a situation, $\psi_\kappa(y)$ is always a subterm of $\psi_\kappa(a)[\kappa \leftarrow a_{\psi\kappa}]$ at ν' while $a_{\psi\kappa}$ need not contain a copy of $\psi_\kappa(y)$ as a subterm at any occurrence.

We shall call $\mathcal{S}(R) = \{ [r : a \rightarrow b]_{\mathcal{AC}} \mid r \in R \}$ the set of *super rules of* R. Now, $\mathcal{S}(R)$ defines a relation $\mathcal{S}(\mathcal{R})$ on $\Gamma/\mathcal{AC}$ as follows: $[s] \overset{\mathcal{S}(\mathcal{R})}{\mapsto} [t]$ if $\exists s' \in [s], \exists t' \in [t]$ and $\exists [r : a \rightarrow b] \in \mathcal{S}(R)$ such that $s'/\nu = \phi(a')$ and $t' = s'[\nu \leftarrow \phi(b')]$ for some $\nu \in \mathcal{D}(s')$, some substitution ϕ, and some equation $a' = b' \in [r : a \rightarrow b]_{\mathcal{AC}}$. In that case, we say $[s]$ is *rewritten to* $[t]$ *by* $[r : a \rightarrow b]_{\mathcal{AC}}$.

Proposition 4.1 *Let* s, t *be terms in* Γ. *Then* $[s] \overset{[\mathcal{R}]_{\mathcal{AC}}}{\mapsto} [t]$ *if and only if* $[s] \overset{\mathcal{S}(\mathcal{R})}{\mapsto} [t]$.

Proof. Since $a = b \in [r : a \rightarrow b]_{\mathcal{AC}}$ for every rule $r : a \rightarrow b$ in R, clearly $[s] \overset{[\mathcal{R}]_{\mathcal{AC}}}{\mapsto} [t]$ implies $[s] \overset{\mathcal{S}(\mathcal{R})}{\mapsto} [t]$.

Conversely, if $[s] \overset{\mathcal{S}(\mathcal{R})}{\mapsto} [t]$, then $\exists s' \in [s], \exists t' \in [t]$ and $\exists [r : a \rightarrow b] \in \mathcal{S}(R)$ such that $s'/\nu = \phi(a')$ and $t' = s'[\nu \leftarrow \phi(b')]$ for some $\nu \in \mathcal{D}(s')$, some substitution ϕ, and some equation $a' = b' \in [r : a \rightarrow b]_{\mathcal{AC}}$. By definition, $a' = \psi_\kappa(a)[\kappa \leftarrow a_{\psi\kappa}] \overset{\mathcal{AC}}{\sim} \psi_\kappa(a)[\kappa \leftarrow \psi_\kappa(a/\kappa)] = \psi_\kappa(a)$ for some exit κ of a and some most general $\mathcal{AC}$-unifier $\psi_\kappa \in \Psi_\kappa(a/\kappa, f_\kappa(v, w))$. Therefore $\phi(a') \overset{\mathcal{AC}}{\sim} \phi \cdot \psi_\kappa(a)$ and $\phi(b') = \phi \cdot \psi_\kappa(b)$. Hence $s'' = s'[\nu \leftarrow \phi \cdot \psi_\kappa(a)] \in [s]$ and $s'[\nu \leftarrow \phi \cdot \psi_\kappa(b)] = s'[\nu \leftarrow \phi(b')] = t'$. Then, clearly, $s'' \overset{\mathcal{R}}{\mapsto} t'$ and it completes the proof. □

Corollary 4.1 *$\mathcal{S}(\mathcal{R}) = [\mathcal{R}]_{\mathcal{AC}}$; and hence, the relation $\mathcal{S}(\mathcal{R})$ is a well-founded reduction on $\Gamma/\mathcal{AC}$.* □

Now we are ready to state the main theorem of this paper.

Theorem 4.1 *Let Γ be a term algebra with some AC-functions and R be a set of rules in Γ. Suppose that R generates a reduction $\mathcal{R}$, which intersects $\mathcal{AC}$ trivially, on Γ; and the induced relation $[\mathcal{R}]_{\mathcal{AC}}$ on $\Gamma/\mathcal{AC}$ is well-founded. Further, we assume that $[\mathcal{R}]_{\mathcal{AC}}$ allows peak-reduction. Then the reduction $[\mathcal{R}]_{\mathcal{AC}}$ on $\Gamma/\mathcal{AC}$ is Church-Rosser.*

Proof. This is a restatement of Theorem 2.2 with the application of Lemma 2.2. □

Our task is now reduced to showing that $[\mathcal{R}]_{\mathcal{AC}}$ allows peak-reduction. However, it is, in general, not clear how to characterize a reduction which allows peak-reduction. On the other hand, it is difficult to check that every possible peak is reducible by the reduction. So one hopes there is a finite set of "special" peaks which "generalize" all the other peaks formed in Γ and then one can effectively check that the set of all the "special" peaks are reducible to deduce that is also the case for the other peaks.

Moreover, since $\mathcal{S}(\mathcal{R}) = [\mathcal{R}]_{\mathcal{AC}}$, it suffices to show that $\mathcal{S}(\mathcal{R})$ allows peak-reduction.

Definition 4.2 Let R be a set of rules in a term algebra Γ. A *critical peak between a pair of super rules $[r : a \to b]_{\mathcal{AC}}$ and $[\ell : c \to d]_{\mathcal{AC}}$ at ν* is a septuple $([s],[t],[u]; r, \ell; \nu; \theta_\nu)$ where s, t, u are terms and θ_ν is a most general AC-unifier in a complete set Θ_ν; and, naturally, its corresponding *critical equation* is defined to be the equation $[s] = [u]$ while its *critical pair* is an equation of the form $s'' = u''$ where $s'' \in [s]$ and $u'' \in [u]$, *i.e.*, a representative of the critical equation. In addition, there are four types of critical peak defined as follows:

Type 1.

(i) $\nu \in \mathcal{G}(a)$

(ii) $a(\nu') \notin \mathcal{F}_{\mathcal{AC}}$ where (ν, ν'), the path consisting of just these two occurrences, is the initial proper segment of the unique path $[\nu, *]_a$

(iii) $a(\nu) = f_\nu \in \mathcal{F} \setminus \mathcal{F}_{\mathcal{AC}}$

(iv) a/ν and c are $\mathcal{AC}$-unifiable

(v) $\exists t', t'' \in [t], \exists u' \in [u]$ and $\exists s' \in [s]$ such that $t' = \theta_\nu(a), t'' = \theta_\nu(a)[\nu \leftarrow \theta_\nu(c)]$, $u' = \theta_\nu(a)[\nu \leftarrow \theta_\nu(d)]$ and $s' = \theta_\nu(b)$ respectively where θ_ν is a most general $\mathcal{AC}$-unifier in the complete set $\Theta_\nu(a/\nu, c)$

Type 2.

(i) $\nu \in \mathcal{G}(a)$

(ii) $a(\nu') \notin \mathcal{F}_{\mathcal{AC}}$ where (ν, ν') is the initial proper segment of $[\nu, *]_a$

(iii) ν is an exit of a

(iv) a/ν and c are $\mathcal{AC}$-unifiable

(v) $\exists t', t'' \in [t], \exists u' \in [u]$ and $\exists s' \in [s]$ such that $t' = \theta_\nu(a), t'' = \theta_\nu(a)[\nu \leftarrow \theta_\nu(c)]$, $u' = \theta_\nu(a)[\nu \leftarrow \theta_\nu(d)]$ and $s' = \theta_\nu(b)$ respectively where θ_ν is a most general $\mathcal{AC}$-unifier in the complete set $\Theta_\nu(a/\nu, c)$

Type 3.

(i) $\nu \in \mathcal{G}(a')$

(ii) $a' = \psi_\kappa(a)[\kappa \leftarrow a_{\psi\kappa}]$ where κ is an exit of a, ψ_κ is a most general $\mathcal{AC}$-unifier in the complete set $\Psi_\kappa(a/\kappa, f_\kappa(v,w))$, $\{v,w\} \subset \mathcal{V} \setminus \mathcal{V}_a$, and $a_{\psi\kappa}$ is the most general unified term defined by ψ_κ (*i.e.* $a' = b' \in [r : a \to b]_{\mathcal{AC}}$ for some b')

(iii) $\nu = \kappa \cdot 1$

(iv) a'/ν and c are $\mathcal{AC}$-unifiable

(v) $\exists t', t'' \in [t], \exists u' \in [u]$ and $\exists s' \in [s]$ such that $t' = \theta_\nu(a'), t'' = \theta_\nu(a')[\nu \leftarrow \theta_\nu(c)]$, $u' = \theta_\nu(a')[\nu \leftarrow \theta_\nu(d)]$ and $s' = \theta_\nu(b')$ respectively where θ_ν is a most general $\mathcal{AC}$-unifier in the complete set $\Theta_\nu(a'/\nu, c)$

Type 4.

(i) $\nu \in \mathcal{G}(a')$

(ii) $a(*) = f_* = c(*) \in \mathcal{F}_{\mathcal{AC}}$. Let $a' = b'$ be an equation in the set $[r : a \to b]^e_{\mathcal{AC}} = \{\psi_e(f_*(v,w)) = \psi_e(f_*(b,y)) \mid \psi_e \in \Psi_e(f_*(a,y), f_*(v,w)); y, v, w \in \mathcal{V} \setminus \mathcal{V}_a\}$

(iii) $\nu = 1$

(iv) a'/ν and c are $\mathcal{AC}$-unifiable

(v) $\exists t', t'' \in [t], \exists u' \in [u]$ and $\exists s' \in [s]$ such that $t' = \theta_\nu(a'), t'' = \theta_\nu(a')[\nu \leftarrow \theta_\nu(c)]$, $u' = \theta_\nu(a')[\nu \leftarrow \theta_\nu(d)]$ and $s' = \theta_\nu(b')$ respectively where θ_ν is a most general $\mathcal{AC}$-unifier in the complete set $\Theta_\nu(a'/\nu, c)$

(vi) it is not necessary to consider the case where the roles of r and ℓ have been interchanged

Note that there are two important points here. Firstly, not every occurrence of a or a' are involved in the definition of critical peak even if the subterms at these occurrences can be unified with the left-hand sides of some rules. It is only those which are exits of a, or where non-AC-functions occur in a or which are $\kappa \cdot 1$ with $\kappa \in \mathcal{K}(a)$ in a'. As a result, this will cut down the amount of computations during the completion procedure.

Secondly, the set of all critical peaks is finite since the complete set of $\mathcal{AC}$-unifiers of any pair of terms is finite; and there are only finitely many super rules, each of which is itself finite. Moreover, suppose that we just consider the triple $([s],[t],[u])$ in the above definition as a $\mathcal{S}(\mathcal{R})$-peak in the usual sense described in Section 2. Then it follows from the definitions of super rule and critical peak that this peak can still be defined uniquely if we replace a (respectively b, c or d) in the definition above by another representative in its $\mathcal{AC}$-equivalence class.

Lemma 4.1 (Peak-Reduction Lemma) *Let Γ be a term algebra which has some AC-functions and let R be a set of rules in Γ. Suppose that R generates a reduction $\mathcal{R}$, which intersects $\mathcal{AC}$ trivially, on Γ; and the induced relation $\mathcal{S}(\mathcal{R})$ on $\Gamma/\mathcal{AC}$ is well-founded. Further, we assume that $\mathcal{S}(\mathcal{R})$ reduces all critical peaks. Then it allows peak-reduction.*

Proof. Let us start by recalling the formal definition of a $\mathcal{S}(\mathcal{R})$-peak in the quotient algebra $\Gamma/\mathcal{AC}$. A $\mathcal{S}(\mathcal{R})$-peak is an triple $([s],[t],[u])$ of elements in $\Gamma/\mathcal{AC}$ such that

(i) there exist $t', t'' \in [t]$; $u' \in [u]$; $s' \in [s]$; and a pair of (not necessarily distinct) super rules $[r : a \to b]_{\mathcal{AC}}$ and $[\ell : c \to d]_{\mathcal{AC}}$ where $r, \ell \in R$

(ii) there exist occurrences $\nu \in \mathcal{D}(t''), \nu' \in \mathcal{D}(t')$; substitutions ϕ, ϕ'; and equations $a' = b' \in [r : a \to b]_{\mathcal{AC}}$, $c' = b' \in [\ell : c \to d]_{\mathcal{AC}}$

(iii) $t''/\nu = \phi(a')$, $t'/\nu' = \phi'(c')$, $s' = t''[\nu \leftarrow \phi(b')]$ and $u' = t'[\nu' \leftarrow \phi'(d')]$

Here, as always, we assume that $\mathcal{V}_{a'} \cap \mathcal{V}_{c'} = \emptyset$ and therefore also that $\phi = \phi'$. Moreover, without loss of generality, we may as well replace t'' by t, a' by a, b' by b, c' by c and d' by d in the definition. However, observe that ν may not be an occurrence of t' and, likewise, ν' may not be an occurrence of t.

Now, it can be shown that there are four outstanding cases to be considered as far as peak reduction is concerned.

Case 1 $t = t'$ and $\nu \perp \nu'$ are occurrences in the same tree (term) t

Case 2 $\nu \geq \nu'$ is a pair of occurrences in the same tree (term) $t[\nu' \leftarrow x]$ where is x is a variable; and $\phi(a)/\nu\backslash\nu' = t/\nu' \overset{AC}{\sim} t'/\nu' = \phi(c)$

Case 3 there exists an occurrence ν'' in t and t' such that $t[\nu'' \leftarrow x] = t'[\nu'' \leftarrow x]$, $t/\nu'' \overset{AC}{\sim} t'/\nu''$ and $\nu \geq$ (in t) $\nu'' >$ (in t') ν'

Case 4 there exists an occurrence ν'' in t and t' such that $t[\nu'' \leftarrow x] = t'[\nu'' \leftarrow x]$, $t/\nu'' \overset{AC}{\sim} t'/\nu''$ and $\nu <$ (in t) $\nu'' >$ (in t') ν'

Now, the reducibility of peaks of Case 1 is trivial. For peaks of Case 2, either it can be shown directly, or by the reducibility of critical peaks of Type 1 or Type 2 or it can be reduced to Case 3.

As far as the reducibility of peaks of Case 3 is concerned, it can be reduced to a special subcase where $\nu = *$ (hence $\phi(a) = t$) and $\nu \geq \nu'' = \kappa$ which is an exit of a. Further, by Proposition 3.1, we can assume that t'/κ has the leftmost tree structure properly containing $\phi(c)$ at $\kappa\backslash\nu'$ with derived length $n \geq 1$. In other words, $t'/\kappa = \rho_0 > \cdots > \rho_n \geq \phi(c) = t'/\nu'$.

Let us firstly consider that $n = 1$ and $\rho_1 = \phi(c)$ (so $\nu' = \kappa{\cdot}1$). It follows from the assumption that $\phi(a/\kappa) \overset{AC}{\sim} t'/\kappa$. Moreover, by extending the substitution ϕ, we can write $t'/\kappa = \phi(f_\kappa(v, w))$ for some distinct pair of variables $v, w \in \mathcal{V}\backslash(\mathcal{V}_t \cup \mathcal{V}_{t'})$ so that $\phi(v) = \phi(c) = t'/\nu'$. This means that the pair a/κ and $f_\kappa(v, w)$ are $\mathcal{AC}$-unifiable by ϕ. Therefore, there exist a most general unifier ψ_κ in the complete set $\Psi_\kappa(\ a/\kappa\ ,\ f_\kappa(v, w)\ )$ and a substitution σ so that $\phi \overset{AC}{\sim} \sigma \cdot \psi_\kappa$ restricted to $\{v, w\} \cup \mathcal{V}_a$. In particular, $\rho_1 = \phi(v) \overset{AC}{\sim} \sigma \cdot \psi_\kappa(v)$. Furthermore, since κ is an exit of a, the most general unified term $a_{\psi\kappa} = \psi_\kappa(f_\kappa(v, w))$ consequently defines an equation $a' = b' \in [r : a \to b]_{\mathcal{AC}}$ where $a' = \psi_\kappa(a)[\kappa \leftarrow a_{\psi\kappa}] \overset{AC}{\sim} \psi_\kappa(a)$ and $b' = \psi_\kappa(b)$. It can, in fact, be seen that $\sigma(a') \overset{AC}{\sim} \sigma \cdot \psi_\kappa(a) \overset{AC}{\sim} \phi(a) = t$ and $\sigma(b') \overset{AC}{\sim} \sigma \cdot \psi_\kappa(b) \overset{AC}{\sim} \phi(b) = s'$. Also $\sigma \cdot \psi_\kappa(v) \overset{AC}{\sim} \phi(v) = \phi(c) = t'/\nu'$ and $\sigma \cdot \psi_\kappa(w) \overset{AC}{\sim} \phi(w) = t'/\kappa{\cdot}2$. Therefore we can assume that $t = \sigma(a')$, $t' = \sigma(a')[\kappa{\cdot}1 \leftarrow \phi(c)]$ since $\kappa{\cdot}1 \perp \kappa{\cdot}2$. It then follows from the reducibility of critical peaks of Type 3 that the peak $(\,[s], [t], [u]\,)$ is reducible.

For all the remaining subcases, it follows either from the reducibility of critical peaks of Type 3 or Type 2; or the induction hypothesis on the derived length n of t'/κ and the special features of leftmost tree structures.

Finally, for peaks of Case 4, either it can be reduced to the above three cases or we can simply employ the reducibility of critical peaks of Type 4. This completes the proof of Peak Reduction Lemma. However, the reader could refer to [Lai 1] for a proper proof. □

5 Application to Free Abelian Group

Let us consider a term algebra Γ with a set of functions $\mathcal{F} = \{0, -, +\}$ with arities 0, 1 and 2 respectively. Further, let $+$ be an AC-function; and so we have an associative equation $x + (y + z) = (x + y) + z$ and an commutative equation $x + y = y + x$. These two equations clearly generate an equivalence relation $\mathcal{AC}$ in Γ. Now, suppose that we are given a set R consisting of the following five rules.

$$\begin{array}{rrcl} r_1 : & x + 0 & \to & x \\ r_2 : & -0 & \to & 0 \\ r_3 : & -(-x) & \to & x \\ r_4 : & -(x + y) & \to & (-x) + (-y) \\ r_5 : & x + (-x) & \to & 0 \end{array}$$

Then the relation $\mathcal{R}$ generated by R is a reduction on Γ and it intersects $\mathcal{AC}$ trivially. So it induces a reduction $[\mathcal{R}]$ on $\Gamma/\mathcal{AC}$. Moreover, $[\mathcal{R}]$ can be shown to be well-founded. Now, note that the left-hand side of r_1 (respectively r_4, or r_5) has only one exit while there are none at all in the left-hand sides of r_2 and r_3. Therefore the set of super rules are given as follows:

(1) $[r_1 : x + 0 \to x]_{\mathcal{AC}}$ consists of

$$\begin{array}{rcll} \underline{x} +_\kappa 0 & = & x & \text{(i)} \\ \underline{0} +_\kappa x & = & x & \text{(ii)} \\ \underline{(0+x)} +_\kappa y & = & x+y & \text{(iii)} \\ \underline{y} +_\kappa (0+x) & = & x+y & \text{(iv)} \end{array}$$

(2) $[r_2 : -0 \to 0]_{\mathcal{AC}}$ consists of

$$-0 = 0 \quad \text{(i)}$$

(3) $[r_3 : -(-x) \to 0]_{\mathcal{AC}}$ consists of

$$(-(-x)) = x \quad \text{(i)}$$

(4) $[r_4 : -(x+y) \to (-x)+(-y)]_{\mathcal{AC}}$ consists of

$$\begin{array}{rcll} -(\underline{x} +_\kappa y) & = & (-x)+(-y) & \text{(i)} \\ -(\underline{y} +_\kappa x) & = & (-x)+(-y) & \text{(ii)} \\ -(\underline{(x+y)} +_\kappa z) & = & (-(x+z))+(-y) & \text{(iii)} \\ -(\underline{z} +_\kappa (x+y)) & = & (-(x+z))+(-y) & \text{(iv)} \\ -(\underline{x} +_\kappa (y+z)) & = & (-(x+y))+(-z) & \text{(v)} \\ -(\underline{(y+z)} +_\kappa x) & = & (-(x+y))+(-z) & \text{(vi)} \\ -(\underline{(x+v)} +_\kappa (y+w)) & = & (-(x+y))+(-(v+w)) & \text{(vii)} \end{array}$$

(5) $[r_5 : x + (-x) \to 0]_{\mathcal{AC}}$ consists of

$$\begin{array}{rcll} \underline{x} +_\kappa (-x) & = & 0 & \text{(i)} \\ \underline{(-x)} +_\kappa x & = & 0 & \text{(ii)} \\ \underline{(-(x+y)+x)} +_\kappa y & = & 0 & \text{(iii)} \\ \underline{y} +_\kappa (-(x+y)+x) & = & 0 & \text{(iv)} \end{array}$$

where if an equation $a' = b'$ belongs to $[r_i : a \to b]_{\mathcal{AC}}$, $i = 1$, 4 or 5, and κ is the exit of a, then $+_\kappa$ is the function $+$ at κ in a' and $\underline{u}$ means u is the subterm $a'/\kappa{\cdot}1$ of a'.

As a result, the set of super rules defines the relation $\mathcal{S}(\mathcal{R})$ which is the same as $[\mathcal{R}]$ by Corollary 4.1. Now, we can write down all the critical peaks of Types 1 to 4 according to Definition 4.2 for this set of super rules. Then one can check that all the critical peaks are indeed reducible by $\mathcal{S}(\mathcal{R})$. Therefore by the Peak-Reduction Lemma (4.1), $\mathcal{S}(\mathcal{R})$ allows peak-reduction and, hence, so does $[\mathcal{R}]$. As a result, rewriting by the set of rules R can solve the word problem in the free abelian group.

6 Effectiveness of the completion algorithm

In this section, we discuss how to cut down the among of computations of critical peaks between rules. All the proofs of the following propositions can be found in [Lai 2].

Proposition 6.1 *Consider critical peaks of* Type 4. *It is enough to compute the critical peaks involving just a subset of* $[r : a \to b]^e_{\mathcal{AC}}$, *namely*

$$\{\psi_e(f_*(v,w)) = \psi_e(f_*(b,y)) \mid \psi_e \in \Psi_e(\, f_*(a,y)\,,\, f_*(v,w)\,); \psi_e(y) \in \mathcal{V}_{\psi_e(v)}; y,v,w \in \mathcal{V} \setminus \mathcal{V}_a\}. \ \square$$

Let us take the set of rules for the free abelian group in Section 5 for example and, particularly, we consider $r_5 : x + (-x) \to 0$. Using the conventional method to compute the critical peaks involving the extension $r_5^e : (x + (-x)) + y \to 0 + y$, we have to unify $(x + (-x)) + y$ with $(x_1 + (-x_1)) + y_1$ to obtain 35 most general unifiers and hence 35 critical peaks and 35 critical pairs.

However, applying Proposition 6.1, the subset of $[r_5 : x + (-x) \to 0]^e_{AC}$ consists of four equations which are

$a'_1 = b'_1$ where $a'_1 = \underline{(x_4 + y_4)} +_* (-x_4)$, $b'_1 = 0 + y_4$ with $\psi_e(x) = x_4$, $\psi_e(y) = y_4$, $\psi_e(v) = x_4 + y_4$ and $\psi_e(w) = -x_4$

$a'_2 = b'_2$ where $a'_2 = \underline{(x_5 + y_5)} +_* (x'_5 + (-(x_5 + x'_5)))$, $b'_2 = 0 + y_5$ with $\psi_e(x) = x_5 + x'_5$, $\psi_e(y) = y_5$, $\psi_e(v) = x_5 + y_5$ and $\psi_e(w) = x'_5 + (-(x_5 + x'_5))$

$a'_3 = b'_3$ where $a'_3 = \underline{(-x_6 + y_6)} +_* (x_6)$, $b'_1 = 0 + y_6$ with $\psi_e(x) = x_6$, $\psi_e(y) = y_6$, $\psi_e(v) = -x_6 + y_6$ and $\psi_e(w) = x_6$

$a'_4 = b'_4$ where $a'_4 = \underline{(-(x_7 + x'_7)) + (x_7 + y_7)} +_* x'_7$, $b'_1 = 0 + y_7$ with $\psi_e(x) = x_7 + x'_7$, $\psi_e(y) = y_7$, $\psi_e(v) = (-(x_7 + x'_7)) + (x_7 + y_7)$ and $\psi_e(w) = x'_7$

Now we are unifying $x + -x$ with the subterm at $\nu = 1$ (*i.e.* the underlined subterm) of each a'_i for $i = 1, 2, 3, 4$ to obtain a total of 14 critical peaks (both $a'_1/1$ and $a'_2/1$ give 4 unifiers respectively while both $a'_3/1$ and $a'_4/1$ give 3 unifiers respectively) and therefore 14 critical pairs. (In fact, by the corresponding propositions of *1* of (6.2) and (6.4), see below, respectively, the equations $a'_2 = b'_2$ and $a'_4 = b'_4$ would not even be considered. Consequently, there are only 7 critical pairs needed to be computed.) As a result, this example shows the number of critical pairs needed to be computed has been cut down by 3/5 (4/5).

The next three propositions are inspired by an educational conversation with Professor Kapur, and his paper [Ka-Mu-Na].

Proposition 6.2 *Consider critical peaks of* Type 3.

1. *Suppose that $a'/\kappa \cdot 1$ is not an AC-instance of c for any super rule $[\ell : c \to d]_{AC}$. If $[a'/\kappa \cdot 1] = [\psi_\kappa(v)]$ is reducible, then we do not have to compute the critical peaks involving the equation $a' = b'$.* □

   An example can be found from equation (iii) of the super rule $[r_5 : x + (-x) \to 0]_{AC}$ in the free abelian example of Section 5.

   Note that there is also a corresponding proposition for critical peaks of Type 4.

2. *Suppose that if $a'/\kappa \cdot 1$ is an AC-instance of c for some super rule $[\ell : c \to d]_{AC}$, then the critical peaks between these two rules have been considered. If $[a'/\kappa \cdot 1] = [\psi_\kappa(v)]$ is reducible, then we do not have to compute the critical peaks involving the equation $a' = b'$ with any other rule different from ℓ.* (We need to be careful when we implement the operational procedure of this particular step in the algorithm here.) □

   An example can be found from equation (iii) of the super rule $[r_1 : x + 0 \to x]_{AC}$ in the free abelian example of Section 5.

   Note that there is also a corresponding proposition for critical peaks of Type 4.

Proposition 6.3 *Consider critical peaks of* Type 3. *Suppose that the variables in a/κ do not appear in the other part of a. Let ψ_κ and ψ'_κ be two most general unifiers of a/κ and $f_\kappa(v, w)$. If $\psi'_\kappa(v)$ and $\psi'_\kappa(w)$ are AC-instances of $\psi_\kappa(v)$ and $\psi_\kappa(w)$ respectively, then we do not have to compute the critical peaks involving the equation defined by ψ'_κ.* □

An example can be found by considering the pair of equations (iii) and (vi) of the super rule $[r_4 : -(x + y) \to (-x) + (-y)]_{AC}$ in the free abelian example of Section 5.

Note that there is also a corresponding proposition for critical peaks of Type 4.

Proposition 6.4 *Consider critical peaks of* Type 3. *If* $[a'/\kappa{\cdot}2] = [\psi_\kappa(w)]$ *is reducible, then we do not have to compute the critical peaks involving the equation* $a' = b'$. □

Note that there is also a corresponding proposition for critical peaks of Type 4. An example can be found by considering the equation $a'_2 = b'_2$ of the set $[r_5 : (x + 0) \rightarrow x]^e_{AC}$ in the example following Proposition (6.1).

Note that with the applications of Propositions 6.1 to 6.4, a total of 48 critical pairs have been computed for proving that $\{r_1, \cdots, r_5\}$ is a complete set of rules for the free abelian group instead of a total of 142 using traditional methods (it is also known that very few are required using the algorithm in [Ka-Mu-Na]). However, without any practical implementation of the described algorithm in this paper at the present stage, it is difficult to make any sort of comparison, in general, with all the existing and well-established implementations yet. Perhaps, it is something the author should address himself to in the future. Finally, useful modifications of some of the crumble-some defintions had be suggested by the referees. Although they have not be enforced here, they will be kept in mind.

Acknowledgement. The author would like to thank the project leader U.Martin for leading him to the door of this wonderful and interesting world of Computer Science. Also, the work would not have been carried out so smoothly without the lively and exciting discussions with P.Watson and D.Cohen in the term.

References

[Ba] L.Bachmair "Proof Methods for Equational Theories" Ph.D. Thesis Univ. of Illinois. (*1987*)

[Hu 1] G.Huet "Confluent Reductions : Abstract Properties and Applications to Term Rewriting Systems" J.ACM 27 797-821 (*1980*)

[Hu 2] G.Huet "A complete Proof of Correctness of the Knuth-Bendix Completion Algorithm" J.Com. and Sys. Sci. 27 (*1981*)

[Jo-Ki] J.Jouannand and H.Kirchner "Completion of A Set of Rules Modulo A Set of Equations" SIAM J. Computing 15 1155-1194 (*1986*)

[Ka-Mu-Na] D.Kapur, D.Musser and P.Narendran "Only Prime Superpositions Need be Considered in the Knuth-Bendix Completion Procedure" J.of Symbolic Computation (*1988*)

[Kn-Be] D.Knuth and P.Bendix "Simple Word Problems in Universal Algebras" In Computational Problems in Abstract Algebra, Pergamon Press 263-297 (*1970*)

[La-Ba 4] D.S.Lankford and A.M.Ballantyne "Decision Procedures for Simple Equational Theories with Permutative Equations : Complete Sets of Permutative Reductions" Tech. Rep., Maths. Dep., Univ. of Texas, April (*1977*)

[La-Ba 8] D.S.Lankford and A.M.Ballantyne "Decision Procedures for Simple Equational Theories with Permutative Equations : Complete Sets of Permutative Reductions" Tech. Rep., Maths. Dep., Univ. of Texas, August (*1977*)

[Lai 1] M.Lai "A Peak Reduction in Rewritings of Term Algebras 'Associatively and Commutatively"' To appear (*1988*)

[Lai 2] M.Lai "A Knuth-Bendix Completion Algorithm Using Peak Reduction Lemma" To appear (*1988*)

[Pe] J.Pedersen "Obtaining Complete Sets of Reductions and Equations Without Using Special Unification Algorithms" Proc. Eurocal 85 Lect. Notes in Comp. Sci. 204 422-423 (*1985*)

[Pe-St] G.Peterson and M.Stickel "Complete Sets of Reductions for Some Equational Theories" J.ACM 28 233-264 (*1981*)

GENERALIZED GRÖBNER BASES: THEORY AND APPLICATIONS. A CONDENSATION.

Dallas Lankford
Mathematics Department
Louisiana Tech University
Ruston, LA 71272

ABSTRACT

Zacharias and Trinks proved that it can be constructively determined whether a finite generating set is a generalized Gröbner basis provided ideals are detachable and syzygies are solvable in the coefficient ring. We develop an abstract rewriting characterization of generalized Gröbner bases and use it to give new proofs of the Spear-Zacharias and Trinks theorems for testing and constructing generalized Gröbner bases. In addition, we use the abstract rewriting characterization to generalize Ayoub's binary approach for testing and constructing Gröbner bases over polynomial rings with Euclidean coefficient rings to arbitrary principal ideal coefficient domains. This also shows that Spear-Zacharias' and Trinks' approach specializes to Ayoub's approach, which was not known before.

1. INTRODUCTION

Many proofs have been omitted from the following condensation. A complete version [1] of this paper is available upon request.

The study of generalized Gröbner bases continues the development of constructive commutative algebra much in the spirit of Noether [2] and Hermann [3], 1923–1925; see also Seidenberg [4], 1974, for a modern development which corrects some earlier errors. These constructions apply mainly to polynomial rings whose coefficient ring is a field. Constructive commutative ring theory can be traced to Szekeres [5], 1952, who constructed canonical bases for ideals in a univariate polynomial ring over the integers. It has been said [12] that Richman [6], 1974, in effect found a univariate construction for coefficient rings in which ideal membership and syzygies are solvable, showed that ideal membership and syzygies are solvable in the polynomial ring, and extended the construction to multivariate polynomial rings by induction using the isomorphism $R[X,Y] \approx \big(R[X]\big)[Y]$. A similar univariate-induction approach has been described in Seidenberg [7], 1975. A direct approach for ideals of $R[X_1,\ldots,X_n]$ was developed for coefficient rings which are fields by Gröbner [8], 1950, and Buchberger [9,10], 1965–1970, and generalized for coefficient rings whose ideals are detachable and whose syzygies are solvable by Spear [11], 1977, Zacharias [12], 1978, and Trinks [13], 1978. Comprehensive collections of background, properties, and applications of Gröbner bases are contained in [12] and Buchberger [14], 1985.

First we establish some basic notation and state some familiar properties. R denotes a *commutative ring with unit*, $R[X_1, \ldots, X_n]$ denotes the *multivariate polynomial ring with indeterminates* $X_1, \ldots, X_n$ and coefficients from R, and *monomials* are finite products of indeterminates. β denotes a *bijection*.

Property 1. If β is a bijection, then each monomial in $R[X_1, \ldots, X_n]$ is equal to a unique monomial $X_{\beta_1}^{m_1} \cdot \ldots \cdot X_{\beta_n}^{m_n}$.

There are $n!$ permutations of $\{1, \ldots, n\}$, and hence $n!$ potential monomial normal forms. Without loss of generality, take $X_1^{m_1} \cdot \ldots \cdot X_n^{m_n}$ as the canonical normal form. $\mathbf{N}$ denotes the *natural numbers*, $\mathbf{N}^n$ denotes the set of all n *tuples* of natural numbers, the *degree* of a monomial $M = X_1^{m_1} \cdot \ldots \cdot X_n^{m_n}$ is $deg(M) = (m_1, \ldots, m_n) \in \mathbf{N}^n$, $\vec{0}$ denotes $(0, \ldots, 0)$, a *partial order* is a transitive relation, monomials are partially ordered by the *monomial* (*semigroup*, or *vector*) *partial order* defined by $(m_1, \ldots, m_n) \lll (p_1, \ldots, p_n)$ iff $m_i \leq p_i$ for $i = 1, \ldots, n$. A *total* (*linear*, *complete*, or *simple*) *order* $\leq$ is a partial order which satisfies (1) if $x \leq y$ and $y \leq x$, then $x = y$, and (2) $x \leq y$ or $y \leq x$ whenever x and y are members of the union of the domain and range of $\leq$, a *well order* is a total order such that every non-empty subset of the union of the domain and range has a least element, and a well order $\leq$ *well orders a set* X when X is a subset of the union of the domain and range of $\leq$. $\leq$ is a *monomial well order* means $\leq$ well orders $\mathbf{N}^n$ for some n, and $\lll$ is a subset of $\leq$. For example, the *tie breaker order*, which is defined by $(m_1, \ldots, m_n) < (p_1, \ldots, p_n)$ iff there exists some $k < n$ such that $m_i = p_i$ for $i = 1, \ldots, k$, and $m_{k+1} < p_{k+1}$, and the *vector lexicographic order*, which is defined by $(m_1, \ldots, m_n) < (p_1, \ldots, p_n)$ iff $m_1 + \cdots + m_n < p_1 + \cdots + p_n$, or $m_1 + \cdots + m_n = p_1 + \cdots + p_n$, and there exists some $k < n$ such that $m_i = p_i$ for $i = 1, \ldots, k$, and $m_{k+1} < p_{k+1}$, are both monomial well orders, where $\vec{m} \leq \vec{p}$ is defined as $\vec{m} < \vec{p}$ or $\vec{m} = \vec{p}$. When $<$ is a monomial well order, *a finite set of monomials* is denoted by $1 = M_0 < M_1 < M_2 < \cdots < M_n$.

Property 2. If monomials are ordered by a monomial well order, then each polynomial P of $R[X_1, \ldots, X_n]$ is equal to a unique $\sum_{i=0}^{d} r_i M_i$ where each $r_i \in R$, and $r_d \neq 0$.

The *head* of P is $head(P) = r_d M_d$, the *tail* of P is $tail(P) = \sum_{i=0}^{d-1} r_i M_i$, the *leading coefficient* of P is $lc(P) = r_d$, the *leading monomial* of P is $lm(P) = M_d$, the *degree* of P is $deg(P) = deg(lm(P))$, the *ideal generated by a set* X is $I(X) = \{\sum_{\text{finite}} r_i \cdot x_i \mid x_i \in X, \text{ and } r_i \in \text{a fixed ring}\}$, and the *least common multiple* of monomials $N_1, \ldots, N_m \in R[X_1, \ldots, X_n]$ is denoted by $\mathrm{LCM}(N_1, \ldots, N_m)$.

2. An Abstract Characterization Of Generalized Gröbner Bases

The definitions of leading immediate reduction, quasi-irreducible, and Gröbner basis below are similar to definitions in Zacharias [12]. The general approach below is also related to combinatorial equivalence methods developed by Newman [15], 1942, Evans [16], 1951, and Knuth and Bendix [17], 1970.

Q is a *leading immediate reduction* of P relative to Γ in $R[X_1, \ldots, X_n]$ means $P, Q \in R[X_1, \ldots, X_n]$, $\Gamma \subseteq R[X_1, \ldots, X_n]$, and there exist a monomial M of $R[X_1, \ldots, X_n]$, $m \in \mathbf{N}$; $\{r_1, \ldots, r_m\} \subseteq R$, and $\{G_1, \ldots, G_m\} \subseteq \{G \in \Gamma \mid deg(G) \lll deg(P)\}$ such that

$$lm(P) = M \cdot \mathrm{LCM}(lm(G_1), \ldots, lm(G_m)), \tag{1}$$

$$lc(P) = \sum_{i=1}^{m} r_i \cdot lc(G_i), \text{ and} \tag{2}$$

$$\begin{aligned} Q &= P - \left(head(P) + M \cdot \Big(\sum_{i=1}^{m} r_i \cdot \frac{\mathrm{LCM}(lm(G_1), \ldots, lm(G_m))}{lm(G_i)} \cdot tail(G_i) \Big) \right) \\ &= P - M \cdot \Big(\sum_{i=1}^{m} r_i \cdot \frac{\mathrm{LCM}(lm(G_1), \ldots, lm(G_m))}{lm(G_i)} \cdot G_i \Big). \end{aligned} \tag{3}$$

$P \longrightarrow_\Gamma Q$ means Q is a leading immediate reduction of P relative to Γ in $R[X_1, \ldots, X_n]$.

A monomial well order *satisfies the multiplicative property* means for any monomials M_1, M_2, and M_3, if $M_1 < M_2$, then $M_1 M_3 < M_2 M_3$. It is easy to show that the tie breaker and vector lexicographic orders satisfy the multiplicative property. Throughout the remainder of this article we assume that a monomial well order satisfies the multiplicative property. To conserve space we omit the assumption of a monomial well order which satisfies the multiplicative property from the hypotheses of lemmas which require that assumption.

Lemma 1. If $P \longrightarrow_\Gamma Q$, then $deg(P) > deg(Q)$ or $Q = 0$.

Proof: If $P \longrightarrow_\Gamma Q$, then $deg(P) > deg(Q)$ or $Q = 0$ follows directly from the definitions of monomial well order which satisfies the multiplicative property and leading immediate reduction.

Lemma 2. If $P \in I(\Gamma)$ and $P \longrightarrow_\Gamma Q$, then $Q \in I(\Gamma)$.

Proof: If $P \in I(\Gamma)$ and $P \longrightarrow_\Gamma Q$, then by (3) of the definition of leading immediate reduction, $Q \in I(\Gamma)$.

P is *quasi-reducible* relative to Γ means

$$lc(P) \in I\big(\{lc(G) \mid G \in \Gamma \text{ and } deg(G) \lll deg(P)\}\big).$$

P is *quasi-irreducible* relative to Γ means P is not quasi-reducible relative to Γ. Γ is a *Gröbner basis* means 0 is the only member of $I(\Gamma)$ which is quasi-irreducible relative to Γ. $\longrightarrow^*_\Gamma$ denotes the *transitive and reflexive closure* of $\longrightarrow_\Gamma$.

Lemma 3. Γ is a Gröbner basis iff for each $P \in I(\Gamma)$, $P \longrightarrow^*_\Gamma 0$.

Proof: ($\Longrightarrow$) Induct on $deg(P)$ after considering $P = 0$. Let $P \in I(\Gamma)$. If $P = 0$, then $P \longrightarrow^*_\Gamma 0$ because $\longrightarrow^*_\Gamma$ is reflexive. If $P \neq 0$, then P is quasi-reducible relative to Γ, so $P \longrightarrow_\Gamma Q$ for some Q. By Lemma 1, $deg(P) > deg(Q)$ or $Q = 0$. If $Q = 0$, then $P \longrightarrow^*_\Gamma 0$ because $\longrightarrow_\Gamma$ is contained in $\longrightarrow^*_\Gamma$. If $deg(P) > deg(Q)$, then $Q \in I(\Gamma)$ by Lemma 2, so by the induction hypothesis, $Q \longrightarrow^*_\Gamma 0$, therefore $P \longrightarrow^*_\Gamma 0$ because $\longrightarrow^*_\Gamma$ is transitive.

($\Longleftarrow$) This case follows immediately from the definitions, and so the proof of the lemma is complete.

Lemma 4. If $P \longrightarrow_\Gamma^* 0$, $Q \longrightarrow_\Gamma^* 0$, and for each $m \in \mathbf{N}$ and each $\{G_1, \ldots, G_m\} \subseteq \Gamma$, there exists a generating set Ω of the R module of solutions of $\sum_{i=1}^m lc(G_i) \cdot x_i = 0$ such that for each $(g_1, \ldots, g_m) \in \Omega$,

$$\sum_{i=1}^m g_i \cdot \frac{\mathrm{LCM}(lm(G_1), \ldots, lm(G_m))}{lm(G_i)} \cdot tail(G_i) \longrightarrow_\Gamma^* 0,$$

then $P + Q \longrightarrow_\Gamma^* 0$.

Proof: The *sum (formal) degree* of $P + Q$ is $sdeg(P + Q) = max\{deg(P), deg(Q)\}$. The proof is by double induction on $sdeg(P+Q)$ and $deg(P+Q)$, after considering the two initial cases $P = 0$ and $Q = 0$.

Lemma 5. If $P \longrightarrow_\Gamma^* 0$, and M is any monomial of $R[X_1, \ldots, X_n]$, then $M \cdot P \longrightarrow_\Gamma^* 0$.

Proof: Observe that if $Q \longrightarrow_\Gamma R$, then $M \cdot Q \longrightarrow_\Gamma M \cdot R$. If $P \longrightarrow_\Gamma^* 0$, then $P = P_0 \longrightarrow_\Gamma P_1 \longrightarrow_\Gamma \cdots \longrightarrow_\Gamma P_m \longrightarrow_\Gamma 0$, so it follows that $M \cdot P = M \cdot P_0 \longrightarrow_\Gamma M \cdot P_1 \longrightarrow_\Gamma \cdots \longrightarrow_\Gamma M \cdot P_m \longrightarrow_\Gamma M \cdot 0 = 0$, therefore $M \cdot P \longrightarrow_\Gamma^* 0$, which completes the proof of the lemma.

Ω is a *generating set* of the R module of solutions of the *linear, homogeneous equation* $\sum_{i=1}^m r_i \cdot x_i = 0$ with coefficients $r_i \in R$ means Ω is a collection of m tuples of elements from R, each $(g_1, \ldots, g_m) \in \Omega$ satisfies $\sum_{i=1}^m r_i \cdot x_i = 0$, and for each m tuple $(s_1, \ldots, s_m)$ of elements from R which satisfies $\sum_{i=1}^m r_i \cdot x_i = 0$, $(s_1, \ldots, s_m) = \sum_{\text{finite}} t_i \cdot (g_{i1}, \ldots, g_{im})$ for some $t_i \in R$ and some $(g_{i1}, \ldots, g_{im}) \in \Omega$.

Lemma 6. If $P \longrightarrow_\Gamma^* 0$, $r \in R$, and for each $m \in \mathbf{N}$ and each $\{G_1, \ldots, G_m\} \subseteq \Gamma$, there exists a generating set Ω of the R module of solutions of $\sum_{i=1}^m lc(G_i) \cdot x_i = 0$ such that for each $(g_1, \ldots, g_m) \in \Omega$,

$$\sum_{i=1}^m g_i \cdot \frac{\mathrm{LCM}(lm(G_1), \ldots, lm(G_m))}{lm(G_i)} \cdot tail(G_i) \longrightarrow_\Gamma^* 0,$$

then $r \cdot P \longrightarrow_\Gamma^* 0$.

Proof: Induct on $deg(P)$, after considering $P = 0$.

Theorem 1. If R is a commutative ring, $\Gamma \subseteq R[X_1, \ldots, X_n]$, and monomials are ordered by a monomial well order which satisfies the multiplicative property, then Γ is a Gröbner basis iff for each $m \in \mathbf{N}$ and each $\{G_1, \ldots, G_m\} \subseteq \Gamma$, there exists a generating set Ω of the R module of solutions of

$$\sum_{i=1}^m lc(G_i) \cdot x_i = 0$$

such that for each $(g_1, \ldots, g_m) \in \Omega$,

$$\sum_{i=1}^m g_i \cdot \frac{\mathrm{LCM}(lm(G_1), \ldots, lm(G_m))}{lm(G_i)} \cdot tail(G_i) \longrightarrow_\Gamma^* 0.$$

Proof: ($\Longrightarrow$) Given $m \in \mathbf{N}$, and $\{G_1, \ldots, G_m\} \subseteq \Gamma$, let Ω be the collection of all solutions of $\sum_{i=1}^{m} lc(G_i) \cdot x_i = 0$, and take any $(g_1, \ldots, g_m) \in \Omega$. Observe that

$$\begin{aligned}
&\sum_{i=1}^{m} g_i \cdot \frac{\mathrm{LCM}(lm(G_1), \ldots, lm(G_m))}{lm(G_i)} \cdot tail(G_i) \\
&\qquad = \mathrm{LCM}(lm(G_1), \ldots, lm(G_m)) \cdot \sum_{i=1}^{m} lc(G_i) \cdot g_i \\
&\qquad\qquad + \sum_{i=1}^{m} g_i \cdot \frac{\mathrm{LCM}(lm(G_1), \ldots, lm(G_m))}{lm(G_i)} \cdot tail(G_i) \\
&\qquad = \sum_{i=1}^{m} g_i \cdot \frac{\mathrm{LCM}(lm(G_1), \ldots, lm(G_m))}{lm(G_i)} \cdot \Big(lc(G_i) \cdot lm(G_i) + tail(G_i)\Big) \\
&\qquad = \sum_{i=1}^{m} g_i \cdot \frac{\mathrm{LCM}(lm(G_1), \ldots, lm(G_m))}{lm(G_i)} \cdot G_i \in I(\Gamma),
\end{aligned}$$

so by Lemma 3,

$$\sum_{i=1}^{m} g_i \cdot \frac{\mathrm{LCM}(lm(G_1), \ldots, lm(G_m))}{lm(G_i)} \cdot tail(G_i) \longrightarrow_{\Gamma}^{*} 0.$$

($\Longleftarrow$) Given $P \in I(\Gamma)$, there exist $m \in \mathbf{N}$, $\{G_1 \ldots, G_m\} \subseteq \Gamma$, and $\{P_1, \ldots, P_m\} \subseteq R[X_1, \ldots, X_n]$ such that

$$P = \sum_{i=1}^{m} P_i \cdot G_i.$$

The *representation* (or *formal*) *degree* of P relative to $\sum_{i=1}^{m} P_i \cdot G_i$ is denoted by $rdeg(P) = max\{deg(lm(P_i) \cdot lm(G_i)) \mid i = 1, \ldots, m\}$. It will be shown below using double induction on $deg(P)$ and $rdeg(P)$ that $P \longrightarrow_{\Gamma}^{*} 0$, from which it follows by Lemma 3 that Γ is a Gröbner basis.

If $deg(P) = rdeg(P) = \vec{0}$, then it follows directly from the definitions that $P \longrightarrow_{\Gamma} 0$, hence $P \longrightarrow_{\Gamma}^{*} 0$.

Observe that $rdeg(P) \geq deg(P)$, so for the induction step it suffices to consider the two cases $rdeg(P) = deg(P)$, and $rdeg(P) > deg(P)$. Without loss of generality, it is assumed that $deg(lm(P_i) \cdot lm(G_i)) = rdeg(P)$ iff $1 \leq i \leq p$.

If $rdeg(P) = deg(P)$, it follows that there exists a monomial M of $R[X_1, \ldots, X_n]$ such that $lm(P) = M \cdot \mathrm{LCM}(lm(G_1), \ldots, lm(G_p))$. Taking

$$Q = P - \left(head(P) + M \cdot \Big(\sum_{i=1}^{p} lc(P_i) \cdot \frac{\mathrm{LCM}(lm(G_1), \ldots, lm(G_p))}{lm(G_i)} \cdot tail(G_i) \Big) \right)$$

it follows that $P \longrightarrow_{\Gamma} Q$. By Lemma 1 $deg(P) > deg(Q)$ or $Q = 0$. The case $Q = 0$ is similar to the ground case. If $deg(P) > deg(Q)$, then by Lemma 2, $Q \in I(\Gamma)$, so by the induction hypothesis, $Q \longrightarrow_{\Gamma}^{*} 0$, and hence $P \longrightarrow_{\Gamma}^{*} 0$ by the definition of $\longrightarrow_{\Gamma}^{*}$.

If $rdeg(P) > deg(P)$, let M be a monomial of $R[X_1, \ldots, X_n]$ such that $deg(M) = rdeg(P)$, and observe that

$$\begin{aligned} P &= \sum_{i=1}^{p} P_i \cdot G_i + \sum_{i=p+1}^{m} P_i \cdot G_i \\ &= \sum_{i=1}^{p} \Big(lc(P_i) \cdot lc(G_i) \cdot M + head(P_i) \cdot tail(G_i) + tail(P_i) \cdot G_i \Big) + \sum_{i=p+1}^{m} P_i \cdot G_i \\ &= \sum_{i=1}^{p} \Big(head(P_i) \cdot tail(G_i) + tail(P_i) \cdot G_i \Big) + \sum_{i=p+1}^{m} P_i \cdot G_i. \end{aligned}$$

It is given that there exists a generating set Ω of the solutions of $\sum_{i=1}^{p} lc(G_i) \cdot x_i = 0$ such that for each $(g_1, \ldots, g_p) \in \Omega$,

$$\sum_{i=1}^{p} g_i \cdot \frac{\mathrm{LCM}(lm(G_1), \ldots, lm(G_m))}{lm(G_i)} \cdot tail(G_i) \longrightarrow_{\Gamma}^{*} 0.$$

By construction $(lc(P_1), \ldots, lc(P_p))$ is a solution of $\sum_{i=1}^{p} lc(G_i) \cdot x_i = 0$, so there exist $q \in \mathbf{N}$, $r_i \in R$, and $(g_{i1}, \ldots, g_{ip}) \in \Omega$ such that

$$\big(lc(P_1), \ldots, lc(P_p)\big) = \sum_{i=1}^{q} r_i \cdot (g_{i1}, \ldots, g_{ip}).$$

Let M be the monomial such that $lm(P_i) \cdot lm(G_i) = M \cdot \mathrm{LCM}(lm(G_1), \ldots, lm(G_p))$ for $1 \le i \le p$, and it follows that

$$\begin{aligned} \sum_{i=1}^{p} head(P_i) \cdot tail(G_i) &= \sum_{i=1}^{p} lm(P_i) \cdot lc(P_i) \cdot tail(G_i) \\ &= \sum_{i=1}^{p} lm(P_i) \cdot \Big(\sum_{j=1}^{q} r_j \cdot g_{ji} \Big) \cdot tail(G_i) \\ &= \sum_{j=1}^{q} r_j \cdot \Big(\sum_{i=1}^{p} lm(P_i) \cdot g_{ji} \cdot tail(G_i) \Big) \\ &= \sum_{j=1}^{q} r_j \cdot \Big(\sum_{i=1}^{p} M \cdot \frac{\mathrm{LCM}(lm(G_1), \ldots, lm(G_m))}{lm(G_i)} \cdot g_{ji} \cdot tail(G_i) \Big) \\ &= \sum_{j=1}^{q} r_j \cdot M \cdot \Big(\sum_{i=1}^{p} g_{ji} \cdot \frac{\mathrm{LCM}(lm(G_1), \ldots, lm(G_m))}{lm(G_i)} \cdot tail(G_i) \Big). \end{aligned}$$

So by hypothesis

$$\sum_{i=1}^{p} g_{ji} \cdot \frac{\mathrm{LCM}(lm(G_1), \ldots, lm(G_m))}{lm(G_i)} \cdot tail(G_i) \longrightarrow_{\Gamma}^{*} 0 \quad \text{for } j = 1, \ldots, q.$$

Because

$$rdeg\Big(\sum_{i=1}^{p} tail(P_i)\cdot G_i + \sum_{i=p+1}^{m} P_i\cdot G_i\Big) < rdeg(P),$$

by the induction hypothesis

$$\sum_{i=1}^{p} tail(P_i)\cdot G_i + \sum_{i=p+1}^{m} P_i\cdot G_i \longrightarrow_\Gamma^* 0.$$

By Lemmas 4, 5, and 6 it now follows that

$$\sum_{i=1}^{p}\Big(head(P_i)\cdot tail(G_i) + tail(P_i)\cdot G_i\Big) + \sum_{i=p+1}^{m} P_i\cdot G_i \longrightarrow_\Gamma^* 0,$$

hence $P \longrightarrow_\Gamma^* 0$, and so the proof is complete.

3. Abstract Algorithms

Consider using Theorem 1 to decide whether a given Γ is a Gröbner basis. *A monomial well order is computable* means there is an a lgorithm which, given monomials M and N, decides whether $M < N$. *A ring R is computable* means there exist algorithms which compute $-r$, $r+s$, and $r\cdot s$ for any $r,s\in R$. *Ideals of R are detachable* means there exists an algorithm which, given $r\in R$, and $\{r_1,\ldots,r_n\}\subseteq R$, decides whether $r\in I(\{r_1,\ldots,r_n\})$, and if so, then outputs $\{s_1,\ldots,s_n\}\subseteq R$ such that $r=\sum_{i=1}^n s_i\cdot r_i$. To decide irreducibility, or compute $\longrightarrow_\Gamma$, it suffices that Γ be finite, $<$ be a computable monomial well order, R be computable, and ideals of R be detachable. *Syzygies of R are solvable* means there exists an algorithm which, given $\{r_1,\ldots,r_n\}\subseteq R$, computes a finite generating set Ω of the R module of solutions of $\sum_{i=1}^n r_i\cdot x_i = 0$. Thus the following has been proved.

Theorem 2 (Spear-Zacharias, Trinks). There is an algorithm which, given a computable, commutative ring R whose ideals are detachable, and whose syzygies are solvable, a finite $\Gamma\subseteq R[X_1,\ldots,X_n]$, and a computable monomial well order satisfying the multiplicative property, decides whether Γ is a Gröbner basis, and if not, then outputs a $P\in I(\Gamma)$ which is quasi-irreducible relative to Γ.

A more elaborate problem is to construct Gröbner bases from given finite Γ. Γ is *quasi-irredundant* means each $G\in\Gamma$ is quasi-irreducible relative to $\Gamma-\{G\}$. If Γ is finite, the monomial well order and R are computable, and ideals of R are detachable, then it can be shown there is an algorithm which computes a finite, quasi-irredundant $Q(\Gamma)$ such that $I(Q(\Gamma)) = I(\Gamma)$. Define $\Gamma_0=\Gamma$, and $\Gamma_{i+1} = Q(\Gamma_i\cup\{P\})$ when the Gröbner basis algorithm generates a $P\in I(\Gamma_i)$ which is quasi-irreducible relative to Γ_i, otherwise $\Gamma_{i+1}=\Gamma_i$. If the sequence $\Gamma_0,\Gamma_1,\Gamma_2,\cdots$ becomes constant, $\cdots,\Gamma_i=\Gamma_{i+1}=\Gamma_{i+2}=\cdots$, then Γ_i is a Gröbner basis. It will be shown that for the sequence to become constant, it suffices that R be Noetherian. It may be observed that the procedure which generates the Γ_i sequence is similar to *completion procedures* which have been used by Evans [16], 1951, Knuth and Bendix [17], 1970, Bergman [18], 1978, Ballantyne and Lankford [19], 1981, and others to solve word problems and uniform word problems. R is *Noetherian* means each ideal of R is finitely generated (equivalently, each ascending chain of ideals of R is finite).

Lemma 7. If R is Noetherian, and monomials are ordered by a monomial well order, then there is no infinite, quasi-irredundant $\Gamma \subseteq R[X_1, \ldots, X_n]$.

Proof: Suppose Γ is infinite and quasi-irredundant. Let $N_1 < N_2 < N_3 < \cdots$, where $\{N_1, N_2, N_3, \cdots\} = \{lm(P) \mid P \in \Gamma\}$. Next suppose there is no infinite subsequence which satisfies $N_{i_1} \ll N_{i_2} \ll N_{i_3} \ll \cdots$, and consider the collection of all maximal $\ll$ chains $N_{i_1} \ll \cdots \ll N_{i_j}$, including singleton chains.The collection of all such N_{i_j} cannot be finite, otherwise $N_1, N_2, N_3, \ldots$ would be finite, which would imply Γ is finite, because R is Noetherian and Γ is quasi-irredundant. But then the collection of all such N_{i_j} would be an infinite collection of mutually incomparable monomials, contradicting Dickson [20], 1913. Thus there is an infinite subsequence which satisfies $N_{i_1} \ll N_{i_2} \ll N_{i_3} \ll \cdots$. Let $x_i \in I(\{lc(P) \mid P \in \Gamma \text{ and } lm(P) = N_{i_i}\}$. But then because Γ is quasi-irredundant, it follows that $I(\{x_1\}) \subset I(\{x_1, x_2\}) \subset I(\{x_1, x_2, x_3\}) \subset \cdots$ is an infinite ascending chain of ideals, contradicting that R is Noetherian.

Theorem 3 (Spear-Zacharias, Trinks). There is an algorithm which, given a computable, commutative, Noetherian ring R whose ideals are detachable, and whose syzygies are solvable, a finite $\Gamma \subseteq R[X_1, \ldots, X_n]$, and a computable monomial well order satisfying the multiplicative property, decides whether Γ is a Gröbner basis, and if not then computes Γ' such that Γ' is a Gröbner basis, and $I(\Gamma') = I(\Gamma)$.

Proof: Suppose the Γ_i sequence does not become constant. Let N_1 be the least monomial which is eventually a leading monomial of each Γ_i, i.e., there exists some $j \in \mathbf{N}$ such that for each $k \in \mathbf{N}$, $j \leq k$, and each $M \in \{lm(P) \mid P \in \Gamma_k\}$, $N_1 = min\{lm(P) \mid P \in \Gamma_j\} \leq M$, and there exists some $N \in \{lm(P) \mid P \in \Gamma_k\}$ such that $N = N_1$. Let $J_i = I(\{lc(P) \mid P \in \Gamma_i \text{ and } lm(P) = N_1\}$. It follows that $J_j \subseteq J_{j+1} \subseteq J_{j+2} \subseteq \cdots$, and because R is Noetherian, there is some $k_1 \in \mathbf{N}$, $j \leq k_1$, such that $J_{k_1} = J_{k_1+1} = J_{k_1+2} = \cdots$. Next let N_2 be the next least monomial which is eventually a leading monomial of each Γ_i, $k_1 \leq i$, and so on. Continuing in this manner, construct sequences $N_1 < N_2 < N_3 \cdots$ and J_{k_1}, J_{k_2}, J_{k_3}, $\cdots$, and let $\bigcup_{i=1}^{\infty}\{P \in \Gamma_{k_i} \mid lm(P) = N_i\}$. It follows that Γ is an infinite, quasi-irredundant subset of $R[X_1, \ldots, X_n]$, contradicting Lemma 7, which completes the proof.

4. Principal Ideal Domains

Generating and testing generalized Gröbner bases using Theorems 1, 2, and 3 is bounded below by the exponential function 2^n, where n is the number of elements of Γ because of the subset test in Theorem 1. However, the **Z**-Gröbner basis work of Butler and Lankford [21], 1984, and remarks by Ayoub [22,23], 1982, 1983, about D-Gröbner basis methods when D is a computable principal ideal domain suggest that an improved test may be developed when the coefficient ring R is a computable principal ideal domain. The immediate goal of this section is to show that if R is a principal ideal domain, then only binary subsets of Γ are required when applying the subset test of Theorem 1.

A *domain* (or *integral domain*) is a commutative ring R with unit which contains no *zero divisors*, i.e., for each $r, s \in R$, if $r \cdot s = 0$, and $r \neq 0$, then $s = 0$. A *principal ideal domain* (or *PID*) is a domain R such that for each ideal I of R, there exists some $r \in R$ such that $I = r \cdot R$. The notation $f: X \to Y$ means f is a function, X is the domain of f, and the range of f is a subset of Y. Let $Choice(\cdot)$ be an arbitrary but fixed choice function, and define $\pi: R^m \to R$ by $\pi(a_1, \dots, a_m) = Choice(\{x \mid x \cdot R = I(\{a_1, \dots, a_m\})\})$.

Property 3. If R is a PID, and $m \in \mathbf{N}$, then $\pi(a_1, \dots, a_m) \cdot R = I(\{a_1, \dots, a_m\})$ for each $a_1, \dots, a_m \in R$.

Property 4 below follows directly from Property 3 and the definitions, and shows that $\pi(a_1, \dots, a_m)$ satisfies the basic properties of a gcd for $a_1, \dots, a_m$.

Property 4. If R is a PID, then for each $a_1, \dots, a_m \in R$, there exist $r_1, \dots, r_m \in R$ and $s_1, \dots, s_m \in R$ such that

$$\pi(a_1, \dots, a_m) = \sum_{i=1}^{m} r_i \cdot a_i \quad \text{and} \quad a_i = s_i \cdot \pi(a_1, \dots, a_m) \quad \text{for } i = 1, \cdots, m.$$

An element $r \in R$ *divides* an element $s \in R$, denoted $r|s$, means there exists $t \in R$ such that $s = r \cdot t$. A *greatest common divisor* (or *gcd*) of $r_1, \dots, r_m \in R$, is an element $g \in R$ which satisfies $g|r_i$, for $i = 1, \dots, m$, and for each $h \in R$, if $h|r_i$, for $i = 1, \dots, m$, then $h|g$. The next property follows immediately from Property 4 and the definitions.

Property 5. If R is a PID, then $\pi(a_1, \dots, a_m)$ is a gcd of $a_1, \dots, a_m$.

The *quotient* of r and s, denoted

$$\frac{r}{s} \quad \text{or} \quad r/s,$$

is the unique x which satisfies $r = sx$ if such an x exists, and is undefined otherwise.

Property 6. If R is a domain, $r, s, t \in R$, and the quotients below exist, then

$$-\frac{r}{s} = \frac{-r}{s} \quad \text{and} \quad r \cdot \frac{s}{t} = \frac{r \cdot s}{t}.$$

Each integral domain can be imbedded in some field using a standard construction called the *field of quotients* (or *fractions*), see, for example, Herstein [24], 1964. However, a complete development of the field of fractions is not necessary because only the identities of Property 6 are required to derive the main results for abstract rewriting in Gröbner bases over PID's.

The next three lemmas establish three kinds of generating sets for R modules of solutions of syzygies. Lemma 9 is a finitely generated variant of the familiar fact that if R is a PID, then submodules of free modules are free, and the submodule dimensions are less than or equal to the module dimension, cf., for example, Lang [25], 1965. The generating sets of Lemmas 8 and 10 are not necessarily linearly independent and are generally not triangular.

Lemma 8. If R is a PID, $a_i, r_i \in R$, for $i = 1, \ldots, m$, and

$$\pi(a_1, \cdots, a_m) = \sum_{i=1}^{m} r_i \cdot a_i,$$

then

$$\Omega = \left\{ \left(\frac{-r_1 \cdot a_i}{\pi(a_1, \ldots, a_m)}, \ldots, \underbrace{1 - \frac{r_i \cdot a_i}{\pi(a_1, \ldots, a_m)}}_{i^{th}\ coordinate}, \ldots, \frac{-r_m \cdot a_i}{\pi(a_1, \ldots, a_m)} \right) \middle| \ i = 1, \ldots, m \right\}$$

generates the R module of solutions for

$$\sum_{i=1}^{m} a_i \cdot x_i = 0.$$

Proof: It is easy to show that each element of Ω is a solution. If $(s_1, \ldots, s_m)$ is a solution of

$$\sum_{i=1}^{m} a_i \cdot x_i = 0,$$

then

$$\pi(a_1, \ldots, a_m) \cdot \sum_{i=1}^{m} a_i / \pi(a_1, \ldots, a_m) \cdot s_i = \sum_{i=1}^{m} a_i \cdot s_i = 0,$$

so it follows that

$$\sum_{i=1}^{m} a_i / \pi(a_1, \ldots, a_m) \cdot s_i = 0,$$

and therefore

$$(s_1, \ldots, s_m) = \sum_{i=1}^{m} s_i \cdot \left(\frac{-r_1 \cdot a_i}{\pi(a_1, \ldots, a_m)}, \ldots, 1 - \frac{r_i \cdot a_i}{\pi(a_1, \ldots, a_m)}, \ldots, \frac{-r_m \cdot a_i}{\pi(a_1, \ldots, a_m)} \right),$$

which completes the proof of the lemma.

Lemma 9. If R is a PID, $a_i, r_{jk} \in R$, for $i = 1, \ldots, m$ and $j, k = 1, \ldots m - 1$, and

$$\pi(a_1, \ldots, a_i) = \sum_{j=1}^{i} r_{ij} \cdot a_j \quad \text{for } i = 1, \ldots, m-1,$$

then

$$\Omega = \left\{ \left(\frac{a_2}{\pi(a_1, a_2)}, \frac{-a_1}{\pi(a_1, a_2)}, 0, \ldots, 0 \right), \right.$$

$$\left(r_{21} \cdot \frac{a_3}{\pi(a_1, a_2, a_3)}, r_{22} \cdot \frac{a_3}{\pi(a_1, a_2, a_3)}, \frac{-\pi(a_1, a_2)}{\pi(a_1, a_2, a_3)}, 0, \ldots, 0 \right),$$

$$\vdots$$

$$\left. \left(r_{m-11} \cdot \frac{a_m}{\pi(a_1, \ldots, a_m)}, \ldots, r_{m-1m-1} \cdot \frac{a_m}{\pi(a_1, \ldots a_m)}, \frac{-\pi(a_1, \ldots, a_{m-1})}{\pi(a_1, \ldots, a_m)} \right) \right\}$$

generates the R module of solutions for

$$\sum_{i=1}^{m} a_i \cdot x_i = 0.$$

Proof: It is easy to show that all members of Ω are solutions. Induct on m. For the ground case, it has been shown in Lemma 8 that if $\pi(a_1, a_2) = r_1 \cdot a_1 + r_2 \cdot a_2$, then

$$\Omega = \left\{ \left(1 - r_1 \cdot \frac{a_1}{\pi(a_1, a_2)}, -r_2 \cdot \frac{a_1}{\pi(a_1, a_2)} \right), \left(-r_1 \cdot \frac{a_2}{\pi(a_1, a_2)}, 1 - r_2 \cdot \frac{a_2}{\pi(a_1, a_2)} \right) \right\}$$

generates the R module of solutions of

$$\sum_{i=1}^{2} a_i \cdot x_i = 0.$$

From

$$1 = r_1 \cdot \frac{a_1}{\pi(a_1, a_2)} + r_2 \cdot \frac{a_2}{\pi(a_1, a_2)}$$

it follows that

$$\left(1 - r_1 \cdot \frac{a_1}{\pi(a_1, a_2)}, -r_2 \cdot \frac{a_1}{\pi(a_1, a_2)} \right) = r_2 \cdot \left(\frac{a_2}{\pi(a_1, a_2)}, \frac{-a_1}{\pi(a_1, a_2)} \right) \quad \text{and}$$

$$\left(-r_1 \cdot \frac{a_2}{\pi(a_1, a_2)}, 1 - r_2 \cdot \frac{a_2}{\pi(a_1, a_2)} \right) = -r_1 \cdot \left(\frac{a_2}{\pi(a_1, a_2)}, \frac{-a_1}{\pi(a_1, a_2)} \right).$$

For the induction step, suppose Lemma 9 holds for $k < m$, and let $(s_1, \ldots, a_m)$ be a solution of

$$\sum_{i=1}^{m} a_i \cdot x_i = 0.$$

It is easy to show that

$$\sum_{i=1}^{m-1} a_i \cdot s_i = r \cdot \pi(a_1, \ldots, a_{m-1}) \quad \text{for some } r \in R,$$

therefore (r, s_m) is a solution of

$$\pi(a_1, \ldots, a_{m-1}) \cdot x_1 + a_m \cdot x_2 = 0.$$

By the ground case of Lemma 9, it follows that

$$(r, s_m) = t \cdot \left(\frac{a_m}{\pi(a_1, \ldots, a_m)}, \frac{-\pi(a_1, \ldots, a_{m-1})}{\pi(a_1, \ldots, a_m)} \right) \qquad \text{for some } t \in R.$$

Clearly

$$\begin{aligned}
&(s_1, \ldots, s_m) = \\
&\quad t \cdot \left(r_{m-1 1} \cdot \frac{a_m}{\pi(a_1, \ldots, a_m)}, \ldots, r_{m-1 m-1} \cdot \frac{a_m}{\pi(a_1, \ldots a_m)}, \frac{-\pi(a_1, \ldots, a_{m-1})}{\pi(a_1, \ldots, a_m)} \right) \\
&\quad + \left(s_1 - t \cdot r_{m-1 1} \cdot \frac{a_m}{\pi(a_1, \ldots, a_m)}, \ldots, s_{m-1} - t \cdot r_{m-1 m-1} \cdot \frac{a_m}{\pi(a_1, \ldots a_m)}, 0 \right),
\end{aligned}$$

and it can be shown that

$$\begin{aligned}
&\sum_{i=1}^{m-1} a_i \cdot \left(s_i - t \cdot r_{m-1 i} \cdot \frac{a_m}{\pi(a_1, \ldots, a_m)} \right) = \\
&\qquad r \cdot \pi(a_1, \ldots, a_{m-1}) - t \cdot \frac{a_m}{\pi(a_1, \ldots, a_m)} \cdot \pi(a_1, \ldots, a_{m-1}) = 0,
\end{aligned}$$

so by the induction hypothesis, $(s_1, \ldots, s_m)$ is a linear combination of the required form.

Lemma 10. If R is a PID, then

$$\Omega = \left\{ \left(\ldots, 0, \underbrace{\frac{a_j}{\pi(a_i, a_j)}}_{i^{th} \text{ coordinate}}, 0, \ldots, 0, \underbrace{\frac{-a_i}{\pi(a_i, a_j)}}_{j^{th} \text{ coordinate}}, 0, \ldots \right) \middle| \, 1 \le i < j \le m \right\}$$

generates the R module of solutions for

$$\sum_{i=1}^{m} a_i \cdot x_i = 0.$$

Proof: Clearly all members of Ω are solutions. Induct on m. The ground case, $m = 2$, is the $m = 2$ case of Lemma 9. For the induction step, the last element $\vec{L}$ of the generating set of Lemma 9 has the form

$$\left(r_1 \cdot \frac{a_m}{\pi(a_1, \ldots, a_m)}, \ldots, r_{m-1} \cdot \frac{a_m}{\pi(a_1, \ldots, a_m)}, \frac{-\pi(a_1, \ldots, a_{m-1})}{\pi(a_1, \ldots, a_m)} \right),$$

and because $\pi(a_1, a_m) = r \cdot \pi(a_1, \ldots, a_m)$ for some $r \in R$, it follows that

$$\left(r_2 \cdot \frac{a_m}{\pi(a_1, \ldots, a_m)}, \ldots, r_{m-1} \cdot \frac{a_m}{\pi(a_1, \ldots a_m)}, r_1 \cdot \frac{r \cdot a_1}{\pi(a_1, a_m)} - \frac{\pi(a_1, \ldots, a_{m-1})}{\pi(a_1, \ldots, a_m)} \right)$$

is a solution of $a_2 \cdot x_1 + \cdots + a_m \cdot x_{m-1} = 0$, and therefore by the induction hypothesis, $\vec{L}$ is a linear combination of elements of Ω. The remaining elements of the generating set of Lemma 9 are linear combinations of elements of Ω because of the induction hypothesis.

Lemma 10 permits the abstract rewriting characterization of Gröbner bases, Theorem 1, to be specialized for PID coefficient rings as follows.

Lemma 11. If $P \longrightarrow_\Gamma^* 0$, $r \in R$, and for each $\{G_1, G_2\} \subseteq \Gamma$,

$$g_1 \cdot \frac{\text{LCM}(lm(G_1), lm(G_2))}{lm(G_1)} \cdot tail(G_1) + g_2 \cdot \frac{\text{LCM}(lm(G_1), lm(G_2))}{lm(G_2)} \cdot tail(G_2) \longrightarrow_\Gamma^* 0,$$

$$\text{where } g_1 = lc(G_2)/\pi(lc(G_1), lc(G_2)) \text{ and } g_2 = -lc(G_1)/\pi(lc(G_1), lc(G_2)),$$

then $r \cdot P \longrightarrow_\Gamma^* 0$.

Proof: Modify the proof of Lemma 6 using Lemma 10.

Theorem 4. If R is a PID, $\Gamma \subseteq R[X_1, \ldots, X_m]$, and monomials are ordered by a monomial well order which satisfies the multiplicative property, then Γ is a Gröbner basis iff for each $\{G_1, G_2\} \subseteq \Gamma$,

$$g_1 \cdot \frac{\text{LCM}(lm(G_1), lm(G_2))}{lm(G_1)} \cdot tail(G_1) + g_2 \cdot \frac{\text{LCM}(lm(G_1), lm(G_2))}{lm(G_2)} \cdot tail(G_2) \longrightarrow_\Gamma^* 0,$$

where $g_1 = lc(G_2)/\pi(lc(G_1), lc(G_2))$ and $g_2 = -lc(G_1)/\pi(lc(G_1), lc(G_2))$.

Proof: Alter the proof of Theorem 1 using Lemmas 1 – 5 and 11.

Theorem 4 shows that testing for Gröbner bases can be considerably more practical in PID's than in general, because the power set test of Theorem 1 can be restricted to binary subsets of Γ. Theorem 4 can be used in PID's to develop algorithms for deciding whether a given Γ is a Gröbner basis, and for generating Gröbner bases from given Γ analogous to the algorithms described in Section 3, see Theorems 1, 2, and 3. A *computable PID* is a computable ring R which is a PID, together with an algorithm which, given any $r_1, r_2 \in R$, computes $\pi(r_1, r_2)$ satisfying $\pi(r_1, r_2) \cdot R = I(\{r_1, r_2\}$. *Quotients of R are computable* means there is an algorithm which, given any $r, s \in R$, determines whether r/s exists, and if so, computes r/s. It follows from Theorem 4 that if R is a computable PID, ideals of R are detachable, and quotients of R are computable, then there is an algorithm which decides whether a given Γ is a Gröbner basis, and if not, generates a Gröbner basis using the binary syzygy test.

Testing for Gröbner bases also includes computing leading immediate reductions, which can be another substantial source of computational difficulty. Again, it is known, for example, from the work of Ayoub [22], Buchberger [26], 1984, Kandri-Rody and Kapur [27], 1984, and Butler and Lankford [21], that in computable Euclidean rings, the leading immediate reduction calculations can be simplified by modifying the definition of leading immediate reduction, although the modification generally leads to increases in the size of Γ. It is shown below that leading immediate reduction calculations can also be simplified in PID's by similarly modifying the definition of leading immediate reduction.

Q is a π *leading immediate reduction* of R relative to Γ in $R[X_1, \ldots, X_m]$ means R is a PID, $P, Q \in R[X_1, \ldots, X_m]$, $\Gamma \subseteq R[X_1, \ldots, x_m]$, and there exist a monomial M of $R[X_1, \ldots, X_m]$, $r \in R$, and $G \in \Gamma$ such that

$$lm(P) = M \cdot lm(G), \tag{1}$$

$$lc(P) = r \cdot lc(G), \text{ and} \tag{2}$$

$$\begin{aligned} Q &= P - (head(P) + M \cdot r \cdot tail(G)) \\ &= P - M \cdot r \cdot G. \end{aligned} \tag{3}$$

$P \, {}_{\pi}\!\!\longrightarrow_{\Gamma} Q$ denotes that Q is a π leading immediate reduction of P relative to Γ, and ${}_{\pi}\!\!\longrightarrow^{*}_{\Gamma}$ denotes the *transitive and reflexive closure* of ${}_{\pi}\!\!\longrightarrow_{\Gamma}$. The next two lemmas are PID variants of Lemmas 1 and 2, and their proofs follow directly from the definitions.

Lemma 12. If $P \, {}_{\pi}\!\!\longrightarrow_{\Gamma} Q$, then $deg(P) > deg(Q)$ or $Q = 0$.

Lemma 13. If $P \in I(\Gamma)$ and $P \, {}_{\pi}\!\!\longrightarrow_{\Gamma} Q$, then $Q \in I(\Gamma)$.

When the coefficient ring is a PID, there is no direct analogue of Lemma 3. However, if Γ satisfies a natural property, namely that superpositions of Γ are π-reducible, then the analogue of Lemma 3 holds.

Lemma 14. If R is a PID and for each $\{G_1, G_2\} \subseteq \Gamma$, there is some $P \in R[X_1, \ldots, X_n]$ such that

$$\pi(lc(G_1), lc(G_2)) \cdot \mathrm{LCM}(lm(G_1), lm(G_2)) \, {}_{\pi}\!\!\longrightarrow_{\Gamma} P,$$

then for each $m \in \mathbf{N}$ and each $\{H_1, \ldots, H_m\} \subseteq \Gamma$ there exists some $Q \in R[X_1, \ldots, X_n]$ such that

$$\pi(lc(H_1), \ldots, lc(H_m)) \cdot \mathrm{LCM}(lm(H_1), \ldots, lm(H_m)) \, {}_{\pi}\!\!\longrightarrow_{\Gamma} Q.$$

Proof: Induct on m. The ground case, $m = 2$, obviously holds. For the induction step, suppose Lemma 14 holds for $k < m$. By the induction hypothesis, there exists some S such that

$$\pi(lc(H_1), \ldots, lc(H_{m-1})) \cdot \mathrm{LCM}(lm(H_1), \ldots, lm(H_{m-1})) \, {}_{\pi}\!\!\longrightarrow_{\Gamma} S,$$

say by H, i.e.,

$$\begin{aligned} r \cdot lc(H) &= \pi(lc(H_1), \ldots, lc(H_{m-1})), \text{ and} \\ L \cdot lm(H) &= \mathrm{LCM}(lm(H_1), \ldots, lm(H_{m-1})) \end{aligned}$$

for some $r \in R$ and some monomial L of $R[X_1, \ldots, X_n]$. By hypothesis, there exists some T such that

$$\pi(lc(H), lc(H_m)) \cdot \mathrm{LCM}(lm(H), lm(H_m)) \, {}_{\pi}\!\!\longrightarrow_{\Gamma} T,$$

say by J, i.e.,

$$s \cdot lc(J) = \pi(lc(H), lc(H_m)), \text{ and}$$
$$M \cdot lm(J) = \mathrm{LCM}(lm(H), lm(H_m))$$

for some $s \in R$ and some monomial M of $R[X_1, \ldots, X_n]$. It is not difficult to show that there is some monomial N of $R[X_1, \cdots, X_n]$ such that

$$N \cdot \mathrm{GCD}\left(\frac{\mathrm{LCM}(lm(H_1), \ldots, lm(H_{m-1})}{L}, lm(H_m)\right)$$
$$= \mathrm{GCD}(\mathrm{LCM}(lm(H_1), \ldots, lm(H_{m-1})), lm(H_m)),$$

so it follows that

$$M \cdot N \cdot lm(J) = \mathrm{LCM}(\mathrm{LCM}(lm(H_1), \ldots, lm(H_{m-1})), lm(H_m))$$
$$= \mathrm{LCM}(lm(H_1), \ldots, lm(H_m)).$$

And it can be shown that

$$t \cdot \pi(lc(H), lc(H_m)) = \pi(lc(H_1), \ldots, lc(H_m))$$

for some $t \in R$, so it follows that

$$s \cdot t \cdot lc(J) = \pi(lc(H_1), \ldots, lc(H_m)).$$

Let

$$Q = \pi(lc(H_1), \ldots, lc(H_m)) \cdot \mathrm{LCM}(lm(H_1), \cdots, lm(H_m)) - s \cdot t \cdot M \cdot N \cdot J,$$

which completes the proof of Lemma 14.

Superpositions of Γ *are* π*-reducible* means for each $G_1, G_2 \in \Gamma$ there exists some $P \in R[X_1, \ldots, X_m]$ such that

$$\pi(lc(G_1), lc(G_2)) \cdot \mathrm{LCM}(lm(G_1), lm(G_2))\, {}_{\pi}\!\!\longrightarrow_{\Gamma} P.$$

Lemma 15. If R is a PID and superpositions of Γ are π-reducible, then Γ is a Gröbner basis iff for each $P \in I(\Gamma)$, $P\, {}_{\pi}\!\!\longrightarrow^{*}_{\Gamma} 0$.

Proof: ($\Longleftarrow$) Easy; like Lemma 3. ($\Longrightarrow$) Induct on $deg(P)$. By Lemma 3, $P \longrightarrow^{*}_{\Gamma} 0$, so WOLOG let $P \longrightarrow_{\Gamma} Q$ for some $Q \in R[X_1, \ldots, X_n]$. It follows from the definition of leading immediate reduction that

$$head(P) = r \cdot M \cdot \pi(lc(G_1), \ldots, lc(G_m)) \cdot \mathrm{LCM}(lm(G_1), \ldots, lm(G_m))$$

for some $r \in R$, some monomial M of $R[X_1, \ldots, X_n]$, and some $\{G_1, \ldots, G_m\} \subseteq \Gamma$. By Lemma 14, $P\, {}_{\pi}\!\!\longrightarrow_{\Gamma} S$ for some $S \in R[X_1, \ldots, X_n]$. By Lemmas 12 and 13, and the induction hypothesis, $S\, {}_{\pi}\!\!\longrightarrow^{*}_{\Gamma} 0$, so it follows that $P\, {}_{\pi}\!\!\longrightarrow^{*}_{\Gamma} 0$.

Lemma 16. If $P_{\pi}\!\longrightarrow^{*}_{\Gamma} 0$, $Q_{\pi}\!\longrightarrow^{*}_{\Gamma} 0$, and for each $\{G_1, G_2\} \subseteq \Gamma$,

$$g_1 \cdot \frac{\mathrm{LCM}(lm(G_1), lm(G_2))}{lm(G_1)} \cdot tail(G_1) + g_2 \cdot \frac{\mathrm{LCM}(lm(G_1), lm(G_2))}{lm(G_2)} \cdot tail(G_2)_{\pi}\!\longrightarrow^{*}_{\Gamma} 0,$$

where $g_1 = lc(G_2)/\pi(lc(G_1), lc(G_2))$ and $g_2 = -lc(G_1)/\pi(lc(G_1), lc(G_2))$, then $P + Q_{\pi}\!\longrightarrow^{*}_{\Gamma} 0$.

Proof: Like the proof of Lemma 4.

Lemma 17. If $P_{\pi}\!\longrightarrow^{*}_{\Gamma} 0$ and M is any monomial of $R[X_1, \ldots, X_n]$, then $M \cdot P_{\pi}\!\longrightarrow^{*}_{\Gamma} 0$.

Proof: Like the proof of Lemma 5.

Lemma 18. If $P_{\pi}\!\longrightarrow^{*}_{\Gamma} 0$, and $r \in R$, then $r \cdot P_{\pi}\!\longrightarrow^{*}_{\Gamma} 0$.

Proof: Simpilfy the proof of Lemma 6.

Theorem 5. If R is a PID, $\Gamma \subseteq R[X_1, \ldots, X_m]$, superpositions of Γ are π-reducible, and monomials are ordered by a monomial well order which satisfies the multiplicative property, then Γ is a Gröbner basis iff for each $\{G_1, G_2\} \subseteq \Gamma$,

$$g_1 \cdot \frac{\mathrm{LCM}(lm(G_1), lm(G_2))}{lm(G_1)} \cdot tail(G_1) + g_2 \cdot \frac{\mathrm{LCM}(lm(G_1), lm(G_2))}{lm(G_2)} \cdot tail(G_2)_{\pi}\!\longrightarrow^{*}_{\Gamma} 0,$$

where $g_1 = lc(G_2)/\pi(lc(G_1), lc(G_2))$ and $g_2 = -lc(G_1)/\pi(lc(G_1), lc(G_2))$.

Proof: Modify the proof of Theorem 1 using Lemmas 12 – 18.

It follows from Theorem 5 that if R is a computable PID, ideals generated by two elements are detachable, and quotients of R are computable, then there is an algorithm which decides whether a given Γ is a Gröbner basis, and if not, generates a Gröbner basis using the binary syzygy test and π-immediate reductions. In general, when some

$$\pi(lc(G_1), lc(G_2)) \cdot \mathrm{LCM}(lm(G_1), lm(G_2))$$

is not π-reducible, a new polynomial

$$\begin{aligned}\pi(lc(G_1), lc(G_2)) &\cdot \mathrm{LCM}(lm(G_1), lm(G_2)) \\ &+ r_1 \cdot \frac{\mathrm{LCM}(lm(G_1), lm(G_2))}{lm(G_1)} \cdot tail(G_1) + r_2 \cdot \frac{\mathrm{LCM}(lm(G_1), lm(G_2))}{lm(G_2)} \cdot tail(G_2),\end{aligned}$$

where $\pi(lc(G_1), lc(G_2)) = r_1 \cdot lc(G_1) + r_2 \cdot lc(G_2)$, called the *superposition* of G_1 and G_2, is added to Γ. To illustrate such an algorithm when the coefficient ring is the ring of integers $\mathbf{Z}$, consider the following variation of an example from Butler and Lankford [21].

Let

$$\Gamma = \{7XY^2 - 2Y, 5X^2Y + 3X\}.$$

It is convenient for reference purposes to consider the elements of Γ as numbered rules, as is done below.

$7XY^2 - 2Y$	given	(1)
$5X^2Y + 3X$	given	(2)

The head term X^2Y^2 of the superposition of rules 1 and 2 is not π-reducible in **Z** because

$$\pi(7,5) = gcd(7,5) = 1 = (-2)(7) + (3)(5),$$

so a new rule is added to Γ.

$$X^2Y^2 + 13XY \quad \text{superposition of 1 and 2} \tag{3}$$

Now superpositions of Γ are π-reducible, so the binary syzygy test of Theorem 5 is applied. The syzygy induced linear combination of rules 1 and 2 reduces to $31XY$, so another new rule is added to Γ.

$$31XY \quad \text{syzygy of 1 and 2} \tag{4}$$

The new Γ is π-reducible, so the binary syzygy test is applied again.

$$XY^2 - 18Y \quad \text{syzygy of 1 and 4} \tag{5}$$

Next, the importance of completion methods is illustrated: rule 5 reduces rule 1, and rules 5 and 3 reduce rule 4 to 0, so rule 4 is deleted.

$$124Y \quad \text{reduction of 1 by 5} \tag{6}$$

The new Γ is π-reducible, so the binary syzygy test is applied again.

$$62Y \quad \text{syzygy of 4 and 5} \tag{7}$$

Rule 7 reduces rule 6 to 0, and so rule 6 is deleted.

$$X^2Y - 18X \quad \text{syzygy of 2 and 4} \tag{8}$$

$$93Y \quad \text{reduction of 3 by 8} \tag{9}$$

Now Γ, which consists of rules 4, 5, 7, 8, and 9, is a Gröbner basis.

Theorem 5 is applicable to any *computable Euclidean domain*, i.e., any computable ring for which there is a computable function d: $R-\{0\} \to \mathbf{N}$ such that for each $r, s \in R$, $d(r) \leq d(r \cdot s)$, and there exist $t, u \in R$ such that $r = t \cdot s + u$, and $u = 0$ or $d(u) < d(s)$. In that case $\pi(lc(G_1), lc(G_2))$ is merely the usual Euclidean domain GCD of $lc(G_1)$ and $lc(G_2)$. For example it is known that the quadratic integer domains $Q(\sqrt{m})$ are computable Euclidean domains for $m = -11, -7, -3, -2, -1, 2, 3, 5, 6, 7, 11, 13, 17, 19, 21, 29, 33, 37, 41, 57$, and 73, and are not Euclidean otherwise, cf. Motzkin [28], 1949, and Ribenboim [29], 1972. So an algorithm using Theorem 5 applies to each of those computable Euclidean domains. It is also known that if $m = -163, -67, -43$, and -19, then $Q(\sqrt{m})$ is a PID which is not Euclidean, and that $Q(\sqrt{m})$ is not a PID for any negative values of m other than $m = -163, -67, -43, -19, -11, -7, -3, -2$, and -1, cf. Heilbromm and Linfoot [30], 1934 and Stark [31], 1967. According to Ribenboim [29] it is an open question whether there are infinitely many positive m for which $Q(\sqrt{m})$ is a PID. Some additional facts relating to this question are known, such as $Q(\sqrt{23})$ is principal, cf. Hardy and Wright [32], 1938. Even when it is known that $Q(\sqrt{m})$ is a PID, it is still frequently non-trivial to determine whether Theorem 5 applies. For example, Wilson [33], 1977 provides details omitted by Motzkin [28] required to prove that $Q(\sqrt{-19})$ is a PID and from which it follows that Theorem 5 is applicable to $Q(\sqrt{-19})$. On the other hand, it is presently an open question whether Theorem 5 can in fact be applied to $Q(\sqrt{m})$ when $m = -163, -67, -43$, and 23, though it is reasonable to conjecture an affirmative answer. These considerations serve to illustrate that much work remains to be done before it can be determined whether Gröbner basis methods in general and algorithms using Theorem 5 in particular provide tractable solutions for many of the fundamental basis problems in classical constructive algebra.

ACKNOWLEDGEMENTS

I would like to express my appreciation to G. Butler, N. Dershowitz, H. Edwards, J. Meseguer, M. Sweedler, and RTA-89 referees for help with this work.

REFERENCES

1. D. Lankford, "Generalized Gröbner bases: theory and applications," Math. Dept., Louisiana Tech University, Ruston, LA 71272, Dec. 1986.

2. E. Noether, "Eliminationstheorie und allgemeine Idealtheorie," *Math. Ann.* **90** (1923), 229–261.

3. G. Hermann, "Die Frage der endlich vielen Schritte in der Theorie der Polynomideale," *Math. Ann.* **95** (1925), 736–783.

4. A. Seidenberg, "Constructions in algebra," *Trans. Amer. Math. Soc.* **197** (1974), 273–313.

5. A. Szekeres, "A canonical basis for the ideals of a polynomial domain," *Amer. Math. Monthly* **59** (1952), 379-386.

6. F. Richman, "Constructive aspects of Noetherian rings," *Proc. Amer. Math. Soc.* **44** (1974), 436–441.

7. A. Seidenberg, "What is Noetherian?" *Rend. Sem. Mat. Fis. Milano* **XLIV** (1974), 55–61.

8. W. Gröbner, "Über die Eliminationstheorie," *Monatshefte für Mathematik* **54** (1950), 71–78.

9. B. Buchberger, "Ein algorithmus zum Auffinden der Basiselemente des Restklassen ringes nach einem nulldimensionalen Polynomideal," Dissertation, Universität Innsbruck, 1965.

10. B. Buchberger, "Ein Algorithmisches Kriterium fur die Losbarkeit eines Algebraischen Gleichungssystems," *Aeq. Math.* **4** (1970), 374–383.

11. D. Spear, "A constructive approach to commutative ring theory," *Proc. 1977 MACSYMA User's Conf.* (NASA CP-2012, 1977), 369–376.

12. G. Zacharias, "Generalized Gröbner bases in commutative polynomial rings." B.Sc. thesis, M.I.T., Cambridge, June 1978.

13. W. Trinks, "Über B. Buchbergers Verfahren, Systeme algebraischer Gleichungen zu lösen," *J. Numb. Th.* **10** (1978), 475–488.

14. B. Buchberger, "Basic features and development of the critical-pair/completion procedure," *Lecture Notes in Computer Science* **202**, Rewriting Techniques and Applications, Springer-Verlag, Berlin, 1985, 1–45.

15. M. Newman, "On thories with a combinatorial definition of equivalence," *Ann. Math.* **43** (1942), 223–243.

16. T. Evans, "On multiplicative systems defined by generators and relations I. Normal form theorems," *Proc. Camb. Philos. Soc.* **47** (1951), 637–649.

17. D. Knuth and P. Bendix, "Simple word problems in universal algebras," *Computational Problems In Abstract Algebras*, ed. J. Leech, Pergamon Press, Oxford, 1970, 376–390.

18. G. Bergman, "The diamond lemma for ring theory," *Adv. Math.* **29** (1978), 178–218.

19. A. Ballantyne and D. Lankford, "New decision algorithms for finitely presented commutative semigroups," *J. Comput. Math. Appls.* **7** (1981), 159–165.

20. L. Dickson, "Finiteness of the odd perfect and primitive abundant numbers with n distinct prime factors," *Amer. J. Math.* **35** (1913), 413–426.

21. G. Butler and D. Lankford, "Dickson's lemma, Hilbert's basis theorem, and applications to completion in commutative, Noetherian rings," dept. rpt., Math. Dept., Louisiana Tech U., Ruston, LA 71272, June 1984.

22. C. Ayoub, "The decomposition theorem for ideals in polynomial rings over a domain," *J. Alg.* **76**, 1 (May 1982), 99–110.

23. C. Ayoub, "On constructing bases for ideals in polynomial rings over the integers," *J. Numb. Th.* **17**, 2 (Oct. 1983), 204–225.

24. I. Herstein, *Topics In Algebra*, Blaisdell Pub. Co., New York, 1964.

25. S. Lang, *Algebra*, Addison-Wesley Pub. Co., Inc., Reading, 1965.

26. B. Buchberger, "A critical-pair/completion algorithm for finitely generated ideals in rings," *Lecture Notes in Computer Science* **171**, Logic and Machines: Decision Problems and Complexity, Springer-Verlag, Berlin, 1984, 137–161.

27. A. Kandri-Rody and D. Kapur, "Algorithms for computing Gröbner bases of polynomial ideals over various Euclidean rings," *Lecture Notes in Computer Science* **174** (1984), EUROSAM 84, International Symposium on Symbolic and Algebraic Computation, Cambridge, England, July 9–11, 1984, 195–208.

28. T. Motzkin, "The Euclidean algorithm," *Bull. Amer. Math. Soc.* **55** (1949), 1142–1146.

29. P. Ribenboim, **Algebraic Numbers**, *Pure And Applied Mathematics* **XXVII**, Wiley-Interscience, New York, 1972.

30. H. Heilbromm and E. Linfoot, "On the imaginary quadratic corpora of class-number one," *Quarterly J. Math. (Oxford)* **5** (1934), 293–301.

31. H. Stark, "A complete determination of the complex quadratic fields of class-number one," *Michigan Math. J.* **14** (1967), 1–27; **MR 36**, Pt. 2 (Oct.–Dec. 1968), #5102, 993.

32. G. Hardy and E. Wright, *An Introduction To The Theory Of Numbers*, Oxford University Press, London, 1938.

33. J. Wilson, "A principal ideal ring that is not a Euclidean ring," *Selected Papers On Algebra*, Vol. 3, The Mathematical Association Of America, 1977, 79–82.

A Local Termination Property for Term Rewriting Systems

Dana May Latch *
Department of Mathematics
North Carolina State University
Raleigh, NC 27695-8205

Ron Sigal †
Dipartimento di Matematica
Università di Catania
Viale A. Doria 6
95125 Catania, Italy

January 18, 1989

Abstract

We describe a desirable property, *local termination*, of rewrite systems which provide an operational semantics for formal functional programming (FFP) languages, and we give a multiset ordering which can be used to show that the property holds.

1 Introduction

In this paper we describe a desirable property, *local termination*, of rewrite systems which provide an operational semantics for FFP languages ([Ba78]), and we give a multiset ordering which can be used to show that the property holds. In [LatSi88], we describe how this property plays a central role in an automated verification and analysis system for functional languages such as *FP* [Ba78], *KRC* [Tu82], *FL* [BaW^{2}86], and *HOPE* [Ei87].

The application of a FFP primitive (that is, primitive function or combining form) results in some transformational effect on its operand(s). The performance of this transformation

*On leave 1985-1988 at CUNY; Work partially supported by NSF Grants #MCS81-04217; #DCR83-02879 and by CUNY Grants #PSC-CUNY-667920; #PSC-CUNY-668293.

†Work supported by the AXL project of Enidata.

might be considered a "macro-operation," or machine instruction, of the interpreter, occurring in a single "instruction cycle." An operational semantics in the form of rewrite rules describes the "micro-operations" that implement a language primitive, where a basic "machine cycle" is occupied by a single rewrite operation. One aspect of verifying the correctness of an operational semantics is the determination that each instruction cycle is finite [Fr79]; that is, each application of a language primitive requires only a finite number of rewrite cycles. Note that this will not guarantee termination of programs. Termination of programs can only be assured by restricting the expressive power of the language sufficiently so that some normalization property holds for the functional rewrite system (for example, see [HaW³85]). Such a restriction produces a programming language that is not Turing complete ([CoBo72]).

On the other hand, the lack of a general termination property need not restrict our ambition in evaluating the correctness of a functional language interpreter. In this paper we formalize the notion of a finite instruction cycle as the *local termination property* and give a multiset ordering $\gg_0$ which, for appropriate sets of rewrite rules, can be used to demonstrate local termination. Since we are interested in program verification by symbolic evaluation, $\gg_0$ is defined for non-ground terms, and our results apply equally to ground and non-ground rewrite sequences.

In section 2, we characterize a class of "good" rewrite rule sets to which we restrict our attention. Sections 3 and 4 define the local termination property and the ordering $\gg_0$ for FFP terms, and in section 5 we prove that $\gg_0$ can be used to demonstrate that good rewrite rule sets are locally terminating.

2 Good Sets of Rewrite Rules for FFP Languages.

An FFP language $\mathcal{L}$ contains three basic kinds of expressions: atoms, applications, and sequences. *Atomic expressions* $A(\mathcal{L})$ include numbers and boolean constants, as well as a collection $F(\mathcal{L})$ of "names" for the language primitives. An expression $(f : \mathrm{e})$ is called an *application*. It represents the application of the function (or program component) f to the operand (or input) component e; both f and e can be any expressions in $\mathcal{L}$. *Sequences* are defined recursively as lists of FFP expressions bracketed by '<' and '>'. No unbracketed list is a well-formed expression in $\mathcal{L}$.

We define a term algebra $\mathbf{T}(\mathcal{L})$ whose ground terms are the expressions of $\mathcal{L}$.

Definition 2.1:

Suppose $\mathcal{L}$ is a FFP language. We define by simultaneous induction the algebras $\mathbf{T}(\mathcal{L})$ and $\mathbf{L}(\mathcal{L})$, starting with two infinite collections of variables: term variables $\{ x_i \}$ ranging over $\mathbf{T}(\mathcal{L})$, and list variables $\{ l_j \}$ ranging over $\mathbf{L}(\mathcal{L})$. The elements of $\mathbf{T}(\mathcal{L})$ are *terms.*

- $\mathbf{L}(\mathcal{L})$ is the free monoid on $\mathbf{T}(\mathcal{L}) \cup \{l_j\}$.
- Each variable x_i is a term.
- Each atom $a \in A(\mathcal{L})$ is a term.
- The binary operator ':' : $\mathbf{T}(\mathcal{L}) \times \mathbf{T}(\mathcal{L}) \rightarrow \mathbf{T}(\mathcal{L})$ constructs applicative terms; ':'$(t_1\ t_2)$ will be written $(t_1 : t_2)$.
- The variadic operator '<>' : $\mathbf{L}(\mathcal{L}) \rightarrow \mathbf{T}(\mathcal{L})$ represents the recursive construction of *sequences*; '<>'$(t_1 \ldots t_n)$ will be written $< t_1, \ldots, t_n >$.

Finally, the *objects* of $\mathbf{T}(\mathcal{L})$ are the application-free terms and the *concrete terms* are terms without list variables. □

With each FFP language $\mathcal{L}$ we can associate a set $\mathbf{RR}(\mathcal{L})$ of rewrite rules that specifies an operational semantics for $\mathcal{L}$. For each $f \in F(\mathcal{L})$, the corresponding rewrite rule set $\mathbf{RR}(f)$ specifies how each f-application $(f : e)$ can be transformed using rewrite evaluation. Utilization of the rules in $\mathbf{RR}(f)$ may be nondeterministic in that an f-application may match both left-recursive and right-recursive rules. If a ground f-application does not match any rewrite rule, then it is assigned the value $\perp$.

Definition 2.2:

(a) Suppose that f is a language primitive in $F(\mathcal{L})$. A *primitive rewrite rule* is of the form

$$(f : \mathrm{u}) \rightarrow rhs(f)$$

where u and $rhs(f)$ are terms in $\mathbf{T}(\mathcal{L})$ and where the set $v(rhs(f))$ of variables occurring in the $rhs(f)$ is a subset of $v(\mathrm{u})$, the set of variables occurring in u.

(b) The set $\mathbf{RR}(meta)$ of meta-composition rewrite rules consists of:

- $\mathrm{RR}_0(meta) = [(< > : x_1) \rightarrow x_1]$

- $\mathrm{RR}_L(meta) = [(< x_1,\ l_1 >: x_2) \ \rightarrow\ (x_1 :<< x_1,\ l_1 >,\ x_2 >)]$.

(c) An operational semantics $\mathbf{RR}(\mathcal{L})$ is *good* whenever $\mathbf{RR}(meta) \subseteq \mathbf{RR}(\mathcal{L})$ and the following conditions hold.

(1) In the left-hand-side $(f : \mathrm{u})$, the term u is an object term.

(2) If f appears in the function component of an application $(\mathrm{t}_1 : \mathrm{t}_2)$ in *rhs*(f), then $\mathrm{t}_1 = f$ and t_2 is an object term (that is, $(f : \mathrm{t}_2)$ is *innermost*).

(3) If the concrete application $(f : \mathrm{t})$ matches an application $(f : \mathrm{v})$ occurring in *rhs*(f), then $(f : \mathrm{t})$ can be rewritten by some f-rewrite rule. □

Example 2.3: The following together with **RR**(*meta*) is a good set of rewrite rules for the primitive functions *append-left* and *append-right* and for the combining forms *composition*, *apply*, and *construction*. In [Wi81] it was shown that any FFP language that contains these combining forms is Turing complete.

Composition:

- $\mathrm{RR}_0(comp) \;=\; [(comp :<< comp >,\ x_1 >) \ \rightarrow\ x_1]$
- $\mathrm{RR}_L(comp) \;=\; [(comp :<< comp,\ x_2,\ l_1 >,\ x_1 >) \ \rightarrow$
  $(x_2 : (comp :<< comp,\ l_1 >,\ x_1 >))]$

Apply:

- $\mathrm{RR}(app) \;=\; [(app :< x_2,\ x_1 >) \ \rightarrow\ (x_2 : x_1)]$

Construction:

- $\mathrm{RR}_0(con) \;=\; [(con :<< con >,\ x_1 >) \rightarrow < >]$
- $\mathrm{RR}_L(con) \;=\; [(con :<< con,\ x_2,\ l_1 >,\ x_1 >) \ \rightarrow$
  $(apnd_L :< (x_2 : x_1),\ (con :<< con,\ l_1 >,\ x_1 >) >)]$
- $\mathrm{RR}_R(con) \;=\; [(con :<< con,\ l_2,\ x_3 >,\ x_1 >) \ \rightarrow$
  $(apnd_R :< (con :<< con,\ l_1 >,\ x_1 >),\ (x_2 : x_1) >)]$

Append-left:

- $\mathrm{RR}(apnd_L) \;=\; [(apnd_L : < x_1, < l_1 >>) \to < x_1, l_1 >]$

Append-right:

- $\mathrm{RR}(apnd_R) \;=\; [(apnd_R : << l_1 >, x_1 >) \to < l_1, x_1 >]$

The rewrite rules $\mathrm{RR}_L(con)$ and $\mathrm{RR}_R(con)$, for example, are well-formed (recursive) rules because the rhs's contain no new variables; the *con*-applications that occur on the rhs's are of the correct format and are innermost; and any concrete application (*con* : t), that matches a *con*-application occurring in a *con*-rhs, matches a *con*-rule.

Note that the collection **RR**(*con*) is nondeterministic. For example, the meta-composition rewrite:

$$(< con,\ and,\ or > : < \text{true},\ \text{false} >) \to_{\mathrm{RR}_L(meta)} (con : << con,\ and,\ or >,\ < \text{true},\ \text{false} >>),$$

results in an application which matches both $\mathrm{RR}_L(con)$ and $\mathrm{RR}_R(con)$. On the other hand, the set $\mathbf{RR}(con) \setminus \{ \mathrm{RR}_R(con) \}$ is deterministic; that is, any application (*con* : t) can match at most one rewrite rule, because in each term $(con : << con, \text{list} >, t_1 >)$, that matches a *con*-rule, 'list' has either zero or more than zero top-level components. □

3 Local Termination.

Definition 3.1:

(a) The notation $t_1[t_2]$ indicates that t_2 is a subterm of t_1.

(b) For concrete term t, rewrite rule $\mathrm{RR}(f) = [(f : \mathrm{u}) \to rhs(f)]$, and substitution θ, let

$$t_1 \to_{\mathrm{RR}(f)} t_2$$

denote the *rewrite* of t_1 to t_2 by replacing a single occurrence of $(f : \theta\,\mathrm{u})$ by $\theta\, rhs(f)$.

(c) If $t_1 \to_{\mathrm{RR}(f)} t_2$ by way of $\mathrm{RR}(f) = [(f : \mathrm{u}) \to rhs(f)]$, then an application $(g : \mathrm{u}_2)$ in t_2 is a *direct descendant* of an application $(f : \mathrm{u}_1)$ in t_1 if:

(1) $(f : \mathrm{u}_1)$ is unaffected by the rewrite and $(f : \mathrm{u}_1) = (g : \mathrm{u}_2)$; or

(2) $(f : \mathrm{u}_1)$ matches a variable in u and $(f : \mathrm{u}_1) = (g : \mathrm{u}_2)$ is an instantiation of that variable in $rhs(f)$; or

(3) $(f : \mathrm{u}_1) = (f : \theta\,\mathrm{u})$, $(g : \mathrm{v})$ is an application appearing in *rhs(f)* (where g is any atom in $A(\mathcal{L})$), and $(g : \mathrm{u}_2) = (g : \theta\,\mathrm{v})$.

If $f = g$ (as is trivially true in cases (1) and (2)), then $(g : \mathrm{u}_2)$ is a ***primary direct descendant*** of $(f : \mathrm{u}_1)$. If $(f : \mathrm{t})$ is rewritten by rule RR(f), then PDD($(f : \mathrm{t})$, RR(f)) denotes the resulting set of primary direct descendants of $(f : \mathrm{t})$. We will write PDD($(f : \mathrm{t})$) when RR(f) is understood from the context. In a rewrite sequence *(primary) descendant* is the reflexive transitive closure of (primary) direct descendant. Given

$$\mathrm{rs} = [\,\mathrm{t}_1[(f : \mathrm{t})] \rightarrow_{\mathrm{RR}(f)} \mathrm{t}_1[\theta\, rhs(f)] \;\rightarrow\; \cdots \;\rightarrow\; \mathrm{t}_i \;\rightarrow\; \cdots],$$

PD($(f : \mathrm{t})$, rs, i) denotes the set of primary descendants of $(f : \mathrm{t})$ in t_i.

(d) In a rewrite sequence $[\cdots\ \mathrm{t}_i \rightarrow_{\mathrm{RR}(f)} \mathrm{t}_{i+1} \;\rightarrow\; \cdots]$, the subterm $(f : \mathrm{t})$ of t_i is ***activated*** when it is rewritten by a rule $\mathrm{RR}(f) = [(f : \mathrm{u}) \;\rightarrow\; rhs(f)]$. It is ***deactivated*** in the first t_j, $j > i$, such that it has no primary descendants in t_j.

(e) A rewrite sequence $[\,\mathrm{t}_1 \;\rightarrow\; \mathrm{t}_2 \;\rightarrow\; \cdots]$ is (*primary*) *function-directed* (*fd*) if $\mathrm{t}_i \rightarrow \mathrm{t}_{i+1}$, for all i, by rewriting a (primary) descendant of an activated term, if one exists.

(f) A rewrite strategy $\mathcal{S}$ is (*primary*) *function-directed* if each rewrite sequence generated by $\mathcal{S}$ is (primary) function-directed.

(g) A fd rewrite sequence is *complete* if it is infinite or it ends in an term which cannot be rewritten.

(h) A complete rewrite sequence is *terminating* if it is finite, and a fd rewrite sequence is *locally terminating* if every term activated in the sequence is eventually deactivated.

(i) A set of rewrite rules **RR** is (*locally*) *terminating* with respect to a fd rewrite strategy $\mathcal{S}$ if every complete sequence generated from **RR** by $\mathcal{S}$ is (locally) terminating. □

Note that, by Definition 3.1.b, we permit rewrite sequences to contain ground terms and terms with term variables but not list variables. Note also that, by Definition 2.2.c.2, primary direct descendants are created by rewriting instantiated left-hand-side applications $(f : \theta\mathrm{u})$ without considering the applicative subterms in the operand component $\theta\,\mathrm{u}$. Hence, primary direct descendants, and therefore (by the obvious induction) primary descendants, are never nested.

If an FFP operational semantics $\mathbf{RR}(\mathcal{L})$ is locally terminating, then the primary descendants of an application of a language primitive $f \in \mathcal{F}(\mathcal{L})$ "disappear" in a finite number of steps, so that, in a sense, the work to be performed by f itself is complete. Informally, we consider the lifetime of the primary descendants of f to be the "instruction" cycle of f. We illustrate the concept of local termination in Example 3.2 by giving a primary fd evaluation of a nonterminating application described in [HaW³85].

Example 3.2: If

$$g = < comp,\ app,\ < con,\ id,\ id >>$$

then any evaluation sequence of the innermost application $(g : g)$ is nonterminating.

$$
\begin{array}{lll}
(1) & (g : g) = (< comp,\ app,\ < con,\ id,\ id >> : g) & \\
(2) & \rightarrow_{\mathbf{RR}_L(meta)} & (comp : << comp,\ app,\ < con,\ id,\ id >>,\ g >) \\
(3) & \rightarrow_{\mathbf{RR}_L(comp)} & (app : (comp : << comp,\ < con,\ id,\ id >>,\ g >)) \\
(4) & \rightarrow_{\mathbf{RR}_L(comp)} & (app : (< con,\ id,\ id > : (comp : << comp >,\ g >))) \\
(5) & \rightarrow_{\mathbf{RR}_0(comp)} & (app : (< con,\ id,\ id > : g)) \\
(6) & \rightarrow_{\mathbf{RR}_L(meta)} & (app : (con : << con,\ id,\ id >,\ g >)) \\
(7) & \rightarrow_{\mathbf{RR}_L(con)} & (app : (apnd_L : < (id : g),\ (con : << con,\ id >,\ g >) >)) \\
(8) & \rightarrow_{\mathbf{RR}_R(con)} & (app : (apnd_L : < (id : g),\ (apnd_R : < \alpha,\ (id : g) >) >)) \\
 & & \quad \text{where } \alpha = (con : << con >,\ g >) \\
(9) & \rightarrow_{\mathbf{RR}_0(con)} & (app : (apnd_L : < (id : g),\ (apnd_R : << >,\ (id : g) >) >)) \\
(10) & \rightarrow_{\mathbf{RR}(apnd_R)} & (app : (apnd_L : < (id : g),\ < (id : g) >>)) \\
(11) & \rightarrow_{\mathbf{RR}(apnd_L)} & (app : < (id : g),\ (id : g) >) \\
(12) & \rightarrow_{\mathbf{RR}(app)} & ((id : g) : (id : g)) \\
(13) & \rightarrow_{\mathbf{RR}(id)} & (g : (id : g)) \\
(14) & \rightarrow_{\mathbf{RR}(id)} & (g : g)
\end{array}
$$

The application $(comp : << comp,\ app,\ < con,\ id,\ id >>,\ g >)$ on line (2), for example, has direct descendants:

(a) $(app : (comp : << comp, < con, id, id >>, g >)$

(b) $(comp : << comp, < con, id, id >>, g >)$,

of which (b) is primary. It also has the primary descendant $(comp : << comp >, g >)$ on line (4), and it is deactivated on line (5). Note that a primary fd strategy is not constrained to rewrite innermost terms, as demonstrated, for example, on line (11). □

4 Weak Reduction Rules.

Definition 4.1:

(a) Let $\gg_0$ denote the multiset extension of the empty partial order [DeMa79].

(b) Let $\delta(\mathrm{t})$ denote the multiset of occurrences of variables and atoms in term $\mathrm{t} \in \mathbf{T}(\mathcal{L})$. Extend δ to $\mathbf{L}(\mathcal{L})$ so that $\delta(\mathrm{t}_1 \ldots \mathrm{t}_n) = \delta(\mathrm{t}_1) \cup \cdots \cup \delta(\mathrm{t}_n)$ and to the power set $\mathcal{P}(\mathbf{L}(\mathcal{L}))$ so that $\delta(\{\mathrm{l}_1, \ldots, \mathrm{l}_n\}) = \delta(\mathrm{l}_1) \cup \cdots \cup \delta(\mathrm{l}_n)$. (Here and henceforth we use '$\cup$' and '\' to mean multiset union and difference.) For any atom f, let $\delta_f(\mathrm{t})$ denote $\delta(\{\mathrm{t}_1, \ldots, \mathrm{t}_n\})$, where $\{\mathrm{t}_1, \ldots, \mathrm{t}_n\}$ is the multiset of all occurrences of outermost subterms of t of the form $(f : \mathrm{u})$, and extend δ_f to $\mathbf{L}(\mathcal{L})$ and $\mathcal{P}(\mathbf{L}(\mathcal{L}))$ as δ was extended.

(c) A rewrite rule $\mathrm{RR}(f) = [(f : \mathrm{u}) \rightarrow rhs]$ is a *weak reduction rule* with respect to partial order $>$ on $\mathbf{T}(\mathcal{L})$ if

$$\delta_f((f : \mathrm{u})) > \delta_f(rhs),$$

and is a *reduction rule* with respect to $>$ if, for any substitution θ,

$$\theta\,(f : \mathrm{u}) > \theta\; rhs\,.$$

A set **RR** of rewrite rules is a *(weak) reduction set* with respect to $>$ if every rule $[(f : \mathrm{u}) \rightarrow rhs] \in \mathbf{RR}$ is a (weak) reduction rule.

(d) The partial ordering $>$ on $\mathbf{T}(\mathcal{L})$ is *(weakly) adequate* for a class $\mathcal{R}$ of rewrite rule sets and rewrite strategy $\mathcal{S}$ if, whenever $\mathbf{RR} \in \mathcal{R}$ is a (weak) reduction set with respect to the ordering $>$, then **RR** is (locally) terminating with respect to $\mathcal{S}$. □

Note that δ_f is defined to avoid double counting of occurrences in the case of nested f-applications. For example, $\delta_f((f : (f : x)))$ is $\{ f, f, x \}$ rather than $\{ f, f, x \} \cup \{ f, x \}$.

We show in the next section that $\gg_0$ is weakly adequate for the class of good sets of FFP rewrite rules and any primary fd rewrite strategy.

Example 4.2: Each of the rewrite rules given in Example 2.3 is a weak reduction rule with respect to $\gg_0$. For example, $\mathrm{RR}_L(comp)$

$$[(comp : << comp,\ x_2\ ,\ l_1 >,\ x_1 >) \rightarrow (x_2 : (comp : << comp,\ l_1 >,\ x_1 >))]$$

is a weak reduction rule because:

$$\begin{aligned} &\delta_{comp}((comp : << comp,\ x_2\ ,\ l_1 >,\ x_1 >)) \\ &\quad = \quad \{ comp,\ comp,\ x_2\ ,\ l_1\ ,\ x_1 \} \\ &\quad \gg_0 \quad \{ comp,\ comp,\ l_1,\ x_1 \} \\ &\quad = \quad \delta_{comp}((x_2 : (comp : << comp,\ l_1 >,\ x_1 >))) \quad \Box \end{aligned}$$

5 Local Termination of Good Rewrite Rule Sets.

Throughout this section we assume that $\mathbf{RR}(\mathcal{L})$ is a good set of weak reduction rules with respect to $\gg_0$.

Lemma 5.1: If $(f : \mathrm{t}) = \theta\,(f : \mathrm{u})$ is rewritten by $\mathrm{RR}(f) = [(f : \mathrm{u}) \rightarrow rhs]$ in $\mathbf{RR}(\mathcal{L})$, then

$$\delta_f((f : \mathrm{t})) \gg_0 \delta_f(\mathrm{PDD}((f : \mathrm{t}))).$$

Proof: Let $\delta_f((f : \mathrm{u})) = \{\mathrm{s}_1,\ \ldots,\ \mathrm{s}_m\}$ and $\delta_f(rhs) = \{\mathrm{t}_1,\ \ldots,\ \mathrm{t}_n\}$. Then, since $\mathrm{RR}(f)$ is a weak reduction rule,

$$\{ \mathrm{s}_1\ ,\ \ldots,\ \mathrm{s}_m \} \gg_0 \{ \mathrm{t}_1\ ,\ \ldots,\ \mathrm{t}_n \},$$

and, by definition of $\gg_0$, for every t_j there is an occurrence s_i in $(f : \mathrm{u})$ such that $\mathrm{t}_j = \mathrm{s}_i$. If t_j is an atom it is unaffected by θ. Otherwise, t_j is a term variable or list variable, and each occurrence of an atom, term variable, or list variable in $\theta\,\mathrm{t}_j = \theta\,\mathrm{s}_i$ also appears in $\theta\,(f : \mathrm{u})$, so that

$$\delta_f((f : \mathrm{t})) = \delta_f(\theta\,(f : \mathrm{u})) \gg_0 \delta(\{ \theta\,\mathrm{t}_1\ ,\ \ldots,\ \theta\,\mathrm{t}_n \}).$$

But $\delta(\{\theta\, t_1, \ldots, \theta\, t_n\}) = \delta_f(\mathrm{PDD}((f : t)))$ by the definitions of δ, δ_f, and PDD, and so the proof is complete. □.

Lemma 5.2: For any primitive f, if a concrete subterm $(f : t)$ of term t_1 matches some f-rewrite rule, then any primary descendant of $(f : t)$ in any finite rewrite sequence

$$rs = [\, t_1[(f : t)] \rightarrow \cdots \rightarrow t_n \,]$$

also matches some f-rewrite rule.

Proof: We argue by induction on n, the length of rs. If $n = 1$, the lemma is trivially true. Assume it is true for k and that $n = k + 1$. An application $(f : t')$ in t_2 is a primary direct descendant of $(f : t)$ by way of Definitions 3.1.c.1, 3.1.c.2, or 3.1.c.3. In the first two cases $(f : t') = (f : t)$ and by assumption it is concrete and matches some f-rewrite rule. If Definition 3.1.c.3 applies, then $(f : t')$ is concrete and matches an f-rewrite rule because of the definition of good rewrite rule sets (Definition 2.2.c.3). In all cases the induction hypothesis applies to $(f : t')$ and $rs' = [\, t_2 \rightarrow \cdots \rightarrow t_n \,]$. Since any primary descendant of $(f : t)$ in rs is a primary descendant in rs$'$ of some such $(f : t')$, the proof is complete. □

Lemma 5.3: Let $\mathcal{S}$ be a primary fd strategy. Then, in every complete rewrite sequence

$$rs = [\, t_1[(f : t)] \rightarrow_{\mathrm{RR}(f)} t_1[\theta\, rhs(f)] \rightarrow \cdots t_i \rightarrow \cdots]$$

generated from $\mathbf{RR}(\mathcal{L})$ by $\mathcal{S}$, there exists a t_i in which $(f : t)$ is deactivated.

Proof: The lemma will follow, by the well-foundedness of $\gg_0$, from the statement:

$$\begin{array}{ll} \text{If } \mathrm{PD}((f : t), rs, i) \neq \emptyset & \\ \quad \text{then } \delta_f(\mathrm{PD}((f : t), rs, i)) \gg_0 \delta_f(\mathrm{PD}((f : t), rs, i + 1)), \quad i \geq 1; & (1) \end{array}$$

which we prove by induction on i. If $i = 1$ then (1) is true by Lemma 5.1. Assume, then, that (1) is true for k and that $i = k + 1$. If $\mathrm{PD}((f : t), rs, k) = \emptyset$ we are done, so assume otherwise. By Lemma 5.2 each element of $\mathrm{PD}((f : t), rs, k)$ can be rewritten, and since $\mathcal{S}$ is primary fd it must choose one of them, say $(f : t')$, to rewrite, so that, since primary descendants cannot be nested,

$$\mathrm{PD}((f : t), rs, k + 1) = (\mathrm{PD}((f : t), rs, k) \setminus \{ (f : t') \}) \cup \mathrm{PDD}((f : t')),$$

and again (1) follows from Lemma 5.1. □

The main theorem now follows from Lemma 5.3.

Theorem 5.4: The multiset ordering $\gg_0$ is weakly adequate for the class $\mathcal{GR}(\mathcal{L})$ of good rewrite rule sets $\mathbf{RR}(\mathcal{L})$ and any primary fd strategy $\mathcal{S}$.

Acknowledgments: We would like to thank Geoffrey Frank, Quan Nguyen, and John Cherniavsky for fundamental contributions to this paper. We also extend our appreciation to Ellis Cooper, Melvin Fitting, and Anil Nerode for many useful conversations.

6 References.

Ba78 BACKUS, J., 'Can programming be liberated from the von Neumann style? A functional style and its algebra of programs,' *Communications of the ACM*, 21(1978), 613-641.

BaW²86 BACKUS, J., WILLIAMS, J. H., and WIMMERS, E. L., 'FL Language Manual,' Research Report RJ 5339, IBM Research Laboratory, San Jose, CA, November, 1986.

BiWa88 BIRD, R. and WALDER, P., *Introduction to Functional Programming*, Prentice Hall, New York, 1988.

BoMo79 BOYER, R. S., and MOORE, J. S., *A Computational Logic*, Academic Press, New York, 1979.

CoBo72 CONSTABLE, R. E., and BORODIN, A., 'Subrecursive programming languages. Part 1,' J. Assoc. Comp. Mach., 19(1972), 526-568.

De82 DERSHOWITZ, N., 'Orderings for term rewriting systems,' *J. of Theoretical Comp. Sci.*, 17(1982), 279-310.

De85 DERSHOWITZ, N., 'Termination,' *Proceedings of Rewriting Techniques and Applications*, Dijon, France, May 1985, LNCS 202, Springer-Verlag, New York, 1985, 180-224.

DeMa79 DERSHOWITZ, N., and MANNA, Z., 'Proving termination with multiset orderings,' Communications of the ACM, 22(1979), 465-467.

Ei87 EISENBACH, S., *Functional Programming: Languages, Tools and Architectures*, John Wiley & Sons, New York, 1987.

Fr79 FRANK, G., *Virtual Memory Systems for Closed Applicative Language Interpreters*, Ph.D. Dissertation, University of North Carolina at Chapel Hill, 1979.

HaW³85 HALPERN, J. Y., WILLIAMS, J. H., WIMMERS, E.L. and WINKLER, T. C., 'Denotational semantics and rewrite rules for FP,' *Proceedings of the Twelfth ACM Symposium of Principles of Programming Languages*, January 1985, 108-120.

HoUl79 HOPCROFT, J. E., and ULLMAN, J. D., *Introduction to Automata Theory, Languages, and Computation*, Addison-Wesley, Reading, MA, 1979.

Hu80 HUET, G., 'Confluent reductions: abstract properties and applications to term rewriting systems,' *J. of Assoc. Comp. Mach.*, 27(1980), 797-821.

KnBe70 KNUTH, D. and BENDIX, P., 'Simple word problems in universal algebras,' *Computational Problems in Abstract Algebra*, Leech, J., ed., Pergamon Press, 1970, 263-297.

La75 LANKFORD, D. S., 'Canonical algebraic simplification in computational logic,' *MEM ATP-25, Automated Theorem Proving Project*, University of Texas, Austin, TX, May 1985.

LatSi88 LATCH, D. M. and SIGAL, R., 'Generating evaluation theorems for functional programming languages,' *Proceedings of the Third International Symposium on Methodologies for Intelligent Systems*, Torino, Italy, October, 1988, 47-58.

Tu82 TURNER, D., 'On the Kent recursive calculator,' *Functional Programming and Its Applications*, Darlington, J., ed., Cambridge University Press, New York, 1982.

Wi81 WILLIAMS, J. H., 'Formal representations for recursively defined functional programs,' *Formalization of Programming Concepts*, LNCS 107, Springer-Verlag, New York, 1981, 460-470.

An Equational Logic Sampler

George F. McNulty
Department of Mathematics
University of South Carolina
Columbia, SC 29208
USA
[Bitnet: N410102 at UNIVSCVM]

§0 Introduction

The concepts that can be expressed by means of equations and the kinds of proofs that may be devised using equations are central concerns of equational logic. The concept of a ring is ordinarily presented by saying that a ring is a system $\langle R, +, -, \cdot, 0, 1\rangle$ in which the following equations are true:

$$x + (y + z) \approx (x + y) + z \qquad x \cdot (y \cdot z) \approx (x \cdot y) \cdot z \qquad x \cdot (y + z) \approx x \cdot y + x \cdot z$$

$$x + y \approx y + x \qquad x \cdot 1 \approx x \qquad (x + y) \cdot z \approx x \cdot z + y \cdot z$$

$$-x + x \approx 0 \qquad 1 \cdot x \approx x$$

$$x + 0 \approx x$$

A ring is an algebra—that is, a nonempty set endowed with a system of finitary operations. Equations, on the other hand, are certain strings of formal symbols. The concept of truth establishes a binary relation between equations and algebras: an equation $s \approx t$ *is true* in an algebra $\mathbf{A}$. This relationship underlies virtually all the work in equational logic. By way of this relation each equational theory—that is, each set of equations closed under logical consequence—is associated with a variety of algebras, which is the class of all algebras in which each equation of the theory is true. Through this connection syntactical tools developed for dealing with equations can be brought to bear on algebraic questions about varieties. Conversely, algebraic techniques and concepts from the theory of varieties can be employed on the syntactical side.

This sampler offers those working with term rewriting systems a selection of concepts and conclusions, many recent, from equational logic at large. Compared to the rich array of material in the literature, this sample is rather spare. It is intended to suggest rather than represent the work being done in equational logic.

Among the most glaring omissions are:

- **The theory of the commutator for congruence modular varieties.** This powerful theory, which applies to the most commonly encountered sets of equations, lies too deep to develop here. A comprehensive account can be found in Freese and McKenzie [1987].
- **The tame congruence theory for locally finite varieties.** This deep theory offers far-reaching insights into the structure of finite algebras and the varieties they generate. Hobby and McKenzie [1988] provide the foundations of this theory.
- **The theory of term rewriting systems.** Book [1987], Buchberger [1987], and Dershowitz [1987], all found in Jouannaud [1987], give highly readable surveys of different aspects of this area. Written for those already familiar with term rewriting, the present sampler focusses on *other* parts of equational logic.

- **The theory of clones** Equational logic and the theory of varieties are two facets of the same jewel. The theory of clones (with a distinguished set of generators) is another facet. Roughly, a clone is a collection of finitary operations on an underlying set, which contains all the projection functions and is closed under composition. Given a variety, the clones of term operations on the free algebras in the variety provide a full view of the variety. Szendrei [1986] is devoted to clone theory.

The last section of this sampler is a little lexicon of concepts from algebra and logic. It can be perused by the interested and curious, or referred to at need. Its entries also offer some explanation of the notation used in the sampler.

The account given in this sampler deals with algebras with only one universe. Also all fundamental operations of these algebras are total. From the viewpoint of practice, such restrictions seem too confining. The need to develop the theory of partial algebras and the theory of multisorted algebras is apparent. Such a development is still in its early stages. Perhaps the total one-sorted theory can be a paradigm. In any case, the results in this sampler have a significance just as they are.

§1 Decidability of Theories Axiomatized by Equations

Equations can be used to specify several kinds of deductively closed sets of sentences. Let Σ be a fixed set of equations of a given similarity type σ.

- The **elementary theory** based on Σ is the set of all sentences of first-order logic which are logical consequences of Σ.
- The **implicational theory** based on Σ is the set of formulas which are logical consequences of Σ and which have the form $\phi \Rightarrow s \approx t$ where ϕ is a conjunction of equations—ϕ may be the empty conjunction, in which case the resulting formula is logically equivalent to $s \approx t$.
- The **equational n-variable theory** based on Σ is the set of all equations which are logical consequences of Σ and in which no variables other than $x_0, \ldots, x_{n-1}$ appear. Equations in which no variables appear are called **ground equations**. The equational 0-variable theories are also called **ground theories**.
- A **Σ-presentation** is a system $\langle G|\Gamma\rangle$ where G is a set of new constant symbols and Γ is a set of ground equations in the similarity type obtained from σ by adjoining the members of G. Construing equations as ordered pairs of terms, the ground theory based on $\Sigma \cup \Gamma$ turns out to be a congruence relation on the algebra Mod Σ-freely generated by G; this ground theory is called the **ground theory of the presentation**. $\langle G|\Gamma\rangle$ is said to **present** the quotient algebra formed from this free algebra using this congruence relation and also to present any algebra isomorphic with this quotient algebra. $\langle G|\Gamma\rangle$ is a **finite** presentation provided both G and Γ are finite.

An elementary, implicational, or n-variable theory is **decidable** iff it is a recursive set of sentences. A Σ-presentation $\langle G|\Gamma\rangle$ has a **recursively solvable word problem** iff the ground theory based on $\Sigma \cup \Gamma$ in the type σ augmented by G is decidable. Σ is said to have **recursively solvable word problems** iff every finite Σ-presentation has a recursively solvable word problem. Σ has a **uniformly recursively solvable word problem** iff the set

$$\{\langle \Pi, s \approx t\rangle : \Pi \text{ is a finite } \Sigma\text{-presentation and } s \approx t \text{ belongs to the ground theory of } \Pi\}$$

is recursive. The connections between these concepts are laid out in the diagram below. This diagram refers to any fixed set Σ of equations.

Decidable elementary theory $\Downarrow$		Decidable elementary theory $\Downarrow$
Decidable implicational theory	$\Leftrightarrow$	Uniformly solvable word problem
$\Downarrow$		$\Downarrow$
Decidable equational theory		Solvable word problems
$\Downarrow$		$\Downarrow$
For each n, the equational n-variable theory is decidable		

In general, none of the implications above can be reversed. This follows from the results mentioned next. The most powerful decidability results that fit into this diagram are the ones which assert the decidability of the elementary theory; the most powerful undecidability results are those that assert the undecidability of the equational n-variable theory for some particular small value of n.

THEOREM.

(1) (TARSKI [1949A]) *The elementary theory of Boolean algebras is decidable.*
(2) (SZMIELEW [1954]) *The elementary theory of Abelian groups is decidable.*
(3) (McKINSEY [1943]) *The implicational theory of lattices is decidable.*
(4) (DEHN [1911]) *The equational theory of groups is decidable.*

The results listed in this theorem are classical, the first two being much more difficult. Boolean algebras and Abelian groups are both provided with powerful algebraic structure theories. Using the tools of tame congruence theory and the theory the commutator for congruence modular varieties, McKenzie and Valeriote [1989] have been able to show that *any* locally finite variety with a decidable elementary theory has a powerful algebraic structure theory.

THEOREM. *Let Σ be a finite set of equations in a finite similarity type. If every algebra with a finite Σ-presentation is residually finite, then Σ has a uniformly recursively solvable word problem. If Π is an finite Σ-presentation and the algebra presented by Π is residually finite, then Π has a recursively solvable word problem.*

THEOREM. *If* T *is a finitely based equational theory of finite similarity type and Mod*T *is generated by its finite members, then* T *is a decidable equational theory.*

These two theorems apply to a surprising number of cases. The first was formulated in Maltsev [1958], although it is related to a result in McKinsey [1943] and it also appears in Evans [1969]. The second is closely connected to Evans [1969]. Evans [1978] provides further discussion on this point. Now consider some undecidability results.

THEOREM.

(1) (MARKOV [1947], POST [1947]) *There is a finite semigroup presentation with a recursively unsolvable word problem.*
(2) (BOONE [1954–1957], NOVIKOV [1955]) *There is a finite group presentation with a recursively unsolvable word problem.*
(3) (TARSKI [1953]) *There is a finite relation algebra presentation with a recursively unsolvable word problem.* (See also Tarski and Givant [1987]).

Of course, to say that there is a finite semigroup presentation with a recursively unsolvable word problem is to assert that there is a finitely based undecidable equational ground theory (in a similarity type with one binary operation symbol and several constant symbols). Tarski's result is the most difficult and it produces the strongest result: let T be the underlying ground theory. T is *essentially undecidable* in the sense that any nontrivial equational theory extending T is also undecidable.

THEOREM.

(1) (MALTSEV [1966]) *There are finitely based undecidable equational 1-variable theories of loops and of quasigroups.*

(2) (MURSKIĬ [1968]) *There is a finitely based undecidable equational 2-variable theory of semigroups.*

There are other results of this kind in the literature. The equational theories arising in this theorem are artificial, as are the presentations of semigroups and groups in the preceding theorem—but not the relation algebra presentation.

THEOREM.

(1) (TARSKI [1953]) *The equational theory of relation algebras is undecidable.*

(2) (HERRMANN [1983]) *The equational 4-variable theory of modular lattices is undecidable.*

The result of Herrmann relies heavily on Freese [1980] where it is shown that the equational 5-variable theory of modular lattices is undecidable. Dedekind [1900] described the free modular lattice on three free generators. It has 28 elements. As a consequence, the equational 3-variable theory of modular lattices is decidable. It is interesting to observe that the equational theory of lattices is decidable in polynomial time, that the equational theory of distributive lattices is decidable, but it is $\mathcal{NP}$-complete, and that the equational theory of modular lattices is not decidable. The undecidability of the equational theory of modular lattices implies that the elementary theory of lattices is undecidable—a result announced by Tarski [1949b].

Let us return to our diagram. The results listed above are enough to insure that none of the arrows can be reversed, with the possible exception of two. These are taken care of by the next two theorems.

THEOREM (WELLS [1982]). *There is a finite set Σ of equations of a finite similarity type such that the equational theory based on Σ is undecidable, but the equational n-variable theory based on Σ is decidable, for each natural number n.*

THEOREM (MEKLER, NELSON, AND SHELAH [1989]). *There is a finite set Σ of equations of a finite similarity type such that Σ does not have a uniformily recursively solvable word problem, but every finite Σ-presentation has a recursively solvable word problem.*

These theorems leave open one conceivable implication in the diagram.

PROBLEM. *Is there a finitely based undecidable equational theory with recursively solvable word problems?*

Mekler, Nelson, and Shelah [1989] give an example of a *recursively* based undecidable equational theory with recursively solvable word problems; the similarity type is infinite.

§2 Finite Axiomatizability

Many common equational theories, e.g. the theory of groups, the theory of rings, and the theory of lattices are finitely based almost by fiat: they are given by listing finitely many equations. However, given such a familiar algebra as $\langle \mathbb{N}, +, \cdot, \uparrow, 1\rangle$ (here $n \uparrow m = n^m$ and $\mathbb{N}$ denotes the set of natural numbers) or such a small algebra as the semigroup of all maps from and to a four element set, it may not be so clear whether the equational theory of the algebra is finitely based.

THEOREM (LYNDON [1951]). *Every two element algebra with only finitely many fundamental operations is finitely based.*

This statement is deceptive in its simplicity. There are infinitely many distinct two element algebras. Lyndon's proof depends on the difficult analysis of the lattice of clones on a two-element set accomplished by Emil Post [1941]. Despite the difficulty of this proof, it *seems* obvious that virtually everything about a finite algebra can be determined, since the algebra can be given by means of a finite list of finite tables, each describing how to compute a fundamental operation. Surprisingly, Lyndon [1954] found a nonfinitely based seven element algebra whose only essential fundamental operation is binary. These discoveries of Lyndon led Alfred Tarski to pose the following problem:

TARSKI'S FINITE BASIS PROBLEM. *Is the set of finitely based finite algebras whose universes are subsets of the natural numbers and which have only finitely many fundamental operations recursive?*

This problem is still open. McKenzie [1984] showed that it reduces to the special case of groupoids. It is not even known whether the set of finitely based finite groupoids is recursively enumerable—or whether it is the complement of an r.e. set. Nevertheless, much has been discovered.

Murskiĭ [1965] presents a 3-element groupoid which is not finitely based. Murskiĭ's groupoid is a graph algebra, a notion introduced by C. Shallon [1979]. Let $\langle V, E\rangle$ be a graph—V is a finite set of vertices and E is a set of undirected edges between vertices in V. Let $A = V \cup \{0\}$, where $0 \notin V$, and define the binary operation $\circ$ on A by

$$a \circ b = \begin{cases} a & \text{if } \langle a, b\rangle \in E \\ 0 & \text{otherwise} \end{cases}$$

The algebra $\langle A, \circ\rangle$ is called the **graph algebra** of the graph $\langle V, E\rangle$. Murskiĭ's groupoid is the graph algebra of

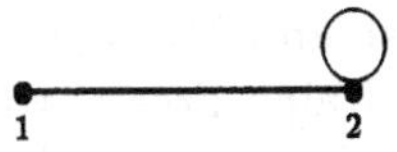

The table for this groupoid is

$\circ$	0	1	2
0	0	0	0
1	0	0	1
2	0	2	2

THEOREM (PERKINS [1966,69]). *Every commutative semigroup is finitely based.*

An equational theory T is **hereditarily finitely based** provided every equational theory (of the same similarity type) extending T is finitely based. Thus, the theory of commutative semigroups is hereditarily finitely based. Many other equational theories of semigroups have this property. Shevrin and Volkov [1985] survey, in depth, the axiomatizability of equational theories of semigroups. Not every finite semigroup is finitely based.

THEOREM (PERKINS [1966,69]). *The multiplicative semigroup consisting of the following six matrices*

$$\begin{pmatrix} 0 & 0 \\ 0 & 0 \end{pmatrix} \quad \begin{pmatrix} 1 & 0 \\ 0 & 1 \end{pmatrix} \quad \begin{pmatrix} 1 & 0 \\ 0 & 0 \end{pmatrix} \quad \begin{pmatrix} 0 & 1 \\ 0 & 0 \end{pmatrix} \quad \begin{pmatrix} 0 & 0 \\ 1 & 0 \end{pmatrix} \quad \begin{pmatrix} 0 & 0 \\ 0 & 1 \end{pmatrix}$$

is not finitely based.

On the other hand, the most familiar kinds of finite algebras *are* finitely based.

THEOREM.

(1) (OATES AND POWELL [1965]) *Every finite group is finitely based.*
(2) (KRUSE [1973] AND L'VOV [1973]) *Every finite ring is finitely based.*
(3) (MCKENZIE [1970]) *Every finite lattice is finitely based.*

This is actually three separate theorems. The proofs of the first two parts have some similarities. The last part is different. In 1972, Kirby Baker announced a considerable extension:

THEOREM (BAKER [1977]). *Every finite algebra with only finitely many fundamental operations which belongs to a congruence distributive variety is finitely based.*

The variety of lattices is congruence distributive, as are most varieties algebras connected with logic (Heyting algebras, cylindric algebras, DeMorgan algebras,...). The variety of groups and the variety of rings are not congruence distributive, but they are congruence modular. Efforts to find the appropriate general setting for the results of Oates and Powell and of Kruse and L'vov have yet to bear a fully satisfying result. The next two theorems are significant steps in the right direction. The theory of the commutator for congruence modular varieties offers the means to frame many of the concepts of group theory in a more general setting. In particular, the notions of an Abelian algebra and of a nilpotent algebra are available. Roughly speaking, an Abelian algebra is little different from a module over some ring and a nilpotent algebra is one that is almost Abelian. (See Freese and McKenzie [1987] for a precise account.) The theorem formulated next is due to Ralph Freese, who generalized a similar result of M. Vaughan-Lee [1983]. The proof can be found in Freese and McKenzie [1987].

THEOREM. *Every finite nilpotent algebra with only finitely many fundamental operations which belongs to a congruence modular variety and which is the direct product of algebras of prime power cardinality is finitely based.*

This theorem does not generalize the results of Oates and Powell (or of Kruse and L'vov) since there are plenty of finite groups which are neither nilpotent nor the direct product of algebras of prime power cardinality

Jónsson [1967] proved that any congruence distributive variety generated by a finite algebra is residually small. So the next theorem generalizes Baker's finite basis theorem above.

THEOREM (MCKENZIE [1988]). *Every finite algebra with finitely many fundamental operations which belongs to a residually small congruence modular variety is finitely based.*

These last two theorems having surprisingly little overlap. This is so since Freese and McKenzie [1981] have shown that a residually small congruence modular variety contains no non-Abelian nilpotent algebra. This observation means that the converse of McKenzie's theorem cannot hold. The inviting prospect of eliminating the hypothesis of residual smallness from this theorem—obtaining a theorem like the one preceding it, with "modular" in place of "distributive"—is closed out by a series of counterexamples. The earliest was a nonfinitely based finite bilinear algebra given by Polin [1976]. Oates-MacDonald and Vaughan-Lee [1978] gave a nonfinitely based finite nonassociative ring, Vaughan-Lee [1979] gave a nonfinitely based finite loop, and R. Bryant [1982] constructed a finite group **G** which has an element p such that when p is given the status of a distinguished element, the resulting "pointed group" $\langle \mathbf{G}, p\rangle$ is not finitely based. In the early 1970's, Bjarni Jónsson speculated that every finite algebra with finitely many fundamental operations which belongs to a residually small variety might be finitely based.

PROBLEM. *Is every finite algebra with only finitely many fundamental operations which belongs to a residually small variety finitely based?*

If Jónsson's speculation turns out to be correct, then the hypothesis of congruence modularity may be eliminated from the theorem above.

The examples of nonfinitely based finite algebras mentioned above, perhaps with the exception of the semigroup of 2×2 matrices, all have an ad hoc quality. This conclusion is reinforced by

THEOREM (MURSKIĬ [1975]). *Denote by $N(k)$ the number of groupoids with universe $\{0, 1, \dots, k-1\}$ and by $Q(k)$ the number of groupoids on the same set which are nonfinitely based. Then*

$$\lim_{k \to \infty} \frac{Q(k)}{N(k)} = 0.$$

In the late 1970's the notion of inherently nonfinitely based finite algebras emerged in the work of Murskiĭ and Perkins—see Murskiĭ [1979] and Perkins [1984]. A finite algebra **A** with finitely many fundamental operations is said to be **inherently nonfinitely based** provided **A** is not a member of any locally finite finitely based variety. Any variety generated by a finite algebra is locally finite—so every inherently nonfinitely based finite algebra is nonfinitely based. Also if an inherently nonfinitely based algebra belongs to the variety generated by a finite algebra **A**, then **A** is itself inherently nonfinitely based: the inherent nonfinite basis property is contagious.

Murskiĭ [1979] proved that the 3-element groupoid given in Murskiĭ [1965] is inherently nonfinitely based. Perkins [1984] gave the first generous sufficient conditions for a finite algebra to be inherently nonfinitely based. McNulty and Shallon [1983] sharpen the results of Murskiĭ and Perkins and analyze most of the known nonfinitely based finite algebras, attempting to discover which ones among them are inherently nonfinitely based. For instance the next theorem extends one from Perkins [1984]. Call an algebra **A collapse-free** provided $t \approx x$ is true in **A** only if t is the variable x. Let **A** be a groupoid. An element $a \in A$ is a **unit** iff $ab = ba = b$ for all $b \in A$; a is a **zero** iff $ab = ba = a$ for all $b \in A$—neither zeros nor units need have the status of fundamental operations.

THEOREM (McNULTY AND SHALLON [1983]). *Every finite nonassociative collapse-free groupoid with a unit and a zero is inherently nonfinitely based.*

Ježek [1985] presents three inherently nonfinitely based 3-element idempotent groupoids. Baker, McNulty, and Werner [1989] provide a very broad but technical sufficient condition for a finite algebra to be inherently nonfinitely based. One consequence is

THEOREM (BAKER, McNULTY, WERNER [1987]). *A graph* **G** *has a finitely based graph algebra if and only if* **G** *has no induced subgraph isomorphic to one of the four graphs below. Every nonfinitely based graph algebra is inherently nonfinitely based.*

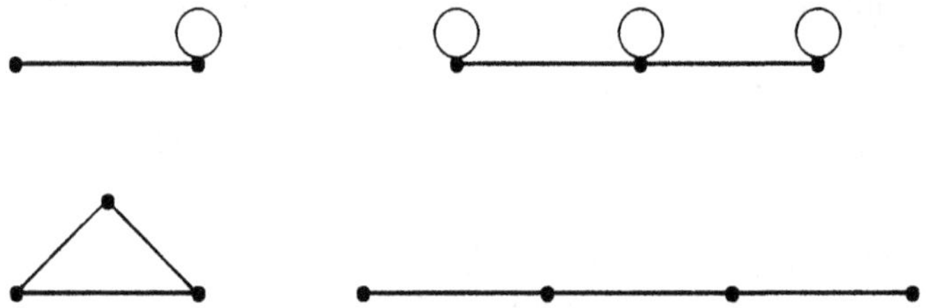

The techniques of Baker, McNulty, and Werner do not seem to apply to algebras with associative operations. Mark Sapir has recently published a breakthrough for semigroups. In view of the associative law, in discussing this result parentheses will be discarded and terms will be taken to be *words*: we would write, for instance, $xyxzx$ instead of $((xy)((xz)x))$. A word w is called an **isoterm** of the semigroup **S** provided w and u are the same whenever the equation $w \approx u$ is true in **S**. The words Z_n are defined by the following recursion scheme:

$$Z_0 = x_0$$
$$Z_{k+1} = Z_k x_{k+1} Z_k \quad \text{for every natural number } k$$

THEOREM (SAPIR [1987A]). *A finite semigroup* **S** *is inherently nonfinitely based if and only if* Z_n *is an isoterm of* **S**, *for every natural number* n.

Sapir has also constructed an algorithm for determining whether a finite semigroup is inherently nonfinitely based, see Sapir [1987b]. Using the theorem above, it is not difficult to prove that Perkins' semigroup of six 2×2 matrices is *inherently* nonfinitely based. It follows that the multiplicative semigroup of all 2×2 matrices over the 2-element field is also inherently nonfinitely based—contrasting with the results of Kruse and L'vov according to which the *ring* of all 2×2 matrices over the 2-element field must be finitely based.

It is not difficult to prove that the algebras $\langle \mathsf{N}, + \rangle$, $\langle \mathsf{N}, \cdot \rangle$, and $\langle \mathsf{N}, +, \cdot \rangle$ are all finitely based (familiar equations suffice). Consider the algebras

$$\mathbf{N} = \langle \mathsf{N}, +, \cdot, \uparrow \rangle$$
$$\mathbf{N}_1 = \langle \mathsf{N}, +, \cdot, \uparrow, 1 \rangle$$

where $n \uparrow m = n^m$ for natural numbers n and m. In his HIGH SCHOOL ALGEBRA PROBLEM, Tarski asked whether these algebras are finitely based, and in particular, whether bases for them could be assembled from the following handful of equations ordinarily taught to high school students.

$$x + y \approx y + x \qquad x \cdot y \approx y \cdot x \qquad x \cdot (y + z) \approx x \cdot y + x \cdot z$$
$$x + (y + z) \approx (x + y) + z \qquad x \cdot (y \cdot z)) \approx (x \cdot y) \cdot z \qquad 1 \cdot x \approx x$$
$$(x \cdot y) \uparrow z \approx (x \uparrow z) \cdot (y \uparrow z) \qquad x \uparrow (y + z) \approx (x \uparrow y) \cdot (x \uparrow z) \qquad x \uparrow (y \cdot z) \approx (x \uparrow y) \uparrow z$$
$$x \uparrow 1 \approx x \qquad 1 \uparrow x \approx x$$

Charles Martin [1973] proved that $\mathbf{N}$ is not finitely based, but he was unable to obtain the same result for $\mathbf{N}_1$. A. Wilkie, in an unpublished manuscript circa 1980, found an equation true in $\mathbf{N}_1$ which cannot be deduced from the eleven equations above. Tarski's High School Algebra Problem has been solved recently. For each natural number $n > 0$, let δ_n denote the equation

$$(A^{2^x} + B_n^{2^x})^x (C_n^x + D_n^x)^{2^x} \approx (A^x + B_n^x)^{2^x} (C_n^{2^x} + D_n^{2^x})^x$$

where $A = 1 + x$, $B_n = \sum_{i<n} x^i$, $C_n = 1 + x^n$, and $D_n = \sum_{i<n} x^{2i}$. Of course, 2 abbreviates $1 + 1$ here. These equations are related to Wilkie's equation and they are all true is $\mathbf{N}_1$.

THEOREM (R. GUREVIČ [1988]). *For every finite set Γ of equations true in the algebra $\mathbf{N}_1$, there is an odd $n \geq 3$ and a finite algebra $\mathbf{G}$ such that Γ is true in $\mathbf{G}$ but δ_n is false in $\mathbf{G}$. Therefore, $\mathbf{N}_1$ is not finitely based.*

§3 LATTICES OF EQUATIONAL THEORIES

Let σ be any similarity type and let L be the set of all equational theories of type σ. Set-inclusion orders the equational theories according to means of proof: the larger theory has more theorems. L is lattice ordered by set-inclusion—the meet in this lattice is just intersection and the join of a collection C of equational theories is just the equational theory based on $\bigcup C$. We use $\mathbf{L}$ to denote this lattice—or $\mathbf{L}(\sigma)$ if we need to emphasize its dependence on σ. $\mathbf{L}$ is an algebraic lattice—its compact members are exactly the finitely based equational theories—and it has the additional property that its largest element is compact, since the largest element is just the theory based on $x \approx y$. Let T be any equational theory. By $\mathbf{L}_{\mathrm{T}}$ we denote the sublattice of $\mathbf{L}$ comprised of all equational theories which include T. Thus if T were the equational theory of all groups, then $\mathbf{L}_T$ would be the lattice of all equational theories of groups and one of the members of $\mathbf{L}_T$ would be the equational theory of Abelian groups.

MALTSEV'S PROBLEM (MALTSEV [1968]). *Find an algebraic characterization of those lattices which can be isomorphic to* $\mathbf{L}_T(\sigma)$ *for some similarity type* σ *and some equational theory* T.

Maltsev's problem is still open, but work on it has led to many of the results described below.

Call σ **bold** iff there is an operation symbol of type σ with rank at least 2 or there are at least two unary operation symbols of type σ; call σ **meek** if it is not bold. We call a term **meek** iff it is a term of some meek similarity type. Terms which are not meek are called **bold**. (Note: bold similarity types may have meek terms.) The lattices $\mathbf{L}(\sigma)$, where σ is meek, have been completely described in Ježek [1970], extending work of Jacobs and Schwabauer [1966] and A.D. Bol'bot [1970]. $\mathbf{L}(\sigma)$, where σ is bold, is very intricate.

THEOREM (MCKENZIE [1971]). *Let* σ *be a bold similarity type and* T *be an equational theory of type* σ. *If* $t \approx s \in T$, *for some terms* $t \neq s$, *then* T *covers at least one member of* $\mathbf{L}(\sigma)$.

THEOREM (MCNULTY [1981]). *Let* σ *be a similarity type and* T *be a finitely based equational theory of type* σ. *If* $t \approx x \in T$ *for some bold term* t, *then* T *covers at least* $2^{\aleph_0}$ *members of* $\mathbf{L}(\sigma)$.

This theorem extends a result in Kalicki [1955]. For bold similarity types, $\mathbf{L}$ is very large.

THEOREM (JEŽEK [1977]). *Let* σ *be any countable similarity type and let* $\mathbf{K}$ *be any algebraic lattice with only countably many compact elements. Then* $\mathbf{K}$ *is isomorphic with some interval in* $\mathbf{L}(\sigma)$.

In particular every finite lattice is an interval in $\mathbf{L}(\sigma)$. This entails that the lattices $\mathbf{L}(\sigma)$ satisfy only those equations of lattice theory which hold in every lattice. This theorem extends earlier work of Burris and Nelson [1971a,1971b].

Fix a similarity type σ. The **multiplicity type** of σ is the sequence $\langle m_0, m_1, m_2, \dots \rangle$, where m_k is the number of distinct operation symbols of similarity type σ with rank k, for each natural number k. Two similarity types which have the same multiplicity type can be regarded as equivalent, since the only distinction between them is the actual "shape" of the operation symbols. Evidently, if σ and τ have the same multiplicity type, then $\mathbf{L}(\sigma)$ is isomorphic with $\mathbf{L}(\tau)$.

THEOREM (MCKENZIE [1971]). *Let* σ *and* τ *be similarity types.* $\mathbf{L}(\sigma)$ *is isomorphic with* $\mathbf{L}(\tau)$ *iff* σ *and* τ *have the same multiplicity type.*

Thus, the multiplicity type of σ can be recovered from the isomorphism type of $\mathbf{L}(\sigma)$, for any similarity type σ.

THEOREM (JEŽEK [1981–86]). *Let* **L** *be the lattice of all equational theories of some fixed similarity type. Each of the following sets is definable in* **L** *by an elementary formula in the language of lattices:*

(1) *The set of all finitely based equational theories.*
(2) *The set of all one-based equational theories.*
(3) $\{\Theta(\mathrm{T}) : \Theta$ *is an automorphism of* $\mathbf{L}\}$, *for any finitely based equational theory* T.
(4) *The set of all equational theories of finite algebras.*
(5) $\{\Theta(\mathrm{T}) : \Theta$ *is an automorphism of* $\mathbf{L}\}$, *where* T *is the equational theory of any finite algebra, provided the similarity type is finite.*

This theorem is a tour de force. It builds on the work of McKenzie [1971], where a number of familiar equational theories were shown to have definable orbits. The fact that many sets are definable supports the view that the structure of **L** is rich. Algebras with intricate structure are expected to have few automorphisms. On the other hand, **L** may well have automorphisms. Suppose that there are two 3-ary operation symbols G and H. Let Φ be a map from terms to terms which interchanges the G's with the H's. So $\Phi(HxGxyzHyxz) = GxHxyzGyxz$. Such a map induces a map of equations to equations, which yields a map that takes equational theories to equational theories. Such a map is an automorphism of **L**. Another kind of automorphism is induced by "permuting positions" within an operation symbol. For example, suppose Ψ is a map from terms to terms defined by $\Psi(Hstr) = H\Psi(t)\Psi(r)\Psi(s)$ and $\Psi(t) = t$ if t is a term not beginning with H. So $\Psi(HxGxyzHyxz) = HGxyzHxzyx$. Ψ also induces an automorphism of **L**. Automorphisms generated by those induced by interchanging symbols and "permuting positions" are called **obvious** automorphisms.

THEOREM (JEŽEK [1981–86]). *Let* **L** *be the lattice of all equational theories of some fixed bold similarity type. Every automorphism of* **L** *is obvious.*

Addressing Maltsev's Problem more directly, we have the following theorem, which was announced independently by Kogalovskiĭ.

THEOREM (PIGOZZI AND TARDOS [1989]). *Let* **K** *be an algebraic lattice whose largest element is compact. If* **K** *has exactly one maximal member, then there is a similarity type* σ *and an equational theory* T *such that* **K** *is isomorphic with* $\mathbf{L}_{\mathrm{T}}(\sigma)$.

However much the theorems above suggest otherwise, there are algebraic lattices with compact largest element—even finite lattices—which are isomorphic to no lattice of the form $\mathbf{L}_{\mathrm{T}}(\sigma)$. The nth **zipper condition** is the following statement:

If $x_0 \vee \ldots \vee x_n \approx 1$ *and* $x_i \wedge y \approx z$ *for all* $i \leq n$, *then* $y \approx z$.

Here 1 denotes the largest element of a lattice. The 0th zipper condition holds in every lattice with a largest element.

THEOREM (LAMPE [1986]). *Let* T *be any equational theory.* $\mathbf{L}_{\mathrm{T}}$ *satisfies the* n^{th} *zipper condition for every natural number* n.

This theorem makes it possible to find finite lattices **K** which are not isomorphic to $\mathbf{L}_{\mathrm{T}}$ for any equational theory T. One such is the lattice $\mathbf{M}_3$ diagrammed below.

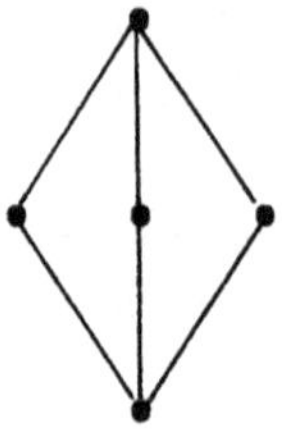

Lampe's Theorem suggests that the class of lattices of the form $\mathbf{L}_{\mathrm{T}}$ might have interesting structural properties.

§4 Undecidable Properties of Finite Sets of Equations

Is the following list of equations a base for the equational theory of Boolean algebras? (Here we construe Boolean algebras using meet and complementation as the only fundamental operations.)

$$x \wedge x \approx x \qquad x \wedge (y \wedge z) \approx (x \wedge y) \wedge z$$
$$x \wedge y \approx y \wedge x \qquad ((x \wedge y)' \wedge (x \wedge y')')' \approx x$$

Does the group with presentation $\langle a, b | a^4 \approx e, a^2 b^{-2} \approx e, abab^{-1} \approx e \rangle$ have more than one element? Questions like these may be addressed individually, but we may wish to have an algorithm for settling them on a more wholesale basis. We will say a property $\mathcal{P}$ of finite sets of equations of similarity type σ is **decidable** iff the set

$$\{\Sigma : \ \Sigma \text{ is a finite set of equations of type } \sigma \text{ which has } \mathcal{P}\}$$

is recursive. There are some interesting positive results, but most properties turn out to be undecidable. The group presented above has 8 elements—it is the group of quaternions. It is unknown whether the listed equations form a base for the equational theory of Boolean algebras—although it is an amusing exercise to prove that any **finite** algebra in which these equations hold is a Boolean algebra.

Perhaps the earliest work in this domain is that of Markov [1951] dealing with properties of finite presentations. Let Σ be any set of equations of some fixed finite similarity type. Let $\mathcal{P}$ be a property of finite Σ-presentations. (Formally, we take $\mathcal{P}$ to be a set of finite presentations.) $\mathcal{P}$ is said to be a **Markov property** over Σ provided each of the following conditions holds

- If Π and Ω are finite Σ-presentations of isomorphic algebras and $\Pi \in \mathcal{P}$, then $\Omega \in \mathcal{P}$.
- There is some finite Σ-presentation in $\mathcal{P}$.
- There is a finite Σ-presentation Π such that if Ω is any finite Σ-presentation such that the algebra presented by Π is embeddable in the algebra presented by Ω, then $\Omega \notin \mathcal{P}$.

In essence, the first condition asserts that $\mathcal{P}$ is almost a property of finitely presented algebras. By an abuse of language, it is common to refer to Markov properties of algebras—for example finiteness is said to be a Markov property of groups. If $\mathcal{P}$ is hereditary, in the sense that any finitely presented subalgebra of an algebra with presentation in $\mathcal{P}$ has a presentation again in $\mathcal{P}$, then the last condition can be simplified to the existence of a finite Σ-presentation not in $\mathcal{P}$. We say that $\mathcal{P}$ is a Markov property of semigroups, groups, etc. when Σ is a base for the theory of semigroups, groups, etc. (and the similarity type is fixed accordingly).

THEOREM.

(1) (MARKOV [1951]) *Every Markov property of semigroups is undecidable.*
(2) (ADDISON [1954] AND FEENEY [1954]) *Every Markov property of cancellation semigroups is undecidable.*
(3) (ADJAN [1958] AND RABIN [1958]) *Every Markov property of groups is undecidable.*

Very many properties of groups interesting from the algebraists' viewpoint are either Markov properties or the complements of Markov properties. Among these are nontriviality (=having more than one element), finiteness, being cyclic, being Abelian, being free in the variety of groups, and solvability. Other properties, while not Markov properties in their own right, can be shown undecidable either through their close connection to Markov properties or directly by using the tools employed in the proofs of this theorem. Simplicity is a property of this kind. In fact, this result is so strong that little work was done on these topics for some twenty years. Recently, however, more work has been directed at the decidability of properties of presentations of monoids—work motivated by developments in string rewriting theory. See Book [1987].

For properties of finite sets of equations in general, there is no single overpowering theorem. There is, however, a pattern of interlocking results according to which many properties of finite sets of equations are seen to be undecidable. The earliest results are those of Perkins [1966–67], many of which were generalized. In particular, Perkins proved that each of the following properties is an undecidable property of finite sets of equations, provided only that the similarty type is rich enough.

- Strong consistency.
- Being the base of a decidable equational theory.
- Being the base of an equationally complete equational theory.
- Being the base of the equational theory of a finite algebra.

Let $\mathcal{P}$ be a property of finite sets of equations of similarity type σ. We say that $\mathcal{P}$ is **closed under logical equivalence** iff whenever $\Gamma \in \mathcal{P}$ and Δ is a finite set of equations such that Γ and Δ are bases of the same equational theory, then $\Delta \in \mathcal{P}$. A property which is closed under logical equivalence might be regarded as a property of finitely based equational theories. We call the similarity type σ **strong** provided there is at least one operation symbol of type σ which has rank at least 2.

Let σ and ρ be two similarity types and let Δ be a set of equations in type σ and Θ be a set of equations in type ρ. We say that Δ is a **term reduct** of Θ provided for each fundamental operation symbol Q of type σ there is a term d_Q of type ρ in variables $x_0, x_1, \ldots, x_{r-1}$, where r is the rank of Q, such that for every infinite model $\mathbf{A} = \langle A, Q^{\mathbf{A}}, \ldots \rangle$ of Δ there is an algebra $\mathbf{B}$, with the same universe as $\mathbf{A}$, which is a model of Θ such that $Q^{\mathbf{A}} = d_Q^{\mathbf{B}}$ for every operation symbol Q of type σ. Roughly speaking, Δ is a term reduct of Θ if all of the infinite models of Δ can be obtained from those of Θ by regarding a certain system of term operations of type ρ has the fundamental operations of type σ.

THEOREM (MCNULTY [1976B]). *Let σ be a strong similarity type and let $\mathcal{P}$ be a property of finite sets of equations of type σ which is closed under logical equivalence. If $\{x \approx y\} \in \mathcal{P}$ and there is a strongly consistent finite set Σ of equations such that Σ is not a term reduct of any finite set of equations with property $\mathcal{P}$, then $\mathcal{P}$ is undecidable.*

The following properties are seen to be undecidable on the basis of this theorem: being the base of a locally finite variety, being the base of a variety which is residually finite (or residually small), having an infinite simple model, and having simple models of arbitrarily large infinite cardinality.

THEOREM (MCNULTY [1976B]). *Let σ be a strong similarity type and let $\mathcal{P}$ be a property of finite sets of equations of type σ which is closed under logical equivalence. If there is some finite set of equations with property $\mathcal{P}$ and for each Γ with $\mathcal{P}$ there is a term t_Γ in which both x and y occur such that $\Gamma \vdash t_\Gamma \approx x$, then $\mathcal{P}$ is undecidable.*

The theory of Maltsev conditions supplies many properties which are undecidable according to this theorem. Among them are being the base of a congruence permutable (or distributive or modular) variety and being the base of congruence regular variety. We also draw the following corollary.

COROLLARY. *Let* T *be a finitely based equational theory in a strong similarity type such that there is a term t in which both x and y occur such that $t \approx x \in$ T. The set of finite bases of* T *is not recursive.*

There are stronger versions of this corollary due to McNulty [1976a] and Murskiĭ [1971]: the corollary remains true even if the similarty type and the term t are only bold. Thus the most commonly encountered finitely based equational theories (groups, rings, lattices, etc.) are *base undecidable* in the sense that the set of finite bases of such a theory will not be recursive. This sort of undecidability sheds light on the difficulty of Tarski's Finite Basis Problem. For a fixed finite algebra **A** one cannot expect to have an algorithm for answering the questions "Is the finite set Σ a base for **A**?" as Σ ranges over finite sets of equations.

The next theorem is an analog of the classical Theorem of Rice [1953].

THEOREM (MURSKIĬ [1971] AND MCNULTY [1976B]). *Let σ be a similarity type with some operation symbol of rank at least 2 or with at least three unary operation symbols. Let Δ be any set of equations of type σ. The set*

$$\{\Sigma : \Sigma \vdash \Delta \text{ and } \Sigma \text{ is a finite strongly consistent set of equations}\}$$

is recursive iff Δ has no finite strongly consistent extensions.

Most of the known undecidable properties of finite sets of equations, which are closed under logical equivalence, can be obtained as consequences of the theorems above, or at any rate from the proofs of these theorems. The original properties proven undecidable by Perkins, under some restrictions on the similarity type, were shown in McNulty [1976a,b] and in Murskiĭ [1971] to be undecidable as long as the similarity type is bold. These extensions are not corollaries of the theorems above, but they rely of techniques underlying those theorems. The next theorem provides a sample of further results of this kind.

THEOREM. *Let σ be similarity type. Each of the following properties of finite sets Σ of equations of type $\sigma$is undecidable.*

(1) (O'DÚNLAING [1983]) *Σ is logically equivalent to a confluent term rewriting system, provided σ is bold.*
(2) (O'DÚNLAING [1983]) *Σ is logically equivalent to a confluent and terminating term rewriting system, provided σ is bold.*
(3) (MURSKIĬ [1971]) *Σ is the base of an hereditarily finitely based equational theory, provided σ is strong.*
(4) (MCNULTY [1976B]) *Σ is a logically irredundant set of equations, provided σ is bold.*
(5) (PIGOZZI, CF. MCNULTY [1976B]) *Σ is logically equivalent to an irredundant set of n equations, provided σ is bold and n is any positive integer.*
(6) (KALFA [1985]) *Σ is a base for a variety with the joint embedding property, provided σ is bold and has at least one constant symbol.*

Many of the results listed above where first proved for various restricted similarity types. See the papers Perkins [1966–67] and Smith [1972]. The results of O'Dúnlaing [1983] were formulated for Thue systems over alphabets with at least four letters. Such results are easy to formulate as statements about sets of equations with unary operation symbols—in this case four of them. In this formulation, they fit into the framework of McNulty [1976b] and can be proved for any bold similarity type.

Unlike the results above, those listed in the theorem below require other techniques. Those due to Pigozzi represent a significant extension of the ideas developed for the theorems above, while McKenzie's results were obtained by substantially different means.

THEOREM. *Let σ be a strong similarity type. Each of the properties below is an undecidable property of finite sets of equations of type σ.*

(1) (MCKENZIE [1975]) *Σ has a finite model with more than one element.*
(2) (PIGOZZI [1976]) *The variety of algebras based on Σ has the amalgamation property.*
(3) (PIGOZZI [1976]) *The variety of algebras based on Σ has the Schreier property.*

Nearly all the properties above are closed under logical equivalence, irredundance being the exception. There are also known undecidable properties of finite sets of equations which are not closed under logical equivalence. These are significant from the perspective of term rewriting systems. Equations and rewrite rules are the same thing—they are just ordered pairs of terms. For properties closed under logical equivalence the leftside-rightside distinction in equations is neglected. For term rewriting theory, this distinction is crucial. Among the undecidable properties of finite sets of equations in any bold similarity type one finds: confluence (Bauer and Otto [1984]), termination (Huet and Lankford [1978], Dershowitz [1987]), and selfembedding (Plaisted [1985]). It would be interesting to see some broad sufficient conditions for the undecidability of properties not closed under logical equivalence.

Finally, we consider some very specialized presentations. A Σ-presentation $\langle G|\Gamma\rangle$ is **full** provided

i. Σ is the empty set,
ii. Every equation in Γ has the form $Qa_0a_1\dots a_{r-1} \approx b$ where Q is an operation symbol (and r is its rank) and $a_0, a_1, \dots, a_{r-1}, b \in G$,
iii. For each operation symbol Q and all $a_0, a_1, \dots, a_{r-1} \in G$, where r is the rank of Q, there is a unique $b \in G$, such that $Qa_0 \dots a_{r-1} \approx b \in \Gamma$.

Of course, the algebra presented by a full presentation $\langle G|\Gamma\rangle$ has universe G and Γ amounts to complete description of each of the operations. So a finite full presentation differs in no essential way from a finite algebra and we will refer to properties of finite algebras instead of properties of finite full presentations.

Most properties of finite algebras which have some natural appeal seem to be decidable. In fact, there do not appear to be any interesting properties of finite algebras which are known to be undecidable. In the theorem below, $\mathbf{A} \equiv_e \mathbf{B}$ means $\mathbf{A}$ and $\mathbf{B}$ are algebras with the same equational theory.

THEOREM. *Fix any finite similarity type. Each of the following sets is recursive.*

(1) (KALICKI [1952]) $\{\langle \mathbf{A}, \mathbf{B}\rangle : \mathbf{A}$ *and* $\mathbf{B}$ *are finite algebras with* $\mathbf{A} \equiv_e \mathbf{B}\}$.
(2) (SCOTT [1956]) $\{\mathbf{A} : \mathbf{A}$ *is a finite equationally complete algebra*$\}$.
(3) $\{\mathbf{A} : \mathbf{A}$ *is a finite algebra which generates a congruence modular variety*$\}$.

In the last item of this theorem, "modular" can be replaced by "permutable" or "distributive" or by a number of other similar properties. Roughly speaking, these positive results flow from the fact that each finitely generated free algebra in the variety generated by a finite algebra $\mathbf{A}$ is finite and can be constructed from $\mathbf{A}$ by effective means. The properties listed can all be determined by examining such free algebras. For instance, Day [1969] proved that $\mathbf{A}$ generates a congruence modular variety iff the free algebra on 4 generators in the variety has a modular congruence lattice.

Although there are no known appealing but undecidable properties of finite algebras, the following questions are open and appear to be very challenging. To be explicit, the type of groupoids is specified, but the problems are interesting in any strong finite similarity type.

DECISION PROBLEMS FOR FINITE ALGEBRAS.

- *Is* $\{\mathbf{A} : \mathbf{A}$ *is a finitely based finite groupoid*$\}$ *recursive?* This is just Tarski's Finite Basis Problem.

- *Is* $\{\mathbf{A} : \mathbf{A}$ *is a finite groupoid which generates a residually large variety*$\}$ *recursive?*

§5 DOING MATHEMATICS VIA EQUATIONS

While many of the most familiar kinds of mathematical structures can be specified by sets of equations, the scope of equational logic seems restricted in both its expressive power and its means of proof. Equational logic is a very simple fragment of first order logic, bereft of both quantifiers and connectives. The straightforward character of both the syntactical elements (terms and proofs) and the semantical elements (free algebras and varieties) of equational logic stem from the simplicity of this fragment.

As indicated by the earliest undecidability results of Post and Markov, equational logic *does* offer the means to capture fully the notion of computability—i.e. Turing machines may be simulated by certain kinds of finite semigroup presentations. So equational logic admits a certain level of complexity. Actually, equational logic embodies the full scope of mathematics, in about the same sense as first-order logic.

Building on the earlier work of C. S. Peirce [1880] and E. Schröder [1890-05], in the 1940's Alfred Tarski formulated the notion of a relation algebra. A **relation algebra** is an algebra $\langle A, \wedge, \vee, \cdot, ^-, ^\smile, 1' \rangle$ such that $\langle A, \wedge, \vee, ^- \rangle$ is a Boolean algebra, $\langle A, \cdot, 1' \rangle$ is a monoid, and each of the following equations holds:

$$x \cdot (y \vee z) \approx (x \cdot y) \vee (x \cdot z) \qquad (x \cdot y)^\smile \approx y^\smile \cdot x^\smile$$

$$(x^\smile)^\smile \approx x \qquad (x^-)^\smile \approx (x^\smile)^-$$

$$(1')^\smile \approx 1' \qquad (x \vee y)^\smile \approx x^\smile \vee y^\smile$$

$$(x^\smile \cdot (x \cdot y)^-) \wedge y \approx x^- \wedge x$$

An example of a relation algebra can be formed by taking A to be the collection of all binary relations on an arbitrary set U, giving A the Boolean operations by regarding A as the power set of U^2 and defining the remaining operations so that

$$\begin{aligned} R \cdot S &= \{\langle a, b\rangle : \langle a, c\rangle \in R \text{ and } \langle c, b\rangle \in S \text{ for some } c \in U\} \\ R^\smile &= \{\langle a, b\rangle : \langle b, a\rangle \in R\} \\ 1' &= \{\langle a, a\rangle : a \in U\} \end{aligned}$$

So $\cdot$ denotes composition, $^\smile$ denotes converse, and 1' denotes the identity relation. The class of relation algebras is axiomatized by equations. Under the intended interpretation just described the Boolean operations can be made to do the work of logical connectives and the operation $\cdot$, intended to denote composition of relations, supplies a quantifier. Using this idea, it is possible to translate sentences of first order logic into equations, so long as the sentences contain no more than three distinct variables (though these may occur any number of times). Roughly speaking, what Tarski was able to do was the following. Suppose that T is any first order theory which is strong enough the define a pairing function, and suppose for simplicity that the only nonlogical symbol of T is a binary relation symbol E. Expand the similarity type of relation algebras by an additional constant symbol e. There is a map * from first order sentences to ground equations in this expanded type such that

$$\mathrm{T} \vdash \Phi \text{ iff } \Sigma \cup \mathrm{T}^* \vdash \Phi^*$$

for every first order sentence Φ, where Σ is the finite set of equations forming a base for the theory of relation algebras. The map * is too technical to describe here, but it is a natural kind of map and it is reasonable to view $\Sigma \cup \mathrm{T}^*$ as just another presentation of T, rather than merely as some coded version of T. The theories which can be taken for T in this context include virtually every version of set theory—including Zermelo-Fraenkel set theory and Gödel-Bernays set theory—as well as elementary Peano arithmetic and even the elementary theory of real closed fields. This theorem of Tarski's is probably the most difficult result mentioned in this sampler. A full exposition of it can be found in Tarski and Givant [1987].

The equational theory of relation algebras is extremely rich. Because it offers the scope in which to view the whole of mathematics via equations, it offers a challenging and worthwhile arena in which to test and develop the tools of equational logic.

§6 WHERE TO FIND IT

The literature in equational logic and its companion, the theory of varieties, is fairly extensive. Here is a guide to that part which may be less familiar to those working in term rewriting.

General References

In the mid-1960's four books where written giving a general overview of universal algebra and the theory of varieties. They are Cohn [1965], Grätzer [1968], Pierce [1968], and Maltsev [1973]. Of these, Grätzer's book is the most comprehensive and it soon became the standard reference. The publication of Maltsev's book was delayed by his untimely death in 1967. The larger portion of work done in the theory of varieties and in equational logic has been accomplished since these books first appeared. The text Burris and Sankappanavar [1981] supplies much of the background necessary to this sampler. R. Freese, R. McKenzie, G. McNulty, and W. Taylor have undertaken a four volume exposition of the field. It is called "Algebras, Lattices, Varieties". The second volume should appear in 1989. The first volume is McKenzie, McNulty, and Taylor [1987].

Specialized Monographs

A deeper understanding of the field is available through monographs of a more specialized kind.

(1) Freese and McKenzie [1987] provide a detailed account of the theory of congruence modular varieties, their principal tool is the commutator operation on congruences. Included among the congruence modular varieties are almost all of the varieties arising from classical mathematical considerations.

(2) Hobby and McKenzie [1988] lays the foundation for a far reaching theory of locally finite varieties, a theory which goes under the name "tame congruence theory". This theory is probably the most significant development in the whole field. Kiss and Pröhle [1988] gives an account of further problems and results concerning tame congruence theory.

(3) McKenzie and Valeriote [1989] establishes that a locally finite variety has a decidable elementary theory iff it has very strong internal structural properties, somewhat like those exhibited by Abelian groups.

(4) Szendrei [1986] gives an account of recent developments in the theory of clones, especially clones of operations on finite sets. While the significance of clone theory in the body of this sampler seems difficult to discern, this is largely due to concerns for the brevity of the sampler. In fact, equational logic, the theory of varieties, and clone theory are really three facets of the same jewel. The developments in clone theory portend surprising results about equational theories and varieties.

(5) Tarski and Givant [1987] gives a full account of the work begun in Tarski [1941] and reported briefly in Tarski [1953]. The richness of equational logic and of the equational theory of relations algebras is made apparent in this monograph. §5 above gave a superficial account of just one aspect of this important work.
(6) Taylor [1986] is likely to be counted the seminal paper in the emerging field of topological universal algebra. In this monograph, the emphasis is on the clone of a topological space.

Other Surveys

Each of the following surveys may also be informative: Tarski [1968], Henkin [1977], Taylor [1979], and McNulty [1983]. Grätzer [1979] is the second edition of Grätzer [1968]. The new edition differs from the earlier work chiefly by the addition of seven appendices, written by leading contributors to the field, each surveying important contributions made in the decade between 1967 and 1978. Jónsson [1980] surveys the work in congruence varieties and Maltsev conditions, giving many proofs. Niether Maltsev conditions nor congruence varieties have been mentioned in this sampler. Roughly speaking, some very powerful algebraic properties of a variety of algebras are reflected as equations holding in the congruence lattices of the algebras in the variety. Indeed, in the sampler hypotheses like congruence modularity and congruence distributivity have made an appearance. This topic would have been the next to include, had this sampler more space.

Journals

A glance through the references of this sampler shows that articles in the field are published in a wide assortment of journals and a number of conference proceedings. There are two journals that include the field among their priorities: Algebra Universalis and Algebra i Logika (translated as Algebra and Logic).

§7 A Little Lexicon of Algebra and Logic

This little lexicon is intended to supply brief definitions or descriptions of the basic notions used in this sampler. A fuller account, a development of most of the relevant underlying theory, and examples for notions mentioned in the lexicon can be found in McKenzie, McNulty, and Taylor [1987]. The lexicon is ordered lexicographically (of course). It can be perused by the curious and referred to at need. It also sets forth the notation used in the sampler.

Algebra: An algebra $\mathbf{A} = \langle A, F \rangle$ is a system such that A is a nonempty set, called the *universe* of $\mathbf{A}$, and F is a system of finitary operations on A, which are called the *fundamental operations* of $\mathbf{A}$. Generally, algebras are denoted by boldface capitals while their universes are denoted by the corresponding italic—for example, A denotes the universe of $\mathbf{A}$. The cardinality of an algebra is taken to be the cardinality of its universe. F is a function who values are operations. The domain of F is called the set of *operation symbols* of $\mathbf{A}$, which we denote be $\mathcal{O}_{\mathbf{A}}$. We always assume that no variables belong to $\mathcal{O}_{\mathbf{A}}$. Usually, we list the values of F if that is convenient: $\langle \mathbb{Z}, +, -, \cdot, 1 \rangle$ denotes the ring of integers.

Algebraic Lattice: An algebraic lattice is one that is complete and generated by its compact elements. According to the Grätzer-Schmidt Theorem (cf. Grätzer [1979]) a lattice is algebraic iff it is isomorphic to the congruence lattice of some algebra.

Amalgamation Property: A variety $\mathcal{V}$ has the amalgamation property provided that whenever $\mathbf{A}, \mathbf{B}, \mathbf{C} \in \mathcal{V}$ with embeddings $f : \mathbf{A} \longrightarrow \mathbf{B}$ and $g : \mathbf{A} \longrightarrow \mathbf{C}$, there exist $\mathbf{D} \in \mathcal{V}$ and embeddings $g' : \mathbf{B} \longrightarrow \mathbf{D}$ and $f' : \mathbf{C} \longrightarrow \mathbf{D}$ such that $g' \circ f = f' \circ g$. A set Σ of equations has the amalgamation property iff the variety of all models of Σ has the amalgamation property.

Automorphism: An automorphism of an algebra $\mathbf{A}$ is a one-to-one map from A onto A which preserves all the fundamental operations of $\mathbf{A}$.

Base: A set Σ of equations is a base for the equational theory T provided $\Sigma \subseteq$ T and every equation in T is a logical consequence of Σ. Σ is said to be a base for the algebra $\mathbf{A}$ or the variety $\mathcal{V}$ provided it is a base of the equational theory of $\mathbf{A}$ or $\mathcal{V}$, as the case may be.

Clone: A clone on the set U is a collection of finitary operations on U which includes all the projection functions and is closed under composition of operations. The collection of all clones on a nonempty set U is lattice ordered under set-inclusion.

Compact Element: An element a of a complete lattice $\mathbf{L}$ is compact provided if $a \leq \bigvee X$, then $a \leq \bigvee F$, for some finite subset $F \subseteq X$.

Complete Lattice: A lattice $\mathbf{L}$ is complete provided every subset of L has both a least upper bound and a greatest lower bound. Note that every complete lattice must have a smallest element—we denote it by 0—and a largest element—we denote it by 1.

Congruence: A congruence of an algebra $\mathbf{A}$ is an equivalence relation on A which is also a subalgebra of $\mathbf{A}^2$; in other words, it is invariant under the coordinatewise application of each fundamental operation of $\mathbf{A}$. The congruences of $\mathbf{A}$ are exactly the kernels of the homomorphisms from $\mathbf{A}$ into other algebras. The congruences of $\mathbf{A}$ constitute an algebraic lattice under set inclusion.

Congruence Modular, Distributive, Permutable, ... : An algebra is said to be congruence modular (distributive) provide its congruence lattice is modular (distributive). An algebra is congruence permutable provided $\phi\circ\theta = \theta\circ\phi$ for any two congruences ϕ and θ, where $\circ$ denotes composition of relations. A variety is said to be congruence modular, distributive, permutable, ..., provided each algebra in the variety has the property.

Constants: Another name for operation symbols of rank 0.

Cover: Let $\mathbf{L}$ be a lattice and let $a, b \in L$. We say that a covers b provided $a > b$ and if $a \geq c \geq b$, then either $a = c$ or $c = b$, for any $c \in L$.

Direct Product: The direct product of a system of similar algebras is obtained by forming the direct (i.e. Cartesian) product of their universes and endowing it with fundamental operations defined coordinatewise.

Distributive: A lattice is distributive if the following equation holds:

$$x \wedge (y \vee z) \approx (x \wedge y) \vee (x \wedge z)$$

Elementary Sentence: This is another name for a sentence of first-order logic. These are the sentences built up from atomic formulas—which denote fundamental finitary relations holding between elements of some universe of discourse—by means of logical connectives: conjunction (and), disjunction (or), implication, and negation, and by means of quantifiers (e.g. $\forall x$ and $\exists x$) which bind variables ranging over elements of the universe of discourse. The word "elementary" refers to this restriction to elements, as opposed, for example, to subsets. The logic of elementary sentences is far from being simple.

Equation: An equation is an ordered pair of terms. Ordinarily we write $s \approx t$ instead of $\langle s, t\rangle$, where s and t are terms. $s \approx t$ is an equation of similarity type σ provided s and t are terms of type σ. Equations are also called rewrite rules.

Equational Completeness: An equational theory is equationally complete if it is a maximal strongly consistent equational theory. An algebra is equationally complete if its equational theory is equationally complete.

Equational Theory: Let σ be a similarity type. An equational theory of type σ is a set T of equations of type σ such that if $s \approx t$ is an equation of type σ and $s \approx t$ is a logical consequence of T, then $s \approx t \in \mathrm{T}$. In other words, equational theories are sets of equations which are closed with respect to logical consequence. The equational theory of an algebra is just the set of all equations which are true in the algebra. The equational theory of a class of similar algebras is the set of all equations true in each member of the class.

Finitary Operation: A finitary operation on the set A is a function from A^n into A, for some natural number n, which is called the *rank* of the operation.

Free Algebra: Let $\mathcal{K}$ be a class of similar algebras. An algebra $\mathbf{A}$ is $\mathcal{K}$-freely generated by G provided $\mathbf{A} \in \mathcal{K}$, G is a set of elements of A which generates A, and if $f : G \longrightarrow B$, then f extends to a homomorphism from $\mathbf{A}$ into $\mathbf{B}$, for every $\mathbf{B} \in \mathcal{K}$. In this circumstance G is said to be a *free generating set*. $\mathbf{A}$ is said to be *free in* $\mathcal{K}$ if it is $\mathcal{K}$-freely generated by some set of elements. $\mathbf{A}$ is a free algebra if it is free in some class of algebras.

Groupoid: A groupoid is an algebra whose only fundamental operation is binary.

Homomorphism: A homomorphism is a function from one algebra into another such that the algebras are similar and the function preserves each of the fundamental operations.

Join: Join, usually denoted by $\vee$, is one of the fundamental lattice operations. $a \vee b$ is the least upper bound of $\{a, b\}$ with respect to the lattice ordering. If X is any set of elements, $\bigvee X$ denotes the least upper bound of X, if it exists, and is called the join of X.

Joint Embedding Property: A variety $\mathcal{V}$ has the joint embedding property provided that whenever $\mathbf{A}, \mathbf{B} \in \mathcal{V}$, then there is $\mathbf{C} \in \mathcal{V}$ such that $\mathbf{A}$ and $\mathbf{B}$ can each be embedded into $\mathbf{C}$. A set Σ of equations has the joint embedding property iff the variety of all models of Σ has the joint embedding property.

Kernel: Let f be a function with domain A. The kernel of f is $\{\langle a, b\rangle : f(a) = f(b), a, b \in A\}$. Evidentally the kernel of f is an equivalence relation on A. The kernels of homomorphisms are exactly the congruences.

Lattice: A lattice is an algebra $\langle L, \wedge, \vee\rangle$ with two fundamental operations, both binary. These operations are commutative, associative, and idempotent ($x \wedge x \approx x$ and $x \vee x \approx x$ hold), and they satisfy the following equation: $x \wedge (y \vee x) \approx x$. A lattice can also be viewed as a partially ordered set in which every finite nonempty set has a least upper bound—called the join of the set—and a greatest lower bound—called the meet.

Locally Finite: An algebra is locally finite if each of its finitely generated subalgebras is finite. A class of algebras is locally finite provided every algebra in the class is locally finite.

Logical Consequence: Let Σ be a set of equations and let $s \approx t$ be an equation. $s \approx t$ is a logical consequence of Σ provided every model of Σ is a model of $s \approx t$. The same concept, after the obvious modifications, applies to sentences of any form, not just equations.

Logically Irredundant: A set Σ of equations is logically irredundant provided $s \approx t$ is not a logical consequence of $\Sigma - \{s \approx t\}$, for all $s \approx t \in \Sigma$.

Meet: Meet, usually denoted by $\wedge$, is one of the fundamental lattice operations. $a \wedge b$ is the greatest lower bound of $\{a, b\}$ with respect to the lattice ordering. If X is any set of elements, $\bigwedge X$ denotes the greatest lower bound of X, if it exists, and is called the meet of X.

Model: An algebra $\mathbf{A}$ is a model of a set Σ of equations, if every equation is Σ is true in $\mathbf{A}$.

Modular: A lattice is modular iff the following equation, which is a weak form of distributivity, holds

$$((x \wedge y) \vee z) \wedge y \approx (x \wedge y) \vee (z \wedge y)$$

$\mathbb{N}$: The set $\{0, 1, 2, \ldots\}$ of natural numbers is denoted by $\mathbb{N}$.

Operation Symbols: See algebra and similarity type.

Permutable: An algebra is called congruence permutable provided $\theta \circ \phi = \phi \circ \theta$ for any two congruences θ and ϕ. (Here $\circ$ denotes composition of relations).

Rank: See finitary operation.

Relation Algebra: See §5.

Residually Finite, Small, Large: An algebra is residually finite if it is a subdirect product of finite algebras. A variety is residually finite if every algebra in the variety is residually finite. A variety is residually small if there is a cardinal κ such that every algebra in the variety is a subdirect product of algebras of cardinality less than κ. Varieties which are not residually small are said to be residually large.

Schreier Property: A variety has the Schreier property if every subalgebra of an algebra free in the variety is again free in the variety. A set Σ of equations has the Schreier property if the variety consisting of all models of Σ has the Schreier property.

Similar: Two algebras are similar if they have the same similarity type.

Similarity Type: Let $\mathbf{A} = \langle A, F \rangle$ be an algebra. The similary type of $\mathbf{A}$ is the function σ with domain $\mathcal{O}_{\mathbf{A}}$ such that

$$\sigma(Q) = \text{the rank of } F(Q)$$

Thus σ is a certain function from operation symbols to natural numbers, assigning each symbol its rank.

Simple: An algebra is simple if it has exactly two congruences.

Strongly Consistent: A set Σ of equations is strongly consistent provided it has a model with more than one element. Equivalently, Σ is strongly consistent iff the equation $x \approx y$ is not a logical consequence of Σ.

Subalgebra: Let $\mathbf{A}$ and $\mathbf{B}$ be similar algebras. $\mathbf{B}$ is a subalgebra of $\mathbf{A}$ provided $B \subseteq A$ and each fundamental operation of $\mathbf{B}$ is the restriction to B of the corresponding fundamental operation of $\mathbf{A}$.

Subdirect Product: An algebra $\mathbf{A}$ is a subdirect product of a system $\langle \mathbf{B}_i : i \in I \rangle$ of algebras provided $\mathbf{A}$ can be embedded in the direct product $\prod_{i \in I} \mathbf{B}_i$ in such a way that the projection functions, restricted to the image of $\mathbf{A}$, are onto the $\mathbf{B}_i$'s.

Term: Fix a similarity type σ. The terms of type σ are certain strings or words made up of variables and operation symbols of type σ. The set of terms is defined by the following recursive scheme:

i Every variable is a term.

ii If Q is an operation symbols and $t_0, t_1, \ldots, t_{r-1}$ are terms, where $r = \sigma(Q)$, then $Qt_0t_1 \ldots t_{r-1}$ is a term.

Term Operation: Let $\mathbf{A} = \langle A, F \rangle$ be an algebra. Each term t will denote a term operation $t^{\mathbf{A}}$ which will be a function from $A^{\mathbb{N}}$ into A—depending on only finitely many arguments.

i $v_i^{\mathbf{A}}$ is the ith projection function. for each natural number i. (v_i is a variable).

ii If $t = Qt_0t_1 \ldots t_{r-1}$, where r is the rank of the operation symbols Q,then $t^{\mathbf{A}} = F(Q)(t_0^{\mathbf{A}}, \ldots, t_{r-1}^{\mathbf{A}})$.

True: Let $\mathbf{A}$ be an algebra and $s \approx t$ be an equation of the same similarity type as $\mathbf{A}$. The equation $s \approx t$ is said to be true in $\mathbf{A}$ provided $s^{\mathbf{A}} = t^{\mathbf{A}}$. Thus $s \approx t$ is true in $\mathbf{A}$ iff s and t denote the same term operation in $\mathbf{A}$.

Universe: See algebra. Algebras are usually denoted by boldface letters, e.g. $\mathbf{A}$, and their universes by the corresponding uppercase italic, e.g A.

Variable: One of the formal symbols $v_0, v_1, v_3, \ldots$. Symbols like x, y and z refer to arbitrary members of this list of formal variables.

Variety: A variety is a class of similar algebras which is closed with respect to the formation of homomorphic images, subalgebras, and arbitrary direct products. Equivalently, a variety is the class of all algebras which are models of some set of equations.

Bibliography

J. W. Addison

[1954] *On some points of the theory of recursive functions*, Ph.D. Thesis, University of Wisconsin, Madison.

S. I. Adjan

[1958] On algorithmic problems in effectively complete classes of groups (Russian), *Doklady Akad. Nauk SSSR* **23**, 13–16.

K. Baker

[1977] Finite equational bases for finite algebras in congruence distributive varieties, *Advances in Mathematics* **24**, 207–243.

K. Baker, G. McNulty, and H. Werner

[1987] The finitely based varieties of graph algebras, *Acta Scient. Math. (Szeged)* **51**, 3–15.

[1989] Shift-automorphism methods for inherently nonfinitely based varieties of algebras, *Czechoslovak Math. J.* (to appear).

G. Bauer and F. Otto

[1984] Finite complete rewriting systems and the complexity of the word problem, *Acta Informatica* **21**, 521–540.

A. D. Bol'bot

[1970] Varieties of Ω-algebras, *Algebra i Logika* **9**, 406–415; English Translation*Algebra and Logic* **9**, 244–248.

R. Book

[1987] Thue systems as rewriting systems, in Jouannaud [1987], pp. 39–68.

W. Boone

[1954] Certain simple unsolvable problems in group theory, *Indag. Math.* **16**, 231–237.

R. Bryant

[1982] The laws of finite pointed groups, *Bull. London Math. Soc.* **14**, 119–123.

B. Buchberger

[1987] History and basic features of the critical-pair/completion procedure, in Jouannaud [1987], pp. 3–38.

S. Burris and E. Nelson

[1971a] Embedding the dual of Π_m in the lattice of equational classes of commutative semigroups, *Proc. Amer. Math. Soc.* **30**, 37–39.

[1971b] Embedding the dual of Π_∞ in the lattice of equational classes of semigroups, *Algebra Universalis* **1**, 248–253.

S. Burris and H. P. Sankappanavar

[1981] *A Course in Universal Algebra*, Springer-Verlag, New York.

P.M. Cohn

[1965] *Universal Algebra*, Harper & Row, New York.

[1981] *Universal Algebra*, second edition, D. Reidel, Dordrecht, Holland.

A. Day

[1969] A characterization of modularity of congruence lattices of algebras, *Canad. Math. Bull.* **12**, 167–173.

R. Dedekind

[1900] Über die von drei Moduln erzeugte Dualgruppe, *Math. Ann.* **53**, 371–403.

M. DEHN

[1911] Über unendlicher diskontinuierliche Gruppen, *Math. Ann.* **71**, 116–144.

N. DERSHOWITZ

[1987] Termination of rewriting, in Jouannaud [1987], pp. 69–116.

T. EVANS

[1969] Some connections between residual finiteness, finite embeddability and the word problem, *J. London Math. Soc.* **1**, 399–403.

[1978] Word problems, *Bull. Amer. Math. Soc.* **84**, 789–802.

W. J. FEENEY

[1954] *Certain Unsolvable Problems in the Theory of Cancellation Semigroups*, Ph.D. Thesis, Catholic University of America.

R. FREESE

[1980] Free modular lattices, *Trans. Amer. Math. Soc.* **261**, 81–91.

R. FREESE AND O. GARCIA, EDS.

[1983] *Universal Algebra and Lattice Theory, Proceedings, Puebla 1982*, Lecture Note in Mathematics, vol. **1004**, Springer-Verlag, Berlin.

R. FREESE AND R. MCKENZIE

[1981] Residually small varieties with modular congruence lattices, *Trans. Amer. Math. Soc.* **264**, 419–430.

[1987] *Commutator Theory for Congruence Modular Varieties*, London Mathematical Society Lecture Note Series vol. **125**, 277pp., Cambridge University Press, Cambridge, England.

R. FREESE, R. MCKENZIE, G. MCNULTY, AND W. TAYLOR

[1989] *Algebras, Lattices, Varieties, Volume II*, Wadsworth & Brook/Cole, Monterey, CA (to appear).

G. GRÄTZER

[1968] *Universal Algebra*, University Series in Higher Mathematics, D. Van Nostrand Company, New York.

[1979] *Universal Algebra*, Second edition of Grätzer [1968], Springer-Verlag, New York, pp. 581.

R. GUREVIČ

[1989] Equational theory of positive numbers with exponentiation is not finitely axiomatizable, *Annals of Pure and Applied Logic* (to appear).

L. HENKIN

[1977] The logic of equality, *Amer. Math. Monthly* **84**, 597–612.

C. HERRMANN

[1983] On the word problem for the modular lattice with four free generators, *Math. Ann.* **256**, 513–527.

D. HOBBY AND R. MCKENZIE

[1988] *The Structure of Finite Algebras (Tame Congruence Theory)*, Contemporary Mathematics, Vol. **76**, American Mathematical Society, Providence, RI.

G. HUET AND D. LANKFORD

[1978] On the uniform halting problem for term rewriting systems, Rapport Laboria 283, Institut de Recherche en Informatique eten Automatique, Le Chesnay, France.

E. JACOBS AND R. SCHWABAUER

[1964] The lattice of equational classes of algebras with one unary operation, *Amer. Math. Monthly* **71**, 151–155.

J. JEŽEK

[1969] Primitive classes of algebras with unary and nullary operations, *Colloq. Math.* **20**, 159–179.

[1976] Intervals in the lattice of varieties, *Algebra Universalis* **6**, 147–158.

[1981] The lattice of equational theories, part I: modular elements, *Czechoslovak Math. J.* **31**, 127–152.

[1981] The lattice of equational theories part II: the lattice of full sets of terms, *Czechoslovak Math. J.* **31**, 573-603.

[1982] The lattice of equational theories part III: definability and automorphisms, *Czechoslovak Math. J.* **32**, 129–164.

[1985] Nonfinitely based three-element groupoids, *Algebra Universalis* **20**, 292–301.

[1986] The lattice of equational theories part IV: equational theories of finite algebras, *Czechoslovak Math. J.* **36**, 331–341.

B. JÓNSSON

[1967] Algebras whose congruence lattices are distributive, *Math. Scand.* **21**, 110–121.

[1980] Congruence varieties, *Algebra Universalis* **10**, 355–394.

J.-P. JOUANNAUD, ED.

[1987] *Rewriting Techniques and Applications*, Academic Press, New York.

C. KALFA

[1986] Decision problems concerning properties of finite sets of equations, *J. Symbolic Logic* **51**, 79–87.

J. KALICKI

[1952] On comparison of finite algebras, *Proc. Amer. Math. Soc.* **3**, 36–40.

[1955] The number of equationally complete classes of equations, *Indag. Math.* **58**, 660–662.

E. KISS AND P. PRŐHLE

[1988] Problems and results in tame congruence theory, Mathematmatical Institute of the Hungarian Academy of Sciences, preprint no. 60/1988.

R. KRUSE

[1973] Identities satisfied by a finite ring, *J. Algebra* **26**, 298–318.

W. LAMPE

[1987] A property of lattices of equational theories, *Algebra Universalis* **24**.

I. V. L'VOV

[1973] Varieities of associative rings,I, *Algebra i Logika* **12**, 269–297; II, *Algebra i Logika* **12**, 667–688, 735.

R. LYNDON

[1951] Identities in two-valued calculi, *Trans. Amer. Math. Soc.* **71**, 457–465.

[1954] Identities in finite algebras, *Proc. Amer. Math. Soc.* **5**, 8–9.

A. I. MALTSEV

[1958] On homomorphisms onto finite groups, *Ucen. Zap. Ivan. Ped. Inst.* **18**, 49–60; English translation, *Amer. Math. Soc. Transl.* **119**, 67–79.

[1966] Identical relations on varieties of quasigroups, *Mat. Sbornik* **69**, 3–12; English translation, (1969), *Amer. Math. Soc. Transl.* **82**, 225–235; Chapter 29 in Maltsev [1971].

[1968] Problems on the borderline between algebra and logic, in: Proc. Int. Congress of Math. (Moscow 1966) Moscow, Mir, 1968 pp.217–231; English translation, (1968), *Amer. Math. Soc. Transl.* **70**, 89–100; Chapter 34 in Maltsev [1971].

[1971] *The Metamathematics of Algebraic Systems*, A selection of Maltsev's papers, edited and translated by B. F. Wells, III, North-Holland Publishing Company, Amsterdam.

[1973] *Algebraic Systems*, Akademie-Verlag, Berlin.

A. MARKOV

[1947] On the impossibility of certain algorithms in the theory of associative systems, *Doklady Akad, Nauk SSSR* **55**, 587–590.

[1951] Impossibility of certain algorithms in the theory of associative systems, *Doklady Akad. Nauk SSSR* **77**, 953–956.

C. MARTIN

[1973] *The Equational Theories of Natural Numbers and Transfinite Ordinals*, Ph.D. Thesis, University of California, Berkeley, CA.

R. MCKENZIE

[1970] Equational bases for lattice theories, *Math. Scand.* **27**, 24–38.

[1971] Definability in lattices of equational theories, *Ann. Math. Logic* **3**, 197–237.

[1975] On spectra, and the negative solution of the decision problem for identities having a finite nontrivial model, *J. Symbolic Logic* **40**, 186–196.

[1984] A new product of algebras and a type reduction theorem, *Algebra Universalis* **18**, 29–69.

[1988] Finite equational bases for congruence modular algebras, *Algebra Universalis* (to appear).

R. MCKENZIE, G. MCNULTY, AND W. TAYLOR

[1987] *Algebras, Lattices, Varieties, Volume I*, Wadsworth & Brooks/Cole, Monterey, CA.

R. MCKENZIE AND M. VALERIOTE

[1989] *The Structure of Decidable Locally Finite Varieties*, Birkhäuser Verlag, Boston, MA (to appear).

J. C. C. MCKINSEY

[1943] The decision problem for some classes of sentences without quantifiers, *J Symbolic Logic* **8**, 61–76.

G. MCNULTY

[1976a] The decision problem for equational bases of algebras, *Ann. Math. Logic* **11**, 193–259.

[1976b] Undecidable properties of finite sets of equations, *J. Symbolic Logic* **41**, 589–604.

[1981] Structural diversity in the lattice of equational theories, *Algebra Universalis* **13**, 271-292.

[1986] Fifteen possible previews in equational logic, in Szabó and Szendrei [1986].

G. MCNULTY AND C. SHALLON

[1983] Inherently nonfinitely based finite algebras, in Freese and Garcia [1983], pp. 205–231.

A. MEKLER, E. NELSON, AND S. SHELAH

[1989] A variety with solvable, but not uniformly solvable, word problem.

V. L. MURSKIĬ

[1965] The existence in three-valued logic of a closed class without a finite complete system of identities, *Doklady Akad. Nauk SSSR* **163**, 815–818.

[1968] Examples of varieties of semigroups, *Mat. Zametki* **3**, 663-670.

[1971] Nondiscernible properties of finite systems of identity relations, *Doklady Akad. Nauk SSSR* **196**, 520-522.

[1975] The existence of a finite basis and some other properties of "almost all" finite algebras, *Problemy Kibernet.* **30**, 43–56.

[1979] Concerning the number of k-element algebras with one binary operation which have no finite basis of identities, *Problemy Kibernet.* **35**, 5–27.

S. OATES AND M. B. POWELL

[1965] Identical relations in finite groups, *J. Algebra* **1**, 11–39.

S. Oates-Williams and M. Vaughan-Lee

[1978] Varieties that make one Cross, *J. Austral. Math. Soc.(A)* **26**, 368-382.

C. O'Dúnlaing

[1983] Undecidable questions of Thue systems, *Theoret. Comp. Sci.* **23**, 339–345.

P. Perkins

[1966] *Decision Problems for Equational Theories of Semigroups and General Algebras*, Ph.D. Thesis, University of California, Berkeley, CA.

[1967] Unsolvable problems for equational theories, *Notre Dame J. of Formal Logic* **8**, 175–185.

[1968] Bases for equational theories of semigroups, *J. Algebra* **11**, 293–314.

[1984] Basis questions for general algebra, *Algebra Universalis* **19**, 16–23.

C. S. Peirce

[1880] On the algebra of logic, *Amer. J. Math.* **3**, 15–57.

R. S. Pierce

[1968] *Introduction to the Theory of Abstract Algebras*, Holt, Rinehart and Winston, New York.

D. Pigozzi

[1976] Base-undecidable properties of universal varieties, *Algebra Universalis* **6**, 193–223.

D. Pigozzi and G. Tardos

[1989] The representation of certain abstract lattices aslattices of subvarieties, (to appear).

D. Plaisted

[1985] The undecidability of self embedding for term rewriting systems, *Inf. Proc. Lett.* **20**, 61–64.

S. V. Polin

[1976] On the identities of finite algebras, *Sib. Mat. J.* **17**, 1356–1366.

E. Post

[1941] *The Two-valued Iterative Systems of Mathematical Logic*, Annals of Math. Studies, No. 5, Princeton University Press, Princeton, NJ.

[1947] Recursive unsolvability of a problem of Thue, *J. Symbolic Logic* **12**, 1–11.

M. Rabin

[1958] Recursive unsolvability of group theoretic problems, *Ann. of Math.* **67**, 172–194.

M. Sapir

[1987a] Problems of Burnside type and the finite basis property in varieties of semigroups, *Izv. Akad. Nauk SSSR* **51**; English translation, *Math. USSR Izvestiya* **30**, 295–314.

[1987b] Inherently nonfinitely based finite semigroups, *Mat. Sbornik* **133**; English Translation, *Math. USSR Sbornik* **61**(1988), 155–166.

E. Schröder

[1890] *Vorlesungen über die Algebra der Logik (exacte Logik)*, edited in part by E. Müller and B .G. Teubner, published in four volumes, Leipzig; *Second edition in three volumes*, (1966), Chelsea Publ. Co., New York.

K. Schütte, ed.

[1968] *Contributions to Mathematical Logic*, North-Holland Publishing Co., Amsterdam.

D. Scott

[1956] Equationally complete extensions of finite algebras, *Indag. Math.* **59**, 35–38.

C. Shallon

[1979] *Nonfinitely Based Binary Algebra Derived for Lattices*, Ph.D. Thesis, University of California, Los Angeles, CA.

L. N. SHEVRIN AND M. V. VOLKOV

[1985] Identities of semigroups, *Izvestiya VUZ Mat.* **29**, 3–47.

D. SMITH

[1972] Non-recursiveness of the set of finite sets of equations whose theories are one based, *Notre Dame J. Formal Logic* **13**, 135–138.

L. SZABÓ AND A. SZENDREI, EDS.

[1986] *Lectures in Universal Algebra*, Colloquia Mathematica Societas János Bolyai vol. **43** pp. 307-331, North-Holland, Amsterdam.

A. SZENDREI

[1986] *Clones in Universal Algebra*, Séminaire de Mathématiques Supérieures vol. 99, Les Presses de l'Université de Montréal, Montréal, Canada.

W. SZMIELEW

[1954] Elementary properties of Abelian groups, *Fund. Math.* **41**, 203–271.

A. TARSKI

[1941] On the calculus of relations, *J. Symbolic Logic* **6**, 73–89.

[1949a] Arithmetical classes and types of Boolean Algebras (preliminary report), *Bull. Amer. Math. Soc.* **55**, p. 64.

[1949b] Undecidability of the theories of lattices and projective geometries, *J. Symbolic Logic* **14**, 77–78.

[1953] A formalization of set theory without variables (abstract), *J. Symbolic Logic* **18**, p. 189.

[1968] Equational logic, in Schütte [1968], pp. 275–288.

A. TARSKI AND S. GIVANT

[1987] *A Formalization of Set Theory Without Variables*, Colloquium Publications, vol **41**, Amer. Math. Soc., Providence, RI.

W. TAYLOR

[1979] Equational logic, *Houston J. Math. (Survey Issue)*.

[1986] *The Clone of a Topological Space*, Research and Exposition in Mathematics, vol. **13**, Heldermann Verlag, Berlin.

M. VAUGHAN-LEE

[1979] Laws in finite loops, *Algebra Universalis* **9**, 269–280.

[1983] Nilpotence in permutable varieties, in Freese and Garcia [1983], pp. 293–308.

B. WELLS

[1982] *Pseudorecursive Varieties and Their Word Problems*, Ph.D. Thesis, University of California, Berkeley, CA.

Modular Aspects of Properties of Term Rewriting Systems Related to Normal Forms

Aart Middeldorp

Department of Mathematics and Computer Science,
Vrije Universiteit, de Boelelaan 1081, 1081 HV Amsterdam.
email: ami@cs.vu.nl

ABSTRACT

In this paper we prove that the property of having unique normal forms is preserved under disjoint union. We show that two related properties do not exhibit this kind of modularity.

Introduction

Recently there has been some interest in establishing the modularity of certain properties of term rewriting systems. The modularity of the Church-Rosser property (CR) was proven by Toyama [4]. He also gave an example which showed that termination is not a modular property. Toyama conjectured the modularity of the completeness (i.e. Church-Rosser and terminating) property but Barendregt and Klop constructed a counterexample (see [5]). In view of this counterexample Hsiang conjectured that completeness is a modular property of reduced term rewriting systems but Toyama falsified that conjecture (see [5]). Toyama, Klop and Barendregt [6] gave an extremely complicated proof showing the modularity of completeness for left-linear term rewriting systems. Rusinowitch [3] gave sufficient conditions under which termination is a modular property. His results were extended by the present author [1].

In this paper we study the modularity of three properties dealing with normal forms. We show that the property of having unique normal forms (UN) is modular. The normal form property (NF) is not a modular property of general term rewriting systems. This is somewhat surprising because NF lies properly between CR and UN. The property $UN^{\rightarrow}$ (unique normal forms with respect to reduction) also is not modular. We give sufficient conditions under which NF and $UN^{\rightarrow}$ are modular.

The remainder of this paper is organized as follows. In the first section we review the basic definitions of term rewriting systems. Section 2 contains the proof of the modularity of UN. In Section 3 we give counterexamples to the modularity of NF and $UN^{\rightarrow}$. We also give some sufficient conditions under which NF and $UN^{\rightarrow}$ become modular. Section 4 contains some concluding remarks and open problems.

1. Term Rewriting Systems: Basic Definitions

Let $\mathcal{V}$ be a countably infinite set of variables. A *term rewriting system* (TRS for short) is a pair $(\mathcal{F},\mathcal{R})$. $\mathcal{F}$ is a set of *function symbols*; associated to every $f \in \mathcal{F}$ is its arity $n \geq 0$. Function symbols of arity 0 are called *constants*. The set of terms built from $\mathcal{F}$ and $\mathcal{V}$, notation $\mathcal{T}(\mathcal{F},\mathcal{V})$, is the smallest set such that $\mathcal{V} \subset \mathcal{T}(\mathcal{F},\mathcal{V})$ and if $t_1,\ldots,t_n \in \mathcal{T}(\mathcal{F},\mathcal{V})$ then $F(t_1,\ldots,t_n) \in \mathcal{T}(\mathcal{F},\mathcal{V})$ for $F \in \mathcal{F}$ with arity n. Terms not containing variables are called *ground* terms or *closed* terms. $\mathcal{R}$ is a set of pairs (l,r) with $l,r \in \mathcal{T}(\mathcal{F},\mathcal{V})$ subject to two constraints: (1) the left-hand side l is not a variable, (2) the variables which occur in the right-hand side r also occur in l. Pairs (l,r) will be called *rewriting rules* or *reduction rules* and will henceforth be written as $l \to r$. We usually write $\mathcal{R}$ instead of $(\mathcal{F},\mathcal{R})$, assuming that $\mathcal{F}$ contains no function symbols which do not occur in the rewriting rules $\mathcal{R}$.

A *substitution* σ is a mapping from $\mathcal{V}$ to $\mathcal{T}(\mathcal{F},\mathcal{V})$. Substitutions are extended to $\mathcal{T}(\mathcal{F},\mathcal{V})$ in the obvious way; we denote by t^σ the term obtained from t by applying the substitution σ. We call t^σ an *instance* of t. An instance of a left-hand side of a rewriting rule is called a *redex* (reducible expression). If $t \in \mathcal{T}(\mathcal{F},\mathcal{V})$ then $V(t)$ is the set of all variables occurring in t. If $s,t \in \mathcal{T}(\mathcal{F},\mathcal{V})$ and $x \in \mathcal{V}$ then $s[x \leftarrow t]$ denotes the term obtained from s by substituting t for every occurrence of x. If $x_1,\ldots,x_n$ are distinct variables and $s,t_1,\ldots,t_n \in \mathcal{T}(\mathcal{F},\mathcal{V})$ we write $s[x_i \leftarrow t_i \mid 1 \leq i \leq n]$ for the simultaneous substitution of the t_i's for every occurrence of the x_i's.

A *context* $C[\,]$ is a 'term' which contains exactly one occurrence of a special symbol $\Box$. Contexts are inductively defined by: $\Box$ is a context, and if $C[\,]$ is a context and $t_1,\ldots,t_{i-1},t_{i+1},\ldots,t_n \in \mathcal{T}(\mathcal{F},\mathcal{V})$ then $F(t_1,\ldots,t_{i-1},C[\,],t_{i+1},\ldots,t_n)$ is a context for every n-ary function symbol F. If $C[\,]$ is a context and $t \in \mathcal{T}(\mathcal{F},\mathcal{V})$ then $C[t]$ is the result of replacing the symbol $\Box$ by t; t is said to be a *subterm* of $C[t]$.

The *rewriting relation* $\to_{\mathcal{R}} \subset \mathcal{T}(\mathcal{F},\mathcal{V}) \times \mathcal{T}(\mathcal{F},\mathcal{V})$ is defined by $s \to_{\mathcal{R}} t$ iff there exists a rewriting rule $l \to r$, a substitution σ and a context $C[\,]$ such that $s \equiv C[l^\sigma]$ and $t \equiv C[r^\sigma]$ (the symbol $\equiv$ stands for syntactic equality). The transitive-reflexive closure of $\to_{\mathcal{R}}$ is denoted by $\twoheadrightarrow_{\mathcal{R}}$; if $s \twoheadrightarrow_{\mathcal{R}} t$ we say that s *reduces* to t. We write $s \leftarrow_{\mathcal{R}} t$ if $t \to_{\mathcal{R}} s$; likewise for $s \twoheadleftarrow_{\mathcal{R}} t$. The symmetric closure of $\to_{\mathcal{R}}$ is denoted by $\leftrightarrow_{\mathcal{R}}$ (so $\leftrightarrow_{\mathcal{R}} = \to_{\mathcal{R}} \cup \leftarrow_{\mathcal{R}}$). The transitive-reflexive closure of $\leftrightarrow_{\mathcal{R}}$ is called *conversion* and denoted by $=_{\mathcal{R}}$. We often omit the subscript $\mathcal{R}$.

DEFINITION 1.1. Let $\mathcal{R}$ be a TRS.

(1) A term s is a *normal form* if there is no term t such that $s \to_{\mathcal{R}} t$. A term s *has a normal form* if there is a normal form t such that $s \twoheadrightarrow_{\mathcal{R}} t$.

(2) $\mathcal{R}$ is *strongly normalizing* (SN) if there are no infinite reduction sequences $t_0 \to_{\mathcal{R}} t_1 \to_{\mathcal{R}} t_2 \to_{\mathcal{R}} \cdots$.

(3) $\mathcal{R}$ is *confluent* or *Church-Rosser* (CR) if for all terms s,t_1,t_2 with $s \twoheadrightarrow_{\mathcal{R}} t_1$ and $s \twoheadrightarrow_{\mathcal{R}} t_2$ we can find a term t_3 such that $t_1 \twoheadrightarrow_{\mathcal{R}} t_3$ and $t_2 \twoheadrightarrow_{\mathcal{R}} t_3$. Such a term t_3 is called a *common reduct* of t_1 and t_2.

(4) $\mathcal{R}$ has *unique normal forms* (is UN) if for all normal forms n_1,n_2 with $n_1 =_{\mathcal{R}} n_2$ we have $n_1 \equiv n_2$.

(5) $\mathcal{R}$ has *unique normal forms with respect to reduction* (is $\mathrm{UN}^{\to}$) if for all terms s,n_1,n_2 such that $s \twoheadrightarrow_{\mathcal{R}} n_1$, $s \twoheadrightarrow_{\mathcal{R}} n_2$ and n_1,n_2 are normal forms we have $n_1 \equiv n_2$.

(6) $\mathcal{R}$ has the *normal form property* (is NF) if for all terms s,t such that $s =_{\mathcal{R}} t$ and t is a normal form we have $s \twoheadrightarrow_{\mathcal{R}} t$.

Often one finds in the literature a different but equivalent formulation of confluence, as suggested by the following proposition whose proof is easy.

PROPOSITION 1.2. *A TRS $\mathcal{R}$ is CR iff for all terms t_1,t_2 with $t_1 =_{\mathcal{R}} t_2$ we can find a term t_3 such that $t_1 \twoheadrightarrow_{\mathcal{R}} t_3 \twoheadleftarrow_{\mathcal{R}} t_2$.* □

The relation between the properties CR, UN, $\mathrm{UN}^{\rightarrow}$ and NF is given in the next lemma.

LEMMA 1.3. *The following implications hold for every TRS $\mathcal{R}$: CR $\Rightarrow$ NF $\Rightarrow$ UN $\Rightarrow$ $\mathrm{UN}^{\rightarrow}$. The reverse implications generally do not hold.*

PROOF.

'CR $\Rightarrow$ NF' Let $\mathcal{R}$ be CR and suppose $s =_{\mathcal{R}} t$ with t in normal form. Because $\mathcal{R}$ is CR s and t have a common reduct. But t is a normal form, so $s \twoheadrightarrow_{\mathcal{R}} t$.

'NF $\Rightarrow$ UN' Let $\mathcal{R}$ be NF. Suppose $n_1 =_{\mathcal{R}} n_2$ is a conversion between normal forms n_1,n_2. Because $\mathcal{R}$ is NF we have $n_1 \twoheadrightarrow_{\mathcal{R}} n_2$ (and $n_2 \twoheadrightarrow_{\mathcal{R}} n_1$). But n_1 is a normal form, therefore $n_1 \equiv n_2$.

'UN $\Rightarrow$ $\mathrm{UN}^{\rightarrow}$' Let $\mathcal{R}$ be UN and suppose $n_1 \twoheadleftarrow_{\mathcal{R}} s \twoheadrightarrow_{\mathcal{R}} n_2$ with n_1,n_2 in normal form. Clearly $n_1 =_{\mathcal{R}} n_2$ and because $\mathcal{R}$ is UN we have $n_1 \equiv n_2$.

'NF $\not\Rightarrow$ CR' Let $\mathcal{R} = \{a \rightarrow b, a \rightarrow c, b \rightarrow b, c \rightarrow c\}$ (see Figure 1(a)). Clearly $\mathcal{R}$ is NF but $\mathcal{R}$ is not CR: $b \leftarrow_{\mathcal{R}} a \rightarrow_{\mathcal{R}} c$ and b,c have no common reduct.

'UN $\not\Rightarrow$ NF' Let $\mathcal{R} = \{a \rightarrow b, a \rightarrow c, b \rightarrow b\}$ (see Figure 1(b)). It is not difficult to see that $\mathcal{R}$ is UN, but $\mathcal{R}$ is not NF: $b =_{\mathcal{R}} c$, c is a normal form and b does not reduce to c.

'$\mathrm{UN}^{\rightarrow}$ $\not\Rightarrow$ UN' Let $\mathcal{R} = \{a \rightarrow b, a \rightarrow c, c \rightarrow c, d \rightarrow c, d \rightarrow e\}$ (see Figure 1(c)). Again it is easy to see that $\mathcal{R}$ is $\mathrm{UN}^{\rightarrow}$, but $\mathcal{R}$ is not UN: $b =_{\mathcal{R}} e$ and b,e are normal forms.

□

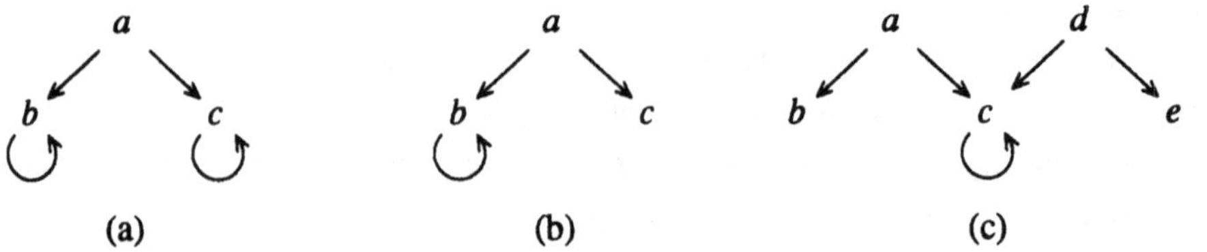

FIGURE 1.

DEFINITION 1.4. Let $(\mathcal{F}_0,\mathcal{R}_0)$ and $(\mathcal{F}_1,\mathcal{R}_1)$ be TRS's. The *direct sum* $\mathcal{R}_0 \oplus \mathcal{R}_1$ of $(\mathcal{F}_0,\mathcal{R}_0)$ and $(\mathcal{F}_1,\mathcal{R}_1)$ is the TRS obtained by taking the disjoint union of $(\mathcal{F}_0,\mathcal{R}_0)$ and $(\mathcal{F}_1,\mathcal{R}_1)$. That is, if $\mathcal{F}_0$ and $\mathcal{F}_1$ are disjoint ($\mathcal{F}_0 \cap \mathcal{F}_1 = \varnothing$) then $\mathcal{R}_0 \oplus \mathcal{R}_1 = (\mathcal{F}_0 \cup \mathcal{F}_1, \mathcal{R}_0 \cup \mathcal{R}_1)$; otherwise we rename the function symbols of $(\mathcal{F}_0,\mathcal{R}_0)$ and $(\mathcal{F}_1,\mathcal{R}_1)$ such that the resulting copies $(\mathcal{F}_0',\mathcal{R}_0')$ and $(\mathcal{F}_1',\mathcal{R}_1')$

have disjoint sets of function symbols, and define $\mathcal{R}_0\oplus\mathcal{R}_1 = (\mathcal{F}_0'\cup\mathcal{F}_1', \mathcal{R}_0'\cup\mathcal{R}_1')$.

DEFINITION 1.5. A property $\mathcal{P}$ of TRS's is called *modular* if for all TRS's $\mathcal{R}_0$, $\mathcal{R}_1$ we have $\mathcal{R}_0\oplus\mathcal{R}_1$ has the property $\mathcal{P}$ iff both $\mathcal{R}_0$ and $\mathcal{R}_1$ have the property $\mathcal{P}$.

2. UN is a Modular Property

In this section we prove that UN is a modular property of TRS's. The key to our proof is the interesting observation that every TRS with the property UN can be extended to a confluent TRS with the same set of normal forms.

Let us first give an intuitive explanation of our construction. We partition the set of all terms in sets of convertible terms. Because the TRS under consideration has the property UN, every equivalence class of convertible terms contains either exactly one normal form or no normal form. If an equivalence class C has a normal form t, we simply add rules which rewrite every term in C different from t to the normal form t. As far as this equivalence class C is concerned, we trivially obtain the Church-Rosser property. However, we have to be careful about adding new rewriting rules. For example, consider the TRS $\mathcal{R} = \{F(x)\to x, F(x)\to 0, 0\to 0\}$ (see Figure 2(a)). At first sight this TRS seems to have the property UN. But then our construction is doomed to fail: the equivalence class $[F(x)]$ $(= \{t \mid F(x) =_{\mathcal{R}} t\})$ contains the normal form x, the term 0 is a member of this equivalence class, but we cannot add the rule $0\to x$ (that would be a violation of the constraint that variables occurring in the right-hand side of a rewriting rule also occur in its left-hand side). Fortunately, a more careful look reveals that $\mathcal{R}$ does not have the property UN; we have the following conversion between different variables x,y: $x \leftarrow F(x) \to 0 \leftarrow F(y) \to y$. In fact, we will prove that if an equivalence class C has a normal form t, then $V(t)\subseteq V(s)$ for every term $s\in C$ (cf. Proposition 2.3).

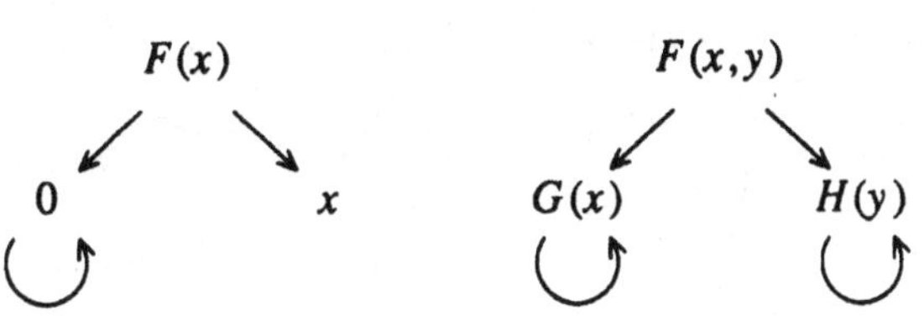

(a) (b)

FIGURE 2.

Difficulties arise if an equivalence class C does not contain a normal form, as will be illustrated for the TRS $\mathcal{R} = \{F(x,y)\to G(x), F(x,y)\to H(y), G(x)\to G(x), H(y)\to H(y)\}$ (see Figure 2(b)). The equivalence class $[F(x,y)]$ does not have a normal form. If we want to extend $\mathcal{R}$ to a confluent TRS, the terms $G(x)$ and $H(y)$ should reduce to a common reduct. A possible common reduct t of $G(x)$ and $H(y)$ cannot have variables (if $G(x) \twoheadrightarrow t$ then $V(t)\subseteq\{x\}$ and if $H(y)\twoheadrightarrow t$ then $V(t)\subseteq\{y\}$), so t must be a closed term. Unfortunately, $\mathcal{R}$ does not have closed terms. Hence, we have to extend the set of function symbols in order to achieve the Church-Rosser property. The simplest way of doing this is adding a single constant $\perp$ and a reduction rule $\perp\to\perp$ (remember that we want to extend a TRS with the property UN to a confluent TRS which has *the same set of normal forms*). Furthermore, adding rules $t \to \perp$ for every term $t \in [F(x,y)]$ results in a TRS $\mathcal{R}_1$ which is Church-Rosser on the equivalence class $[F(x,y)]$. In fact, in this particular case the TRS $\mathcal{R}_1$ as a whole is Church-Rosser.

In general this treatment of equivalence classes without a normal form does not work without further ado. For example, the TRS $\mathcal{R}$ of Figure 3 has at least two equivalence classes

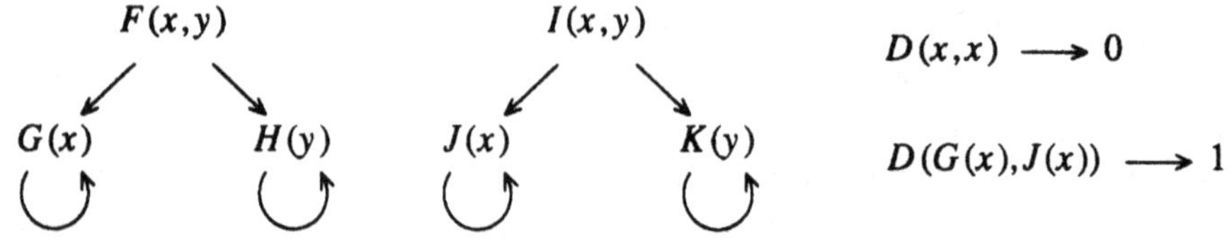

FIGURE 3.

without a normal form: $[F(x,y)]$ and $[I(x,y)]$. The technique illustrated above suggests adding a single constant $\perp$ and reduction rules (among others) $G(x) \to \perp$, $J(x) \to \perp$. However, the resulting TRS $\mathcal{R}_1$ is not Church-Rosser. In fact is is not even UN:

$$0 \leftarrow D(\perp,\perp) \twoheadleftarrow D(G(x),J(x)) \to 1.$$

The problem is that the terms $G(x)$ and $J(x)$, which belonged to different equivalence classes in $\mathcal{R}$, are members of the same equivalence class in $\mathcal{R}_1$ (see Figure 4). In other words, the

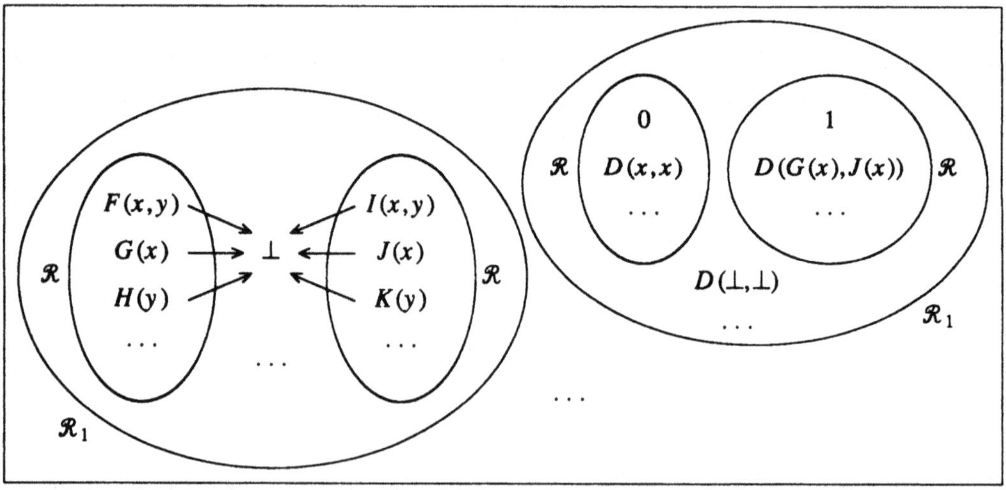

FIGURE 4.

extended TRS $\mathcal{R}_1$ is not a *conservative* extension of the original TRS $\mathcal{R}$. This problem can be solved, at least for this particular $\mathcal{R}$, by using a different $\perp$ for every equivalence class without a normal form (see Figure 5, where $\perp$'s are labeled with associated equivalence classes).

This refinement is still not sufficient. Consider the TRS of Figure 6. The equivalence class $[F(x,y,z)]$ does not have a normal form. According to the above discussion we should

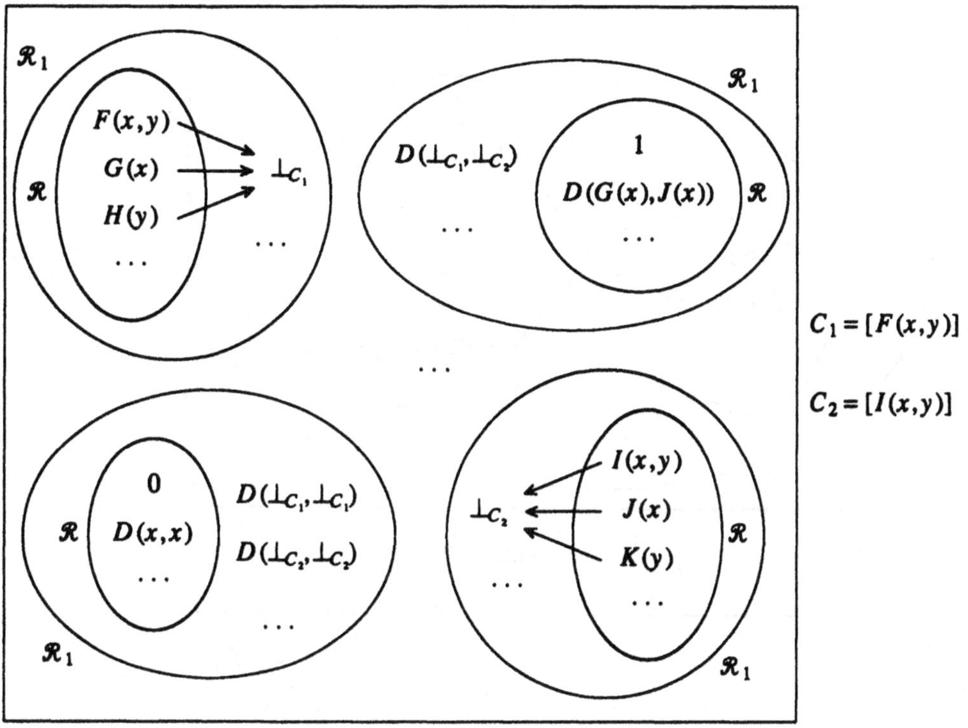

FIGURE 5.

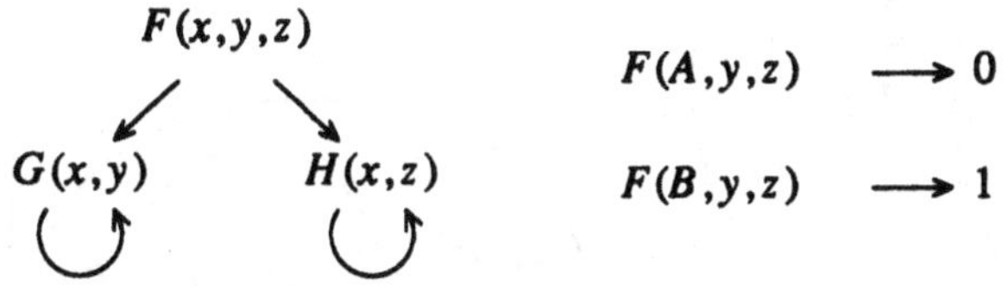

FIGURE 6.

add a constant $\perp_{[F(x,y,z)]}$ and reduction rules

$$\perp_{[F(x,y,z)]} \to \perp_{[F(x,y,z)]},$$
$$t \to \perp_{[F(x,y,z)]} \text{ for every term } t \in [F(x,y,z)].$$

In particular we have the new rule $F(x,y,z) \to \perp_{[F(x,y,z)]}$, which induces a conversion between the normal forms 0, 1: $0 \leftarrow F(A,y,z) \to \perp_{[F(x,y,z)]} \leftarrow F(B,y,z) \to 1$. Again the extended TRS $\mathcal{R}_1$ is not a conservative extension of $\mathcal{R}$ (see Figure 7). If we add to $\mathcal{R}$ a *unary function symbol* $\perp_{[F(x,y,z)]}$ and reduction rules

$$\perp_{[F(x,y,z)]}(x) \to \perp_{[F(x,y,z)]}(x),$$
$$t \to \perp_{[F(x,y,z)]}(x) \text{ for every term } t \in [F(x,y,z)],$$

then we are out of trouble: the extended TRS $\mathcal{R}_1$ preserves the equivalence class structure of $\mathcal{R}$ (see Figure 8).

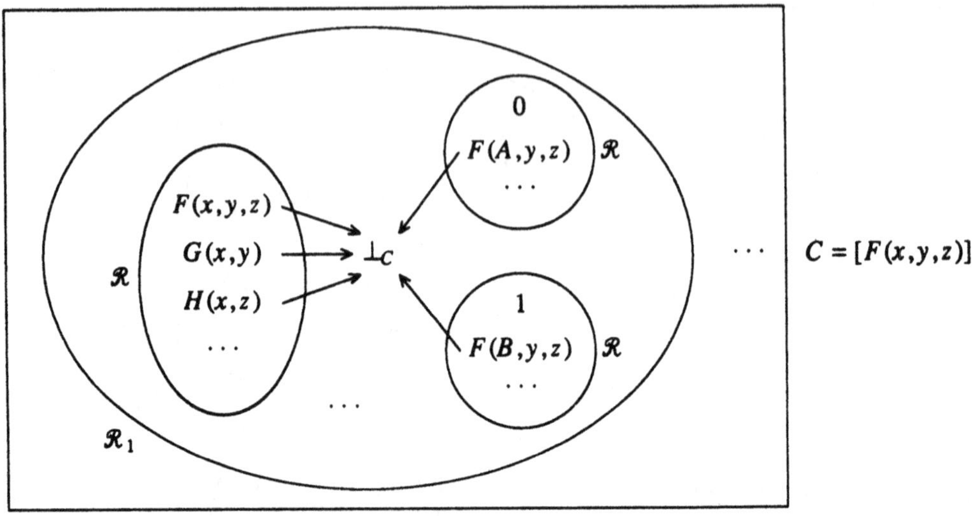

FIGURE 7.

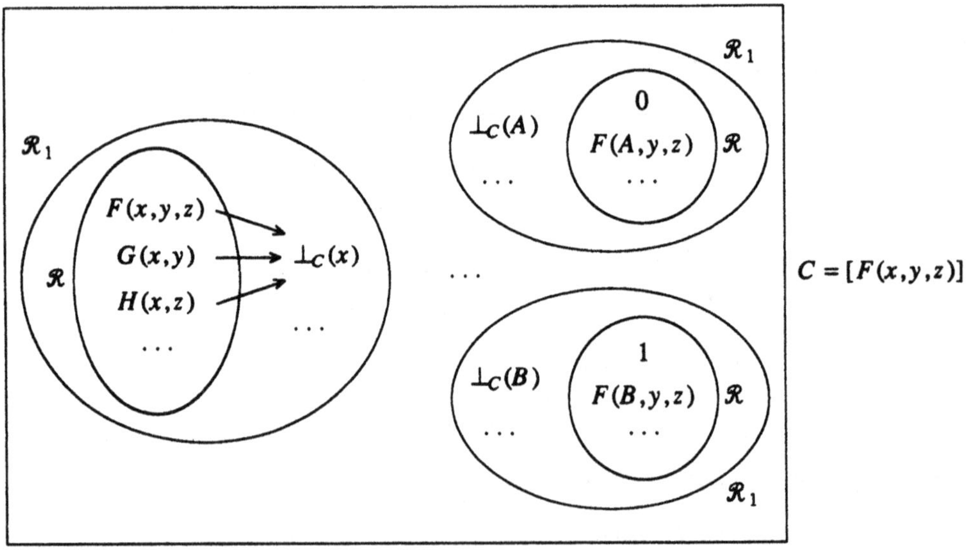

FIGURE 8.

In general the treatment of an equivalence class C without a normal form consists of (1) adding a new n-ary function symbol $\perp_C$, and (2) adding reduction rules

$$\perp_C(x_1,\ldots,x_n) \to \perp_C(x_1,\ldots,x_n),$$

$$t \to \perp_C(x_1,\ldots,x_n) \text{ for every term } t \in C,$$

where $\{x_1,\ldots,x_n\}$ is the set of all variables which occur in every term $t \in C$.

This concludes the informal account of our construction. We proceed with a formal explanation.

DEFINITION 2.1. Let $(\mathcal{F},\mathcal{R})$ be a TRS.

(1) If $t\in\mathcal{T}(\mathcal{F},\mathcal{V})$ then $[t]_{\mathcal{R}}=\{s\in\mathcal{T}(\mathcal{F},\mathcal{V}) \mid s=_{\mathcal{R}} t\}$. When no confusion can arise, we often omit the subscript $\mathcal{R}$.

(2) $\mathcal{C}_{\mathcal{R}}=\{[t]_{\mathcal{R}} \mid t\in\mathcal{T}(\mathcal{F},\mathcal{V})\}$.

(3) $\mathcal{C}_{\mathcal{R}}^{\perp}=\{C\in\mathcal{C}_{\mathcal{R}} \mid C$ does not contain a normal form$\}$.

(4) If $C\in\mathcal{C}_{\mathcal{R}}$ then $V_{fix}(C)=\bigcap_{t\in C} V(t)$.

PROPOSITION 2.2. *Let $(\mathcal{F},\mathcal{R})$ be a TRS. Suppose $t\in C\in\mathcal{C}_{\mathcal{R}}$ and $V(t)-V_{fix}(C)=\{x_1,\ldots,x_n\}$. Then $t[x_i\leftarrow s_i \mid 1\le i\le n]\in C$ for all terms $s_1,\ldots,s_n\in\mathcal{T}(\mathcal{F},\mathcal{V})$.*

PROOF. Let us first prove the statement for all terms $s_1,\ldots,s_n\in\mathcal{T}(\mathcal{F},\mathcal{V})$ such that $V(s_i)\cap\{x_1,\ldots,x_n\}=\varnothing$ $(i=1,\ldots,n)$. Define a sequence of terms $t_0,\ldots,t_n$ as follows:

$$t_0\equiv t,$$
$$t_i\equiv t_{i-1}[x_i\leftarrow s_i] \text{ if } 0<i\le n.$$

We prove that $t_i=_{\mathcal{R}} t$ by induction on i. The case $i=0$ is trivial. Suppose the statement is true for all $i<k$ $(k>0)$. Because $x_k\notin V_{fix}(C)$ there exists a term $u\in C$ such that $x_k\notin V(u)$. The induction hypothesis tells us that $t_{k-1}=_{\mathcal{R}} t$. This implies that $t_k\equiv t_{k-1}[x_k\leftarrow s_k]=_{\mathcal{R}} u[x_k\leftarrow s_k]\equiv u=_{\mathcal{R}} t$. We conclude that $t_n\equiv t[x_1\leftarrow s_1]\ldots[x_n\leftarrow s_n]\equiv t[x_i\leftarrow s_i \mid 1\le i\le n]\in C$. Now let $s_1,\ldots,s_n$ be arbitrary terms of $\mathcal{T}(\mathcal{F},\mathcal{V})$. Choose distinct variables $y_1,\ldots,y_n$ such that $y_i\notin V(t)\cup\bigcup_{i=1}^{n} V(s_i)$ $(i=1,\ldots,n)$. By the above argument we have $t[x_i\leftarrow y_i \mid 1\le i\le n]\in C$ and because $V(t[x_i\leftarrow y_i \mid 1\le i\le n])-V_{fix}(C)=\{y_1,\ldots,y_n\}$ we obtain $t[x_i\leftarrow y_i \mid 1\le i\le n][y_i\leftarrow s_i \mid 1\le i\le n]\equiv t[x_i\leftarrow s_i \mid 1\le i\le n]\in C$. □

In the remainder of this section $(\mathcal{F},\mathcal{R})$ denotes a TRS which has the property UN.

PROPOSITION 2.3. *If $C\in\mathcal{C}_{\mathcal{R}}$ contains the normal form t, then $V_{fix}(C)=V(t)$.*

PROOF. Let $s\in C$. We prove that $V(t)\subseteq V(s)$ by induction on the length of the conversion $s=_{\mathcal{R}} t$. The case of zero length is trivial. Let $s\leftrightarrow_{\mathcal{R}} s_1=_{\mathcal{R}} t$. From the induction hypothesis we obtain $V(t)\subseteq V(s_1)$. If $s\rightarrow_{\mathcal{R}} s_1$ then $V(s_1)\subseteq V(s)$ and we are done. Let $s\leftarrow_{\mathcal{R}} s_1$. We have to show that every variable of t occurs in s. Suppose to the contrary that there is a variable $x\in V(t)$ which does not occur in s. Let $y\in\mathcal{V}$ be any variable which does not occur in the conversion between s_1 and t. If we replace every occurrence of x by y in the conversion between s_1 and t we get a conversion $s_1'=_{\mathcal{R}} t'$. Notice that t' is a normal form of $\mathcal{R}$ different from t. Because x does not occur in s we have $s_1'\rightarrow_{\mathcal{R}} s$. But now we have the following conversion between t and t': $t=_{\mathcal{R}} s_1\rightarrow_{\mathcal{R}} s\leftarrow_{\mathcal{R}} s_1'=_{\mathcal{R}} t'$, which is impossible by the unicity of normal forms of $\mathcal{R}$. We conclude that $V(t)\subseteq V(s)$ for every term $s\in C$, from which we immediately obtain $V_{fix}(C)=V(t)$. □

In the first step of our construction we associate with every equivalence class without a normal form a $\perp$-term.

DEFINITION 2.4. The TRS $(\mathcal{F}_1,\mathcal{R}_1)$ is defined by

- $\mathcal{F}_1 = \mathcal{F} \cup \{\perp_C \mid C \in \mathcal{C}_{\mathcal{R}}^{\perp}\}$ where $\perp_C$ is a n-ary function symbol if the cardinality of $V_{fix}(C)$ equals n. (If $V_{fix}(C) = \varnothing$ then $\perp_C$ is a constant.)
- $\mathcal{R}_1 = \mathcal{R} \cup \{\perp_C(x_1,\ldots,x_n) \to \perp_C(x_1,\ldots,x_n) \mid C \in \mathcal{C}_{\mathcal{R}}^{\perp}$ and $V_{fix}(C) = \{x_1,\ldots,x_n\}^{\dagger}\}$
  $\cup \{t \to \perp_C(x_1,\ldots,x_n) \mid t \in C \in \mathcal{C}_{\mathcal{R}}^{\perp}$ and $V_{fix}(C) = \{x_1,\ldots,x_n\}^{\dagger}\}$.

Notice that $\mathcal{R}_1$ contains only legal rewriting rules because $V_{fix}(C) \subseteq V(t)$ whenever $t \in C$, and if $t \in C \in \mathcal{C}_{\mathcal{R}}^{\perp}$ then t is not a variable (since every variable is a normal form).

NOTATION. We abbreviate $\mathcal{T}(\mathcal{F},\mathcal{V})$ to $\mathcal{T}$ and $\mathcal{T}(\mathcal{F}_1,\mathcal{V})$ to $\mathcal{T}_1$.

PROPOSITION 2.5. $(\mathcal{F},\mathcal{R})$ *and* $(\mathcal{F}_1,\mathcal{R}_1)$ *define the same set of normal forms.*
PROOF. Trivial. □

We want to show that every equivalence class of $\mathcal{R}_1$ contains exactly one equivalence class of $\mathcal{R}$ (this establishes the conservativity of $\mathcal{R}_1$ over $\mathcal{R}$). The usefulness of the next definition will soon become apparent.

DEFINITION 2.6. Let $t \in \mathcal{T}_1$. We define the set $t^{\perp}$ of *lifted terms* of t by induction on the structure of t:

$$x^{\perp} = \{x\},$$
$$F(t_1,\ldots,t_n)^{\perp} = \{F(s_1,\ldots,s_n) \mid s_i \in t_i^{\perp}\ (i=1,\ldots,n)\},$$
$$\perp_C(t_1,\ldots,t_n)^{\perp} = \{s[x_i \leftarrow s_i \mid 1 \le i \le n] \mid V_{fix}(C) = \{x_1,\ldots,x_n\}, s \in C \text{ and } s_i \in t_i^{\perp}\ (i=1,\ldots,n)\}.$$

EXAMPLE 2.7. Let $\mathcal{R} = \{F(x,G(x)) \to H(x),\ H(x) \to H(x),\ F(A(x,y),G(A(x,y))) \to B(y),\ B(y) \to B(y)\}$. We have $[F(x,G(x))] \in \mathcal{C}_{\mathcal{R}}^{\perp}$, $V_{fix}([F(x,G(x))]) = \{x\}$, $[B(y)] \in \mathcal{C}_{\mathcal{R}}^{\perp}$ and $V_{fix}([B(y)]) = \{y\}$. Therefore, in $\mathcal{R}_1$ we have the conversion $\perp_{[B(y)]}(y) \leftarrow_{\mathcal{R}_1} F(A(x,y),G(A(x,y))) \to_{\mathcal{R}_1} \perp_{[F(x,G(x))]}(A(x,y))$. It is not difficult to verify that $\perp_{[B(y)]}(y)^{\perp} = [B(y)]$ and $\perp_{[F(x,G(x))]}(A(x,y))^{\perp} = \{F(A(x,y), G(A(x,y))), H(A(x,y))\}$. Notice that for all $s \in \perp_{[B(y)]}(y)^{\perp}$ and $t \in \perp_{[F(x,G(x))]}(A(x,y))^{\perp}$ we have $s =_{\mathcal{R}} t$.

PROPOSITION 2.8. *Let* $t \in \mathcal{T}_1$.
(1) *The set* $t^{\perp}$ *is a non-empty subset of* $\mathcal{T}$.
(2) *If* $t \in \mathcal{T}$ *then* $t^{\perp} = \{t\}$.
(3) *If* $s \in t^{\perp}$ *then* $s =_{\mathcal{R}_1} t$. *In other words, the set* $t^{\perp}$ *is a subset of* $[t]_{\mathcal{R}_1}$.
(4) *If* $s_1, s_2 \in t^{\perp}$ *then* $s_1 =_{\mathcal{R}} s_2$.
PROOF. Straightforward induction on the structure of t. □

For a proof of the next proposition we refer to the full version of this paper [2].

---

† In this paper we assume that variables are always listed in the same order.

PROPOSITION 2.9. *If* $t_1 =_{\mathcal{R}_1} t_2$ *then* $s_1 =_{\mathcal{R}} s_2$ *for every* $s_1 \in t_1^{\perp}$ *and* $s_2 \in t_2^{\perp}$. □

COROLLARY 2.10. *Every equivalence class of* $\mathcal{R}_1$ *contains exactly one equivalence class of* $\mathcal{R}$.
PROOF. Let $C \in \mathcal{C}_{\mathcal{R}_1}$. By Proposition 2.8(1,3), C contains at least one equivalence class of $\mathcal{R}$. Suppose C contains equivalence classes $C_1, C_2 \in \mathcal{C}_{\mathcal{R}}$. Choose $t_1 \in C_1$ and $t_2 \in C_2$. We have $t_1 =_{\mathcal{R}_1} t_2$. Using Propositions 2.8(2) and 2.9 we obtain $t_1 =_{\mathcal{R}} t_2$. Therefore $C_1 = C_2$ and we are done. □

NOTATION. If $C \in \mathcal{C}_{\mathcal{R}_1}$ then we denote by $\phi(C)$ the unique equivalence class of $\mathcal{R}$ which it contains. Notice that $\phi(C) = C \cap \mathcal{T}$ (by Corollary 2.10).

Clearly, every equivalence class of $\mathcal{R}$ is contained in an equivalence class of $\mathcal{R}_1$. Hence, we have the situation of Figure 9.

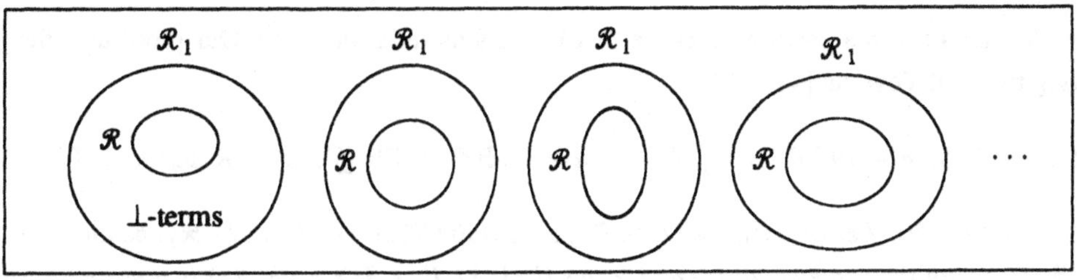

FIGURE 9.

PROPOSITION 2.11. *The* TRS $(\mathcal{F}_1, \mathcal{R}_1)$ *has the property* UN.
PROOF. This is an immediate consequence of Proposition 2.5 and Corollary 2.10. □

DEFINITION 2.12. Let $C \in \mathcal{C}_{\mathcal{R}_1}$. The representative $\pi(C) \in C$ is defined by

$$\pi(C) = \begin{cases} t & \text{if } C \text{ contains the normal form } t, \\ \perp_{\phi(C)}(x_1, \ldots, x_n) & \text{if } C \in \mathcal{C}^{\perp}_{\mathcal{R}_1} \text{ and } V_{fix}(\phi(C)) = \{x_1, \ldots, x_n\}. \end{cases}$$

PROPOSITION 2.13. *If* $C \in \mathcal{C}_{\mathcal{R}_1}$ *then* $V_{fix}(C) = V(\pi(C))$.
PROOF. By definition of $\pi(C)$ and Proposition 2.3, we have $V(\pi(C)) = V_{fix}(\phi(C))$. Suppose there is a term $t \in C$ such that $V_{fix}(\phi(C)) \not\subseteq V(t)$. Let $x \in V_{fix}(\phi(C))$ such that $x \notin V(t)$. Choose a term $s \in \phi(C)$ and let y be any variable different from x. By Proposition 2.2 we have $s[x \leftarrow y] \in C$. From Corollary 2.10 we obtain $s[x \leftarrow y] \in \phi(C)$. But this means that $x \notin V_{fix}(\phi(C))$. Therefore, $V_{fix}(\phi(C)) \subseteq V(t)$ for every term $t \in C$. □

DEFINITION 2.14. The TRS $(\mathcal{F}_2, \mathcal{R}_2)$ is defined by
- $\mathcal{F}_2 = \mathcal{F}_1$,
- $\mathcal{R}_2 = \mathcal{R}_1 \cup \{t \to \pi(C) \mid t \in C \in \mathcal{C}_{\mathcal{R}_1} \text{ and } t \not\equiv \pi(C)\}$.

Notice that according to Proposition 2.13, $\mathcal{R}_2$ contains only legal rewriting rules. The next three propositions state that $\mathcal{R}_2$ is a conservative extension of $\mathcal{R}_1$ which is confluent and preserves the normal forms of $\mathcal{R}_1$. The proofs of Propositions 2.15 and 2.16 are straightforward.

PROPOSITION 2.15. *$(\mathcal{F}_1,\mathcal{R}_1)$ and $(\mathcal{F}_2,\mathcal{R}_2)$ define the same set of normal forms.* □

PROPOSITION 2.16. *For all terms $s,t\in\mathcal{T}_1$ we have $s =_{\mathcal{R}_1} t$ iff $s =_{\mathcal{R}_2} t$.* □

PROPOSITION 2.17. *The TRS $(\mathcal{F}_2,\mathcal{R}_2)$ is confluent.*
PROOF. Let $t_1 =_{\mathcal{R}_2} t_2$. By Proposition 2.16 we have $t_1 =_{\mathcal{R}_1} t_2$. Let $C = [t_1]_{\mathcal{R}_1}$. By construction of $\mathcal{R}_2$, the term $\pi(C)$ is a common reduct of t_1 and t_2. □

COROLLARY 2.18. *Every TRS with the property UN can be conservatively extended to a confluent TRS with the same set of normal forms.* □

We are now in a position to prove that UN is a modular property. Our proof uses the following result of Toyama [4].

THEOREM 2.19. *For all TRS's $\mathcal{R}_0$, $\mathcal{R}_1$ we have $\mathcal{R}_0\oplus\mathcal{R}_1$ is CR iff $\mathcal{R}_0$ is CR and $\mathcal{R}_1$ is CR.* □

PROPOSITION 2.20. *Let $(\mathcal{F}_0,\mathcal{R}_0)$ and $(\mathcal{F}_1,\mathcal{R}_1)$ be disjoint TRS's and let $(\mathcal{F}'_i,\mathcal{R}'_i)$ be an extension of $(\mathcal{F}_i,\mathcal{R}_i)$ with the same set of normal forms $(i=0,1)$. If $t\in\mathcal{T}(\mathcal{F}_0\cup\mathcal{F}_1,\mathcal{V})$ is a normal form of $\mathcal{R}_0\oplus\mathcal{R}_1$, then t is also a normal form of $\mathcal{R}'_0\oplus\mathcal{R}'_1$.*
PROOF. See the full version of this paper [2]. □

THEOREM 2.21. *For all TRS's $\mathcal{R}_0$, $\mathcal{R}_1$ we have $\mathcal{R}_0\oplus\mathcal{R}_1$ is UN iff $\mathcal{R}_0$ is UN and $\mathcal{R}_1$ is UN.*
PROOF.
'⇒' Trivial.
'⇐' By Corollary 2.18 we can extend $\mathcal{R}_i$ to a confluent TRS $\mathcal{R}'_i$ with the same set of normal forms $(i=0,1)$. Without loss of generality we may assume that $\mathcal{R}_0$, $\mathcal{R}_1$ are disjoint and $\mathcal{R}'_0$, $\mathcal{R}'_1$ are disjoint. Let $n_1 =_{\mathcal{R}_0\oplus\mathcal{R}_1} n_2$ be a conversion between normal forms of $\mathcal{R}_0\oplus\mathcal{R}_1$. Because $\mathcal{R}'_0\oplus\mathcal{R}'_1$ is an extension of $\mathcal{R}_0\oplus\mathcal{R}_1$ we have $n_1 =_{\mathcal{R}'_0\oplus\mathcal{R}'_1} n_2$. Furthermore, n_1,n_2 are normal forms of $\mathcal{R}'_0\oplus\mathcal{R}'_1$ by Proposition 2.20, and therefore $n_1\equiv n_2$ by Theorem 2.19.

□

3. Modular Aspects of NF and UN$^{\rightarrow}$

The following examples show that neither NF nor UN$^{\rightarrow}$ is a modular property of TRS's.

EXAMPLE 3.1. Let $\mathcal{R}_0=\{F(x,x)\rightarrow C\}$ and $\mathcal{R}_1=\{a\rightarrow b, a\rightarrow c, b\rightarrow b, c\rightarrow c\}$ (see Figure 1(a)). Both $\mathcal{R}_0$ and $\mathcal{R}_1$ have the property NF. The following conversion shows that $F(b,c)$ is $\mathcal{R}_0\oplus\mathcal{R}_1$-convertible to the normal form C: $F(b,c)\leftarrow F(a,c)\leftarrow F(a,a)\rightarrow C$. However, it is clear that the term $F(b,c)$ does not reduce to C. So $\mathcal{R}_0\oplus\mathcal{R}_1$ is not NF. Note that $\mathcal{R}_1$ refutes the implication NF ⇒ CR (cf. Lemma 1.3).

EXAMPLE 3.2. Let $\mathcal{R}_0 = \{F(x,x) \to C\}$ and $\mathcal{R}_1 = \{a \to b, a \to c, c \to c, d \to c, d \to e\}$ (see Figure 1(c)). Both systems have the property $UN^{\to}$. The term $F(a,d)$ reduces in $\mathcal{R}_0 \oplus \mathcal{R}_1$ to the different normal forms $F(b,e)$ and C: $F(b,e) \leftarrow F(b,d) \leftarrow F(a,d) \to F(c,d) \to F(c,c) \to C$. Therefore $\mathcal{R}_0 \oplus \mathcal{R}_1$ is not $UN^{\to}$. Note that $\mathcal{R}_1$ refutes the implication $UN^{\to} \Rightarrow UN$ (cf. Lemma 1.3).

Both counterexamples involve a non-left-linear TRS:

DEFINITION 3.3. A rewriting rule $l \to r$ is *left-linear* if its left-hand side l does not contain more than one occurrence of the same variable. A TRS $\mathcal{R}$ is *left-linear* if all rewriting rules of $\mathcal{R}$ are left-linear.

It seems reasonable to conjecture that NF and $UN^{\to}$ are modular properties of left-linear TRS's. However, we have not been able to prove this. Complications arise in the presence of collapsing rules (cf. [1], [3], [4], [6] and [5] in which all counterexamples involve a TRS which has collapsing rules):

DEFINITION 3.4. A rewriting rule $l \to r$ is *collapsing* if its right-hand side r is a variable.

In the remainder of this section we will prove that NF and $UN^{\to}$ are modular properties of left-linear TRS's without collapsing rules. Let $(\mathcal{F}_0, \mathcal{R}_0)$ and $(\mathcal{F}_1, \mathcal{R}_1)$ be two disjoint TRS's. A term $t \in \mathcal{T}(\mathcal{F}_0 \cup \mathcal{F}_1, \mathcal{V})$, in its tree notation, is an alternation of $\mathcal{F}_0$-parts and $\mathcal{F}_1$-parts (see Figure 10). To formalize this structure we introduce the following definitions as in [4].

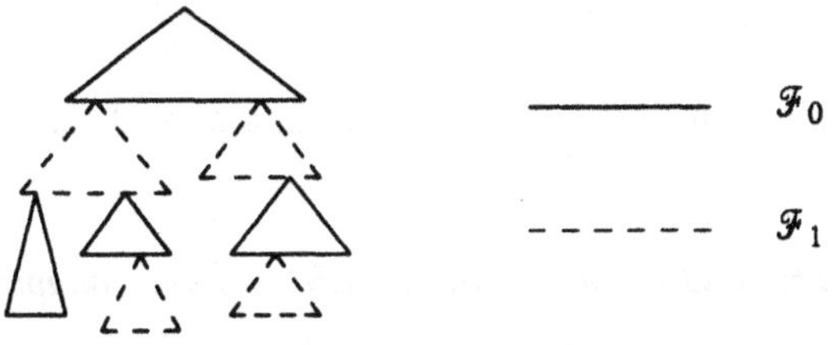

FIGURE 10.

DEFINITION 3.5. Let $t \in \mathcal{T}(\mathcal{F}_0 \cup \mathcal{F}_1, \mathcal{V})$. The *outermost symbol* of t, notation $root(t)$, is defined by $root(t) = F$ if $t \equiv F(t_1, \ldots, t_n)$ and $root(t) = t$ otherwise.

DEFINITION 3.6. Let $t \equiv C[t_1, \ldots, t_n] \in \mathcal{T}(\mathcal{F}_0 \cup \mathcal{F}_1, \mathcal{V})^{\dagger}$ and $C[\,,\ldots,\,] \not\equiv \square$. We write $t \equiv C[\![t_1, \ldots, t_n]\!]$ if $C[\,,\ldots,\,]$ is a $\mathcal{F}_a$-context and $root(t_i) \in \mathcal{F}_{1-a}$ for $i = 1, \ldots, n$ ($a \in \{0,1\}$).

DEFINITION 3.7. Let $t \in \mathcal{T}(\mathcal{F}_0 \cup \mathcal{F}_1, \mathcal{V})$. Then $rank(t) = 1$ if $t \in \mathcal{T}(\mathcal{F}_a, \mathcal{V})$ ($a \in \{0,1\}$) and $rank(t) = 1 + \max_{1 \le i \le n} rank(t_i)$ if $t \equiv C[\![t_1, \ldots, t_n]\!]$ ($n > 0$).

† $C[\,,\ldots,\,]$ is a 'context' containing n occurrences of the symbol $\square$. For notational convenience we allow for $n = 0$.

PROPOSITION 3.8. *If $s \twoheadrightarrow_{\mathcal{R}_0 \oplus \mathcal{R}_1} t$ then rank$(s) \geq$ rank(t).*
PROOF. Routine. □

DEFINITION 3.9. Let $s \equiv C[\![s_1, \ldots, s_n]\!] \rightarrow_{\mathcal{R}_0 \oplus \mathcal{R}_1} t$ by application of the rewriting rule $l \rightarrow r$. We write $s \rightarrow_i t$ if $l \rightarrow r$ is being applied in one of the s_j's; otherwise we write $s \rightarrow_o t$. The relation $\rightarrow_i$ is called *inner* reduction and $\rightarrow_o$ is called *outer* reduction.

The following proposition states that inner reductions can be postponed for left-linear TRS's without collapsing rules.

PROPOSITION 3.10. *Let $\mathcal{R}_0$ and $\mathcal{R}_1$ be left-linear* TRS's *without collapsing rules. If $s \twoheadrightarrow_{\mathcal{R}_0 \oplus \mathcal{R}_1} t$ then there exists a term s' such that $s \twoheadrightarrow_o s' \twoheadrightarrow_i t$.*
PROOF. Straightforward induction on the length of the reduction $s \twoheadrightarrow t$. □

Proposition 3.10 is in general not true for left-linear TRS's with collapsing rules, as can be seen from the next example.

EXAMPLE 3.11. Let $\mathcal{R}_0 = \{F(A) \rightarrow B\}$ and $\mathcal{R}_1 = \{e(x) \rightarrow x\}$. The term $F(e(A))$ is in normal form with respect to $\rightarrow_o$, but we do have the reduction $F(e(A)) \rightarrow_i F(A) \rightarrow_o B$.

The following example shows that Proposition 3.10 is not true if $\twoheadrightarrow_o$ and $\twoheadrightarrow_i$ are interchanged.

EXAMPLE 3.12. Let $\mathcal{R}_0 = \{F(x) \rightarrow G(x,x)\}$ and $\mathcal{R}_1 = \{a \rightarrow b\}$. We have $F(a) \rightarrow_o G(a,a) \rightarrow_i G(a,b)$, but there is no term t such that $F(a) \twoheadrightarrow_i t \twoheadrightarrow_o G(a,b)$.

However, the restriction to reductions ending in a normal form yields postponement of outer reductions.

PROPOSITION 3.13. *Let $\mathcal{R}_0$ and $\mathcal{R}_1$ be left-linear* TRS's *without collapsing rules. If $s \twoheadrightarrow_{\mathcal{R}_0 \oplus \mathcal{R}_1} t$ and t is a normal form, then we can find a term s' such that $s \twoheadrightarrow_i s' \twoheadrightarrow_o t$.*
PROOF. Routine. □

Again the restriction to left-linear TRS's without collapsing rules is essential, as the following example shows.

EXAMPLE 3.14. Let $\mathcal{R}_0 = \{F(x) \rightarrow G(x,x), G(C,x) \rightarrow H(x)\}$ and $\mathcal{R}_1 = \{e(x) \rightarrow x, e(x) \rightarrow f(x)\}$. The term $H(f(C))$ is a normal form of $F(e(C))$: $F(e(C)) \rightarrow_o G(e(C),e(C)) \rightarrow_i G(C,e(C)) \rightarrow_o H(e(C)) \rightarrow_i H(f(C))$. It is not difficult to verify that there exists no term t such that $F(e(C)) \twoheadrightarrow_i t \twoheadrightarrow_o H(f(C))$.

THEOREM 3.15. NF *is a modular property for left-linear* TRS's *without collapsing rules.*
PROOF. Suppose $\mathcal{R}_0$ and $\mathcal{R}_1$ are left-linear TRS's without collapsing rules which have the property NF. Let $s =_{\mathcal{R}_0 \oplus \mathcal{R}_1} t$ with t in normal form. We have to prove that $s \twoheadrightarrow t$. We use

induction on the length of the conversion $s = t$. The case of zero length is trivial. Suppose $s \leftrightarrow u = t$. The induction hypothesis states that $u \twoheadrightarrow t$. If $s \to u$ then we are done. The difficult case is $u \to s$. We use induction on $rank(u)$. If $rank(u) = 1$ then $s \leftarrow u \twoheadrightarrow t$ is either a conversion in $\mathcal{R}_0$ or $\mathcal{R}_1$ (by Proposition 3.8). Hence, $s \twoheadrightarrow t$. Let $rank(u) > 1$. We distinguish two cases: (1) $u \to_i s$ or (2) $u \to_o s$.

(1)   Using Proposition 3.13, we have the situation of Figure 11. Again we consider two

$$u \equiv C[\![u_1, \ldots, u_j, \ldots, u_n]\!] \xrightarrow{\ i\ } C[\![t_1, \ldots, t_j, \ldots, t_n]\!] \xrightarrow{\ o\ } C_1[\![t_{i_1}, \ldots, t_{i_m}]\!] \equiv t$$

$$\Big\downarrow i$$

$$s \equiv C[\![u_1, \ldots, u_j', \ldots, u_n]\!]$$

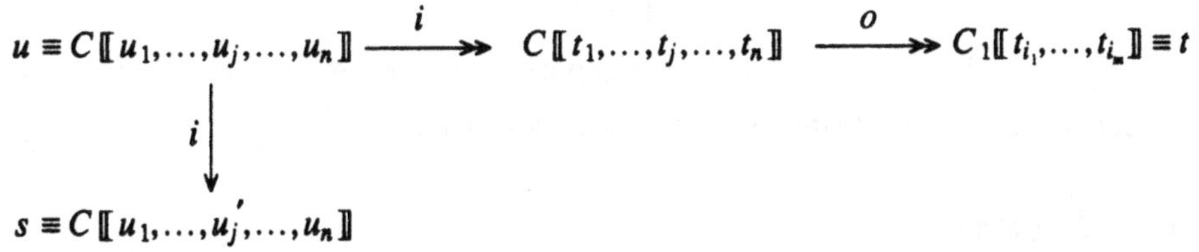

FIGURE 11.

cases: (i) t_j is a normal form or (ii) t_j is not a normal form.

(i)   In this case we have $u_j' \leftarrow u_j \twoheadrightarrow t_j$, $rank(u_j) = rank(u) - 1$ and t_j is a normal form. Applying the induction hypothesis gives us $u_j' \twoheadrightarrow t_j$ and therefore we have the following reduction from s to t: $s \equiv C[\![u_1, \ldots, u_j', \ldots, u_n]\!] \twoheadrightarrow C[\![t_1, \ldots, t_j, \ldots, t_n]\!] \twoheadrightarrow C_1[\![t_{i_1}, \ldots, t_{i_m}]\!] \equiv t$.

(ii) If t_j is not a normal form then $C[\![t_1, \ldots, v, \ldots, t_n]\!] \twoheadrightarrow C_1[\![t_{i_1}, \ldots, t_{i_m}]\!] \equiv t$ for every term v, because t is a normal form. In particular we have $s \equiv C[\![u_1, \ldots, u_j', \ldots, u_n]\!] \twoheadrightarrow C[\![t_1, \ldots, u_j', \ldots, t_n]\!] \twoheadrightarrow C_1[\![t_{i_1}, \ldots, t_{i_m}]\!] \equiv t$.

(2)   We have the situation of Figure 12. If we replace the terms $t_1, \ldots, t_n$ in the conversion

$$u \equiv C[\![u_1, \ldots, u_n]\!] \xrightarrow{\ i\ } C[\![t_1, \ldots, t_n]\!] \xrightarrow{\ o\ } C_1[\![t_{i_1}, \ldots, t_{i_m}]\!] \equiv t$$

$$o\Big\downarrow \qquad\qquad\qquad \Big\downarrow o$$

$$s \equiv C_2[\![u_{j_1}, \ldots, u_{j_k}]\!] \xrightarrow{\ i\ } C_2[\![t_{j_1}, \ldots, t_{j_k}]\!]$$

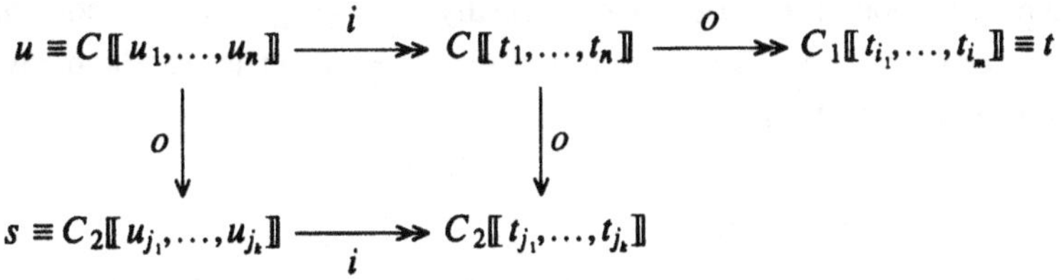

FIGURE 12.

$C_2[\![t_{j_1}, \ldots, t_{j_k}]\!] \leftarrow_o C[\![t_1, \ldots, t_n]\!] \twoheadrightarrow_o C_1[\![t_{i_1}, \ldots, t_{i_m}]\!]$ by distinct fresh variables $x_1, \ldots, x_n$, we get a conversion in either $\mathcal{R}_0$ or $\mathcal{R}_1$: $C_2[x_{j_1}, \ldots, x_{j_k}] \leftarrow C[x_1, \ldots, x_n] \twoheadrightarrow C_1[x_{i_1}, \ldots, x_{i_m}]$. The term $C_1[x_{i_1}, \ldots, x_{i_m}]$ is a normal form, hence $C_2[x_{j_1}, \ldots, x_{j_k}] \twoheadrightarrow C_1[x_{i_1}, \ldots, x_{i_m}]$. Instantiation of this reduction gives us $C_2[t_{j_1}, \ldots, t_{j_k}] \twoheadrightarrow C_2[t_{i_1}, \ldots, t_{i_m}]$. We conclude that $s \twoheadrightarrow t$.

□

THEOREM 3.16. $UN^{\to}$ is a modular property for left-linear TRS's without collapsing rules.
PROOF. Suppose $\mathcal{R}_0$ and $\mathcal{R}_1$ are left-linear TRS's without collapsing rules which have the property $UN^{\to}$. Let $n_1 \twoheadleftarrow_{\mathcal{R}_0 \oplus \mathcal{R}_1} t \twoheadrightarrow_{\mathcal{R}_0 \oplus \mathcal{R}_1} n_2$ with n_1, n_2 in normal form. We have to prove that $n_1 \equiv n_2$. We use induction on $rank(t)$. If $rank(t) = 1$ then $n_1 \twoheadleftarrow t \twoheadrightarrow n_2$ is either a conversion in

$\mathcal{R}_0$ or $\mathcal{R}_1$. Therefore, $n_1 \equiv n_2$. Let $rank(t) > 1$. Using Proposition 3.10, we have the situation of Figure 13. By a similar argument as in case (2) of the proof of Theorem 3.15 we

$$
\begin{array}{ccccc}
 & \xrightarrow{o}\!\!\!\!\rightarrow & C_1\llbracket t_{i_1},\ldots,t_{i_m}\rrbracket & \xrightarrow{i}\!\!\!\!\rightarrow & C_1\llbracket u_{i_1},\ldots,u_{i_m}\rrbracket \equiv n_1 \\
t \equiv C\llbracket t_1,\ldots,t_n\rrbracket & & & & \\
 & \xrightarrow[o]{}\!\!\!\!\rightarrow & C_2\llbracket t_{j_1},\ldots,t_{j_k}\rrbracket & \xrightarrow[i]{}\!\!\!\!\rightarrow & C_2\llbracket v_{j_1},\ldots,v_{j_k}\rrbracket \equiv n_2
\end{array}
$$

FIGURE 13.

have $C_1\llbracket t_{i_1},\ldots,t_{i_m}\rrbracket \equiv C_2\llbracket t_{j_1},\ldots,t_{j_k}\rrbracket$. Furthermore, $v_k \twoheadleftarrow t_k \twoheadrightarrow u_k$ for $k \in \{i_1,\ldots,i_m\}$. The induction hypothesis states that $v_k \equiv u_k$. We conclude that $n_1 \equiv n_2$. □

4. Concluding Remarks

In this paper we have proved the following facts: (1) UN is a modular property, (2) NF and $\mathrm{UN}^{\rightarrow}$ are not modular properties of general TRS's, and (3) NF and $\mathrm{UN}^{\rightarrow}$ are modular properties of left-linear TRS's without collapsing rules. Two problems remain:

CONJECTURE.

(1) NF *is a modular property of left-linear* TRS'*s*.

(2) $\mathrm{UN}^{\rightarrow}$ *is a modular property of left-linear* TRS'*s*.

It is conceivable that some of the proof techniques in [6] are applicable to prove the above conjecture.

Acknowledgements. The author would like to thank Jan Willem Klop for his comments on improving the presentation of this paper and especially for his suggestion of reducing the modularity of UN to Corollary 2.18. The author also thanks Roel de Vrijer for the many discussions leading to improvements of the text.

References

[1] A. Middeldorp, *A Sufficient Condition for the Termination of the Direct Sum of Term Rewriting Systems*, Report IR-150, Vrije Universiteit Amsterdam, 1988. To appear in: Proceedings of the 4th Conference on Logic in Computer Science (Asilomar, 1989).

[2] A. Middeldorp, *Modular Aspects of Properties of Term Rewriting Systems Related to Normal Forms*, Report IR-164, Vrije Universiteit Amsterdam, 1988.

[3] M. Rusinowitch, *On Termination of the Direct Sum of Term Rewriting Systems*, Information Processing Letters **26**, pp. 65-70, 1987.

[4] Y. Toyama, *On the Church-Rosser Property for the Direct Sum of Term Rewriting Systems*, Journal of the ACM **34**(1), pp. 128-143, 1987.

[5] Y. Toyama, *Counterexamples to Termination for the Direct Sum of Term Rewriting Systems*, Information Processing Letters **25**, pp. 141-143, 1987.

[6] Y. Toyama, J.W. Klop and H.P. Barendregt, *Termination for the Direct Sum of Left-Linear Term Rewriting Systems*, in this proceedings.

Priority Rewriting:
Semantics, Confluence, and Conditionals

Chilukuri K. Mohan
School of Computer and Information Science,
Syracuse University, Syracuse, NY 13244-4100

Abstract: Priority rewrite systems (PRS) are partially ordered finite sets of rewrite rules; in this paper, two possible alternative definitions for rewriting with PRS are examined. A logical semantics for priority rewriting is described, using equational formulas obtained from the rules, and inequations which must be assumed to permit rewriting with rules of lower priority. Towards the goal of using PRS to define data type and function specifications, restrictions are given that ensure confluence and encourage modularity. Finally, the relation between priority and conditional rewriting is studied, and a natural combination of these mechanisms is proposed.

1 Introduction

Baeten, Bergstra and Klop[BBK] have proposed the formalism of *Priority Rewrite Systems* (PRS), a powerful and interesting generalization of ordinary Term Rewriting Systems (TRS). PRS consist of rewrite rules augmented by a partial ordering which determines which rule may next be used to rewrite a term. Their expressive power is illustrated in the following PRS, which defines an equality predicate: as [BBK] point out, such a definition cannot be given using ordinary TRS nor conditional TRS formalisms (without negation).

Example 1

$$\Re_1 = \left\{ \Big\downarrow \begin{array}{lcl} r_1 & : & eq(x,x) \to T \\ r_2 & : & eq(x,y) \to F \end{array} \right\}$$

Here, the 'priority' partial order determines that $r_1 > r_2$, hence the second rule is used for reduction only if the first fails to apply.

Rewrite rules without priority have a clear and well-understood equational first order logical semantics. PRS are more powerful and operationally easy to implement, but their logical semantics has not been studied. This is an important concern, because PRS can be difficult to visualize and understand; the "meaning" and applicability of a rewrite rule cannot be grasped in isolation, without looking at other (higher priority) rules.

By imposing restrictions and limiting the amount of non-determinism, we establish a framework in which priorities are useful, and enable compact expression of semantically meaningful specifications. To ensure equational semantics, as in the case with ordinary (unprioritized) term rewriting, we require the properties of

- stability on instantiation (p rewrites to $q \Rightarrow p\sigma$ rewrites to $q\sigma$), and
- stability on replacement (p rewrites to $q \Rightarrow M[...p...]$ rewrites to $M[...q...]$).

Recent years have also seen considerable work on *Conditional* Term Rewriting Systems (CTRS) [BDJ, BK, Re, Ka1, KJ, Ga]. In particular, the work in [MS, Mo1] has emphasized the task of constructing well defined specifications using conditional rewriting systems with negation. In this paper, we present a first attempt to compare and combine these ideas with priority rewriting, yielding a powerful rewriting formalism.

In section 2, we present alternative definitions of priority rewriting, aiming to satisfy the requirements of equational reasoning. In section 3, we give an equational logical semantics for PRS, and criteria for soundness and completeness. Section 4 presents the restrictions on PRS needed for satisfying useful properties like modularity and ground confluence. Finally, we discuss the relation between priority and conditional rewriting mechanisms. Proofs of theorems, omitted for brevity, may be found in [Mo2].

2 Preliminaries

Definition 1 A Priority Rewrite System (PRS) $[R, >]$ consists of an 'underlying' TRS together with a partial ordering among the rewrite rules (called priority). $[R_1, >_1]$ is a sub-PRS of $[R_2, >_2]$ if $R_1 \subset R_2$ and $>_1$ is a restriction of $>_2$ to the rules in R_1.

Notation: We adopt standard notation and definitions for unexplained terms (see [HO, Kl]). "$p \equiv q$" denotes that p and q are syntactically identical. "$\mathcal{V}(t)$" denotes the set of variables in t; a term t is ground if $\mathcal{V}(t)$ is empty. We indicate that p is a subterm of M by writing $M[p]$, and $M[p/q]$ is the term resulting from the replacement of an occurrence of subterm p (in $M[p]$) by q. In the examples, the priority ordering is indicated by a downward arrow: each rule is of 'higher' priority than a lower rule to which it is connected by '$\downarrow$'. Rules may be labelled to allow convenient reference; in example 1 above, r_1 and r_2 are labels identifying the two rules, and $r_1 > r_2$ in the priority ordering ($>$) among these rules.

The priority ordering establishes the precedence in which applicable rules may be used for rewriting. It has been pointed out [BBK] that in the definition of rewriting using PRS, it is necessary to check that 'internal' reductions (of subterms) cannot be performed yielding a term reducible by a higher rule. This restriction is not sufficient to ensure stability on instantiation: it is possible that p reduces to q but $p\sigma$ does not reduce to $q\sigma$ for some substitution σ. For instance, if non-ground terms can be indiscriminately reduced using the rules in example 1 defining *eq*, we may reduce $eq(x, y)$ to F, although its instance $eq(x, x)$ reduces to T instead. Such a reduction must be avoided because $\forall x, y.[x \neq y]$ cannot be a logical consequence of any consistent (non-empty) theory. A default inequational assumption (reflecting that higher priority rules are inapplicable) is often made when a lower priority rule in a PRS is invoked. While ground inequational assumptions (*e.g.*, $a \neq b$) can be semantically meaningful (indicating that a and b denote distinct objects), non-ground assumptions like $x \neq y$ are neither meaningful nor useful. Hence we choose to allow rewriting of non-ground terms using only rules which are of the highest priority, modifying [BBK]'s definition of rewriting using a PRS.

Definition 2 A term p is P-reducible, and P-rewrites (or P-reduces) to $p[m/t\sigma]$ using a PRS $\Re$ (abbreviated $p \rightarrow_{\Re}^{P} p[m/t\sigma]$, with $\rightarrow_{\Re}^{P*}$ denoting reflexive transitive closure) if

- $r : s \rightarrow t$ is a highest priority rule in $\Re$ such that
- p contains a subterm m (the redex), such that $m \equiv s\sigma$ for some substitution σ;
- m is ground if the sub-PRS $\Re_r$ of rules (in $\Re$) higher than r is non-empty, and
- m has no P-reducible proper subterm n such that $n \rightarrow_{\Re}^{P*} n'$ and $p[m[n/n']] \rightarrow_{\Re_r}^{P} p'$ for any terms n', p'.

Using the PRS of example 1, for instance, $eq(F, eq(x, x))$ cannot be P-reduced at the outer occurrence of *eq* by either rule; and the priority ordering ($r_1 > r_2$) prohibits reduction at the inner occurrence of *eq* using rule r_2. The only possible reduction replaces $eq(x, x)$ by T using rule r_1; the resulting term $eq(F, T)$ can then be P-reduced to F using rule r_2.

One major problem with the above definition is that P-rewriting is not always decidable. Since performing a P-rewrite involves checking whether P-reduction of subterms yields a term P-reducible by a higher rule, the original step of P-rewriting may never finish due to non-termination of the P-rewriting sequences issuing from the subterms.

Example 2 An attempt to P-rewrite $f(g(a))$ neither succeeds nor fails, using the following PRS, although $f(g(a))$ matches with the *lhs* of the second rule, because of the need to predetermine that the subterm $g(a)$ does not have a reduct 'a' which makes the first rule applicable instead.

$$\Re_2 = \left\{ \left| \begin{array}{c} f(a) \rightarrow b \\ f(x) \rightarrow c \\ g(a) \rightarrow f(g(a)) \\ g(x) \rightarrow a \end{array} \right. \downarrow \right\}$$

P-rewriting is decidable when the TRS underlying a PRS is 'noetherian' (every rewriting sequence terminates) or 'bounded' (every rewriting sequence consists of finitely many terms) [BBK]. To enable decidability of each rewriting step using a PRS which is not bounded, we explore an alternative definition of rewriting using PRS, enforcing an innermost-first rewrite strategy. Another motivation for such an alternative definition is that P-rewriting is not stable on replacement, as illustrated below, unlike equational systems and conventional TRS.

Example 3

$$\Re_3 = \left\{ \begin{array}{l} r_1 : \ f(g(x)) \rightarrow s(x) \\ r_2 : \ g(h(x)) \rightarrow t(x) \end{array} \right\}$$

Here, $f(g(z)) \rightarrow^P_{\Re} s(z)$ which does not conflict with the result of P-rewriting $f(g(h(a)))$ to $s(h(a))$ (applying r_1); this indicates stability on instantiation. But P-rewriting $g(h(a))$ yields $t(a)$, whereas P-rewriting $f(g(h(a)))$ does not yield $f(t(a))$. This is a result of non-confluence of the underlying TRS: a non-trivial critical pair is obtained by superposing the *lhs*'s of the two rules. Despite non-confluence, whereas $f(g(h(a)))$ would still be reduced to $f(t(a))$ using the second rule in ordinary term rewriting, this is forbidden for P-rewriting by the priority ordering, violating stability on replacement. Another possible reason for lack of stability on replacement could be undecidability of P-rewriting, due to the underlying TRS being non-noetherian, as in example 2.

Proposition 1 Let $\Re$ be a PRS whose underlying TRS R is confluent and noetherian. Then any ground term reducible using R is also P-reducible using $\Re$; and for any p, q, if $p \rightarrow^P_{\Re} q$ then there are converging P-reduction sequences from (every) $M[p]$ and $M[p/q]$.

We thus obtain a sufficient condition for a property very similar to stability on replacement. If *lhs*'s of rules in R do not overlap at non-variable positions, then the lack of non-trivial critical pairs of the underlying TRS ensures that if $p \rightarrow^P_{\Re} q$, then <u>either</u> $M[p] \rightarrow^P_{\Re} M[p/q]$ <u>or</u> $\exists N$ such that $M[p] \rightarrow^P_{\Re} N$ and $M[p/q] \rightarrow^P_{\Re} N$. However, the premises of Proposition 1 are rather strong, and are not satisfied by simple and interesting PRS's like $\Re_1$ in example 1. We now give an alternative definition for rewriting using PRS, which is decidable and stable on replacement, irrespective of whether the underlying TRS is noetherian or confluent.

Definition 3 A term p is <u>I-reducible</u>, and <u>I-rewrites</u> (<u>I-reduces</u>) to q using a PRS $\Re$ if

- p contains a ground subterm (<u>redex</u>) m, no proper subterm of which is I-reducible;
- $\Re$ contains a rule $s \rightarrow t$, such that
- m matches s with a matching substitution σ (*i.e.*, $m \equiv s\sigma$),
- q is the result of replacing m in p by $t\sigma$ (*i.e.*, $q \equiv p[m/t\sigma]$), and
- m is not I-reducible by any rule in $\Re$ of higher priority than $s \rightarrow t$.

Notation: We abbreviate 'p I-rewrites to q using $\Re$' by '$p \rightarrow^I_{\Re} q$'. We denote both '$p \rightarrow^P_{\Re} q$' and '$p \rightarrow^I_{\Re} q$' ambiguously by "priority rewriting" or '$p \rightarrow_{\Re} q$'. The reflexive transitive closure of priority rewriting is indicated by '$\rightarrow^*_{\Re}$'; if $p \rightarrow^*_{\Re} q$, then we say q is a reduct of p. Since $\mathcal{V}(rhs) \subseteq \mathcal{V}(lhs)$ in each rewrite rule, ground terms have only ground reducts.

Unlike P-rewriting, each step of I-rewriting is decidable. Further, this definition enforces an innermost-first reduction strategy which ensures stability on replacement, unlike P-rewriting. However, we observe that the priority ordering on the rules now becomes effective only when comparing rules whose lhs's have identical outermost function symbols. Also, non-ground reductions are practically banned, in the interest of preserving the property of stability on instantiation. With the PRS in example 3, for instance, we have $f(g(h(a))) \rightarrow^I_{\Re_3} f(t(a))$, applying r_2 to the inner subterm $g(h(a))$, in spite of the contrary priority ordering ($r_1 > r_2$). So we refrain from I-rewriting the non-ground term $f(g(z))$ to $s(z)$, to avoid the situation that $p \rightarrow^I q$ but not $p\sigma \rightarrow^I q\sigma$.

While I-rewriting is decidable irrespective of whether the underlying TRS is noetherian, non-termination of reduction sequences can have other consequences. Plausible reductions may not be reachable, because the innermost reduction strategy is not fair, as in the following example.

Example 4

$$\Re_4 = \left\{ \left| \begin{array}{c} f(x) \rightarrow t \\ p \rightarrow q \\ q \rightarrow p \end{array} \right. \right\}$$

We find that $f(p) \rightarrow^I_{\Re_4} f(q) \rightarrow^I_{\Re_4} f(p) \rightarrow^I_{\Re_4} \cdots$, a non-terminating I-rewriting sequence which prohibits the reduction of $f(p)$ to t using the first rule. This contrasts with P-rewriting, where $f(p) \rightarrow^P_{\Re_4} t$, although the P-reduction sequence $p \rightarrow^P_{\Re_4} q \rightarrow^P_{\Re_4} p \rightarrow^P_{\Re_4} \cdots$ is not finite.

3 Semantics

In [BBK], a semantics is given for PRS with respect to a 'sound and complete set of rewrites'. This definition is operational in flavor: a logical semantics is preferable, which characterizes the theory (set of equational first order formulas) described by a PRS, and whose logical consequences we should expect to compute via the priority rewriting mechanism.

Term rewriting systems (without priorities) have an intended equational semantics; for example, $R = \{ p_1 \rightarrow q_1, \cdots, p_n \rightarrow q_n \}$ is intended to implement $\mathcal{E}(R)$ defined as $\{p_1 = q_1 \wedge \cdots \wedge p_n = q_n\}$. Soundness and completeness of rewriting with R is judged by asking whether rewriting sequences yield pairs of terms whose equalities are precisely the logical consequences of $\mathcal{E}(R)$. Similarly, we need to define the semantics of priority rewriting with a PRS $\Re$ by identifying the relevant equational formulas they represent, and finding conditions under which priority rewriting with $\Re$ yields all and only the equations which are logical consequences of these formulas.

We first observe that the equations obtained from rules by syntactically changing '$\rightarrow$' to '$=$' do not give a good description of the PRS, although every priority reduction is then automatically sound. For example, the equations obtained in this manner from

$$\begin{array}{|l} f(a) \rightarrow a \\ f(x) \rightarrow b \end{array}\downarrow$$

are $f(a) = a$ and $\forall x.f(x) = b$, implying $a = b$, which is not intended to be a consequence of this PRS. A better description of intended equational formulas is given below.

From a PRS $\Re$, we obtain a set (conjunction) of first order formulas $\mathcal{E}(\Re)$ as follows:

- If $p \rightarrow q \in \Re$ with no rule $\in \Re$ of higher priority, then $\mathcal{E}(\Re)$ contains the equation $p = q$ (with all variables universally quantified).
- Let $\{ r_1 : f(\tau_1) \rightarrow p_1, \cdots, r_n : f(\tau_n) \rightarrow p_n \}$ be the rules in $\Re$ whose *lhs* has outermost symbol f, and which are also of higher priority than $r_k : f(\tau_k) \rightarrow p_k$ (each τ_i is a term-tuple, and variables in different rules are renamed to be distinct); then, $\mathcal{E}(\Re)$ contains $\forall \mathcal{V}(\tau_k).[f(\tau_k) = p_k \vee [\exists \mathcal{V}(\tau_1).\tau_1 = \tau_k] \vee \cdots \vee [\exists \mathcal{V}(\tau_n).\tau_n = \tau_k]]$.

The first part of the above definition of $\mathcal{E}(\Re)$ reflects the fact that when there are no rules of higher priority, a rule $l \rightarrow r$ can be used for reducing any term matching l, and hence is assumed to denote just $l = r$. The second part of the definition intuitively means that if a term $f(\tau)$, with irreducible arguments, is priority reducible by a rule $f(\cdots) \rightarrow \cdots$, then there are no rules of higher priority defining f whose arguments can be considered equal to τ.

Example 5

$$\text{Let } \Re_5 = \left\{ \begin{array}{|c} f(nil) \rightarrow true \\ f(cons(x,x)) \rightarrow true \\ f(x) \rightarrow false. \end{array}\downarrow \right\}.$$

$$\text{Then } \mathcal{E}(\Re_5) = \left(\begin{array}{c} f(nil) = true \ \wedge \\ \forall x.[f(cons(x,x)) = true] \ \wedge \\ \forall x.[f(x) = false \ \vee \ \exists y.(x = cons(y,y))] \end{array} \right).$$

In addition to the above $\mathcal{E}(\Re)$ formulas, we need to make implicit default assumptions of inequations when performing reduction with a lower priority rule. For instance, in example 5 above, it is natural to reduce $f(cons(0, nil))$ to $false$ using the last rule in $\Re_5$, although $f(cons(0, nil)) = false$ is not a direct consequence of $\mathcal{E}(\Re_5)$, *i.e.*, there are models (E-interpretations) of $\mathcal{E}(\Re_5)$ in which 0 and *nil* are equal. More generally, non-Horn clauses of the kind that may be contained in $\mathcal{E}(\Re)$ do not have initial or minimal models, and using such clauses (*e.g.*, $p = q \vee r = s$) to infer a single literal (like $p = q$) is possible only if we also assume some negative literals (like $r \neq s$).

As in [MS], we seek a set of inequations Ψ with respect to which priority rewriting is "sound" and "complete"; these semantic notions are defined below for ground term reductions, since we prohibit reduction of non-ground terms in most cases.

Definition 4 Priority rewriting with $\Re$ is said to be

- sound w.r.t. Ψ iff $\forall$ ground $p, q.\ p \rightarrow_{\Re} q \ \Rightarrow\ \mathcal{E}(\Re) \cup \Psi \models p = q$; and
- complete w.r.t. Ψ iff $\forall$ ground $p, q.\ \mathcal{E}(\Re) \cup \Psi \models p = q \ \Rightarrow\ \exists t. p \rightarrow^*_{\Re} t \wedge q \rightarrow^*_{\Re} t$.

In other words, priority rewriting is 'complete' if two terms have a common reduct whenever their equality is a logical consequence of the theory (of $\mathcal{E}(\Re)$, together with inequations in Ψ). We now explore conditions under which priority rewriting is sound and complete, following a strategy described in [Mo1]. Note that ordinary (unprioritized) rewriting is always sound, and confluence of a TRS implies completeness w.r.t. any Ψ. In priority rewriting, unlike conditional rewriting, we do not explicitly attempt to prove (by checking possible convergence, not merely matching terms) that the arguments of a lower priority rule cannot equal those of a higher priority rule whose *lhs* has the same outermost function symbol. For example, given the PRS

$$\left\downarrow \begin{array}{l} f(f(x)) \rightarrow b \\ f(b) \rightarrow a \end{array}\right.$$

we find that $f(b)$ priority reduces to a, although $f(b) = a$ is not a logical consequence of

$$\overbrace{f(f(x)) = b \ \wedge [f(b) = a \vee \exists x. f(x) = b]}^{\mathcal{E}(\Re)} \wedge \overbrace{a \neq b \ \wedge \cdots}^{\Psi}$$

The need to avoid disproving existentially quantified disjuncts with non-constructor terms motivates the 'left-normalization' premise in the following theorems. Left-normalization ensures that *lhs*'s of rules do not contain arguments whose reduction may conflict with the inequational assumption (*e.g.*, $\forall x. f(x) \neq b$ above) made during reduction with a lower priority rule.

Definition 5 A PRS $\Re$ is left-normalized if for every rule $r : f(t_1, \ldots, t_n) \rightarrow rhs$ and for every instantiation σ mapping variables to irreducible ground terms, none of the instantiated *lhs*'s arguments $(t_1\sigma, \ldots, t_n\sigma)$ is priority reducible using $\Re$ (*i.e.*, no t_i is 'narrowable').

Theorem 2 Let $\Re$ be a left-normalized PRS such that every priority rewriting sequence from every ground term terminates. Then, priority rewriting with $\Re$ is sound and complete with respect to '$\Psi_{\Re}$', the set of inequations between every pair of irreducible ground terms.

This theorem does need the premise that all ground priority reduction sequences must be finite. For example, let $\Re$ be $\{a \rightarrow b,\ b \rightarrow a,\ p \rightarrow a,\ p \rightarrow c,\ c \rightarrow d,\ d \rightarrow c\}$, a non-terminating TRS; then $\Psi_\Re$ is empty. Although $\mathcal{E}(\Re) \models\ a = c$, terms a and c have no common reduct; hence (ground) reduction is not complete with respect to $\Psi_\Re$.

We also observe that $\mathcal{E}(\Re) \cup \Psi_\Re$ may be inconsistent if $\Re$ is not 'confluent', *i.e.*, if there are non-converging reduction sequences obtainable by reducing the same term. For example, let $\Re$ be the non-confluent TRS $\{r \rightarrow s, r \rightarrow t\}$; then $\mathcal{E}(\Re) \models\ s = t$, but $\Psi_\Re \models s \neq t$ since s, t are irreducible, hence $\Psi_\Re \cup \mathcal{E}(\Re)$ is inconsistent.

In practice, usefulness of the above theorem is also limited by the dependence of the definition of $\Psi_\Re$ on the notion of irreducibility using $\Re$. A more useful characterization of correctness would be with respect to an independently chosen set of inequations Ψ. From a specification point of view, a sublanguage of 'free' constructor terms presents such a useful candidate for Ψ. For example, specifications on a *List* data type assume that *nil* and every instance of $cons(x, y)$ are inequal. Then, Ψ will be the set of non-unifiable terms constructed using *nil* and *cons*. For soundness and completeness of ground priority rewriting with respect to arbitrary Ψ, we need a property similar to 'sufficient completeness' [GH], defined here for priority rewriting.

Definition 6 Let Ψ be a set of inequations between terms, and let $\Re$ be a PRS. Let $\tau(\Psi)$ be any set of terms such that $p \neq q \in \Psi$ for every distinct pair $p, q \in \tau(\Psi)$. For example, if Ψ is $\{0 \neq S^i(0)\} \cup \{0 \neq P^j(0)\} \cup \{P^j(0) \neq S^i(0)\}$, for integer specifications, where $0, P, S$ are constructors, then $\tau(\Psi)$ is $\{0\} \cup \{P^j(0)\} \cup \{S^i(0)\}$.

- Priority rewriting with $\Re$ is said to be sufficiently complete with respect to Ψ if there is some $\tau(\Psi)$ such that every ground term built using the function symbols occurring in $\Re$ has a reduct in $\tau(\Psi)$.

Theorem 3 Priority rewriting with a left-normalized PRS $\Re$ is sound and complete w.r.t. a set of inequations Ψ if priority rewriting with $\Re$ is sufficiently complete with respect to Ψ.

4 Properties of PRS

We now explore conditions which ensure that priority rewriting is sound and complete, in a framework where functions are partitioned into defined functions ($\in D$) and constructors ($\in C$). We say that a rule whose *lhs* has f as outermost symbol defines f; a constructor term is one in which defined functions do not occur; defined functions occur in non-constructor terms. When constructors are unrelated or 'free', Ψ above (for soundness and completeness) is naturally chosen to be Ψ_C, the set of inequations between distinct ground constructor terms. As in the case of conditional specifications [MS], termination of ground reduction, ground confluence and reducibility of every ground non-constructor term are sufficient to ensure that each ground term has a unique constructor term reduct, thereby assuring sufficient completeness, hence soundness and completeness w.r.t. Ψ_C.

Termination (the 'noetherian' property) of the underlying TRS guarantees that priority rewriting also terminates. Methods to show termination of TRS involve proving the necessary and sufficient condition that there is a well-founded, stable (on instantiation and replacement) ordering among terms, in which every instance of the *lhs* of each rule is greater than the corresponding instance of the *rhs* of the rule [De]. We note that this is not a necessary condition for priority rewriting to terminate; for example, priority rewriting using

$$\Re_6 = \left\{ \begin{array}{|l} p \rightarrow q \\ p \rightarrow p \end{array} \right\}$$

terminates, since the first rule is of higher priority than the second, although the underlying TRS is not noetherian. The following is an appropriate modification to the termination criterion: priority rewriting with a PRS terminates if there is a well-founded stable ordering ">" among terms such that for every rule $l \rightarrow r$ and every instantiation σ, we have either $l\sigma > r\sigma$ or there is a higher priority rule $L \rightarrow R$ such that $l\sigma$ is an instance of L.

Ground confluence can be assured by avoiding 'top-level' overlaps between *lhs*'s of rules, giving higher priority to one member of each subset of rules with unifiable *lhs*'s, and ensuring that non-unifiable *lhs*'s of rules cannot be equal. These conditions also enforce modularity of function definitions and specifications in a natural way. For modularity, a PRS specification should consist of disjoint sub-PRS's separately defining each function for all possible constructor term arguments.

Definition 7 A Definitional PRS (or DPRS) is a PRS in which the *lhs* of each rule is of the form "$f(\tau)$", where f is not a constructor and τ is a tuple of constructor terms.

The sets of ground irreducible terms ("normal forms") obtained using a DPRS $\Re$ coincide for both definitions ($\rightarrow^I, \rightarrow^P$) of priority rewriting, if the TRS underlying $\Re$ is noetherian. The following proposition also holds for either definition of priority rewriting.

Proposition 4 A DPRS $\Re$ is ground confluent if the following premise is satisfied: (Premise:) for every pair of rules $r_1 : f(\tau_1) \rightarrow t_1$ and $r_2 : f(\tau_2) \rightarrow t_2$ whose *lhs*'s unify with m.g.u. σ, one of the following conditions holds:
- one of the rules r_1 or r_2 has a higher priority than the other;
- $f(\tau_1)\sigma$ is an instance of the *lhs* of another rule of higher priority than r_1 (or r_2);
- $t_1\sigma$ and $t_2\sigma$ are identical.

For sufficient completeness w.r.t. Ψ_C, the final requirement is that every ground non-constructor term must be reducible (cf. 'inductive reducibility', 'totality' [JK, Ko, MS]). The simplest way to ensure that a function is defined for every possible constructor term argument is to have an 'otherwise'-case, a rule whose *lhs* has only variable arguments. Unlike ordinary TRS with such rules, we can avoid disturbing or complicating the ground confluence requirement: a rule like $f(x_1, \ldots, x_n) \to t$ is ordered to have a priority lower than every other rule defining the same function f. In the absence of such a rule, it is necessary to use techniques similar to those of [Th, Ko] to ensure that each function definition is total.

Since every DPRS is left-normalized, a combination of the above conditions assures soundness and completeness w.r.t. the set of inequations between ground constructor terms.

Proposition 5 If a DPRS $\Re$ satisfies the premise of proposition 4, its underlying TRS is noetherian, and for each defined function f, $\Re$ contains a rule whose *lhs* is $f(x_1, \ldots, x_n)$, then priority rewriting with $\Re$ is sound and complete w.r.t. Ψ_C.

It is relatively easy to account for PRS specifications with non-free constructors. For example, we can give specifications using integers (with constructors $0, P, S$) which contain the rules $S(P(x)) \to x$, $P(S(x)) \to x$ with priority higher than every other rule; for the semantics, we choose Ψ to be $\{0 \neq S^i(0)\} \cup \{0 \neq P^j(0)\} \cup \{P^j(0) \neq S^i(0)\}$, instead of the set of inequations between ground constructor terms.

Proposition 6 Let $\Re_1$ be a DPRS which satisfies the premise of proposition 4. Let $\Re_0$ be a PRS in which both *lhs* and *rhs* of each rule are constructor terms, using which every ground constructor term t can reduce to a unique irreducible reduct or *normal form* $\mathcal{N}(t)$. Let $\Re = \Re_0 \uplus \Re_1$, extending the priority orderings of $\Re_0$ and $\Re_1$, with each rule in $\Re_0$ having priority higher than each rule in $\Re_1$. **Then:**

- $\Re$ is ground confluent; and
- **if** every ground non-constructor term is reducible, and every priority reduction sequence terminates, **then**
  - every ground term has a unique reduct $\in \aleph$,
  the set of all normal forms of ground constructor terms; and
  - priority reduction using $\Re$ is sound and complete with respect to $\Psi_\aleph$,
  the set of all inequations between distinct terms $\in \aleph$.

5 Priority and Conditional Systems

Since PRS do not have initial algebra semantics, [BBK] point out that PRS are a strict generalization of Conditional TRS whose rules are accompanied by antecedents consisting of conjunctions of equations. PRS also appear to generalize 'Equational-Inequational-Conditional TRS' (EI-CTRS) [MS, Ka2] in which the antecedent of each rewrite rule consists of a conjunction of equations and inequations. The operational use of PRS implicitly uses negated preconditions in rules of lower priority: $\Re_1$ of example 1 is similar to the unordered CTRS $\{ eq(x,x) \rightarrow T, \quad x \neq y \Rightarrow eq(x,y) \rightarrow F \}$. By using $\Re_1$, together with the PRS defining an '*if*' predicate, it is possible to avoid conditionals in rules altogether. For example, a conditional rule like

$$p_1 = q_1 \wedge \cdots \wedge p_m = q_m \wedge r_1 \neq s_1 \wedge \cdots \wedge r_n \neq s_n \Rightarrow \lambda \rightarrow \rho$$

can be simulated by the following PRS, with each rule r_i having a higher priority than r_{i+1}:

$$\begin{array}{ll} r_1 : not(T) \rightarrow F & r_5 : neq(x,y) \rightarrow not(eq(x,y)) \\ r_2 : not(F) \rightarrow T & r_6 : and(T,T) \rightarrow T \\ r_3 : eq(x,x) \rightarrow T & r_7 : and(x,y) \rightarrow F \\ r_4 : eq(x,y) \rightarrow F & r_8 : if(T,x) \rightarrow x \end{array}$$

$r_9 : \lambda \rightarrow if(and[eq(p_1,q_1), and(eq(p_2,q_2) \cdots and(neq(r_1,s_1), \cdots, neq(r_n,s_n) \cdots)], \rho)$

In general, this is not an accurate simulation of the original conditional rewrite rule in the following sense: the last (least priority) rule would apply (λ can always be rewritten), even when the antecedent or if-condition does not hold. Whereas we would not reduce a term using the corresponding conditional rewrite rule, the PRS yields an irreducible term "$if(\cdots,\rho)$". Hence CTRS are not always accurately translatable to PRS. However, if an EI-CTRS R satisfies the property that every reduction sequence from any given ground term t is finite and yields the same irreducible constructor term $\mathcal{N}_R(t)$, then a PRS $\mathcal{P}(R)$ can be constructed which is equivalent to R in the sense that priority reduction sequences from any ground term t using $\mathcal{P}(R)$ terminate yielding the same normal form $\mathcal{N}_R(t)$ as produced using R.

Conversely, priority rewrite systems do not (in general) have equivalent unprioritized conditional rewrite systems. For example, the following PRS

$$\downarrow \; \begin{array}{l} f(c(x,y)) \rightarrow s \\ f(z) \rightarrow t \end{array}$$

cannot be easily specified using a CTRS, since we cannot have a rule to help us check and decide whether $\exists x,y.[c(x,y) = p]$ and accordingly reduce $f(p)$. This is because, traditionally, 'logical' variables not occurring in the *lhs* of a conditional rule are not allowed to occur in the antecedent. However, equivalent EI-CTRS formulations can be obtained for those PRS which satisfy the stringent restriction that whenever we have $r_1 : f(\overline{s}) \rightarrow p \; > \; r_2 : g(\overline{t}) \rightarrow q$ in the priority ordering for any two rules, we must have $f \equiv g$, and $\overline{s}$ must contain only variables and ground terms.

Hence, and for conciseness of specification, we propose to incorporate both priorities and conditionals into a new formalism, *viz., Priority Conditional Rewriting Systems (PCRS).* The associated rewriting mechanism is a combination of priority rewriting and conditional rewriting. Due to the fact that $p \neq q$ can be represented by $eq(p,q) = F$ using a sub-PRS ($\Re_1$ of example 1), antecedents of rules in PCRS need contain only equations. We assume then that the rules defining eq have higher priority than all other rules in the PCRS.

Definition 8 A PCRS is a set of conditional rewrite rules

$$\{r:\ p_1 = q_1 \wedge \cdots \wedge p_n = q_n \Rightarrow L \rightarrow R \mid \mathcal{V}(\{p_1, q_1, \cdots, p_n, q_n, R\}) \subseteq \mathcal{V}(L)\}$$

with a priority ordering among the rules. A term p rewrites to q using a PCRS $\Re$ if

- $\Re$ contains a rule $r:\ p_1 = q_1 \wedge \cdots \wedge p_n = q_n \Rightarrow L \rightarrow R$ such that
- p has a subterm M which matches L with a substitution σ, and $q \equiv p[L\sigma \ / \ R\sigma]$; and
- there are finite converging reduction sequences (using $\Re$) from the instantiated terms $(p_i\sigma, q_i\sigma)$ in each conjunct of the antecedent;
- and {we distinguish between the two possible definitions}
  let $\Re_r \subset \Re$ be the sub-PCRS consisting only of rules in $\Re$ higher than r; then
  - Case $(\rightarrow^P_\Re)$: if $\Re_r$ is non-empty, then M is a ground term with no proper subterm N such that $N \rightarrow^{P*}_{\Re_r} N'$ and $p[M[N/N']] \rightarrow^P_{\Re_r} p'$ for any p', N';
  - Case $(\rightarrow^I_\Re)$: no proper subterm of M is reducible using $\Re$, and M is ground;
- $\Re$ contains no rule of higher priority than r which satisfies the above conditions.

The semantics of PCRS can be given in a manner closely following that given for PRS in section 3. Note that $[eq(x,x) = T]$ and $[x = y \vee eq(x,y) = F]$ need to be included in $\mathcal{E}(\Re)$, as also the antecedent of each formula denoting a rule. As before, in a language with functions partitioned into free constructors and defined functions of fixed arity, the semantics of PCRS specifications may be determined with respect to Ψ_C, which stipulates that non-unifiable constructor terms cannot be equal. From Ψ_C, it follows that a pair of terms $f(s_1, \cdots, s_n)$, $g(t_1, \cdots, t_m)$ are inequal if they are constructor-wise distinct, *i.e.*, if

- f and g are distinct constructors; or
- f and g are identical constructors, but $\exists i.\ s_i$ and t_i are constructor-wise distinct.

Priority conditional rewriting terminates if conditional rewriting with the underlying CTRS terminates, *e.g.*, if terms in the antecedent and *rhs* of each rule are smaller than the *lhs* in some well-founded stable ordering.

Proposition 7 A PCRS is ground confluent if the outermost symbol is the only constructor in the *lhs* of each rule, and for every pair of rules $r_1 :\ C_1 \Rightarrow L_1 \rightarrow R_1$, $r_2 :\ C_2 \Rightarrow L_2 \rightarrow R_2$ whose *lhs*'s unify with m.g.u. σ, one of the following conditions holds:

- $r_1 > r_2$ or $r_2 > r_1$ in the priority ordering;

- $L_1\sigma$ is an instance of the *lhs* of another rule of higher priority than r_1 (or r_2);
- $R_1\sigma$ and $R_2\sigma$ are identical;
- $\exists eq(p,q) = F \in C_1 \cup C_2$ such that $p\sigma, q\sigma$ are identical;
- $\exists eq(p,q) = T$ or $p = q \in C_1 \cup C_2$ such that $p\sigma$ and $q\sigma$ are constructor-wise distinct;
- $\exists eq(p,q) = T$ or $p = q$ as well as $eq(P,Q) = F \in C_1 \cup C_2$ such that $P\sigma, Q\sigma$ are identical upto mutual replacement of some occurrences of their subterms $p\sigma, q\sigma$;
- $C_1 \cup C_2$ contains $eq(p,q) = T$ or $p = q$, as well as $eq(P,Q) = T$ or $P = Q$ or $eq(Q,P) = T$ or $Q = P$, such that $P\sigma, p\sigma$ are identical, but $Q\sigma, q\sigma$ are constructor-wise distinct.

Similarly, conditions may be given ensuring that each function is defined by a PCRS for every possible ground constructor term tuple of arguments; these would comprise those given for CTRS [MS], with a possible "otherwise" rule for each defined function f, which is a lowermost priority rule of the form $f(x_1 \cdots x_n) \rightarrow rhs$.

6 Summary

In this paper, we have studied priority rewriting, *i.e.*, term rewriting with rewrite rules partially ordered by a 'priority' ordering. We have suggested modifications to the definition of priority rewriting, aimed at making possible a logical semantics following the traditional idea that rewriting is essentially simplification in a theory described by formulas in first order logic with equality. We have given such a logical semantics, which resembles a similar approach to the semantics of conditional rewriting with negation. We briefly discussed restrictions needed for soundness and completeness of priority rewriting, focusing on a sufficient condition for ground confluence. Finally, we have compared priority rewriting and conditional rewriting, and suggested a new combination (PCRS), a powerful formalism which extends PRS and CTRS.

References

[BBK] J.C.M.Baeten, J.A.Bergstra, J.W.Klop, *Term Rewriting Systems with Priorities*, Proc. II Conf. Rewriting Techniques and Applications, France, Springer-Verlag LNCS 256, 1987, pp83-94.

[BK] J.A.Bergstra, J.W.Klop, *Conditional Rewrite Rules : Confluency and Termination* Res. Rep. IW 198/82, Center for Math. and Comp. Sci., Amsterdam, 1982 (also in JCSS 32, pp.323-362, 1986).

[BDJ] D.Brand, J.A.Darringer, W.Joyner, *Completeness of Conditional Reductions*, Res. Rep. RC7404, IBM T.J.Watson Res. Center, Yorktown Heights (NY), 1978.

[De] N. Dershowitz, *Termination of Rewriting,* Proc. First Int'l. Conf. on Rewriting Techniques and Applications, Dijon (France), 1985, Springer-Verlag LNCS 202, pp180-224.

[Ga] H.Ganzinger, *Ground term confluence in parametric conditional equational specifications,* Proc. STACS 87, pp286-298, 1987.

[GH] J.V.Guttag, J.J.Horning, *The Algebraic Specification of Abstract Data Types,* Acta Informatica **10,** 1978, pp27-52.

[HO] G.Huet, D.S.Oppen, *Equations and Rewrite Rules: A Survey,* in **Formal Languages: Perspectives and Open Problems,** R.Book (ed.), Academic Press, 1980, pp349-405.

[JK] J.-P.Jouannaud, E.Kounalis, *Proofs by Induction in Equational Theories without Constructors,* Symp. on Logic in C.S., Cambridge (Mass.), USA, 1986, pp358-366.

[Ka1] S.Kaplan, *Conditional Rewrite Rules,* Theoretical Computer Science **33**, 1984, pp175-193.

[Ka2] S.Kaplan, *Positive/Negative Conditional Rewriting,* Proc. First Int'l. Workshop on Conditional Term Rewriting Systems, Orsay (France), Springer-Verlag LNCS 308, 1988, pp129-143.

[KJ] S.Kaplan, J.-P.Jouannaud (eds.), *Proc. First Int'l. Workshop on Conditional Term Rewriting Systems,* Orsay (France), Springer-Verlag LNCS 308, 1988.

[Kl] J.W. Klop, *Term Rewriting Systems: a tutorial,* Rep. IR 126, Centre for Math. & C.S., Amsterdam, 1987.

[Ko] E.Kounalis, *Completeness in Data Type Specifications,* Res. Rep. 84-R-92, C.R.I.N., Nancy (France), 1984.

[Mo1] C.K.Mohan, *Negation in Equational Reasoning and Conditional Specifications,* Ph.D. Thesis, State University of New York at Stony Brook, 1988.

[Mo2] C.K.Mohan, *Priority Rewriting: Semantics, Confluence and Conditionals,* Tech.Rep. CIS-88-6, Syracuse University, Nov. 1988.

[MS] C.K.Mohan, M.K.Srivas, *Conditional Specifications with Inequational Assumptions,* Proc. First Int'l. Workshop on Conditional Term Rewriting Systems, Orsay (France), Springer-Verlag LNCS 308, 1988, pp161-178.

[Re] J.-L.Remy, *Etudes des systemes de reecriture conditionnels et application aux types abstraits algebriques,* These d'Etat, Nancy (France), 1982.

[Th] J.J.Thiel, *Stop Losing Sleep over Incomplete Specifications,* Proc. 11th ACM Symp. on Princ. of Prog. Lang., 1983.

Negation with Logical Variables in Conditional Rewriting

Chilukuri K. Mohan
School of Computer & Info. Sci.
Syracuse University
Syracuse, NY 13244-4100
mohan@top.cis.syr.edu

Mandayam K. Srivas
Odyssey Research Associates, Inc.
301A Harris B. Dates Drive
Ithaca, NY 14850-1313
oravax!srivas@cu-arpa.cs.cornell.edu

ABSTRACT: We give a general formalism for conditional rewriting, with systems containing conditional rules whose antecedents contain literals to be shown satisfiable and/or unsatisfiable. We explore semantic issues, addressing when the associated operational rewriting mechanism is sound and complete. We then give restrictions on the formalism which enable us to construct useful and meaningful specifications using the proposed operational mechanism.

1. Introduction

Conditional Term Rewriting Systems (CTRS) are a generalization of term rewriting systems (TRS), proposed as a means of having conditionals while preserving the localized declarative semantics. In CTRS, each rewrite rule is associated with an *antecedent*, which has to be shown to 'hold' before rewriting can be performed. A CTRS formalism in which antecedents of rules contain '=' and '≠' (equality and its negation) has been explored in [MS2, Ka2]. In another direction extending classical rewriting, it is illustrated in [DP] that conditional rewriting can be a meaningful notion even when the *lhs*'s of rules do not exclusively determine the bindings for all variables in the rule. Such 'logical' variables distinguish relational logic programs from functional and rewrite programs. Combining both these extensions of traditional rewriting is very useful, and yields a powerful language for writing concise and meaningful specifications. This paper is a detailed study of such an extension to CTRS, which we call *Sat-Unsat-CTRS* (SU-CTRS).

In the rest of this section, we present motivation and discuss related work. SU-CTRS and the associated operational mechanisms are described and illustrated in section 2. Section 3 contains a study of the semantics of these mechanisms. Section 4 gives restrictions to help construct meaningful 'fully-defined' specifications using SU-CTRS. Concluding remarks are given in section 5. Omitted proofs of theorems may be found in [Mo].

1.1. Motivation

Recently, there have been several attempts [DL, DP, GM] made to combine term rewriting techniques (viewed as adhering to a 'functional' programming style) with 'relational' logic programming (*a la* Prolog [Kow]). SU-CTRS, the formalism we propose, is an exploration in this direction; we include conditionals, negation, as well as 'logical variables' which distinguish relational from functional programs. In this paper, we focus our attention to investigating the use of SU-CTRS for building complete function and data type specifications. We show that when certain sufficient conditions are satisfied, it is possible to build usable and meaningful specifications with conditional rewrite rules in which literals in the antecedent are to be proved satisfiable and/or unsatisfiable. In general, checking satisfiability is semi-decidable while unsatisfiability is even more intractable; hence the need to impose restrictions which ensure that rewriting is decidable, and that satisfiability or unification w.r.t. an equational theory can be accomplished by finding instances of terms with converging reduction sequences.

The motivation for the formalism is illustrated by the following example. Suppose we wish to specify a function Filter which must pass on certain lists (whose first element is also a list) unaltered, but map objects of every other kind or sort to the constant "**err**". The first requirement of the function can be easily expressed by the following rewrite rule:

$$Filter(cons(cons(u,v),z)) \rightarrow cons(cons(u,v),z) \qquad \text{.......................(1)}$$

But in earlier CTRS formalisms without logical variables, the formulation of conditional rewrite rules complementing the above is not straightforward when there are many other possible constructors different from *nil* and *cons* (not known when *Filter* is specified). We overcome this problem by using a rule like

$$\textbf{unsat}\ [x=cons(cons(u,v),z)]::Filter(x) \rightarrow \textbf{err} \qquad \text{.......................(2)}$$

which falls within the scope of the new formalism. The intention of this rule is to say that *Filter* evaluates to **err** if its argument does not equal (evaluate to) any term with the structure $cons(cons(\_,\_),\_)$. The new rule (2) above is to be interpreted to mean:

$$\forall x.\left[\textbf{if}\ [x=cons(cons(u,v),z)\ \textit{is unsatisfiable}\ ]\ \ \textbf{then}\ \ [Filter(x)\ \textit{rewrites to}\ \textbf{err}\ ]\right].$$

The original rule (1) can be applied when, for some u,v,z, the argument of *Filter* is reducible or provably equal to $cons(cons(u,v),z)$. The new rule (2) is useful in the opposite case, when there is no $cons(cons(u,v),z)$ to which the argument of *Filter* is reducible or provably equal. The above specification of *Filter* continues to be applicable even when new data types are added to the language. Unlike priority rewrite systems (PRS) [BBK], it is possible to have partial specifications; for instance, an SU-CTRS may consist of rule (2) alone, which cannot be separately specified using an unconditional PRS. Also, when antecedents of rules contain more complex non-constructor terms, there is no straightforward way to obtain an equivalent PRS. Conditional rewriting occurs only after ensuring that the antecedent of a rule holds, while rewriting to some intermediate term occurs for unconditional rewriting using a defined *if–then* predicate.

Example 1: The two rules (1,2) mentioned above constitute an SU-CTRS. To 'SU-rewrite' a term like $Filter(cons(a,cons(0,nil)))$ which unifies with the *lhs* of rule (2) (but not (1)), we need to show that $cons(a,cons(0,nil))=cons(cons(u,v),z)$ is unsatisfiable. In general, this is done by repeatedly 'narrowing' this pair of terms (using all applicable rewrite rules) to show that no unifiable pair results. In the present example, neither of the terms is narrowable, and they do not unify, allowing us to conclude that there is no substitution which can instantiate $cons(a, cons(0, nil))$ and $cons(cons(u,v),z)$ to equal terms. Hence SU-rewriting does occur, replacing the instance of the *lhs* by the corresponding *rhs* ('**err**' in this case). The conclusion $[Filter(cons(a, cons(0, nil)))=\mathbf{err}]$ is a logical consequence of the following formulas obtained from the rewrite rules

$$\forall u \forall v \forall z. Filter(cons(cons(u,v),z))=cons(cons(u,v),z),$$
$$\forall x. \left[\forall u \forall v \forall z.\ x \neq cons(cons(u,v),z)\right] \supset Filter(x)=\mathbf{err},$$

<u>together with</u> (universally quantified) inequational assumptions

$$\{a \neq cons(u,v),\ 0 \neq cons(u,v),\ a \neq 0,\ nil \neq a,\ nil \neq cons(u,v), \cdots \}.$$

This illustrates the semantics of SU-rewriting: we deduce equations which are logical consequences of the given rules supplemented with a certain set of inequational assumptions.

The above SU-CTRS is a 'fully-defined' specification of *Filter*, *i.e.*, the following useful properties are satisfied, ensuring that every invocation of *Filter* on a ground term argument t reduces to a term in which '*Filter*' does not occur (t or **err** in this case):

(a) Reduction of each ground term $Filter(t)$ terminates, despite the presence of logical variables u, v, z in the antecedent.

(b) The two rules defining *Filter* never match the same redex: although the *lhs*'s unify, the instantiated antecedent of the second rule does not hold for such overlapping cases.

(c) Every ground term $Filter(t)$ is reducible by one of the two rules.

1.2. Related Work

Conditional Term Rewriting Systems (CTRS) were first studied in [BDJ, La2, BK, PEE] where some important issues were raised. Remy [Rem], Zhang [ZR] and Navarro [Na] developed the theory and techniques of contextual rewriting and hierarchical CTRS. For CTRS in which antecedents of rules are conjunctions of equations, Kaplan [Ka1] and Jouannaud & Waldmann [JW] have imposed restrictions under which conditional rewriting is "well-behaved". Various recent results were reported in the First International Workshop on CTRS [KJ], including 'Equational-Inequational-CTRS' [MS2, Ka2], which introduce the negation of equality into conditional rules, and are generalized in this paper. SU-CTRS were first proposed in [MS1], where the expressive power of such a formalism was compared to unconditional term rewriting and applicative programming, and the notion of a specification being 'fully-defined' was informally described.

Most formalisms combining functional and relational logic programming [DL, DP, Fr, GM, Red, Li] are more restrictive than SU-CTRS in that positive literals can occur in antecedents of rules, but negative literals are generally not allowed. Our treatment of negation resembles that of "negation as failure" in Prolog [Rei, Cl] rather than explicit deduction of inequations. This is because we are concerned with the [un]satisfiability of equations w.r.t. an equational theory, rather than just showing that two terms have syntactically distinct instances [Co]; proving/solving inequations in a theory containing equations and inequations is in general inefficient and semi-decidable [MSK]. An additional reason is that we may intend to have infinitely many inequational assumptions (*e.g.*, all ground constructor terms are inequal) which are not axiomatized and used explicitly in proofs. If efficient and available, direct methods to prove literals are preferable, and may be used in conjunction with other rewriting mechanisms.

Operationally, we use 'narrowing' and 'T-unification' techniques, *a la* [DP], to check antecedents of conditional rules. The T-Unification of a pair of terms s, t is the problem of finding a substitution σ such that $T \models (s\sigma = t\sigma)$, where T is a theory or set of formulas. Narrowing [Sl, La1] can be (semi-decidably) used to T-unify terms when the theory is described by a confluent and noetherian TRS [Fa, Hul]. For termination of narrowing, a sufficient condition is the basic narrowing criterion [Hul], refined in [Ret], which is satisfied by a TRS iff it is canonical and every *basic* narrowing sequence from the *rhs* of every rule terminates. Intuitively, a narrowing is 'basic' if it affects terms other than those which can be introduced by mere instantiation, rather than narrowing. Examples that satisfy this criterion are canonical TRS's in which all subterms of the *rhs* are variables, ground terms, and terms whose function symbols occur in the *lhs*'s of only those rules that also satisfy this criterion. We use this criterion to ensure that antecedent checking is decidable; any other criterion for decidability of T-unification can equally well be used.

Although the examples given in this paper look simple enough, SU-CTRS cannot easily be subsumed by other mechanisms. The presence of logical variables is not redundant: for example, the formalisms of [MS2, Ka2] cannot easily express the equivalent of **unsat** $[x=cons(cons(u,v),z)]::Filter(x) \rightarrow$ **err** . In fact, if non-equality literals are allowed in SU-CTRS, the SU-rule mentioned above is to be interpreted differently from **sat** $[x \neq cons(cons(u,v),z)]::Filter(x) \rightarrow$ **err** (where some substitution has to be found for u,v,z to make the inequation hold). On a different note, we also observe that CTRS cannot easily be replaced by equivalent priority rewrite systems [BBK]. Unlike CTRS, priorities cannot always ensure that the antecedent is evaluated before the rest of the rule; with 'equivalent' unconditional systems, rewriting to an intermediate form occurs even if the antecedent does not evaluate to "true" [MS1, Mo2].

2. Sat-Unsat-Conditional Term Rewriting Systems

For standard notation and definitions, the reader is referred to [HO, CL]. By $s \equiv t$, we mean that s and t are syntactically identical; conversely, $s \not\equiv t$ denotes that s,t are not identical. The set of variables occurring in t is denoted by $V(t)$.

Definition: A *Sat-Unsat-Conditional Term Rewriting System (SU-CTRS)* is a finite set of 'Sat-Unsat-Conditional rules' (SU-rules), of the form

$$\textbf{sat } [p_1=q_1, \ldots, p_m=q_m] \textbf{unsat } [s_1=t_1, \ldots, s_n=t_n]::lhs \rightarrow rhs, \quad \text{where}$$

(a) each of p_i, q_i, s_j, t_j, rhs may contain variables not occurring in lhs; but
(b) every variable in rhs must occur either in lhs or the **sat** part of the antecedent, i.e., $V(rhs) \subseteq V(\{lhs, p_1, \cdots, p_m, q_1, \cdots, q_m\})$.

For example, **sat** $[f(x)=g(z)]$**unsat** $[f(y)=h(x,z)]$ :: $l(x) \rightarrow r(z)$ is an SU-rule according to the above definition, but not **unsat** $[f(y)=h(x)]$ :: $l(x) \rightarrow r(y)$. In the operational use of SU-CTRS, the term to be rewritten instantiates variables of the lhs. During reduction, the intention is to solve for variables in the **sat** -part of the antecedent, such that the **unsat** -part is unsatisfiable. So the variables in the **sat** -part are instantiated during reduction and may be used in the rhs to yield the final result. But variables which occur only in the **unsat** -part are not instantiated during reduction, hence these cannot occur in the rhs.

The rule **sat** $[P_1, \ldots, P_m]$**unsat** $[N_1, \ldots, N_n]$::$lhs \rightarrow rhs$ is intended to denote

$$\forall \bar{z}. \forall \bar{x}. \Big[[(P_1 \wedge \cdots \wedge P_m) \wedge \forall \bar{y}. (\neg N_1 \wedge \cdots \wedge \neg N_n)] \supset (lhs=rhs) \Big]$$

where $\bar{z} \equiv V(lhs)$, $\bar{x} \equiv V(\bigcup_i \{P_i\}) - V(lhs)$, and $\bar{y} \equiv V(\bigcup_j \{N_j\}) - V(lhs) - V(\bigcup_i \{P_i\})$.

We refer to the variables $\bar{x}, \bar{y} \notin V(lhs)$ as logical variables; $\bar{x}$ as **sat** -variables; and $\bar{y}$ as **unsat** -variables.

The universal quantification ($\forall \bar{x}$) of the **sat** -variables may seem counter-intuitive to the idea of solving for (satisfying) them. But the formula, taken as a whole, is an assertion that for any instantiation (of $\bar{z}, \bar{x}$), if the antecedent holds, then so does the consequent ($lhs=rhs$). When the conditional rule is used for simplifying terms, however, we have to find some instantiation for the **sat** -variables $\in \bar{x}$ such that the antecedent holds. This is the reason for using the mnemonic '**sat**' in the rule. But none of the literals in the **unsat** -part should be satisfiable: $\forall \bar{y}. (\neg N_1 \wedge \cdots \wedge \neg N_n)$ is equivalent to $\neg \exists \bar{y}. (N_1 \vee \cdots \vee N_n)$.

It is easy to build complete specifications, where desirable, using rules of the form given above, with complementary antecedents. For example, the three rules

$$\begin{array}{rcl} \mathbf{sat}\ [P_1, \ldots, P_m]\mathbf{unsat}\ [N_1, \ldots, N_n] & :: & lhs \rightarrow rhs_0 \\ \mathbf{unsat}\ [P_1, \ldots, P_m] & :: & lhs \rightarrow rhs_1 \\ \mathbf{sat}\ [P_1, \ldots, P_m, N_1, \ldots, N_n] & :: & lhs \rightarrow rhs_2 \end{array}$$

span all possible cases of the antecedent, and every ground instance of lhs should be reducible using one of these rules, assuming we can decide whether each antecedent holds.

2.1. SU-Rewriting

For rewriting with an SU-CTRS, we check whether or not literals (in the antecedent of a rule) are satisfiable using an *SU-Unification* procedure, a modified version (cf. [Kal, Hus]) of Hullot's T-Unification procedure [Hul]. To SU-unify s,t, we examine narrowing sequences from s,t to determine whether there is a unifiable pair of terms to which s,t

narrow using compatible substitutions. SU-Unification is described in section 2.2; we now describe <u>*SU–rewriting*</u> (or <u>SU-reduction</u>).

If a term Q is to be SU-Reduced, we must first find a subterm M of Q that matches with the *lhs* of some rule of the SU-CTRS. Then, to check if the instantiated antecedent **sat** $[P_1 \cdots P_m]$**unsat** $[N_1 \cdots N_n]$ holds, we first attempt to satisfy literals in **sat** $[P_1 \cdots P_m]$ by jointly SU-unifying terms equated by each P_i, without instantiating variables in M. If this succeeds, each literal in **unsat** $[N_1 \cdots N_n]$ must then be proved unsatisfiable, by showing that there is no joint SU-unifier for the terms equated by each N_i. If this can be done, SU-rewriting is completed by replacing (in Q, the given term) the subterm M, which matched with the *lhs* of the reducing rule, by the appropriately instantiated *rhs*. Variables in M are not instantiated while checking the **sat** -part, but their instantiations are considered while checking the **unsat** -part, because if M can be reduced, then so must every instance of M. A non-ground term may result from SU-rewriting a ground term, because the *rhs* of a rule may contain logical variables.

A more precise description of SU-Reduction is given below. We abbreviate 's SU-Reduces to t using R' by $s \underset{R}{\rightarrow} t$, and denote by $\underset{R}{\rightarrow}^*$ the relexive transitive closure of $\underset{R}{\rightarrow}$. For convenience, we assume that variables of different rules are first renamed to be distinct from each other and from those in the term to be reduced.

(1) A rule **sat** $[P_1, \ldots, P_m]$**unsat** $[N_1, \ldots, N_n]$::$lhs \rightarrow rhs$ is chosen so that *lhs* matches with some subterm t of the given term, i.e., $t \equiv (lhs)\sigma_0$, for some substitution σ_0.

- If no such rule is found, conclude that the given term is irreducible.
- If found, apply matching substitution σ_0 to the rest of the rule.

(2) Pairs of terms s_i, t_i equated in **sat** $[P_1\sigma_0, \ldots, P_n\sigma_0]$ are jointly SU-Unified, finding an instantiation σ for variables in $[\bigcup_i V(P_i\sigma_0) - V(t)]$ such that $\forall i.\ [s_i\sigma \underset{R}{\rightarrow}^* n_i,\ t_i\sigma \underset{R}{\rightarrow}^* n_i]$.

- If no such SU-unifier is found, we conclude that this rule cannot be used to SU-Reduce the given term, and attempt SU-Reduction using some other rule.
- If found, the SU-Unifier σ is applied to the rest of the rule.

(3) Terms s_i, t_i equated by each $N_i\sigma_0\sigma$ from the instantiated **unsat** -part of the rule are now jointly SU-Unified, attempting to find some term n_i and substitution ρ for variables in $\left[\bigcup_i V(N_i\sigma_0\sigma) - \bigcup_j V(P_j\sigma_0)\right] \cup V(t)$, such that $s_i\rho \rightarrow^* n_i$ and $t_i\rho \rightarrow^* n_i$.

- If no such SU-Unifier ρ is found, we conclude that SU-Reduction can occur: in the given term, $[(rhs)\sigma_0\sigma]$ replaces the subterm t.
- If any SU-unifier ρ is found, we conclude that for substitution σ for variables in the **sat** -part, this rule cannot be used to SU-Reduce the given term: we go back to step (2) and try to find another σ (a new substitution for **sat** -variables).

Non-termination is possible in every invocation of steps (2) and (3) above. For example, we do not succeed in reducing $f(x)$ to c using the rule **sat** $[f(y)=c]::f(x)\rightarrow c$, nor can we clearly terminate with the conclusion that $f(x)$ does not SU-reduce to c. Following Fitting [Fi], to account for such cases when we cannot correctly decide whether a term reduces to another, SU-rewriting can be given a more precise definition using 3-valued logic (as in [MS2] for 'EI-rewriting'). We have $s \underset{R}{\rightarrow} t \in \{\mathbf{T}, \mathbf{F}, \perp\}$ for any terms s, t and SU-CTRS R, where $s \underset{R}{\rightarrow} t$ is undefined ("$\perp$") when there is some SU-reduction step from s which does not terminate. A term p is SU-reducible (to q) if $(p \underset{R}{\rightarrow} q) \in \{\mathbf{T}\}$ for some term q; in what follows, we abbreviate $(p \underset{R}{\rightarrow} q) \in \{\mathbf{T}\}$ as "$p \underset{R}{\rightarrow} q$". Otherwise, if $\forall q.(p \underset{R}{\rightarrow} q) \in \{\mathbf{F}\}$, then p is irreducible.

The definition of SU-rewriting allows considerable non-determinism, especially since the process of satisfying literals in the antecedent can yield more than one possible instantiation of the *rhs* of the same rule. Since our emphasis is on rewriting in a functional style, we impose restrictions that suppress non-determinism; a relational logic programming approach may instead make full use of the potential for non-determinism in SU-CTRS.

Example 2 (illustrating SU-reduction): The term $f(cons(a,cons(b,cons(b,nil))))$ is to be SU-Reduced using the SU-rule

$$\textbf{sat}\ [z=cons(x,cons(v,y))]\textbf{unsat}\ [x=v] :: f(z)\rightarrow cons(x,f(cons(v,y))).$$

The matching substitution $[z \leftarrow cons(a,cons(b,cons(b,nil)))]$ matches the given term with the *lhs* of the rule and instantiates the antecedent to

$$\textbf{sat}\ [cons(a,cons(b,cons(b,nil)))=cons(x,cons(v,y))]\textbf{unsat}\ [x=v].$$

The unifier $[x\leftarrow a,\ v\leftarrow b,\ y\leftarrow cons(b,nil)]$ helps satisfy the first equality, and the antecedent then simplifies to just **unsat** $[a=b]$. Non-unifiability of the non-narrowable constants a and b implies the unsatisfiability of $(a=b)$. Hence SU-reduction of the given term succeeds, resulting in the reduct $cons(a,f(cons(b,cons(b,nil))))$.

We now attempt to apply the same rule again to this term, instantiating the variable z to $cons(b,cons(b,nil))$. The substitution $[x\leftarrow b,\ v\leftarrow b,\ y\leftarrow nil]$ instantiates $[x=v]$ to $[b=b]$ and satisfies $[cons(b,cons(b,nil))=cons(x,cons(v,y))]$. Since $b\equiv b$, the **unsat**-literal $(b=b)$ is trivially satisfiable; hence the term $cons(a,f(cons(b,cons(b,nil))))$ cannot be SU-reduced.

For reduction of non-ground terms, the definition of SU-reduction involves 'freezing' variables of the given term while checking satisfiability, but checking possible instantiations of these variables when checking unsatisfiability. We consider $f(z)$ as well as $f(cons(x,cons(y,z)))$ irreducible using the above SU-rule, because these terms have irreducible instances. On the other hand, $f(cons(a,cons(b,z)))$ can be safely SU-rewritten to $cons(a,f(cons(b,z)))$.

2.2. SU-Unification

T-unification of a pair of terms p,q is the task of solving $p=q$ in the theory T, *i.e.*, finding a substitution σ such that $T \models (p\sigma = q\sigma)$. For the theory described by a SU-CTRS, we decompose T-unification into two sub-problems:

(a) finding substitutions instantiating variables in p, q such that they have converging SU-reduction sequences using the SU-CTRS; and

(b) showing that the existence of finite converging reduction sequences is a sound and complete method for proving equations in the theory T (cf. section 3).

SU-Unification is the former task: two terms p,q are SU-unifiable using a SU-CTRS R if there is some substitution σ such that $p\sigma$ and $q\sigma$ have converging SU-reduction sequences using R. We now describe a procedure for SU-unification, mutually recursive with SU-rewriting.

Definition : A term s SU-narrows to t using the SU-CTRS R under the substitution σ (denoted $s \underset{\sigma, R}{\rightsquigarrow} t$), if it has an instance $s\sigma$ which SU-reduces to t, *i.e.*, if there is a non-variable subterm u of s, and a rule $[C::l \rightarrow r] \in R$, such that

(i) l and u unify with m.g.u. θ,

(ii) ρ is a (most general) substitution for $V(C\theta)$ such that $C\theta\rho$ is shown to 'hold',

(iii) σ is the restriction of $\theta\rho$ to $V(s)$, and

(iv) t is the result of replacing in $s\sigma$ an occurrence of $u\sigma$ by $r\theta\rho$.

If there is any (possibly empty) finite sequence of narrowing steps $s_0 \underset{\sigma_1, R}{\rightsquigarrow} s_1 \underset{\sigma_2, R}{\rightsquigarrow} s_2 \rightsquigarrow \cdots \underset{\sigma_m, R}{\rightsquigarrow} s_m$, and if $\sigma \equiv (\sigma_1.\sigma_2 \cdots \sigma_m)$ is the composition of all the narrowing substitutions, we say $s_0 \underset{\sigma, R}{\rightsquigarrow}^{*} s_m$. Subscripts σ and R may be omitted from '$\rightsquigarrow$' when irrelevant or when they can be clearly inferred from the context.

We assume that rules can be and are normalized, *i.e.*, no subterm of the *lhs*, *rhs*, or antecedent of a rule is reducible using some other rule. To avoid some unnecessary and irrelevant attempts to SU-narrow terms, we also assume that the rules are in a standard form in which the *lhs*'s of rules have no redundant variables, *e.g.*, any SU-rule of the form **sat** $[x=t, S]$**unsat** $[U]$ :: $lhs(...x...) \rightarrow rhs$ is transformed by eliminating the redundant variable x and using the equivalent SU-rule **sat** $[S]$**unsat** $[U]$::$lhs(...t...) \rightarrow rhs$.

Two terms are SU-Unifiable if unifiable terms are obtained from them by (0 or finitely many) SU-narrowing steps, with substitutions that are 'compatible' — the same variable is not instantiated to non-unifiable terms in the two narrowing sequences. For example, $f(x)$ and $g(x)$ are not SU-unifiable using $\{f(a) \rightarrow c,\ g(b) \rightarrow c\}$, although $f(x)$ and $g(y)$ are SU-unifiable.

The SU-Unification procedure, illustrated using example 3 below, consists of repeatedly SU-narrowing the given pair of terms, exploring all possible narrowing sequences, until some unifiable terms are obtained. The SU-Unifier is the composition of the SU-narrowing substitutions and the m.g.u. in the last step of the procedure. SU-Unification fails if the procedure terminates without yielding a unifiable pair; in general, the procedure may not terminate. The following results relate SU-Rewriting, SU-Narrowing and SU-Unification.

Theorem 1:

(1) If $s \underset{\sigma, R}{\twoheadrightarrow}^{*} t$ then $s\sigma \underset{R}{\rightarrow}^{*} t$.

(2) If SU-Unification of s,t succeeds with SU-Unifier ρ, then $s\rho$, $t\rho$ have a common reduct.

(3) If $\exists\rho[s\rho \underset{R}{\rightarrow} t]$ then <u>either</u> $\exists\sigma[s\sigma \equiv t]$ <u>or</u> $\left[\exists t_1 \exists\sigma[s \underset{\sigma, R}{\twoheadrightarrow} t_1] \text{ such that } \exists\theta[t_1\theta \equiv t]\right]$.

(4) If $\exists\sigma\exists t[p\sigma \underset{R}{\rightarrow} t$ and $q\sigma \underset{R}{\rightarrow} t]$, then $\exists s\exists s'\exists\theta[p \underset{R}{\twoheadrightarrow} s$ and $q \underset{R}{\twoheadrightarrow} s'$ and $s\theta \equiv s'\theta]$.

Example: We attempt to SU-Unify $g(w)$ and $h(u)$ using the following SU-rules:

(1) **sat** $[x{=}nil] :: f(x) \rightarrow nil$

(2) **unsat** $[x{=}nil] :: f(x) \rightarrow cons(x,x)$

(3) **sat** $[f(x){=}cons(y,z)] :: g(cons(x,y)) \rightarrow r(z)$

(4) **unsat** $[f(y_1){=}nil] :: h(cons(y_1,y_2)) \rightarrow r(y_1)$

[because
$g(w)$ *SU-Unifies with* $h(u)$
$\downarrow$ $\downarrow$
$r(cons(x_1, x_2))$ *unifies with* $r(cons(z_1, z_2))$]

We have $f(x) \underset{[x \leftarrow cons(x_1, x_2)]}{\overset{(2)}{\rightsquigarrow}} cons(x,x)$. Therefore, **sat** $[f(cons(x_1,x_2)){=}cons(y,z)]$ holds, which implies $g(w) \underset{(3)}{\twoheadrightarrow} r(cons(x_1,x_2))$. Similarly, $f(y_1) \underset{[y_1 \leftarrow cons(z_1, z_2)]}{\overset{(2)}{\rightsquigarrow}} cons(y_1,y_1)$, hence **unsat** $[f(cons(z_1,z_2)){=}nil]$ which implies $h(u) \underset{(3)}{\twoheadrightarrow} r(cons(z_1, z_2))$. These terms $r(cons(x_1, x_2))$ and $r(cons(z_1, z_2))$ obtained by narrowing $g(w)$ and $h(u)$ unify, with the m.g.u. $[x_1 \leftarrow z_1, x_2 \leftarrow z_2]$. Hence we conclude that $g(w)$ and $h(u)$ SU-unify with the substitution $[w \leftarrow cons(cons(z_1, z_2), cons(z_1, z_2)), u \leftarrow cons(cons(z_1, z_2), y_2)]$.

3. Semantics of SU-Rewriting

Rewrite systems with negation do not have initial or minimal model semantics; in this section, we show that it is nevertheless possible to ascribe meaningful logical semantics to SU-CTRS. From each SU-CTRS R, we directly obtain a set (conjunction) of formulas $\Xi(R)$, by replacing '**sat**', '**unsat**' with quantifiers, and by syntactically changing the symbols '::', '$\rightarrow$' in each rule to '$\supset$' and '=' respectively. For example,

if R is $\{$**sat** $[p(x){=}q(z)]$**unsat** $[t(y){=}s(x,z)]::l(z) \rightarrow r(x,z)\}$,

then $\Xi(R)$ is $\{\forall z.\forall x.\left[p(x){=}q(z) \wedge [\forall y.t(y) \neq s(x,z)] \supset l(z){=}r(x,z)\right]\}$.

Since SU-CTRS allow some form of negation in the antecedents of rules, their semantics also needs to include additional negative formulas. For example, when reducing $f(nil)$ using the SU-rule **unsat** $[x{=}cons(y,z)]::f(x)\rightarrow rhs$, SU-reduction assumes that $nil{=}cons(y,z)$ is unsatisfiable when the narrowing sequences issuing from nil and $cons(y,z)$ terminate without yielding unifiable terms. Note that $nil{\neq}cons(y,z)$ is not a direct logical consequence of $\Xi(R)$; thus we need to assume some additional set of inequations (Ψ) in order to perform reductions using SU-rules with **unsat** -parts. Ψ contains default assumptions like $nil{\neq}cons(y,z)$ which must be included along with $\Xi(R)$ in defining the notions of soundness and completeness.

Definition: For any set of inequations Ψ, SU-Rewriting with an SU-CTRS R is

- *sound* w.r.t. Ψ if $\forall$ ground p, q. $[p\underset{R}{\rightarrow}q] \Rightarrow [\ \Xi(R)\cup\Psi \models p{=}q\ ]$, and
- *complete* w.r.t. Ψ if $\forall$ ground p, q. $[\ \Xi(R)\cup\Psi\models p{=}q\ ] \Rightarrow \exists r.\ [p\underset{R}{\rightarrow}^{*} r$ and $q\underset{R}{\rightarrow}^{*} r]$.

Theorems 2, 3 below answer the following questions regarding the relationship between an SU-CTRS R and the set of inequations Ψ with respect to which SU-Rewriting (with R) is sound and complete (for proofs, see [Mo]):

(1) Given R, what Ψ helps define the semantics for SU-reduction?

(2) Given Ψ, when is SU-rewriting with R sound and complete w.r.t. Ψ?

Theorem 2 : Let R be a given SU-CTRS. Let Ψ_R be the set of inequations between distinct ground terms irreducible by R. If every ground SU-Rewriting sequence is finitely terminating, then ground SU-Rewriting is sound and complete with respect to Ψ_R.

Definition: Let $\tau(\Psi)$ be a set of ground terms such that if two distinct terms $s, t\in\tau(\Psi)$, then $s{\neq}t\in\Psi$. An SU-CTRS R is said to be *sufficiently complete with respect to* Ψ if every ground term t formed using symbols occurring in R has a reduct in $\tau(\Psi)$, i.e., $\exists t'\in\tau(\Psi).[t\underset{R}{\rightarrow}^{*} t']$.

Theorem 3: Ground SU-Rewriting with R is sound and complete with respect to Ψ if R is sufficiently complete with respect to Ψ.

4. Fully-Defined SU-CTRS

In this section, we explore the issue of constructing specifications using SU-CTRS which satisfy the semantic requirements described above. We identify conditions under which functions are 'fully-defined' by SU-CTRS. These conditions are useful for writing data type and function specifications in a two-tiered framework with function symbols $\in F$ separated into a set C of *constructors* and a set D of *defined functions* being specified using functions in C. A term in which the only function symbols are those $\in C$ is said to be a constructor term; terms in which defined function symbols also occur are *defined terms*. Every rule of the form $C::f(...)\rightarrow rhs$ is said to be a *defining* rule of f.

The central assumption in this framework is that every ground constructor term represents a distinct semantic object. It is then natural to establish the semantics of EI-rewriting with respect to the set of inequations between non-unifiable constructor terms (Ψ_C); a natural choice for $\tau(\Psi_C)$ is the set of ground constructor terms. Two terms $c(s_1, \ldots, s_m)$ and $d(t_1, \ldots, t_n)$ are said to be constructor-wise distinct iff their outermost symbols c, d are constructors, and either c, d are distinct symbols, or they have corresponding arguments s_i, t_i that are constructor-wise distinct. Then, the definition of Ψ_C ensures that $\Xi(R)\cup\Psi_C \models c(s_1, \ldots, s_m) \neq d(t_1, \ldots, t_n)$.

If two terms are constructor-wise distinct, every possible replacement of their non-constructor subterms by constructor terms makes them instances of non-unifiable constructor terms. For example, *nil* and $cons(x,y)$ are constructor-wise distinct; hence the pair $cons(nil, cons(x, f(x)))$ and $cons(cons(u,v), cons(w,y))$ are also constructor-wise distinct, although defined function symbols occur in them. It can be safely assumed that equality between such terms is unsatisfiable. The definition of SU-rewriting hence can be further refined: we may suspend all attempts to show that constructor-wise distinct terms are T-unifiable, and this can result in avoiding the useless exploration of narrowing sequences issuing from such terms. For example, with an SU-rule like **unsat** $[x=cons(y,z)]::g(x,y)\rightarrow q$, it is legitimate to SU-reduce $g(nil, f(w))$ to q, because $[nil=cons(f(w),z)]$ is deemed unsatisfiable, even though infinite narrowings ensue from $f(w)$, where function f is defined by other rules in the SU-CTRS. Conversely, we may decide that it is not possible to SU-reduce $g(nil, f(w))$ using **sat** $[x=cons(y,z)]::g(x,y)\rightarrow q$ whereas the original definition of SU-rewriting (in section 2) may lead to an infinite narrowing sequence from $f(w)$.

Definition : A function f is said to be *fully-defined* by a CTRS R iff every ground term of the form $f(t_1, \ldots, t_n)$ has a unique constructor term reduct (using R).

Theorem 4: If every defined function is fully-defined by R, then ground SU-rewriting is sound and complete w.r.t. $\Psi_C \equiv$ {inequations between non-unifiable constructor terms}.

Definition : An SU-CTRS R is said to be ground confluent if $\forall$ ground terms p, q, r, whenever $p \xrightarrow[R]{*} q$ and $p \xrightarrow[R]{*} r$, then there is some term m such that $q \xrightarrow[R]{*} m$, and $r \xrightarrow[R]{*} m$.

Definition : An SU-CTRS R is said to be ground terminating if every SU-reduction sequence beginning from every ground term finitely terminates.

Theorem 5: Every defined function is fully-defined by R if R is ground confluent, ground terminating, and every ground defined term is SU-reducible.

The development of syntactic restrictions for fully-definedness is firstly motivated by the absence of decision procedures for checking any of the above three criteria (in theorem 5). We can only develop separate conditions sufficient to ensure that each of the three criteria is satisfied. For modularity, the conditions are developed in such a way that we can assert that an entire specification (*i.e.*, every defined function) is fully-defined by separately verifying the sufficient conditions for subsets of rules defining each function. Furthermore, the sufficient conditions developed must be statically checkable and intuitive enough to serve as useful guidelines in writing specifications. In the rest of this section, we describe and illustrate the following sufficient conditions (ensuring ground termination, confluence and reducibility, respectively): *halting*, *unambiguity*, and *totality*. It follows that:

Theorem 6: If R is halting, unambiguous and total, then every function defined by R is fully-defined; hence ground SU-rewriting with R is sound and complete w.r.t. Ψ_C.

4.1. Halting

There are three possible causes for non-termination of SU-rewriting, for which we suggest solutions below.

(1) The reduction sequences may yield progressively "bigger" terms. For this, the traditional solution has been to find a stable and monotonic well-founded term (*s.m.w.t.*) ordering in which terms in the antecedent and *rhs* are smaller than the *lhs* [Ka1, JW, MS2]. This condition must be relaxed because it prevents the use of logical variables in antecedents of rules. For example, we would like to use (terminating) SU-rules like **sat** $[f(x)=cons(x,y)]::l(x)\rightarrow r(y)$; since $l(x)$ is a subterm of $cons(x,l(x))$, an instance of a term in the antecedent, there is no *s.m.w.t.* ordering in which $l(x)>cons(x,y)$ (nor $l(x)>r(y)$, similarly). We enforce termination by allowing logical variables in terms on which other restrictions, given below, are imposed, as well as requiring the existence of an *s.m.w.t.* ordering such that in every rule, *lhs* > every term in *rhs* and *antecedent* without logical variables.

(2) Backtracking among **sat** and **unsat** parts of a rule (to find suitable substitutions) may not terminate, *e.g.*, when a substitution satisfying one literal in the **sat** -part of the antecedent does not satisfy another literal, and it is necessary to look for another substitution satisfying both literals. To avoid this possibility, we impose a 'linearity' restriction: logical variables may occur no more than once in the antecedent of a rule.

(3) The process of SU-narrowing terms in antecedents may not terminate. To allow logical variables while simultaneously imposing a termination requirement, we need to ensure that logical variables cannot be instantiated to any terms bigger than *lhs* while attempting to SU-rewrite any term. For this, we impose a two-level hierarchical structure. At the 'lower' level we have functions defined using TRS which satisfy Hullot's basic narrowing criterion (or any similar condition on TRS which ensures

decidability of T-Unification), assuring termination of narrowing sequences from terms containing only these functions and constructors. By restricting logical variables to occur only in such terms, we assure that all narrowing sequences triggered during SU-reduction terminate.

Definition: Let R_0 be a TRS which satisfies the basic narrowing criterion. Then $R_0 \cup R_1$ is *halting* iff for each rule $C::l \rightarrow r$ in R_1 (an SU-CTRS), the following conditions hold:

(a) every logical variable occurs at most once in the antecedent C;

(b) every logical variable occurs only in terms whose leading symbols are constructors or functions fully-defined by R_0; and

(c) $l >$ every subterm in C and r with no logical variables, in some *s.m.w.t.* ordering.

The above definition ensures that narrowing is carried out only on terms which satisfy Hullot's basic narrowing criterion, hence is guaranteed to terminate. This property, together with the well-foundedness of the s.m.w.t. ordering, ensures the following:

Theorem 7: Every halting SU-CTRS is ground terminating.

For example, if f is fully-defined by R_0 (a TRS satisfying the basic narrowing criterion), c, d are constructors, and R_1 also contains a halting definition of the function g where $h >_F g$ according to an ordering "$>_F$" on function symbols, then the following rule in R_1 conforms to the halting condition:

$$\textbf{sat}\ [f(g(x),\ c(y, g(x))) = d(z)]\textbf{unsat}\ [x = d(u)] ::\ h(x) \rightarrow c(x, y).$$

Note that y, z, u occur no more than once in the antecedent, hence no backtracking is required. The terms in which these logical variables occur are $f(\_, c(y, \_))$, $d(z)$, $d(u)$, where the functions f, c, d satisfy the condition of being either constructors or fully-defined by R_0; hence narrowing these terms terminates, given that x is already ground instantiated. Condition (c) in the above definition is satisfied using the ordering $>_F$ among outermost function symbols. Since $h >_F g$, we have $h(x) > g(x)$; finally, although the *rhs* contains a logical variable, it is a constructor term.

4.2. Unambiguity

We now formulate a sufficient syntactic condition to ensure ground-confluence of SU-CTRS. We assume that variables in different rules are first renamed to be distinct.

Definition: An SU-CTRS is *unambiguous* if each of the following conditions hold:

(1) All and only proper subterms of *lhs* of each rule must be constructor terms.

(2) Every logical variable z occurring in the *rhs* of each rule occurs in an equation $t = c[z]$ in the **sat**-part of the rule, where $V(t) \subseteq V(lhs)$, and the leading symbol of every subterm of $c[z]$ in which z occurs is a constructor, *e.g.*, **sat** $[g(x) = cons(z, cons(u, g(y)))] :: f(x, y) \rightarrow f(z, u)$.

(3) Let **sat** $[P_1]$**unsat** $[N_1]::l_1 \to r_1$ and **sat** $[P_2]$**unsat** $[N_2]::l_2 \to r_2$ be any two rules in R whose *lhs*'s unify with m.g.u. σ, where P_1, P_2, N_1, N_2 are sets of equations.
Let P denote $P_1\sigma \cup P_2\sigma$ and let N denote $N_1\sigma \cup N_2\sigma$; then
(a) either $r_1\sigma \equiv r_2\sigma$;
(b) or $\exists [m_1 = m_2] \in N$ and substitution ρ for the **unsat**-variables, such that $m_1\rho \equiv m_2\rho$;
(c) or $\exists [n_1 = n_2] \in P$ such that n_1, n_2 are constructor-wise distinct;
(d) or $\exists p = q \in P \wedge \exists p' = q' \in N$ and some substitution ρ for the **unsat** -variables, such that $p'\rho$ and $q'\rho$ are identical upto mutual replacements of occurrences of p, q in them;
(e) or $\exists p = q,\ p' = q' \in P$ such that $p \equiv p'$, but q, q' are constructor-wise distinct.

Each of the following pairs of SU-rules is unambiguous, illustrating the above conditions:

$or(x, \mathbf{true})$	$\to$	**true**
$or(\mathbf{true}, x)$	$\to$	**true** , by (3)(a);
$mem(x, cons(x, y))$	$\to$	**true**
unsat $[z = cons(u, v)]::mem(w, z)$	$\to$	**false** , by (3)(b);
$Isempty(nil)$	$\to$	**true**
sat $[x = cons(u, v)]::Isempty(x)$	$\to$	**false** , by (3)(c);
sat $[x = cons(y, y)]::Duplist(x, y)$	$\to$	**true**
unsat $[u = cons(v, w)]::Duplist(u, z)$	$\to$	**err** , by (3)(d);
sat $[x = cons(u, v)]::First(x)$	$\to$	u
sat $[z = succ(y)]::First(z)$	$\to$	y, by (3)(e).

Theorem 8: Every unambiguous SU-CTRS is ground confluent.

4.3. Totality

Finally, we give a sufficient condition for checking that every ground term with defined function symbols is reducible using a given SU-CTRS (cf. [Th, Kou, JK]). Unlike unconditional TRS, we observe that the condition for termination (halting) is also required to ensure that ground terms are reducible. For example, due to non-termination, $f(a)$ is not reducible using $\{$**sat** $[f(x) = c]::f(x) \to c$, **unsat** $[f(x) = c]::f(x) \to c\}$, although $f(a) = c$ is either satisfiable or unsatisfiable in every model.

We give a procedure which checks whether subsets of rules with unifying *lhs*'s (in a given SU-CTRS) have instantiated antecedents one of which always holds. When the procedure *SU–AnteTotal* (given below) returns with success, it ensures that the antecedent of one of the rules holds, whenever a term to be reduced matches with the *lhs* of the rules. We can then use known procedures for showing that every possible ground constructor term argument-tuple is 'covered' by the instantiated *lhs*'s of rules [Th] for which procedure *SU–AnteTotal* revealed that the antecedents imply a tautology. Some preprocessing is needed, generating equivalent formulations of rules, and identifying variables in corresponding positions in different rules. For example, we consider **sat** $[P]$**unsat** $[N]::f(\ldots t \ldots) \rightarrow rhs$ and **sat** $[x=t, P]$**unsat** $[N]::f(\ldots x \ldots) \rightarrow rhs$ to be equivalent.

procedure *SU–AnteTotal*(R);
  If R contains an unconditional rule, **return** success;
  If R is empty, **return** failure;
  Let $m=n$ be an equation in the **unsat** -part of a rule in R;
  Let R_0 be $\{(C::l \rightarrow r) \mid [(C::l \rightarrow r) \in R] \wedge (m=n \notin C)\}$;
  Let R_+ be $R_0 \cup \{(C::l \rightarrow r) \mid (\textbf{sat}\ [m=n] \cup C::l \rightarrow r) \in R\}$;
  Let R_- be $R_0 \cup \{(C::l \rightarrow r) \mid (C \cup \textbf{unsat}\ [m=n]::l \rightarrow r) \in R\}$;
  return success if *SU–AnteTotal*(R_+) and *SU–AnteTotal*(R_-) return success;
end procedure .

Theorem 9: Let λ be the (identical) *lhs* of a set R of SU-rules, obtained as instances of rules with unifying *lhs*'s from a ground terminating SU-CTRS. If *SU–AnteTotal*(R) returns success, then every ground instance of λ is reducible.

Example 4 : Function *mem* is specified below to check membership in a list polymorphically w.r.t. the types of elements. This specification allows for lists whose elements are any (unspecified) objects, and returns **false** if the second argument is not a nonempty list.

(1) $mem(x_1,\ x_1{:}x_2) \rightarrow \textbf{true}$

(2) **unsat** $[y_1=y_2]$ :: $mem(y_1,\ y_2{:}y_3) \rightarrow mem(y_1,\ y_3)$

(3) **unsat** $[z_2=u{:}v]$ :: $mem(z_1,\ z_2) \rightarrow \textbf{false}$

The specification is ground terminating and ground confluent because the rules satisfy the halting and unambiguity conditions. The specification is fully-defined, as we show below that it is total. We can view rules (1,2) as equivalent to the following rules (1′,2′):

$$(1')\ \textbf{sat}\ [z_2=u{:}v,\ z_1=u]::\ mem(z_1,z_2) \rightarrow \textbf{true}\ ,$$

$$(2')\ \textbf{sat}\ [z_2=u{:}v]\textbf{unsat}\ [z_1=u]\ ::\ mem(z_1,z_2) \rightarrow mem(z_1,\ v).$$

As per procedure *SU–AnteTotal*, we can 'separate' rule (3) from (1′,2′) using the **unsat** -literal $(z_2=u{:}v)$, which is stripped away on separating the rules into $R_+ \equiv \{\ \textbf{sat}\ [z_1=u]::\ mem(z_1,\ z_2) \rightarrow \textbf{true}\ ,\quad \textbf{unsat}\ [z_1=u]\ ::\ mem(z_1,z_2) \rightarrow mem(z_1,\ v)\}$, and $R_- \equiv \{mem(z_1,\ z_2) \rightarrow \textbf{false}\ \}$. We now invoke procedure *SU–AnteTotal* on both these rule-sets. Invoking *SU−AnteTotal*(R_-) returns with success since R_- contains an unconditional rule. When *SU−AnteTotal*(R_+) is invoked, the two rules in R_+ are separated by $(z_1{=}u)$, into $\{mem(z_1, z_2)\rightarrow \mathbf{true}\}$ and $\{mem(z_1, z_2)\rightarrow mem(z_1, v)\}$, after deleting the separating literal from the antecedents. These are unconditional rules, for which invocations of *SU−AnteTotal* return success. Hence the original invocation of *SU−AnteTotal* on all three rules returns with success. Since we started from rules (1′,2′,3) with only variables as arguments to the *lhs*, the above definition of *mem* is total, and every ground term $mem(t_1, t_2)$ is reducible.

5. Concluding Remarks

In this paper, we studied a new and powerful conditional rewriting framework and mechanism. SU-CTRS differ from previously studied CTRS, in that this formalism allows negation with logical variables in the antecedents of rewrite rules. This feature makes it easier to write complete, concise and declarative specifications. SU-CTRS can also be viewed as a language for combining functional and logic programming.

We developed a mechanism for performing reductions using SU-CTRS which uses narrowing for instantiating the logical variables in the rules. The rewriting mechanism is operationally complex, a price paid for increased expressiveness of the formalism. We developed a logical semantics for SU-CTRS, based on equational formulas obtained from the rewrite rules, augmented by a set of default inequational assumptions. We developed a set of restrictions under which ground SU-rewriting terminates, is confluent, and specifications define total functions.

Further study of SU-CTRS and SU-rewriting can follow two distinct directions. One possibility is a closer study of restrictions needed to ensure that SU-CTRS specifications are fully-defined. We need to build and examine larger SU-CTRS specifications to check if the restrictions given in this paper are realistic. It may be possible to extend various known techniques for checking termination, confluence and reducibility, obtaining weaker sets of restrictions despite inherent undecidability of checking these properties. This approach may even rely on an available theorem-prover to accomplish some of the above tasks.

Another possibility is to change the emphasis from the specification-oriented approach towards 'general logic programming'. Instead of attempting the formulation of complete SU-CTRS specifications which necessitate restrictions like decidability and sufficient completeness, we may use the expressiveness of the language to write complex programs. Such an approach would proceed towards combining equational and relational logic programming, with narrowing as the chief operational mechanism. SU-CTRS may allow further generalization, to rules whose antecedents contain arbitrary first order formulas involving equality, although it is presently unclear how such complexity (*e.g.*, nested quantifiers) can be handled operationally.

References

[BBK] J.C.M.Baeten, J.A.Bergstra, J.W.Klop, *Term Rewriting Systems with Priorities,* Proc. 2nd RTA Conf., Bourdeaux, Springer-Verlag LNCS 256, pp83-94, 1987.

[BDJ] D.Brand, J.A.Darringer, W.Joyner, *Completeness of Conditional Reductions,* IBM Res. Rep. RC7404, 1978.

[BK] J.A.Bergstra, J.W.Klop, *Conditional Rewrite Rules : Confluency and Termination,* JCSS 32, pp.323-362, 1986.

[Cl] K.L.Clark, *Negation as Failure,* in **Logic and Databases**, H.Gallaire & J.Minker (eds.), Plenum, NY, pp293-322, 1978.

[CL] C.L. Chang, R.C. Lee, *Symbolic Logic and Mechanical Theorem Proving,* Academic Press, New York, 1973.

[Co] H.Comon, *Sufficient Completeness, Term Rewriting Systems, and "Anti-Unification",* Proc. 8th CADE, Springer-Verlag LNCS 230, pp128-140, 1986.

[De] N.Dershowitz, *Termination of Rewriting,* Proc. First RTA Conf., Dijon, Springer-Verlag LNCS 202, pp180-224, 1985.

[DL] D. DeGroot, G. Lindstrom, *Logic Programming: Functions, Relations and Equations,* Prentice Hall (NJ), 1986.

[DP] N. Dershowitz, D.A. Plaisted, *Logic Programming cum Applicative Programming,* Proc. 1985 Symp. on Logic Programming, Boston, 1985.

[Fa] M. Fay, *First-Order Unification in an Equational Theory,* Proc. 4th CADE, Austin (Texas), 1979, pp161-167.

[Fi] M. Fitting, *A Kripke-Kleene Semantics for Logic Programs,* The Journal of Logic Programming, vol.4, Elsevier Science Pub., 1985, pp295-312.

[Fr] L. Fribourg, *Oriented Equational Clauses as a Programming Language,* Proc. 11th ICALP, Antwerp (Belgium), July 1984.

[GM] J.A. Goguen, J. Meseguer, *Equality, Types, Modules and Generics for Logic Programming,* J. of Logic Programming, Vol. 1, No.2, 1984, pp179-210.

[HO] G.Huet, D.S.Oppen, *Equations and Rewrite Rules: A Survey,* Tech.Rep. CSL-111, SRI Int'l., California, 1980.

[Hul] J.M. Hullot, *Canonical Forms and Unification,* Proc. 5th CADE, Springer-Verlag LNCS 87, April 1980.

[Hus] H. Hussmann, *Unification in Conditional-Equational Theories,* Proc. EUROCAL Conf. at Linz (Austria), Springer-Verlag, LNCS 204, Apr. 1985, pp543-553.

[JK] J.-P.Jouannaud, E.Kounalis, *Proofs by Induction in Equational Theories without Constructors,* Symp. on Logic in C.S., Cambridge (Mass.), USA, 1986, pp358-366.

[JW] J.-P. Jouannaud, B. Waldmann, *Reductive Conditional Term Rewriting Systems,* Proc. Third TC2 Working Conf. on the Formal Description of Prog. Concepts, Ebberup, Denmark, Aug. 1986.

[Ka1] S.Kaplan, *Fair Conditional Term Rewriting Systems: Unification, Termination and Confluence,* Res.Rep. 194, Universite de Paris-Sud, Orsay, Nov. 1984.

[Ka2] S.Kaplan, *Positive/Negative Conditional Rewriting,* Proc. First Int'l. Workshop on Conditional Term Rewriting Systems, Orsay, Springer-Verlag LNCS 308, 1988, pp129-143.

[KJ] S.Kaplan, J.-P.Jouannaud (eds.), *Proc. First Int'l. Workshop on Conditional Term Rewriting Systems,* Orsay, Springer-Verlag LNCS 308, 1988.

[Kou] E.Kounalis, *Completeness in Data Type Specifications,* Res. Rep. 84-R-92, C.R.I.N., Nancy (France) 1984.

[Kow] R.A. Kowalski, *Predicate Logic as a Programming Language,* Information Processing 74, North Holland, Apr. 1974, pp556-574.

[La1] D.S.Lankford, *Canonical Inference,* Rep. ATP-32, Dept. of Math. & C.S., Univ. of Texas at Austin, Dec.1975.

[La2] D.S. Lankford, *Some New Approaches to the Theory and Applications of Conditional Term Rewriting Systems,* Research Report MTP-6, Univ. of Louisiana, Ruston, 1979.

[Li] G. Lindstrom, *Functional Programming and the Logical Variable,* Proc. ACM POPL Symp., Jan. 1985.

[Mo] C.K.Mohan, *Negation in Equational Reasoning and Conditional Specifications,* Ph.D. Thesis, State University of New York at Stony Brook, 1988.

[Mo2] C.K.Mohan, *Priority Rewriting: Semantics, Confluence and Conditionals,* Proc. 3rd RTA Conf., Chapel Hill, 1989.

[MS1] C.K. Mohan, M.K. Srivas, *Function Definitions in Term Rewriting and Applicative Programming,* Information and Control, Academic Press, New York, Vol.71, No.3, Dec. 1986, pp186-217.

[MS2] C.K.Mohan, M.K.Srivas, *Conditional Specifications with Inequational Assumptions,* Proc. First Int'l. Workshop on Conditional Term Rewriting Systems, Orsay, Springer-Verlag LNCS 308, 1988, pp161-178.

[MSK] C.K.Mohan, M.K.Srivas, D.Kapur, *Inference Rules and Proof Procedures for Inequations,* (to appear) Journal of Logic Programming, 1989.

[Na] M.L. Navarro, *Tecnicas de Reescritura para especificaciones condicionales,* These Doctorale de l'Universite Polytechnique de Catalunya, Barcelona (Spain), 1987.

[PEE] U. Pletat, G. Engels, H.D. Ehrich, *Operational Semantics of Algebraic Specifications with Conditional Equations,* 7eme C.A.A.P., Lille (France), 1982.

[Red] U. Reddy, *Narrowing as the Operational Semantics of Functional Languages,* Proc. IEEE Logic Programming Symp., Boston, 1985.

[Rei] R.Reiter, *On Closed World Data Bases,* in **Logic and Data Bases,** H.Gallaire & J.Minker (eds.), Plenum, NY, 1978.

[Rem] J.-L.Remy, *Etudes des systemes de reecriture conditionnels et application aux types abstraits algebriques,* These d'Etat, Nancy (France), 1982.

[Ret] P.Rety, *Improving Basic Narrowing Techniques,* Proc. 2nd RTA Conf., Bourdeaux, Springer-Verlag LNCS 256, pp228-241, 1987.

[Sl] J.R. Slagle, *Automated Theorem-Proving for Theories with Simplifiers, Commutativity and Associativity,* J. ACM, Vol.21, No.4, 1974, pp622-642.

[Th] J.J.Thiel, *Stop Losing Sleep over Incomplete Specifications,* Proc. 11th ACM POPL Symp., 1983.

[ZR] H. Zhang, J.L. Remy, *Contextual Rewriting,* Proc. Conf. on Rewriting Techniques and Applications, Dijon (France), Springer-Verlag, LNCS 202, June 1985.

Algebraic Semantics and Complexity of Term Rewriting Systems

Tohru Naoi and Yasuyoshi Inagaki
Faculty of Engineering, Nagoya University
Furo-cho, Chikusa-ku, Nagoya, JAPAN

Abstract: The present paper studies the semantics of linear and non-overlapping TRSs. To treat possibly non-terminating reduction, the limit of such a reduction is formalized using Scott's order-theoretic approach. An interpretation of the function symbols of a TRS as a continuous algebra, namely, continuous functions on a cpo, is given, and universality properties of this interpretation are discussed. Also a measure for computational complexity of possibly non-terminating reduction is proposed. The space of complexity forms a cpo and function symbols can be interpreted as monotone functions on it.

0. Introduction

We study, in this paper, the semantics of term rewriting systems (TRSs) as a model of computation. The classical semantics of equational logic is not sufficient for our purpose, since it gives the semantics of TRSs on the condition that they are terminating and confluent, and the termination property restricts the computational power of TRSs [2]. We can observe an obvious difference between these two kinds of formal systems: Rewrite rules express more procedural information by their rewriting direction. We should, therefore, also consider such information when formalizing the semantics of TRSs. In addition, we can not place the termination assumption.

To give the semantics, we use Scott's order theoretic approach [19, 20] and define the limit of (possibly) non-terminating reduction. Furthermore, we give an interpretation of function symbols using the notion of continuous algebra [1] (complete magma in [3-7, 15]), which is a natural combination of algebraic and order-theoretic methods.

In this paper we treat *regular* (i.e., *left-linear* and *non-overlapping*) TRSs which has been studied intensively in O'Donnel [17], and Huet and Lévy [9]. As the reasons for our choice, we point out the following facts. First, every computable function can be described by such a TRS. Hence, in spite of the syntactical restrictions, they constitute a quite general mathematical tool for analyzing computation like the lambda calculus. Second, they allow several rewriting strategies that ensure computers to obtain the normal form (if any) of a term without an explosion of the search space [17, 9]. Hence, an implementation of such a TRS can be free from extra control descriptions followed by complicated semantics. In this respect, they seem hopeful as a real programming language and, in particular, a tool for automated program development.

We introduce the preliminary definitions and notations in Chapter 1. In Chapter 2, we formalize the limit of possibly infinite reduction, using the notion of approximate normal forms; the TRS version of this notion is due to Huet and Lévy [9] (see Chapter 5, for its origin). This notion of the limit generalizes that of the ordinary normal forms. In general, the limit is an infinite tree. We obtain a retraction (i.e., a continuous and idempotent function) on the ordered set of infinite trees, extending the correspondence between terms and their limits. It is the denotation of the TRS in question. Then, we define a preorder over infinite trees, as follows: A tree T *is less defined than* T' if the limit of T has less information than that of T'. The preorder induces a continuous quotient algebra and trees can be interpreted as functions on its domain. The universalities and the characterizations of this interpretation are discussed in Chapter 3. In Chapter 4, we give a complexity measure for TRSs by formalizing *speed* of the convergence of infinite reduction. This measure generalizes the conventional time complexity and is compatible with our denotation of trees. Then, we refine the preorder described above in such a way that a tree T *is less defined than* T' if (1) the limit of T has less information than that of T', and (2) T converges more slowly than T'. It is shown that this preorder induces an ordered quotient algebra, and the function symbols are interpreted as monotone functions. Based on this observation, we propose:

$$Semantics = Information \times Complexity.$$

Finally, related work is discussed in Chapter 5.

1. Preliminaries

1.1 Continuous Algebra

Several preliminary notions on continuous algebra will be given here. See [1, 4-7] for more details.

Let F be the *set of function symbols* and X be the *set of variable symbols*. We assume that F contains a special constant Ω. An *ordered F-algebra* is a pair $A = \ll <D_A, \sqsubseteq_A>, < f_A | f \in F > \gg$ such that (1) $< D_A, \sqsubseteq_A >$ is a partial order with a least element $\perp_A$, and (2) for $f \in F$ with arity n, f_A is a monotone map from D_A^n to D_A; in particular, $\Omega_A = \perp_A$. The order $< D_A, \sqsubseteq_A >$ is said to be the *domain of A*. For ordered F-algebras A and B, an *homomorphism* $\eta: A \to B$ is a monotone map from D_A to D_B such that:

$$\eta(f_A(d_1, ..., d_n)) = f_B(\eta(d_1), ..., \eta(d_n)) \text{ for } f \text{ in } F.$$

A *continuous F-algebra* A is an ordered F-algebra such that the domain is a cpo (i.e., a (directed-) complete partial order) and each f_A is continuous (w.r.t. Scott topology). An *homomorphism* between continuous algebras is a continuous homomorphism between ordered algebras.

Let A be an ordered F-algebra and Γ be a class of ordered F-algebras. A is said to be *initial* (resp. *final*) *in* Γ if (1) A belongs to Γ and (2) for any algebra B in Γ, a unique homomorphism $\eta: A \to B$ (resp. $\eta: B \to A$) exists. A is said to be *free over X in Γ* if (1) A is in Γ, and (2) there is a map $\iota: X \to D_A$ satisfying the following condition: for any B in Γ and any map $h: X \to D_B$, there is a unique homomorphism $\eta: A \to B$ such that $h = \eta \circ \iota$. These properties are similarly defined for continuous algebras.

Similar to F-algebras, we also define $F \cup X$-algebras by taking variable symbols as constant symbols. An F-algebra is obtained from an $F \cup X$-algebra when the interpretation of variables is forgotten.

1.2. Infinite Trees

Here, we informally introduce the basic definitions and notations for infinite trees. See [1, 5] for a precise treatment.

An *(F, X)-tree* is a finite or infinite tree having nodes labeled with function or variable symbols such that the number of sons of a node equals the arity of the label. Figure. 1 shows an example. We denote the *set of (F, X)-trees* (resp. *the set of finite (F, X)-trees*) by $T^\infty(F, X)$ (resp. $T(F, X)$). A finite (F, X)-tree is identified with a *well-formed term*. We use letters $S, T, U, ...$ to denote (F, X)- trees and $s, t, u, ...$ to denote finite (F, X)-trees. We write $Node(T)$ for the set of nodes in T. For p in $Node(T)$, the label on p is denoted by $T(p)$; T/p is the subtree of T with root p; $T[p \leftarrow T']$ is the tree obtained from T by replacing the subtree occurred at p by T'. $T[T_1/x_1, T_2/x_2, ...]$ denotes the tree obtained from T by substituting T_i for x_i for all i. We sometimes write $T[T_1, T_2, ...]$ for short.

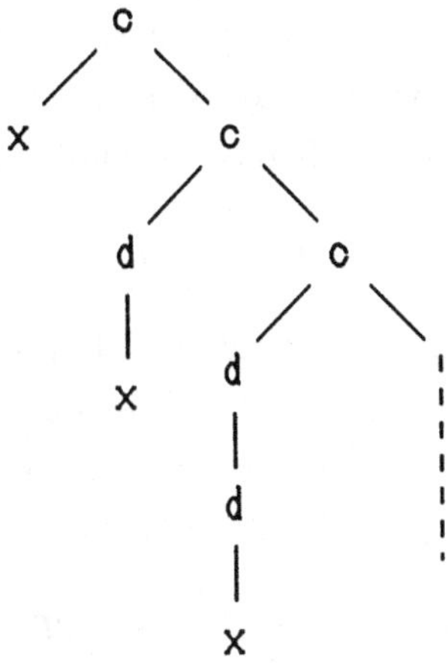

Fig. 1.

Consider a cpo $< F \cup X, \sqsubseteq >$ defined by: $f \sqsubseteq g$ iff $f = g$ or $f = \Omega$. This order induces an order $\preceq$ on $T^\infty(F, X)$, as follows:

$$(*) \qquad T \preceq T' \text{ iff } Node(T) \subseteq Node(T') \text{ and } T(p) \sqsubseteq T'(p) \text{ for all } p \in Node(T).$$

Proposition 1.1. *(ADJ [1]) $T^\infty(F, X)$ is a bounded-complete algebraic cpo with the base $T(F, X)$. The least element is Ω.*

For instance, the lub of the following infinite chain is the tree shown in Fig. 1:

$$\Omega \preceq c(x, \Omega) \preceq c(x, c(d(x), \Omega)) \preceq c(x, c(d(x), c(d(d(x)), \Omega))) \preceq \ldots$$

We can interpret f in $F \cup X$ with arity n as a continuous n-ary function on $T^\infty(F, X)$ that sends $T_1, \ldots, T_n$ to $f(T_1, \ldots, T_n)$. This defines a continuous $F \cup X$-algebra with the domain $T^\infty(F, X)$. It is denoted by $\mathbf{T}^\infty(F \cup X)$. $\mathbf{T}^\infty(F, X)$ denotes the F-algebra obtained from $\mathbf{T}^\infty(F \cup X)$ by forgetting the interpretation of variable symbols.

Proposition 1.2. *(ADJ [1])*
(1) $\mathbf{T}^\infty(F \cup X)$ *is initial in the class of all continuous* $F \cup X$*-algebras.*
(2) $\mathbf{T}^\infty(F, X)$ *is free over* X *in the class of all continuous* F*-algebras.*

Let A be a continuous $F \cup X$-algebra with the domain $<< D_A, \sqsubseteq_A >$. Proposition 1.2 (1) claims that a tree T represents the element T^A in D_A given by:

$$T^A = v_A(T),$$

where $v_A : \mathbf{T}^\infty(F \cup X) \to A$ is the unique homomorphism.

Now, let A be a continuous F-algebra. A function $v: X \to D_A$ is called a *valuation of variables into* D_A. Using (2) in the proposition, a tree T can be interpreted as a continuous map $T_A: (X \to D_A) \to D_A$ defined as follows. For $v: X \to D_A$,

$$T_A(v) = \nu(T),$$

where ν is the unique homomorphism extending v. Hence, the set $\{T_A | T \in T^\infty(F, X)\}$ is the set of "representable" functions on D_A. It is ordered pointwisely and this order is also denoted by $\sqsubseteq_A$ without fear of confusion. We define an ordered $F \cup X$-algebra $\mathbf{T}^\infty(F, X)_A$ with the domain $\{T_A | T \in T^\infty(F, X)\}$, as follows [3]:

$$f_{\mathbf{T}^\infty(F,X)_A}((T_1)_A, \dots, (T_n)_A) = (f(T_1, \dots, T_n))_A$$

$\mathbf{T}^\infty(F, X)_A$ is not continuous in general [3, 7].

Let π be a preorder on $T^\infty(F, X)$. We write $T^\infty(F, X)/\pi$ for the quotient set of $T^\infty(F, X)$ by the equivalence relation $\pi \cap \pi^{-1}$. It is equipped with an induced order $\pi/\pi \cap \pi^{-1}$. We also write $[T]_\pi$ for the equivalence class of T. We say π is a *precongruence* if (1) $\preceq \subseteq \pi$, and (2) $T_i \pi U_i$ for all i implies $T_0[T_1, T_2, \dots] \pi U_0[U_1, U_2, \dots]$ (compatible with tree composition). A precongruence π induces an ordered $F \cup X$-algebra $\mathbf{T}^\infty(F, X)_\pi$ with a domain $T^\infty(F, X)/\pi$, as follows:

$$f_{\mathbf{T}^\infty(F,X)_\pi}([T_1]_\pi, \dots, [T_n]_\pi) = [f(T_1, \dots, T_n)]_\pi.$$

Also $\mathbf{T}^\infty(F, X)_\pi$ need not be continuous [7].

1.3. Term Rewriting Systems

Let R be a TRS. The notions of *reduction in R*, *redexes of R*, and *R-normal forms* are assumed to be defined in the standard manner (see, e.g. [8]). We denote the *reduction relation in R* (on $T(F, X)$) by $\to_R$, the *set of redexes of R* by Red_R ($\subseteq T(F, X)$), the *set of the redex occurrences of R in t* by $Redocc_R(t)$ ($\subseteq Node(t)$), and the *set of R-normal forms* by NF_R. For any binary relation $\to$, we write $\to^n$, $\to^*$, and $\leftrightarrow$ for the n-fold composition, the reflexive-transitive closure, and the symmetric closure of $\to$, respectively.

A term t is *Ω-free* if Ω does not occur in t, and R is *Ω-free* if for all $< t \to u >$ in R, t is Ω-free. In this paper, we assume that each TRS R is left-linear, non-overlapping (non-ambiguous) and Ω-free. It is well-known that such an R is confluent (see, e.g. [8]).

For any subset P of $Redocc_R(t)$, simultaneous rewriting specified by P is called *development of P in t* (for the precise definition, see [3, 14, 17]). We write $t \to_{\delta R} u$ if t

$$\begin{aligned} R = \{ \; & a && \to b, \\ & f(b,x) && \to g(x), \\ & g(x) && \to c(x, g(d(x))), \\ & h(x) && \to c(x, c(d(x), h(d(d(x))))) \} \end{aligned}$$

Fig. 2.

reduces to u by some development. $Dev_R(t)$ denotes the set of development sequences starting at t, i.e., $< t_0, t_1, \ldots > \in Dev_R(t_0)$ iff $t_i \to_{\delta R} t_{i+1}$ for all i. Development of $Redocc_R(t)$ is called *full development in t*. The unique full development sequence starting at t is denoted by $FDS_R(t)$. Remark that $\to^*_{\delta R} = \to^*_R$.

2. Continuous Algebra Semantics

2.1. Approximate Normal Forms

From the conventional viewpoint, an interpreter of programs can give output only when the evaluation terminates. On the other hand, Scott's framework [20] insists that an interpreter could output *partial* results, that is, incomplete pieces of information, while running. If we regard TRSs as interpreters, such partial results can be formalized as approximate normal forms. First, we prepare some useful notions.

The *set of candidates for the redexes of R*, denoted by $Cand_R$, is defined inductively as: (1) $t \in Red_R$ implies $t \in Cand_R$, and (2) $t, t' \in Cand_R$ and $p \in Node(t)$ implies $t[p \leftarrow t'] \in Cand_R$.

Example. *For the TRS R shown in Fig. 2, $Cand_R$ contains $a, f(b,x), f(a,x), f(g(b),a)$, etc.*

Let $\downarrow Cand_R = \{t | \exists u \in Cand_R \;\; t \preceq u\}$. We define the *set of the candidates occurrences of R in t*, denoted by $Candocc_R(t)$, as follows:

$$Candocc_R(t) = \{p \in Node(t) | t/p \in \downarrow Cand_R\}.$$

The *set of approximate normal forms of R*, denoted by ANF_R, is defined by

$$ANF_R = \{ t \,| p \in Candocc_R(t) \text{ implies } t/p = \Omega \}.$$

Remark that $ANF_R \subseteq NF_R \cup \{\Omega\}$ holds since $Red_R \subseteq Cand_R \subseteq \downarrow Cand_R$. On the other hand, a normal form containing Ω need not be in ANF_R. All Ω-free normal forms belong to ANF_R.

For any set P of nodes, $min(P)$ denotes the set of nodes with no proper ancestors in P. We define a function ω_R on $T(F,X)$ as below:

$$\omega_R(t) = t[p \leftarrow \Omega | p \in min(Candocc_R(t))].$$

Note that ω_R is idempotent, and $\omega_R(t)$ is in ANF_R for any t. If $t \rightarrow_R^* t'$, we call $\omega_R(t')$ an *approximate normal form of* t. The relation $\rightarrow_R^* \cdot \omega_R$ is called the *approximate reduction relation in* R.

Proposition 2.1. *(Huet and Lévy [9])*
(1) ω_R *is monotone, and*
(2) $t \rightarrow_R^* t'$ *implies* $\omega_R(t) \preceq \omega_R(t')$.

Example 2.2. *Using R shown in Fig. 2, we have a reduction sequence*

$$f(a,x) \rightarrow_R f(b,x) \rightarrow_R g(x) \rightarrow_R c(x, g(d(x)))$$

$$\rightarrow_R c(x, c(d(x), g(d(d(x))))) \rightarrow_R \cdots$$

Applying ω_R *to each terms in the sequence, we obtain the following monotone sequence of approximate normal forms:*

$$< \Omega,\ \Omega,\ \Omega,\ c(x,\Omega),\ c(x, c(d(x),\Omega)), \ldots >$$

The next lemma is one of the most important properties of approximate reduction in a left-linear and non-overlapping TRS; it need not hold without these assumptions.

Lemma 2.3. *For any* s, t *and* u *such that* $s[t] \rightarrow_{\delta R}^n u$,
(1) there are t' *and* u' *such that* $t \rightarrow_{\delta R}^n t'$, $s[\omega_R(t')] \rightarrow_{\delta R}^n u'$ *and* $\omega_R(u) \preceq \omega_R(u')$,
(2) there are s' *and* u' *such that* $s \rightarrow_{\delta R}^n s'$, $\omega_R(s')[t] \rightarrow_{\delta R}^n u'$ *and* $\omega_R(u) \preceq \omega_R(u')$.

We obtain a weaker form of this lemma by replacing $\rightarrow_{\delta R}^n$ by $\rightarrow_R^*$ (see Chapter 5). It expresses *the principle of lazy evaluation*. Namely, when we evaluate a program $s[t]$, we do not require the normal form of the subprogram t, which may not exist; we only need some approximate normal form $\omega_R(t')$ of t, depending on the approximate normal form $\omega_R(u)$ of the whole program $s[t]$. The weaker version of the lemma is sufficient to prove the main result of this chapter, Theorem 2.8. However, the original form is required in Chapter 4.

2.2. The Limit of Infinite Reduction

For a term t, the set $\{\omega_R(t')|t \to_R^* t'\}$ is directed, since R is confluent and ω_R is monotone increasing along $\to_R^*$ (Proposition 2.1 (2)). Intuitively, the partial outputs from the interpreter evaluating t are consistent as a whole. Thus, the *eventual* output can be defined by summing them up. We define the *symbolic value of t in R*, denoted by $Val_R(t)$, as follows:

$$Val_R(t) = lub\{\omega_R(t')|t \to_R^* t'\}.$$

Example 2.2. *(continued)* $Val_R(f(a,x))$ *is the tree given in Fig. 1.*

It is easy to see that the function ω_R is computable (in fact, it is primitive recursive). Thus, Val_R actually defines a new kind of interpreter (for a rewriting strategy of this interpreter, see Proposition 4.2 in Chapter 4). From Proposition 2.1 (1) and on the assumption that R is left-linear and Ω-free, it can be shown that Val_R is monotone. Hence, it is uniquely extended into a continuous function on $T^\infty(F,X)$ by:

$$Val_R(T) = lub\{Val_R(t)|t \preceq T\}.$$

Let $ANF_R^\infty = \{lub\Delta|\Delta \subseteq ANF_R \text{ directed }\}$. Then, we have:

Proposition 2.4. Val_R *is a retraction on* $T^\infty(F,X)$ *and its range is* ANF_R^∞.

The output domain of the interpreter forms a cpo, as does the input domain $T^\infty(F,X)$. That is,

Proposition 2.5. ANF_R^∞ *is an algebraic cpo with the base* ANF_R.

By Lemma 2.3 and the continuity of Val_R, we have:

Lemma 2.6. *(Parallelization of execution)*

$$Val_R(T_0[T_1, T_2 ...,]) = Val_R(Val_R(T_0)[Val_R(T_1), Val_R(T_2), ...]).$$

We define a preorder $\precsim_R$ on $T^\infty(F,X)$ by

$$T \precsim_R T' \quad \text{iff} \quad Val_R(T) \preceq Val_R(T'),$$

and an equivalence relation $\approx_R$ by

$$T \approx_R T' \quad \text{iff} \quad T \precsim_R T' \quad \text{and} \quad T' \precsim_R T.$$

Corollary 2.7. $\precsim_R$ *is a precongruence on* $T^\infty(F,X)$.

From Propositions 2.4 and 2.5, and Corollary 2.7, we obtain the main result of this chapter:

Theorem 2.8. *An $F \cup X$-algebra $\mathbf{T}^\infty(F,X)_{\precsim_R}$ is continuous algebra. Moreover, it is isomorphic to the $F \cup X$-algebra $\mathbf{A}_R = << ANF_R^\infty, \preceq>, < f_{\mathbf{A}_R} | f \in F \cup X >>$ defined by:*

$$f_{\mathbf{A}_R}(T_1, ..., T_n) = Val_R(f(T_1, ..., T_n)).$$

The isomorphism maps $[T]_{\precsim_R}$ to $Val_R(T)$.

Thus, each symbol is interpreted as a continuous function, and so are trees. Since we have obtained Corollary 2.7 from Lemma 2.3, we *did* use our assumptions on R to prove the theorem.

The next proposition relates the algebra $\mathbf{T}^\infty(F,X)_{\precsim_R}$ to the domain theory approach [19, 20]. That is, it claims that we can solve the *domain equation*

$$D \cong (X \to D) \to D,$$

if we take the right-hand side as the set of continuous and representable functions.

Proposition 2.9. *(Extensionality) $\mathbf{T}^\infty(F,X)_{\precsim_R}$ is isomorphic to $\mathbf{T}^\infty(F,X)_{\mathbf{T}^\infty(F,X)_{\precsim_R}}$. The isomorphism sends $[T]_{\precsim_R}$ to $T_{\mathbf{T}^\infty(F,X)_{\precsim_R}}$.*

In the proposition, $\mathbf{T}^\infty(F,X)_{\mathbf{T}^\infty(F,X)_{\precsim_R}}$ denotes the algebra of representable functions over $\mathbf{T}^\infty(F,X)_{\precsim_R}$, where the latter is considered to be an F-algebra by forgetting the interpretation of variable symbols. The *solution* $\mathbf{T}^\infty(F,X)_{\precsim_R}$ we have obtained is a model of R as an equational theory. See the following result:

Proposition 2.10. *$t \leftrightarrow_R^* t'$ implies $t \eqsim_R t'$.*

The above implication is proper and it is a gap between the meaning of equations and rewriting rules. For a counterexample, consider $g(x)$ and $h(x)$ with R as in Fig. 2. We do not have $g(x) \leftrightarrow_R^* h(x)$, though $g(x) \eqsim_R h(x)$ holds.

Lemma 2.11. *For a tree T and an approximate normal form u,*

$$u \precsim_R T \text{ iff } u(\preceq \cup \leftarrow_R)^* t \preceq T \text{ for some } t.$$

Corollary 2.12. *For a term t and an Ω-free normal form u,*

$$u \eqsim_R t \text{ iff } t \to_R^* u \text{ iff } Val_R(t) = u.$$

The corollary gives a case that the converse of Proposition 2.10 holds. It also states that Val_R generalizes the conventional interpreter.

3. Universality of the Continuous Algebra Semantics

In this chapter, we discuss universality properties and characterizations of the interpretation $\mathbf{T}^\infty(F,X)_{\precsim_R}$.

We start with initiality. Let Γ_R be the class of continuous F-algebras $\{A | T \precsim_R U$ implies $T_A \sqsubseteq_A U_A\}$, and let $\Phi_R = \{\mathbf{T}^\infty(F,X)_A | A$ is in $\Gamma_R\}$.

Theorem 3.1. $\mathbf{T}^\infty(F,X)_{\precsim_R}$ *is*

(1) free over X in Γ_R, when the interpretation of variable symbols is forgotten, and

(2) initial in Φ_R with continuous homomorphisms.

Corollary 3.2. *$T \precsim_R U$ iff $T_A \sqsubseteq_A U_A$ for any A in Γ_R.*

With respect to finality we have stronger results. A precongruence π is said to be a *continuous implementation of R* iff

(1) $\mathbf{T}^\infty(F,X)_\pi$ is continuous,

(2) $R^{-1} \subseteq \pi$,

(3) $u\pi T$ iff $u \ (\preceq \cup \leftarrow R)^* \cdot \preceq T$, for any u in ANF_R and any T, and

(4) $[u]_\pi$ is compact in $T^\infty(F,X)/\pi$, for any u in ANF_R.

Proposition 3.3. *$\precsim_R$ is the greatest continuous implementation of R.*

Let $\Psi_R = \{\mathbf{T}^\infty(F,X)_\pi | \ \pi$ is a continuous implementation of $R\}$.

Theorem 3.4. $\mathbf{T}^\infty(F,X)_{\precsim_R}$ *is final in* Ψ_R.

Corollary 3.5. *$T \precsim_R U$ iff $T^A \sqsubseteq_A U^A$ for some A in Ψ_R.*

By Theorems 3.1 and 3.4, we have:

Corollary 3.6. *For any two algebras A in Ψ_R and B in Γ_R, there exists a unique continuous homomorphism from A to $\mathbf{T}^\infty(F,X)_B$.*

In other words, an element of A in Ψ_R can be interpreted as a continuous function on the domain of B in Γ_R.

4. Computational Cost

In this chapter, we give a measure of computational complexity to explain the cost spent by the Val_R interpreter. Using the interpreter, we can discuss in a precise manner,

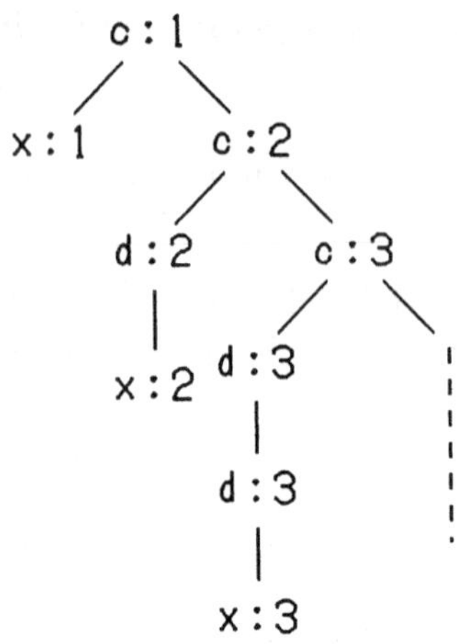

Fig. 3.

for example, programs that produce *infinite lists.* Then, the question arises as to how efficient the programs are. Recall that two non-terminating terms with the same limits can differ in their efficiency. For instance, with R as in Fig. 2, both $g(x)$ and $h(x)$ has the limit shown in Fig. 1. However, $h(x)$ converges faster than $g(x)$. This will become clear by counting, for each node p in the limit, the number of rewriting steps required to obtain p. To do this, we can associate a counter with each node of the tree as illustrated in Fig. 3.

Let $\mathbf{N}$ be the set of non-negative integers. We denote $\mathbf{N} \cup \{\infty\}$ by $\mathbf{N}^\infty$, and extend the usual order $\leq$ on $\mathbf{N}$ with $n \leq \infty$ for any n. Then, the inverse order $< \mathbf{N}^\infty, \leq^{-1} >$ is a bounded-complete algebraic cpo with a *least* element ∞ (in fact, it is an algebraic complete lattice also with a *greatest* element zero). An element $< f, n >$ in $(F \cup X) \times \mathbf{N}^\infty$ is called a *costed symbol*, and it is denoted by $f\colon n$. Similar to (F, X)-trees, we consider well-formed infinite trees whose nodes are labeled with costed symbols. We call such trees *costed* (F, X)*-trees*, and denote the set of those trees by $CT^\infty(F, X)$. We write $CT(F, X)$ for the *set of finite costed* (F, X)*-trees*; they are identified with *costed terms* such as $f\colon 1(a\colon 3,\ \Omega\colon \infty)$. Define an order $\sqsubseteq$ on $(F \cup X) \times \mathbf{N}^\infty$ by: $f\colon n \sqsubseteq g\colon m$ iff $f \sqsubseteq g$ and $n \geq m$. Then, an order $\preceq$ on $CT^\infty(F, X)$ is defined using (*) in Section 1.2 again.

Proposition 4.1. *([14]) $CT^\infty(F, X)$ is a bounded-complete algebraic cpo with the base $CT(F, X)$. The least element is $\Omega\colon \infty$.*

The least element represents both inexistence of information and the worst of computation cost.

For T in $T^\infty(F,X)$ and n in $\mathbf{N}$, $T{:}\,n$ denotes the costed tree obtained from T by replacing any label f ($\neq \Omega$) by $f{:}\,n$, and Ω by $\Omega{:}\,\infty$. Let γ be a monotone function on $\mathbf{N}$, and an (infinite) development sequence $\rho = < t_0, t_1, \ldots >$. We define the *symbolic value of* ρ *under the efficiency* γ, denoted by $Val_R^\gamma(\rho)$, as follows:

$$Val_R^\gamma(\rho) = lub\{\omega_R(t_n){:}\,\gamma(n) | n \geq 0\}.$$

In the definition, $\gamma(n)$ represents the cost to obtain the n-th output $\omega_R(t_n)$. $Val_R^\gamma(\rho)$ is defined for any ρ since $\{\omega_R(t_n){:}\,\gamma(n) | n \geq 0\}$ is bounded by $Val_R(t_0){:}\,0$.

Example. *Consider R in Fig. 2 again. Let* $\gamma(n) = n$ *for any* n*, and let*

$$\rho = < g(x),\ c(x, g(d(x))),\ c(x, c(d(x), g(d(d(x))))), \ldots > .$$

Then, by the definition, we have

$$Val_R^\gamma(\rho) = lub\{\Omega{:}\,\infty,\ c{:}\,1(x{:}\,1, \Omega{:}\,\infty),\ c{:}\,2(x{:}\,2, c{:}\,2(d{:}\,2(x{:}\,2), \Omega{:}\,\infty)), \ldots\}.$$

This is the tree shown in Fig. 3.

Now, we define the *symbolic value of t under the efficiency* γ by:

$$Val_R^\gamma(t) = lub\{Val_R^\gamma(\rho) | \rho \in Dev_R(t)\}.$$

We denote the projection from $CT^\infty(F,X)$ to $T^\infty(F,X)$ by Sym. It is easy to see that $Val_R(t) = Sym(Val_R^\gamma(t))$.

Proposition 4.2. $Val_R^\gamma(t) = Val_R^\gamma(FDS_R(t))$.

In other words, the full development sequence $FDS_R(t)$ gives the optimal reduction whether or not the computation terminates. The terminating case is treated as follows:

Corollary 4.3. *Suppose that t has an* Ω*-free normal form u. Let* $\bar{u} = Val_R^\gamma(t)$*, and K be the length of the shortest development sequence from t to u. Then,*

$$\gamma(K) = Max\{n | p \in Node(\bar{u}), \bar{u}(p) = f{:}\,n\}.$$

Since Val_R^γ is monotone, we can uniquely extend it into a continuous function on $CT^\infty(F,X)$. Define a preorder $\sqsubseteq_R^\gamma$ by:

$$T \sqsubseteq_R^\gamma T' \quad \text{iff} \quad Val_R^\gamma(T) \preceq Val_R^\gamma(T').$$

Proposition 4.4. $t \rightarrow_R^* t'$ *implies* $t \sqsubseteq_R^\gamma t'$.

Reduction may, therefore, improve the complexity of a term, while it preserves the symbolic information (see Corollary 2.10). Using Lemma 2.3, we can show that:

Theorem 4.5. $\sqsubseteq_R^\gamma$ *is a precongruence.*

That is, $\sqsubseteq_R^\gamma$ is compatible with program composition. Note that Lemma 2.3 does not hold if we replace $\rightarrow_{\delta R}^n$ by $\rightarrow_R^n$. This is why we count the number of rewriting steps by $\rightarrow_{\delta R}$, instead of $\rightarrow_R$. A TRS is considered to be a model of parallel computation in this formulation. As a consequence of the theorem, we obtain an ordered algebra $\mathbf{T}^\infty(F, X)_{\sqsubseteq_R^\gamma}$. This gives an interpretation of symbols where we take the complexity of programs as a part of the semantics. It is open whether or not the algebra is continuous. In addition,

5. Related work

The semantics treated in this paper is based on the notion of continuous algebra and the limit of a non-terminating reduction. Such a framework was first introduced by Nivat [15] for the *algebraic semantics* of *recursive program schemes (RPSs)*, and it has been developed by Courcelle, Guessarian and many other researchers (see [3-7, 15, 16, 18]). The present paper generalizes some of their results and adds some new discoveries. See also [13].

On the semantics of the lambda calculus, the notion of approximate normal forms is found in Lévy [11] and Wadsworth [20]. Our key lemma, Lemma 2.3., is a slightly stronger counterpart of those in the above papers. In fact, it is proved by *labeled reduction* technique due to [11] (The proof has been omitted in the present paper). Wadsworth utilized approximate normal forms, together with *head normal forms,* to characterize Scott's D_∞- model from the viewpoint of computational behavior of the lambda terms. A similar result on TRSs has been obtained by the authors [12], which parallels the result on RPSs due to Raoult and Vuillemin [18].

For the semantics of TRSs, Courcelle [4] and Boudol [3] have also taken approaches based on the notion of continuous algebra. The former studied an initial continuous algebra determined by confluent, terminating and monotone TRSs. Similar to the present paper, a retraction induced by rewriting played an important role in the construction (see also [10]). The latter studied a wider class of TRSs than ours, though the results obtained are weaker. It treated nondeterministic computation which does not allow the straightforward construction using retraction.

Finally, as far as the authors know, concrete measures of computational complexity from the viewpoint of order-theoretic approach was first given by the authors [14]. Chapter 4 of the present paper is a summerization of [14] with some additional results.

Acknowledgment

The authors wish to thank Professors Namio Honda and Teruo Fukumura for their encouragements to conduct this work. They also thank Professor Bruno Courcelle for his stimulative comments at the starting point of this research; Professors Toshiki Sakabe and Tomio Hirata, and Mr. Ken Mano for their helpful discussions; and Mr. Olav Stål Aandraa for his suggestions for polishing the manuscript of this paper. Their work is partly supported by the Grants from Ministry of the Education, Science and Culture of Japan (Developmental Scientific Research No. 62880007, Co-operation Research No.62302032 and Scientific Research on Priority Areas No. 63633008)

References

[1] ADJ, "Initial Algebra Semantics and Continuous Algebras," J.ACM 24, pp. 68-95(1977).

[2] J. Avenhaus, "On the Descriptive Power of Term Rewriting Systems," J. of Symbolic Computation 2, pp. 109-122 (1986).

[3] G. Boudol, "Computational Semantics of Term Rewriting Systems," [16, pp. 167-236].

[4] B. Courcelle, "Infinite Trees in Normal Form and Recursive Equations Having a Unique Solution," Math. Systems Theory 13, pp. 131-180 (1979).

[5] B. Courcelle, "Fundamental Properties of Infinite Trees," Theor. Comput. Sci. 25, pp. 95-169(1983).

[6] I. Guessarian, "Algebraic Semantics," LNCS 99, Springer-Verlag(1978).

[7] I. Guessarian, "Survey on Classes of Interpretations and Some of Their Applications," [16, pp. 383-410].

[8] G. Huet, "Confluent Reductions: Abstract Properties and Applications to Term Rewriting Systems," J. ACM 27, pp. 797-821 (1980).

[9] G. Huet and J.-J. Lévy, "Call by Need Computations in Non-Ambiguous Linear Term Rewriting Systems," Raport Laboria 359, IRIA (1979).

[10] M.R. Levy and T.S.E. Maibaum, "Continuous data types," SIAM J. Comput. 11, pp. 201-216(1982).

[11] J.-J.Lévy, "An Algebraic Interpretation of the $\lambda\beta K$-calculus and a Labeled λ-calculus," Theor. Comp. Sci. 2, pp. 97-114(1976).

[12] T. Naoi and Y. Inagaki, "The Conservative Extension and Behavioral Semantics of Term Rewriting Systems," Technical Research Report No.8604, Department of Information Science, Nagoya University(1986).

[13] T. Naoi and Y. Inagaki, "The Relation between Algebraic and Fixedpoint Semantics of Term Rewriting Systems," Technical Report COMP86-37, IEICE (1986).

[14] T. Naoi and Y. Inagaki, "Time-Cost of Infinite Computation in Term Rewriting Systems", Technical Report COMP88-34, IEICE(1988).

[15] M. Nivat, "On the Interpretation of Recursive Polyadic Program Schemes," Symposia Mathematics 15, Rome, pp. 255-281(1975).

[16] M. Nivat and J.C. Reynolds, Eds., "Algebraic Methods in Semantics," Cambridge University Press (1985).

[17] M. J. O'Donnell, "Computing in Systems Described by Equations," LNCS 58, Springer-Verlag (1977).

[18] J.-C. Raoult and J. Vuillemin, "Operational and Semantic Equivalence Between Recursive Programs," J.ACM 27, pp.772-796(1980).

[19] D. Scott, "Data Types as Lattices," SIAM J. Comput 5, pp. 522-587(1976).

[20] D. Scott, "Domains for Denotational Semantics," in: "ICALP 82," M. Nielson and E.M. Schmidt, Eds., LNCS 140, pp. 577-613, Springer-Verlag (1982).

[21] C.P. Wadsworth, "The Relation between Computational and Denotational Properties for Scott's D_∞-Models of the Lambda-Calculus," SIAM J. Comput. 5, pp. 488-521(1976).

Optimization by non-deterministic, lazy rewriting

Sanjai Narain
RAND Corporation
1700 Main Street
Santa Monica, CA 90406
narain@rand.org

ABSTRACT

Given a set S and a condition C we address the problem of determining which members of S satisfy C. One useful approach is to set up the generation of S as a tree, where each node represents a subset of S. If from the information available at a node, we can determine that no members of the subset it represents satisfy C, then the subtree rooted at it can be pruned, i.e. its generation suppressed. Thus, large subsets of S can be quickly eliminated from consideration. We show how such a tree can be simulated by interpretation of non-deterministic rewrite rules, and its pruning simulated by lazy evaluation.

1.0 INTRODUCTION

Given a condition C, and a set S, the problem is to compute those members of S which satisfy C. The obvious way of solving it is to simply generate all members of S, and test which of these satisfy C. However, a much more effective approach can be the following: set up the generation of S as a tree, each node in which represents a subset of S. The subset represented by an internal node is the union of the subsets represented by each of its immediate descendants. External, or leaf nodes represent either empty sets, or singleton sets consisting of individual members of S. From the information available at a node N, check if any members of the set represented by N satisfy C. If not, prune, i.e. do not generate, the subtree rooted at N. Thus, large numbers of members in S can quickly be eliminated from consideration, particularly if trees are deep and pruning occurs near the root.

For example, S could be the set of all lists of the form [X1,..,Xm], where each Xi is in {0,1,..,9}. C could be the condition that the sum of all numbers in a list be equal to some fixed number K. We could set up the generation of S as a tree where each node is of the form [A1,A2,..,Ak], 0=<k=<m, each Ai in {0,..,9}, and its immediate descendants are the nodes of the form [A1,..,Ak,Ak+1], Ak+1 in {0,..,9}. The subset of S it represents is all lists whose first k elements are, respectively, A1,..,Ak. Clearly, if it can be determined that the sum of A1,..,Ak is greater than K, then the tree rooted at [A1,..,Ak] need not be generated. In particular, none of the 10**(m-k) subsets that [A1,..,Ak] represents, need be generated.

A convenient way to generate sets is via non-deterministic algorithms e.g. [Sterling & Shapiro 1986]. A non-deterministic algorithm SA for generating S is, typically, a definition of what it means for an object to belong to S. From it, an interpreter for SA can automatically compute all members of S. Interpretation of SA can usually be laid out in the form of a tree. Nodes in this tree represent unfinished computations. Information available at these is that gathered in the computation along the branch from the root to these.

Now, we can try to write SA in such a way that the tree resulting from its interpretation is similar in structure, and information content at nodes, to the original tree for generating S. For the purpose of tree-pruning we can try to use this tree in place of the original one. To ensure that whenever a new node is created in this tree, the condition C is evaluated, if possible, using the information available at this node, we need to suitably interleave steps in SA with those of CA, the algorithm implementing C.

For example, the set of all lists of the form [X1,..,Xm], each Xi in {0,..,9}, can be generated by the following non-deterministic Prolog program:

```
tuple([X1,X2,..,Xm]):-d(X1),d(X2),..,d(Xm).

d(0). d(1). d(2). d(3). d(4). d(5). d(6). d(7). d(8). d(9).
```

Now the query tuple([Y1,..,Ym]), Y1,..,Ym variables, will halt with each instantiation of Y1,..,Ym to members of {0,1,..,9}. Moreover, it can be easily seen that the structure of the SLD-search tree for this query, and information available at its nodes (in the form of partial answer substitutions [Lloyd 1984]), are similar to those with the tree defined in the second paragraph above.

A Prolog program for checking whether the sum of all members of a list is equal to K is:

```
sum([],S,S).
sum([U|V],S,K):-A is U+S,A=<K,sum(V,A,K).
```

Now, the query sum(L,0,K) will succeed if the sum of members of L is equal to K, otherwise it will fail. The second rule says that the sum of members of [U|V], given S, is K provided U+S=<K, and the sum of members of V, given U+S, is K. Thus, if U+S>K, no further members of V need be considered.

Tree-pruning would occur if we could suitably interleave steps in these two programs. In particular, we need to ensure that whenever the tuple program instantiated a new variable in [X1,..,Xm], the variables already instantiated are immediately tested by sum, and moreover, that sum never instantiates any variables. Of course, it is clear that pruning would not occur with the simple-minded Prolog query:

```
tuple(X),sum(X,0,K).
```

as tuple would generate an entire tuple of m digits, before passing it on to sum.

There are three possible approaches for interleaving steps in SA with those in CA. First, we can combine SA and CA into a single algorithm, in particular, by explicitly programming the appropriate interleaving. However, it is highly desirable to keep SA and CA separate, i.e. think about them, and develop them independently of each other. If not, the combined algorithm can become quite complex, especially if SA and CA are complex.

Second, we can keep SA and CA separate but interleave their steps by connecting them via a facility for coroutining. For example, one can use the variable annotations (?,^) of IC-Prolog [Clark & McCabe 1979]. However, if SA and CA are complex, connecting them together may not be simple, as intricate knowledge of their execution may be required. Also, there do not seem to be efficient enough implementations of languages such as IC-Prolog, in which one can write non-deterministic programs such as tuple above, and also do coroutining.

Third, we can develop SA and CA separately, but in such a way that the interleaving is accomplished transparently, in the natural course of interpreting SA and CA, without our having to program it. This seems possible in languages such as LEAF [Barbuti et al. 1986], or FUNLOG [Subrahmanyam & You 1984] which combine logic programming with lazy rewriting. A similar possibility also exists in Prologs which perform intelligent backtracking [Kumar & Lin 1988, Bruynooghe & Pereira 1984, Chang & Despain 1984]. However, implementations of these languages, efficient enough for practical programming, seem to be still under development.

We propose a realization of the third approach within the framework of rewriting. Specifically, we propose F*, a first-order, lazy non-deterministic rewrite rule system [Narain 1988]. We show how SA and CA can be developed independently of each other in F*, yet steps in them interleaved by non-deterministic, lazy rewriting in such a way that tree-pruning is accomplished.

F* programs can be compiled into Horn clauses in such a way that when SLD-resolution interprets these it directly simulates the behavior of the lazy F* interpreter. In particular, the non-determinism of F* is mapped on to the non-determinism of Horn clauses. Due to the nature of the clauses obtained by compilation [Narain 1988], we effectively obtain a lazy interpreter which operates at roughly the same speed as does Prolog. For problems in which lazy evaluation does not reduce lengths of computation, F* is somewhat slower than Prolog. Otherwise, F* is faster than Prolog by unbounded, even infinite amounts. Thus, F* seems to be efficient enough for practical programming.

Intuitively, tree-pruning is achieved in F* as follows: Let E be a ground term. In F*, as with other rewrite rule systems, there can be more than one reduction starting at E. However, in F* there can also be more than one *lazy* reduction starting at E. Due to non-determinism, E can possess more than one normal form. A *reduction-completeness* theorem states that each of these normal forms can be computed by generating all the lazy reductions starting at E. These can all be laid out in the form of a *lazy reduction-tree*. This tree is analogous to an SLD-search tree [Lloyd 1984], while the lazy reduction strategy is analogous to SLD-resolution.

Given the set S, we can define a non-deterministic algorithm in F* such that some ground term E possesses as normal forms, all the members of S. Moreover, we define this algorithm in such a way that the lazy reduction-tree rooted at E is similar to that we have in mind for generating S, and upon which we wish to perform the pruning.

Now, for conditions such as C, we take the point of view that *when they hold for objects, they return, not the truth-value true, but the objects themselves. Furthermore, if they do not hold, they do not return anything, but simply fail.* Thus, where CA is the function symbol defining condition C in F* with this point of view, the normal forms of the term CA(E) represent the members of S which satisfy C.

Given CA(E), the F* interpreter generates the lazy reduction-tree rooted at CA(E). *Due to laziness, reduction of* E *is interleaved with evaluation of* CA. In particular, whenever new information about E becomes available, an attempt is automatically made to evaluate CA. If at a node it is discovered that CA does not hold, no further descendants of it are generated. Thus, the tree-pruning we seek is automatically achieved. In particular, the lazy reduction tree rooted at CA(E) is, usually, much smaller than that rooted at E.

This paper discussess solutions to four problems illustrating the above idea. These have all been implemented and tested in the LOG(F) system, which is simply a logic programming system augmented with an F* compiler. Performance figures obtained for these seem to compare favorably with those obtained by languages such as CHIP [Dincbas et al. 1988, van Hentenryck & Dincbas 1987], which contain specialized capabilities for solving combinatorial problems.

F* can be viewed as a way of bringing to rewriting, one of the most powerful features of logic programming: the ability to develop non-deterministic algorithms. Due to the compilation of F* in Prolog, it can also be viewed as a means of doing lazy evaluation in Prolog. Before we discuss the problems solved by our approach, we first define the F* system.

2.0 DEFINITION OF F*

F* is intended mainly for lazy reduction of ground terms. Nevertheless, it forms a sufficient basis for functional programming.

Function symbols are partitioned, *in advance*, into constructors and non-constructors. For example, 0, 1, 2, 3.1415,..., true, false, [] are 0-ary constructor symbols and | a 2-ary constructor symbol.

A **term** is either a variable, or an expression of the form f(t1,..tn) where f is an n-ary function symbol, and each ti is a term. A ground term E is said to **match** another term F, with substitution α, if E=Fα. A **reduction rule** is of the form LHS=>RHS, where LHS and RHS are terms, satisfying the following restrictions:

(a) LHS is of the form f(L1,..,Lm), f an m-ary non-constructor function symbol, and each Li either a variable, or a term of the form c(X1,..,Xn), c an n-ary constructor symbol, and each Xi a *variable.*

(b) A variable occurs at most once in LHS.

(c) All variables of RHS occur in LHS.

Note that reduction rules with left-hand-sides of arbitrary depth can easily be expressed in terms of rules with left-hand-sides of depth at most two, as required by (a). For example, fib(s(s(X)))=>plus(fib(X),fib(s(X))) can be expressed as fib(s(A))=>g(A), g(s(X))=>plus(fib(X),fib(s(X))). An **F* program** is a set of reduction rules. Some examples of F* programs are:

```
append([],X)=>X
append([U|V],W)=>[U|append(V,W)]
int(N)=>[N|int(s(N))].
merge([A|B],[C|D])=>if(lesseq(A,C),[A|merge(B,[C|D]),[C|merge([A|B],D)]).
if(true,X,Y)=>X.
if(false,X,Y)=>Y.
```

Note that there is no restriction that F* programs be *Noetherian*, or even *confluent*. In particular, infinite structures can be freely defined and manipulated in F*. Also, terms can be simplified in more than one way, a fact which we exploit for implementing tree-pruning.

Let P be an F* program and E and E1 be ground terms. We say **E=>$_P$E1** if there is a rule LHS=>RHS in P such that E matches LHS with substitution σ and E1 is RHSσ. Where E,F,G,H, are ground terms, let F be the result of replacing an occurrence of G in E by H. Then we say **F=E[G/H]**. Let P be an F* program and E a ground term. Let G be a subterm of E such that G=>$_P$H. Let E1=E[G/H]. Then we say that **E->$_P$E1**. **-*>$_P$** is the reflexive-transitive closure of ->$_P$. The subscript P is dropped if clear from context.

A ground term is said to be in **simplified form,** or simplified, if it is of the form c(t1,..,tn) where c is an n-ary constructor symbol, n>=0, and each ti is a ground term. F is called a simplified form of E if E-*>F and F is in simplified form. Simplified forms can be used to represent finite approximations to infinite structures. For example, [0|int(s(0))] is a simplified form, and is a finite approximation to [0,s(0),s(s(0)),..].

A ground term is said to be in **normal form** if each function symbol in it is a constructor symbol. F is called a normal form of E if E-*>F and F is in normal form. Note that this notion of normal form is different from the usual one which has to do with non-reducibility. This does not lead to any loss of generality, at least for programming purposes. Moreover, it allows us to define the notion of *failure form* below, which is useful for tree-pruning.

Let P be an F* program. A **reduction** in P is a, possibly infinite, sequence E1,E2,E3... such that for each i, Ei->$_P$Ei+1. A **successful reduction** in P is a reduction E0,..,En, n>=0, in P, such that En is simplified.

Let P be an F* program. We now define a reduction strategy, **select$_P$** for P. Informally, given a ground term E it will select that subterm of E whose reduction is necessary in order that some => rule in P apply to the whole of E. *In this, is implicit its laziness.* The relation select$_P$, whose second argument is the subterm selected from E, is defined by the following pseudo-Horn clauses:

select$_P$(E,E) if E=>$_P$X.
select$_P$(E,X) if
    E=f(T1,..,Ti,..,Tn), and
    there is a rule f(L1,..,Li,..,Ln)=>RHS in P, and
    Ti does not match Li, and
    select$_P$(Ti,X).

The first rule states that if E is the given ground term, and there exists another term X such that E=>X, then E itself can be selected from E.

In the second rule, E=f(T1,..,Tn), and there is some rule f(L1,..,Ln)=>RHS, such that for some i, Ti in T1,..,Tn does not match Li in L1,..,Ln. In order to reduce E by this rule, it is necessary to reduce Ti. Thus, a term in Ti must be recursively selected for reduction. This rule is a schema, so that an instance of it is assumed written for each 1=<i=<n, and each non-constructor function symbol f. For example, where P is the set of reduction rules which appear above, we have the following:

select(merge(int(1),int(2)),int(1)).
select(merge(int(1),int(2)),int(2)).
select(merge([1,3],int(2)),int(2)).
select(merge([1,2],[3,4]),merge([1,2],[3,4])).
If E=[1|merge(int(1),int(2))] then select is undefined for E.

Note that select is non-deterministic, in that given E, there can be more than one F, such that select(E,F). Thus, starting at E, there can be more than one reduction computed by select. Also, select is more general than a leftmost-outermost strategy. For example, from the rules:

f(X,[])=>[].
a=>a.
b=>[].

the only leftmost-outermost reduction of f(a,b) is f(a,b),f(a,b),f(a,b),... However, repeated application of select yields only the finite reduction f(a,b),f(a,[]),[].

Let P be an F* program and E,G,H be ground terms. Suppose select_P(E,G) and G$=>_P$H. Let E1 be the result of replacing G by H in E. Then we say that E reduces to E1 in an **N-step** in P. The prefix N in N-step is intended to connote normal order. Let P be an F* program. An **N-reduction** in P is a reduction E1,E2,.... in P such that for each i, Ei reduces to Ei+1 in an N-step in P. We now have:

Theorem A. Reduction-completeness of F* for simplified forms. Let P be an F* program and D0 a ground term. Let D0,D1,..,Dn be a successful reduction in P. Then there is a successful N-reduction D0,E1,..,Em in P, such that Em-*>Dn.

A sketch of its proof can be found in the Appendix. It states that to reduce a ground term to a simplified form it is sufficient to generate only the N-reductions starting at it.

Note that if a term E is already simplified, e.g. E=[1|append([],[])], then select is undefined for E. Thus, an N-reduction ending at E cannot be extended further. If we wish to compute normal forms of E, we need a reduction strategy more general than select. It turns out that this can be based upon repeated application of select. Specifically, where P is an F* program, we define a reduction strategy **select-r$_P$**, where r stands for repeated or recursive, by the following pseudo-Horn clauses:

select-r_P(E,F) if select_P(E,F).
select-r_P(c(T1,..,Ti,..,Tm),F) if
c is a constructor symbol, and
select-r_P(Ti,F).

Thus, select-r is like select except that if a ground term is in simplified form, it recursively calls select on one of the arguments of the outermost constructor symbol. So, its repeated use can yield normal-forms of ground terms. Again, the second rule is a schema so that an instance of it is assumed written for each 1=<i=<m, and each constructor symbol c.

Let P be an F* program and E,G,H be ground terms. Suppose select-r_P(E,G) and G$=>_P$H. Let

E1 be the result of replacing G by H in E. Then we say that E reduces to E1 in an **NR-step** in P. NR is intended to connote normal-repeated.

Let P be an F* program. An **NR-reduction** in P is a reduction E1,E2,.... in P such that for each i, Ei reduces to Ei+1 in an NR-step in P. We now have:

Theorem B. Reduction-completeness of F* for normal forms. Let P be an F* program and D0 a ground term. Let D0,D1,..,Dn be a reduction in P, where Dn is in normal form. Then there is an NR-reduction D0,E1,..,Em=Dn, in P.

Again, a sketch of its proof can be found in the Appendix. It states that to compute normal forms of a term, it is sufficient to generate only the NR-reductions starting at it. Of course, not all N- or NR-reductions are finite, even when normal forms exist. For example, with the program:

```
a=>a.
a=>[].
f([])=>[].
```

There is an infinite N-reduction f(a),f(a),f(a),.... However, there is also a finite N-reduction f(a),f([]),[]. Thus, some searching among alternative N- or NR-reductions may be required to compute simplified, or normal forms. However, laziness ensures that search paths are cutoff as soon as possible. In [Narain 1988] is studied D(eterministic)F*, a restriction of F*, in which simplified, or normal forms may be computed without any search. However, in DF*, every term has at most one normal form, and so it is not relevant to this paper.

Let P be an F* program, and E a ground term. An **NR-tree** rooted at E is constructed as follows: E is the root node. The immediate descendants of any node Q in the tree are Q0,Q1,..,Qk, where Q reduces to each Qi in an NR-step. NR-trees are also called **lazy reduction-trees**.

Let P be an F* program and E a ground term. E is said to be a **failure form** if E is not a normal form, and E cannot be reduced to another term in an NR-step in P. Note that E may still be reducible. Failure forms are analogous to failure nodes in SLD-search trees [Lloyd 1984], while normal forms are analogous to success nodes (or empty goals). If while generating a lazy reduction-tree, a failure form is encountered, the interpreter backs up and generates other parts of the tree. *It is for this reason that normal forms are not defined in terms of non-reducibility.*

3.0 TUPLE SUM

We now show how the example in Section 1.0 can be programmed in F* to achieve the type of tree-pruning we desire. An F* program to compute the set S of all lists [X1,..,Xm], each Xi in

{0,..,9} can be:

```
tuple=>[d,d,d,..,d].

d=>c(0). d=>c(1). d=>c(2). d=>c(3). d=>c(4).
d=>c(5). d=>c(6). d=>c(7). d=>c(8). d=>c(9).
```

Here c, |, [], and 0,1,..,9 are constructor symbols, while d and tuple are non-constructor symbols. Now, the term tuple, whose right hand side contains m ds, possesses 10**m normal forms, each of the form [c(X1),c(X2),..,c(Xm)], Xi in {0,..,9}. Each normal form can be taken to represent a member of S. Assuming that m=2, a subtree of the lazy reduction-tree rooted at tuple is:

```
                tuple
                  |
                  |
                [d,d]
                / | \
               /  |  \
        [c(0),d]...[c(9),d]
           / | \      / | \
          /  |  \      ...
         /   |   \
[c(0),c(0)] ... [c(0),c(9)]
```

Each internal node, except the root, has ten descendants. Each tip is of the form [c(X1),c(X2)], Xi in {0,..,9}. This tree can be seen to be similar to that defined in the second paragraph of Section 1.0.

We now define the condition that the sum of all members of a list be equal to K. *Again, we do this in such a way that when it hold for a list, it returns, not true, but the list itself. If it does not hold, it does not return anything, but simply fails.* A satisfactory definition is:

```
sum_eq([],c(S),c(K))=>cond(equal(S,K),[]).
sum_eq([c(U)|V],c(S),c(K))=>cond((U+S)=<K,[c(U)|sum_eq(V,c(S+U),c(K))]).

cond(true,X)=>X.
```

These rules exactly parallel the two Prolog clauses for sum, in Section 1.0. Here, true is a constructor symbol (along with [], |, and c), and sum_eq, cond, equal, +, and =< are non-constructor symbols. The last three are F* primitives and are evaluated eagerly. Now, where m=2, the term sum_eq(tuple,c(0),c(0)) possesses as the only normal form, the term [c(0),c(0)]. Effectively, sum_eq is evaluated as soon as enough information becomes available about its first argument. If it does not hold, then the rest of the argument is not evaluated further. In particular, one branch in the lazy reduction-tree rooted at sum_eq(tuple,c(0),c(0)) is:

```
sum_eq(tuple,c(0),c(0))
       |
sum_eq([d,d],c(0),c(0))
       |
sum_eq([c(0),d],c(0),c(0))
       |
cond(0=<0,[c(0)|sum_eq([d],c(0),c(0))])
       |
cond(true,[c(0)|sum_eq([d],c(0),c(0))])
       |
[c(0)|sum_eq([d],c(0),c(0))]
       |
[c(0)|cond(0=<0,[c(0)|sum_eq([],c(0),c(0))])]
       |
[c(0)|cond(true,[c(0)|sum_eq([],c(0),c(0))])]
       |
[c(0),c(0)|sum_eq([],c(0),c(0))]
       |
[c(0),c(0)|cond(equal(0,0),[])]
       |
[c(0),c(0)|cond(true,[])]
       |
[c(0),c(0)]
```

Some, minor steps involving + have been omitted. However, all other branches in this tree rapidly reach failure forms. For example:

```
sum_eq(tuple,c(0),c(0))
       |
sum_eq([d,d],c(0),c(0))
       |
sum_eq([c(1),d],c(0),c(0))
       |
cond(1=<0,[c(1)|sum_eq([d],c(0),c(0))])
       |
cond(false,[c(1)|sum_eq([d],c(0),c(0))])
```

Thus, the lazy reduction tree T2 rooted at sum_eq(tuple,c(0),c(0)) is much smaller than the tree T1, rooted at tuple, as was our objective. In particular, while T1 had 100 leaves, T2 has only

19. The gap widens as larger tuples are considered (for the same value of K).

Moreover, we are also able to define the generation algorithm, tuple independently of the testing algorithm, sum_eq. The interleaving between these two is naturally accomplished in the process of generating the lazy reduction-tree.

A similar program has been developed for computing all the ways in which a dollar can be changed using half-dollars, quarters, dimes, nickels, and pennies [Polya 1973].

4.0 SUBSET SUM

Subset sum is problem SP13 in Appendix: A list of NP-complete problems, [Garey & Johnson 1979]. Given a finite set A, and size $s(a) \in Z^+$ for each $a \in A$, and a positive integer K, the problem is to determine whether there is a subset $A1 \subseteq A$ such that the sum of the sizes of the elements in A1 is exactly K. We consider a special case of this problem where A is a list of positive integers, and s is the identity function. Sets are represented as lists. Subsets can be computed non-deterministically by the following F* program:

```
subset([])=>[].
subset([U|V])=>[U|subset(V)].
subset([U|V])=>subset(V).
```

Now the term subset([c(1),c(2),c(3)]) possesses as normal forms each of the eight subsets of [c(1),c(2),c(3)]. Each node Q in the lazy reduction-tree rooted at subset([a1,a2,..,am]) is of the form [U1,..,Un|subset([A1,..,Ap])], where each Ui and Aj is in {a1,..,am}, 0=<n,p=<m. The number of normal forms appearing at the leaves of the subtree rooted at Q is 2**p. The program to test whether sum of numbers in a subset is equal to K, is exactly the same as in the previous section.

In the lazy reduction-tree for sum_eq(subset([a1,a2,..,am]),c(0),c(K)), subset([a1,..,am]) is reduced to [U1,..,Un|subset([A1,..,Ap])], and whenever the sum of U1,..,Un is greater than K, none of the 2**p subsets of the form [U1,..,Un|X] are generated.

5.0 N-QUEENS

The problem is to place N queens on an NxN chess board so that no two queens attack each other. It is easily seen that each queen must be in a distinct row and column, so that candidates for solutions can be represented by permutations of the list [1,2,..,N]. The position of the ith queen in a permutation p is [i,q] where q is is the ith element of p. The problem now reduces to generating all permutations of [1,2,..,N] and testing whether they are safe, or represent a solution. Permutations can be generated by the following F* program:

```
perm([])=>[].
perm([U|V])=>insert(U,perm(V)).
insert(U,X)=>[U|X].
insert(U,[A|B])=>[A|insert(U,B)].
```

In the lazy reduction-tree rooted at perm([1,..,N]) if there is a node Q of the form [U1,..,Up|Z], p>=1, each Ui in {1,..,N}, and Z unsimplified, then the subtree rooted at Q contains (N-p)! leaves, each representing a permutation of the form [U1,..,Up|_]. If at Q it is determined that [U1,..,Up] already form an unsafe configuration, then none of these (N-p)! permutations need be generated.

The condition safe, as usual, is defined in such a way that if it holds for a list, it returns that list itself:

```
safe([])=>[].
safe([U|V])=>[U|safe(nodiagonal(U,V,1))].
nodiagonal(U,[],N)=>[].
nodiagonal(U,[A|B],N)=>cond(noattack(U,A,N),[A|nodiagonal(U,B,N+1)]).
noattack(U,A,N)=>neg(equal(abs(U-A),N)).
cond(true,X)=>X.
```

Here neg, equal, and abs are F* primitives computing, respectively, logical inversion, syntactic equality, and real number modulus. Finally, queens can be computed by:

```
queens(X)=>safe(perm(X)).
```

In particular, queens([1,2,3,4]) yields [2,4,1,3], and [3,1,4,2].

6.0 SEND+MORE=MONEY

The problem is to assign the variables S,E,N,D,M,O,R,Y to distinct values in {0,1,..,9} so that SEND+MORE=MONEY is a correct equation, where for example, SEND is interpreted as 1000*S+100*E+10*N+D. Clearly, there are 10p8 assignments, where NpR is the number of permutations of N distinct things taken R at a time.

The main challenge here is designing an appropriate condition which, given an assignment A to the first k variables, can determine that A *cannot* be part of any correct assignment to all 8 variables. If the condition is simply to find the values of SEND, MORE, and MONEY, and check the addition, then it requires complete assignment before evaluation. This degenerates to checking each of the 10p8 assignments, and there is no tree-pruning.

A suitable condition is based upon checking the addition from right to left, the way a child would do it. In particular, given an assignment, check whether the list [D,E,Y,N,R,E,E,O,N,S,M,O,0,0,M] represents a correct sum, in that D+E yields Y as sum and the carry+N+R yields E as sum, and so on. Now, for example, all assignments with D=0,E=1,Y=2 can be discarded, and substantial pruning can take place. The following four F* rules construct the above list:

```
form_1([c(D),c(E),c(Y)|Z])=>[D,E,Y|form_2(D,E,Y,Z)].
form_2(D,E,Y,[c(N),c(R)|Z])=>[N,R,E|form_3(D,E,Y,N,R,Z)].
form_3(D,E,Y,N,R,[c(O)|Z])=>[E,O,N|form_4(D,E,Y,N,R,O,Z)].
form_4(D,E,Y,N,R,O,[c(S),c(M)|Z])=>[S,M,O,0,0,M].
```

Function admit below, takes as input, an initial carry of 0, and the above list, checks whether they represent a correct sum, and if so, return the list as output. Functions sum and carry are F* primitives yielding the obvious results:

```
admit(_,[])=>[].
admit(C,[A,B,D|Z])=>cond(equal(sum(C,A,B),D),[A,B,D|admit(carry(C,A,B),Z)]).

cond(true,X)=>X.
```

The program below computes permutations of a list of L items, taken N at a time, such that each item in a permutation is distinct. If N is zero, there is a single permutation []. Otherwise, some element from L is removed, and put at the front of a permutation of the resulting list of items, taken N-1 at a time. Note that the conditional function, if, is 3-ary, in contrast to cond, which was binary. The symbol pair is a binary constructor.

```
npr(L,N)=>if(equal(N,0),[],npr_aux(remove(L),N)).
npr_aux(pair(U,V),N)=>[U|npr(V,N-1)].
remove([U|V])=>pair(U,V).
remove([U|V])=>remove_aux(U,remove(V)).
remove_aux(U,pair(A,B))=>pair(A,[U|B]).
if(true,X,Y)=>X.
if(false,X,Y)=>Y.
```

Finally, instantiations of [D,E,Y,N,R,E,E,O,N,S,M,O,0,0,M] such that it is a correct sum are obtained by computing normal forms of soln:

```
soln=>admit(0,form_1(npr(digits,8))).

digits=>[c(0),c(1),c(2),c(3),c(4),c(5),c(6),c(7),c(8),c(9)].
```

7.0 PERFORMANCE FIGURES

These figures provide some idea of the performance we have been able to obtain with the above algorithms using our current implementation of F*. If we try to solve these problems in Prolog, using unintellligent generate-and-test, the time taken is far too much to be practical.

Time in seconds on SUN-3/50 with 4 MB main memory	
Changing dollar: all 292 solutions	230
Zebra Puzzle: Only solution	35
SEND+MORE=MONEY: All 25 solutions	20
8-Queens: All solutions	20
9-Queens: All solutions	97
15-Queens: First solution	37

8.0 SUMMARY

A great deal of work in rewriting has focussed on *confluent* rewrite rule systems. Of course, these are very important because of their close relationship to equality theories. In particular, they can be used to solve the word problem in a computationally feasible manner, e.g. [Knuth & Bendix 1970].

We have tried to show that non-confluent rewrite rule systems such as F* can also exhibit interesting behavior. In particular, we can use these to generate sets using non-deterministic algorithms, which can be very compact. Moreover, if rewrite rules are interpreted lazily, we can achieve substantial reductions of search spaces of problems of the form "given set S, and condition C, find those members of S which satisfy C".

Whenever critical, algorithms achieving such reductions could always be developed, and in any language, declarative, or procedural. What we have proposed is a technique for simplifying their development. In particular, we can separate development of algorithms for generating search spaces, from development of algorithms for testing candidates, yet achieve the search space reduction transparently. Due to the efficient implementation of F*, and as the performance figures show, our technique could be viable for important cases of real problems.

REFERENCES

Barbuti, R., Bellia, M., Levi, G. [1986]. LEAF: A language which integrates logic, equations, and functions. In *Logic programming: functions, relations and equations* (eds.) D. DeGroot, G. Lindstrom, Prentice Hall, N.J.

Bruynooghe, M., Pereira, L.M. [1984]. Deduction revision by intelligent backtracking. In,

Implementations of Prolog, ed. J.A. Campbell, Ellis Horwood.

Chang, J.-H., Despain, A.M. [1984]. Semi-intelligent backtracking of Prolog based on a static data dependency analysis. *Proceedings of IEEE symposium on logic programming*, Boston, MA.

Clark, K.L., McCabe F. [1979]. Programmer's guide to IC-Prolog. *CCD Report 79/7*, London: Imperial College, University of London.

Dincbas, M., Simonis, H., van Hentenryck, P. [1988]. Solving a cutting-stock problem in constraint logic programming. *Proceedings of fifth international conference and symposium on logic programming*, eds. R. Kowalski, K. Bowen, MIT Press, Cambridge, MA.

Garey, M.R., Johnson, D.S. [1979]. *Computers and intractability. A guide to the theory of NP-completeness.* W.H. Freeman & Co. New York, N.Y.

Huet, G., Levy, J.-J. [1979]. Call by need computations in non-ambiguous linear term rewriting systems. IRIA technical report 359.

Kahn, G., MacQueen, D. [1977]. Coroutines and Networks of Parallel Processes. *Information Processing-77*, North-Holland, Amsterdam.

Knuth, D.E., Bendix, P.B. [1970]. Simple word problems in universal algebras. *Computational problems in abstract algebra*, ed. J. Leech, Pergamon Press.

Kumar, V., Lin, Y.-J. [1988]. A data-dependency-based intelligent backtracking scheme for Prolog. *Journal of Logic Programming*, vol. 5, No. 2, June.

Lloyd, J. [1984]. *Foundations of logic programming.* Springer Verlag, New York.

Narain, S. [1986]. A Technique for Doing Lazy Evaluation in Logic. *Journal of Logic Programming*, vol. 3, no. 3, October.

Narain, S. [1988]. LOG(F): An optimal combination of logic programming, rewriting and lazy evaluation. Ph.D. Thesis, Department of Computer Science, University of California, Los Angeles.

O'Donnell, M.J. [1985]. Equational logic as a programming language. MIT Press, Cambridge, MA.

Polya, G. [1973]. *How to solve it.* Princeton university press. Princeton, N.J.

Sterling, L., Shapiro, E. [1986]. *The art of Prolog.* MIT Press, Cambridge, MA.

Subrahmanyam, P.A. and You J.-H. [1984]. Conceptual Basis and Evaluation Strategies for Integrating Functional and Logic Programming. *Proceedings of IEEE Logic Programming Symposium,* Atlantic City, N.J.

van Hentenryck, P., Dincbas, M. [1987]. Forward checking in logic programming. *Proceedings of fourth international conference on logic programming,* ed. J.-L. Lassez, MIT Press, Cambridge, MA.

APPENDIX: Sketches of proofs of Theorems A and B

Full proofs of these theorems can be found in [Narain 1988]. Copies of it can be obtained upon request from the author. We first define:

R_P(G,H,A,B). Let P be an F* program. Where G,H,A,B are ground terms, R_P(G,H,A,B) if (a) G=>H, and (b) B is identical with A except that zero or more occurrences of G in A are *simultaneously* replaced by H. Note that A and G can be identical. Again, if P is clear from context, the subscript P is dropped.

Lemma 1. Let P be an F* program. If A->B and B is simplified but A is not, then A=>B.

Proof. Easy, by restriction (a). **QED.**

Lemma 2. Let P be an F* program. Let X1,..,Xn be variables, G,H,t1,..,tn,t1*,..,tn* be ground terms such that for each i, R(G,H,ti,ti*). Let σ={<X1,t1>,..,<Xn,tn>} and τ={<X1,t1*>,..,<Xn,tn*>} be substitutions. Let M be a term, possibly containing variables, but only from {X1,..,Xn}. Then R(G,H,Mσ,Mτ).

Proof. By induction on length of M. **QED.**

Lemma 3. Let P be an F* program. If (1) G, H, E1=f(t1,..,tn) and F1=f(t1*,..,tn*) are ground terms, and (2) for every i in 1,..,n, R(G,H,ti,ti*), and (3) B=f(L1,..,Ln) is the head of some rule in P, and (4) E1=Bσ for some substitution σ

Then there exists a substitution τ such that (1) F1=Bτ, and (2) σ and τ define exactly the same variables, i.e. only those occurring in B, and (3) If pair <X,s> occurs in σ and <X,s*> occurs in τ then R(G,H,s,s*).

Proof. By restrictions (a), (b), (c) on F* rules. **QED..**

Lemma 4. Let P be an F* program. If (1) f(t1,..,ti,..,tn) is a ground term, and (2) f(L1,..,Li-1,c(X1,..,Xm),Li+1,..,Ln)=>RHS is a rule in P, and (3) ti=d1,d2,d3,..,dr, r>0, is an N-reduction.

Then, f(t1,..,ti-1,d1,ti+1,..,tn), f(t1,..,ti-1,d2,ti+1,..,tn), .., f(t1,..,ti-1,dr,ti+1,..,tn) is also an N-reduction.

Proof. By restrictions (a), (b), (c). **QED.**

Theorem 1. Let P be an F* program. Let E1,F1,F2,G,H be ground terms such that (1) R(G,H,E1,F1), and (2) F1 reduces to F2 in an N-step

Then there is an N-reduction E1,..,E2 in P such that R(G,H,E2,F2).

Proof. By induction on length of E1.

Lemma 5. Let P be an F* program. Let R(G,H,E0,F0) and F0,F1,..,Fn be an N-reduction. Then there is an N-reduction E0,..,En such that R(G,H,En,Fn).

Proof. By induction on length n of F0,F1,..,Fn, and using Theorem 1. **QED.**

Theorem A. Reduction-completeness of F* for simplified forms. Let P be an F* program and D0 a ground term. Let D0,D1,..,Dn, n>=0, be a successful reduction in P. Then there is a successful N-reduction D0,E1,..,Em in P, such that Em-*>Dn.

Proof. By induction on length n of D0,D1,..,Dn, and using Lemma 5. Note that if R(G,H,A,B), and B is simplified, then either A is simplified, or A=>B. **QED.**

Theorem 3. Let P be an F* program. Let E1,F1,F2,G,H be ground terms such that (1) R(G,H,E1,F1), and (2) F1 reduces to F2 in an NR-step.

Then there is an NR-reduction E1,..,E2 in P such that R(G,H,E2,F2).

Proof. By induction on length of E1, and using Theorem 1. **QED.**

Lemma 6. Let P be an F* program. Let R(G,H,E0,F0) and F0,..,Fn, n>=0, be an NR-reduction. Then there is an NR-reduction E0,..,Ek such that R(G,H,Ek,Fn).

Proof. Similar to that of Lemma 5. **QED.**

Theorem B. Reduction-completeness of F* for normal forms. Let P be an F* program and D0 a ground term. Let D0,D1,..,Dn, n>=0, be a reduction in P, where Dn is in normal form. Then there is an NR-reduction D0,E1,..,Em=Dn, m>=0, in P.

Proof. By induction on length n of D0,D1,..,Dn, and using Lemma 6. **QED.**

Combining Matching Algorithms: The Regular Case*

Tobias Nipkow
Laboratory for Computer Science
MIT
545 Technology Square
Cambridge MA 02139

Abstract

The problem of combining matching algorithms for equational theories with disjoint signatures is studied. It is shown that the combined matching problem is in general undecidable but that it becomes decidable if all theories are regular. For the case of regular theories an efficient combination algorithm is developed. As part of that development we present a simple algorithm for solving the word problem in the combination of arbitrary equational theories with disjoint signatures.

1 Introduction

One all-pervasive concern in computing science is *compositionality*: the ability of composing the solution to a problem from the solutions to its subproblems. In unification theory this is known as the *combination problem*: how do unification algorithms for disjoint equational theories compose to yield a unification algorithm for the combined theory. In recent years a number of papers [19, 11, 17, 6, 15] have addressed this problem quite successfully. Because matching can be viewed as a special case of unification (where one side of the equation is variable-free), it is possible to obtain a matching algorithm for a combination of theories by combining unification algorithms for each theory using one of the algorithms in [19, 11, 17, 6, 15]. However, there are a number of problems with this simple-minded approach. Because it requires a unification rather than just a matching algorithm for each theory, it is not always applicable.

Example 1 There are equational theories which have an undecidable unification but a decidable matching problem. In [20] it is shown that the following equational theory DA has an undecidable unification problem:

$$DA = \left\{ \begin{array}{rcl} x * (y + z) & = & (x * y) + (x * z) \\ (x + y) * z & = & (x * z) + (y * z) \\ (x + y) + z & = & x + (y + z) \end{array} \right.$$

On the other hand, Szabó also shows that any *permutative* theory, i.e. where all equivalence classes are finite, has a decidable matching problem. Since DA is permutative, DA is an example of a theory with decidable matching but undecidable unification problem.

It may be interesting to note that the converse phenomenon can also occur: [1] presents an example of an equational theory with decidable unification but undecidable matching problem.

*This research was supported in part by NYNEX, NSF grant CCR-8706652, and by the Advanced Research Projects Agency of the DoD, monitored by the ONR under contract N00014-83-K-0125.

However, this pathological situation can only arise if by "unification" we mean "unification of terms without free constants".

Even if the required unification algorithms exist, we would lose efficiency by turning a matching into a unification problem. Not only is unification in many theories more complex than matching, but the same is also true for the combination: the algorithms in [19, 11, 17, 6] do not cover collapsing theories (which are covered by the current paper) and the algorithm in [15], which does, is extremely complex due to a number of nondeterministic choice points. However, efficiency is crucial for the most important application of matching: rewriting modulo equations, the basic inference mechanism in systems like OBJ [4], REVE [13], and LP [5]. Especially in a context like OBJ, where rewriting is used as a computational model, efficiency is very important.

Once we have convinced ourselves there is some merit in studying the combination of matching algorithms, the next question concerns the decidability of this problem. For unification algorithms this is still open. It has been shown to be undecidable if the individual unification algorithms can only deal with terms over variables and function symbols from their theory, but not free constants [1]. On the other hand, it is decidable if the individual unification algorithms can deal with terms with free constants and if in addition the so called "constant elimination problem" can be solved for each theory [15]. In contrast, matching problems don't combine that nicely, due to the asymmetric nature of matching.

Example 2 Let $E_1 = \{f(f(x)) = a\}$ and $E_2 = \{g(x, x) = x\}$. E_1 is linear but non-regular and E_2 is non-linear but regular. Both E_1 and E_2 have a decidable matching problem, even with free constants. For E_1 this is not difficult to work out and for E_2 it follows from the decidability of unification under idempotence as detailed for example in [7]. Now let Σ_3 be disjoint from $\Sigma_1 = \{f, a\}$ and $\Sigma_2 = \{g\}$. For $s, t \in \mathcal{T}(\Sigma_3, V)$ we have the following equivalences:

$$
\begin{array}{cll}
 & f(g(f(s), f(t))) = a & \text{is solvable (has a matcher)} \\
\Leftrightarrow & g(f(s), f(t)) = f(x) & \text{is solvable} \\
\Leftrightarrow & f(s) = f(t) & \text{is solvable} \\
\Leftrightarrow & s = t & \text{is solvable (has a unifier)}
\end{array}
$$

Thus we have shown that a matching problem in the combined theory is equivalent to a *unification* problem in one of the constituent theories. In particular, if unification in E_3 is undecidable, matching in the combination of the three theories is undecidable.

This example shows that matching in a combination of theories, where one is non-regular and one is non-linear, can be undecidable. The example can be simplified and weakened by looking at a theory which is both non-regular and non-linear, for example $E = \{x + x = 0\}$. The matching problem is easily seen to be decidable and it follows that $s + t = 0$ has a matcher iff $s = t$ has a solution.

Example 2 clearly is an upper bound on how far we can go in combining matching algorithms. It can be shown that it is also a tight upper bound in the following sense: if either all theories are regular or all theories are linear, we can provide a combination algorithm. The rest of the paper spells out how to combine regular theories. The linear case will be the subject of a forthcoming paper. In a sense the regular case is the important one since the main application of matching is in term rewriting. Matching modulo non-regular theories can result in substitutions which are not ground - an extremely undesirable property in the context of term rewriting.

The rest of the paper is organized as follows. Section 2 introduces the basic definitions and notations connected with equational logic. Section 3 establishes the main theorems about the combination of disjoint equational theories, in particular a simple solution to the word problem which is an important part of our matching algorithms. Section 4 finally presents 2 variants of a combined matching algorithm based on the principles of variable and constant "abstraction".

2 Preliminaries

Although this section reviews most of the notions used in the paper it is helpful if the reader is familiar with standard terminology concerning terms, term algebras, and equational theories as for example in [8, 9, 16].

A *signature* Σ is a set of function symbols with fixed arity, V is an infinite set of *variables* and $\mathcal{T}(\Sigma, V)$ denotes the free Σ-algebra over V. A term is called *proper* if it is not a variable, and *linear* if no variable occurs in it twice. The set of variables in a term t is denoted by $\mathcal{V}(t)$. *Substitutions* are functions from V to $\mathcal{T}(\Sigma, V)$ which differ from the identity only in finitely many places and which automatically extend to endomorphisms on $\mathcal{T}(\Sigma, V)$. Substitutions can be written as finite mappings of the form $\sigma = \{x_1 \mapsto t_1, \ldots, x_n \mapsto t_n\}$ with $t_i \neq x_i$ for all i. The domain and range are defined as $dom(\sigma) = \{x_1, \ldots, x_n\}$ and $ran(\sigma) = \{t_1, \ldots, t_n\}$. Composition is written as $\sigma; \tau = \lambda x. \tau\sigma x$, restriction to a set of variables W as $\sigma|_W$. A substitution σ is called *idempotent* iff $\sigma; \sigma = \sigma$ iff $dom(\sigma) \cap \mathcal{V}(ran(\sigma)) = \{\}$.

To get a graphical representation of terms and their subterms, we introduce the notion of a *context* which should be thought of as a λ-term of the form $F = \lambda x_1, \ldots, x_n.s$, where $s \in \mathcal{T}(\Sigma, V)$ and $\{x_1, \ldots, x_n\} \subseteq \mathcal{V}(s)$. Now F can be viewed as an n-ary function

$$F(t_1, \ldots, t_n) = \{x_1 \mapsto t_1, \ldots, x_n \mapsto t_n\}(s)$$

or simply as a term. In the latter case we have to remember the conventions of the λ-calculus which imply for example $\mathcal{V}(F) = \mathcal{V}(s) - \{x_1, \ldots, x_n\}$ and

$$\sigma(F) = \lambda x_1, \ldots, x_n. \sigma|_W(s) \quad \text{where} \quad W = \mathcal{V}(s) - \{x_1, \ldots, x_n\}$$

The set of contexts over a signature Σ is denoted by Σ^*, Σ^+ denotes proper contexts, i.e. those not of the form $\lambda x.x$. It will sometimes be convenient to just write $F(s_I)$ where I is some finite linearly ordered index set. In certain cases s_I will be identified with $\{s_i \mid i \in I\}$. If $I = J \dot\cup K$, we take the liberty to write $F(s_J, s_K)$ for $F(s_I)$.

To distinguish syntactic equality of terms from a number of other equalities that are used, the former is often denoted by $\equiv$.

In the sequel Σ stands for signatures, r, s, t stand for arbitrary terms, x, y, z for variables, f and g for function symbols in Σ, F and G for contexts, $\sigma, \tau, \theta, \delta$ are reserved for substitutions, I, J, K, L, M, N for index sets, and W for a set of variables.

Functions $f : D \to \mathcal{P}(R)$ extend automatically to $\mathcal{P}(D) \to \mathcal{P}(R)$ with $f(S) = \bigcup_{s \in S} f(s)$ for $S \subseteq D$. In addition we write $f(s_1, \ldots, s_n)$ instead of $f(\{s_1, \ldots, s_n\})$. This convention implies for example that $\mathcal{V}(s,t) = \mathcal{V}(\{s,t\}) = \mathcal{V}(s) \cup \mathcal{V}(t)$.

2.1 Equational Theories, Unification and Matching

A set of equations $E \subseteq \mathcal{T}(\Sigma, V) \times \mathcal{T}(\Sigma, V)$ over a signature Σ induces an *equational theory* $=_E$ which is the closure of E under the laws of equational logic, namely symmetry, reflexivity, transitivity, substitutivity and congruence. An equational theory $=_E$ is called *consistent* iff $x =_E y$ implies $x \equiv y$ for all $x, y \in V$, *collapse-free* iff $t =_E x$ implies $t \equiv x$ for all $t \in \mathcal{T}(\Sigma, V)$ and $x \in V$, and *regular* iff $s =_E t$ implies $\mathcal{V}(s) = \mathcal{V}(t)$ for all $s, t \in \mathcal{T}(\Sigma, V)$. A theory $=_E$ is collapse-free or regular iff E is.

Two substitutions σ and τ are E-equal w.r.t. a set of variables W, written as $\sigma =_E \tau\ [W]$ iff $\sigma x =_E \tau x$ for all $x \in W$; $\sigma =_E \tau$ is short for $\sigma =_E \tau\ [V]$. τ is an *instance* of σ w.r.t. W, written as $\sigma \leq_E \tau\ [W]$, iff there is a θ such that $\sigma; \theta =_E \tau\ [W]$. The set of *E-unifiers* and the set of *complete sets of E-unifiers* of a set of equations Γ away from W are defined as

$$\begin{aligned} \mathcal{U}_E(\Gamma, W) &= \{\sigma \mid \mathcal{V}(ran(\sigma)) \cap W = \{\} \wedge \forall s = t \in \Gamma.\, \sigma(s) =_E \sigma(t)\} \\ \mathcal{CSU}_E(\Gamma, W) &= \{\Theta \subseteq \mathcal{U}_E(\Gamma, W) \mid \forall \sigma \in \mathcal{U}_E(\Gamma) \exists \tau \in \Theta.\, \tau \leq_E \sigma\ [\mathcal{V}(\Gamma)]\} \end{aligned}$$

$\mathcal{U}_E(\Gamma)$ and $\mathcal{CSU}_E(\Gamma)$ are short for $\mathcal{U}_E(\Gamma, \{\})$ and $\mathcal{CSU}_E(\Gamma, \{\})$.

The literature contains two different definitions of what a matcher of two terms s and t should be. However, the comparison in [2] shows that if $\mathcal{V}(s)$ and $\mathcal{V}(t)$ are disjoint, both definitions coincide. In particular, matching can be viewed as a special case of unification if the variables in one term are considered as constants. Hence we restrict ourselves to matching problems of the form $s = t$, where t is ground, and use unification terminology.

3 Combinations of Equational Theories

In the sequel let the Σ_i be disjoint signatures and the E_i sets of equations over Σ_i, for all i in some index set $\mathcal{I}$. In particular we assume $0 \in \mathcal{I}$, $E_0 = \{\}$, and $\Sigma_0 = C \cup V$, where C is an infinite set of (free) *constants*; $C(t)$ denotes the set of constants in t. Let $\Sigma = \bigcup_{i \in \mathcal{I}} \Sigma_i$ and $E = \bigcup_{i \in \mathcal{I}} E_i$ for the rest of the paper unless stated otherwise. *Notice that as a consequence $\mathcal{T}(\Sigma)$ is the set of* all *terms and only $\mathcal{T}(\Sigma \backslash V)$ is the set of* ground *terms.* Instead of $=_{E_i}$ we will often just write $=_i$. The *theory* of a term $t \in \mathcal{T}(\Sigma)$, denoted by $\mathcal{T}(t)$, is i if the top function symbol of t comes from Σ_i. Terms in $\mathcal{T}(\Sigma_i, V)$ are called *i-pure*, those in $\mathcal{T}(\Sigma)$ *mixed*.

The notion of a context is particularly useful in separating a term into a pure top-layer and subterms from different theories. If $F \in \Sigma_i^*$ we write $F[t_1, \ldots, t_n]$ only if $\mathcal{T}(t_j) \neq i$ and $t_j \notin V$ for all j. Notice that *any* term $t \in \mathcal{T}(\Sigma)$ can be written as $F[t_1, \ldots, t_n]$ with $F \in \Sigma_i^*$ for any $i \in \mathcal{I}$: if $\mathcal{T}(t) \neq i$, let $F = \lambda x.x$, $n = 1$, and $t_1 = t$. Note also that any constant or variable a can be written as $a[]$.

In the sequel the context notation is often used as a pattern matching device, i.e. given a term t we write "let t be of the form $F[t_I]$". Even if we insist that F is a proper context, neither F nor the t_I are uniquely defined: the term $f(a, a)$ can be written as both $F[a, a]$ and $G[a]$, where F and G are the contexts $\lambda x, y.f(x, y)$ and $\lambda x.f(x, x)$ respectively. However, this ambiguity is of no consequence in this paper and simply reflects a certain degree of freedom in dealing with mixed terms.

In order to decompose mixed terms we introduce the following terminology:

Definition 1 *The* immediate alien subterms, immediate i-alien subterms, $i \in \mathcal{I}$, proper alien subterms *and* alien subterms *are defined as*

$$\begin{aligned}
IAST(F[s_1, \ldots, s_n]) &= \{s_1, \ldots, s_n\} \cup \mathcal{V}(F) \\
IAST_i(t) &= \begin{cases} IAST(t) & \text{if } \mathcal{T}(t) = i \\ \{t\} & \text{otherwise} \end{cases} \\
PAST(F[s_1, \ldots, s_n]) &= \{s_1, \ldots, s_n\} \cup PAST(s_1) \cup \ldots \cup PAST(s_n) \\
AST(t) &= \{t\} \cup PAST(t)
\end{aligned}$$

where $F \in \Sigma_i^+$. The alternation depth *or* theory height *of a term is defined as*

$$Alt(t) = 1 + \max_{s \in IAST(t)} Alt(s)$$

where $\max_{i \in \{\}} . = 0$.

The most important notational devices in this paper are contexts as defined above, and abstraction of terms w.r.t. equivalence relations. Given a term $t \in \mathcal{T}(\Sigma)$ and an equivalence $\approx$ on $\mathcal{T}(\Sigma)$, $[t]_\approx$ is the equivalence class of t w.r.t. $\approx$. Equivalence classes are treated as constants in C. The reverse translation from a constant which denotes an equivalence class to some member of that class is denoted by a partial function $\pi : C \to \mathcal{T}(\Sigma)$ such that $\pi([t]_\approx) \approx t$. Note that π is well defined up to $\approx$. In the sequel we sometimes take the liberty to treat π as a substitution in the obvious way. All the equivalence relations we deal with are consistent, i.e. on V they coincide with $\equiv$. Hence we can identify the singleton equivalence class $[x]_\approx$ with x, a convention that will be useful in certain places.

In many cases we only want to abstract the alien subterms of a term. Given a term $t \equiv F[t_1, \ldots, t_n]$, $F \in \Sigma_j^+$, and a theory $i \in \mathcal{I}$, we define

$$[t]^i_\approx = \begin{cases} F[[t_1]_\approx, \ldots, [t_n]_\approx] & \text{if } i = j \\ [t]_\approx & \text{otherwise} \end{cases}$$

Instead of $F[[s_1]_\approx, \ldots, [s_n]_\approx]$ we write $F[[s_I]_\approx]$ where $I = \{1, \ldots, n\}$. If $\approx$ is a congruence, $\pi([t]^i_\approx) \approx t$. For substitutions define $[\sigma]^i_\approx = \{x \mapsto [\sigma(x)]^i_\approx \mid x \in dom(\sigma)\}$.

Here are some useful lemmas about equivalence class manipulations:

Lemma 1 *Let $\approx_1$ and $\approx_2$ be two equivalence relations, $=_E$ some equational theory, $F, G \in \Sigma^*$, such that $r \approx_1 r' \Rightarrow r \approx_2 r'$ for all $r, r' \in s_I \cup t_J$. Then*

$$F[[s_I]_{\approx_1}] =_E G[[t_J]_{\approx_1}] \Rightarrow F[[s_I]_{\approx_2}] =_E G[[t_J]_{\approx_2}]$$

Proof The claim follows because equational validity is preserved under homomorphisms. By definition there is a surjection from $\approx_1$ to $\approx_2$-equivalence classes on terms in $s_I \cup t_J$, inducing an endomorphism on the term algebra. □

As a corollary we obtain the same lemma, but with $\Leftrightarrow$ instead of $\Rightarrow$.

Lemma 2 *For any equational theory $=_E$*

$$F[[s_I]_{=_E}] =_E G[[t_J]_{=_E}] \;\Rightarrow\; F(s_I) =_E G(t_J)$$

Proof $F[[s_I]_{=_E}] =_E G[[t_J]_{=_E}]$ implies $F(s_I) =_E \pi(F[[s_I]_{=_E}]) =_E \pi(G[[t_J]_{=_E}]) =_E G(t_J)$ because π is an endomorphism. □

3.1 Equality

The main objective of this section is to develop a construction which facilitates equational reasoning in $\mathcal{T}(\Sigma)/_{=_E}$.

The intuitive idea about $=_E$ is as follows: $s =_E t$ holds iff s and t can be "collapsed" to two "normal forms" s' and t' which are equal without further collapses. The point of collapsing s and t first is that, in the comparison of s' and t', alien subterms can be treated as constants. If terms are not fully collapsed, subterms may appear alien which are in fact of the same theory.

First we define the collapse-free equality $\cong$ which compares terms as if no collapse could occur:

$$s \cong t \;\Leftrightarrow\; \exists i.\, \mathcal{T}(s) = \mathcal{T}(t) = i \wedge [s]^i_\cong =_i [t]^i_\cong$$

This recursive definition defines $\cong$ uniquely because the $\cong$-equivalence class on the RHS are on terms of strictly lower theory height than those of the LHS. A more programming oriented formulation should make this point clearer:

$s \cong t$ = **if** $\mathcal{T}(s) \neq \mathcal{T}(t)$ **then false**
else let $F[s_I] = s$ **where** $F \in \Sigma^+_{\mathcal{T}(s)}$
$G[t_J] = t$ **where** $G \in \Sigma^+_{\mathcal{T}(t)}$
in $F[[s_I]_\cong] =_{\mathcal{T}(s)} G[[t_J]_\cong]$

Notice that on C and V, $\cong$ coincides with $\equiv$.

If all $=_i$ are computable, $\cong$ is a total computable function. In most papers on combining unification algorithms $\cong$ appears in some form or other, though it is usually not part of the actual algorithm but a device used in the proofs. For collapse-free theories $\cong$ coincides with $=_E$, as fact 2 below shows (see also [19, 7]).

Next we show that $\cong$ has a number of useful properties.

Lemma 3 *$s \cong t \Rightarrow [s]^i_{\cong} =_i [t]^i_{\cong}$ holds for any s, t and i.*

Proof Assume $s \cong t$. If $\mathcal{T}(s) = \mathcal{T}(t) = i$ it follows by definition of $\cong$. If $\mathcal{T}(s) = \mathcal{T}(t) \neq i$ we get $[s]^i_{\cong} = [s]_{\cong} = [t]_{\cong} = [t]^i_{\cong}$. □

Lemma 4 *$\cong$ is a congruence on $\mathcal{T}(\Sigma)$.*

Proof From the definition $\cong$ is easily seen to be an equivalence. Now let $f \in \Sigma_i$ and let $s_j \cong t_j$ for all $j \in J$:

$$\begin{array}{llll} \Rightarrow & \forall j \in J.\, [s_j]^i_{\cong} =_i [t_j]^i_{\cong} & \Rightarrow & f([s_J]^i_{\cong}) =_i f([t_J]^i_{\cong}) \\ \Rightarrow & [f(s_J)]^i_{\cong} =_i [f(t_J)]^i_{\cong} & \Rightarrow & f(s_J) \cong f(t_J) \end{array}$$

□

Lemma 5 $\cong\, \subseteq\, =_E$

Proof by induction on terms. Let $s \equiv F[s_I] \cong G[t_J] \equiv t$, and hence $\mathcal{T}(s) = \mathcal{T}(t) = i$, for some $F, G \in \Sigma_i^+$, and assume the claim holds for all terms in $s_I \cup t_J$.

$$\begin{array}{rlll} s \cong t & \Rightarrow & F[[s_I]_{\cong}] =_i G[[t_J]_{\cong}] & \\ & \Rightarrow & F[[s_I]_{=_E}] =_i G[[t_J]_{=_E}] & \text{by ind. hyp. and lemma 1} \\ & \Rightarrow & F[[s_I]_{=_E}] =_E G[[t_J]_{=_E}] & \\ & \Rightarrow & F[s_I] =_E G[t_J] & \text{by lemma 2} \end{array}$$

□

These lemmas are all we need to know about $\cong$. Now we turn to the task of collapsing terms. The following little lemma explains why collapsing a term is equivalent to equating it to one of its alien subterms:

Lemma 6 *If $s \in \mathcal{T}(\Sigma_i, V)$ and $=_i$ is consistent, $(\exists x \in V.\, s =_i x) \Leftrightarrow (\exists x \in \mathcal{V}(s).\, s =_i x)$.*

Proof The $\Leftarrow$-direction is immediate. For the $\Rightarrow$-direction let $s =_i x \in V$. If $x \notin \mathcal{V}(s)$ it follows that $y = \{x \mapsto y\}x =_i \{x \mapsto y\}s = s =_i x$ which contradicts consistency. □

Mutatis mutandis, the same lemma holds for free constants instead of variables.

For a given term there is in general no unique fully collapsed equivalent. Therefore the collapsing process takes the form of a mapping from terms to equivalence classes of terms, more precisely elements in $A = \mathcal{T}(\Sigma)/_{\cong}$. To this end A is turned into an algebra $\underline{A}$ which is different from the factor algebra $\mathcal{T}(\Sigma)/_{\cong}$. This construction is inspired by the one given in chapter 3 of [15]. For each n-ary function symbol $f \in \Sigma_i$ define an n-ary function $f^A : A^n \rightarrow A$:

$$f^A([t_I]_{\cong}) = \begin{cases} [t]_{\cong} & \text{if } t \in \mathit{IAST}_i(f(t_I)) \text{ and } [f(t_I)]^i_{\cong} =_i [t]_{\cong} \\ [f(t_I)]_{\cong} & \text{otherwise} \end{cases}$$

Lemma 7 *f^A is uniquely defined.*

Proof Let $s_j \cong t_j$ for all $j \in J = \{1, \ldots, n\}$. Because $\cong$ is a congruence, $f(s_J) \cong f(t_J)$ follows. Thus the second case in the definition of f^A is well defined.

If there are $s, s' \in \mathit{IAST}_i(f(s_J))$ such that $[f(s_J)]^i_{\cong} =_i [s]_{\cong}$ and $[f(s_J)]^i_{\cong} =_i [s']_{\cong}$ then $s \cong s'$. Furthermore, $f(s_J) \cong f(t_J)$ implies $[f(s_J)]^i_{\cong} =_i [f(t_J)]^i_{\cong}$. By lemma 6 it follows that $s \in \mathit{IAST}_i(f(t_J))$ and hence $f^A([s_J]_{\cong}) = f^A([t_J]_{\cong}) = [s]_{\cong}$. □

Since $\cong$ is a congruence, the above definition extends verbatim to F^A, $F \in \Sigma_i^*$, instead of f^A.

The homomorphic extension of f^A from function symbols to terms yields an interpretation t^A for every term $t \in \mathcal{T}(\Sigma)$. This interpretation is the mapping referred to above that collapses a term to some $\cong$-equivalence class of collapse-free terms. Note that this interpretation is only a surjection if $=_E$ is collapse-free.

To see that collapsing acts reasonable, we show that it preserves E-equalities:

Lemma 8 $s^A = [t]_{\cong} \Rightarrow s =_E t$

Proof by induction on terms. Let $s \equiv f(s_J)$, $f \in \Sigma_i$, and assume the induction hypothesis holds for all s_j, i.e $s_j^A = [t_j]_{\cong}$ implies $s_j =_E t_j$. Therefore $s^A = f^A(s_J^A) = [f(t_J)]_{\cong}$ implies $s \equiv f(s_J) =_E f(t_J)$.

If there is a $t \in IAST_i(f(s_J))$ such that $[f(s_I)]^i_{\cong} =_i [t]_{\cong} = [t]^i_{\cong}$ then $s^A = [t]_{\cong}$. Now lemma 5 implies $[f(s_I)]^i_{=_E} =_E [t]^i_{=_E}$ and with lemma 2 $s \equiv f(s_J) =_E t$ follows. □

Next we need two technical lemmas relating $\cong$ and $\underline{A}$.

Lemma 9 *Let $F, G \in \Sigma_k^*$. Then $F[[s_I]_{\cong}] =_k G[[t_J]_{\cong}]$ implies $[F(s_I)]^k_{\cong} =_k [G(t_J)]^k_{\cong}$.*

Proof

$$\begin{array}{rrcll} & F[[s_I]_{\cong}] & =_k & G[[t_J]_{\cong}] & \\ \Rightarrow & F[[[s_I]^k_{\cong}]_{=_k}] & =_k & G[[[t_J]^k_{\cong}]_{=_k}] & \text{by lemmas 1 and 3} \\ \Rightarrow & F[[s_I]^k_{\cong}] & =_k & G[[t_J]^k_{\cong}] & \text{by lemma 2} \\ \Rightarrow & [F(s_I)]^k_{\cong} & =_k & [G(t_J)]^k_{\cong} & \end{array}$$

□

Lemma 10 *Let $F, G \in \Sigma_i^*$. Then $F[[s_I]_{\cong}] =_i G[[t_J]_{\cong}]$ implies $F^A[[s_I]_{\cong}] = G^A[[t_J]_{\cong}]$.*

Proof From lemma 9 it follows that $F[[s_I]_{\cong}] =_i G[[t_J]_{\cong}]$ implies $[F(s_I)]^i_{\cong} =_i [G(t_J)]^i_{\cong}$.

If there is a $t \in IAST_i(F(s_I))$ with $[F(s_I)]^i_{\cong} =_i [t]_{\cong}$, it follows from the above that $G[[t_J]_{\cong}] =_i [t]_{\cong}$ and hence that $F^A[[s_I]_{\cong}] = G^A[[t_J]_{\cong}] = [t]_{\cong}$.

If there is no such t in $IAST_i(F(s_I))$, there cannot exist a $t \in IAST_i(G(t_J))$ with $G[[s_I]_{\cong}] =_i [t]_{\cong}$ either (lemma 6). This means in particular that $F, G \in \Sigma_i^+$ and thus $F(s_I) \cong G(t_J)$. Hence $F^A[[s_I]_{\cong}] = [F(s_I)]_{\cong} = [G(s_J)]_{\cong} = G^A[[t_J]_{\cong}]$. □

Now we can show that $\underline{A}$ is a model of E or, equivalently, that E is valid in $\underline{A}$.

Lemma 11 $\underline{A} \models E$

Proof Let $F = G \in E_i$ and $\sigma : V \to A$ be a substitution. From lemma 10 it follows that, since $\sigma F =_i \sigma G$, $(\sigma F)^A = (\sigma G)^A$, i.e. $A \models F = G$. The proposition follows immediately. □

Finally we have

Theorem 1 $s =_E t \Leftrightarrow s^A = t^A$

Proof

- $\Rightarrow$ follows directly from lemma 11.
- $\Leftarrow$ $s^A = t^A$ implies that there is an r such that $s^A = [r]_{\cong} = t^A$. Thus $s =_E r =_E t$ follows by lemma 8. □

It remains to be shown that this theorem is effective, i.e. that s^A can be computed. Above we have shown (by giving an algorithm) that $\cong$ is decidable provided each $=_i$ is. Below we define a function $\Downarrow$ which, given s, computes some representative $s\Downarrow$ of s^A. By theorem 1 it follows that $s =_E t$ holds iff $s\Downarrow \cong t\Downarrow$. It is not difficult to extract the following bit of "code" from the definition of f^A:

$s\Downarrow$ = **let** $F[s_1,\ldots,s_n] = s$ **where** $F \in \Sigma^+_{\mathcal{T}(s)}$
$G[t_I] = F(s_1\Downarrow,\ldots,s_n\Downarrow)$ **where** $G \in \Sigma^+_{\mathcal{T}(s)}$
in if $\exists i \in I.\ G[[t_I]_\cong] =_{\mathcal{T}(s)} [t_i]_\cong$ **then** t_i **else** $G[t_I]$

Notice that $x\Downarrow = x$ and that $s\Downarrow$ is uniquely defined up to $\cong$. In fact we have

Fact 1 *$[s\Downarrow]_\cong = s^A$ and hence $s =_E t \Leftrightarrow s\Downarrow \cong t\Downarrow$.*

Fact 2 *If $=_E$ is collapse-free, $s\Downarrow = s$ and hence $s =_E t \Leftrightarrow s \cong t$.*

Fact 1 shows that the word problem in the combination ($=_E$) is decidable if it is decidable in each theory ($=_i$) and fact 2 strengthens the result for collapse-free theories. Both [18] and [15] prove the decidability of the combined word problem using similar methods. However, I would claim that the separation of the collapse-free equality $\cong$ from the collapsing process $\Downarrow$ provides a better understanding of the computational content. This is important because both $\Downarrow$ and $\cong$ are part of the matching algorithm below. The correctness proof of that algorithm requires some more definitions and lemmas.

Definition 2 *The* normalized theory *of a term is defined as $\mathcal{NT}(s) = \mathcal{T}(s\Downarrow)$. A term t is in* collapse-free normal form *or* CNF *iff $\mathcal{NT}(s) = \mathcal{T}(s)$ for all alien subterms s of t. A substitution is in CNF if every term in its range is; $\sigma\Downarrow$ denotes the substitution obtained by applying $\Downarrow$ to every term in the range.*

A simple inductive consequence of this definition is

Fact 3 *If s is in CNF and $s \cong t$ then t is in CNF.*

The following lemma is purely technical.

Lemma 12 *Let $F, G \in \Sigma^*_n$, $n \in \mathcal{I}$, and $s_I \cup t_J \subseteq \mathcal{T}(\Sigma)$ such that $\mathcal{T}(r) \neq n \neq \mathcal{NT}(r)$ for all $r \in s_I \cup t_J$. Then*

$$F[s_I^A] =_n G[t_J^A] \;\Leftrightarrow\; F^A[s_I^A] = G^A[t_J^A]$$

Proof Let $s_i^A = [s_i']_\cong$ and $t_j^A = [t_j']_\cong$ for all $i \in I$ and $j \in J$. The important point is that $\mathcal{T}(r) = \mathcal{NT}(r) \neq n$ for all $r \in s_I' \cup t_J'$.

- $\Rightarrow$ Follows from lemma 10.
- $\Leftarrow$ If there is an $i \in I$ such that $F[[s_I']_\cong] =_n [s_i']_\cong$, then $G^A[t_J^A] = F^A[s_I^A] = [s_i']_\cong$ implies $G[[t_J']_\cong] =_n [s_i']_\cong$, proving the claim. Otherwise, $G^A[t_J^A] = F^A[s_I^A] = [F[s_I']]_\cong$, implying $G^A[t_J^A] = [G[t_J']]_\cong$ and hence $F[[s_I']_\cong] =_n G[[t_J']_\cong]$, again proving the claim. □

The next lemma justifies abstracting alien subterms to constants in equational derivations, provided certain conditions are met.

Lemma 13 *Let $s, t \in \mathcal{T}(\Sigma)$ and $i \in \mathcal{I}$ such that the normalized theory of all immediate i-alien subterms of s and t is different from i, and let $\approx\ \subseteq\ =_E$ be an equivalence such that $r_1 \approx r_2 \Leftrightarrow r_1 =_E r_2$ for all $r_1, r_2 \in IAST_i(s,t)$. Then*

$$s =_E t \;\Leftrightarrow\; [s]^i_\approx =_i [t]^i_\approx$$

Proof Let $s \equiv F[s_I]$ and $t \equiv G[t_J]$ with $F, G \in \Sigma_i^*$ and $\mathcal{T}(r) \neq i \neq \mathcal{NT}(r)$ for all $r \in s_I \cup t_J$.

$$\begin{array}{rrcll}
 & [s]^i_\approx & =_i & [t]^i_\approx & \\
\Leftrightarrow & F[[s_I]_\approx] & =_i & G[[t_J]_\approx] & \\
\Leftrightarrow & F[[s_I]_{=_E}] & =_i & G[[t_J]_{=_E}] & \text{by assumption and lemma 1} \\
\Leftrightarrow & F[s_I^A] & =_i & G[t_J^A] & \text{by lemma 1 and theorem 1} \\
\Leftrightarrow & F^A[s_I^A] & =_i & G^A[t_J^A] & \text{by lemma 12} \\
\Leftrightarrow & s & =_E & t & \text{by theorem 1}
\end{array}$$

□

In a slightly different form this lemma can also be found in [15].

3.2 Matching

This section provides a partial solution to the combination problem based on the principle of "constant abstraction" used in [6, 15] to combine unification algorithms. The idea is to divide matching into 3 steps. First all alien subterms are consistently replaced by free constants. The resulting terms are matched using the appropriate basic matching algorithm. Finally all occurrences of the newly introduced constants in a matcher are replaced by the subterms they stand for. As we shall see, this works fine for variable free subterms in CNF.

In the sequel we assume that each theory i comes with a correct and complete matching algorithm $\mathcal{M}^i$: given a finite set of equations $\Gamma \subseteq \mathcal{T}(\Sigma_i, V \cup C) \times \mathcal{T}(\Sigma_i, C)$, and a set of variables W, $\mathcal{M}^i(\Gamma, W) \in \mathcal{CSU}_i(\Gamma, W)$ is a (potentially infinite) set of substitutions. Without loss of generality we assume $dom(\sigma) \subseteq \mathcal{V}(\Gamma)$ for all $\sigma \in \mathcal{M}^i(\Gamma, W)$. $\mathcal{M}^i(\Gamma)$ is short for $\mathcal{M}^i(\Gamma, \{\})$. Note that $\mathcal{M}^0(s = t, W)$ returns $\{\{\}\}$ if $s \equiv t$ and $\{\}$ otherwise.

Intuitively speaking, the constant abstraction process described above is not really necessary. Each $\mathcal{M}^i$ only needs the ability to test alien subterms for equality. In fact, explicit substitution by constant may well be much less efficient. Consider the matching problem $f(x, y) = f(s, t)$ where s and t are alien ground terms and f is a free function symbol. The most general unifier of this problem is obvious and can be computed without abstraction. Even more, in this situation it is completely unnecessary to check s and t for equality. Hence we could call the traditional constant abstraction "eager" and the idea of passing an equality test to the matching algorithm "lazy". The following extension of $\mathcal{M}^i$ w.r.t. a congruence $\approx$ tries to formalize this intuition:

$$\mathcal{M}^i_\approx(s = t, W) = \{(\sigma; \pi)|_{dom(\sigma)} \mid \sigma \in \mathcal{M}^i([s]^i_\approx = [t]^i_\approx, W)\} \tag{1}$$

$\mathcal{M}^i_\approx$ combines constant abstraction, matching, and reversal of constant abstraction. The next lemma shows that it works as intended.

Lemma 14 *Let $s \in \mathcal{T}(\Sigma_i, V \cup \mathcal{T}(\Sigma \backslash V))$, $t \in \mathcal{T}(\Sigma \backslash V)$, and let the congruence $\approx$ be such that the preconditions of lemma 13 are met. Then*

$$\mathcal{M}^i_\approx(s = t) \in \mathcal{CSU}_E(s = t).$$

Proof W.l.o.g. let $s \equiv F[s_I]$ where $F \in \Sigma_i^*$, and $s_I \subseteq \mathcal{T}(\Sigma \backslash V)$, i.e $\rho s_i = s_i$ and hence $\rho s = (\rho F)(s_I)$ for all substitutions ρ.

For correctness let $\tau \in \mathcal{M}^i_\approx(s = t)$, i.e. there is a σ such that $\tau = \sigma; \pi$ $[\mathcal{V}(F)]$. Hence $\tau s = (\tau\sigma F)(s_I) \approx (\pi\sigma F)(\pi([s_I]_\approx)) = \pi\sigma(F[[s_I]_\approx]) = \pi\sigma([s]^i_\approx) =_i \pi([t]^i_\approx) \approx t$ and thus $\tau s =_E t$.

For completeness let τ such that $\tau s =_E t$. W.l.o.g. let τ be in CNF – otherwise use $\tau\Downarrow =_E \tau$. Now τ can be decomposed as $\tau = \tau_1; \tau_2$ $[dom(\tau)]$ such that $ran(\tau_1) \subseteq \mathcal{T}(\Sigma_i, V)$, $dom(\tau_2) \subseteq \mathcal{V}(ran(\tau_1))$, and $\mathcal{T}(t) = \mathcal{NT}(t) \neq i$ for all $t \in ran(\tau_2)$. Let $\tau_2' = [\tau_2]^i_\approx$. Therefore $\tau_1; \tau_2' = [\tau]^i_\approx$ $[\mathcal{V}(F)]$. Hence $[\tau]^i_\approx([s]^i_\approx) = ([\tau]^i_\approx F)([s_I]_\approx) = ([\tau F]^i_\approx)([s_I]_\approx) = [\tau(F[s_I])]^i_\approx = [\tau s]^i_\approx =_E [t]^i_\approx$ – the last step is an application of lemma 13. Thus there is a $\sigma \in \mathcal{M}^i([s]^i_\approx = [t]^i_\approx)$ such that $\sigma \leq_{E_i} [\tau]^i_\approx = \tau_1; \tau_2'$ $[\mathcal{V}(F)]$ and therefore $\sigma; \pi \leq_{E_i} \tau_1; \tau_2'; \pi =_E \tau_1; \tau_2 = \tau$ $[\mathcal{V}(F)]$. This shows that τ is an instance of $\sigma; \pi \in \mathcal{M}^i_\approx(s = t)$. □

Armed with this lemma we are finally in a position to tackle the problem of combining matching algorithms.

4 Combining Regular Theories

In this section we assume that all theories $=_i$ are regular. As a crucial consequence all substitutions returned from any of the $\mathcal{M}^i$ are ground (because the RHS's of all equations are) and the second parameter to $\mathcal{M}^i$ is superfluous. Not surprisingly this carries over to $\mathcal{M}^i_\approx$.

Lemma 15 *If the equivalence $\approx$ is regular, all substitutions returned from $\mathcal{M}^{i}_{\approx}$ are ground.*

Proof Any $\tau \in \mathcal{M}^{i}_{\approx}(s = t)$ is of the form $(\sigma;\pi)|_{dom(\sigma)}$ for some $\sigma \in \mathcal{M}^{i}([s]^{i}_{\approx} = [t]^{i}_{\approx})$. Thus we have $\mathcal{V}(ran(\tau)) = \mathcal{V}(ran(\sigma)) \cup \mathcal{V}(\pi(\mathcal{C}(ran(\sigma))))$. From $\sigma([s]^{i}_{\approx}) =_i [t]^{i}_{\approx}$ and the facts that $=_i$ is regular and $[t]^{i}_{\approx}$ is ground, it follows that σ is ground too. Again by regularity of $=_i$ we have $\mathcal{C}(ran(\sigma)) \subseteq \mathcal{C}([t]^{i}_{\approx}) = \{[r]_{\approx} \mid r \in IAST_i(t)\}$ and because $\approx$ is regular $\mathcal{V}(\pi(\mathcal{C}(ran(\sigma)))) = \mathcal{V}(\{\pi([r]_{\approx}) \mid r \in IAST_i(t)\}) = \mathcal{V}(IAST_i(t)) = \{\}$. □

Note that ground substitutions are idempotent, $\sigma = \sigma;\tau\ [dom(\sigma)]$ if σ is ground, and $\sigma \cup \tau = \sigma;\tau = \tau;\sigma$ if σ and τ are ground and their domains are disjoint.

4.1 Transformations of Equations

Since our matching algorithms are expressed as transformations on sets of equations, we first look at some general principles connected with transforming sets of equations. The idea is to simplify the equations until we eventually arrive at a set which is in "solved form", i.e. where the solution is readily apparent.

Definition 3 *A set of equations $\Gamma = \{s_1 = t_1, \ldots, s_n = t_n\}$ is in* substitution form *if $s_i \in V$ and $s_i \neq s_j$ for $i \neq j$. In this case $\vec{\Gamma}$ denotes $\{s_1 \mapsto t_1, \ldots, s_n \mapsto t_n\}$. If $\vec{\Gamma}$ is idempotent, Γ is in* solved form. *Given a substitution σ, $\sigma_=$ denotes the obvious translation into a set of equations in substitution form.*

The following simple lemma justifies the name solved form.

Lemma 16 *If Γ in solved form and all RHS's are ground then $\{\vec{\Gamma}\} \in \mathcal{CSU}_E(\Gamma)$, and $\vec{\Gamma} =_E \tau\ [\mathcal{V}(\Gamma)]$ for all $\tau \in \mathcal{U}_E(\Gamma)$.*

For a set of transformations to be useful we need the following properties.

Definition 4 *A rewrite relation $\Longrightarrow$ on sets of equations is called*

correct iff $\Gamma \Longrightarrow \Delta$ implies $\mathcal{U}_E(\Gamma) \supseteq \mathcal{U}_E(\Delta)$.

complete iff $\mathcal{U}_E(\Gamma)|_{\mathcal{V}(\Gamma)} \subseteq \bigcup\{\mathcal{U}_E(\Delta)|_{\mathcal{V}(\Gamma)} \mid \Gamma \Longrightarrow \Delta\}$ for all Γ not in normal for w.r.t. $\Longrightarrow$.

non-blocking iff $\mathcal{U}_E(\Gamma) \neq \{\}$ and Γ is not in solved form imply that Γ is not in normal form (w.r.t. $\Longrightarrow$).

Correctness means that transformations don't introduce new unifiers, completeness that unifiers don't get lost, and non-blocking that rewriting of solvable equation sets cannot get stuck before a solved form is reached.

The following theorem spells out the conditions under which rewriting on equation sets constitutes a unification algorithm.

Theorem 2 *Let $\Longrightarrow$ be a correct, complete, non-blocking, and terminating rewrite relation on sets of equations. Then*

$$\{\vec{\Delta} \mid \Gamma \Longrightarrow^{*} \Delta \wedge \Delta \text{ is in normal and solved form}\} \in \mathcal{CSU}_E(\Gamma)$$

If $\Longrightarrow$ is finitely branching and r.e. and normal forms are decidable, then unification in E is decidable.

Proof *Correctness:* Let $\Gamma \Longrightarrow^{*} \Delta$ such that Δ is in normal and solved form. Then $\vec{\Delta} \in \mathcal{U}_E(\Delta) \subseteq \mathcal{U}_E(\Gamma)$. *Completeness:* Let $\sigma \in \mathcal{U}_E(\Gamma)$. The combination of completeness and termination yields that there are Δ and τ such that $\Gamma \Longrightarrow^{*} \Delta$, Δ is in normal form, $\tau \in \mathcal{U}_E(\Delta)$, and $\sigma = \tau\ [\mathcal{V}(\Gamma)]$. Because $\Longrightarrow$ is non-blocking, Δ is also in solved form and the claim follows directly from lemma 16. *Decidability:* By assumption the set of normal forms is r.e. and finite, and it is decidable if an equation system is in solved form. □

If we think of $\Longrightarrow$ as being generated by a set of rules, correctness and completeness are properties which are local to each rule, whereas termination and non-blocking depend on the interaction of the rules, i.e they are global properties.

Here are some useful facts about manipulating unifiers of equation sets.

Fact 4 $\mathcal{U}_E(\Gamma \cup \Delta) = \{\sigma_1;\sigma_2 \mid \sigma_1 \in \mathcal{U}_E(\Gamma) \wedge \sigma_2 \in \mathcal{U}_E(\sigma_1\Delta)\}$

Fact 5 *If* $|x_I| = |I|$, $x_I \cap W = \{\}$ *and* $x_I \cap \mathcal{V}(F(s_I), t) = \{\}$ *then*

$$\mathcal{U}_E(\{F(s_I) = t\})|_W = \mathcal{U}_E(\{F(x_I) = t\} \cup \{s_i = x_i \mid i \in I\})|_W$$

In addition we frequently use that $\tau \in \mathcal{U}_E(\sigma\Gamma) \Leftrightarrow \sigma;\tau \in \mathcal{U}_E(\Gamma)$.

4.2 The Matching Algorithms

This section presents two related algorithms for solving the combination problem. The first one, based on "variable abstraction", is simpler to understand and prove correct but potentially less efficient. In one step it is transformed into a more efficient version oriented towards "constant abstraction". Finally a number of further optimizations are discussed.

The following transformation rules constitute a correct, complete and, if each E_i has a finitary matching problem, terminating matching algorithm for the combined theory E.

$$\frac{x \in \mathcal{V}(\Gamma)}{\Gamma \,\dot{\cup}\, \{x = s\} \Longrightarrow \{x \mapsto s\}\Gamma \cup \{x = s\}} \qquad \text{(R)}$$

$$\frac{F \in \Sigma_i^+ \quad F \notin V \quad \sigma \in \mathcal{M}^{\dot{i}}_{\cong}(F(x_I) = t\Downarrow) \quad |x_I| = |I| \quad x_I \cap (\mathcal{V}(\Gamma) \cup \mathcal{V}(F[r_I])) = \{\}}{\Gamma \,\dot{\cup}\, \{F[r_I] = t\} \Longrightarrow \sigma(\Gamma \cup \{r_i = x_i \mid i \in I\}) \cup \sigma_=} \qquad \text{(M)}$$

The underlying idea is the same as in all algorithms for combined unification: alien subterms are replaced by variables or constants and matching is performed on the pure term. On the RHS of the equation all terms are ground and $\mathcal{M}^{\dot{i}}_{\cong}$ automatically performs "constant abstraction". On the LHS rule (M) uses "variable abstraction": all proper alien subterms (r_I) are replaced by new variables (x_I) and the r_I are subsequently equated to the x_I.

The one important invariant these rules maintain is that all RHS's of equations are ground. For (R) this is obvious, for (M) it follows from regularity of $=_E$ and the fact that $\sigma(F(x_I)) =_E t\Downarrow =_E t$ and $x_I \subseteq dom(\sigma)$. Thus we assume in the sequel that all equations have ground RHS's.

Lemma 17 *Rules* (R) *and* (M) *are correct.*

Proof For (R) this is obvious. Let τ be a solution to the RHS of the conclusion of (M), implying $\tau \in \mathcal{U}_E(\Gamma)$, $\tau\sigma r_i =_E \sigma x_i$ for $i \in I$, and $\tau =_E \sigma\ [dom(\sigma)]$. From the latter condition it follows that $\tau = \sigma;\tau$. By lemma 14 it follows that $\tau(F[r_I]) = (\tau\sigma F)(\tau\sigma r_I) =_E (\tau\sigma F)(\tau\sigma x_I) = \tau\sigma(F(x_I)) =_E \tau t\Downarrow = t\Downarrow =_E t$. Hence τ is a solution to the LHS of the conclusion of (M). □

Lemma 18 *Rules* (R) *and* (M) *are complete.*

Proof For (R) this is obvious. Let $\Gamma_0 = \Gamma \,\dot{\cup}\, \{F[r_I] = t\}$ and $\tau \in \mathcal{U}_E(\Gamma_0)$.

$$\begin{array}{ll}
\Rightarrow \exists\tau_1 \in \mathcal{U}_E(F[r_I] = t), \tau_2 \in \mathcal{U}_E(\tau_1\Gamma).\ \tau = \tau_1;\tau_2 & \text{by fact 4} \\
\Rightarrow \exists\theta \in \mathcal{U}_E(\{F(x_I) = t\} \cup \{r_i = x_i \mid i \in I\}).\ \tau_1 = \theta\ [\mathcal{V}(\Gamma_0)] & \text{by fact 5} \\
\Rightarrow \exists\theta_1 \in \mathcal{U}_E(F(x_I) = t), \theta_2 \in \mathcal{U}_E(\theta_1\{r_i = x_i \mid i \in I\}).\ \theta = \theta_1;\theta_2 & \text{by fact 4}
\end{array}$$

Because of lemma 14, fact 2, and the fact that all substitutions are ground, there is some $\sigma \in \mathcal{M}^{\dot{i}}_{\cong}(F(x_I) = t\Downarrow)$ such that $\sigma =_E \theta_1\ [\mathcal{V}(F(x_I))]$. Thus $\Gamma_0 \Longrightarrow \Delta = \Gamma \cup \{r_i = \sigma x_i \mid i \in I\} \cup \sigma_=$. We claim that $\theta;\tau_2 = \tau\ [\mathcal{V}(\Gamma_0)]$ and $\theta;\tau_2 \in \mathcal{U}_E(\Delta)$. The former follows because $\tau_1 = \theta\ [\mathcal{V}(\Gamma_0)]$. Thus $\theta;\tau_2 \in \mathcal{U}_E(\Gamma)$ because $\tau \in \mathcal{U}_E(\Gamma)$. Because $\theta;\tau_2 = \theta_1 =_E \sigma\ [\mathcal{V}(F(x_I))]$, $\theta;\tau_2 \in \mathcal{U}_E(\sigma_=)$. Finally we have $\tau_2\theta r_i =_E \tau_2\theta x_i =_E \sigma x_i$, and hence $\theta;\tau_2 \in \mathcal{U}_E(\{r_i = \sigma x_i \mid i \in I\})$. □

Lemma 19 *Rules* (R) *and* (M) *are terminating.*

Proof The complexity of a set of equations Γ is a pair. The first component is the number of variables which appear more than once in Γ. Obviously it is decreased by (R). The second component is the *multiset* $\{Alt(s) \mid s = t \in \Gamma\}$ together with the usual extension of $\leq$ from integers to multisets of integers (see for example [3]). For (M) we need to distinguish 2 cases. If $dom(\sigma)$ and $\mathcal{V}(\Gamma) \cup \mathcal{V}(r_I)$ are not disjoint, the first component of the complexity decreases. Otherwise the first component cannot increase and the second one decreases. Thus the lexicographic ordering on these pairs (which is well-founded) decreases with every rule application. □

Lemma 20 *Rules* (R) *and* (M) *are non-blocking.*

Proof Assume $\mathcal{U}_E(\Gamma) \neq \{\}$ and Γ is not in solved form. We can distinguish 3 cases.

If there are two equations $x = s, x = t \in \Gamma$ with $s \neq t$, then rule (R) is applicable.

If there is an equation $s = t \in \Gamma$ with $s \notin V$, i.e. $s \equiv F[r_I]$, then lemma 14 tells us that $\mathcal{M}^{i}_{\cong}(F(x_I) = t\Downarrow) \neq \{\}$ because $\mathcal{U}_E(F(x_I) = t\Downarrow) \supseteq \mathcal{U}_E(s = t\Downarrow) = \mathcal{U}_E(s = t) \neq \{\}$. Hence rule (M) is applicable.

Finally, Γ may be in substitution form but $\vec{\Gamma}$ is not idempotent. This is impossible because of the invariant that all RHS's are ground.

Thus Γ is not in normal form either. □

The combination of lemmas 17 - 20 with theorem 2 yields

Theorem 3 *Rules* (R) *and* (M) *enumerate a complete set of matchers of a set of equations with ground RHS's. If each $\mathcal{M}_i$ is finitary, the enumeration is finite and matching is decidable.*

Comparing rule (M) with lemma 14 we notice that we haven't used the full generality of that lemma: abstracting alien subterms which are ground is not necessary. This is intuitively obvious because ground subterms act like constants. The following modified rule (N) uses this additional freedom of applying constant abstraction on the LHS too: only non-ground alien subterms on the LHS are replaced by new variables. This has certain advantages and disadvantages.

$$\frac{\begin{array}{c}\forall j \in J.\, \mathcal{V}(r_j) = \{\} \wedge s_j = r_j\Downarrow \\ \forall k \in K.\, \mathcal{V}(r_k) \neq \{\} \wedge s_m \in V - (\mathcal{V}(\Gamma) \cup \mathcal{V}(F[r_I])) \\ F \in \Sigma_i^+ \quad F \notin V \quad \sigma \in \mathcal{M}^{i}_{\cong}(F(s_I) = t\Downarrow) \quad I = J \mathbin{\dot\cup} K \quad |s_K| = |K|\end{array}}{\Gamma \mathbin{\dot\cup} \{F[r_I] = t\} \Longrightarrow \sigma(\Gamma \cup \{r_k = s_k \mid k \in K\}) \cup \sigma_=} \tag{N}$$

The motivation for constant abstraction is to minimize the number of substitutions returned from $\mathcal{M}^{i}_{\cong}$. The price we pay for that is an additional collapsing step for all r_J — otherwise $\mathcal{M}^{i}_{\cong}(F(s_I) = t\Downarrow)$ is not complete. Section 4.3 discusses whether collapsing can be avoided.

Instead of going through the same correctness, completeness, etc. routine as above, we show that rules (M) and (N) are closely related, using the following lemma:

Lemma 21 *Let $\Longrightarrow_1$ and $\Longrightarrow_2$ be two rewrite relations such that for all $i, j \in \{1, 2\}$:*

$$(\Gamma \Longrightarrow_i \Delta \wedge \delta \in \mathcal{U}_E(\Delta)) \Rightarrow \exists \Delta', \delta' \in \mathcal{U}_E(\Delta').\, \Gamma \Longrightarrow_j \Delta' \wedge \delta =_E \delta'\ [\mathcal{V}(\Gamma)]. \tag{2}$$

Then $\Longrightarrow_1$ is correct, complete, and non-blocking iff $\Longrightarrow_2$ is.

Proof By symmetry it suffices to show only one direction of the "iff". *Correctness*: If $\Gamma \Longrightarrow_2 \Delta$ and $\delta \in \mathcal{U}_E(\Delta)$, there are Δ' and $\delta' \in \mathcal{U}_E(\Delta')$ such that $\Gamma \Longrightarrow_1 \Delta'$ and $\delta =_E \delta'$ $[\mathcal{V}(\Gamma)]$. By correctness of $\Longrightarrow_1$ it follows that $\delta' \in \mathcal{U}_E(\Gamma)$ and hence $\delta \in \mathcal{U}_E(\Gamma)$, i.e. $\Longrightarrow_2$ is also correct. *Completeness*: From (2) it follows that $\bigcup\{\mathcal{U}_E(\Delta)|_{\mathcal{V}(\Gamma)} \mid \Gamma \Longrightarrow_1 \Delta\} = \bigcup\{\mathcal{U}_E(\Delta)|_{\mathcal{V}(\Gamma)} \mid \Gamma \Longrightarrow_2 \Delta\}$

and thus the correctness case is immediate. *Non-Blocking*: Let $\Longrightarrow_1$ be complete and non-blocking. If $\mathcal{U}_E(\Gamma) \neq \{\}$ and Γ is not in solved form, Γ cannot be in $\Longrightarrow_1$-normal form. By completeness there must be a Δ such that $\Gamma \Longrightarrow_1 \Delta$ and $\mathcal{U}_E(\Delta) \neq \{\}$. By assumption it follows that Γ cannot be in $\Longrightarrow_2$-normal form either. □

Using this lemma we show

Theorem 4 *Rules* (R) *and* (N) *are correct, complete and non-blocking.*

Proof We apply lemma 21 by showing that the relation generated by (R) and (M) and the one generated by (R) and (N) correspond as required. For (R)-transitions this is obvious. For (M) and (N) we distinguish 2 cases.

First we show that (N) simulates (M). Assume that the premise of (M) holds and that $\Gamma_0 \Longrightarrow \Delta$ matches the conclusion of (M), and let $\delta \in \mathcal{U}_E(\Delta)$. We have to exhibit a corresponding (N) rewrite. Let $I = J \dot\cup K$ and s_J as required by (N), and let $s_k = x_k$ for $k \in K$. Because $\delta \in \mathcal{U}_E(\Delta)$ and the r_j and σx_j are ground, it follows that $\sigma s_j = \sigma(r_j{\Downarrow}) =_E \sigma r_j = \delta\sigma r_j =_E \delta\sigma x_j = \sigma x_j$. Thus $\sigma(F(s_I)) = (\sigma F)(\sigma s_J, \sigma s_K) =_E (\sigma F)(\sigma x_J, \sigma x_K) = \sigma(F(x_I)) =_E t{\Downarrow}$. Hence there is a $\tau \in \mathcal{M}^i_{\cong}(F(s_I) = t{\Downarrow})$ such that $\tau =_E \sigma\ [\mathcal{V}(F) \cup x_K]$. Therefore $\Gamma_0 \Longrightarrow_{(\mathrm{N})} \Delta' = \tau(\Gamma \cup \{r_k = t_k \mid k \in K\}) \cup \tau_=$. Now $\delta \in \mathcal{U}_E(\Delta')$ follows easily: $\delta \in \mathcal{U}_E(\sigma(\Gamma \cup \{r_k = s_k \mid k \in K\})) = \mathcal{U}_E(\sigma|_{\mathcal{V}(F)\cup x_K}(\Gamma \cup \{r_k = s_k \mid k \in K\})) = \mathcal{U}_E(\tau(\Gamma \cup \{r_k = s_k \mid k \in K\}))$ and $\delta \in \mathcal{U}_E(\sigma_=) \subseteq \mathcal{U}_E(\tau_=)$.

Now we show that (M) simulates (N). This time assume that the premise of (N) holds and that $\Gamma_0 \Longrightarrow \Delta$ matches the conclusion of (N), and let $\delta \in \mathcal{U}_E(\Delta)$. Now let $x_k = s_k$ for all $k \in K$, $\sigma' = \{x_j \mapsto s_j \mid j \in J\}$, and $\theta = \sigma \cup \sigma'$. Thus $\theta(F(x_I)) = (\theta F)(\theta x_J, \theta x_K) = (\sigma F)(s_J, \sigma x_K) = (\sigma F)(\sigma s_J, \sigma s_K) = \sigma(F(s_I)) =_E t{\Downarrow}$. Hence there is a $\tau \in \mathcal{M}^i_{\cong}(F(x_I) = t{\Downarrow})$ such that $\tau =_E \theta\ [F(x_I)]$. Therefore $\Gamma_0 \Longrightarrow_{(\mathrm{M})} \Delta' = \tau(\Gamma \cup \{r_i = s_i \mid i \in I\}) \cup \tau_=$. Let $\delta' = \delta \cup \sigma'$. Obviously $\delta = \delta'\ [\mathcal{V}(\Gamma_0)]$. We claim that $\delta' \in \mathcal{U}_E(\Delta')$ and distinguish several cases.

Since $\delta \in \mathcal{U}_E(\sigma(\Gamma \cup \{r_k = s_k \mid k \in K\})) = \mathcal{U}_E(\theta(\Gamma \cup \{r_k = s_k \mid k \in K\})) = \mathcal{U}_E(\tau(\Gamma \cup \{r_k = s_k \mid k \in K\}))$, $\delta' \in \mathcal{U}_E(\tau(\Gamma \cup \{r_k = x_k \mid k \in K\}))$ follows.

From $\delta'\tau r_j = r_j =_E r_j{\Downarrow} = s_j = \theta x_j = \tau x_j$ it follows that $\delta' \in \mathcal{U}_E(\tau\{r_j = x_j \mid j \in J\})$.

From $\delta \in \mathcal{U}_E(\sigma_=)$ it follows that $\delta =_E \sigma\ [\mathcal{V}(F) \cup s_K]$. Thus $\delta' x_k = \delta x_k =_E \sigma x_k = \theta x_k =_E \tau x_k$ holds for all $k \in K$. Together with $\delta' x_j = \sigma' x_j = \theta x_j =_E \tau x_j$ this implies $\delta \in \mathcal{U}_E(\tau_=)$. □

The termination proof is identical to the one for lemma 19:

Lemma 22 *Rules* (R) *and* (M) *are terminating.*

As for (R) and (M), we conclude that (R) and (N) yield a complete and terminating matching algorithm for the combined theory, provided all $\mathcal{M}^i$ are finitary.

4.3 Optimizations

In this section we discuss a number of optimizations of the matching algorithms presented so far. Most of them could have been incorporated into the algorithms from the start but would have complicated the presentation and the proofs.

4.3.1 Removing ⇓

One obvious source of inefficiency is the collapse in each matching step $\mathcal{M}^i_{\cong}(F[\ldots] = t{\Downarrow})$ which can in fact be avoided. We show that under certain assumptions the rules keep all RHS's of equations in CNF. In order to achieve this the individual matching algorithms must not return substitutions which can be collapsed or which contain subterms from other theories.

Definition 5 *Let $i \in \mathcal{I}$ and $\Gamma \subseteq \mathcal{T}(\Sigma_i, V \cup C) \times \mathcal{T}(\Sigma_i, C)$. $\mathcal{M}^i$ is called* pure *if $ran(\sigma) \subseteq \mathcal{T}(\Sigma_i, V \cup C)$ for all $\sigma \in \mathcal{M}^i(\Gamma)$, and* collapse-free *if all $\sigma \in \mathcal{M}^i(\Gamma)$ are in CNF.*

Because of regularity we immediately obtain

Fact 6 *All $\mathcal{M}^i$ are pure.*

Lemma 23 *If $\mathcal{M}^i$ is collapse-free, $s \in \mathcal{T}(\Sigma)$ and $t \in \mathcal{T}(\Sigma\backslash V)$ such that all immediate i-alien subterms of t are in CNF, then all substitutions in $\mathcal{M}^i_{\cong}(s = t)$ are in CNF.*

Proof Let $\sigma \in \mathcal{M}^i_{\cong}(s=t)$ and $r \in ran(\sigma)$. By definition of $\mathcal{M}^i_{\cong}$ there is a $\tau \in \mathcal{M}^i([s]^i_{\cong} = [t]^i_{\cong})$ such that $\sigma = (\tau;\pi)|_{dom(\sigma)}$. Thus there is an $r' \in ran(\tau)$ with $r = \pi r'$. W.l.o.g. $r' \equiv F[r'_J]$ for some $F \in \Sigma^*_i$. Because $\mathcal{M}^i$ is pure and $=_i$ is regular, we have $IAST_i(ran(\tau)) = \mathcal{C}(ran(\tau)) \subseteq \mathcal{C}([t]^i_{\cong}) = \{[t']_{\cong} \mid t' \in IAST_i(t)\}$. Therefore $r'_j = [t_j]_{\cong}$ for some $t_j \in IAST_i(t)$. Hence $r = F[r_J]$ where $r_j \cong t_j$ and hence $\mathcal{T}(r_j) = \mathcal{T}(t_j) \neq i$ and r_j is in CNF because t_j is (lemma 3). Now we distinguish 2 cases.

If F is the empty context, then $J = \{j\}$ and $r = r_j$ and hence r is in CNF.

Now let $F \in \Sigma^+_i$ and assume r is not in CNF, i.e. there is a j such that $F[r'_J] = F[[r_J\Downarrow]_{\cong}] =_i [r_j\Downarrow]_{\cong} = r'_j$, contradicting the collapse-freeness of $=_i$. Hence r must be in CNF. □

Theorem 5 *$\mathcal{M}^i_{\cong}(F[\ldots] = t\Downarrow)$ in (M) and (N) can be replaced by $\mathcal{M}^i_{\cong}(F[\ldots] = t)$, provided all $\mathcal{M}^i$ are collapse-free and all RHS's of the initial system of equations are in CNF.*

Proof Let (M') and (N') be the rules (M) and (N) without $\Downarrow$. If t is in CNF, $t\Downarrow = t$ and thus the primed rules behave like the original ones. It remains to be shown that all RHS's stay in CNF. Rule (R) doesn't change RHS's because they are ground. If $\Gamma \Longrightarrow \Delta$ via one of the primed rules, then the RHS's of all equations in $\Gamma - \Delta$ are of the form σv, where $\sigma \in \mathcal{M}^i_{\cong}(F[\ldots] = t)$ and $v \in dom(\sigma)$. Since t is in CNF, it follows by lemma 23 that σv is in CNF. □

Note that the $\mathcal{M}^i$ can be made collapse-free by composing them with $\Downarrow$.

The occurrence of $r_j\Downarrow$ in (N) cannot be removed that easily. The problem is that even if a term s and a substitution σ are both in CNF, σs need not be. Thus the LHS's of the system $\sigma(\Gamma \cup \{r_k = s_k \mid k \in K\})$ on the RHS of the conclusion of (N) need not be in CNF even if Γ, the r_K and σ are. It depends very much on the representation of variables, terms, and substitutions whether the collapsing process can take advantage of the fact that the terms and the substitution on their own are in CNF.

4.3.2 Removing (R)

We can dispose of rule (R) provided (R) is inapplicable to the initial system. The reason is that (M) and (N) preserve the the property that (R) is inapplicable. Hence it suffices to apply (R) to the initial system as long as possible and then forget about it. This strategy preserves completeness because (R) is a complete transformation.

4.3.3 Taking Advantage of Collapse-Free Theories

The one major optimization we haven't mentioned so far is taking collapse-freeness into account, the reason being that the resulting changes are pretty obvious.

The last line in the definition of $s\Downarrow$ can be changed to

in if $E_{\mathcal{T}(s)}$ *is collapse-free* **then** $G[t_I]$
else if $\exists i \in I.\ G[[t_I]_{\cong}] =_{\mathcal{T}(s)} [t_i]_{\cong}$ **then** t_i **else** $G[t_I]$

The justification is that if $\mathcal{T}(s) \neq 0$ and $E_{\mathcal{T}(s)}$ is collapse-free then $s =_{\mathcal{T}(s)} c$ does not hold for any constant $c \in C$. As a consequence we have that $\mathcal{NT}(\sigma s) = \mathcal{T}((\sigma s)\Downarrow) = \mathcal{T}(s)$ holds for any σ and any $s \notin V$ such that $E_{\mathcal{T}(s)}$ is collapse-free.

More importantly, a *failure condition* can be added to the matching rules. Failure conditions identify certain easily recognizable cases of unsolvable equation systems. If there is an equation $s = t$ in the current set of equations Γ (hence t is ground) such that $E_{\mathcal{T}(s)}$ is collapse free and $\mathcal{T}(s) \neq \mathcal{T}(t\Downarrow)$ then $s = t$ and hence Γ are unsolvable: $\mathcal{NT}(\sigma s) = \mathcal{T}(s) \neq \mathcal{T}(t) = \mathcal{NT}(t) \Rightarrow (\sigma s)\Downarrow \not\cong t \Rightarrow \sigma s \neq_E t$.

Even for collapsing theories efficiency can potentially be increased by using a special "collapsing algorithm" rather than collapsing by equality test as in the current definition of $\Downarrow$.

4.3.4 Miscellany

Matching ground terms can be replaced by equality test. More precisely we get a rule

$$\frac{s \in \mathcal{T}(\Sigma \backslash V) \quad s\Downarrow \cong t}{\Gamma \dot{\cup} \{s = t\} \Longrightarrow \Gamma} \tag{E}$$

and a failure condition if a system contains an equation $s = t$ where $s \in \mathcal{T}(\Sigma \backslash V)$ and $s\Downarrow \not\cong t$. More generally, failure occurs if in trying to apply one of the rules (M) or (N) the set $\mathcal{M}^{i}_{\cong}(F[\ldots] = t\Downarrow)$ is empty.

Finally, it should be noted that although, because of notational reasons, we have used *sets* of equations, *everything* (including the termination proofs!) works just as well with *multisets*.

5 Conclusion

Although I am not aware of any theoretical treatment of the combination of matching problems, there are several implementations for particular theories, notably in REVE, LP, and OBJ. The implementation in REVE and LP is based on Kathy Yelick's work on combining unification algorithms in [19]. As a consequence it is restricted to collapse-free theories. OBJ on the other hand provides rewriting modulo some special theories, in particular associativity with a 1, i.e. where $x * 1 = 1 * x = x$, a collapsing theory. However, according to [12], the collapsing axioms are not integrated into the matching algorithm but are added as new rewrite rules. In order to obtain a canonical set of rewrite rules, a simple form of completion modulo equations (e.g. [10]) is applied. In general the resulting set of rules is however not canonical.

Acknowledgements

Many thanks go to Behnam Banieqbal for some early collaboration, to John Guttag for providing the stimulating environment, to Daniel Jackson for his cappuccino, to Claude Kirchner for detailed comments, to Kathy Yelick for innumerable patient technical discussions, and to all the other people who have previously worked on combination problems.

References

[1] H.-J. Bürckert: *Some Relationships Between Unification, Restricted Unification and Matching*, in: Proc. CADE-8, LNCS 230 (1986), 514-524.

[2] H.-J. Bürckert: *Matching - A Special Case of Unification?*, Tech. Rep., Universität Kaiserslautern (1987).

[3] N. Dershowitz, Z. Manna: *Proving Termination with Multiset Orderings*, CACM 22 (1979), 465-476.

[4] K. Futatsugi, J.A. Goguen, J.-P. Jouannaud, J. Meseguer: *Principles of OBJ2*, in: Proc. 12th POPL (1985), 52-66.

[5] S.J. Garland, J.V. Guttag: *An Overview of LP, The Larch Prover*, in: Proc. RTA-89, LNCS (1989).

[6] A. Herold: *Combination of Unification Algorithms*, in: Proc. CADE-8, LNCS 230 (1986), 450-469.

[7] A. Herold: *Combination of Unification Algorithms in Equational Theories*, Ph.D. Thesis, Fachbereich Informatik, Universität Kaiserslautern (1987).

[8] G. Huet: *Confluent Reductions: Abstract properties and Applications to Term Rewriting Systems*, JACM 27, 4 (1980), 797-821.

[9] G. Huet, D.C. Oppen: *Equations and Rewrite Rules - A Survey*, in: Formal Languages: Perspectives and Open Problems, R. Book (ed.), Academic Press (1982).

[10] J.-P. Jouannaud, H. Kirchner: *Completion of a Set of Rules Modulo a Set of Equations*, SIAM Journal of Computing 15 (1986), 1155-1194.

[11] C. Kirchner: *Méthodes et outils de conception systématique d'algorithmes d'unification dans les théories équationnelles*, Thèse d'état de l'Université de Nancy I (1985).

[12] C. Kirchner, personal communication, June 1988.

[13] P. Lescanne: *REVE: A Rewrite Rule Laboratory*, in: Proc. CADE-8, LNCS 230 (1986), 695-696.

[14] G.E. Peterson, M.E. Stickel: *Complete Sets of Reductions for Some Equational Theories*, JACM, 28, 2 (1981), 233-264.

[15] M. Schmidt-Schauß: *Unification in a Combination of Arbitrary Disjoint Equational Theories*, Tech. Rep., Universität Kaiserslautern (1987).

[16] J.H. Siekmann: *Universal Unification*, in: Proc. CADE-7, LNCS 170 (1984).

[17] E. Tidén: *Unification in Combinations of Collapse-Free Theories with Disjoint Sets of Function Symbols*, in: Proc. CADE-8, LNCS 230 (1986), 431-449.

[18] E. Tidén: *First-Order Unification in Combinations of Equational Theories*, PhD Thesis, Department of Numerical Analysis and Computing Science, The Royal Institute of Technology, Stockholm (1986).

[19] K. Yelick: *Unification in Combinations of Collapse-Free Regular Theories*, JSC 3 (1987), 153-181.

[20] P. Szabó: *Unifikationstheorie erster Ordnung*, Ph.D. Thesis, Universität Karlsruhe (1982).

RESTRICTIONS OF CONGRUENCES GENERATED BY FINITE CANONICAL STRING-REWRITING SYSTEMS*

Friedrich Otto

Department of Computer Science, State University of New York
Albany, New York 12222, U.S.A.

currently visiting at

Fachbereich Informatik, Universität Kaiserslautern
6750 Kaiserslautern, Fed. Rep. Germany

* This research was partially supported by a Faculty Research Award from SUNY Albany.

Abstract
Let Σ_1 be a subalphabet of Σ_2, and let R_1 and R_2 be finite string-rewriting systems on Σ_1 and Σ_2, respectively. If the congruence $\overset{*}{\longleftrightarrow}_{R_1}$ generated by R_1 and the congruence $\overset{*}{\longleftrightarrow}_{R_2}$ generated by R_2 coincide on Σ_1^*, then R_1 can be seen as representing the restriction of the system R_2 to the subalphabet Σ_1. Is this property decidable ? This question is investigated for several classes of finite canonical string-rewriting systems.

I. INTRODUCTION

For algebraic structures like monoids and groups, that are given through finite presentations, the word problem is one of the most fundamental decision problems. It is well-known that there are finite presentations for which the word problem is undecidable. In addition, it is even undecidable in general, whether or not the word problem for a given finite presentation is decidable. However, if an algebraic structure is given through a finite canonical rewriting system, then its word problem is necessarily decidable. In fact, a finite canonical rewriting system allows to effectively determine a normal form for each given element, and by comparing the normal forms u_0 and v_0 of two elements u and v, respectively, one can determine whether or not u and v represent the same element of the given structure. Although this algorithm may be of a much higher degree of complexity than the word problem itself [3], it is employed usually whenever a finite canonical rewriting system is given because of its simple syntactic structure.

Here we are interested in the case of string-rewriting systems, i.e., term-rewriting systems involving unary function symbols only, and hence, the algebraic structures considered are monoids. In order to solve the word problem for a monoid that is given through a finite non-canonical string-rewriting system S one often tries to determine a finite canonical string-rewriting system R that is equivalent to S by way of completion [12]. But even if the word problem for S is decidable, there may not exist a finite canonical system R that is equivalent to S [9,10,21], i.e., completion cannot possibly succeed in this situation.

Introducing additional generators does not always help either, since as shown by Squier [22] there exist finitely presented monoids with decidable word problem that cannot be presented by any finite canonical string-rewriting system.

This leaves the question of whether every monoid with a decidable word problem can at least be embedded into a monoid that is presented by a finite canonical string-rewriting system. This would improve upon the fact that every monoid with a decidable word problem can be embedded into a finitely presented monoid with a decidable word problem [1]. Bauer has investigated this question as part of his doctoral dissertation [2], but he only succeeded in giving a restricted result: Every monoid with a decidable word problem can be embedded into a monoid that is presented by a finite string-rewriting system that is canonical on the embedded monoid. So the above question is still open. However, it should be mentioned that by introducing additional functions a finite canonical term-rewriting can be obtained such that the restriction of this system to the original letters still presents the monoid we started out with [4].

Here we are concerned with a decision problem that is related to the above question. Let Σ_1 and Σ_2 be two alphabets such that $\Sigma_1 \subseteq \Sigma_2$, and let R_1 and R_2 be two string-rewriting systems on Σ_1 and Σ_2, respectively. Under what conditions is it decidable whether or not the congruence $\overset{*}{\longleftrightarrow}_{R_1}$ on Σ_1^* generated by R_1 is the restriction of the congruence $\overset{*}{\longleftrightarrow}_{R_2}$ generated by R_2 to Σ_1^* ? To phrase it differently we may ask under which conditions the congruence $\overset{*}{\longleftrightarrow}_{R_2}$ is an extension of the congruence $\overset{*}{\longleftrightarrow}_{R_1}$. If $\Sigma_1 = \Sigma_2$, this problem is nothing but the equivalence problem for string-rewriting systems. Thus, here we are dealing with a generalization of this problem. The equivalence problem for string-rewriting systems is known to be undecidable in general. If, however, R_1 and R_2 are finite and canonical, then their equivalence is decidable [7].

After establishing notation in Section 2 we shall prove in Section 3 that the above problem is undecidable in general, even if R_1 is the empty system, and R_2 is finite, length-reducing, and confluent. The same is true if R_1 is empty, and R_2 is finite and monadic. However, if R_2 is finite, monadic, and confluent, then this problem becomes decidable for all finite canonical systems R_1. Finally, a particularly simple decision procedure is obtained if both, R_1 and R_2, are finite, special, reduced, and confluent string-rewriting systems.

2. THUE CONGRUENCES

In this section we only introduce the minimum number of definitions and results that will be needed throughout this paper. For additional information regarding canonical string-rewriting systems the reader is asked to consult Book's excellent overview paper [6].

Let Σ be a finite alphabet. Then Σ^*, the free monoid generated by Σ, is the set of words (or strings) over Σ including the empty word e. The concatenation of two words u and v is simply written as uv. The length $|w|$ of a word w is defined by: $|e| = 0$, $|wa| = |w|+1$ for all $w \in \Sigma^*$, and $a \in \Sigma$.

A string-rewriting system R on Σ is a subset of $\Sigma^* \times \Sigma^*$. Its elements are called (rewrite) rules . Let $\mathrm{dom}(R) = \{l \mid \exists r \in \Sigma^*: (l,r) \in R\}$, and $\mathrm{range}(R) = \{r \mid \exists l \in \Sigma^*: (l,r) \in R\}$. The string-rewriting system R induces a number of binary relations on Σ^*, the most fundamental one being the single-step reduction relation $\longrightarrow_R$:

$$u \longrightarrow_R v \text{ iff } \exists x,y \in \Sigma^* \ \exists (l,r) \in R\colon u = xly \text{ and } v = xry.$$

The reflexive, transitive closure $\xrightarrow{*}_R$ of $\longrightarrow_R$ is the reduction relation induced by R, and the reflexive, symmetric, and transitive closure $\xleftrightarrow{*}_R$ of $\longrightarrow_R$ is a congruence relation on Σ^*, the Thue congruence generated by R. For $w \in \Sigma^*$, $[w]_R = \{u \in \Sigma^* \mid u \xleftrightarrow{*}_R w\}$ is the congruence class of w modulo R. Since $\xleftrightarrow{*}_R$ is a congruence relation, the set $\{[w]_R \mid w \in \Sigma^*\}$ of congruence classes forms a monoid M_R under the operation $[u]_R \circ [v]_R = [uv]_R$ with identity $[e]_R$. This is the factor monoid $\Sigma^*/\xleftrightarrow{*}_R$ of the free monoid Σ^* modulo the congruence $\xleftrightarrow{*}_R$, and it is uniquely determined (up to isomorphism) by Σ and R. Therefore, whenever a monoid M is isomorphic to M_R, the ordered pair $(\Sigma;R)$ is called a (monoid-)presentation of M with generators Σ and defining relations R.

Two string-rewriting systems R_1 and R_2 on Σ are called equivalent if they generate the same congruence relation, i.e., $\xleftrightarrow{*}_{R_1} = \xleftrightarrow{*}_{R_2}$. This condition is much stronger than that of isomorphism, where two string-rewriting systems R_1 on Σ_1 and R_2 on Σ_2 are called isomorphic if the factor monoids $M_{R_1} = \Sigma_1^*/\xleftrightarrow{*}_{R_1}$ and $M_{R_2} = \Sigma_2^*/\xleftrightarrow{*}_{R_2}$ are isomorphic.

The problem of deciding whether two given finite string-rewriting systems are equivalent, and the problem of deciding whether two given finite string-rewriting systems are isomorphic are both undecidable in general [13,14,17]. However, if R_1 and R_2 are isomorphic, then the presentation $(\Sigma_1;R_1)$ can be transformed into the presentation $(\Sigma_2;R_2)$ by a finite sequence of Tietze-transformations [23, c.f.,e.g., 20].

Here we are interested in the following generalization of the equivalence problem.

The PROBLEM OF RESTRICTION :

INSTANCE : A finite string-rewriting system R_1 on alphabet Σ_1, and a finite string-rewriting system R_2 on alphabet $\Sigma_2 \supseteq \Sigma_1$.

QUESTION : Is $\overset{*}{\longleftrightarrow}_{R_1} = \overset{*}{\longleftrightarrow}_{R_2}\big|_{\Sigma_1^*\times\Sigma_1^*}$, i.e., is the congruence $\overset{*}{\longleftrightarrow}_{R_1}$ the restriction of the congruence $\overset{*}{\longleftrightarrow}_{R_2}$ to Σ_1^* ?

Since the problem of restriction is a generalization of the equivalence problem, it is also undecidable in general. So the best we can hope for is that we may be able to solve it for some restricted classes of finite string-rewriting systems. Of particular interest are the finite canonical string-rewriting systems, since for them many problems like the word problem or the equivalence problem, that are undecidable in general, become decidable [7].

Let R be a string-rewriting system on Σ. The system R is called Noetherian if there is no infinite reduction sequence of the form $x_0 \longrightarrow_R x_1 \longrightarrow_R x_2 \longrightarrow_R \ldots$ A word $w \in \Sigma^*$ is called reducible (modulo R), if there exists a word x such that $w \longrightarrow_R x$; otherwise, w is called irreducible (modulo R). By IRR(R) we denote the set of all irreducible words (modulo R). If R is Noetherian, then each congruence class $[w]_R$ contains one or more irreducible words.

The system R is called confluent if, for all $u,v,w \in \Sigma^*$, $u \overset{*}{\longrightarrow}_R v$ and $u \overset{*}{\longrightarrow}_R w$ imply that $v \overset{*}{\longrightarrow}_R z$ and $w \overset{*}{\longrightarrow}_R z$ for some word $z \in \Sigma^*$. If R is confluent, then no congruence class $[w]_R$ contains more than one irreducible word.

Finally, if R is Noetherian and confluent, then each congruence class $[w]_R$ contains exactly one irreducible word w_0, which can then be taken as the normal form for this class, and w_0 can be obtained from w by a finite sequence of reduction steps. A rewriting system that is both Noetherian and confluent is therefore called canonical or complete.

If R is finite, then the process of reduction $\longrightarrow_R$ is effective. Thus, the word problem for a finite canonical string-rewriting system is decidable. Further, if $R_1 = \{(l_i,r_i) \mid i = 1,2,\ldots,n\}$ and $R_2 = \{(u_j,v_j) \mid j = 1,2,\ldots,m\}$ are two finite string-rewriting systems on Σ, then R_1 and R_2 are equivalent if and only if $l_i \overset{*}{\longleftrightarrow}_{R_2} r_i$, $i = 1,2,\ldots,n$, and $u_j \overset{*}{\longleftrightarrow}_{R_1} v_j$, $j = 1,2,\ldots,m$. Hence, if R_1 and R_2 are canonical, the equivalence of R_1 and R_2 is decidable.

Finally, we want to introduce some syntactic restrictions for string-rewriting systems. A system $R = \{(l_i,r_i) \mid i \in I\}$ on Σ is called length-reducing if $|l_i| > |r_i|$ holds for all $i \in I$. It is called monadic if it is length-reducing, and range(R) $\subseteq \Sigma \cup \{e\}$, and it is called special if it is length-reducing and range(R) $= \{e\}$.

3. UNDECIDABLE CASES

In this section we prove that the PROBLEM OF RESTRICTION remains undecidable even if $R_1 = \emptyset$, and R_2 is also fairly restricted.

Theorem 3.1. The following restricted version of the PROBLEM OF RESTRICTION is undecidable:

INSTANCE : A finite, length-reducing, and confluent string-rewriting system R on alphabet Σ_2, and a subalphabet $\Sigma_1 \subsetneq \Sigma_2$.

QUESTION : Is $\overset{*}{\longleftrightarrow}_R|_{\Sigma_1^* \times \Sigma_1^*} = id_{\Sigma_1^*}$, i.e., is the restriction of the congruence $\overset{*}{\longleftrightarrow}_R$ generated by R to Σ_1^* the identity on Σ_1^* ?

Proof. The proof consists of a reduction from the halting problem for Turing machines. Let Σ be a finite alphabet, let $L \subseteq \Sigma^*$ be recursively enumerable, but non-recursive, and let $M = (Q, \Sigma_b, q_o, q_n, \delta)$ be a single-tape Turing machine that accepts L. Here $Q = \{q_o, q_1, \ldots, q_n\}$ is the set of states, $\Sigma_b = \Sigma \cup \{b\}$, where b denotes the blank symbol, is the tape alphabet, q_o is the initial state, q_n is the halting state, and $\delta: (Q-\{q_n\}) \times \Sigma_b \longrightarrow Q \times (\Sigma_b \cup \{L,R\})$ is the transition function of M. Observe that M halts if and only if it enters state q_n.

Without loss of generality we may assume that M never gets into a loop, i.e., for all $u,v \in \Sigma_b^*$ and $q \in Q$, $uqv \not\overset{+}{\longrightarrow}_M uqv$. This is true, since we may simulate the machine M by another Turing machine M_o, where M_o simulates the stepwise behavior of M, but while doing so it counts M's steps on a special track. Then, for $x \in \Sigma^*$, if M does not halt on input x, then neither does M_o, and in addition, the tape inscription of M_o on input x grows arbitrarily long. Hence, M_o cannot get into a loop. Since we will only be interested in computations that start with a proper initial configuration, we don't have to worry about configurations that cannot be reached from any initial configuration.

Let $\bar{\Sigma}_b$ be an alphabet in one-to-one correspondence to Σ_b such that $\Sigma_b \cap \bar{\Sigma}_b = \emptyset$, let \$, ¢, d be three additional symbols, and let $\Gamma := \Sigma_b \cup \bar{\Sigma}_b \cup Q \cup \{\$, ¢, d\}$. On Γ we now define a string-rewriting system R(M) that simulates the stepwise behavior of M. R(M) consists of the following three groups of rules:

(1)	$q_i a_k dd$	$\longrightarrow$	$q_j a_l$	if $\delta(q_i, a_k) = (q_j, a_l)$	
	$q_i ¢ dd$	$\longrightarrow$	$q_j a_l ¢$	if $\delta(q_i, b) = (q_j, a_l)$	
	$q_i a_k dd$	$\longrightarrow$	$\bar{a}_k q_j$	if $\delta(q_i, a_k) = (q_j, R)$	
	$q_i ¢ dd$	$\longrightarrow$	$\bar{b} q_j ¢$	if $\delta(q_i, b) = (q_j, R)$	
	$\bar{a}_l q_i a_k dd$	$\longrightarrow$	$q_j a_l a_k$	if $\delta(q_i, a_k) = (q_j, L)$	for all $\bar{a}_l \in \bar{\Sigma}_b$
	$\bar{a}_l q_i ¢ dd$	$\longrightarrow$	$q_j a_l ¢$	if $\delta(q_i, b) = (q_j, L)$	for all $\bar{a}_l \in \bar{\Sigma}_b$
	$\$ q_i a_k dd$	$\longrightarrow$	$\$ q_j b a_k$	if $\delta(q_i, a_k) = (q_j, L)$	
	$\$ q_i ¢ dd$	$\longrightarrow$	$\$ q_j b ¢$	if $\delta(q_i, b) = (q_j, L)$	

(2) $a_ia_jdd \longrightarrow a_ida_j$
$a_ida_jdd \longrightarrow a_idda_j$ } for all $a_i \in \Sigma_b$, $a_j \in \Sigma_b \cup \{¢\}$

(3) $q_na_idd \longrightarrow q_n$ for all $a_i \in \Sigma_b$

$\bar{a}_iq_n¢dd \longrightarrow q_n¢$ for all $\bar{a}_i \in \bar{\Sigma}_b$

The rules of group (1) simulate the stepwise behavior of M, where the symbol d is used as "enable symbol", while the symbols \$ and ¢ serve as left and right end marker, respectively. The rules of group (2) shift occurrences of the symbol d to the left, and the rules of group (3) erase the tape inscription (or rather its description) of halting configurations.

From the given construction it can easily be seen that R(M) has the following properties.

(1.) R(M) is finite, length-reducing, and confluent. (Recall that M is a deterministic machine !)

(2.) For all $x \in \Sigma^*$ the following two properties are equivalent:

(i) $x \in L$,

(ii) $\exists m \in \mathbf{N} : \$q_ox¢d^m \xrightarrow{*}_{R(M)} \$q_n¢$.

Using R(M) we now define instances of the above restricted version of the PROBLEM OF RESTRICTION. Let $\Sigma_1 := \{\$,¢,q_n,d\}$, $\Sigma_2 := \Gamma$, and, for some $x \in \Sigma^*$, $R(x) := R(M) \cup \{\$d^{|x|+2}¢ \longrightarrow \$q_ox¢\}$. Then R(x) is finite, length-reducing, and confluent, i.e., $(R(x),\Sigma_2,\Sigma_1)$ is a proper instance of the above problem. Given $x \in \Sigma^*$, this instance is easily obtainable.

Claim 1. If $x \in L$, then there exists an integer $m \in \mathbf{N}$ such that $\$d^{|x|+2}¢d^m \xleftrightarrow{*}_{R(x)} \$q_n¢$.

Proof. Let $x \in L$. Then by condition (2.) there exists an integer $m \in \mathbf{N}$ such that $\$q_ox¢d^m \xrightarrow{*}_{R(M)} \$q_n¢$. Thus, since $R(M) \subseteq R(x)$, we have $\$d^{|x|+2}¢d^m \longrightarrow_{R(x)} \$\, q_ox¢d^m \xrightarrow{*}_{R(x)} \$q_n¢$.

Hence, if $x \in L$, then $\xleftrightarrow{*}_{R(x)}\big|_{\Sigma_1^* \times \Sigma_1^*} \neq id_{\Sigma_1^*}$.

Claim 2. If $x \notin L$, then, for all $u,v \in \Sigma_1^*$, $u \xleftrightarrow{*}_{R(x)} v$ implies $u = v$.

Proof. Let $x \in \Sigma^*$ such that $x \notin L$. Then starting from the configuration q_ox the Turing machine M will never reach state q_n. Let $u,v \in \Sigma_1^*$ such that $u \xleftrightarrow{*}_{R(x)} v$. Since $dom(R(M)) \cap \Sigma_1^* = \emptyset$, since R(x) is confluent, and since $\$d^{|x|+2}¢ \longrightarrow \$q_ox¢$ is the only

rule in R(x) - R(M), we see that u and v can be factored as follows:
$u = u_0\$d^{|x|+2}¢d^{i_1}u_1\$ d^{|x|+2}¢d^{i_2}u_2 \ldots \$d^{|x|+2}¢d^{i_r}u_r$, and
$v = v_0\$d^{|x|+2}¢ d^{j_1}v_1\$d^{|x|+2}¢d^{j_2}v_2 \ldots \$d^{|x|+2}¢d^{j_s}v_s$, where $r,s \geq 0$,

$u_0,u_1,\ldots,u_r,v_0,v_1,\ldots,v_s \in IRR(R(x)), u_1,\ldots,u_r,v_1,\ldots,v_s$ do not begin with an occurrence of the letter d, and $i_1,i_2,\ldots,i_r,j_1,j_2,\ldots,j_s \in \mathbf{N}$.

Since occurrences of the end markers \$ and ¢ can neither be generated nor deleted, all applications of rules of R(x) to u and v and their descendants do involve only the factors of the form $\$ d^{|x|+2}¢ d^{i}\lambda$ or $\$ d^{|x|+2}¢ d^{j}\lambda$, respectively, and their descendants. Hence, $u \xrightarrow{*}_{R(x)} u_0\$ x_1q_{\lambda_1}y_1¢ d^{k_1}u_1 \ldots \$ x_rq_{\lambda_r}y_r¢d^{k_r}u_r$, and $v \xrightarrow{*}_{R(x)} v_0\$w_1q_{\mu_1}z_1¢ d^{l_1}v_1 \ldots \$w_sq_{\mu_s} z_s¢d^{l_s}v_s$, where $x_1,\ldots,x_r, w_1,\ldots,w_s \in \overline{\Sigma}_b^*$, $y_1,\ldots,y_r, z_1,\ldots,z_s \in (\Sigma_b \cup \{d\})^*$, $q_{\lambda_1},\ldots,q_{\lambda_r}, q_{\mu_1},\ldots,q_{\mu_s} \in (Q-\{q_n\})$, and $k_1,\ldots,k_r, l_1,\ldots,l_s \in \{0,1\}$ (Recall that $x \notin L$), such that these words are irreducible (modulo R(x)).

Now $u \xleftrightarrow{*}_{R(x)} v$ and R(x) being confluent imply that these two irreducible words coincide. Thus, $r = s$, $u_i = v_i$, $x_i = w_i$, $y_i = z_i$, $q_{\lambda_i} = q_{\mu_i}$, and $k_i = l_i$, all i.

Let $v \in \{1,2,\ldots,r\}$. Then $\$d^{|x|+2}¢d^{i_v} \longrightarrow_{R(x)} \$ q_0x¢ d^{i_v} \xrightarrow{*}_{R(x)} \$ x_vq_{\lambda_v}y_v¢d^{k_v} = \$ w_vq_{\mu_v}z_v¢ d^{l_v} \xleftarrow{*}_{R(x)} \$ q_0x¢ d^{j_v} \longleftarrow_{R(x)} \$ d^{|x|+2}¢ d^{j_v}$, i.e., both these reductions describe computations of the Turing machine M that start with the initial configuration q_0x and end with the configuration $x_vq_{\lambda_v}y_v'$, where y_v' is obtained from y_v by deleting all occurrences of the letter d. Since M does never get into a loop, these two computations coincide, and hence, $i_v = j_v$. This holds for all $i = 1,2,\ldots,r$, and so we can conclude that $u = v$.

Hence, if $x \notin L$, then $\xleftrightarrow{*}_{R(x)}|_{\Sigma_1^*\times\Sigma_1^*} = id_{\Sigma_1^*}$. Together Claims 1 and 2 yield that, for $x \in \Sigma_1^*$, $\xleftrightarrow{*}_{R(x)}|_{\Sigma_1^*\times\Sigma_1^*} = id_{\Sigma_1^*}$ if and only if $x \notin L$. By choice of L, this is undecidable, and so it is undecidable whether $\xleftrightarrow{*}_{R(x)}|_{\Sigma_1^*\times\Sigma_1^*} = id_{\Sigma_1^*}$ holds or not. This completes the proof of Theorem 3.1. □

The next lemma shows that each finite string-rewriting system can be embedded into a finite monadic string-rewriting system. This then allows to carry over Theorem 3.1 to finite monadic systems that are not confluent.

<u>Lemma 3.2.</u> Let R be a finite string-rewriting system on Σ such that no rule of R is of the form (a,b) with $a,b \in \Sigma$. Then one can effectively construct a finite monadic system S on alphabet $\Delta \supseteq \Sigma$ such that $\xleftrightarrow{*}_R = \xleftrightarrow{*}_S|_{\Sigma^*\times\Sigma^*}$.

Proof. Let $R = \{(l_i,r_i) \mid i = 1,2,\dots,n\}$ be a finite string-rewriting system on Σ such that no rule of R is of the form (a,b) with $a,b \in \Sigma$. For each rule $(l_i,r_i) \in R$ with $|l_i| > 1$ and $|r_i| > 1$, we introduce a new letter b_i: $\Delta := \Sigma \cup \{b_i \mid i \in \{1,2,\dots,n\}, |l_i| > 1, \text{ and } |r_i| > 1\}$. The string-rewriting system S now contains the following rules:

$$S := \{l_i \longrightarrow b_i,\ r_i \longrightarrow b_i \mid i \in \{1,2,\dots,n\},\ |l_i| > 1, \text{ and } |r_i| > 1\}$$
$$\cup \{l_i \longrightarrow r_i \mid |l_i| \geq 1, \text{ and } |r_i| \leq 1\}$$
$$\cup \{r_i \longrightarrow l_i \mid |r_i| \geq 1, \text{ and } |l_i| \leq 1\}.$$

Then S is a finite monadic string-rewriting system on Δ that can easily be obtained from R. Let $\psi: \Delta^* \longrightarrow \Sigma^*$ be the homomorphism generated by $a \longmapsto a$ $(a \in \Sigma)$ and $b_i \longmapsto l_i$ $(b_i \in \Delta\text{-}\Sigma)$. Then ψ induces an isomorphism between the factor monoids $\Sigma^*/\overset{*}{\longleftrightarrow}_R$ and $\Delta^*/\overset{*}{\longleftrightarrow}_S$, i.e. for all $u,v \in \Delta^*$, $u \overset{*}{\longleftrightarrow}_S v$ if and only if $\psi(u) \overset{*}{\longleftrightarrow}_R \psi(v)$. Since $\psi|_{\Sigma^*} = id_{\Sigma^*}$, this implies that $\overset{*}{\longleftrightarrow}_R = \overset{*}{\longleftrightarrow}_S|_{\Sigma^* \times \Sigma^*}$. □

Lemma 3.2 establishes an effective reduction of the decision problem of Theorem 3.1 to the one in the following corollary, therewith giving the undecidability of this problem as well.

Corollary 3.3. The following restricted version of the PROBLEM OF RESTRICTION is undecidable:

INSTANCE: A finite, monadic string-rewriting system R on alphabet Σ_2, and a subalphabet $\Sigma_1 \subsetneq \Sigma_2$.

QUESTION: Is $\overset{*}{\longleftrightarrow}_R|_{\Sigma_1^* \times \Sigma_1^*} = id_{\Sigma_1^*}$?

4. DECIDABLE CASES

As we have just seen the PROBLEM OF RESTRICTION is undecidable in general for finite, non-monadic, length-reducing systems R_2 that are confluent as well as for finite monadic systems R_2 that are not confluent. If, however, we only deal with finite monadic systems R_2 that are confluent, then we get the following decidability result.

Theorem 4.1. The following restricted version of the PROBLEM OF RESTRICTION is decidable:

INSTANCE: A finite canonical string-rewriting system R_1 on Σ_1, and a finite, monadic, and confluent string-rewriting system R_2 on alphabet Σ_2 containing Σ_1.

QUESTION: Is $\overset{*}{\longleftrightarrow}_{R_1} = \overset{*}{\longleftrightarrow}_{R_2}|_{\Sigma_1^* \times \Sigma_1^*}$?

Proof. Let R_1 be a finite canonical string-rewriting system on alphabet Σ_1, and let R_2 be a finite, monadic, and confluent string-rewriting system on alphabet Σ_2, where $\Sigma_1 \subseteq \Sigma_2$. If $\Sigma_1 = \Sigma_2$, we are faced with the equivalence problem for finite canonical string-rewriting systems, which is decidable [7]. Hence, we may assume without loss of generality that Σ_1 is a proper subalphabet of Σ_2.
Let $\Sigma_3 := \{a \in \Sigma_2 \mid \exists w \in \Sigma_1^*: a \overset{*}{\longleftrightarrow}_{R_2} w\}$. Then $\Sigma_3 = (\Delta_{R_2}^*(\Sigma_1^*) \cap \Sigma_2) \cup ([e]_{R_2} \cap \Sigma_2)$, where $\Delta_{R_2}^*(L) = \{u \in \Sigma_2^* \mid \exists w \in L: w \overset{*}{\longrightarrow}_{R_2} u\}$ is the set of descendants of L modulo R_2. Hence, Σ_3 is computable in polynomial time from R_2 and Σ_1 [cf., e.g., 16]. Further, let $R_3 := R_2 \cap (\Sigma_3^* \times \Sigma_3^*)$. Then R_3 is a finite monadic string-rewriting system on Σ_3 that can easily be obtained from R_2 and Σ_3.

Claim 1. $\overset{*}{\longleftrightarrow}_{R_2}\big|_{\Sigma_3^* \times \Sigma_3^*} = \overset{*}{\longleftrightarrow}_{R_3}\big|_{\Sigma_3^* \times \Sigma_3^*}$.

Proof. Let $(l,r) \in R_2$ with $r \in \Sigma_2$. If $l \in \Sigma_3^*$, then there exists a word $w \in \Sigma_1^*$ such that $w \overset{*}{\longleftrightarrow}_{R_2} l \longrightarrow_{R_2} r$ implying that $r \in \Sigma_3$. Thus, for all $u \in \Sigma_3^*$, if $u \overset{*}{\longrightarrow}_{R_2} v$, then $v \in \Sigma_3^*$, and $u \overset{*}{\longrightarrow}_{R_3} v$.
Let $u,v \in \Sigma_3^*$ such that $u \overset{*}{\longleftrightarrow}_{R_2} v$. Then $u \overset{*}{\longrightarrow}_{R_2} w \overset{*}{\longleftarrow}_{R_2} v$ for some $w \in \Sigma_2^*$, since R_2 is canonical. So, from the above observation, $w \in \Sigma_3^*$ and $u \overset{*}{\longrightarrow}_{R_3} w \overset{*}{\longleftarrow}_{R_3} v$, i.e., $u \overset{*}{\longleftrightarrow}_{R_3} v$. On the other hand, if $u \overset{*}{\longleftrightarrow}_{R_3} v$, then $u \overset{*}{\longleftrightarrow}_{R_2} v$, since R_3 is a subsystem of R_2. []

Thus, when comparing $\overset{*}{\longleftrightarrow}_{R_1}$ to $\overset{*}{\longleftrightarrow}_{R_2}\big|_{\Sigma_1^* \times \Sigma_1^*}$, we can forget about R_2 and take R_3 instead.

The following claim shows that R_3 inherits all the restrictions from R_2 that are stated in the formulation of Theorem 4.1.

Claim 2. R_3 is confluent.

Proof. Let (u,v) be a critical pair of R_3. Then $u,v \in \Sigma_3^*$ and $u \overset{*}{\longleftrightarrow}_{R_3} v$. Since $R_3 \subseteq R_2$, and since R_2 is confluent, there exists a word $w \in \Sigma_2^*$ such that $u \overset{*}{\longrightarrow}_{R_2} w$ and $v \overset{*}{\longrightarrow}_{R_2} w$. Again, by the above observation, $w \in \Sigma_3^*$ and $u \overset{*}{\longrightarrow}_{R_3} w \overset{*}{\longleftarrow}_{R_3} v$, i.e., the critical pair (u,v) resolves modulo R_3. []

For each $a \in \Sigma_3 - \Sigma_1$, let $u_a \in \Sigma_1^*$ be a word such that $u_a \overset{*}{\longleftrightarrow}_{R_3} a$. Given $a \in \Sigma_3 - \Sigma_1$, such a word $u_a \in \Sigma_1^*$ can be computed effectively. Now we define a homomorphism $\psi: \Sigma_3^* \longrightarrow \Sigma_1^*$ through $a \longmapsto a$ $(a \in \Sigma_1)$ and $a \longmapsto u_a$ $(a \in \Sigma_3 - \Sigma_1)$. Furthermore, let $R_4 := \{(\psi(l),\psi(r)) \mid$

$(l,r) \in R_3\}$. Then R_4 is a finite string-rewriting system on Σ_1, and ψ induces an isomorphism between the two factor monoids $\Sigma_3^*/\overset{*}{\longleftrightarrow}_{R_3}$ and $\Sigma_1^*/\overset{*}{\longleftrightarrow}_{R_4}$. Since $\psi|_{\Sigma_1^*} = id_{\Sigma_1^*}$, this implies that

$\overset{*}{\longleftrightarrow}_{R_4} = \overset{*}{\longleftrightarrow}_{R_3}|_{\Sigma_1^* \times \Sigma_1^*} = \overset{*}{\longleftrightarrow}_{R_2}|_{\Sigma_1^* \times \Sigma_1^*}$. Thus, $\overset{*}{\longleftrightarrow}_{R_1} = \overset{*}{\longleftrightarrow}_{R_2}|_{\Sigma_1^* \times \Sigma_1^*}$ if and only if $\overset{*}{\longleftrightarrow}_{R_1} = \overset{*}{\longleftrightarrow}_{R_4}$. Now this holds if and only if the following two conditions are satisfied:

(a) $\forall (l,r) \in R_1$: $l \overset{*}{\longleftrightarrow}_{R_3} r$, and

(b) $\forall (l,r) \in R_3$: $\psi(l) \overset{*}{\longleftrightarrow}_{R_1} \psi(r)$.

These two conditions are decidable, since R_1 and R_3 are canonical. In fact, Condition (a) can be verified in polynomial time, while the complexity of verifying Condition (b) depends on the form of the system R_1. This completes the proof of Theorem 4.1.

Theorems 3.1 and 4.1 give another example of a decision problem that distinguishes between the computational power of finite, length-reducing, and confluent string-rewriting systems that are non-monadic on the one hand and that of finite, monadic, and confluent string-rewriting systems on the other hand. For more results about this distinction see [15,18,19].

Since the words u_a used in the proof of Theorem 4.1 may be of exponential length, the algorithm given in that proof is in general of exponential time complexity only even if the system R_1 is length-reducing. Of course, if the word problem for R_1 is of a higher degree of complexity [3], then also this algorithm is of a higher degree of complexity. However, if R_1 and R_2 are both special systems that are confluent and reduced, then, because of the following theorem, the problem of whether or not $\overset{*}{\longleftrightarrow}_{R_1} = \overset{*}{\longleftrightarrow}_{R_2}|_{\Sigma_1^* \times \Sigma_1^*}$ is easily decidable. Here a special system $R = \{(l_i,e) \mid i \in I\}$ is called <u>reduced</u> if, for each $i \in I$, $l_i \in IRR(R-\{(l_i,e)\})$, i.e., no left-hand side of a rule contains another left-hand side as a factor. Given a finite, special, and confluent string-rewriting system R on Σ, an equivalent finite, special, and confluent system R_o that is reduced can be constructed in polynomial time by just deleting those rules that violate the above condition [11].

<u>Theorem 4.2</u>. Let R_1 be a special, reduced, and confluent string-rewriting system on Σ_1, and let R_2 be a special, reduced, and confluent string-rewriting system on $\Sigma_2 \supsetneq \Sigma_1$.

Then $\overset{*}{\longleftrightarrow}_{R_1} = \overset{*}{\longleftrightarrow}_{R_2}|_{\Sigma_1^* \times \Sigma_1^*}$ if and only if $R_1 = R_2 \cap (\Sigma_1^* \times \{e\})$.

<u>Proof.</u> <u>"===>"</u>: Assume that $\overset{*}{\longleftrightarrow}_{R_1} = \overset{*}{\longleftrightarrow}_{R_2}|_{\Sigma_1^* \times \Sigma_1^*}$, and let $u,v \in \Sigma_1^*$. Since R_2 is canonical, $u \overset{*}{\longrightarrow}_{R_2} u_o \in IRR(R_2)$ and $v \overset{*}{\longrightarrow}_{R_2} v_o \in IRR(R_2)$. Since $u,v \in \Sigma_1^*$, and since R_2 is special, $u_o,v_o \in \Sigma_1^*$. Let $R_3 := R_2 \cap (\Sigma_1^* \times \{e\})$. Then $u \overset{*}{\longrightarrow}_{R_3} u_o$ and $v \overset{*}{\longrightarrow}_{R_3} v_o$.

Let (l,e) $\in R_3$. Since $\overset{*}{\longleftrightarrow}_{R_2}\big|_{\Sigma_1^*\times\Sigma_1^*} = \overset{*}{\longleftrightarrow}_{R_1}$, and since R_1 is also canonical, we have $l \overset{*}{\longrightarrow}_{R_1} e$. Hence, $l = xl_1y$, where $x,y \in \Sigma_1^*$ and $(l_1,e) \in R_1$. Since R_2 is canonical, and $\overset{*}{\longleftrightarrow}_{R_1} = \overset{*}{\longleftrightarrow}_{R_2}\big|_{\Sigma_1^*\times\Sigma_1^*}$, this means that $l_1 \overset{*}{\longrightarrow}_{R_2} e$. But R_2 being reduced yields that

$l = l_1$, i.e. $(l,e) \in R_1$. Hence, $R_3 \subseteq R_1$.
Conversely, let $(l,e) \in R_1$. Then $l \overset{*}{\longrightarrow}_{R_3} e$, and so $l = xl_1y$, where $x,y \in \Sigma_1^*$ and $(l_1,e) \in R_3$. Again this implies that $l_1 \overset{*}{\longrightarrow}_{R_1} e$, and since R_1 is reduced, this yields $l = l_1$, i.e., $(l,e) \in R_3$. Thus, $R_1 = R_3$.

"<===": Assume that $R_1 = R_2 \cap (\Sigma_1^* \times \{e\})$.

Then obviously, $\overset{*}{\longleftrightarrow}_{R_1} \subseteq \overset{*}{\longleftrightarrow}_{R_2}\big|_{\Sigma_1^*\times\Sigma_1^*}$.

Further, $IRR(R_1) = IRR(R_2) \cap \Sigma_1^*$, and so $\overset{*}{\longleftrightarrow}_{R_1} = \overset{*}{\longleftrightarrow}_{R_2}\big|_{\Sigma_1^*\times\Sigma_1^*}$.

Acknowledgement

The author gratefully acknowledges several fruitful discussions that he had with Geraud Senizergues about the problems dealt with in this paper. Actually, it was Geraud Senizergues, who gave the first proof of Theorem 4.1. His proof, which is more complicated than the one presented here, is based on results about congruential languages and reduced context-free grammars [5,8].

REFERENCES

1. **J. Avenhaus, K. Madlener,** Subrekursive Komplexität bei Gruppen II. Der Einbettungssat von Higman für entscheidbare Gruppen; Acta Informatica 9 (1978), 183-193.
2. **G. Bauer,** Zur Darstellung von Monoiden durch Regelsysteme; Ph.D. Dissertation, Fachbereich Informatik, Universität Kaiserslautern, 1981.
3. **G. Bauer, F. Otto,** Finite complete rewriting systems and the complexity of the word problem; Acta Inf. 21 (1984), 521-540.
4. **J. A. Bergstra, J.V. Tucker,** A characterization of computable data types by means of a finite, equational specification method; Technical Report, Mathematical Centrum, Amsterdam, Holland, 1979. Preprint IW 124/79, 1979.
5. **L. Boasson,** Grammaires a non-terminaire separes, 7th ICALP, Lecture Notes in Compute Science 85 (1980), 105-118.
6. **R.V. Book,** Thue systems as rewriting systems; J. Symbolic Computation 3 (1987), 39-68.
7. **R.V. Book, C. O'Dunlaing,** Thue congruences and the Church-Rosser property; Semigroup Forum 22 (1981), 325-331.

8. **C. Frougny**, Une famille de langages algebriques congruentiels; les langages a non-terminaire separes; These de 3eme cycle de l'universite de Paris VII, 1980.

9. **M. Jantzen**, A note on a special one-rule semi-Thue system, Inf. Proc. Letters 21 (1985), 135-140.

10. **D. Kapur, P. Narendran**, A finite Thue system with decidable word problem and without equivalent finite canonical system; Theoretical Computer Science 35 (1985), 337-344.

11. **D. Kapur, P. Narendran**, The Knuth-Bendix completion procedure and Thue systems; SIAM J. on Computing 14 (1985), 1052-1072.

12. **D. Knuth, P. Bendix**, Simple word problems in universal algebras; in: J. Leech (ed.), Computational Problems in Abstract Algebra, Oxford: Pergamon Press, 1970, pp. 263-297.

13. **A.A. Markov**, Impossibility of algorithms for recognizing some properties of associative systems; Dokl. Akad. Nauk SSSR 77 (1951), 953-956.

14. **A. Mostowski**, Review of [13]; J. of Symbolic Logic 17 (1952), 151-152.

15. **P. Narendran, C. O'Dunlaing**, Cancellativity in finitely presented semigroups; Journal of Symbolic Computation, to appear.

16. **P. Narendran, F. Otto**, Some polynomial-time algorithms for finite monadic Church-Rosser Thue systems, Technical Report (87-20), Department of Computer Science, State University of New York, Albany, 1987, also: Theoretical Computer Science, to appear.

17. **C. O'Dunlaing**, Undecidable questions of Thue systems; Theoretical Computer Science 23 (1983), 339-345.

18. **F. Otto**, Some undecidability results for non-monadic Church-Rosser Thue systems; Theoretical Computer Science 33 (1984), 261-278.

19. **F. Otto**, On two problems related to cancellativity, Semigroup Forum 33 (1986), 331-356.

20. **F. Otto**, Deciding algebraic properties of finitely presented monoids; in: B. Benninghofen, S. Kemmerich, M.M. Richter, Systems of Reductions, Lecture Notes in Computer Science 277 (1987), 218-255.

21. **F. Otto**, Finite canonical rewriting systems for congruences generated by concurrency relations; Math. Systems Theory 20 (1987), 253-260.

22. **C. Squier**, Word problems and a homological finiteness condition for monoids; Journal of Pure and Applied Algebra 49 (1987), 201-217.

23. **H. Tietze**, Über die topologischen Invarianten mehrdimensionaler Mannigfaltigkeiten; Monatsheft Math. Physik 19 (1908), 1-118.

Embedding with Patterns and Associated Recursive Path Ordering

Laurence Puel
LIENS, URA 1327 CNRS
Ecole Normale Supérieure
45 rue d'Ulm
75005 PARIS FRANCE

Introduction

Termination is an important property for term rewriting systems. For proving this Dershowitz [1] introduces quasi-simplification orderings that are monotonic extensions of the embedding relation. He proves their well-foundedness as a consequence of Kruskal's theorem [5]. Dershowitz's method is powerful but cannot be used on rules whose left hand side is embedded in the right hand side. The purpose of this paper is to override this constraint and thus to define a quasi-ordering on trees strictly included in the embedding one but which is still a well quasi-ordering.

In [6] we give a first generalization of Kruskal's theorem using the unavoidability concept like Erhenfeucht et al.[3] have done for the words. Here, we define a new relation, whose definition is close to the classical definition of embedding. Usually, we say that a tree t is embedded in a tree t' if we can get t from t' by erasing some symbols. We wish to erase patterns belonging to a given set S instead of symbols. Here is the first problem: it is not obvious to decide which pattern to erase. There is not always a pattern at the root of the tree and even if there is one, it is not always the only one. So, we define a decomposition operation that associates to each tree an unique pattern belonging to a given set S. Thus we associate to this kind of decomposition an embedding where the part played usually by the head-symbol is now played by the selected pattern. We define the S-embedding extending a quasi-ordering and prove it to be a well quasi-ordering provided the underlying qo is a well quasi-ordering. This result generalizes Kruskal's theorem. Then we define RPO-like orderings extending S-embedding. We use them for proving termination of term rewriting systems. Their use is not as easy as RPO's one because they are not compatible with term structure (also called monotonic). We give sufficient conditions for proving termination. This method has been implemented in CAML, the language developed at INRIA and ENS (FRANCE) by FORMEL team. The implementation leads to technical algorithms as unification with constraints on terms and terms decomposition. As the RPO is a subcase of the SRPO, we keep the same results and we add, for example, the termination proof for some systems in which the left hand side of a rule is embedded in the right one.

1 Preliminaries

We use classical notations and vocabulary on words and trees.

Definition 1 (Lexicographic Ordering) *Let Σ be an alphabet, Σ^* the set of words on Σ and $\leq$ an ordering on Σ. The relation $\leq_{lexico}$ on Σ^* is defined by $u = u_1 \dots u_n \leq_{lexico} v = v_1 \dots v_p$ if and only if either $u = v$ or there exists an integer k $(1 \leq k \leq p)$ such that for every $l < k$, $u_l = v_l$ and $u_k < v_k$. This relation is an ordering.*

Definition 2 (Hierarchical Ordering) *Let Σ be an alphabet, Σ^* the set of words on Σ and $\preceq$ an ordering on Σ. The relation $\preceq_{hier}$ on Σ^* is defined by $u \preceq_{hier} v$ if and only if $|u| < |v|$ or $|u| = |v|$ and $u \preceq_{lexico} v$. This relation is an ordering.*

Let Σ be an alphabet and ω a symbol with null arity which does not belong to Σ. A tree t on Σ is defined by a subset, $Vertex(t)$, of N^*, the set of words on N, and a map, named also t, from $Vertex(t)$ into Σ such that:

1. $\forall u \in Vertex(t) \;\; \forall v \leq_{prefix} u \quad v \in Vertex(t)$

2. $\forall u \in Vertex(t) \quad ui \in Vertex(t) \Rightarrow \forall j < i \quad uj \in Vertex(t)$

Let t be a tree. We define depth(t) as the maximum of $\{|u| \; |u \in Vertex(t)\}$ where $|u|$ denotes the length of the finite sequence $u \in N^*$. When Σ contains variables a substitution is an application that associates to each variable a tree. The set of trees build on Σ (resp. Σ and ω) is denoted $T(\Sigma)$(resp. $T(\Sigma, \omega)$). Let S be a subset of $T(\Sigma, \omega)$. We say that a tree s is a factor of t if t can be obtained from s in two stages:

1. Substitute trees from $T(\Sigma, \omega)$for terminal ω-vertices of s.

2. Substitute the preceding result for a terminal ω-vertex of a tree from $T(\Sigma, \omega)$.

A subset S is said to be $factor - unavoidable$ or unavoidable in short, if there exists an integer k such that for every tree $t \in T(\Sigma, \omega)$ with $depth(t) > k$ there exists $s \in S$ factor of t. We call k the avoidance bound. Given a set A and a quasi-ordering $\leq$ on A, $\leq$ and $<$ are both called well-founded if and only if each strictly descending sequence is finite. $\leq$ is a well quasi-ordering if $\leq$ is well-founded and each set of pairwise incomparable elements is finite.

2 Embedding with Patterns

2.1 Classical Embedding and Method of Generalization

We recall the definition of the embedding relation extending a quasi-ordering given by Kamin-Levy [4]. Let $\preceq$ be a quasi-ordering on trees. The embedding extending $\preceq$, denoted $TEO(\preceq)$, is defined by

$$t = f(t_1, \ldots, t_n) \quad TEO(\preceq) \; t' = g(t'_1, \ldots, t'_p)$$

if one of the following properties is satisfied:

1. There exists an integer i $(1 \leq i \leq p)$ such that $t \;\; TEO(\preceq) \; t'_i$.

2. $t \preceq t'$ and for every integer i $(1 \leq i \leq n)$ $t_i \;\; TEO(\preceq) \; t'_{j_i}$ with $1 \leq j_1 < \ldots < j_n \leq p$.

We wish to generalize the embedding relation in such a way that the part played by symbols will be played by patterns. Instead of removing the head-symbol, we want to remove a pattern belonging to a given subset S. But the concept of head-pattern is not well defined. We need a method to select, in an unique way, a pattern in a tree.

2.2 Tree Decomposition

Let t be a tree in $T(\Sigma)$. Either t has no factor in S, or t has a factor in S and there exist a tree R, a terminal vertex u in R, an element s in S and n trees $t_i (1 \leq i \leq n)$ such that $t = R[u \leftarrow s[u_i \leftarrow t_i | (1 \leq i \leq n), s(u_i) = \omega]]$. This remark leads to a decomposition of t in three parts which are respectively R in $T(\Sigma, \omega)$, s in S and $(t_1, \ldots, t_n)$in $T(\Sigma, \omega)^*$, the set of words on $T(\Sigma, \omega)$. Such a decomposition is not unique as we can see on the following example: Let $\Sigma = \{f, g, a\}$ and $S = \{g(f(\omega, \omega))\}$. Let $t = f(g(f(g(a), a)), g(f(a, g(a))))$. There are two decompositions of t. In one of them $R = f(\omega, g(f(a, g(a))))$, $t_1 = g(a)$, $t_2 = a$ and in the other one $R' = f(g(f(g(a), a)), \omega)$, $t'_1 = a$, $t'_2 = g(a)$. To get uniqueness, we have to specify the choice of the vertex u and of the pattern s in S. Let us describe our choice. We choose the vertex u as the smallest for the hierarchical ordering on N^* the set of words on integers.

Definition 3 (Decomposition of a tree t) *Let t be an element of $T(\Sigma)$, S a subset of $T(\Sigma, \omega)$and $<_p$ a strict ordering on S such that two elements of S with the same head symbol are comparable. The decomposition of t, denoted decomp(t), belongs to $S \times T(\Sigma, \omega) \times N^* \times T(\Sigma, \omega)^*$ and is specified by*

1. *decomp(t) $=(s, R[u \leftarrow \omega], u, (t_1, \ldots, t_n))$ if there exists s in S such that*

$$t = R[u \leftarrow s[u_i \leftarrow t_i | (1 \leq i \leq n), s(u_i) = \omega]]$$

with the following property: If t can also be written

$$t = R'[u' \leftarrow s'[u'_i \leftarrow t'_i | (1 \leq i \leq n'), s'(u'_i) = \omega]]$$

then $u <_{hier} u'$ or $u = u'$ and $s' <_p s$;

2. decomp(t) $=(t, \omega, \epsilon, \emptyset)$ *if not.*

Let us call *decomposition pattern* (resp. *decomposition vertex*)of a tree t the first (resp. second) component of *decomp*(t).

remarks

1. decomposition pattern of t either belongs to S or is t itself.
2. It is not necessary for $<_p$ to be a total ordering on S. This relation is used only when there are two patterns at the same selected vertex.

example

Let $\Sigma = \{*, f, a, b, c, d\}$. Let $S = \{s_1 = *(*(\omega,\omega),\omega), s_2 = *(\omega, *(\omega,\omega))\}$ let us suppose $s_2 <_p s_1$.

- $decomp(f(a)) = (f(a), \omega, \epsilon, \emptyset)$
- $decomp(f(*(*(a,b),*(c,d)))) = (*(*(\omega,\omega),\omega), f(\omega), 1, (a, b, *(c,d)))$
- $decomp(\omega) = (\omega, \omega, \epsilon, \emptyset)$

Proposition 1 *For every tree t, the decomposition decomp(t) is unique.*

2.3 S-Embedding

Definition 4 (S-Embedding) *Let $\preceq$ be a quasi-ordering, called the underlying qo, on $T(\Sigma,\omega)$and S a subset of $T(\Sigma,\omega)$. Let t and t' be two trees from $T(\Sigma,\omega)$such that*

$$decomp(t) = (s, R[u \leftarrow \omega], u, (t_1, \ldots, t_n))$$

$$decomp(t') = (s', R'[u' \leftarrow \omega], u', (t'_1, \ldots, t'_p))$$

t $STEO(S,\preceq)$ t' if and only if one of the following properties is satisfied:

1. $\exists\, i\ (1 \le i \le p)$ *such that* $t\ STEO(S,\preceq)\ R'[u' \leftarrow t'_i]$.
2. $t \preceq t'$ *and* $\forall\, i\ (1 \le i \le p)\ R[u \leftarrow t_i]\ STEO(S,\preceq)\ R'[u' \leftarrow t'_{j_i}]$ *with* $1 \le j_1 < \ldots < j_n \le p$.

remarks

1. In order to use condition 1 in the above definition, t' must have a factor in S. Thus we say that t $STEO(S,\preceq)$ t' by pattern erasing or erasing in short.

2. When S is the set of trees whose depth is equal to one, $STEO(S,\preceq)$ is equal to $TEO(\preceq)$.

Proposition 2 *The relation* $STEO(S,\preceq)$ *is a quasi-ordering on* $T(\Sigma,\omega)$.

The relation $STEO(S,\preceq)$ can be seen as a kind of generalized patterns insertion. t $STEO(S,\preceq)$ t' if t' is built from t by insertion of patterns belonging to S. This insertion can be done directly in t or after erasing patterns in t, insertion of new patterns and finally reinsertion of erased patterns. We show that on the following example where the underlying qo is defined by $t \preceq t'$ if and only if the *decomposition patterns* are the same for both trees. Let $S = \{g(g(\omega)), h(h(\omega))\}$. Let $t = f(g(g(a)), x)$ and $t' = f(h(g(g(h(a)))), x)$. In order to go from t to t', erase $g(g(\omega))$, insert $h(h(\omega))$ and insert again $g(g(\omega))$.

2.4 Comparison with Embedding

We have already seen that, when S is the set of trees whose depth is equal to one, $TEO(\preceq)$ and $STEO(S,\preceq)$ are equal.

Proposition 3 $STEO(S,TEO(\preceq))$ *is included in* $TEO(\preceq)$.

The proof is done by induction on the number of vertices of the terms.

In general this inclusion is strict as we can see on the following example. Let $\Sigma = \{f, g\}$, $S = \{f(f(\omega,\omega),\omega), g(\omega)\}$ and $\preceq$ be the head symbol equality.

$$f(g(x), f(x,x)) \;\; TEO(\preceq) \;\; f(f(x, g(x)), f(x,x))$$

but the relation is false with respect to $STEO(S,TEO(\preceq))$ because the term $f(x, f(x,x))$ is not comparable with x or $g(x)$ with respect to this relation.

Let S be a subset of $T(\Sigma,\omega)$ and Res be the subset of $T(\Sigma,\omega)$whose elements have no factor in S. We get a natural extension of Kruskal embedding when the underlying qo is defined by $t \preceq t'$ if and only if the *decomposition patterns* s and s' satisfies:

1. either $s \in S$, $s' \in S$ and $s \;\; TEO(\preceq)\; s'$.
2. or $t \in Res$, $t' \in Res$ and $t \;\; TEO(\preceq)\; t'$.
3. or $t \in Res$, $s' \in S$ and $t \;\; TEO(\preceq)\; s'$.

When S is the set of trees of depth one, we get exactly Kruskal embedding.

2.5 Well Quasi-Ordering

Theorem 1 *Let S be a subset of $T(\Sigma,\omega)$and $\preceq$ a well quasi-ordering on $T(\Sigma,\omega)$. Then $STEO(S,\preceq)$ is a well quasi-ordering on $T(\Sigma,\omega)$.*

The proof is like the Kruskal's theorem one but the *decomposition patterns* play the part of the head symbols.

Corollary 1 *Let S be an unavoidable subset of $T(\Sigma,\omega)$and $\preceq_1$ a quasi-ordering on the trees whose depth is less than the avoidance bound. Let $\preceq$ be the quasi-ordering on $T(\Sigma,\omega)$defined by $t \preceq t'$ if and only if the decomposition patterns s and s' satisfy $s \preceq_1 s'$. If $\preceq_1$ is a well quasi-ordering then $STEO(S,\preceq)$ is a well quasi-ordering on $T(\Sigma,\omega)$.*

In the above corollary, S's unavoidability hypotheses is interesting because it allows us to extend of a well quasi-ordering on a proper subset of trees into a well quasi-ordering on trees. When Σ is a finite ranked alphabet, we can take a finite unavoidable set and the set of trees whose depth is less than the avoidance bound is finite. In that case, every quasi-ordering on these finite sets is a well quasi-ordering. There is another reason to use unavoidable sets. These quasi-orderings are fastened to tree's decomposition which is, most of the time, not trivial when S is unavoidable.

3 Recursive Path Ordering

Let us recall Recursive Path Ordering definition given by Dershowitz[1] and extensions defined by Kamin-Levy[4]. Let $\preceq$ a relation on $T(\Sigma, \omega)$. In the following, $O(\preceq)$ denotes either the multiset extension or the lexicographic extension of $\preceq$.

$$f(t_1, \ldots, t_n) O(\prec) f(t'_1, \ldots, t'_n) \text{ if } (t_1, \ldots, t_n) \prec_{lexico} (t'_1, \ldots, t'_n)$$

$$f(t_1, \ldots, t_n) O(\prec) f(t'_1, \ldots, t'_n) \text{ if } (t_1, \ldots, t_n) \prec_{mult} \{t'_1, \ldots, t'_n\}$$

Definition 5 (RPO) *Let $\prec$ be an irreflexive and transitive relation, named precedence, on Σ. The relation $RPO(\prec)$ is defined by*

$$t = f(t_1, \ldots, t_n) RPO(\prec) t' = g(t'_1, \ldots, t'_p)$$

if and only if one of the following property is satisfied:

1. *$f \prec g$ and for every integer i $(1 \leq i \leq n)$ t_i $RPO(\prec)$ t'.*
2. *there exists an integer i $(1 \leq i \leq p)$ such that t $RPO(\prec)$ t'_i or $t = t'_i$.*
3. *$f = g$ and $tO($ $RPO(\prec)$ $)t'$ and for every integer i $(1 \leq i \leq n)$ t_i $RPO(\prec)$ t'.*

Now, we extend these definitions like we have done for the embedding. The part played by the head-symbol is now played by the *decomposition pattern*.

3.1 Recursive Path Ordering Related to a Subset S

Definition 6 (SRPO) *Let S be a subset of $T(\Sigma, \omega)$, $\preceq$ a quasi-ordering on $T(\Sigma, \omega)$and $\approx$ the associated equivalence. The relation $SRPO(\prec)$ is defined by t $SRPO(\prec)$ t' if and only if*

$$decomp(t) = (s, R[u \leftarrow \omega], u, (t_1, \ldots, t_n))$$

$$decomp(t') = (s', R'[u' \leftarrow \omega], u', (t'_1, \ldots, t'_p))$$

and one of the following properties is satisfied:

1. *$t \prec t'$ and for every integer i $(1 \le i \le n)$ $R[u \leftarrow t_i]$ $SRPO(\prec)$ t'.*
2. *there exists an integer i $(1 \le i \le p)$ such that t $SRPO(\prec)$ $R'[u' \leftarrow t'_i]$ or $t \equiv R'[u' \leftarrow t'_i]$.*
3. *$t \approx t'$, $tO($ $SRPO(\prec)$ $)t'$ and for every integer i $(1 \le i \le n)$ $R[u \leftarrow t_i]$ $SRPO(\prec)$ t'.*

with $t \equiv t'$ if and only if $t \approx t'$ and $R[u \leftarrow t_i] \equiv_{O(SRPO(\prec))} R'[u' \leftarrow t'_i]$.

Proposition 4 *If $\preceq$ is a well quasi-ordering on $T(\Sigma, \omega)$, then the relation $SRPO(\prec)$ defined above is a well quasi-ordering.*

To prove this property, we prove that $STEO(S, \preceq)$ is included in $SRPO(\prec$ $)$ by induction on the number of vertices in the trees.

In [6] we give more general version of this RPO, like Kamin-Levy[4], which keeps the property to be a well quasi-ordering.

3.2 Examples

Let S be a subset of $T(\Sigma, \omega)$and Res be the subset of $T(\Sigma, \omega)$with no factor in S. Let $\prec$ be an irreflexive and transitive relation on $S \bigcup Res$. Let $\preceq$ be a relation defined on $T(\Sigma, \omega)$by $t \preceq t'$ if and only if $s = s'$ or $s \prec s'$ where s (resp.s') are the *decomposition pattern* of t (resp.t'). When S is finite unavoidable $\preceq$ is a well quasi-ordering. We use such an ordering in examples below.

1. Let $\Sigma = \{f, p, s, *\}$ and $S = \{*(\omega, \omega), s(\omega), p(s(\omega)), p(\omega), f(\omega)\}$. The precedence $\prec_S$ on S used to decompose a tree is defined by $p(\omega) \prec_S p(s(\omega))$. The precedence on S is defined by $*(\omega, \omega) \prec f(\omega), p(s(\omega)) \prec p(\omega), p(s(\omega)) \prec s(\omega)$. Let $t = *(s(x), f(p(s(x))))$ and $t' = f(s(x))$.

$$decomp(t) = (*(\omega, \omega), \omega, \epsilon, (s(x), f(p(s(x))))))$$

$$decomp(t') = (f(\omega), \omega, \epsilon, (s(x))))$$

decomposition patterns satisfies $*(\omega, \omega) \prec f(\omega)$

   (a) $s(x)$ $SRPO(\prec)$ $f(s(x))$by 2

(b) $f(p(s(x)))\ \ SRPO(\prec)\ f(s(x))$ because $p(s(x))\ \ SRPO(\prec)\ s(x)$

The last relation is a consequence of

(a) $decomp(p(s(x))) = (p(s(\omega)), \omega, \epsilon, (x)))$

(b) $decomp(s(x)) = (s(\omega), \omega, \epsilon, (x)))$

(c) $p(s(\omega)) \prec s(\omega)$ and $x\ \ SRPO(\prec)\ s(x)$ by 2.

Thus we conclude $t\ \ SRPO(\prec)\ t'$. This inequality cannot be proved by a classical RPO because t' is embedded in t.

2. Let $\Sigma = \{f, g, a\}$. We wish to compare $f(x, g(a))$ and $f(a, x)$. Let $S = \{g(g(\omega)), f(a, \omega), f(\omega, g(\omega)), f(\omega, \omega)\}$ given in a decreasing ordering for tree decomposition. The precedence on $S \bigcup Res$ is defined by

$$f(\omega, \omega) \prec f(\omega, g(\omega)) \prec f(a, \omega)$$

$$g(a) \prec a \prec f(a, \omega)$$

The decomposition of $f(x, g(a))$ depends on the value given to the variable x.

(a) $x \neq a$. *Decomposition patterns* satisfy $f(\omega, g(\omega)) \prec f(a, \omega)$ and $x\ \ SRPO(\prec)\ f(a, x)$ (case 2), $a\ \ SRPO(\prec)\ f(a, x)$ (case 1). Thus $f(x, g(a))\ \ SRPO(\prec)\ f(a, x)$.

(b) $x = a$. *Decomposition patterns* are both equal to $f(a, \omega)$ and $g(a)\ \ SRPO(\prec)\ a$(case 1). Thus $f(a, g(a))\ \ SRPO(\prec)\ f(a, a)$

4 Termination of Rewriting Systems

A rewriting system terminates for a set of terms T if it does not exist an infinite sequence (t_i) in T such that $t_1 \to t_2 \to t_3 \to \ldots$. A system does not terminate if such a sequence exists. To determine whether a rewriting system terminates is an undecidable problem. We refer to Dershowitz survey on termination [2] for all the results on this problem. A classical way for proving termination is to use Dershowitz's simplification orderings. But these orderings cannot be applied to rules in which the left hand side is included in the right one though embedding is a necessary but not sufficient condition for terminating. Another argument to look for stronger orderings is given by the following example:

$$f(a) \to f(b)$$

$$g(b) \to g(a)$$

In that case RPO is useless to prove termination. It can be proved by SRPO when $S = \{f(a), f(b), g(a), g(b), f(\omega), g(\omega)\}$.

4.1 SRPO and Termination

The SRPO is not compatible with term structure. So, its use for proving termination of rewriting systems is based on the following theorem:

Theorem 2 (Manna-Ness, Lankford) *A rewriting system R over a set of terms T is terminating, if and only if, there exists a well-founded ordering $\prec$ over T such that*

$$t \to^{+} u \textit{ implies } u \prec t$$

for all terms t and u in T.

Thus a rewriting system terminates if for every term C, for every vertex u of C, and for every substitution σ, $\sigma(C[u \leftarrow d])$ $SRPO(\prec)$ $\sigma(C[u \leftarrow g])$ where $SRPO(\prec)$ is defined in a previous section. When S is any set the SRPO is not the right tool to prove termination because the number of cases to consider is not finite. When S is unavoidable, the number of contexts to consider is finite.

Definition 7 *Let C be a tree and $dom(C)$ its set of vertices. C is a ridge if and only if for every $u \in dom(C)$ either u is a terminal vertex or there exists at most one integer i such that ui is not a terminal vertex.*

Definition 8 *A context is a tree with a distinguished vertex. A ridge context is a ridge in which the distinguished vertex is a vertex of maximal length. When it is not necessary, we don't mention the vertex.*

Lemma 1 *For every context (C,u) there exist a ridge context (C',u') and a substitution σ such that $\sigma(C')=C$ and $\sigma(u')=u$.*

Theorem 3 *Let Σ be a finite ranked alphabet, S an unavoidable subset of $T(\Sigma,\omega)$ and R a rewriting system. There exists a finite set of contexts C such that R is terminating if for every context $(C,u)\in C$, for every substitution σ, $\sigma(C[u\leftarrow d])$ $SRPO(\prec)$ $\sigma(C[u\leftarrow g])$ for every rule $g\rightarrow d$ in R.*

To prove the theorem, we show that the only useful contexts are those whose depth is less than $k+sup\{depth(s)|s\in S\}$ where k is the avoidance bound.

When Σ is a finite alphabet with some symbols of variable arity, we get analogous results if these symbols occur as labels in S only at vertices just above terminal vertices.

Now we have to take care of substitutions. We show that it is possible to decompose termination proof in finite steps.

Definition 9 (variant of a decomposition) *Let t and t' be two terms in $T(\Sigma,\omega)$ and S an unavoidable subset of $T(\Sigma,\omega)$. The decomposition of t' is a variant of t's decomposition if and only if*

$$decomp(t)=(s,R[u\leftarrow\omega],u,(t_1,\ldots,t_n))$$

$$decomp(t')=(s',R'[u'\leftarrow\omega],u',(t'_1,\ldots,t'_p))$$

and there exists a substitution ρ such that $R'=\rho(R)$, $s=s'$, $u=u'$ and for every integer i $(1\leq i\leq p)$, $t'_i=\rho(t_i)$.

Lemma 2 *Let Σ be a finite ranked alphabet and S an unavoidable subset of $T(\Sigma,\omega)$ whose avoidance bound is equal to k. There exists a finite set E of substitutions σ such that , for every substitution σ', there exists an element $\sigma\in E$ such that for every term $t\in T(\Sigma)$, the decomposition of $\sigma'(t)$ is a variant of the decomposition of $\sigma(t)$.*

In the following E denotes this particular set of substitutions.

Definition 10 *A substitution ρ is a variant of σ with respect to a term t if the decomposition of $\rho(t)$ is a variant of the decomposition of $\sigma(t)$.*

Definition 11 *Let F be a set of substitutions. $t <_F t'$ if and only if for every substitution $\sigma \in F$, $\sigma(t)\ \ SRPO(\prec)\ \sigma(t')$.*

Let us denote

$$F_\sigma(t,t') = \{\nu | \nu \circ \sigma \in F \text{ and is a variant of } \sigma \ \text{ w.r.t. } t \text{ and } t'\}$$

Proposition 5 *Let $\prec$ be a precedence on S. Let $SRPO(\prec)$ the quasi-ordering extending $\preceq$ in which we take the multiset extension, denoted $\ll$. Let F be a set of substitutions and t and t' be two terms. $t <_F t'$ if and only if for every substitution $\sigma \in E$,*

$$decomp(\sigma(t)) = (s, R[u \leftarrow \omega], u, (t_1, \ldots, t_n))$$

$$decomp(\sigma(t')) = (s', R'[u' \leftarrow \omega], u', (t'_1, \ldots, t'_p))$$

and one of the following properties is satisfied:

1. *$s \prec s'$ and for every integer i $(1 \leq i \leq n)$ $R[u \leftarrow t_i] <_{F_\sigma(t,t')} \sigma(t')$.*
2. *For every substitution $\nu \in F_\sigma(t,t')$, there exists an integer i $(1 \leq i \leq p)$ such that*

$$\nu(\sigma(t))\ \ SRPO(\prec)\ \nu(R')[u' \leftarrow \nu(t'_i)]$$

3. *$s = s'$ and $\{R[u \leftarrow t_i]\} \ll_{F_\sigma(t,t')} \{R'[u' \leftarrow t'_i]\}$.*

To prove this property, we remark that $t <_F t'$ if and only if for every substitution $\sigma \in E$ and for every substitution $\nu \in F_\sigma(t,t')$, $\nu \circ \sigma(t)\ \ SRPO(\prec)\ \nu \circ \sigma(t')$. Then we scan the different cases. When we consider the lexicographic extension we only get a sufficient condition. We give below a sufficient condition for $t <_F t'$ that allows a comparison of t and t' by steps. This condition is equivalent to the first one in case of monadic patterns or of ordering compatible with term structure.

Proposition 6 *Let $\prec$ be a precedence on S. Let $SRPO(\prec)$ the quasi-ordering extending $\preceq$ in which we take the multiset extension, denoted $\ll$.*

Let F be a set of substitutions and t and t' be two terms. $t <_F t'$ if for every substitution $\sigma \in E$,

$$decomp(\sigma(t)) = (s, R[u \leftarrow \omega], u, (t_1, \ldots, t_n))$$

$$decomp(\sigma(t')) = (s', R'[u' \leftarrow \omega], u', (t'_1, \ldots, t'_p))$$

and one of the following properties is satisfied:

1. *$s \prec s'$ and for every integer i $(1 \leq i \leq n)$ $R[u \leftarrow t_i] <_{F_\sigma(t,t')} \sigma(t')$.*
2. *There exists an integer i $(1 \leq i \leq p)$ such that*

$$\nu(\sigma(t)) <_{F_\sigma(t,t')} R'[u' \leftarrow t'_i]$$

3. *$s = s'$ and $\{R[u \leftarrow t_i]\} \ll_{F_\sigma(t,t')} \{R'[u' \leftarrow t'_i]\}$.*

The difference between the two properties above is that in the first one for every substitution there exists a subterm of t' comparable with t, and in the second one there exists a subterm that allows the comparison for every substitution. To prove termination of a rewriting system using this condition with F equal to the set of all substitutions, we have to calculate

- the finite set $\mathcal{C}$ of useful contexts C.
- for every pair $(C[u \leftarrow d], C[u \leftarrow g])$, a finite set of substitutions $\mathcal{E}_C$ included in E such that every substitution σ' is a variant of a substitution $\sigma \in \mathcal{E}$ w.r. to $C[u \leftarrow d]$ and $C[u \leftarrow g]$.

and then to compare for every $C \in \mathcal{C}$ and for every related substitution $\sigma \in \mathcal{E}_C$ $\sigma(C[u \leftarrow d])$ and $\sigma(C[u \leftarrow g])$ *decomposition patterns.*

4.2 Particular Cases and Examples

We study some particular cases in which the set of contexts to consider can be defined more efficiently.

4.2.1 Words on a Finite Alphabet

Let R a rewriting system on Σ^* and S an unavoidable set. On words insertion operation is defined as concatenation and the decomposition of a word w has only three components, the pattern s, the prefix P and the suffix U with $w = PsU$.

Definition 12 *A word w is a residue if and only if there exists a word w' such that $decomp(ww') = (s, P, U)$ and w is either the empty word ϵ or a proper prefix of Ps.*

Proposition 7 *A rewriting system R is terminating if for every rule $g \rightarrow d$, for every residue w and for every word v, $wdv \ \ SRPO(\prec) \ wgv$.*

Let $\Sigma = \{a, b\}$, $R = \{aa \rightarrow aba\}$ and $S = \{a, ba, bb\}$ in decreasing ordering for words decomposition. The precedence on S is defined by $ba < a$.The only *residues* are ϵ and$b$, thus we have to show $abau \ \ SRPO(\prec) \ aau$ and $babau \ \ SRPO(\prec) \ baau$ for every word u.

1. $abau \ \ SRPO(\prec) \ aau$ if $bau \ \ SRPO(\prec) \ au$ and thus, as $ba < a$, if $u \ \ SRPO(\prec) \ au$ that is clear.

2. $babau \ \ SRPO(\prec) \ baau$ if $bau \ \ SRPO(\prec) \ au$ that has been already proved.

4.2.2 Unavoidable Pattern at the Root

Let S be an unavoidable set such that any tree from $T(\Sigma, \omega)$has a factor from S inserted at its root. Its avoidance bound is null. Let C_0 be the set of terms strictly overlapping elements from S.

$$C_0 = \bigcap_{s \in S} \{\tau \in T(\Sigma, \omega) | \exists \text{ a substitution } \rho \neq Identity \text{ s.t. } \rho(\tau) = s\}$$

Proposition 8 *Let Σ be a finite ranked alphabet, S an unavoidable set with a null avoidance bound. Let C_0 be the set defined above and R a rewriting system. R is terminating if for every context C belonging to C_0, for every substitution σ and for every rule $g \rightarrow d \in R$*

$$\sigma(C[u \leftarrow d]) \ \ SRPO(\prec) \ \sigma(C[u \leftarrow g])$$

Let $\Sigma = \{f, p, s, *\}$ and $R = \{f(s(x)) \rightarrow *(s(x), f(p(s(x)))); p(s(x)) \rightarrow x\}$. We take S as $S = \{*(\omega,\omega), s(\omega), p(s(\omega)), p(\omega), f(\omega)\}$. Thus the set of contexts $\mathcal{C}_0 = \{\omega, p(\omega)\}$. The precedence $\prec_S$ on S used to decompose a tree is defined by $p(\omega) \prec_S p(s(\omega))$. The precedence on S is defined by $*(\omega,\omega) \prec f(\omega), p(s(\omega)) \prec p(\omega), p(s(\omega)) \prec s(\omega)$. In order to prove termination for R, considering the contexts the only inequalities to check are

1. $*(s(x), f(p(s(x)))) \;\; SRPO(\prec) \; f(s(x))$

2. $x \;\; SRPO(\prec) \; p(s(x))$

3. $p(x) \;\; SRPO(\prec) \; p(p(s(x)))$

The first inequality has been already proved in section 3.2. The second one is clear by erasing. To prove the third one we have to consider two cases: either x is instanced in $s(y)$ or x is not instanced or is instanced in something that does not begin by s. In the last case *decomposition pattern* of both sides is $p(\omega)$ and we get the result by erasing. In the other one we have to prove $p(s(y)) \;\; SRPO(\prec) \; p(p(s(s(y))))$ that is satisfied if $p(s(y)) \;\; SRPO(\prec) \; p(s(s(y)))$ and thus if $y \;\; SRPO(\prec) \; s(y)$.

5 Implementation and Conclusion

We have implemented, in collaboration with A.Suarez, this method for proving termination in CAML, the language developed at INRIA and ENS (FRANCE) inside projet FORMEL. The difficulty is due to the non compatibility of our ordering with term structure. Thus we define a tree decomposition procedure that takes account the different possible substitutions for the variables and that assures a finite number of possibilities. For avoiding the variants of already considered substitutions in the decomposition of a given tree, we associate constraints to the variables of this tree. These constraints consist in a list of terms such that their assignation is prohibited for the variable. We have been led to write an unification algorithm that takes into account these constraints on variables.

In conclusion there are two major problems with the SRPO. The first one is the non compatibility with term structure, the second one is the choice of the unavoidable set S. In the example treated with this method, we don't look for minimal sets. We always use sets that are trivially unavoidable because they contain the trees of depth equal to one. Instead of these problem this ordering remains stronger than the usual RPO that is only a particular case of SRPO. It allows termination proof for rewriting systems containing a rule whose left hand side is embedded in the right one.

References

[1] N. Dershowitz. Ordering for Term Rewriting Systems. Theoretical Computer Science 17 (1982) 279-300

[2] N. Dershowitz. Termination of Rewriting J.of Symbolic Computation (1987)

[3] A.Erhenfeucht, D.Haussler, G.Rozenberg. On Regularity of Context-Free Languages. Theoretical Computer Science 27 (1983) 311-332

[4] S.Kamin, J.J.Levy. Two Generalizations of the Recursive Path Ordering. Unpublished note, Department of computer science, University of Illinois, Urbana,IL, February 1980

[5] J.B.Kruskal. Well Quasi-Ordering, the Tree Theorem and Vazsonyi's Conjecture. Trans. AMS 95 (1960) 210-225

[6] L.Puel. Bons Preordres sur les Arbres Associes a des Ensembles Inevitables et Preuves de Terminaison de Systemes de Reecriture. These d'Etat, Paris 1987.

[7] L.Puel. Using Unavoidable Sets of Trees to Generalize Kruskal's Theorem. To appear J. of Symbolic Computation

Rewriting Techniques for Program Synthesis

Uday S. Reddy[1]
Department of Computer Science
University of Illinois at Urbana-Champaign
Net address: reddy@a.cs.uiuc.edu
January 20, 1989

Abstract

We present here a completion-like procedure for program synthesis from specifications. A specification is expressed as a set of equations and the program is a Noetherian set of rewrite rules that is efficient for computation. We show that the optimizations applicable for proving inductive theorems are useful for program synthesis. This improves on the use of general completion procedure for program synthesis, reported by Dershowitz, in that it generates fewer rules and terminates more often. However, there is a qualitative difference between this procedure and completion, as superposition is used not for eliminating critical overlaps but to find a complete set of cases for an inductive theorem.

1 Introduction

In [Der82a, Der85] Dershowitz proposed the Knuth-Bendix completion procedure as an approach to program synthesis. The initial program specification and the definitions of auxiliary functions are given as a set of equations and rewrite rules. The completion procedure then generates new rules comprising the program for the specified functions by superposing the specifications with the definitions of auxiliary functions.

There are two problems in using the completion procedure for synthesis. Firstly, the procedure generates too many rules only a few of which comprise the program. It is not easy to decipher which rules are essential program rules and which are not. Secondly, the completion procedure can go on infinitely (producing an infinite number of rules) even though there exists a finite program. It is not clear when enough rules have been generated to comprise a valid program. In [Der88], Dershowitz remarks that the test of *inductive reducibility* (called *ground reducibility* in this paper) can be used to determine when enough rules have been generated to constitute a program. This test is useful for checking *sufficient completeness* which is somewhat orthogonal to the problem of synthesis. (For instance, we may want to synthesize a program that is not sufficiently complete). In any case, this test is made after the fact and does not guide the generation of rules itself. The conventional techniques for program synthesis/transformation, such as the unfold-fold technique [BD77] do not appear to have these problems even though they are based on the inference rules of equational logic without the concept of reduction.

The problem of generating too many rules (critical pairs) exists for inductive theorem proving as well, and Fribourg [Fri86] and Küchlin[Kuc88] give efficient procedures for inductive completion by explicitly using the fact the resulting system only needs to be ground confluent, not necessarily confluent. This paper similarly attempts a direct

[1]This research was supported in part by NASA grant NAG-1-613 and NSF grant CCR-87-00988..

approach to the problem of synthesis instead of using the general completion procedure. We show, through examples, that this procedure has better termination properties than the general completion procedure, and its performance is comparable to conventional unfold-fold techniques for program transformation [BD77].

2 Definitions

We assume the general familiarity with term rewriting systems. Our notation is borrowed from Bachmair, Dershowitz and Hsiang [Bac87, BDH86, Der88]. If E is a set of equations, $\leftrightarrow_E$ is its symmetric congruence closure. An equational proof is a sequence of proof steps $t_1 \leftrightarrow_E t_2 \leftrightarrow_E \cdots \leftrightarrow_E t_n$, which is also denoted $t_1 \leftrightarrow_E^* t_n$. An equality is *valid* in E, $E \vdash s = t$ if there exists a proof $s \leftrightarrow_E^* t$. We also say $s = t$ is an *equational consequence* of E. By Birkhoff's completeness theorem [Bir35], this is true iff $s = t$ holds in all models of E.

If an equality holds in the initial algebra of E, we say that $s = t$ is an *inductive consequence* of E and write $E \vdash_{ind} s = t$. This holds iff every ground instance of $s = t$ is an equational consequence of E. For most practical applications, the notion of inductive consequence requires *many-sorted equational logic*, *i.e.*, there is a set of sort symbols, all function symbols have signatures over the sort symbols, and an equality is an inductive consequence iff every *well-sorted* ground instance is an equational consequence.

If R is a set of rewrite rules written as $l \to r$, then $\to_R$ is the congruence closure, $\leftarrow_R$ its inverse, and $\leftrightarrow_R$ is its symmetric closure.

The set of function symbols in an equational system or rewrite system E is denoted F_E. $G(F)$ is the set of all ground (variable-free) terms over the function symbols F. $T(F)$ is the set of all ground and nonground terms (with variables implicitly drawn from a set X) over F. We use the following notion of a program:

Definition 1 (program) A *program* is a ground confluent and Noetherian set of rewrite rules.[2]

The rationale for this definition is that a *computation* is the rewriting of a ground term and the *value* of the term is its normal form. So, a well-defined program should give a unique value for each ground term in a finite amount of time. There are of course a variety of extensions of the above definition which would still meet this criterion (*e.g.*, ground confluence modulo a congruence, or weak termination instead of Noetherian condition etc.). Our definition identifies the assumptions we make in this paper.

Definition 2 (program equivalence) Two programs P and P' are *equivalent* with respect to a set of function symbols F if, for all $t, u \in G(F)$, $t \leftrightarrow_P^* u$ iff $t \leftrightarrow_{P'}^* u$.

---

[2] A rewrite system R is *confluent* if, for all terms a, b, c, whenever $a \to_R^* b$ and $a \to_R^* c$, there exists a term d such that $b \to_R^* d$ and $c \to_R^* d$. It is *ground confluent* if this property holds for all ground terms a, b, c. It is *Noetherian* if every rewriting sequence is finite.

Example 3 The following is a program for Fibonacci numbers. (We have a sort *nat* and function symbols $0 : nat$, $1 : nat$, $+ : nat \times nat \to nat$ and $fib : nat \to nat$).

$$\begin{array}{lrcl}
n1: & 0+x & \to & x \\
n2: & x+0 & \to & x \\
n3: & x+(y+z) & \to & x+y+z \\
f1: & fib(0) & \to & 0 \\
f2: & fib(1) & \to & 1 \\
f3: & fib(1+1) & \to & 1 \\
f4: & fib(x+1+1)) & \to & fib(x+1)+fib(x)
\end{array}$$

A derived program for this may be the following. (We have an additional sort *natpair* and new function symbols $\langle\rangle : nat \times nat \to natpair$, $sp : natpair \to nat$, $np : natpair \to natpair$, and $g : nat \to natpair$).

$$\begin{array}{lrcll}
n1: & 0+x & \to & x & \\
n2: & x+0 & \to & x & \\
n3: & x+(y+z) & \to & x+y+z & \\
f1: & fib(0) & \to & 0 & \\
f2: & fib(1) & \to & 1 & \\
f3: & fib(1+1) & \to & 1 & \\
f4': & fib(x+1+1) & \to & sp(g(x)) & \\
s1: & sp\langle x,y\rangle & \to & x+y & \text{– “sum pair”} \\
s2: & np\langle x,y\rangle & \to & \langle x+y,x\rangle & \text{– “new pair”} \\
g1: & g(0) & \to & \langle 1,0\rangle & \\
g2: & g(1) & \to & \langle 1,1\rangle & \\
g3: & g(x+1) & \to & np(g(x)) &
\end{array}$$

In most functional languages, functions like *sp* and *np* can be expressed using a **let** or **where** construct without introducing new function symbols. For instance:

$$\begin{array}{rcl}
fib(x+1+1) & \to & \textbf{let } \langle p,q\rangle = g(x) \textbf{ in } p+q \\
g(x+1) & \to & \textbf{let } \langle p,q\rangle = g(x) \textbf{ in } \langle p+q,p\rangle
\end{array}$$

Burstall and Darlington [BD77] use an inference rule called *abstraction* to introduce such constructs.

It may be seen intuitively that the two programs are equivalent with respect to $\{fib, 0, 1, +\}$, but the formal details are cumbersome. We introduce some more terminology and return to the problem of their equivalence. □

Definition 4 (consistent enrichment) Let D be a program, and S a set of equations. S is a *consistent enrichment* of D iff, for all $t, u \in G(F_D)$, $t \leftrightarrow^*_{D\cup S} u$ iff $t \leftrightarrow^*_D u$.

In other words, a consistent enrichment preserves the set of ground equivalence classes of D.

Definition 5 (specification) A *specification* is a pair (D, S) where D is a program called the *definition* and S is a consistent enrichment of it. When D is clear from context, we often refer to S itself as the specification.

D is supposed to contain the definitions of all the "primitive" functions used in S, together with any properties required of them. S then defines some new functions without modifying the congruence $\leftrightarrow_D$. This notion is convenient to identify what should be preserved by the program synthesis process. Note that it can be trivially satisfied by choosing D to be empty.

Definition 6 (rewrite-consistent enrichment) Let D and P be rewrite systems. P is a *rewrite-consistent enrichment* of D if P is a consistent enrichment and, further, P does not reduce any irreducible ground term of D.

Definition 7 (derived program) A *derived program* for a specification (D, S) is a rewrite-consistent enrichment P of D so that

1. $D \cup P$ is a program (ground confluent and terminating),

2. $D \cup S \vdash_{ind} P$,

3. $D \cup P \vdash_{ind} S$.

Immediate from the definitions is the following result:

Theorem 8 If P is a derived program for a specification (D, S), then $D \cup S$ and $D \cup P$ are equivalent. (We also say that S and P are equivalent in the context of D).

Example 9 To prove the equivalence of the two Fibonacci number programs of Example 3, we need to first postulate the "specification" of the function g, viz.,

$$s3: \qquad g(n) = \langle fib(n+1), fib(n)\rangle$$

Now, let

$$\begin{aligned} D_1 &= \{n1, n2, n3, f1, f2, f3, f4\} \\ S_1 &= \{s1, s2, s3\} \\ P_1 &= \{s1, s2, g1, g2, g3\} \end{aligned}$$

We use definition 7 to show that P_1 is a correct derived program for (D_1, S_1). The rules $s1$ and $s2$ are repeated in the program. For $s3$, we can show that it is an inductive consequence of $D_1 \cup P_1$ using the following proof schemes:

$$\begin{aligned} &g(0) \rightarrow_{P_1} \langle 1, 0\rangle \leftarrow^{+}_{D_1} \langle fib(1), fib(0)\rangle \\ &g(1) \rightarrow_{P_1} \langle 1, 1\rangle \leftarrow^{+}_{D_1} \langle fib(1+1), fib(1)\rangle \\ &g(x+1) \rightarrow_{P_1} np(g(x)) \leftrightarrow_{S_1} np\langle fib(x+1), fib(x)\rangle \\ &\qquad \rightarrow_{P_1} \langle fib(x+1) + fib(x), fib(x+1)\rangle \leftarrow_{D_1} \langle fib(x+1+1), fib(x+1)\rangle \end{aligned}$$

It is also clear that $g1$-$g3$ are equational consequences of $D_1 \cup S_1$. Thus, P_1 is a derived program for (D_1, S_1). (We can say that $g1$-$g3$ is a program that "meets the specification" $s3$).

To get the final program, we have to do this trick once more. Let

$$\begin{aligned} D_2 &= \{n1, n2, n3, s1, s2, g1, g2, g3\} \\ S_2 &= \{f1, f2, f3, f4\} \\ P_2 &= \{f1, f2, f3, f4'\} \end{aligned}$$

First note that $D_2 \cup S_2 \vdash_{ind} s3$. So, for $D_2 \cup S_2 \vdash_{ind} f4'$, we have the proof

$$fib(x+1+1) \rightarrow_{S_2} fib(x+1) + fib(x) \leftarrow_{D_2} sp\langle fib(x+1), fib(x)\rangle \leftrightarrow_{s3} sp(g(x))$$

Now, $s3$ is an inductive consequence of $D_2 \cup P_2$ as well. So, a rearrangement of the above proof shows $D_2 \cup P_2 \vdash_{ind} f4$. □

The problem of program synthesis is a generalization of inductive proof (program verification). Instead of proving an equation to be an inductive consequence of a program (as done above), we want to produce a program for which the given equation is an inductive consequence.

Efficiency

Central to the problem of program synthesis is the issue of efficiency. If we disregard efficiency, then specification equations can often be trivially oriented as program rules so that they make a rewrite-consistent enrichment of the definition rules. The problem of synthesis arises mainly because the specifications are often *inefficient* programs and we are interested in deriving efficient programs which meet such specifications. In the framework of term rewriting, the notion of efficiency can be captured by insisting that the derived program be *contained* in a given reduction ordering.

Definition 10 (reduction ordering, Dershowitz 1987) A reduction ordering $>$ is a well-founded ordering over $T(F)$ which is *stable* (preserved under instantiation) and *monotonic* ($a > b$ implies $t[a] > t[b]$).

Even though reduction orderings were proposed for proving termination, we find that they are also useful for qualitatively comparing relative efficiencies of terms. The stability axiom consistently generalizes the efficiency of ground terms to nonground terms, and the monotonicity axiom is consistent with efficiency under *innermost* evaluation.

Even more useful for the purposes of program synthesis are path orderings [Der82b] which are generated from precedence orderings on function symbols. In synthesizing a program for a new function, we often have an idea of which existing functions can be used in its program. With reference to the first derivation in Example 9, we expect that an efficient program can be derived for g without using fib. So, we can choose for the reduction ordering, the recursive path ordering generated by a precedence relation in which $fib > g$. Now, the specification equation $g(n) = \langle fib(n+1), fib(n)\rangle$ has to be oriented backwards as $\langle fib(n+1), fib(n)\rangle \rightarrow g(n)$. It cannot be directly added as a program rule as it violates the ground confluence requirement. If, on the other hand, it is acceptable to use fib in the derived program for g, then the precedence relation $g > fib$ allows the specification to be oriented as $g(n) \rightarrow \langle fib(n+1), fib(n)\rangle$, and the synthesis task is trivially completed.

The problem of program synthesis, then, is to derive a program which satisfies the correctness condition of Definition 7 and is contained in the given reduction ordering $>$.

3 Synthesis

We formulate the basic synthesis procedure below as a state transformation procedure in the notation of [BDH86, Bac87, Der88]. A state is a triple of the form (R, S, P). The component S contains the current *Specification equations*, and P contains the current *Program rules*. The R component contains what we call *Recursion rules*. They are the rules which were part of S in some earlier state, but need not be in S any more since all their ground instances are covered by other rules in $S \cup P$. The problem of synthesis now is to remove rules from S and add new ones to P maintaining the relation $\leftrightarrow^*_{D\cup S\cup P}$ invariant on ground terms. The notion of progress is that every ground proof in $D \cup S \cup P$ is eventually transformed into one in $D \cup P$.

We use the following concept from [Der82a] and [JK86]:

Definition 11 (ground reducibility) A term t is ground reducible (also called "quasi-reducible") by a rewrite system R, if every ground instance $t\sigma$, such that σ is irreducible, is reducible by R.

Methods for testing for ground reducibility may be found in [JK86, KNZ86].

The inference rules of the basic synthesis procedure are as follows:

expand :	$\dfrac{(R, S \cup \{t = u\}, P)}{(R \cup \{t \to u\}, S \cup S_C, P)}$	if $t > u$, t is ground reducible by $D \cup P$, and $S_C = CP(D \cup P \cup R, \{t \to u\})$
accept :	$\dfrac{(R, S \cup \{t = u\}, P)}{(R, S, P \cup \{t \to u\})}$	if $t > u$, t is not ground reducible by $D \cup P$ and $t \notin T(F_D)$
simplify :	$\dfrac{(R, S \cup \{t = u\}, P)}{(R, S \cup \{t' = u\}, P)}$	if $t \to_{D\cup R\cup P} t'$
delete :	$\dfrac{(R, S \cup \{t = t\}, P)}{(R, S, P)}$	

where $CP(R_1, R_2)$ is the set of all critical pairs between R_1 and R_2.[3] If $t = u$ is an equation in some S_i such that none of the inference rules is applicable, then either the equation is unorientable or there exists no program for the specification included in $>$. (Cf. Theorem 24).

Example 12 As an example of the basic synthesis procedure, consider the following starting state in the context of the rules in Example 3:

$$(\emptyset,\ \{g(n) = \langle fib(n+1), fib(n)\rangle\},\ \emptyset)$$

and assume that $D = \{n1, n2, n3, f1, f2, f3, f4, s1, s2\}$ and the reduction ordering is the recursive path ordering generated by

$$fib > g > np > sp > \langle\rangle > + > 1 > 0$$

[3]The set of *critical pairs* between rewrite systems R_1 and R_2, denoted $CP(R_1, R_2)$, is defined as follows. Let $s \to t$ be a rule in R_1 and $l \to r$ be a rule in R_2 (or *vice versa*) with no variables in common. If, for some position p, s/p is not a variable and is unifiable with l, σ being the most general unifier, then the equation $t\sigma = s\sigma[p \leftarrow r\sigma]$ is a critical pair in $CP(R_1, R_2)$.

An application of *expand* yields:

$$
\begin{aligned}
R_1 &= \{\langle fib(n+1), fib(n)\rangle \rightarrow g(n)\} \\
S_1 &= \{\langle fib(1), fib(0)\rangle = g(0),\ \ \langle 1, fib(1)\rangle = g(1), \\
&\quad \langle fib(x+1) + fib(x), fib(x+1)\rangle = g(x+1), \\
&\quad \langle fib(0+1), 0\rangle = g(0),\ \ \langle fib(1+1), 1\rangle = g(1),\ \ \langle fib(1+1+1), 1\rangle = g(1+1), \\
&\quad \langle fib(x+1+1+1), fib(x+1) + fib(x)\rangle = g(x+1+1), \\
&\quad \langle fib(x+1) + fib(x), fib(x+1)\rangle = np(g(x)), \\
&\quad fib(x+1) + fib(x) = sp(g(x))\} \\
P_1 &= \emptyset
\end{aligned}
$$

The equations in S_1 get simplified to

$$
\begin{aligned}
S_2 = \{&\langle 1, 0\rangle = g(0),\ \ \langle 1, 1\rangle = g(1),\ \ \langle fib(x+1) + fib(x), fib(x+1)\rangle = g(x+1), \\
&\langle 1+1, 1\rangle = g(1+1), \\
&\langle fib(x+1) + fib(x) + fib(x+1), fib(x+1) + fib(x)\rangle = g(x+1+1), \\
&\langle fib(x+1) + fib(x), fib(x+1)\rangle = np(g(x)), \\
&fib(x+1) + fib(x) = sp(g(x))\}
\end{aligned}
$$

Picking the last equation for expansion, we get a set of equations S_{sp} of critical pairs. We omit listing them as they all eventually reduce to identities. After adding the equation to R and using it to simplify the remaining equations in S_2, we obtain the state:

$$
\begin{aligned}
R_3 = \{&\langle fib(n+1), fib(n)\rangle \rightarrow g(n), \\
&fib(x+1) + fib(x) \rightarrow sp(g(x))\} \\
S_3 = \{&\langle 1, 0\rangle = g(0),\ \ \langle 1, 1\rangle = g(1),\ \ \langle sp(g(x)), fib(x+1)\rangle = g(x+1), \\
&\langle 1+1, 1\rangle = g(1+1),\ \ \langle sp(g(x)) + fib(x+1), sp(g(x))\rangle = g(x+1+1), \\
&\langle sp(g(x)), fib(x+1)\rangle = np(g(x)) \\
&\} \cup S_{sp}
\end{aligned}
$$

Again picking the last equation for expansion, we get another set of equations S_{np} and the state:

$$
\begin{aligned}
R_4 = \{&\langle fib(n+1), fib(n)\rangle \rightarrow g(n), \\
&fib(x+1) + fib(x) \rightarrow sp(g(x)), \\
&\langle sp(g(x)), fib(x+1)\rangle \rightarrow np(g(x))\} \\
S_4 = \{&\langle 1, 0\rangle = g(0),\ \ \langle 1, 1\rangle = g(1),\ \ np(g(x)) = g(x+1), \\
&\langle 1+1, 1\rangle = g(1+1),\ \ \langle sp(g(x)) + fib(x+1), sp(g(x))\rangle = g(x+1+1) \\
&\} \cup S_{sp} \cup S_{np}
\end{aligned}
$$

The first three equations can now be accepted as the program:

$$P_5 = \{\ g(0) \rightarrow \langle 1, 0\rangle,\ \ g(1) \rightarrow \langle 1, 1\rangle,\ \ g(x+1) \rightarrow np(g(x))\}$$

After further superposition and simplification, all the remaining equations in S_4 reduce to identities and get deleted. □

When the derivation is complete, all the rules left in the R component are inductive theorems of the newly derived program. Even though they are not necessary for performing computations, they are useful properties for further program synthesis activities. For instance, the rule $fib(x+1) + fib(x) \rightarrow sp(g(x))$ in R_4 is useful for simplifying the original rule $f4$ of fib. Thus, we always maintain, in addition to the definition rules D, a set of known theorems T. These can be used in the inference rule *simplify*.

3.1 Correctness

To show that the program derived by the synthesis procedure is correct, we need to show that it satisfies the correctness condition of Definition 7. A ***basic synthesis derivation*** is a sequence of states $\{R_i, S_i, P_i\}_{i\geq 0}$ such that $R_0 = P_0 = \emptyset$ and every $(i+1)$th state is obtained from the previous one by one of the basic synthesis inference rules.

Lemma 13 The ground proof relation $\leftrightarrow_{D \cup R \cup S \cup P}$ is invariant in a basic synthesis derivation.

Lemma 14 The following assertion is invariant in a basic synthesis derivation: For every ground rewrite step $a \rightarrow_R b$, there is a proof $a \leftrightarrow^*_{D \cup S \cup P} b$ such that every $a' \leftrightarrow_S b'$ step in the latter proof satisfies $\{a, b\} \gg \{a', b'\}$.[4]

Proof: Since $R_0 = P_0 = \emptyset$, the assertion is vacuously true for the initial state. Suppose it holds for (R_i, S_i, P_i). The next state $(R_{i+1}, S_{i+1}, P_{i+1})$ is generated by one of the inference rules. Consider each rule:

expand Let $a \rightarrow b$ be a ground instance of $(t \rightarrow u) \in R_{i+1}$, i.e., $a = t\theta$ and $b = u\theta$ for some ground substitution θ. If θ is a reducible substitution with $\theta \rightarrow^+_{D \cup P \cup R} \theta'$, then the proof

$$t\theta \rightarrow^+_{D \cup P \cup R} t\theta' \rightarrow_{R_{i+1}} u\theta' \leftarrow^*_{D \cup P \cup R} u\theta$$

contains a smaller R_{i+1} step. So, it suffices to consider irreducible ground substitutions for θ. Since t is ground reducible by $D \cup P$, there is a critical peak proof $v\sigma \leftarrow_{D \cup P} t\sigma \rightarrow u\sigma$ such that a is an instance of $t\sigma$. The equation $v\sigma = u\sigma$ is in S_C. So, we have a proof

$$t\sigma \rightarrow_{D \cup P} v\sigma \leftrightarrow_{S_{i+1}} u\sigma$$

of which the proof for $a = b$ is an instance. Note that $\{t\sigma, u\sigma\} \gg \{v\sigma, u\sigma\}$.

accept Trivial.

simplify For each ground proof $t\theta \leftrightarrow_{S_i} u\theta$, there is a proof

$$t\theta \rightarrow_{D \cup R \cup P} t'\theta \leftrightarrow_{S_{i+1}} u\theta$$

Since $t > t'$, $\{t\theta, u\theta\} \gg \{t'\theta, u\theta\}$. The invariant applied to the ith state shows that every R_i step in $t\theta \rightarrow t'\theta$ has a smaller S_i step. So, the inference step either leaves an $\leftrightarrow_{S_i}$ step untouched, or it replaces it by a proof which has only a smaller $\leftrightarrow_{S_{i+1}}$ step.

delete Trivial.

Hence, the assertion holds in $(R_{i+1}, S_{i+1}, P_{i+1})$. □

[4] $\gg$ is the multiset ordering [DM79] generated by $>$. It is defined by $M \cup \{x\} \gg M \cup \{y_1, \ldots, y_n\}$ whenever $x > y_i$ for all y_i.

Theorem 15 Suppose (D, S_0) is a specification and there is a basic synthesis derivation starting from a state $(\emptyset, S_0, \emptyset)$ and ending in a state $(R_f, \emptyset, P_f)$. If $D \cup P_f$ is ground confluent, then P_f is a correct derived program for (D, S_0).

Proof: It is clear that $R_f \cup P_f$ only contains equational consequences of $D \cup S_0$. Since S_0 is a consistent enrichment of D, $R_f \cup P_f$ is also a consistent enrichment. Further, since the left hand side of every P_f-rule is not in $T(F_D)$, it cannot reduce any irreducible ground term of D. Hence P_f is a rewrite-consistent enrichment of D.

By lemma 13, every ground proof $a \leftrightarrow^*_{D \cup S_0} b$ has a proof of the form $a \leftrightarrow^*_{D \cup R_f \cup P_f} b$. By lemma 14, there is also a proof of the form $a \leftrightarrow^*_{D \cup P_f} b$. Hence, $D \cup P_f \vdash_{ind} S_0$. □

The derived program P_f is often ground confluent (together with D). But, it may not always be. Since we do not yet have a general ground completion procedure for arbitrary rewriting systems, with the current state of the art, we can only use a general completion procedure for making $D \cup P_f$ ground confluent.

4 Inductive expansion

It is now well-known that the Knuth-Bendix completion procedure performs many inessential deductions when used for proving inductive theorems [Fri86, Kuc88]. Since program synthesis attempts to satisfy inductive theorems, it is not surprising that the same kind of redundancy is present in the above program synthesis procedure based on completion. For instance, in the state (R_1, S_1, P_1) of Example 12, the first three equations of S_1 essentially subsume the next four equations. In fact, the first and the fourth equation reduce to the same equation after simplification by D rules. The references [Fri86, Kuc88] give examples where the redundant work may be infinite. Example 20 below illustrates a similar situation in the context of program synthesis leading to an infinite number of inessential deductions.

Definition 16 (Inductively complete set of positions, Küchlin 1987) A set of non-variable positions P of a term t is called an *inductively complete set of positions using* $D' \subseteq D$ iff, for every normalized ground substitution σ, $t\sigma$ is reducible at a position $p \in P$ by a rule in D'. If $P = \{p\}$ then p is called an *inductively complete position*, and t/p is called an *induction subterm*.

Example 17 Consider the term $\langle fib(n+1), fib(n) \rangle$. Since every ground term of type *nat* reduces to an instance of $\{0, 1, x+1\}$, and the three instantiations of the term (for n) are reducible at positions 1 and 11, $\{1, 11\}$ is an inductively complete set of positions. Similarly, every ground term of type *nat* is also reducible to an instance of $\{0, 1, 1+1, x+1+1\}$. Since all these instantiations of the term (for n) are reducible at position 2, the singleton set $\{2\}$ is also an inductively complete set. □

When there exists a single inductively complete position, the ground reducibility test at that position is enough to show this fact. However, when multiple positions are needed to achieve inductive completeness, a more sophisticated test is needed. We find the set of critical pairs at the chosen set of positions (as defined below) and check that the instantiations used in them cover all the irreducible ground terms in an appropriate test set [KNZ86].

Definition 18 (Inductively complete critical pairs) A set of *inductively complete critical pairs* $ICP(R, \{l \to r\})$, is obtained by choosing an inductively complete set of positions P of l using some $R' \subseteq R$ and then finding the critical pairs of R' into $l \to r$ at only the positions P. (This is abuse of notation since ICP is not a function).

Using these concepts, we can modify the rule *expand* to use ICP instead of CP.

$$expand: \quad \frac{(R, S \cup \{l = r\}, P)}{(R \cup \{l \to r\}, S \cup S_C, P)} \quad \begin{array}{l} \text{if } l > r,\ l \text{ is ground reducible by } D \cup P, \\ ICP(D \cup P, \{l \to r\}) \subseteq S_C, \text{ and} \\ S_C \subseteq CP(D \cup R \cup P, \{l \to r\}) \end{array}$$

Notice that the lemma 14 still holds because it only depends on the presence of an inductively complete set of critical pairs in S_C. Thus, the derivations made using the modified *expand* rule are still correct.

The synthesis procedure is now *nondeterministic* just as Küchlin's inductive completion procedure [Kuc88]. Any one choice of the inductively complete set of positions may lead to successful synthesis. It cannot be predicted which choice would. So, it is necessary to try all of them in parallel. However, this is not any more work than the general completion procedure which computes all the critical pairs, but does not recognize the choice involved in them. In our implementation, called the Focus program derivation system [Red88], the user can interactively suggest the induction terms to be expanded.

We permit the set S_C computed in an expansion to be any set of critical pairs containing an inductively complete set. The additional critical pairs included in it may be useful as "lemmas" for performing simplifications that are otherwise impossible. (Cf. Example 19 below). In practice, it is useful to compute all the additional critical pairs in each expansion and keep them in a separate component of the state (say H). Whenever such a critical pair is useful for simplification, it can be moved to S, expanded, and then used for simplification. It should be noted that the additional critical pairs are merely inductive hypotheses generated during completion. They need to be "proved" for the proof of the original hypothesis to be valid. It is somewhat tempting to add them directly to R. This was done by Küchlin in an early version of his paper [Kuc87]. But, adding such hypotheses to R is not, in general, valid. Even though they are inductive consequences of $D \cup S \cup P$, they may not have equivalent simpler (by $\gg$) proofs in $D \cup S \cup P$. So, the invariant in lemma 14 is not maintained. (The error was independently detected by Küchlin and corrected in [Kuc88]). When a derivation is complete, however, we have a proof $D \cup P \vdash_{ind} S_0$ and, since the equations in H are equational consequences of $D \cup P \cup S_0$, we also have $D \cup P \vdash_{ind} H$. So, at this stage, all the equations in H can be oriented as rules and used as inductive theorems T.

Example 19 Let us redo the derivation of Example 12 using inductive expansion. The rules generated are shown in Figure 1 as a tree. The rules labeled as Cases are the essential critical pairs obtained by expanding the induction terms shown enclosed in boxes. The rules $5'$ and $6'$ labeled as Lemmas are inessential critical pairs obtained by superposing $1'$ with the definition rules of *sp* and *np*. The proofs of the lemmas are intricately interdependent, so much so that, if we omit $5'$, the proof of $6'$ goes on *ad infinitum*. The final program is the set of non-identity leaf rules of the tree, viz., $\{2', 3', 4'\}$. All the internal rules of the tree, viz., $\{1', 5', 9', 6', 12'\}$ are recursion rules which can be accepted as inductive theorems.

1. $g(n) = \langle \mathit{fib}(n+1), \mathit{fib}(n)\rangle$

1′. $\langle \boxed{\mathrm{fib}(\boxed{\mathrm{n+1}})}, \mathit{fib}(n)\rangle \rightarrow g(n)$

Cases:

2. $\langle \mathit{fib}(1), \mathit{fib}(0)\rangle = g(0)$ 2′. $g(0) \rightarrow \langle 1, 0\rangle$

3. $\langle 1, \mathit{fib}(1)\rangle = g(1)$ 3′. $g(1) \rightarrow \langle 1, 1\rangle$

4. $\langle \mathit{fib}(x+1) + \mathit{fib}(x), \mathit{fib}(x+1)\rangle = g(x+1)$ 4′. $g(x+1) \rightarrow np(g(x))$

Lemmas:

5′. $\boxed{\mathrm{fib}(\boxed{\mathrm{n+1}})} + \mathit{fib}(n) \rightarrow sp(g(n))$

Cases: ...

9. $\mathit{fib}(x+1) + \mathit{fib}(x) + \mathit{fib}(x+1) = sp(g(x+1))$

9′. $sp(\boxed{\mathrm{g(x)}}) + f(x+1) \rightarrow sp(np(g(x)))$

Cases: ...

15. $sp(np(g(y))) + \mathit{fib}(y+1+1) = sp(np(g(y+1)))$

15′. $sp(np(g(y))) + sp(g(y)) = sp(np(g(y))) + sp(g(y))$

6. $\langle \mathit{fib}(n+1) + \mathit{fib}(n), \mathit{fib}(n+1)\rangle = np(g(n))$

6′. $\langle sp(\boxed{\mathrm{g(n)}}), \mathit{fib}(n+1)\rangle \rightarrow np(g(n))$

Cases: ...

12. $\langle sp(np(g(x))), \mathit{fib}(x+1+1)\rangle = np(g(x+1))$

12′. $np(np(\boxed{\mathrm{g(x)}})) \rightarrow \langle sp(np(g(x))), sp(g(x))\rangle$

Cases: ...

18. $np(np(np(g(y)))) = \langle sp(np(g(y+1))), sp(g(y+1))\rangle$

18′. $\langle sp(np(g(y))) + sp(g(y)),\ sp(np(g(y)))\rangle$
$= \langle sp(np(g(y))) + sp(g(y)),\ sp(np(g(y)))\rangle$

Figure 1: Derivation tree for g

Recall from Example 17 that the subterm $\mathit{fib}(n)$ in 1′ is also an induction term. If we choose to expand that instead, we would obtain a similar derivation yielding the program

$$\begin{aligned} g(0) &\rightarrow \langle 1, 0\rangle \\ g(1) &\rightarrow \langle 1, 1\rangle \\ g(1+1) &\rightarrow \langle 1+1, 1\rangle \\ g(x+1+1) &\rightarrow np(np(g(x))) \end{aligned}$$

which is different, but equivalent to, the program derived in Figure 1. □

Example 20 The following is a nontrivial example of program synthesis using inductive expansion. The basic synthesis procedure does not terminate for this problem.

Given the sorts $atom, tree, list, bool$ with the constructors $tip : atom \rightarrow tree$, $\langle\rangle :$ $tree \times tree \rightarrow tree$, $nil : list$, "." : $atom \times list \rightarrow list$, and $true, false : bool$, let

$$\begin{array}{rcl} D & = & \{fringe(tip(a)) \rightarrow a.nil \\ & & fringe(\langle tip(a), t\rangle) \rightarrow a.fringe(t) \\ & & fringe(\langle\langle s,t\rangle, u\rangle) \rightarrow fringe(\langle s, \langle t,u\rangle\rangle) \\ & & leq(nil,nil) \rightarrow true,\ \ leq(nil, a.x) \rightarrow false,\ \ leq(a.x, nil) \rightarrow false, \\ & & leq(a.x, b.y) \rightarrow eq(a,b) \wedge leq(x,y) \\ & & false \wedge x \rightarrow false,\ \ x \wedge false \rightarrow false,\ \ true \wedge x \rightarrow x,\ \ x \wedge true \rightarrow x\ \} \\ S & = & \{feq(s,t) = leq(fringe(s), fringe(t))\} \end{array}$$

For the reduction ordering, use a combination of recursive path ordering and lexicographic path ordering with the precedence relation:

$$fringe > leq > feq > \wedge > \langle\rangle > tip > \text{`.'} > nil > true > false$$

and a left-to-right status for $\langle\rangle$. Figure 2 shows the derivation of a program for feq included in this reduction ordering.

The leaf rules of the tree, viz., $\{5', 6', 11'\text{-}13', 7', 14'\text{-}16', 9', 10', 4'\}$ constitute the derived program. The large size of the program is due to the fact that no useful properties of $fringe$ were supplied. If we add lemmas like $leq(nil, fringe(t)) = false$, then all the rules at level 4 can be eliminated. (These rules essentially contain the proofs of the lemmas).

The basic synthesis procedure (based on general completion) does not terminate for this example. Notice that the rule 3 has two possible induction terms, $fringe(s)$ and $fringe(t)$. If we expand the former, we would obtain a new rule of the form

$$leq(a.b.fringe(s), fringe(t)) \rightarrow feq(\langle tip(a), \langle tip(b), s\rangle\rangle, fringe(t))$$

which in turn has a similar induction term $fringe(s)$. Since the basic synthesis procedure insists on expanding all the induction terms, the $fringe(s)$ subterm in each such rule is expanded producing every larger rules. The derivation sequence for the basic synthesis procedure is thus infinite. □

5 Completeness

In this section, we show that the synthesis procedures using the inference rules of section 3 are complete in that they always succeed in generating a program (albeit an infinite one). The proof uses the techniques of of *proof orderings* [BDH86].

First we define a rewrite relation on ground equational proofs denoted by $\Rightarrow$. This is expressed as the least monotonic and stable extension of a set of *elimination patterns* of the form $p \Rightarrow p'$. $\Rightarrow^+$ is shown to be well-founded by defining a *complexity measure* $C(p)$ for proofs and by showing that, if $p_1 \Rightarrow p_2$, then $C(p_1) \gg C(p_2)$. $C(p)$ is defined to be the multiset of the complexity measures of its individual proof steps where $C(s \rightarrow_{D\cup P} t) = \{s\}$, $C(s \leftrightarrow_{R\cup S} t) = \{s,t\}$. The elimination patterns, together with proofs of their wellfoundedness, are the following:

1. $s \leftrightarrow_S t$	$\Rightarrow$	$s \rightarrow_{D\cup P'} u \leftrightarrow_{S'} t$	$\{\{s,t\}\}$	$\gg$	$\{\{s\},\{u,t\}\}$	since $s > u$
2. $s \leftrightarrow_S t$	$\Rightarrow$	$s \rightarrow_{R'} u \leftrightarrow_{S'} t$	$\{\{s,t\}\}$	$\gg$	$\{\{s,u\},\{u,t\}\}$	since $s > u$
3. $s \leftrightarrow_S s$	$\Rightarrow$		$\{\{s,s\}\}$	$\gg$	$\{\emptyset\}$	
4. $s \leftrightarrow_S t$	$\Rightarrow$	$s \rightarrow_{P'} t$	$\{\{s,t\}\}$	$\gg$	$\{\{s\}\}$	

1. $feq(s,t) = leq(fringe(s), fringe(t))$
1'. $leq(\boxed{\text{fringe(s)}}, fringe(t)) \rightarrow feq(s,t)$
  2. $leq(a.nil, \boxed{\text{fringe(t)}}) \rightarrow feq(tip(a),\ t)$
    5. $leq(a.nil, b.nil) = feq(tip(a),\ tip(b))$
    5'. $feq(tip(a),\ tip(b)) \rightarrow eq(a,b)$
    6. $leq(a.nil, b.fringe(t)) = feq(tip(a),\ \langle tip(b), t\rangle)$
    6'. $eq(a,b) \wedge leq(nil, \boxed{\text{fringe(t)}}) \rightarrow feq(tip(a),\ \langle tip(b), t\rangle)$
      11. $eq(a,b) \wedge leq(nil, b1.nil) = feq(tip(a),\ \langle tip(b), tip(b1)\rangle)$
      11'. $feq(tip(a),\ \langle tip(b), tip(b1)\rangle) \rightarrow false$
      12. $eq(a,b) \wedge leq(nil, b1.fringe(t1)) = feq(tip(a),\ \langle tip(b), \langle tip(b1), t1\rangle\rangle)$
      12'. $feq(tip(a),\ \langle tip(b), \langle tip(b1), t1\rangle\rangle) \rightarrow false$
      13. $eq(a,b) \wedge leq(nil, fringe(\langle t1, \langle t2, t3\rangle\rangle)) = feq(tip(a),\ \langle\langle t1, t2\rangle, t3\rangle)$
      13'. $feq(tip(a),\ \langle\langle t1, t2\rangle, t3\rangle) \rightarrow feq(tip(a),\ \langle t1, \langle t2, t3\rangle\rangle)$
    7. $leq(a.nil, fringe(\langle t1, \langle t2, t3\rangle\rangle)) \rightarrow feq(tip(a),\ \langle\langle t1, t2\rangle, t3\rangle)$
    7'. $feq(tip(a),\ \langle\langle t1, t2\rangle, t3\rangle) \rightarrow feq(tip(a),\ \langle t1, \langle t2, t3\rangle\rangle))$
  3. $leq(a.fringe(s), \boxed{\text{fringe(t)}}) \rightarrow feq(\langle tip(a), s\rangle,\ t)$
    8. $eq(a,b) \wedge leq(fringe(s), nil) = feq(\langle tip(a), s\rangle,\ tip(b))$
    8'. $eq(a,b) \wedge leq(\boxed{\text{fringe(s)}}, nil) \rightarrow feq(\langle tip(a), s\rangle,\ tip(b))$
      14. $eq(a,b) \wedge leq(tip(a1), nil) = feq(\langle tip(a), tip(a1)\rangle,\ tip(b))$
      14'. $feq(\langle tip(a), tip(a1)\rangle,\ tip(b)) \rightarrow false$
      15. $eq(a,b) \wedge leq(a1.s1, nil) = feq(\langle tip(a), \langle tip(a1), s1\rangle\rangle,\ tip(b))$
      15'. $feq(\langle tip(a), \langle tip(a1), s1\rangle\rangle,\ tip(b)) \rightarrow false$
      16. $eq(a,b) \wedge leq(fringe(\langle s1, \langle s2, s3\rangle\rangle), nil) = feq(\langle tip(a), \langle\langle s1, s2\rangle, s3\rangle\rangle,\ tip(b))$
      16'. $feq(\langle tip(a), \langle\langle s1, s2\rangle, s3\rangle\rangle,\ tip(b)) \rightarrow feq(\langle tip(a), \langle s1, \langle s2, s3\rangle\rangle\rangle,\ tip(b))$
    9. $eq(a,b) \wedge leq(fringe(s), fringe(t)) = feq(\langle tip(a), s\rangle,\ \langle tip(b), t\rangle)$
    9'. $feq(\langle tip(a), s\rangle,\ \langle tip(b), t\rangle) \rightarrow eq(a,b) \wedge feq(s,t)$
    10. $leq(a.fringe(s), fringe(\langle t1, \langle t2, t3\rangle\rangle)) = feq(\langle tip(a), s\rangle,\ \langle\langle t1, t2\rangle, t3\rangle)$
    10'. $feq(\langle tip(a), s\rangle,\ \langle\langle t1, t2\rangle, t3\rangle) \rightarrow feq(\langle tip(a), s\rangle,\ \langle t1, \langle t2, t3\rangle\rangle)$
  4. $leq(fringe(\langle s1, \langle s2, s3\rangle\rangle), fringe(t)) = feq(\langle\langle s1, s2\rangle, s3\rangle,\ t)$
  4'. $feq(\langle\langle s1, s2\rangle, s3\rangle,\ t) \rightarrow feq(\langle s1, \langle s2, s3\rangle\rangle,\ t)$

Figure 2: Derivation tree for feq

Next we show that each inference rule reduces ground equational proofs, i.e., if $(R, S, P) \vdash (R', S', P')$ then for every proof p in $D \cup R \cup S \cup P$, there is a proof p' in $D \cup R' \cup S' \cup P'$ such that $p \Rightarrow^* p'$. The inference rule *expand* and *simplify* reduce proofs via the elimination patterns 1 and 2, *accept* via 4, and *delete* via 3.

If $\{R_i\}_i$ is a sequence of sets, then its *persisting set* is $R^\infty = \bigcup_{i \geq 0} \bigcap_{j \geq i} R_j$.

Definition 21 A derivation (R_i, S_i, P_i) is *fair* if S^∞ does not contain any rewrite rules to which one of the inference rules is applicable (in particular *expand* and *accept*).

Theorem 22 If a derivation is fair with $S^\infty = \emptyset$, then for every ground proof p in $D \cup S_i \cup P_i$ there is a proof p' in $D \cup P_j$ for some $j \geq i$ such that $p \Rightarrow^* p'$. (That is, such a derivation eliminates all the S steps in all ground proofs).

Lemma 23 If a derivation is fair with $S^\infty = \emptyset$, then for every ground proof p in $D \cup S_i \cup P_i$ containing a $\leftrightarrow_{S_i}$ step, there is a proof p' in $D \cup S_j \cup P_j$ for some $j \geq i$, such that $p \Rightarrow p'$.

Proof: If p contains a proof step $q \equiv s \rightarrow_{S_i} s'$ by a rule $r \in S_i$, then since $S^\infty = \emptyset$, there exists $j \geq i$ so that $r \notin S_j$. Since the inference rules are compatible with the elimination patterns there must be a proof $q' \equiv s \rightarrow_{D \cup S_j \cup P_j} s'$ so that $q \Rightarrow^* q'$. But $q \neq q'$ since it cannot have a use of r. Hence $q \Rightarrow^+ q'$, and $p \Rightarrow p'$ for some p'. □
Proof of Theorem 22: By induction on $\Rightarrow$ (since it is well-founded). If p contains a proof step involving S_i, by the lemma, there is a proof p' at a later state so that $p \Rightarrow p'$. The induction hypothesis may then be applied to p'. □

The overall effect of the inductive synthesis procedure is stated as follows:

Theorem 24 Let (D, S_0) be a candidate specification, and $\{(R_i, S_i, P_i)\}_i$ be a fair derivation starting from the state $(\emptyset, S_0, \emptyset)$ and using a reduction ordering $>$. Assume that S^∞ does not contain an unorientable equation, and let G^∞ be the ground completion of $D \cup P^\infty$. Then the following statements hold:

1. If $S^\infty = \emptyset$ and G^∞ is a rewrite-consistent enrichment of D, then G^∞ contains a possibly infinite program for (D, S_0) included in $>$.

2. If $S^\infty \cup G^\infty$ contains an equation $t = u$ such that $t, u \in T(F_D)$, $t > u$ and t is not ground reducible by D, then (D, S_0) is not a consistent specification.

3. If $S^\infty \cup G^\infty$ contains an equation $t = u$ such that $t \in T(F_D)$, $u \in T(F) - T(F_D)$, $t > u$ and t is not ground reducible by D, then there does not exist a program for (D, S_0) contained in $>$.

Equivalently, we can state that $S^\infty = \emptyset$, G^∞ is a rewrite-consistent enrichment of D, and G^∞ contains a program for (D, S_0) *if and only if* (D, S_0) is a consistent specification and there is a program for it contained in $>$.

6 Discussion

We have presented here a completion-based procedure for program synthesis from equational specifications, whose performance and termination behavior are comparable to the conventional unfold-fold transformation procedures [BD77, Hog76, KS85, TS84, Tur86]. The *expand* rule achieves the combined effect of instantiation and unfolding, except that it also finds the required substitutions by unification. For instance, in Figure 1, the application of *expand* to $1'$ instantiates the specification of g with $n = 0$, $n = 1$ and $n = x + 1$, simultaneously unfolding at least one rewritable subterm. This simultaneous instantiation and unfolding is crucial to ensure that the applications of the inductive hypothesis (recursion rule) are correct. The use of unification with unfolding is also found in the approaches of [Hog76, KS85, TS84, Tur86]). *Simplify* using a definition rule or a program rule is like plain unfolding.

On the other hand, our completion-based procedure does not use *folding*. Folding can use arbitrary rewrite rules backwards, resulting in huge search spaces. It can also introduce new variables [KS85]. The effect of folding is achieved in our procedure by two means. Firstly, specification equations, unless accepted as program rules, are oriented backwards using the efficiency information captured in the reduction ordering. Simplification using such back-oriented recursion rules achieves the most common effect of folding in a "controlled" manner. This is similar to the restricted use of folding in [TS84] and "reduction to a basic configuration" in [Tur86]. However, simplification achieves only some of the rewritings that can be done by folding. The others are achieved by the formation of lemmas or inductive theorems via inessential critical pairs. For instance, the lemma 5′ in Figure 1 achieves the combined effect of a folding step followed by a simplification:

$$fib(n+1) + fib(n) \leftarrow_D sp\langle fib(n+1), fib(n)\rangle \rightarrow_R sp(g(n))$$

But, it should be pointed out that the critical pair generation is over-eager in its attempt to obviate all possible foldings. An infinite number of crtical pairs may be generated in this fashion, as illustrated by Example 20. We are investigating methods to generate critical pairs in a goal-directed fashion.

Another important issue that is relevant to both inductive proofs and program synthesis is deciding whether an inessential critical pair is a lemma (to be proved) or an inductive theorem. The rules 5′ and 6′ in Figure 1, for instance, need not be proved, even though our method requires them to be proved. That is because their use in simplifying equation 4 is for a strictly smaller instance of the original specification 1. (See the proof schemes in Example 9). Once they are used for producing the rule 4′, the specification is proved and hence they are automatically inductive theorems. Thus, even with the inductively complete optimization, our method performs more work than necessary. The additional work can also be infinite. (If the lemma 5′ is removed from Figure 1, then the proof of 6 goes on forever). It should be possible to keep track of the instances of the hypothesis used in an inessential critical pair, and decide whether it can be directly used for simplification.

Finally, methods for achieving ground confluence of arbitrary rewriting systems need to be investigated. This is crucially required for the final product of our procedure, viz., the ground completion of $D \cup P^{\infty}$.

References

[Bac87] L. Bachmair. *Proof Methods for Equational Theories.* PhD thesis, Univ. Illinois at Urbana-Champaign, 1987.

[BD77] R. M. Burstall and J. Darlington. A transformation system for developing recursive programs. *Journal of the ACM*, 24(1):44–67, January 1977.

[BDH86] L. Bachmair, N. Dershowitz, and J. Hsiang. Orderings for equational proofs. In *Symp. on Logic in Computer Science*, pages 346–357, IEEE, 1986.

[Bir35] G. Birkhoff. On the structure of abstract algebras. *Proc. of the Cambridge Philosophical Society*, 31:433–454, 1935.

[Der82a] N. Dershowitz. Applications of the Knuth–Bendix completion procedure. In *Proc. of the Seminaire d'Informatique Theorique, Paris*, pages 95–111, December 1982.

[Der82b] N. Dershowitz. Orderings for term-rewriting systems. *Theoretical Computer Science*, 17(3):279–301, 1982.

[Der83] N. Dershowitz. *Computing with rewrite systems.* Technical Report ATR-83(8478)-1, Information Sciences Research Office, The Aerospace Corp., El Segundo, CA., January 1983.

[Der85] N. Dershowitz. Synthesis by completion. In *IJCAI*, pages 208–214, Los Angeles, 1985.

[Der88] N. Dershowitz. Completion and its applications. In *1987 MCC Colloq. on Resolution of Equations in Algebraic Structures*, Austin, Texas, 1988.

[DM79] N. Dershowitz and Z. Manna. Proving termination with multiset orderings. *Communications of the ACM*, 22(8):465–476, August 1979.

[Fri86] L. Fribourg. A strong restriction of the inductive completion procedure. In *Intern. Conf. Aut., Lang. and Program.*, pages 105–115, Rennes, France, July 1986. (Springer Lecture Notes in Computer Science, Vol. 226).

[Ham88] D. Hammerslag. *Treemacs Manual.* Technical Report UIUCDCS-R-88-1427, Univ. Illinois at Urbana-Champaign, May 1988.

[Hog76] C. J. Hogger. Derivation of logic programs. *Journal of the ACM*, 23(4), 1976.

[JK86] J.-P. Jouannaud and E. Kounalis. Automatic proofs by induction in equational theories without constructors. In *Symp. on Logic in Computer Science*, pages 358–366, IEEE, Cambridge, MA., June 1986.

[KNZ86] D. Kapur, P. Narendran, and H. Zhang. Proof by induction using test sets. In *Conf. on Automated Deduction*, Oxford, U.K., 1986.

[KS85] D. Kapur and M. Srivas. A rewrite rule based approach for synthesizing data types. In *TAPSOFT 85*, pages 188–207, Springer-Verlag, 1985.

[Kuc87] W. Küchlin. *Inductive completion by ground proof transformation.* Technical Report 87-08, Department of Computer Science, Univ. of Deleware, Newark, February 1987.

[Kuc88] W. Küchlin. Inductive completion by ground proof transformation. In *Proc. 1987 MCC Colloquium on Resolution of Equations in Algebraic Structures*, MCC, Austin, Texas, 1988.

[Red88] U. S. Reddy. Transformational derivation of programs using the FOCUS system. In *Symp. Software Development Environments*, pages 163–172, ACM, December 1988.

[TS84] H. Tamaki and T. Sato. Unfold/fold transformation of logic programs. In *Intern. Conf. on Logic Program.*, pages 127–138, Uppsala, 1984.

[Tur86] V. F. Turchin. The concept of a supercompiler. *ACM Transactions on Programming Languages and Systems*, 8(3):292–325, 1986.

Transforming Strongly Sequential Rewrite Systems with Constructors for Efficient Parallel Execution

R.C. Sekar, Shaunak Pawagi and I.V. Ramakrishnan*

Department of Computer Science

State University of New York at Stony Brook

Stony Brook, NY 11794

Abstract

Strongly sequential systems, developed by Huet and Levy [2], has formed the basis of equational programming languages. Experience with such languages so far suggests that even complex equational programs are based only on strongly sequential systems with constructors. However, these programs are not readily amenable for efficient parallel execution. This paper introduces a class of strongly sequential systems called *path sequential systems*. Equational programs based on path sequential systems are more natural for parallel evaluation. An algorithm for transforming any strongly sequential system with constructors into an equivalent path sequential system is described.

1 Introduction

Term rewriting systems (TRS) play a fundamental role in applications involving computing with equations. Huet and Levy [2] and O'Donnell [6] laid the theoretical foundations for implementing equational logic based on rewriting. This was followed by the development of an equational programming language by Hoffman and O'Donnell [1] and Futatsugi et al [4]. Unlike OBJ2 (the language described in [4]), O'Donnell [7] provided a complete implementation for a restricted class of equational programs called *strongly left sequential systems*, without any compromise on semantics. However, both these efforts were directed towards efficient sequential implementations.

Only recently Goguen et al [8] and Pawagi et al [11] have begun investigating specialized parallel architectures for general term rewriting. However, execution of equational programs does not require the full generality of term rewriting and efficient techniques for parallel evaluation that exploit the special properties of the rewrite systems used in programs are not known.

An equational program is evaluated by repeatedly replacing *redices* (a redex is an instance of a left-hand side) in the input term by the corresponding right-hand sides. This process is known as *normalization* and the *normal form* of a term contains no redices. There may be several redices in any term and only some of them need to be rewritten

*Partially supported by NSF grant CCR-8805734

in every reduction sequence to normal form. A *lazy* evaluation procedure replaces such *needed* redices only. Lazy evaluation results in an efficient and complete normalization algorithm (i.e., it will reduce a term to its normal form whenever one exists).

An intuitive way of normalizing a term on parallel computers is to search for needed redices in parallel along different paths in the term and rewriting them. For efficiency reasons, it is critical that this search and reduction be done without excessive communication and synchronization overhead.

A major source of inter-processor communication is in the matching phase of normalization. Specifically, the nodes of the *subject* (i.e., the input term) to be matched against a pattern (the left-hand side of a rule) may be split among several processors. An obvious approach to handle this would be to copy all the nodes to the processor holding the node at which the match was initiated and then perform the match locally. As observed by Goguen [8], handling *split matches* in this manner could generate a large volume of communication. Ideally, we would like to handle each split using a single communication step.

The matching process also gives rise to synchronized waiting. Specifically, for a match initiated at a node to progress, the subterms rooted at some of the descendants of that node should either be reduced or regarded as substitutions for variables. Lazy evaluation requires that we do not reduce those subterms that become variable substitutions. This choice at a descendant may depend on the outcome of reductions performed at another descendant. Such a dependence requires synchronized waiting, since the match in progress must be suspended. In general, there could be a chain of such suspended matches, thereby greatly reducing the concurrency in evaluation. Resuming these waits also requires additional communication.

This paper is a preliminary investigation into efficient methods for parallel evaluation of equational programs. The following summarizes our results.

1.1 Summary and Significance of our Results

We propose a class of systems known as *path sequential systems* that are natural for parallel evaluation. Some important advantages of using path sequential systems are as follows:

1. They readily support the intuitive approach of searching and rewriting redices in parallel along different paths.
2. Split matches require minimal communication.
3. The need for synchronized waits is completely eliminated.
4. Finally, two processors communicate only when one processor holds a node that is the parent of a node in the other. This results in a highly organized and regular pattern of communication.

It may appear that path sequential systems restrict the expressive power of the equational language. We show that they are no less "powerful" than a more general class known as *strongly sequential systems*. (Experience so far suggests that even complex equational programs are based on strongly sequential constructor systems [7]). We describe a technique to transform a strongly sequential constructor system into an equivalent path sequential constructor system.

The transformation procedure given in this paper is significant in the sense that the path sequentiality restriction is transparent to the user, i.e., the user can write strongly sequential programs which may not be path sequential. In contrast, O'Donnell's equational language explicitly enforces strong left sequentiality on the programs. Since path sequential systems are also strongly left sequential, our technique can be used to repair violations of strong left sequentiality also.

The transformed system may contain at most thrice the number of rules in the original system. In practice, we find that very few rules are added and that the transformation procedure is very efficient.

The rest of this paper is organized as follows. In the following section, we present the necessary notations and definitions. In section 3 we give an overview of how path sequential systems satisfactorily handle the synchronization and communication problems mentioned above. Section 4 deals with the transformation process. We illustrate the transformation using an example followed by a detailed description and proof of correctness of the transformation algorithm. In section 5 bounds on the time complexity of the algorithm and on the size of the transformed systems are established. We also briefly discuss the practical implications of our algorithm. Finally, concluding remarks appear in section 6.

2 Preliminaries

In this section we introduce the terminology commonly used in term rewriting systems. We also develop the notations that will be used throughout the rest of this paper.

All the symbols used in a term rewriting system are drawn from a non-empty *ranked alphabet* Σ. In general, the Σ-terms may contain *variables*. A Σ-term is said to be linear iff no variable in it occurs more than once. Henceforth, a *term* always refers to a Σ-term. We use the obvious tree representation for terms, and use *terms* and *trees* interchangeably.

A *path* in a term is a (possibly empty) sequence of integers. The empty sequence Λ reaches the root of the term itself, the sequence consisting of the integer k reaches the kth argument of the root, and the sequence $k \cdot m$ reaches the mth argument of the kth argument of the root and so on. The concatenation of paths p and q is denoted by $p \cdot q$. We use t/p to refer to the descendant of t reached by p. For a path $p = i_1.i_2....i_k$, let $s_0, s_1, ..., s_k$ be the symbols on this path in a term t (listed in the order from root to the node at t/p). The path string for a term t and a path p, denoted by $pathstr(t,p)$, refers to the string $s_0.i_1.s_1.i_2....i_k.s_k$. In fig.1, the paths 2.1 and 1 reach nodes 5 and 2 respectively, and $pathstr(l_1, 1.1) = f.1.b.1.e$.

A *substitution* maps variables to terms. An *instance* $t\beta$ of a term t is obtained by replacing x by $\beta(x)$ for each variable x in t. For two terms t and u, $t \leq u$ means that u is an instance of t. Two terms t_1 and t_2 are said to *overlap* if there exists a nonvariable subterm u of t_1 and substitutions α and β such that $u\alpha = t_2\beta$.

A term rewriting system $\mathcal{R}$ over an alphabet Σ is a set of *rewrite rules* of the form $[l_i \rightarrow r_i]$. $\mathcal{L}$ is the set $\{l_i\}$ of left hand sides (lhs). In a *regular term rewriting system* [2], the left hand sides are *linear* and *non-overlapping.* We further restrict $\mathcal{L}$ such that $\mathcal{R}$ is a *constructor system*, as defined in [9]. In a constructor system, the nonvariable symbols in Σ can be partitioned into two disjoint sets Δ and Θ, such that the outermost symbol of each lhs belongs to Δ and all other nonvariable symbols belong to Θ.

A *redex* in term u is an occurrence of an instance of an lhs. A term t is in *normal form* if it contains no redices. A reduction $t \rightarrow^{!}_{l} u$ means that u is obtained by replacing the

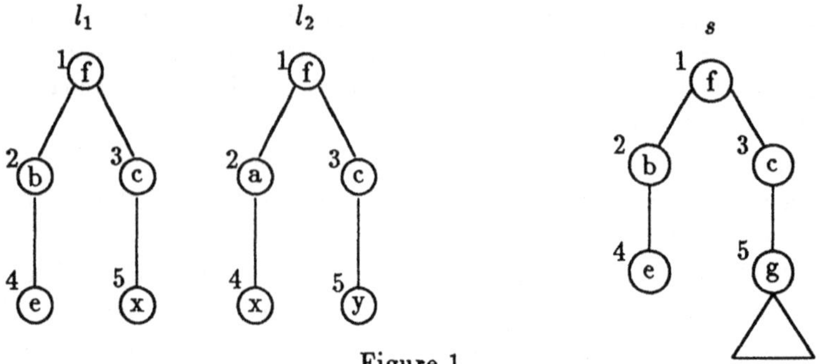

Figure 1

redex $l\beta$ in t at path s by $r\beta$, where $[l \rightarrow r]$ is a rewrite rule. A redex in a term t is called a *needed redex* [3] if it is rewritten in every reduction of t to normal form. A *maximal redex* in t is a redex that is not a subterm of any other redex in t.

An important theorem by Huet and Levy [2] states that in a regular system the normal form can be reached by repeated rewriting of needed redices. To ensure lazy evaluation it is necessary to identify at least one needed redex at every reduction step. By restricting $\mathcal{R}$ to be *strongly sequential*, we can compute at least one maximal needed redex at every step. Henceforth $\mathcal{R}$, $\mathcal{R}^N$, $\mathcal{R}'$ and $\mathcal{R}_i$ will denote rewriting systems, and $\mathcal{L}$, $\mathcal{L}^N$, $\mathcal{L}'$ and $\mathcal{L}_i$ will denote the corresponding sets of lhs.

There is a *block* in the set $\mathcal{L}$ of left hand sides at $p.i$, if there are two lhs l_1 and l_2 in $\mathcal{L}$ such that $pathstr(l_1, p) = pathstr(l_2, p)$ and $l_1/p.i$ is a variable (here p is a path and i is an integer). If there is a block at $p.i$ and there is no block at p or any prefix of p, we refer to $p.i$ as a *blocking path*, and to p as a *maximal block-free* path. The nodes at $p.i$ are called *blocking nodes*. In fig.2, node 4 is a blocking node. Path 1.1 is a blocking path and 1 and 2 are maximal block-free paths.

$Top(\mathcal{L}, l)$ is the structure obtained from l by deleting all subtrees rooted at blocking nodes in l. $Roof(\mathcal{L}, l)$ is the structure obtained by replacing each subtree rooted at a blocking node by a new variable. We use $top(l)$ and $roof(l)$ where $\mathcal{L}$ is understood. Note that $top(l)$ can be obtained from $roof(l)$ by dropping the variables. In fig.2, $top(l_1) = f(a, c)$, $top(l_2) = f(a, d)$ and $roof(l_1) = f(a(x), c(y))$.

A constructor system $\mathcal{R}$ is *path sequential* iff there are no two lhs l_1 and l_2, a path p and an integer i such that $pathstr(l_1, p) = pathstr(l_2, p)$, $l_1/p.i$ is a variable and $l_2/p.i$ is a nonvariable. In other words, the only blocks that are permitted in a path sequential systems are on paths p such that l/p is a variable for all the blocking rules. The patterns in fig.2 are not path sequential, whereas the ones in fig.1 are path sequential. For a set of lhs $\mathcal{L}$ and a term t, let $\mathcal{L}_t = \{l \in \mathcal{L} \mid t \leq l\}$. Let $\mathcal{P} = \{p \mid t/p \text{ is a variable}\}$. A constructor system $\mathcal{R}$ is *strongly sequential* iff

$$\forall t \; (\mathcal{P} = \phi) \vee \exists p \in \mathcal{P} \text{ such that there is no block in } \mathcal{L}_t \text{ at } p$$

This is an operational definition of strong sequentiality that has been shown in [9] to be equivalent to the original definition of strong sequentiality by Huet and Levy [2] for constructor systems. Note that path sequential systems are also strongly sequential.

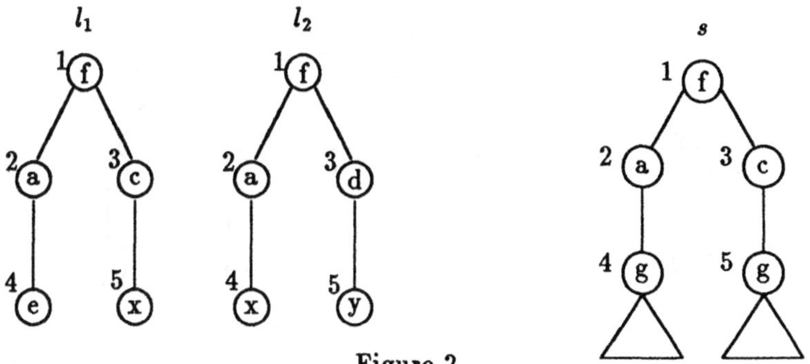

Figure 2

3 Matching in Path Sequential Systems

We give an overview of matching for path sequential systems on parallel processors. This will illustrate the appropriateness of this class of equational programs for parallel evaluation. We assume that the processors are connected through an interconnection network. The term to be normalized is distributed among them and each processor has a copy of the program.

Matching in path sequential systems is based on path strings, as illustrated through the following example (see fig.1). In the figure, l_1 and l_2 are two lhs and s is the term to be normalized. Assume (for simplicity of explanation) that s is completely distributed, with nodes 1, 2, 3, 4 and 5 in processors p_1, p_2, p_3, p_4 and p_5. p_1 initiates a match at the root of s and succeeds in matching the root symbol of s with the root of l_1 and l_2. It sends a message to p_2 and p_3 asking them to continue the match for both patterns. p_2 succeeds in matching node 2 with the corresponding node of l_1, and asks p_4 to continue the match for l_1. Now p_4 also succeeds in matching node 4 with that of l_1 and reports success to p_2, which then passes this on to p_1. Similarly, p_3 succeeds in matching node 3 with that of l_1 and l_2 and reports it to p_1. Based on the information received from p_2 and p_3, p_1 concludes that there is a match for l_1 at node 1. Note that only one communication step per split was required for matching.

In contrast, if the program is not path sequential then the matching process gives rise to synchronization waits and additional communication steps. To see this, consider the example shown in fig.2. Here again p_1 initiates a match at node 1 of s and succeeds in matching the symbol on node 1 with the root symbol of l_1 and l_2. It therefore asks p_2 and p_3 to continue the match initiated at node 1. p_2 succeeds in matching symbol a in node 2 with the symbols in node 2 in both l_1 and l_2 and then asks p_4 to continue the match further. Now p_4 does not know whether to reduce the subtree rooted under its node (labelled g) hoping to match the symbol e on node 4 of l_1 or to regard it as a variable substitution for l_2. This decision can be made only by looking at node 3 which is in p_3. Hence p_4 must *wait* for p_3 to complete its match. The outcome of this match is then reported to p_4 by p_3. (This obviously needs additional communication.) We can completely eliminate such waits and communication steps only if each processor is able to decide whether the subtree rooted under its node is to be rewritten or regarded as substitution for a variable based entirely on the information sent to it by its parent. This is indeed possible for path sequential systems and they are therefore appropriate for efficient parallel evaluation of equational programs. In order to make path sequentiality transparent to the user, we present a procedure for

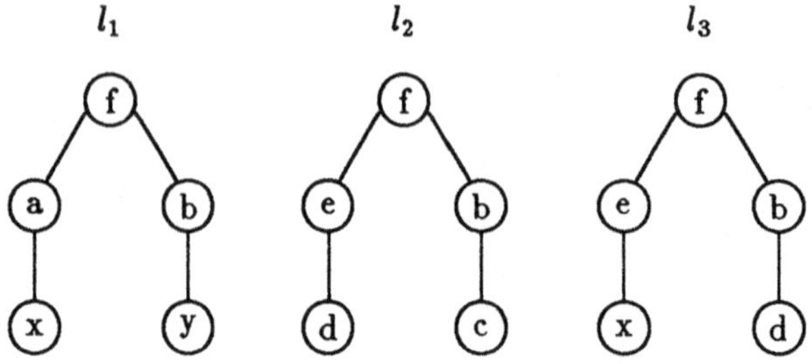

Figure 3

transforming a strongly sequential constructor system into an equivalent path sequential system.

4 Transformation Algorithms

In this section we present algorithms for transforming a strongly sequential constructor system into a path sequential system. Our algorithm reports failure when the input system is not strongly sequential. We illustrate the algorithm with an example and prove its correctness. Tight bounds on the size of the transformed system are then given and the time complexity of our transformation algorithm is established.

Note that in a constructor system, *strong sequentiality* or *path sequentiality* is determined independently for each set of rules that define a function. Hence we assume without any loss of generality that all the lhs in $\mathcal{L}$ have the same root symbol f. The bulk of the transformation work is done by two procedures – *BuildTree* and *Transform*. *BuildTree* constructs a tree which is essentially an automaton to recognize the maximal block-free path strings in $\mathcal{L}$. *Transform* makes use of this tree to replace $\mathcal{R}$ by $\mathcal{R}^N = \mathcal{R}' \cup (\bigcup_i \mathcal{R}_i)$. $\mathcal{R}'$ is path sequential, and the $\mathcal{R}_i$'s need to be processed recursively by *transform*. Before proceeding with the details of the algorithm we first illustrate the transformation process on the following example. We will show the effect of *buildtree* and *transform* after each step of the transformation.

Let $\mathcal{R} = \{[f(a(x), b(y)) \rightarrow r_1(x, y)],\ [f(e(d), b(c)) \rightarrow r_2],\ [f(e(x), b(d)) \rightarrow r_3(x)]\}$. The lhs of these rules are illustrated in fig.3. Invocation of *buildtree* on these patterns results in the tree shown in fig.4. Observe that there are blocks among these patterns at 1.1 (between l_2 and l_3) and 2.1 (between l_1 and l_2). Thus, the maximal block-free path strings are $f.1.a$, $f.1.e$ and $f.2.b$.

We now show how *transform* generates a new set of rules. Based on these blocks, the *top*'s of patterns l_1, l_2 and l_3 are $f(a, b), f(e, b)$ and $f(e, b)$ respectively. We use a new function symbol g_1 to abbreviate $top(l_1)$ and g_2 to abbreviate $top(l_2)$ and $top(l_3)$. Next we introduce the following new equations in place of those in $\mathcal{R}$:

$$f(a(x), b(y)) \rightarrow g_1(x, y) \qquad \cdots (1)$$
$$g_1(x, y) \rightarrow r_1(x, y) \qquad \cdots (2)$$
$$f(e(x), b(y)) \rightarrow g_2(x, y) \qquad \cdots (3)$$
$$g_2(d, c) \rightarrow r_2 \qquad \cdots (4)$$
$$g_2(x, d) \rightarrow r_3(x) \qquad \cdots (5)$$

Note that in rule (3) the lhs is the same as $roof(l_2)$ (or $roof(l_3)$), and the rhs is obtained by making each variable in $roof(l_2)$ an argument to g_2. The lhs of (4) (or (5)) is obtained

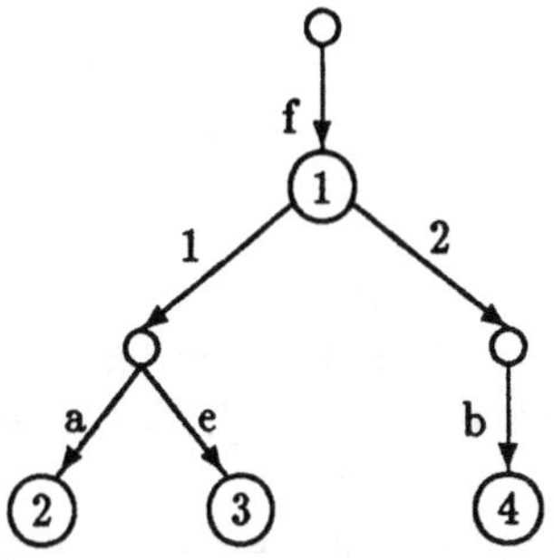

Figure 4

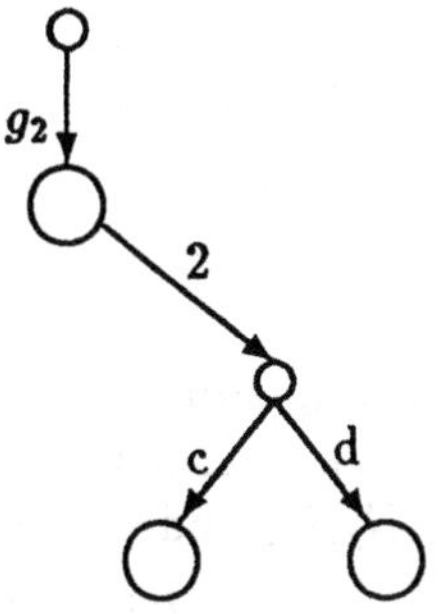

Figure 5

by making each blocking node in pattern l_2(or l_3) an argument to g_2.

Since there is only one rule with g_1 on lhs it is trivially path sequential. However, the rules (4) and (5) defining g_2 need to be processed further. *Buildtree* is called again on the lhs of (4) and (5), resulting in the tree shown in fig.5. Based on this, *transform* generates the following rules to replace (4) and (5):

$$g_2(x,c) \rightarrow g_3(x) \qquad \cdots (6)$$
$$g_3(d) \rightarrow r_2 \qquad \cdots (7)$$
$$g_2(x,d) \rightarrow g_4(x) \qquad \cdots (8)$$
$$g_4(x) \rightarrow r_3(x) \qquad \cdots (9)$$

Note that these rules are again path sequential. Thus, the transformed system consists of rules (1), (2), (3), (6), (7), (8) and (9).

We now provide the algorithmic details of the process described above. We begin with *BuildTree*.

procedure *BuildTree*($\mathcal{L}, p, v$)
  $\mathcal{L}$ is a list of lhs
  p is a path
  v is a dummy node in the tree. This invocation of *Buildtree*
    constructs the subtree rooted at v
1. Partition $\mathcal{L}$ into
    $\mathcal{L}_x \leftarrow \{l \in \mathcal{L} \mid l/p$ is a variable$\}$ and
    $\mathcal{L}_c \leftarrow \{l \in \mathcal{L} \mid l/p = c\}, \forall c \in \Sigma$
2. If $\mathcal{L}_x \neq \phi$ then {there is a block at this point in this path}
3.   return(null)
  else
4.   for each $\mathcal{L}_c$ do
5.   create a new node v_c {The v_c's are *state nodes* or *states*}
      create an edge from v to v_c, labelled c
      $poss(v_c) \leftarrow \mathcal{L}_c$ { $poss(v_c)$ is the set of lhs s.t.}
        $\{pathstr(l,p) =$ path in tree from root to $v_c\}$
6.   for $i \leftarrow 1$ to $rank(c)$ do
7.   create a new node v_{ci} { The v_{ci}'s are the *dummy nodes*}
        create an edge from v_c to v_{ci}, labelled i
8.   *buildtree*($\mathcal{L}_c, p.i, v_{ci}$)
      end

```
            end
        end
        return(v)
    end
```

On invoking *buildtree* on the rules in fig.3, the tree shown in fig.4 is obtained. The possibility sets associated with state nodes are: $poss(1) = \{l_1, l_2, l_3\}$, $poss(2) = \{l_1\}$, $poss(3) = \{l_2, l_3\}$ and $poss(4) = \{l_1, l_2, l_3\}$.

T refers to the tree constructed by this algorithm. For a state node v in T, define $trace(v)$ to be the sequence of labels on the edges from the root to v. The correctness of our algorithm depends on the following properties of T.

Lemma 1 *For each state v in T, $poss(v)$ is precisely the set* $\{l \in \mathcal{L} \mid \exists p\; pathstr(l,p) = trace(v)\}$

Proof: Omitted. ∎

Lemma 2 *For any state v in T, let v' be any child of v. For any two distinct children v_1 and v_2 of v', $poss(v_1) \cap poss(v_2) = \phi$. If v' has children $v_1, \ldots, v_k$, $poss(v) = poss(v_1) \cup \cdots \cup poss(v_k)$.*

Proof: This is obvious, since the $\mathcal{L}_c$'s are distinct and $\bigcup_c \mathcal{L}_c = \mathcal{L}$. ∎

Lemma 3 *If $p.i$ is a blocking path in l, then there is a state node v in T such that $trace(v) = pathstr(l,p)$.*

Proof: We need to prove that there is a state node v in T corresponding to every block-free path in $\mathcal{L}$. This is proved by induction. . ∎

Note that in our transformation process we have to identify the *top*'s of lhs from T. This is done using the algorithm *prune* below.

```
    procedure Prune(v, l)
        v is a dummy node in the tree T
        l is a pattern in L
1.      for each child v_c of v do
2.          if l ∉ poss(v_c)
3.              then remove the edge from v to v_c
            else for each child v_ci of v_c do
4.              prune(v_ci, l)
        end
        return(v)
    end
```

Invoking *prune* on the tree shown in fig.4 with respect to patterns l_1 and l_2, results in the trees shown in figs.6 and 7. The corresponding *top*'s are shown in figs.8 and 9.

Now we state and prove some important properties of *prune* and show how $top(l)$ can be constructed from $prune(T,l)$. In the following T_l refers to $Prune(T,l)$.

Lemma 4 *T_l contains exactly the states v in T such that $poss(v)$ includes l.*

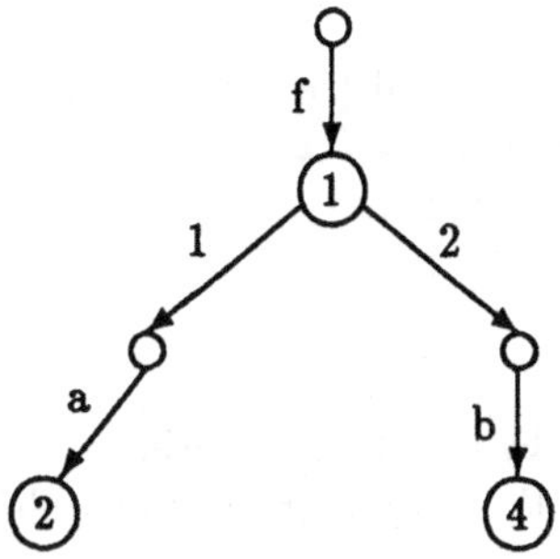

Figure 6

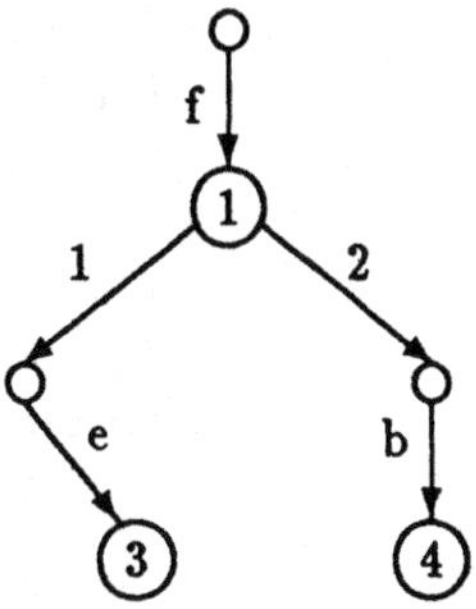

Figure 7

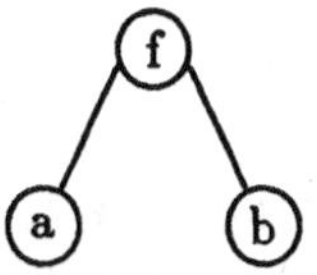

Figure 8

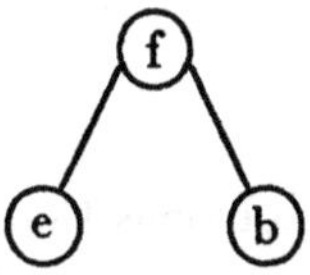

Figure 9

Proof: The proof follows immediately from the definition of *prune* and Lemma 2. ∎

It is clear from this and lemma 3 that T_l contains all and only maximal block-free paths in l. By definition of $top(\mathcal{L}, l)$, it follows that it also contains all and only maximal block-free paths of l. This establishes an obvious correspondence between the state nodes in T_l and $top(\mathcal{L}, l)$. An algorithm to compute *top* from *prune* would first merge each dummy node with its parent. This removes all integer labelled edges in T_l. Now the label on each edge is moved to the state node it points to. Dropping the root from this structure we get $top(\mathcal{L}, l)$. This procedure is illustrated in figs. 8 and 9. *Roof* can be obtained from *top* by adding variables at appropriate places.

In the following let $\mathcal{V}_l = \{v \mid v \text{ is a leaf of } prune(T, l)\}$.

Lemma 5 $l_1 \in \bigcap_{v \in \mathcal{V}_l} poss(v)$ *iff* $Prune(T, l) = Prune(T, l_1)$

Proof: (only if) From lemma 2 it is clear that $l_1 \in poss(v)$ for all ancestors of the leaves of $prune(T, l)$. Thus, all the states v in $prune(T, l)$ contain l_1 in $poss(v)$. From lemma 2, for every state v in T such that $l \in poss(v)$, exactly one of the children v' of each child v_i of v has $l \in poss(v')$. Since v' appears in $prune(T, l)$, $l_1 \in poss(v')$. By lemma 2 no other child v'' of v_i can have $l_1 \in poss(v'')$. This implies that all v in T such that $l_1 \in poss(v)$ are in $Prune(T, l)$. Hence $Prune(T, l) = Prune(T, l_1)$ by lemma 4. We omit the proof of the *if* part. ∎

We now present details of procedure *transform* that generates a new set of rules after a single step of transformation.

```
procedure Transform(ℒ)
        ℒ is the set of lhs of the input equations
        ℒ' ← φ; i ← 1
        T ← buildtree(ℒ, ∧, root)
        while ℒ ≠ φ do
                ℒ_i ← φ
```

3. choose an $l \in \mathcal{L}$

4. Let e be the equation $[roof(\mathcal{L}, l) \rightarrow g(\overline{x})]$ { e is the equation *generating* g }
   where $\overline{x}$ is the list of variables in $roof(\mathcal{L}, l)$
   and g is a new functor symbol.

5. $\mathcal{L}' \leftarrow \mathcal{L}' \cup e$

6. $\mathcal{L}_g = \bigcap_{v \in \mathcal{V}_l} poss(v)$

7. for each $l_1 \in \mathcal{L}_g$ do

8. Let t be the term obtained by rewriting[1] l_1 using e

9. $\mathcal{L}_i \leftarrow \mathcal{L}_i \cup [t \rightarrow r_1]$, where $[l_1 \rightarrow r_1]$ is the original rule.

   end {$[t \rightarrow r_1]$ is an equation *defining* g}

10. $\mathcal{L} \leftarrow \mathcal{L} - \mathcal{L}_g$

   end

11. if $|\mathcal{L}'| = 1$ return(*failed*)

end

It is obvious that the new rules generated are left linear. It is also clear that the non-variable symbols in the new system can be partitioned into two disjoint sets – constructors and functors. Because of this any overlap of lhs, if it exists, should be at the root.

Theorem 1 *The transformed system* $\mathcal{R}^N = \mathcal{R}' \cup (\bigcup_i \mathcal{R}_i)$ *is a constructor system.*

Proof: Note that all l' (in $\mathcal{L}'$) have f as the root symbol and l_i (in $\mathcal{L}_i$) have a root function symbol distinct from f and that of any other l_j (in $\mathcal{L}_j$). So overlaps can occur only within $\mathcal{L}'$ or an $\mathcal{L}_i$. Hence it suffices to show that $\mathcal{R}'$ and each $\mathcal{R}_i$ are constructor systems.

Assume that there are two distinct lhs l'_1 and l'_2 in $\mathcal{L}'$ that overlap. Let l'_1 be generated from $top(\mathcal{L}, l_1)$ and l'_2 from $top(\mathcal{L}, l_2)$. As mentioned before l'_1 and l'_2 can overlap only at the root, implying that $top(\mathcal{L}, l_1) = top(\mathcal{L}, l_2)$. This means that $prune(T, l_1) = prune(T, l_2)$. From lemma 5, for all leaves v of $prune(T, l_1)$, $l_2 \in poss(v)$. But then from steps 6 and 10, only one of l_1 or l_2 will generate an lhs in $\mathcal{L}'$, which contradicts the assumption $l'_1 \neq l'_2$. Hence there is no overlap in $\mathcal{L}'$.

Now suppose that $l_{i1}, l_{i2} \in \mathcal{L}_i$ overlap. Let l_{i1} and l_{i2} be generated from $l_1, l_2 \in \mathcal{L}$. Clearly $top(\mathcal{L}, l_1) = top(\mathcal{L}, l_2)$ since l_{i1} and l_{i2} belong to the same $\mathcal{L}_i$. Since the *top*'s as well as l_{i1} and l_{i2} overlap, l_1 and l_2 must also overlap – a contradiction. Therefore the new system is regular. It has already been mentioned that the symbols in the new system can be partitioned into constructors and functors. Thus the new system is constructor based. ∎

Now we establish that $\mathcal{R}$ and $\mathcal{R}^N$ are equivalent. In the succeeding lemmas we show that every reduction sequence in $\mathcal{R}$ can be simulated by a corresponding sequence in $\mathcal{R}^N$ and vice versa. Note that the alphabet of $\mathcal{R}^N$'s contains some g's that are not present in the alphabet of $\mathcal{R}$ and hence $\mathcal{R}$ cannot directly simulate $\mathcal{R}^N$. Therefore we define a function h that maps a term with g's into one that does not contain g's. Under this mapping we show that the two systems can simulate each other.

Let Σ^N and Σ be the alphabets of $\mathcal{R}^N$ and $\mathcal{R}$ and $l_g \rightarrow g(x_1, ..., x_{rank(g)})$ be the rule generating g for every $g \in (\Sigma^N - \Sigma)$. $t[t_1 \leftarrow t_2]$ denotes the term obtained from t by replacing one of its subterms t_1 by t_2. We define a function h that maps a Σ^N-term to a

[1] In rewriting, the variables in l_1 are treated like constants

Σ-term as follows.

$h(g(t_1, ..., t_{rank(g)})) = g(h(t_1), ..., h(t_{rank(g)})),\ g \in \Sigma$

$h(g(t_1, ..., t_{rank(g)})) = l_g[x_1 \leftarrow h(t_1), ..., x_{rank(g)} \leftarrow h(t_{rank(g)})],\ g \in (\Sigma^N - \Sigma)$

Note that the equation generating g does not contain g in its lhs and hence $h(t)$ is indeed a Σ-term. Intuitively, h "replaces" each g by the *top* it corresponds to.

Fact 1: $h(t[u \leftarrow v]) = h(t)[h(u) \leftarrow h(v)]$

Lemma 6 *If* $t_1 \rightarrow^* t_2$ *in* $\mathcal{R}^N$, *then* $h(t_1) \rightarrow^* h(t_2)$ *in* $\mathcal{R}$.

Proof: By induction on the length of the reduction sequence $t_1 \rightarrow \cdots \rightarrow t_2$. The base case is for a sequence of length zero, whence $t_1 = t_2$ and $h(t_1) = h(t_2)$ and thus holds. Suppose $t_1 \rightarrow^* t'$ in n steps and $t' \rightarrow^s_{l''} t_2$. By induction hypothesis $h(t_1) \rightarrow^* h(t')$. There are two cases to consider:

1. $l'' \in \mathcal{L}'$: the rule used in reduction is of the form $l_g \rightarrow g(x_1, ..., x_{rank(g)})$. From the definition of h, it is clear that $h(t') = h(t_2)$ and thus the lemma holds.

2. $l'' \in \mathcal{L}_i$: In this case the rule used in reduction is of the form $g(u_1, ..., u_{rank(g)}) \rightarrow r$. Let $l \rightarrow r$ be the rule in $\mathcal{L}$ such that $l \rightarrow_{l_g} g(u_1, ..., u_{rank(g)})$. Let v' be the subterm of t' at s and v_2 be the corresponding subterm of t_2. By fact 1 it suffices to show that $h(v') \rightarrow h(v_2)$ in $\mathcal{R}$. It is clear by definition of h that $h(v') = l\sigma$. Thus $h(v') \rightarrow r\sigma = h(v_2)$ and this completes the proof. ∎

Lemma 7 *If* $t_1 \rightarrow^* t_2$ *in* $\mathcal{R}$, *then* $t_1 \rightarrow^* t_2$ *in* $\mathcal{R}^N$.

Proof: The proof is again by induction on the length of the reduction sequence in $\mathcal{R}$. The base case is once again for sequence length of zero whence it holds vacuously. For the induction step, assume that the lemma holds whenever $t_1 \rightarrow^* t'$ in $\mathcal{R}$ in n steps or less. Let $[l \rightarrow r]$ be the rule in reducing t' to t_2 in $\mathcal{R}$ and $[l' \rightarrow g(x_1, \ldots, x_{rank(g)})]$ and $[l_g \rightarrow r]$ be the rules (in $\mathcal{R}^N$) generated from l. Let $l\beta$ be the subterm that matches l in t' at path p. Since $l = l'\beta'$ we have $l\beta = l'\beta'\beta$. Hence t' can be reduced to t_2 in $\mathcal{R}^N$ by replacing $l'\beta'\beta$ by $g(x_1, \ldots, x_k)\beta'\beta = l_g\beta$. This can be further reduced by replacing $l_g\beta$ by $r\beta$, obtaining t_2. ∎

Theorem 2 $\mathcal{R}^N$ *is equivalent to* $\mathcal{R}$.

Proof: Straight forward from lemmas 6 and 7. ∎

Lemma 8 $\mathcal{R}'$ *is path sequential.*

Proof: Assume to the contrary that $\mathcal{R}'$ is not path sequential. Then, by definition there exists an l'_1 and l'_2 in $\mathcal{R}'$ and a path p such that $pathstr(l'_1, p) = pathstr(l'_2, p)$, and there is an i such that $l'_1/p.i$ is a variable and $l'_2/p.i$ is a nonvariable. From the fact that $pathstr(l'_1, p) = pathstr(l'_2, p)$ and lemma 1, it follows that there is a single state node v in T corresponding to l'_1/p and l'_2/p. Since $l'_1/p.i$ is a variable v cannot have an ith child. If $l'_2/p.i$ were to be a nonvariable then v must have to have an ith child – a contradiction. Therefore $\mathcal{R}'$ is path sequential. ∎

Lemma 9 *If* $\mathcal{R}$ *is strongly sequential then each of the* $\mathcal{R}_i$*'s is strongly sequential.*

Proof: Assume that some $\mathcal{R}_i$ is not strongly sequential. By definition,

$$\exists\, t'\ \forall\, p \in \mathcal{P}' \text{ there is a block in } \mathcal{L}_{it} \text{ at } p$$

where $\mathcal{L}_{it} = \{l \in \mathcal{L}_i \mid t' \leq l\}$. Let g be the root symbol of rules in $\mathcal{L}_{it}$. Let l be the rule in $\mathcal{L}$ that generated a rule in $\mathcal{L}_{it}$. Consider $t = h(t')$ and the set $\mathcal{L}_t = \{h(l) \mid l \in \mathcal{L}_{it}\}$. Since all the rules in $\mathcal{L}_{it}$ have the same outer function symbol it follows that all the blocks that existed in $\mathcal{L}_{it}$ at variable nodes in t' would exist in $\mathcal{L}_t$ at the variable nodes in t. This means that $\mathcal{R}$ was not strongly sequential – a contradiction. ∎

Lemma 10 *If the algorithm Transform fails then the input system $\mathcal{R}$ is not strongly sequential.*

Proof: By definition $\mathcal{R}$ will not be strongly sequential if

$$\exists t\ \forall p \in \mathcal{P} \text{ there is a block in } \mathcal{L}_t$$

When *transform* reports failure at step 11, $\mid \mathcal{L}' \mid = 1$. Let t be the only lhs in $\mathcal{L}'$. In this case $\mathcal{L}_t = \mathcal{L}$ since t is the roof of all rules in $\mathcal{L}$. It remains to be proved that there is a block on all paths in $\mathcal{P}$. Suppose there was no block on a path p. In that case *buildtree* would have added at least one more state node to T. This would have resulted in a member $l' \in \mathcal{L}'$ which is different from t – a contradiction. Therefore $\mathcal{R}$ is not strongly sequential. ∎

The following procedure operates recursively on the $\mathcal{R}_i$'s to construct the equivalent path sequential system.

procedure *PathSeq*($\mathcal{R}$)
    $\mathcal{R}$ is the set of input equations
    if [*transform*($\mathcal{R}$) ≠*failed*] then
        for each system $\mathcal{R}_i$ generated by *transform* do
            if $\mid \mathcal{R}_i \mid > 1$ then *PathSeq*($\mathcal{R}_i$)
    end
end

Theorem 3 *PathSeq($\mathcal{R}$) returns a new path sequential system if $\mathcal{R}$ is strongly sequential.*

Proof: The algorithm *transform* returns sets of equations $\mathcal{R}'$ that is path sequential and $\mathcal{R}_i$'s that are strongly sequential. From theorems 1 and 2, R^N is a constructor system equivalent to $\mathcal{R}$. If any $\mathcal{R}_i$ contains just one equation then it is path sequential. Otherwise *transform* is applied on $\mathcal{R}_i$. Hence when *PathSeq* terminates it has converted the original system into a path sequential system. It remains to prove that *PathSeq* terminates.

Note that when $\mathcal{R}$ is transformed it is split into $\mathcal{R}'$ and $\mathcal{R}_i$. Only $\mathcal{R}_i$'s need to be processed recursively. Since the number of $\mathcal{R}_i$'s is more than 1 (or else *transform* reports failure), $\forall i \ \mid \mathcal{R}_i \mid < \mid \mathcal{R} \mid$. Thus after each call of *transform* we are left with smaller sets of equations that need to be processed. Since the processing stops when $\mathcal{R}_i = 1$, this process terminates. ∎

5 Complexity Issues

In this section we present tight bounds on the size (i.e., the number of rules) of the transformed system in terms of the size of the original system, and the time complexity of the transformation algorithm. Following this is a discussion related to the practical aspects of our algorithm.

5.1 Size Bounds for Transformed System

Let $|\mathcal{R}'| = k$ after one step of transformation. There are exactly k new function symbols introduced and let $|\mathcal{R}_1| = n_1, \ldots, |\mathcal{R}_k| = n_k$. It is clear that $n_1 + \cdots + n_k = n$. Representing the number of rules in the final system by $N(n)$, we have:

$$N(n) = k + \sum_{i=1}^{k} N(n_i) \qquad \text{such that } \sum_{i=1}^{k} n_i = n$$

We will prove by induction that $N(n) \leq 3n - 2$. It is clear that this holds for the base case $n = 1$, since no further transformation is attempted by *PathSeq* at step 3. For the induction step, since all n_i's are less than n, we can assume $N(n_i) \leq 3n_i - 2$.

$$N(n) \leq k + \textstyle\sum_{i=1}^{k}(3n_i - 2) \qquad \textit{such that } \textstyle\sum_{i=1}^{k} n_i = n$$

$$N(n) \leq k + 3n - 2k = 3n - k$$

By step 11 of *transform*, $k > 1$, thus establishing the upper bound.

Now we show that this is a tight bound, in the sense that there exist systems of equations that expand to size $3n - 2$ when transformed. Assume inductively that $N(n) \geq 3n - 2$. This obviously holds for $n = 1$. The worst case of $3n - 2$ occurs when $k = 2$, $n_1 = 1$ and $n_2 = n - 1$. In this case,

$$N(n) \geq 2 + N(1) + N(n-1) = 2 + 1 + 3n - 5 = 3n - 2$$

It remains to show that such a case actually occurs. An example of this worst case behaviour is given below:

$$\mathcal{L} = \{f(a, b, x), f(b, x, t_1), \ldots, f(b, x, t_{n-1}))\}$$

where

$$t_1 = e(a, b, x),\ t_2 = e(b, x, t_2'), \ldots, t_{n-1} = e(b, x, t_{n-1}')$$

and so on. After one step of transformation,

$$\mathcal{L}' = \{f(a, x, y), f(b, x, y)\}$$

$$\mathcal{L}_1 = \{g_1(b, x)\}$$

$$\mathcal{L}_2 = \{g_2(x, e(a, b, y)), g_2(x, e(b, y, t_2')), \ldots, g_2(x, e(b, y, t_{n-1}'))\}$$

It is easy to see that $\mathcal{L}_2$ is like $\mathcal{L}$, and hence in the recursive invocations of *transform*, the same splitting of groups takes place. This establishes that in the worst case the transformed system can have as many as $3n - 2$ rules.

5.2 Time Complexity

Let s be the sum of the sizes (i.e., the number of symbols) of all the lhs in $\mathcal{L}$ and $|\mathcal{L}| = n$. It is clear that *buildtree* looks at each node in every pattern in $\mathcal{L}$ at most once. All the operations involving a node take constant time. Hence *buildtree* takes $O(s)$ time. With a slight modification *prune* can be made to generate the *tops* in $O(s)$ time. Since

$| roof(l) | < | l |$ and $top(l)$ is already computed, computing the $roofs$ take at most $O(s)$ time. The rewriting at step 8 in $transform$ takes time proportional to the size of the rule being rewritten, and hence takes at most $O(s)$ time. Thus the total time taken at step 1 of *PathSeq* is $O(s)$. The recurrence relation for the time taken by *PathSeq* is therefore:

$$T(n, s) = c.s + \sum_{i=1}^{k} T(n_i, s_i)$$

where c is a constant. Noting that $s_i \leq s$ for all i, we get

$$T(n, s) \leq c.s + \sum_{i=1}^{k} T(n_i, s)$$

whose solution is $O(ns)$.

It can be shown that this bound of $O(ns)$ cannot be improved, as long as the variables are explicitly represented. The worst case example can be obtained by modifying the previous example slightly. Consider

$$\mathcal{L} = \{f(x_1, a, b, x), f(x_1, b, x, t_1), \ldots, f(x_1, b, x, t_{n-1}))\}$$

where

$$t_1 = e(x_2, a, b, x),\ t_2 = e(x_2, b, x, t_2'), \ldots, t_{n-1} = e(x_2, b, x, t_{n-1}')$$

and so on. It can be verified that the transformation process takes $O(ns)$ time.

5.3 Discussion

From the above analysis it may appear that the size of the transformed system can become considerably larger than the original system. However, in practice we find that the size of the system increases very little. The transformation procedure takes time almost linear in the sum of the sizes of the lhs. The worst case behaviour occurs only in the case of contrived examples.

Even though the number of pattern matches and rewrites have increased, the complexity of each reduction is reduced. Because of the special nature of the reductions performed using the rules $l_g \rightarrow g(x_1, ..., x_k)$, it can be shown that the computational effort involved will be the same for the original and the transformed system. Actually, parallel evaluation of the transformed system will exhibit better performance because of the reduction in communication overhead.

The algorithms presented may generate 'unnecessary' equations (such as equation (9) generated in the transformation of the system in Fig. 3) of the form $g(x_1, ..., x_k) = r(x_1, ..., x_k)$. Such cases can be detected and eliminated by a slight modification of $transform$. Whenever the transformation procedure reports failure, the input system is not strongly sequential. Hence we require a parallel evaluation strategy similar to the one needed for handling *parallel-or* [7] type of constructs. $Transform$ can be modified so that at step 11 it reports that such a strategy needs to be applied in matching the rules in $\mathcal{L}_1$.

6 Conclusions

In this paper we introduced a new class of strongly sequential systems called path sequential systems. Equational programs based on path sequential constructor systems are natural for parallel evaluation. In such systems parallel search for needed redices is very much simplified. Interprocess communication is reduced and is also uniform and localized between a parent and child processors. Path sequential systems are natural in situations wherein different processors search different parts of a term being normalized. We believe that the research reported in this paper is a first step towards efficient parallel implementation of equational programming languages.

References

[1] HOFFMAN, C. AND O'DONNELL, M., Programming with Equations, *ACM Transactions on Programming Languages and Systems (1982) pp. 83-112.*

[2] HUET, G. AND LEVY, J.J., Computations in Nonambiguous Linear Term Rewriting Systems, *Tech. Rep. No. 359(1979), INRIA, Le Chesney, France.*

[3] KLOP, J.W. AND MIDDELDORP, A, Strongly Sequential Term Rewriting Systems, *Report CS-R8730, Centre for Mathematics and Computer Science, Amsterdam.*

[4] KOKICHI FUTATSUGI, JOSEPH A. GOGUEN, JEAN-PIERRE JOUANNAUD, AND JOSE MESEGUER, Principles of OBJ2, *Proc. 12th ACM Symposium on Principles of Programming Languages (1985).*

[5] O'DONNELL, M.J., Term-Rewriting Implementation of Equational Logic Programming, *Rewriting Techniques and Applications (1987) pp. 1-12.*

[6] O'DONNELL, M.J., Computing in Systems described by Equations, *Springer LNCS 58 (1977).*

[7] O'DONNELL. M.J., Equational Logic as a Programming Language, *MIT Press (1985).*

[8] JOSEPH GOGUEN, CLAUDE KIRCHNER AND JOSE MESEGUER, A Rewrite Rule Machine: Models of Computation for the Rewrite Rule Machine, *SRI International, July 1986.*

[9] SATISH THATTE, A Refinement of Strong Sequentiality for Term Rewriting With Constructors, *Information and Computing 72(1), 1987.*

[10] SATISH THATTE, On the correspondence between two classes of Reduction systems, *Information Processing Letters 20 (2), pp. 83-85 (1985).*

[11] SHAUNAK PAWAGI, R. RAMESH AND I.V. RAMAKRISHNAN, R^2M : A Reconfigurable Rewrite Machine, *Second International Workshop on Unification, June 1988.*

EFFICIENT GROUND COMPLETION:
An $O(n \log n)$ Algorithm for Generating Reduced Sets of Ground Rewrite Rules Equivalent to a Set of Ground Equations E

Wayne Snyder
Boston University
Department of Computer Science
111 Cummington St.
Boston, MA 02215

Abstract. We give a fast method for generating reduced sets of rewrite rules equivalent to a given set of ground equations. Since, as we show, reduced ground rewrite systems are in fact canonical, this is essentially an efficient Knuth-Bendix procedure for the ground case. The method runs in $O(n \log n)$, where n is the number of occurrences of symbols in E. We also show how our method provides a precise characterization of the (finite) collection of all reduced sets of rewrite rules equivalent to a given ground set of equations E, and prove that our algorithm is complete in that it can enumerate every member of this collection. Finally, we show how to modify the method so that it takes as input E and a total precedence ordering on the symbols in E, and returns a reduced rewrite system contained in the lexicographic path ordering generated by the precedence.

1 Introduction

The theory of ground equality systems has developed in two rather distinct strands. On the one hand, a number of researchers [9,20,23,26] studied word problems in ground equational theories as the problem of calculating the *congruence closure* of a relation on a graph. On the other, researchers in term rewriting theory [5,7,18,21,25,24] were interested in ground rewriting systems as a simple special case of the general theory. Few attempts have been made to connect the two approaches (but see [7]). Recently, this author and four others needed an algorithm to 'compile' a set of ground equations E into a canonical set of rewrite rules R in polynomial time; this was a crucial step in showing the NP-completeness of a new form of E-Unification called *Rigid E-Unification*, invented by Jean Gallier [11, 12, 13, 14]. The paper [15] presented an algorithm which took as input some E and a reduction ordering $\succ$ total on ground terms, and returned an equivalent R canonical under $\succ$ in $O(n^3)$, where n is the size of E.

The current paper is an attempt to extend these results and, more generally, to investigate reduced ground rewriting systems. We present an algorithm which compiles a set of ground equations E into a reduced set of rewrite rules R in $O(n \log n)$, where no reduction ordering is necessary. The algorithm works by generating a graph representation for the set E and using two passes of the congruence closure algorithm to reduce the graph representation itself into a certain minimal form, from which a set of reduced rules may be extracted. It will turn out that this method can find all such sets, and in fact provides a rather nice combinatorial characterization of the (finite) set of all reduced sets of rules equivalent to E. Finally, we show how this algorithm can be modified to incorporate the lexicographic path ordering.

2 Preliminaries

We present in this section a brief review of the necessary definitions; for a more detailed presentation, see [10], [17], or [23].

Definition 2.1 For a given signature Σ, let the set of function symbols of arity n be denoted by Σ_n and the set of all ground terms (hereafter simply called *terms*) on Σ be T_Σ. For any terms s, t and tree address α in s, let s/α denote the subterm of s rooted at α, and let $s[\alpha \leftarrow t]$ denote the the result of replacing the subterm s/α in s by t.

Definition 2.2 Let $R \subseteq T_\Sigma \times T_\Sigma$ be a binary relation on terms. Given any $s, t \in T_\Sigma$, we say that s *rewrites to* t, denoted by $s \longrightarrow_R t$, iff there is some $(l, r) \in R$ and some address α in s such that $s/\alpha = l$ and $t = s[\alpha \leftarrow r]$. (Such a *rewrite step* is fully specified by $s \longrightarrow_{[\alpha, l \dot{\rightarrow} r]} t$.) We denote the reflexive and transitive closure, the symmetric closure, and the reflexive, symmetric, and transitive closure by $\overset{*}{\longrightarrow}_R$, $\longleftrightarrow_R$, and $\overset{*}{\longleftrightarrow}_R$, respectively. When we focus on the non-symmetric relations $\longrightarrow_R$ and $\overset{*}{\longrightarrow}_R$ (or their converses $\longleftarrow_R$ and $\overset{*}{\longleftarrow}_R$), emphasizing the *oriented* use of pairs in R, we call (l, r) a *rewrite rule* and use the notation $l \dot{\rightarrow} r$. A set R of such rules is called a *ground term rewriting system* (abbreviated *gtrs*).

On the other hand, when we wish to emphasize the symmetric relations $\longleftrightarrow_R$ and $\overset{*}{\longleftrightarrow}_R$, which correspond to the unoriented use of pairs, we denote (s, t) by $s \doteq t$ and call it an *equation*; we shall use E (rather than R) to denote a set of equations. Let the *size* of a set E be defined as the number of occurrences of function and constant symbols in E. It is well known that $\overset{*}{\longleftrightarrow}_E$ is the least congruence on T_Σ containing E.

The word problem for (ground) E (i.e., to decide for arbitrary terms s and t whether $s \overset{*}{\longleftrightarrow}_E t$) was shown to be decidable by [1], but a practical decision procedure was not given until a number of researchers [9,20,23,26] studied the problem and independently invented the *congruence closure* method. In this approach, to decide if $s \overset{*}{\longleftrightarrow}_E t$, we construct a graph encoding the subterm relationships of s and t and all the terms in E, and then propagate the equational consequences of E over the vertices of the graph. This motivates the following definition.

Definition 2.3 A *subterm graph* is a directed, labeled, acyclic multigraph $G = (V, A, \lambda)$, (hereafter denoted simply (V, A)) where V is the set of vertices, $A \subseteq V \times V$ is the multiset of (directed) arcs, and $\lambda : V \to \Sigma$ is a function which assigns a label to each vertex such that if $outdegree(v) = n$, then $\lambda(v) \in \Sigma_n$. The multiset of arcs leaving a vertex are ordered; we use v/i to denote the ith *successor* of vertex v, i.e., the target of the ith arc leaving v, and $succ(v)$ is defined as the ordered list $\langle v/1, \ldots, v/n \rangle$ of the successors of v.

The interpretation of the vertices as terms is given by the function $\widehat{\ } : V \longrightarrow T_\Sigma$, which is defined by recursion on the (well-founded) successor relation on V. For any $v \in V$, if $outdegree(v) = 0$, then (by abuse of notation) $\widehat{v} = \lambda(v)$; if $outdegree(v) = n > 0$ and $\lambda(v) = f$, then $\widehat{v} = f(\widehat{v/1}, \ldots, \widehat{v/n})$. This function is extended to apply to subsets of V and even to relations on V. Let $ST(E)$ be defined as the set of all subterms (proper and otherwise) of terms occurring in E, and extend this notation to individual terms. (Note that the cardinality of $ST(E)$ is no greater than the size of E.) If $\widehat{V} = ST(E)$ and each vertex has a unique denotation in $ST(E)$ (i.e., the function $\widehat{\ }$ is bijective) then G is called a *subterm graph for* E, denoted by G_E.

Thus, a subterm graph for E is just a *structure sharing* diagram of the terms in E, where structure sharing arises because we defined $ST(E)$ as a *set* and because $\widehat{\ }$ is a bijection. For example, if $E = \{f^3a \doteq a, f^2a \doteq a, gc \doteq fa, gha \doteq gc, c \doteq ha, b \doteq mfa\}$, then $ST(E) = \{a, b, c, fa, f^2a, f^3a, gc, ha, gha, mfa\}$, and G_E is represented pictorially as

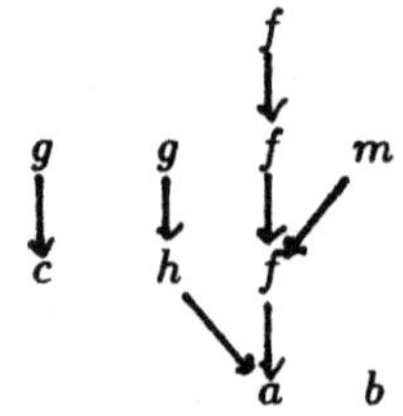

(For the rest of this paper, we shall use this set E as a running example.)

Definition 2.4 Let $G = (V, A)$ be a subterm graph. An equivalence relation Π over V is a *congruence on* V iff $(u, v) \in \Pi$ whenever $\lambda(u) = \lambda(v) \in \Sigma_n$ and $(u/i, v/i) \in \Pi$ for $1 \leq i \leq n$. The *congruence closure* of a given relation Π on V is the finest congruence on V containing Π.

It is well-known that for any relation Π on a graph G, there exists a unique congruence closure Π^* of Π, and congruence closure algorithms are given in [23] and [9].

The latter can be implemented with a worst-case time of $O(|A|\,log\,|A|)$; the former runs in $O(|A|^2)$ but seems to be faster in practice. The motivation for this is that a word problem $s \overset{*}{\longleftrightarrow}_E t$ can be decided by finding the congruence closure Π of the relation

$$\Pi_E \;=\; \{(u,v) \,|\, (\widehat{u},\widehat{v}) \in E\}$$

induced by E on the subterm graph for E, s, and t (i.e., where $\widehat{V} = ST(E,s,t)$); since $|A| \leq n$, where n is the combined size of E, s, and t, this can be done with a worst case time of $O(n\,log\,n)$.

In fact we shall be interested in this paper in simply calculating the congruence closure of E over $ST(E)$, so this means the relation $\overset{*}{\longleftrightarrow}_E$ restricted to $ST(E)$ can be calculated in $O(n\,log\,n)$, where n is the size of E. We thus assume we have available a procedure (such as detailed in [9]) which, given a set of equations E, returns a subterm graph G_E and associated congruence Π, which is the congruence closure of Π_E on V; this may be interpreted as a congruence $\widehat{\Pi}$ on $ST(E)$, which we shall represent by a partition of the set $ST(E)$. For example, using the E of our running example above, we would obtain the following congruence on G_E:

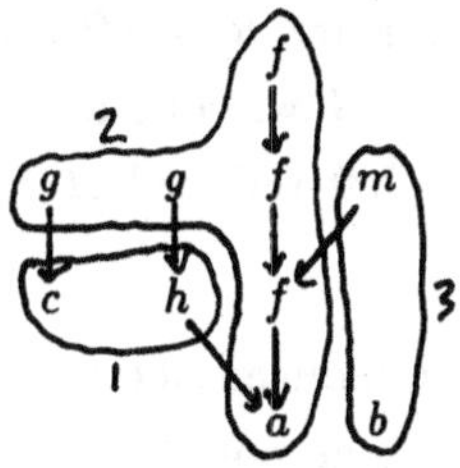

The associated partition is $\widehat{\Pi} = \{[c,ha]_1, [a,fa,f^2a,f^3a,gc,gha]_2, [b,mfa]_3\}$. The essential fact about G_E and Π is that they are logically equivalent to E in the following sense.

Lemma 2.5 For a given set E, let Π be the congruence closure of Π_E on G_E. For any $s,t \in T_\Sigma$, let $s \longleftrightarrow_{G_E,\Pi} t$ iff there is some address α in s and some $u \in ST(E)$, where $[s/\alpha]_{\widehat{\Pi}} = [u]_{\widehat{\Pi}}$, such that $t = s[\alpha \leftarrow u]$. Let $\overset{*}{\longleftrightarrow}_{G_E,\Pi}$ be the reflexive and transitive closure of this relation. Then $\overset{*}{\longleftrightarrow}_{G_E,\Pi} \;=\; \overset{*}{\longleftrightarrow}_E$.

The proof is not hard and is omitted. The relation $\overset{*}{\longleftrightarrow}_{G,\Pi}$ for a subterm graph G and associated congruence Π (which will generally be denoted $\overset{*}{\longleftrightarrow}_G$, leaving the congruence implicit) will be central in showing the soundness of our method.

3 Generating Reduced Sets of Rewrite Rules

Before we present our method, we review some basic definitions and then prove a useful sufficient condition for a *gtrs* to be canonical.

Definition 3.1 A *gtrs* R is *noetherian* iff there exists no infinite sequence of terms $t_1, t_2, t_3, \ldots$ such that $t_1 \longrightarrow_R t_2 \longrightarrow_R t_3 \longrightarrow_R \ldots$, and it is *confluent* iff whenever $t_2 \overset{*}{\longleftarrow}_R t_1 \overset{*}{\longrightarrow}_R t_3$, there exists a term t_4 such that $t_2 \overset{*}{\longrightarrow}_R t_4 \overset{*}{\longleftarrow}_R t_3$. R is *canonical* iff it is noetherian and confluent. A term t is *irreducible* by R (or in *normal form*) if there exists no t' such that $t \longrightarrow_R t'$. A system R is *left-reduced* iff for every $l \dot{\rightarrow} r \in R$, l is irreducible by $R - \{l \dot{\rightarrow} r\}$; R is *right-reduced* iff for every $l \dot{\rightarrow} r \in R$, r is irreducible by R. R is called *reduced* iff it is left-reduced and right-reduced.

Theorem 3.2 If R is a reduced set of ground rewrite rules, then R is canonical.

Proof. Observe that if $t \longrightarrow_{[\alpha, l \dot{\rightarrow} r]} t'$, then t' can not be rewritten at or below α, since $t'/\alpha = r$ and R is reduced. Thus the set of potential rewrite addresses in t' is smaller than in t. So, if $t_1 \longrightarrow_{[\alpha_1, l_1 \dot{\rightarrow} r_1]} t_2 \longrightarrow_{[\alpha_2, l_2 \dot{\rightarrow} r_2]} \ldots$ is some sequence of rewrite steps, inductively define a sequence of sets $A_1 \supseteq A_2 \supseteq \ldots$ by letting A_1 be the set of all tree addresses in t_1, and, for $i \geq 1$, let $A_{i+1} = A_i - \{\beta \mid \beta \text{ is at or below } \alpha_i\}$. But then $|A_1| > |A_2| > \ldots$, because $\{\alpha_i\} \subseteq A_i - A_{i+1}$ for each i, and so the length of the rewrite sequence is no greater than the size of the (finite) term t_1. Thus if R is right-reduced then it must be noetherian. Now, since R is left-reduced, it has no critical pairs, and thus is locally confluent, so that by Newman's lemma (see [16]), it must be confluent. (Alternately, we could show that a left-reduced *gtrs* is confluent.) Thus R is canonical. □

It is always possible to find reduced sets of ground rewrite rules equivalent to a given set of ground equations E, as noted in [7], where the result is attributed to [21]. This is because the Knuth-Bendix procedure (e.g. in the abstract approach of [7] and [3,4]) restricted to ground rules does not make use of the overlap rule, and so only reductions, orientation of rules, and removal of rules are necessary. Since it is always possible to find reduction orderings total on ground terms [6], rules can always be oriented and failure is impossible. Unfortunately, a naive approach to this method can result in exponentially long reduction sequences; in [15] we presented an algorithm using an initial application of congruence closure in addition to a more sophisticated approach to finding redexes to reduce the time bound to $O(n^3)$ to find a *gtrs* canonical under a given reduction ordering total on ground terms. Our approach here is different from [15] in two respects: (1) it avoids standard rewriting techniques by using a graph representation and integrating the use of congruence closure more essentially into the reduction process, and (2) using the previous theorem, we can dispense with term orderings entirely, since we need only find a reduced system (but see Theorem 3.16). The result is that by a structural manipulation of the graph G_E and *two* applications of congruence closure, we can find a reduced, and hence canonical, *gtrs* in $O(n \log n)$.

Definition 3.3 (The Reduction Procedure) The basic method consists of four steps:

(1) Given a set of ground equations E, create the subterm graph $G_E = (V, A)$ and find the congruence closure Π of the relation Π_E on V induced by E.

(2) For each class in $\Pi = \{C_1, \ldots, C_n\}$, find a ***representative*** vertex for the class and for each class C_i in Π, repoint the arcs in A so that every arc pointing to a vertex in C_i now points to the representative for C_i. Call this new graph G_n.

(3) Find the congruence closure Id^* of the identity relation on V in G_n; this will be a congruence whose classes consist of vertices interpreted as identical (and hence redundant) terms. Produce the quotient graph $H = (V', A') = G_n/Id^*$ by removing redundant vertices to 'merge' the classes of Id^*. Let Π' be the restriction of Π to the new set of vertices V'.

(4) Generate a relation T on V' which consists of pairs (u, v) of distinct but congruent vertices u and v where v is a representative and u is not. Interprete the corresponding relation $R = \widehat{T}$ as a *gtrs*.

For the remainder of this section we discuss each step of the reduction procedure in detail and justify the time bound $O(n \log n)$ and the ***soundness*** of the method, i.e., that R is reduced and equivalent to E. Then we show a rather stronger ***completeness*** result, namely, that ***every*** reduced ***gtrs*** equivalent to E can be found by choosing the appropriate sequence of representatives in Step Two. Finally, we show how the method can be modified to make use of term orderings.

Step One: We have discussed this step in detail in the previous section. In [8] it is shown that a subterm graph $G_E = (V, A)$ for E can be produced in $O(n)$, where n is the size of E. Thus the total cost, including generating the associated congruence closure Π, is $O(n \log n)$. As shown above, this step is sound in the sense that $\overset{*}{\longleftrightarrow}_{G_E} = \overset{*}{\longleftrightarrow}_E$.

Step Two: First, we discuss the notion of a representative for a class.

Definition 3.4 For any subterm graph G_E with associated congruence Π, a ***sequence of representatives for*** Π is a sequence $\langle v_1, \ldots, v_n \rangle$ of vertices, exactly one from each class of Π, which defines an ordering $\{[v_1] < [v_2] < \ldots < [v_n]\}$ on the classes of Π such that for each v_i, $1 \leq i \leq n$, if $succ(v_i) = \langle v_i/1, \ldots, v_i/k \rangle$, then $[v_i/j] < [v_i]$ for each j, $1 \leq j \leq k$.

Lemma 3.5 For any subterm graph $G = (V, A)$ and relation Π on V, a sequence of representatives for Π can be found in $O(|V|)$.

Proof. (Sketch) The basic idea is to topologically sort the vertices in V and extract the least member of each class with respect to this ordering. It is not hard to show that this is a sequence of representatives, and that this can be done in $O(|V|)$ worst case time. □

In fact, not all sequences of representatives can be found using a topological sort, but the important fact for now is that we can find at least one. For instance, in our running example, denoting a vertex v by its term $\widehat{v}$, for the graph

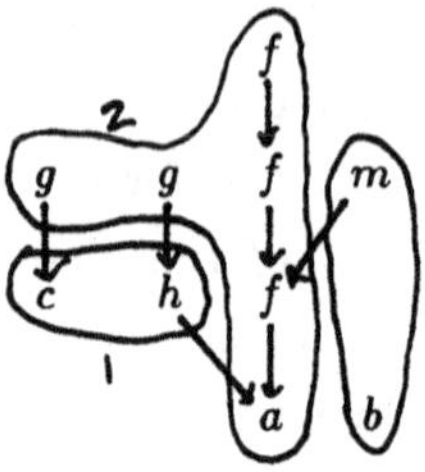

we have (among others) the sequences of representatives $\langle c, a, b\rangle$, $\langle a, ha, mfa\rangle$, and $\langle c, gha, b\rangle$, but this last can not be found by the method suggested above.

Now we modify G_E by 'instantiating' each class in Π by its representative.

Definition 3.6 For any $G_E = (V, A)$ with associated Π and sequence of representatives $\langle v_1, \ldots, v_n\rangle$, inductively define a sequence $A_0, \ldots, A_n$ as follows. Let $A_0 = A$, and, for $0 \leq i < n$, let $B_i = \{(u, w) | w \in [v_i]\}$ be the multiset of arcs with targets in $[v_i]$, and then define $A_{i+1} = (A_i - B_i) \cup \{(u, v_i) | (u, v) \in B_i\}$. Finally, let $G_n = (V, A_n)$.

For instance, using $\langle c, a, b\rangle$ on our example, we have G_n as follows:

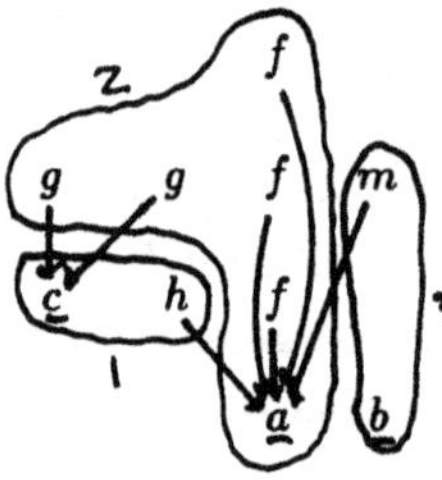

Note that Π is still a congruence on V, but now the relation $\widehat{\Pi}$ has changed; that this does not affect the soundness of the method is shown by

Lemma 3.7 The graph G_n as defined above is an acyclic graph, and the relation $\overset{*}{\longleftrightarrow}_{G_n}$ on T_Σ is such that $\overset{*}{\longleftrightarrow}_{G_n} = \overset{*}{\longleftrightarrow}_{G_E} = \overset{*}{\longleftrightarrow}_E$.

Proof. (Sketch) If G_n had a cycle, it would occur among the representatives, since only representatives are now targets of arcs in A_n; but it is easy to check that this violates the ordering condition of Definition 3.4. The rest of the proof consists of a verification using Lemma 2.5 that moving a single arc to another, congruent target does not change the global congruence on T_Σ, and concludes by induction on the number of arcs moved.
□

Another way of thinking about this step is that we 'inter-rewrite' the equations of E to obtain a sequence of equivalent sets of equations $E_0, E_1, \ldots, E_n$ where each G_i

would then (roughly) be the result of performing Step One on E_i. (The only difference is that, strictly speaking, the G_i may not be subterm graphs for the E_i, since $\widehat{}$ may not be injective.) For instance, in our running example, 'instantiating' the representative c to obtain G_1 corresponds to using the equation $c \doteq ha$ to rewrite $gha \doteq gc$ in E_0 to get $E_1 = \{f^3a \doteq a, f^2a \doteq a, gc \doteq fa, gc \doteq gc, c \doteq ha, b \doteq mfa\}$.

Since each arc is modified at most once, the cost of this step is $O(|A|)$.

Step Three: Note that in the set E_1 above, we have a redundant equation $gc \doteq gc$. The general problem is that in the graph G_n of our example, we have two distinct vertices with the same interpretation gc, i.e., the function $\widehat{}$ is not injective. Clearly, such redundant vertices (which will always be congruent) contribute no essential information to the congruence $\stackrel{*}{\longleftrightarrow}_{G_n}$, which is defined in terms of $\widehat{}$. Finding such redundant vertices is an example of the *common subexpression* problem [9], and can be solved using congruence closure. The formal justification for this is the following lemma, which is proved by an easy induction on the successor relation on the graph.

Lemma 3.8 Let Id^* be the congruence closure of the identity relation Id on V in any subterm graph G. Then for any two distinct vertices $u, v \in V$, $\widehat{u} = \widehat{v}$ iff $(u,v) \in Id^*$.

(Another way to think of Id^* is as the kernel relation of the function $\widehat{}$ viewed as a homomorphism.) Now we produce the graph H, which is effectively the quotient graph G_n/Id^*, by simply running through V and removing the redundant vertices and arcs.

Definition 3.9 For $G_n = (V, A_n)$ as defined above, for every $u \in V$ define $A_u = \{(u, u/i) | 1 \leq i \leq outdegree(u)\}$ as the set of arcs leaving u. Let $V' = V$ and $A' = A_n$ initially and then apply the following algorithm to (V', A'):

for each $u \in V'$ **do**
  if u is not a representative and $[u]_{Id^*}$ is not a singleton in V'
    then begin
      $V' := V' - \{u\}$;
      $A' := A' - A_u$
    end;

Now let H be the resultant graph (V', A'). (This is the quotient graph G_n/Id^*.)

In our running example we would have:

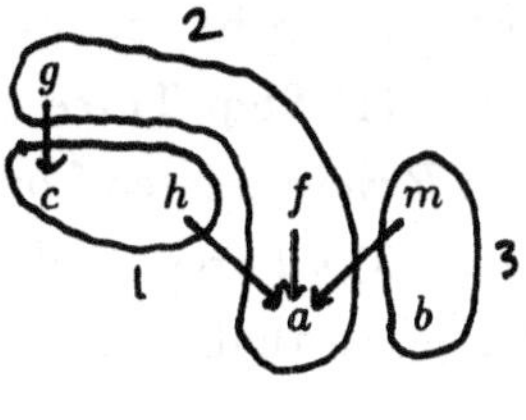

Clearly, the transformation of G_n into H can be done in $O(|A_n| \log |A_n| + |A_n| + |V|)$. Soundness is shown by

Lemma 3.10 Let $H = (V', A')$ be as in the previous definition, and let Π' be the restriction of Π to V'. Then $\overset{*}{\longleftrightarrow}_{H,\Pi'} = \overset{*}{\longleftrightarrow}_{G_n,\Pi}$.

Proof. We simply use Lemma 3.8 to show that $\widehat{\Pi'} = \widehat{\Pi}$. □

Step Four: It seems intuitively plausible that the graph H is unique up to the picking of representatives and that it represents the 'information content' of E in a certain minimal form. Thus is is not surprizing that we can extract from this minimal representation a minimal set of rewrite rules, i.e., a reduced *gtrs*.

Definition 3.11 For H and Π' as above, define a subset of Π'

$$T = \{(u,v) \in \Pi' \mid v \text{ is a representative and } u \text{ is not}\}.$$

In other words, T connects the non-representatives in non-trivial classes with their representatives. (Note that the trivial classes of Π' will contribute nothing to T.) Finally, interprete the relation $R = \widehat{T}$ as a set of ground rewrite rules.

For instance, our running example produces the set of rules

$$R = \{fa \dot{\rightarrow} a,\ gc \dot{\rightarrow} a,\ ma \dot{\rightarrow} b,\ ha \dot{\rightarrow} c\}.$$

Clearly the extraction of T from H and Π' can be done in time $O(|V'|)$, as we can simply run through all the non-representatives residing in non-trivial classes of Π' and pair them up with the representative of their class. Soundness is shown by

Lemma 3.12 For H, Π', and R as above, we have $\overset{*}{\longleftrightarrow}_R = \overset{*}{\longleftrightarrow}_{H,\Pi'}$.

Proof. This follows from the fact that $(u,v) \in \Pi'$ iff $(u,v) \in (Id \cup T \cup T^{-1} \cup T \circ T^{-1})$, where Id is the identity relation on V'. □

This leads us to state our major soundness result.

Theorem 3.13 (Soundness) For any set of ground equations E, the method specified in Steps One through Four above results in a reduced *gtrs* R such that $\overset{*}{\longleftrightarrow}_R = \overset{*}{\longleftrightarrow}_E$.

Proof. Given the cascade of partial soundness results above, the only difficulty is in showing that R is in fact reduced. Let S be the multiset of all lhs's and rhs's of rules in R. The essential facts are that (1) Step Three assures us that every term in S is unique, (2) all the lhs's of rules in R were created from non-representatives, and all rhs's were created from representatives, (3) *all* arcs in A' point to representatives (i.e. rhs's) by virtue of Steps Two and Three, so that every term in S is either a constant or a compound term, all of whose subterms are rhs's of rules. Now suppose one rule reduces another as specified in Definition 3.1. This reduction can not take place at the root, by (1), so it must take place at a proper subterm. But this would imply that the lhs of the reducing rule occurs as a proper subterm of some term in S, which contradicts (2) and (3). Thus R must be reduced. □

Finally, it should be clear by collecting together the separate complexity results (and noting that both the size of the vertex set and the number of arcs are initially bounded by the size of E, and throughout the algorithm can only decrease, so that neither can exceed the size of the original E) that the dominating cost is the double application of congruence closure, so the cost of the whole algorithm is $O(n \log n)$, where n is the size of the original E. Interestingly, there are some sets of equations E of size n for which there are reduced equivalent *gtrs* of size $O(n^2)$, but it is possible to show (see [14]) that there are always reduced systems no larger than the original set. Thus our result does not mean that any reduced system can be literally *printed out* in $O(n \log n)$, but simply that we can find a representation T for a reduced system. This is similar to the situation in fast unification algorithms, where a 'triangular form' unifier can be found in $O(n)$, but the explicit unifier might be exponentially large. In both cases, the seeming contradiction arises out of the size of non-structure-sharing representations of solutions.

The second major result of this paper is a completeness theorem which states that our method is powerful enough to generate *every* reduced system equivalent to E.

Theorem 3.14 (Completeness) For any set E, let H and Π' be the result of Steps One through Three above. For any reduced *gtrs* R equivalent to E, there exists a sequence of representatives for Π' such that we can obtain R by reapplying Steps Two and Four above.

Proof. (Sketch) The complete proof is given in detail in the full paper. Let S' be the original sequence of representatives used to create H, and let R' be the result of applying Step Four to H and Π'. The proof proceeds in two parts to show that we can convert R' into R and thereby derive the appropriate sequence of representatives S which generate R. The first part shows that for any reduced *gtrs* R there exists a simplification ordering $\succ$ total on R-equivalent ground terms which contains it. Then we show that the following transformation:

$$\{l \dot{\rightarrow} r\} \cup R \implies \{r \dot{\rightarrow} l\} \cup R[r := l],$$

where r is not a subterm of l and where $R[r := l]$ denotes the result of replacing every occurrence of r by l in R (i.e., a single round of reductions using $r \dot{\rightarrow} l$), is sufficient to imitate the way the Knuth-Bendix procedure (restricted to ground equations) would proceed to convert R' into a reduced system R'' canonical under $\succ$. But then it can be shown that any two equivalent, reduced systems contained in the same simplification ordering are unique (see [22]), and so $R'' = R$. Thus R' can be converted into R by essentially 'flipping' the orientation of rules and doing single rounds of replacements. The second part of the proof shows how each single transformation in this process corresponds to changing one of the representatives in S', and concludes by induction on the length of the sequence of transformations to show that we can define the sequence S which corresponds to the reduced *gtrs* R. □

Corollary 3.15 For any set E of ground equations, there are at most 2^k reduced *gtrs* equivalent to E, where k is the number of equations in E. Furthermore, each such reduced system has the same number of rules.

Proof. The upper bound follows from the fact that the KB algorithm (in the ground case) never adds rules to a ground system while converting it into a reduced *gtrs*, and so the resultant system has no more than k rules. Since one reduced system can always be converted into another, equivalent one by the transformation given above, we can show that no trivial rules ever arise, and obviously no new rules are ever added, so all reduced sets equivalent to E have the same number of rules. The number of reduced systems is thus bounded by the number of possible orientations of the rules in a reduced R equivalent to E, i.e., by the number of possible sequences of representatives. □

In the full paper (in preparation) we give an algorithm for enumerating all possible sequences of representatives for a subterm graph, and although there is not room to present it here, it should be intuitively clear that such an algorithm exists (for example, one can simply enumerate all possible sequences consisting of exactly one vertex from each class, and check whether it is a sequence of representatives by looking for cycles 'modulo the congruence') and so our completeness result shows that our method can in fact enumerate the complete set of all reduced *gtrs* equivalent to a given set E.

These results give us a nice combinatorial view of the set of all reduced *gtrs* equivalent to a given set of ground equations E. Each intuitively consists of a different way of 'centering' the congruence classes of the minimal graph H around normal forms (i.e., the rhs's of the rules), and we can move from one system to another by switching the normal forms (i.e., the representatives) by moving pointers, which corresponds to flipping a rule and performing a single round of reductions. For instance, our running example resulted in the graph H given above which has the following allowable sequences of representatives (where we have eliminated redundant sequences arising from inessential permutations of the same representatives, e.g. $\langle a, b, c\rangle$ and $\langle b, c, a\rangle$):

$(i)\,\langle a,b,c\rangle$, $(ii)\langle a,b,ha\rangle$, $(iii)\,\langle a,c,ma\rangle$, $(iv)\,\langle a,ma,ha\rangle$, $(v)\,\langle b,c,gc\rangle$, $(vi)\,\langle c,gc,ma\rangle$, which correspond (after applying Steps Two and Four to each of these) to the following complete set of reduced *gtrs* equivalent to our original E:

(v)		(i)		(ii)
$fgc \dot{\rightarrow} gc$ $a \dot{\rightarrow} gc$ $mgc \dot{\rightarrow} b$ $hgc \dot{\rightarrow} c$	$\overset{2}{\Longleftrightarrow}$	$fa \dot{\rightarrow} a$ $gc \dot{\rightarrow} a$ $ma \dot{\rightarrow} b$ $ha \dot{\rightarrow} c$	$\overset{4}{\Longleftrightarrow}$	$fa \dot{\rightarrow} a$ $gha \dot{\rightarrow} a$ $ma \dot{\rightarrow} b$ $c \dot{\rightarrow} ha$
$\Updownarrow 3$		$\Updownarrow 3$		$\Updownarrow 3$
(vi)		(iii)		(iv)
$fgc \dot{\rightarrow} gc$ $a \dot{\rightarrow} gc$ $b \dot{\rightarrow} mgc$ $hgc \dot{\rightarrow} c$	$\overset{2}{\Longleftrightarrow}$	$fa \dot{\rightarrow} a$ $gc \dot{\rightarrow} a$ $b \dot{\rightarrow} ma$ $ha \dot{\rightarrow} c$	$\overset{4}{\Longleftrightarrow}$	$fa \dot{\rightarrow} a$ $gha \dot{\rightarrow} a$ $b \dot{\rightarrow} ma$ $c \dot{\rightarrow} ha$

The numbered arrows indicate which rule (numbered top to bottom) was flipped in the transition from one system to another. Note that collapse rules (where the rhs is a subterm of the lhs) can never be flipped, as the resulting system would not be reduced; sometimes, a given rule can be a collapse rule in one system and not in another, while some rules (e.g., $fa \dot{\rightarrow} a$) are always collapse rules and can never be reoriented. This is all due to the cycles among the classes of the partition Π' and provides another intuition for the ordering conditions placed on sequences of representatives in Definition 3.4.

It is interesting to note that for any reduction ordering total on ground terms, one of these systems will be canonical under the ordering. In the full paper, we present a modified version of our original method which incorporates term orderings. Unfortunately there is not room to pursue this here, and so we sketch this result as follows.

Theorem 3.16 For any set of ground equations E and total precedence ordering $\prec$ on the symbols of E, an equivalent reduced *gtrs* R contained in the *lexicographic path ordering* [19, 6] can be found in a worst-case time of $O(n \log n)$.

Proof. (Sketch) By the results presented in [15], if we select our sequence of representatives in Step Two such that $r_1 \prec_{lpo} r_2 \prec_{lpo} \cdots \prec_{lpo} r_n$, then R will be contained in the *lpo*. As discussed in [15], this selection process works by picking, at each stage, the least term (i.e., vertex) among all terms in the graph residing in non-trivial classes for which representatives have yet to be picked. Now, let a term residing in such a class, all of whose immediate subterms are representatives previously chosen, be called a *candidate*. It can be shown that, at each stage, the next representative must be among the candidates. We therefore use a 2-3 tree as a *priority queue* to store the ordered list of all candidates, and show how to suitably implement the ordering so that insertions can be done efficiently. We start by inserting all constants into the queue, selecting the least constant as the first representative. Thereafter, as each representative is chosen and 'repointing' is performed, we remove all the other terms in the new representative's class from the queue, and insert into the queue all of the terms which just became candidates. The representative at each stage is just the first term in the queue, and so can be found in $O(1)$. Now each term is inserted and deleted from the queue at most once, and it can be shown that a sequence of $|V|$ insertions and $|V|$ deletions can be done in $O(n \log n)$, where n is the size of E (and $n \geq |V|$). □

4 Conclusion

We have presented an algorithm which takes as input a set of ground equations E and produces a reduced (and thus canonical) set of ground rewrite rules R equivalent to E in a worst case time of $O(n \log n)$, where n is the size of E. Furthermore, we have proved that this algorithm can enumerate the complete set of all reduced *gtrs* equivalent to E, a result which tells us much about the structure of this set. Finally, we have shown that this algorithm can be modified to find reduced *gtrs* contained in a given *lpo* in the same worst-case time.

Although we hope this research contributes to the theoretical understanding of rewrite and equality systems, by relating the fundamental concepts of minimality, termination, and confluence in the ground case, we also hope the methods developed here might have practical application in term rewriting and theorem proving. For example, reduction procedures for ground equations are of great importance to the concept of ***Rigid E-Unification*** which was invented by Jean Gallier in extending the method of *matings* [2] to first-order languages with equality (see [11, 12, 13, 14]). One possible line of future research is to investigate the use of this new method for decision procedures for Rigid E-Unification.

The algorithm suggested in the last theorem (which we present in detail in the full paper) is perhaps more in the spirit of a ground Knuth-Bendix procedure than the original method, since in some cases, we may be concerned with producing a system noetherian under a specific ordering; it seems possible to adapt the method to work with orderings other than the *lpo*, but this is left for further research. It would be interesting to investigate whether this last algorithm might have some application as a subsystem within an implementation of the Knuth-Bendix completion procedure for the general case. In any case, the last result shows that the method can be used to reduce ground terms to particular normal forms which are of interest, if these can be specified via the *lpo* generated by the given precedence; note that the first method (inspired by Theorem 3.2) is perhaps too general in that it produces *gtrs* with arbitrary normal forms. It would be interesting to explore this approach in the more general context of rewrite (equational) programs where the normal forms must be in a specific domain (e.g., built from symbols considered to be constructors in the larger context, so that normal forms don't contain 'defined' symbols); the algorithm could be modified in various ways depending on the specification of this domain to generate *gtrs* which produce normal forms with the desired property.

Remark: One of the anonymous referees commented that the graph-based approach used here for generating reduced systems bears some resemblence to the details of the implementation of congruence closure in Greg Nelson's thesis. Although I was not able to examine this work at the time of submitting this report, a close comparison of the two approaches will be made and reported on by the time of the conference and in the full paper in preparation.

5 References

[1] Ackermann, W., *Solvable Cases of the Decision Problem.* North-Holland, Amsterdam (1954).

[2] Andrews, P.B., "Theorem Proving via General Matings," JACM 28:2 (1981) 193-214.

[3] Bachmair, L. *Proof Methods for Equational Theories*, Ph.D thesis, University of Illinois, Urbana Champaign, Illinois (1987).

[4] Bachmair, L., Dershowitz, N., and Hsiang, J., "Orderings for Equational Proofs," In *Proc. Symp. Logic in Computer Science*, Boston, Mass. (1986) 346-357.

[5] Dauchet, M., Tison, S., Heuillard, T., and Lescanne, P., "Decidability of the Confluence of Ground Term Rewriting Systems," *LICS'87*, Ithaca, New York (1987) 353-359.

[6] Dershowitz, N,. "Termination of Rewriting," *Journal of Symbolic Computation* 3 (1987) 69-116.

[7] Dershowitz, N,. "Completion and its Applications," Proceedings of CREAS, Lakeway, Texas (May 1987).

[8] Downey, P.J., Samet, H., and Sethi, R., "Off-line and On-line Algorithms for Deducing Equalities," POPL-5, Tucson, Arizona (1978).

[9] Downey, P.J., Sethi, R., and Tarjan, E.R., "Variations on the Common Subexpressions Problem," JACM 27:4 (1980) 758-771.

[10] Gallier, J.H. *Logic for Computer Science: Foundations of Automatic Theorem Proving*, Harper and Row, New York (1986).

[11] Gallier, J.H., Raatz, S., and Snyder, W., "Theorem Proving using Rigid *E*-Unification: Equational Matings," *LICS'87*, Ithaca, New York (1987) 338-346.

[12] Gallier, J., Narendran, P., Raatz, S., and Snyder, W., "Theorem Proving using Equational Matings and Rigid E-Unification," submitted to JACM (1988).

[13] Gallier, J.H., Narendran, P., Plaisted, D., and Snyder, W., "Rigid E-Unification is NP-complete," *LICS'88*, Edinburgh, Scotland (July 1988)

[14] Gallier, J.H., Narendran, P., Plaisted, D., and Snyder, W., "Rigid E-Unification: NP-Completeness and Applications to Equational Matings," submitted to Information and Computation (1988).

[15] Gallier, J.H., Narendran, P., Plaisted, D., Raatz, S., and Snyder, W., "Finding Canonical Rewriting Systems Equivalent to a Finite Set of Ground Equations in Polynomial Time," CADE-9, Argonne, Ill. (1988) (Journal version submitted to JACM (1988).)

[16] Huet, G., "Confluent Reductions: Abstract Properties and Applications to Term Rewriting Systems," JACM 27:4 (1980) 797-821.

[17] Huet, G. and Oppen, D. C., "Equations and Rewrite Rules: A Survey," in *Formal Languages: Perspectives and Open Problems*, R. V. Book (ed.), Academic Press, NY (1982).

[18] Huet, G., Lankford, D., "On the Uniform Halting Problem for Term Rewriting Systems," Rapport de Recherche 283 (March 1978).

[19] Kamin, S., and Levy, J.-J., "Two Generalizations of the Recursive Path Ordering," unpublished note, Department of Computer Science, University of Illinois, Urbana, IL.

[20] Kozen, D., "Complexity of Finitely Presented Algebras," Technical Report TR 76-294, Department of Computer Science, Cornell University, Ithaca, NY (1976).

[21] Lankford, D.S., "Canonical Inference," Report ATP-32, University of Texas (1975)

[22] Metivier, Y., "About the Rewriting Systems Produced by the Knuth-Bendix Completion Algorithm," Information Processing Letters 16 (1983) 31-34.

[23] Nelson G., and Oppen, D. C., "Fast Decision Procedures Based on Congruence Closure," JACM 27:2 (1980) 356-364.

[24] Oyamaguchi, M., "The Church-Rosser Property for Ground Term-Rewriting Systems is Decidable," TCS 49 (1987) 43-79.

[25] Rosen, B., "Tree-Manipulating Systems and Church-Rosser Theorems," JACM 20 (1973) 160-187.

[26] Shostak, R., "An Algorithm for Reasoning about Equality," CACM 21:7 (July 1978) 583-585.

Extensions and Comparison of Simplification orderings

Joachim Steinbach
Universität Kaiserslautern, Fachbereich Informatik
Postfach 3049, D-6750 Kaiserslautern, FRG

The effective calculation with term rewriting systems presumes termination. Orderings on terms are able to guarantee termination. This report deals with some of those term orderings : Several path and decomposition orderings and the Knuth-Bendix ordering. We pursue three aims :

- Known orderings will be newly defined.
- New ordering methods will be introduced : We will extend existing orderings by adding the principle of status (see [KL80]).
- The comparison of the power as well as the time behaviour of all orderings will be discussed.

1 Introduction

Term rewriting systems (TRS, for short) gain more and more importance because they are a useful model for non-deterministic computations (since they are based on directed equations with no explicit control), with various applications in many areas of computer science and mathematics. Automatic theorem proving and program verification, abstract data type specifications and algebraic simplification, to name a few, are based on this concept.

The simplicity of the semantic of a TRS is guaranteed whenever the result of such a computation does not depend on the choice of the rules to be applied. This property is called *confluence* and is related with the so-called *Church-Rosser* property that justifies the possible solution of the word problem by checking the equality of *normal forms* (irreducible terms). If a TRS is not confluent, it can sometimes be transformed into a confluent one using the *Knuth-Bendix completion* procedure (cf. [KB70]) which adds new rules (non-convergent *critical pairs* that are derived from the *overlappings* of two left members of rules) to the initial rewrite system. Unfortunately, the successful use of this process crucially depends on the ability of proving the termination of a TRS. In general, the termination of an arbitrary TRS is an *undecidable* property, even if the number of rules is bounded by 1 ([Da88], [De85]). Thus, the best we can hope for are different strategies which are together able to cope with many rewrite rule systems occurring in practice. These methods are based on verifying that the rewrite relation $\Rightarrow_{\mathfrak{R}}$ is included in an ordering on terms. Such an ordering must be *well-founded* to prevent derivations of infinite length. To check the inclusion '$\Rightarrow_{\mathfrak{R}} \subseteq >$' all (infinitely many) possible derivations must be tested. The key idea is to restrict this infinite test to a finite one. For that purpose we have to require a *reduction ordering* > (a well-founded ordering that is *compatible* with the structure of terms) *stabilized* with respect to (w.r.t.) substitutions (cf. [La77]). Guaranteeing these properties is very difficult. This fact leads to the basic idea of characterizing classes of orderings for which there is no need to prove these conditions. One possible solution is represented by the class of *simplification orderings* ([De82]) which are at least reduction orderings. A partial ordering is a simplification ordering if it has two characteristics: The compatibility and the *subterm property* (any term is greater than any of its proper subterms). A great number of simplification orderings has been defined. Most of them are *precedence orderings* using a special ordering on operators (called precedence).

After giving some basic notations in section 2, we will deal with the definitions of well-known and new simplification orderings. All orderings presented are connected by an essential characteristic : Each operator f has a *status* $\tau(f)$ that determines the order according to which the subterms of f are compared ([KL80]). Formally, status is a function which maps the set of operators into the set {mult , left , right}. Therefore, a function symbol can have one of the following three statuses : *Mult* (the arguments will be compared as multisets), *left* (lexicographical comparison from left to right) and *right* (the arguments will lexicographically be compared from right to left).

Each definition of the orderings in chapter 3 will be preceded by an abstract verbal description of the way two terms are compared. The list of orderings consists of the following well-known ones : The recursive path ordering with status of Dershowitz, Kamin and Lévy ([De82], [KL80]), the path of subterms ordering of Plaisted and Rusinowitch ([Pl78], [Ru87]), the path ordering with status of Kapur, Narendran and Sivakumar ([KNS85]). A new recursive decomposition ordering with status will lead off the catalogue of orderings introduced in this paper. It is stronger than the one of [Le84]. [St88b] contains an ordering on decompositions equivalent to the path of subterms ordering. The ordering on decompositions has an advantage over the corresponding ordering on paths : The combination with the principle of status is much easier. This extension (by incorporating status) will be presented in this paper. We have also added the principle of status to the improved recursive decomposition ordering of Rusinowitch ([Ru87]). Furthermore, we will deal with the weight oriented ordering of Knuth and Bendix by incorporating status.

The orderings based on decompositions will be presented in a new and simple style : The decomposition of a term consists of terms only. The original definitions take tuples composed of three (or four) components. At the end of chapter 3, important properties of the newly introduced orderings will be listed, including simplification ordering and stability w.r.t. substitutions.

Besides the introduction of new orderings, the second main point of this paper is the comparison of the power of the given orderings (see chapter 4), i.e. we will examine the sets of comparable terms for each combination of two orderings. This will be done w.r.t. an underlying fixed total precedence and irrespective (unrelated to) the precedence.

An implementation of the orderings is integrated into our rewrite rule laboratories *TRSPEC* (a **t**erm **r**ewriting based system for algebraic **spec**ifications) and *COMTES* (**com**pletion of **te**rm rewriting **s**ystems). A series of experiments has been conducted to study the time behaviour of the orderings. An evaluation of these results concludes the report.

2 Preliminary notations

In order to mark definitions they will be printed in italics. A *term rewriting system* $\mathfrak{R}$ over a set of terms Γ is a finite or countably infinite set of rules, each of the form $l \to_{\mathfrak{R}} r$, where l and r are terms in Γ, such that every variable that occurs in r also occurs in l. The *set Γ of all terms* is constructed from elements of a set $\mathcal{F}$ of *operators* (or *function symbols*) and some denumerably infinite set $\mathcal{X}$ of *variables*. The set of *ground terms* (terms without variables) is denoted by Γ_G. The leading function symbol and the tuple of the (direct) arguments of a term t are referred to by *top*(t) and *args*(t), respectively.

A *substitution* σ is defined as an endomorphism on Γ with the finite domain $\{x \mid \sigma(x) \neq x\}$. The structure of a term is partially altered by rule application. Consequently, it is advantageous to have a precise scheme for specifying how and what particular part of it is to be changed. For this, we use the formalism of labelling terms with positions which are sequences of non-negative integers. The set of all positions of a term t is called the set of *occurrences* and its abbreviation is *O*(t). *Ot*(t) denotes the set of all *terminal occurrences* (occurrences of the leaves) of the term t.

A *precedence* is a partially ordered set $(\mathcal{F}, \triangleright)$ consisting of the set $\mathcal{F}$ of operators and an irreflexive and transitive binary relation $\triangleright$ defined on elements of $\mathcal{F}$. We consider the precedence p to be a parameter of an ordering >, denoted by *>(p)*. If there is no ambiguity, we will use the notation > instead of >(p). A partial ordering > is said to be *total* if for any two distinct elements s and t, either $s > t$ or $t > s$ holds.

An ordering $\blacktriangleright$ on any set M is an *extension* of > if and only if $s > t$ implies $s \blacktriangleright t$ for all $s, t \in M$: We write $> \subseteq \blacktriangleright$.

Note that a partial ordering > is used to compare elements of any set M. Since operators have terms as arguments we define an extension of >, called *lexicographically greater* ($>^{lex}$), on tuples of elements as follows :

$$
\begin{array}{lll}
(m_1,m_2,\ldots,m_p) >^{lex} (n_1,n_2,\ldots,n_q) & & \\
\text{if either } p>0 & \wedge & q=0 \\
\quad \text{or } \quad m_1 > n_1 & & \\
\quad \text{or } \quad m_1 = n_1 & \wedge & (m_2,\ldots,m_p) >^{lex} (n_2,\ldots,n_q).
\end{array}
$$

If there is no order among the elements of such tuples then the structures over M^n are called multisets. *Multisets* resemble sets, but allow multiple occurrences of identical elements. The extension of > on multisets of elements is defined as follows : A multiset M_1 is greater than a multiset M_2 over M, denoted by

$$
\begin{array}{ll}
M_1 \gg M_2 & \\
\text{iff } \bullet\ M_1 \neq M_2 & \wedge \\
\quad\ \bullet\ (\forall y \in M_2 \backslash M_1)\,(\exists x \in M_1 \backslash M_2)\ x > y. &
\end{array}
$$

The result of an application of the function args to a term $f(t_1,\ldots,t_n)$ depends on the status of f: If $\tau(f) = \text{mult}$, then the multiset $\{t_1,\ldots,t_n\}$ is returned, otherwise args delivers the tuple $(t_1,\ldots,t_n)$.

To define the orderings, we need some kind of formalism. A *path* of a term is a sequence of terms starting with the whole term followed by a path of one of its arguments :

- $path_\varepsilon(\Delta) = \Delta$ if Δ is a constant symbol or a variable,
- $path_{i.u}(\ f(t_1,...,t_n)\ ) = f(t_1,...,t_n)\ ;\ path_u(t_i)$ if $u \in Ot(t_i)$.

Moreover, $path(\{t_1,...,t_n\}) = \{\ path_u(t_i) \mid i \in [1,n]\ ,\ u \in Ot(t_i)\ \}$ is the multiset of all paths of the specified terms $t_1,...,t_n$. A path will be enclosed in square brackets. For a path $p = [t_1;t_2;...;t_n]$, *set*(p) denotes the set $\{t_1,...,t_n\}$ of all terms in p. This set is also called *path-decomposition* and its abbreviation is $dec_u(t)$ (= $set(path_u(t))$). An element (i.e. a term) of a path-decomposition is named *elementary decomposition.* Analogous with paths, the *decomposition* $dec(\{t_1,...,t_n\}) = \{\ dec_u(t_i) \mid i \in [1,n]\ ,\ u \in Ot(t_i)\ \}$ is the multiset of all path-decompositions of the terms $t_1,...,t_n$. There are two operations on a path-decomposition $P \subseteq \Gamma$ to describe. The set of subterms and the set of superterms of P relative to a term t are defined as

- $sub(P\ ,\ t) = \{s \in P \mid (\exists u \neq \varepsilon)\ t/u = s\}$ and
- $sup(P\ ,\ t) = \{s \in P \mid (\exists u \neq \varepsilon)\ s/u = t\}$, respectively.

Analogous with decompositions we use sub and sup to denote subsequences of paths. Suppose $t = (x + y) * z$, then $path_{12}(t) = [t\ ;\ x + y\ ;\ y]$, $path(\{t\}) = \{path_{11}(t)\ ,\ path_{12}(t)\ ,\ path_2(t)\}$, $dec_2(t) = \{t\ ,\ z\}$, $sub(dec_2(t)\ ,\ z) = \emptyset$, and $sup(path_{11}(t)\ ,\ x + y) = [t]$.

When writing s, t and $\triangleright$ we will always assume that s and t are terms over Γ and $\triangleright$ is a precedence on the set $\mathcal{F}$ of operators. Moreover, we use $>_{ord}$ with ord to denote a representative of an ordering (e.g., ord = RPOS). The index $\tau(f)$ of $>_{ord,\tau(f)}$ marks the extension of $>_{ord}$ w.r.t. the status of the operator f :

$$
\begin{array}{llllll}
(s_1,...,s_m) & >_{ord,\tau(f)} & (t_1,...,t_n) & & & \\
\text{iff} & \tau(f) = \text{mult} & \wedge & \{s_1,...,s_m\} & \gg_{ord} & \{t_1,...,t_n\} \\
 & \text{or } \tau(f) = \text{left} & \wedge & (s_1,...,s_m) & >_{ord}^{lex} & (t_1,...,t_n) \\
 & \text{or } \tau(f) = \text{right} & \wedge & (s_m,...,s_1) & >_{ord}^{lex} & (t_n,...,t_1).
\end{array}
$$

Permitting variables, we have to consider each and every one of them as an additional constant symbol uncomparable (w.r.t. $\triangleright$) to all the other operators in $\mathcal{F}$.

Every ordering of this report defines a congruence $\sim$ depending on $\mathcal{F}$ and τ. All these congruences are equivalent : $f(s_1,...,s_m) \sim g(t_1,...,t_n)$ iff $f = g$ and $m = n$ and i) $\tau(f) = \text{mult}$ and there is a permutation π of the set $\{1,...,n\}$ such that $s_i \sim t_{\pi(i)}$, for all $i \in [1,n]$ or ii) $\tau(f) \neq \text{mult}$ $\wedge$ $s_i \sim t_i$, for all $i \in [1,n]$.

Most of the orderings are based on the principle of root ordering, i.e. two terms are compared depending on their leading function symbols. This or other kinds of case distinctions will be represented as the union of conditions that will each be marked by Roman numerals i), ii), and so on. The lexicographical performance of conditions will each be indicated by hyphens, i.e.

$$
\begin{array}{lll}
s > t & \text{iff} & -\ s >_1 t \\
 & & -\ s >_2 t
\end{array}
$$

stands for: $s > t$ iff $s >_1 t$ or $(s =_1 t \wedge s >_2 t)$. The equality sign $=_1$ is the congruence relation induced by the quasi-ordering $\geq_1$. A *quasi-ordering* $\geq$ is a transitive and reflexive binary relation which defines an equivalence relation $=$ as both $\geq$ and $\leq$, and a partial ordering $>$ as $\geq$ but not $\leq$.

3 Definitions of the orderings

All orderings described in this chapter are recursively defined simplification orderings. Most of them are well-known : A version with status exists for the recursive path ordering (RPO), the recursive decomposition ordering (RDO, a new and more powerful version of the RDO will be presented) and the path ordering of Kapur, Narendran and Sivakumar (KNS). We have added the principle of status to the others. The main point of this chapter is the description of all these orderings with status. For a better understanding, some of these methods of comparing terms will be demonstrated by an example at the end of the chapter. The orderings described satisfy properties that qualify them for proving termination of term rewriting systems : Well-foundedness, *stability w.r.t. substitutions* ($\forall \sigma$, s, t: $s > t$ implies $\sigma(s) > \sigma(t)$) and *monotony w.r.t. the precedence* ($\forall$p, q precedences : $p \subseteq q$ implies $>(p) \subseteq >(q)$, i.e. if the precedence is increased the ordering becomes stronger).

3.1 Path orderings

i) Recursive path ordering with status ([KL80] , [De82])

A comparison of terms with respect to the recursive path ordering with status (RPOS, for short) is based on the following idea : A term is decreased by replacing a subterm with any number of smaller terms which are connected by any structure of operators smaller (w.r.t. $\triangleright$) than the leading function symbol of the replaced subterm. The method of comparing two terms depends on their leading function symbols. The relationship w.r.t. $\triangleright$ between these operators and the status τ is responsible for decreasing one of the (or both) terms in the recursive definition of the RPOS. If one of the terms is 'empty' (i.e. totally decreased) then the other one must be greater.

Definition : RPOS

$s >_{RPOS} t$

iff i) $top(s) \triangleright top(t) \quad \wedge \quad \{s\} \gg_{RPOS} args(t)$

ii) $top(s) = top(t) \quad \wedge \quad \tau(top(s)) = mult \quad \wedge \quad args(s) \gg_{RPOS} args(t)$

iii) $top(s) = top(t) \quad \wedge \quad \tau(top(s)) \neq mult \quad \wedge \quad args(s) >_{RPOS,\tau(top(s))} args(t)$

$\wedge \quad \{s\} \gg_{RPOS} args(t)$

iv) $args(s) \succeq\!\!\succeq_{RPOS} \{t\}$

ii) Path of subterms ordering ([Ru87] , [Pl78] , [St88b])

Plaisted's path of subterms ordering (PSO, for short) is a predecessor of the RPO and compares two terms by comparing all their paths. A slightly modified version (equivalent to the original one) of Rusinowitch is given next.

Definition : PSO

$s >_{PSO} t$
iff $path(\{s\}) \gg_{PO} path(\{t\})$
with $p >_{PO} q$
iff $set(p) \gg_{T} set(q)$
with $s >_{T} t$
iff - $top(s) \triangleright top(t)$
- $path(args(s)) \gg_{PO} path(args(t))$

iii) Path ordering with status of Kapur, Narendran and Sivakumar ([KNS85])

Like the PSO, the KNSS (KNS with status) is an ordering which compares terms using their paths. It has been devised by Kapur, Narendran and Sivakumar. They have implemented the RPO within their rewrite rule laboratory and have found it weak in handling terms which should intuitively be comparable. The KNSS is a consequence of these experiments and it extends the RPOS.

Definition : KNSS

$s >_{KNSS} t$
iff $path(\{s\}) \gg_{LK} path(\{t\})$
with $p >_{LK} q$
iff $(\forall t' \in q)\ (\exists s' \in p)\ \ s' >_{LT} t'$
with $p \ni s >_{LT} t \in q$
iff i) $top(s) \triangleright top(t)$
ii) $top(s) = top(t) \quad \wedge \quad \tau(top(s)) = mult \quad \wedge$
- $sub(p , s) >_{LK} sub(q , t)$
- $path(args(s)) \gg_{LK} path(args(t))$
- $sup(p , s) >_{LK} sup(q , t)$
iii) $top(s) = top(t) \quad \wedge \quad \tau(top(s)) \neq mult \quad \wedge$
- $args(s) >_{KNSS,\tau(top(s))} args(t) \quad \wedge$
$\{s\} \gg_{KNSS} args(t)$
- $sup(p , s) >_{LK} sup(q , t)$

3.2 Decomposition orderings

i) Recursive decomposition ordering with status (based on [JLR82] , [Ru87])

Like the KNSS, the recursive decomposition ordering with status (RDOS, for short) has been developed from the RPO. One of the important differences to the RPO is the fact that the RDOS stops a comparison as soon as it has to compare incomparable operators. The RDOS defined here is different from that described in [Le84] : Both orderings are based on [JLR82] but the incorporations of status are different. Moreover, our various decomposition orderings (see the next three definitions) are founded on another decomposition : We use terms (cf. the definitions of chapter 2) instead of triples or even quadruplets. The original decomposition ordering works on quadruplets

which divide a term into the following parts : The leading function symbol, any selected immediate subterm, the rest of the immediate subterms, the context (which marks the occurrence where the decomposition takes place).
A term s is greater than a term t (w.r.t. RDOS) if the decomposition of s is greater than the decomposition of t. The ordering on these multisets ($\gg\gg_{LD}$) is an extension of the basic ordering on terms ($>_{LD}$) to multisets of multisets.

Definition : RDOS

$s >_{RDOS} t$
iff $dec(\{s\}) \gg\gg_{LD} dec(\{t\})$
with $dec_u(s') \ni s >_{LD} t \in dec_v(t')$
iff i) $top(s) \triangleright top(t)$
ii) $top(s) = top(t) \quad \wedge \quad \tau(top(s)) = mult \quad \wedge$
- $sub(dec_u(s'), s) \quad \gg_{LD} \quad sub(dec_v(t'), t)$
- $args(s) \quad \gg_{RDOS} \quad args(t)$
iii) $top(s) = top(t) \quad \wedge \quad \tau(top(s)) \neq mult \quad \wedge$
$args(s) >_{RDOS,\tau(top(s))} args(t) \quad \wedge \quad \{s\} \gg_{RDOS} args(t)$

ii) Path of subterms ordering on decompositions and with status ([St88a] , [St88b])
Another ordering based on decompositions results from the PSO. It is remarkable that the PSO is an extremely recursive ordering which takes three suborderings ($>_{PO}$, $>_T$ and $\triangleright$) into account. We have succeeded in redefining this path ordering in such a way that the result, called PSD, provides a much simpler method of using decompositions (cf. [St88b], [St88a]). The PSD has another advantage over the PSO : The combination with the concept of status is much easier. The PSD with status (PSDS, for short) as well as the PSD depends on the fact that a path is an ordered path-decomposition.

Definition : PSDS

$s >_{PSDS} t$
iff $dec(\{s\}) \gg\gg_{LP} dec(\{t\})$
with $s >_{LP} t$
iff i) $top(s) \triangleright top(t)$
ii) $top(s) = top(t) \quad \wedge \quad \tau(top(s)) = mult \quad \wedge \quad dec(args(s)) \gg\gg_{LP} dec(args(t))$
iii) $top(s) = top(t) \quad \wedge \quad \tau(top(s)) \neq mult \quad \wedge$
$args(s) >_{PSDS,\tau(top(s))} args(t) \quad \wedge \quad \{s\} \gg_{PSDS} args(t)$

iii) Improved recursive decomposition ordering with status ([St88a], based on [Ru87])
The relatively simple definition (PSDS) of the complicated PSO has two further advantages. In the first place, we can compare both implementations according to their efficiency (see paragraph 5.1). Secondly, it is easy to compare the PSO with the decomposition orderings, e.g. with the improved recursive decomposition ordering of Rusinowitch (so-called IRD, see next definition). The essential

difference between the PSO (= PSD) and the IRD lies in the way by which a comparison is processed. While the PSD works according to the principle of 'breadth-first', the IRD follows the principle of 'depth-first' : If the leading function symbols of the terms to compare are identical, the IRD chooses only one subterm. On the other hand, the PSD proceeds by simultaneously considering the multiset of the decompositions of all subterms.

In addition to the definition of Rusinowitch (see 5.2), we have incorporated status to the IRD (IRDS, for short), so that it is equivalent to the path ordering with status of Kapur, Narendran and Sivakumar (see chapter 4).

Definition : IRDS

$s >_{IRDS} t$

iff $dec(\{s\}) \ggg_{EL} dec(\{t\})$

with $dec_u(s') \ni s \;\; >_{EL} \;\; t \in dec_v(t')$

iff i) $top(s) \triangleright top(t)$

ii) $top(s) = top(t) \quad \wedge \quad \tau(top(s)) = mult \quad \wedge$

- $sub(dec_u(s'), s) \gg_{EL} sub(dec_v(t'), t)$

- $dec(args(s)) \ggg_{EL} dec(args(t))$

iii) $top(s) = top(t) \quad \wedge \quad \tau(top(s)) \neq mult \quad \wedge$

$args(s) >_{IRDS,\tau(top(s))} args(t) \quad \wedge \quad \{s\} \gg_{IRDS} args(t)$

3.3 Knuth-Bendix ordering with status

The ordering of Knuth and Bendix (KBO, for short) assigns natural (or possibly real) numbers to the function symbols and then to terms by adding the numbers of the operators they contain. Two terms are compared by comparing their weights (the sum of the numbers of the operators), and, if the weights are equal, by lexicographically comparing the subterms. In [La79] (and other reports), a generalization of this ordering is described: The comparison of terms depends on polynomials instead of weights.

Analogous with the path and decomposition orderings, we succeeded in adding the idea of status to the KBO and therefore, in extending the method of comparing the arguments of two equivalent function symbols. To describe this strategy, called Knuth-Bendix ordering with status (KBOS, for short), we need some prerequisites and helpful definitions.

If x is a variable and t is a term we denote the number of occurrences of x in t by $\#_x(t)$. We assign a non-negative integer $\varphi(f)$ (the *weight* of f) to each operator in $\mathcal{F}$ and a positive integer φ_0 to each variable such that

$$\varphi(c) \geq \varphi_0 \qquad \text{if } c \text{ is a constant and}$$
$$\varphi(f) > 0 \qquad \text{if } f \text{ has one argument.}$$

Now we extend the weight function to terms. For any term $t = f(t_1,...,t_n)$ let

$$\varphi(t) = \varphi(f) + \sum \varphi(t_i).$$

Definition : KBOS (based on [Ma87] , [KB70])

$s >_{KBOS} t$

iff $(\forall x \in \mathcal{X})\ \#_x(s) \geq \#_x(t) \quad \wedge$

- $\varphi(s) > \varphi(t)$
- $top(s) \triangleright top(t)$
- $args(s) >_{KBOS,\tau(top(s))} args(t)$

This definition is slightly different from that given in [KB70]. Knuth and Bendix require that the precedence is total and that $\#_x(s) = \#_x(t)$ if $\varphi(s) = \varphi(t)$. The version given here is from [Ma87] and is an extension of the original definition. This claim can be sustained with the help of the rule $(y \supset x) \vee x \rightarrow (y \vee 0) \supset x$ which can only be oriented with the new definition ($\#_x(s) \geq \#_x(t)$ instead of $\#_x(s) = \#_x(t)$) if $\vee \triangleright \supset$ is presumed.
There exists another slight improvement of the ordering that allows one unary operator f with weight zero (see [KB70]), at most. To conserve the well-foundedness all other operators in $\mathcal{F}$ have to be smaller than f (w.r.t. the precedence).
Using a quasi-ordering on the function symbols instead of a partial ordering, we may permit more than one unary operator with weight zero (see [St88a]). On the premise that all these operators are equivalent w.r.t. the precedence, the induced KBOS also is a simplification ordering stabilized w.r.t. substitutions. For example, the termination of $(x*y)^2 \rightarrow x^2*(-y)^2$ can be shown with the improved (quasi-ordered) version but cannot be shown with the partial version of the KBOS (if $^2 = - \triangleright *$).

3.4 Properties of the orderings

Lemma All orderings described are

- well-founded,
- stable w.r.t. substitutions and
- monotonous w.r.t. the precedence.

Proof: All proofs (respectively the references to proofs) are given in [St88a]. ❑

In addition, all orderings are simplification orderings on Γ_G.

3.5 Example

In order to illustrate how terms are compared w.r.t. different orderings, we want to prove that $s = (x + y) + z >_{ord} x + (y + z) = t$ on the premises $ord \in \{RPOS, IRDS, KBOS\}$ and $\tau(+) = left$.

- $s >_{RPOS} t$ since α) $args(s) >_{RPOS,\,left} args(t)$ and β) $\{s\} \gg_{RPOS} args(t)$:
  α) $x + y >_{RPOS} x$ (because $x \in args(x + y)$), β) $s >_{RPOS} x$ (since $x \in args(s)$) and $s >_{RPOS} y + z$ (because $x + y >_{RPOS} y$ and $\{s\} \gg_{RPOS} \{y, z\}$).

- $s >_{IRDS} t$ iff $\{ \{s,x+y,x\}, \{s,x+y,y\}, \{s,z\} \} \gg\gg_{EL} \{ \{t,x\}, \{t,y+z,y\}, \{t,y+z,z\} \}$:
  - $\{s, x+y, x\} \gg_{EL} \{t, x\}$ since $dec_{11}(s) \ni s >_{EL} t \in dec_1(t)$:
    This is true because $args(s) = (x+y, z) >_{IRDS,\,left} (x, y+z) = args(t)$ (since $x \in args(x+y)$) $\wedge$ $\{s\} \gg_{IRDS} \{x, y+z\}$ (because $x \in args(s)$ and $y \in args(x+y)$).
  - $\{s, x+y, y\} \gg_{EL} \{t, y+z, y\}$ since $dec_{12}(s) \ni s >_{EL} t \in dec_{21}(t)$ (analogous with the previous case) $\wedge$ $dec_{12}(s) \ni s >_{EL} y+z \in dec_{21}(t)$:
    This is true because $args(s) = (x+y, z) >_{IRDS,left} (y, z) = args(y+z)$ (since $y \in args(x+y)$) $\wedge$ $\{s\} \gg_{IRDS} \{y, z\}$ (because $y, z \in args(s)$).
  - $\{s, z\} \gg_{EL} \{t, y+z, z\}$ since $dec_2(s) \ni s >_{EL} t \in dec_{22}(t)$ and $dec_2(s) \ni s >_{EL} y+z \in dec_{22}(t)$ (analogous with the previous cases).

- $s >_{KBOS} t$ since $\varphi(s) = \varphi(t)$, $top(s) = top(t)$ and $args(s) = (x+y, z) >_{KBOS,\,left} (x, y+z) = args(t)$: This is true because $\varphi(x+y) > \varphi(x)$ (since $\varphi(y) = \varphi_0 > 0$).

4 Comparison

In this chapter we compare the power of the presented orderings. We are additionally including the basic orderings restricted to multiset status. The power of an ordering is represented by the set of comparable terms. We do not compare the size of these sets but examine the <u>relation</u> between two sets. There are three possible relations : Two orderings can be *equivalent* ($> = \blacktriangleright$), one ordering can be *properly included* in the other ($> \subset \blacktriangleright$) or they *overlap* ($> \# \blacktriangleright$). The orderings $>$ and $\blacktriangleright$ overlap if there exist some terms such that $s_1 > t_1 \wedge s_1 \not\blacktriangleright t_1$ and $s_2 \blacktriangleright t_2 \wedge s_2 \not> t_2$. Consequently, the proof of such an overlapping is composed by specifying two counter-examples (see 4.2).

Note that the orderings described depend on a parameter : The precedence. This parameter may be either partial or total. Our results would be more general if we could give some information about the comparisons of orderings <u>unrelated to the precedence</u>. Unfortunately, the proofs of such comparisons are very complex. Therefore, we are only interested in total precedences since this kind of comparison in conjunction with the following proposition leads to the desired result.

4.1 Proposition Let $>$ and $\blacktriangleright$ be orderings which are monotonous w.r.t. the precedence. Furthermore, let $>$ be included in $\blacktriangleright$ w.r.t. <u>total</u> precedences. Then, if $s >(p)\ t$ holds for some precedence p, there exists a precedence q (possibly different from p) so that $s \blacktriangleright(q)\ t$.

Proof : q may be any total extension of p (see [St88b]). ❑

This statement will be of practical importance if we consider it together with the relations between orderings with an underlying total precedence : Only two (either IRDS (= KNSS) or KBOS) of the thirteen orderings collected in the diagram of 4.2 are needed to cover the union of comparable terms of all the orderings presented here. In other words, if terms can be oriented with any ordering (of the figure) there exists a precedence such that the terms are also comparable with either the

IRDS (KNSS) or the KBOS. Consequently, if you are implementing a system where the termination of a rewriting system must be guaranteed, only two of the thirteen orderings will have to be made available for the user. The cause of it is that the IRDS (resp. the KNSS) is stronger than all other path and decomposition orderings irrespective of the precedence.

The relations $\subset$, $=$ and $\#$ w.r.t. a total precedence have the following meanings : Let p (resp. p' and p") be a total precedence, τ (resp. τ' and τ'') a status, s and t terms.

- $> \subset \blacktriangleright$ iff $s >_{(p,\tau)} t \rightarrow s \blacktriangleright_{(p,\tau)} t \;\wedge$
  $(\exists p',\tau')\,(\nexists p'',\tau'')\; s \blacktriangleright_{(p',\tau')} t \;\wedge\; s >_{(p'',\tau'')} t$
- $> = \blacktriangleright$ iff $s >_{(p,\tau)} t \leftrightarrow s \blacktriangleright_{(p,\tau)} t$
- $> \# \blacktriangleright$ iff $(\exists p',\tau')\,(\nexists p'',\tau'')\; s >_{(p',\tau')} t \;\wedge\; s \blacktriangleright_{(p'',\tau'')} t \;\wedge$
  $(\exists p',\tau')\,(\nexists p'',\tau'')\; s \blacktriangleright_{(p',\tau')} t \;\wedge\; s >_{(p'',\tau'')} t$

We use a kind of *Hasse diagrams* to represent the relations. If $> \subset \blacktriangleright$ then we arrange $\blacktriangleright$ above $>$ joining them with a thick arrow :

4.2 Total precedence and arbitrary terms

Theorem Assuming total precedence and terms with variables, the following diagram holds :

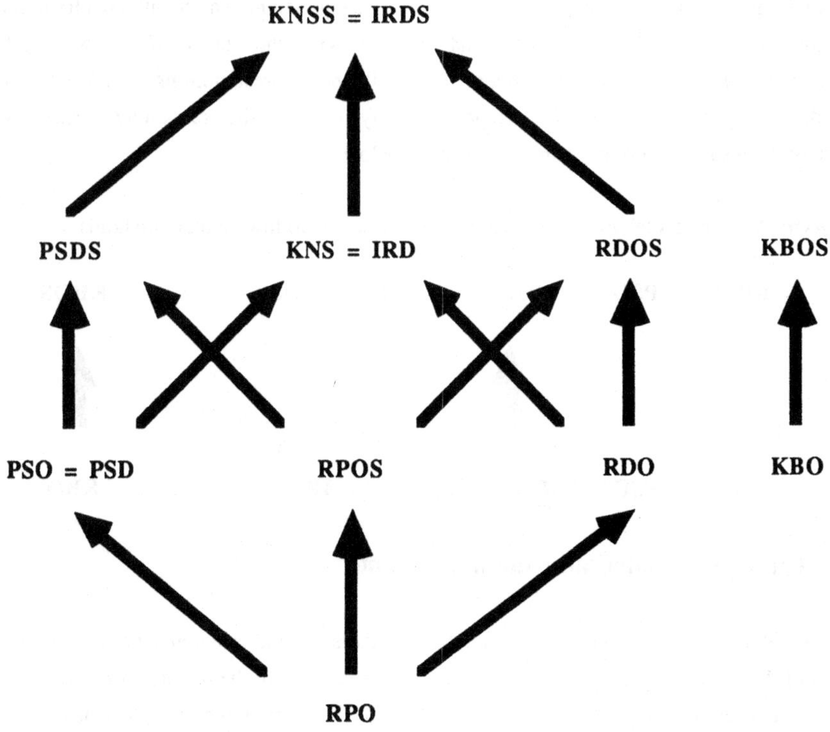

The diagram of 4.2 summarizes the relations between the orderings. The proofs are given in [St88a].

The following counter-examples account for the inclusions and overlappings of the orderings. Every pair of terms is comparable w.r.t. more than one of the orderings. Due to lack of space, we only specified the weakest (w.r.t. the theorem on the previous page) ones. For example, ⑥ is only comparable with the PSD, PSO, PSDS, KNS, IRD, KNSS and the IRDS.

①	$(-x - (-x)) - (-y - (-y)) >_{KBO,RDO} (x - y) - (x - y)$	with $\tau(-) = mult$ and $\varphi(-) > 0$
②	$x * ((-y) * y) >_{KBOS,RPO} (-y * y) * x$	with $* \triangleright -$ and $\tau(*) = mult$
③	$(x + y) + z >_{KBO,RPOS} x + (y + z)$	with $\tau(+) = left$
④	$\neg x \supset (y \supset z) >_{RDO,PSD,KBO} y \supset (x \vee z)$	with $\neg \triangleright \supset \triangleright \vee$ and $\varphi(\neg) > \varphi(\vee)$
⑤	$(\neg x \supset y) \vee z >_{KBO} (y \vee z) \vee x$	with $\varphi(\supset) > \varphi(\vee)$
⑥	$and(not(not(x)),y,not(z)) >_{PSD} and(y,nand(x,z),x)$	with $not \triangleright nand$ and $\tau(and) = mult$
⑦	$x^2 >_{RPO} x * x$	with $^2 \triangleright *$

Up to now we have compared the orderings w.r.t. arbitrary terms. In the following, we will investigate the restrictions to ground terms and to monadic terms.

4.3 Total precedence and ground terms

If we consider Γ_G instead of Γ and take a total precedence $\triangleright$, then the path and decomposition orderings are total on $\Gamma_G/\sim$. The orderings with multiset status are equivalent and are included in the orderings on arbitrary status which themselves are equivalent. The Knuth-Bendix ordering is also total on $\Gamma_G/\sim$ (KBO ⊂ KBOS) and overlaps with the others. $f(g(a)) >_{KBO} g(f(f(a)))$ and $g(f(f(a))) \blacktriangleright f(g(a))$ hold under the following presumptions : $\varphi(f) = 0$, the total precedence $f \triangleright g \triangleright a$ and $\blacktriangleright$ represents any one of the remaining orderings. The graph of the results on ground terms can be found below.

Theorem Assuming total precedence and ground terms, the following diagram holds :

RPOS = PSDS = RDOS = KNSS = IRDS **KBOS**

↑ ↑

RPO = PSO = PSD = RDO = KNS = IRD **KBO**

4.4 Total precedence and monadic terms

A *monadic term* only contains unary function symbols and either a constant or a variable. The subset of the monadic terms without constants can unequivocally be transformed into strings and vice versa. Therefore, the subsequent result refers to the corresponding orderings on string

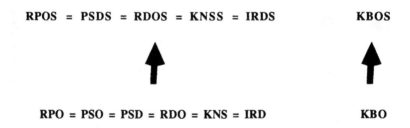

rewriting systems. Note that, on monadic terms, an ordering with status and the version without status coincide. On condition of a total precedence, all orderings with the exception of the KBO(S) are equivalent. The counter-example mentioned in 4.3 accounts for the overlapping of the KBO(S) with the others. Due to lack of space, the proofs cannot explicitly be given here but may be found in [St86]. Analogous with arbitrary terms, a synopsis in the form of a diagram is presented below.
Note that the class of path and decomposition orderings on monadic terms is equivalent to the *syllabled* or collected ordering $>_{CO}$ (see [Si87]) on the reverse words : u $>_{RPO,...}$ v iff reverse(u) $>_{CO}$ reverse(v).

Theorem Assuming total precedence and monadic terms, the following diagram holds :

$$\begin{array}{ccc} \mathbf{RPO = KNS = PSO = PSD = RDO = IRD} & & \\ \parallel \qquad\qquad\qquad\qquad\qquad \parallel & & \mathbf{KBOS = KBO} \\ \mathbf{RPOS = KNSS = PSDS = RDOS = IRDS} & & \end{array}$$

5 Conclusion

TRSPEC is a system for algebraic specifications based on term rewriting techniques and has been developed at the university of Kaiserslautern from the research group PROGRESS (**Pro**jekt**g**ruppe **Re**duktion**ss**ysteme), cf. [AGGMS87]. It is implemented in Common-Lisp and is currently running on Apollo Domain Systems. It consists of approximately 8000 lines of source code (without documentation) corresponding to 400 KB of compiled code. The kernel of the TRSPEC-system is COMTES, an extended Knuth-Bendix completion procedure. COMTES has been designed as a parametric system that is particularly suited for efficiency experiments. Besides different reduction strategies, the parameters include all orderings presented.

5.1 Time behaviour

A series of experiments has been conducted to study the time behaviour of the orderings. We have chosen 18 pairs of terms. Each pair is comparable w.r.t. all the orderings implemented. The diagram on the following page represents the average factors of all tests related to the RPOS.
Compared to the other orderings, the RPOS and the KBOS show a good result : The KBOS is indeed slower than the RPOS (by factor 1,7) but all other orderings are slower by a minimum factor of 4. The probable reason is the time for constructing the paths and the decompositions, respectively. In general, it is more expensive to build up a path-decomposition than the corresponding path (since the data structure of a path-decomposition is more complex than that of a path). The time difference between the KNSS (resp. the PSO) and the IRDS (resp. the PSDS) confirms our observation. Generally, the PSDS and the IRDS approximately have the same time behaviour. The reason why the IRDS is much better than the PSDS in some experiments could be the following : The selection of a direct subterm is immediately successful.

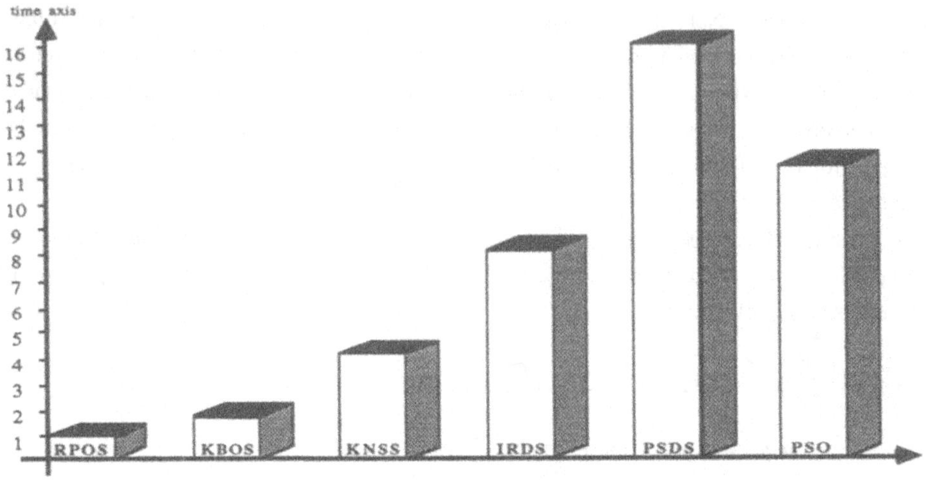

Considering the remarks of this section we come to the conclusion that the implementation of the RPOS, the KNSS and the KBOS is the most favourable choice. The RPOS serves as a pre-processor for the KNSS since the RPOS is faster than (and properly included in) the KNSS.

5.2 Concluding remarks

Various methods of proving the termination of term rewriting systems have been suggested. Most of them are based on the following notion of a simplification ordering : Any term that is syntactically simpler than an other one is smaller than the other one. In this paper, a collection of simplification orderings has been pointed out, including the well-known recursive path and decomposition ordering, the improved decomposition ordering of Rusinowitch, the path of subterms ordering (and an equivalent version on decompositions), the path ordering of Kapur/Narendran/Sivakumar and the Knuth-Bendix ordering. Most of the original definitions of those orderings cannot, for example, orient the associative laws. However, this is possible by using the principle of status (cf. [KL80]) in conjunction with the orderings. A variant with status exists as RPOS ([KL80]), RDOS (two versions : [Le84] and [Ru87]) and KNSS ([KNS85]). A substantial part of our work consisted of adding the principle of status to existing orderings and therefore, of creating the following new orderings : A new RDOS , PSDS (PSO with status) , IRDS (IRD of Rusinowitch with status) and KBOS. It turned out that the IRDS (= KNSS) is the most powerful ordering of the class of path and decomposition orderings presented in this paper. Since the KBOS and the IRDS overlap only two (resp. three) of the thirteen orderings must be emphasized : **IRDS = KNSS** and **KBOS**. The definition of the KNSS is more suitable for an implementation than the IRDS. However, we are of the opinion that the technique of comparing terms w.r.t. the IRDS is much easier to twig. The cause of it is that the definitions of the various orderings based on decompositions have a new look : The used decompositions only contain terms instead of vectors with three or even four components. This report extends the considerable set of different decomposition orderings. In short, the relations between the orderings presented in this paper and other well-known orderings (from different sources) can be described in the following way :

- The RDOS of [Le84] is included in our RDOS.
- The RDS of [Ru87] is a technique which connects the notions of the PSO (the principle of breadth-first), the IRD (the principle of depth-first) and lexicographical status. We suppose that the RDS and the IRDS overlap.
- The PSDS is included in the RDS of [Ru87].

A closer examination of these conjectures will be part of future plans.

[AGGMS87] J. Avenhaus, R. Göbel, B. Gramlich, K. Madlener, J. Steinbach : *TRSPEC: A Term Rewriting Based System for Algebraic Specifications*, Proc. 1st Int. Workshop on Conditional Rewriting Systems, LNCS 308, Orsay, 245 - 248

[Da88] M. Dauchet : *Termination of rewriting is undecidable in the one rule case*, Proc. 14th Mathematical foundations of computer science, LNCS 324, Carlsbad, 262 - 268

[De85] N. Dershowitz : *Termination* , Proc. 1st RTA, LNCS 202, Dijon, 180 - 224

[De82] N. Dershowitz : *Orderings for term rewriting systems* , J. Theor. Comp. Sci., Vol. 17, No. 3, 279 - 301

[JLR82] J.-P. Jouannaud, P. Lescanne, F. Reinig : *Recursive decomposition ordering*, I.F.I.P. Working Conf. on Formal Description of Programming Concepts II, North Holland, Garmisch Partenkirchen, 331 - 348

[KB70] D. E. Knuth, P. B. Bendix : *Simple word problems in universal algebras*, Computational problems in abstract algebra, Pergamon Press, 263 - 297

[KL80] S. Kamin, J.-J. Lévy : *Attempts for generalizing the recursive path orderings* , Unpublished manuscript, Dept. of Computer Science, Univ. of Illinois

[KNS85] D. Kapur, P. Narendran, G. Sivakumar : *A path ordering for proving termination of term rewriting systems* , Proc. 10th CAAP, LNCS 185, 173 - 187

[La79] D. S. Lankford : *On proving term rewriting systems are noetherian* , Memo MTP-3, Mathematics Dept., Louisiana Tech. Univ., Ruston

[La77] D. S. Lankford : *Some approaches to equality for computational logic : A survey and assessment* , Report ATP-36, Depts. of Mathematics and Computer Science, Univ. of Texas, Austin

[Le84] P. Lescanne : *Uniform termination of term rewriting systems - the recursive decomposition ordering with status* , Proc. 9th CAAP, Cambridge University Press, Bordeaux, 182 - 194

[Ma87] U. Martin : *How to choose the weights in the Knuth-Bendix ordering* , Proc. 2nd RTA, LNCS 256, Bordeaux, 42 - 53

[Pl78] D. A. Plaisted : *A recursively defined ordering for proving termination of term rewriting systems* , Report UIUCDCS-R-78-943, Dept. of Computer Science, Univ. of Illinois

[Ru87] M. Rusinowitch : *Path of subterms ordering and recursive decomposition ordering revisited* , J. Symbolic Computation 3 (1 & 2), 117 - 131

[Si87] C. C. Sims : *Verifying nilpotence* , J. Symbolic Computation 3 (1 & 2), 231 - 247

[St88a] J. Steinbach : *Term orderings with status* , SEKI REPORT SR-88-12, Artificial Intelligence Laboratories, Dept. of Computer Science, Univ. of Kaiserslautern

[St88b] J. Steinbach : *Comparison of simplification orderings* , SEKI REPORT SR-88-02, Artificial Intelligence Laboratories, Dept. of Computer Science, Univ. of Kaiserslautern

[St86] J. Steinbach : *Orderings for term rewriting systems* , M.S. thesis, Dept. of Computer Science, Univ. of Kaiserslautern (in German)

Classes of Equational Programs that Compile into Efficient Machine Code

Robert I. Strandh
Laboratoire Bordelais de Recherche En Informatique,
351, Cours de la Libération, Bordeaux, France

Abstract

Huet and Lévy [HL79] showed that, if an equational program E is strongly sequential, there exists an automaton that, given a term in the language L(E), finds a redex in that term.

The most serious problem with their approach becomes evident when one tries to use their result in a programming system. Once a redex has been found, it must be replaced by a term built from the structure of the right-hand side corresponding to the redex, and from parts of the old term. Then, the reduction process must be restarted so that other redexes can be found. With their approach, a large part of the term tree may have to be rescanned.

Hoffmann and O'Donnell [HO82a] improved the situation by defining the class of strongly left-sequential programs. For this class, a particularly simple reduction algorithm exists. A stack is used to hold information about the state of the reduction process. When a redex has been found and replaced by the corresponding right-hand side, the stack holds all the relevant information needed to restart the reduction process in a well defined state such that no unnecessary rescanning of the term is done.

However, it turns out that the approach of Hoffmann and O'Donnell is unnecessarily restrictive. In this paper, we define a new class of Equational Programs, called the forward branching programs. This class is much larger than the class of strongly left-sequential programs. Together with a new reduction algorithm, briefly discussed in this paper, our approach allows us to use the hardware stack to hold reduction information in a way similar to the way a block structured programming language uses the stack to hold local variables. In effect, our approach allows us to use innermost stabilization, while preserving the overall outermost reduction strategy.

1 Introduction

Equational programming [HO82b] [HOS85] uses lazy evaluation uniformly. The simple semantics make equational programming attractive from a theoretical point of view. In order for equational programming to be attractive also from a practical point of view, it needs to be efficient. Most importantly, it should compile into efficient machine code comparable to code generated by compiled Lisp [McC60] or C [KR78].

Huet and Lévy [HL79] showed that for every *strongly sequential* equational program E, there exists an automaton (called a matching DAG in [HL79]) that, given a term T in the language $L(E)$, finds an outermost redex in T if one exists. Huet and Lévy also give procedures to produce the automaton for two subsets of the strongly sequential programs.

In order to produce an efficient programming language based on reduction, we not only need to find redexes. We also need to replace the redexes with corresponding right-hand sides and then repeat the process until no redex can be found. The process of replacing a redex has been found to take substantial resources, and is discussed in [Str88]. Finally, after replacement has taken place, the reduction process needs to be restarted in a well defined state. The naïve approach is to restart completely, ignoring the fact that a large part of the term T may be unchanged, and thereby repeating substantial parts of the matching process.

Hoffmann and O'Donnell [HO82a] improved the situation by defining a more restricted class of equational programs, the *strongly left-sequential* programs. Equational programs in this class permit an efficient matching algorithm that makes use of a *stack* to hold relevant information on the matching process. When a redex has been found and replaced, the matching process can be restarted in a state that avoids unnecessary rescanning of the term T. The method of Hoffmann and O'Donnell uses an automaton based on string pattern matching by Knuth, Morris and Pratt [KMP77] improved by Aho and Corasick to handle multiple patterns [AC75].

In this paper, we show that the method of Hoffman and O'Donnell is unnecessarily restrictive. We define a new class of equational programs, called the *forward branching* equational programs. This class is much larger than the class of strongly left-sequential programs, but still a subset of the strongly sequential programs. For this class, we design a new reduction algorithm. The algorithm is based on *strong stabilization* [HL79]. Surprisingly, it is possible to stabilize in an *innermost* order while still preserving the overall *outermost* reduction order. This permits us to use a hardware stack to hold information on the reduction process in much the same way a traditional programming language [JW76] uses the stack to hold *local variables*.

2 Terminology

We adopt the terminology of [Str88]

Definition 2.1 *A symbol σ is a pair, (n, a), where n is the* name *used to refer to σ and a is a nonnegative integer that defines the number of places, or the* arity *of σ. We write $ARITY(\sigma)$ for a. We denote the set of all symbols in a program E by $\Sigma(E)$. We leave out the program, and simply write Σ, whenever E is understood.*

Definition 2.2 *A term T is a rooted ordered DAG (Directed Acyclic Graph) with a set $N(T)$ of nodes and a set $A(T)$ of arcs. Each node $\mathbf{n}$ in $N(T)$ has assigned to it a* label, *$LABEL(\mathbf{n}) \in \Sigma$. A subterm U rooted at node $\mathbf{n}$ of a term T is the subgraph of T rooted at $\mathbf{n}$.*

Notice that other authors sometimes call a term a *ground term*, since it cannot have any variables. Also notice that we allow for a term to be a DAG, and not just a tree. We shall have little or no reason, however, to use this fact. The set of all possible terms that can be constructed from symbols in a set Σ is called $\Theta(\Sigma)$. We write $\Theta(E)$ as an abbreviation for $\Theta(\Sigma(E))$, or simply Θ, whenever E is understood. The set $\Theta(E)$ is sometimes called the *language* of E, and written $L(E)$.

Definition 2.3 *A path is a sequence of positive integers, $i_1, i_2, \ldots, i_k$. A path defines a unique position in a term by giving a sequence of child numbers, in a left to right order, to follow, starting at the root. We represent a path with numbers separated by periods, e.g.* 2.3.1 *for the first child of the third child of the second child of the root. The empty path to the root node is represented by the symbol ϵ.*

The notation $\mathbf{n} = NODE(p,T)$, where p is a path and T is a term, means that $\mathbf{n}$ is the node in the term T that you find by following the path p starting at the root of T. Notice that, although a path defines a unique node in a term, there may be more than one path that defines the same node.

Using paths, we can define a natural equivalence relation on terms. We consider two terms T and U equivalent if and only if $\forall p, LABEL(NODE(p,T)) = LABEL(NODE(p,U))$.

2.1 Patterns and Templates

Definition 2.4 *A variable is a zeroary symbol that cannot appear in a term. We write $\Psi(E)$ when we mean the set of variables of E. As usual, E is left out if it is understood.*

Definition 2.5 *A pattern π is a rooted ordered tree with a set $N(\pi)$ of nodes and a set $A(\pi)$ of arcs. Each node $\mathbf{n} \in N(\pi)$ has assigned to it a* label, *$LABEL(\mathbf{n}) \in \Sigma \cup \Psi$. We require that a variable appear at most once in a pattern.*

The set of all possible patterns that can be constructed from a set $\Sigma \cup \Psi$ of symbols and variables is called $\bar{\Theta}(\Sigma \cup \Psi)$, The bar indicates that we are restricted to a tree, rather than the more general DAG. The set of all patterns actually appearing in some equational program E is written $\Pi_p(E)$. (See definition 2.9, Equational Program .) Again we leave out E whenever it is understood. Thus, $\Pi_p(E) \subseteq \bar{\Theta}(\Sigma(E) \cup \Psi(E))$.

In this paper, we use Lisp M-notation to represent terms and patterns, such as f[g[x;(1 2 3)];h[y]]. This representation does not capture the DAG structure of a term, but always shows an equivalent term tree. Lists are conveniently abbreviated using S-notation, and we mix the two notations freely [McC60].

A term is said to *match* a pattern if and only if for every path p in the pattern, either the pattern has a variable at p, or both the pattern and the term have the same symbol at p.

Definition 2.6 *A node $\mathbf{n} \in N(T)$ for some term T is a* redex *if and only if $\exists \pi \in \Pi_p$ such that the subterm rooted at $\mathbf{n}$ in T matches π. A term T* is a *redex if and only if $NODE(\epsilon, T)$ is a redex. A term T* has a *redex if and only if $\exists \mathbf{n} \in N(T)$ such that $\mathbf{n}$ is a redex. A term T is in* normal form *if and only if no $\mathbf{n} \in N(T)$ is a redex.*

The right-hand sides of the equations consist of *templates*. A template looks just like a pattern, but the variables serve a different purpose. Also, the same variable may appear twice in a template, but not in a pattern. Finally, a template is a DAG, whereas a pattern is a tree.

We frequently want to talk about terms about which we have incomplete knowledge, *i.e.*, we know only part of the structure of the term. In order to avoid confusion with patterns and templates, we define a separate item for this purpose. Its structure is very similar to that of a pattern.

Definition 2.7 *A* skeleton *α is a rooted ordered tree with a set $N(\alpha)$ of nodes and a set $A(\alpha)$ of arcs. Each node $\mathbf{n} \in N(\alpha)$ has assigned to it a* label, $LABEL(\mathbf{n}) \in \Sigma \cup \{?\}$.

The set of all skeletons that can be constructed from a set Σ of symbols is called $\bar{\Theta}(\Sigma \cup \{?\})$. Just as with terms, we use paths (see definition 2.3, Path) to identify nodes in a pattern or a skeleton. Notice, however, that since patterns and skeletons are trees, every node has a unique path.

2.2 Equations and Equational Programs

Definition 2.8 *A* rule, *r, is a pair $(\pi, \tau) \in \bar{\Theta}(\Sigma \cup \Psi) \times \Theta(\Sigma \cup \Psi)$. For some set of symbols Σ and some set of variables Ψ. We require that $LABEL(NODE(\epsilon, \pi)) \notin \Psi$, and $\forall \mathbf{n} \in N(\tau), LABEL(\mathbf{n}) \in \Psi \Longrightarrow \exists \mathbf{m} \in N(\pi)$ such that $LABEL(\mathbf{m}) = LABEL(\mathbf{n})$ The pattern π is referred to as the* left-hand side *of the rule, and the template τ is the* right-hand side.

In this paper we use the term *equation* as a synonym for rule. The set of equations defined in an equational program E is written $\Gamma(E)$.

Definition 2.9 *An* equational program *E is a triple (Σ, Ψ, Γ), where Σ is a set of symbols (see definition 2.1, Symbol), Ψ is a set of variables (see definition 2.4, Variable), and Γ is a set of rules constructed from Σ and Ψ (see definition 2.8, Rule). An equational program in our definition is sometimes called* left-linear *when a variable may appear at most once in a pattern.*

Reduction is defined in the usual way [ODo85]. We use the arrow $\rightarrow$ to mean reduction in one step, and $\rightarrow^*$ to mean the reflexive and transitive closure of $\rightarrow$. A subscript on an arrow as in $\rightarrow_E$ and $\rightarrow_E^*$ means reduction with respect to the equational program E. It is left out whenever E is understood.

3 Reduction Algorithms for Equational Programming

3.1 Reduction Strategy vs. Output Strategy

One of the major advantages of equational programming over other programming languages is the fact that we use *lazy evaluation*, also called *outermost evaluation*. Outermost evaluation implies that the reduction strategy must choose an *outermost redex* to reduce before any other redexes are considered. An outermost redex is a redex that is closest to the root in the natural partial ordering of the term. In general, there can be many outermost redexes. The strategy used to choose the next one is called a ***reduction strategy*** [Str87]. A popular strategy is the *leftmost-outermost* strategy. It always chooses the outermost redex that is *furthest to the left* in the term.

Equational programming uses a slightly different way to choose an outermost redex. In order to explain the strategy, we need some more terminology. Most of the terminology is adapted from [HL79] and [ODo85].

Definition 3.1 *Let U be the subterm of the term T rooted at some node $\mathbf{n}$ in T. The node $\mathbf{n}$ is said to be* stable *if and only if $\forall V$ such that $U \rightarrow^*_E V$, V is not a redex. A term T is* root stable *if and only if $NODE(\epsilon, T)$ is stable.*

The fact that $\mathbf{n}$ is stable implies that the label of $\mathbf{n}$ cannot change as a result of a sequence of reductions in the subterm rooted at $\mathbf{n}$. Stability is undecidable in general.

In order to characterize the behavior of the equational programming system, we distinguish between the ***output strategy*** and the ***reduction strategy***. A discussion on ***output strategy*** is, however, outside the scope of this paper. Thus, most of this paper is concerned with the reduction strategy. (See [Str87] and [Str88] for a discussion on output strategy.) The reduction strategy can be characterized as ***stabilizing***. That term is explained in greater detail below.

3.2 Regular Equational Programs

Both Huet and Lévy [HL79] and Hoffmann and O'Donnell [HO82a] require that the equational program be ***regular***. The same holds for the equational programs discussed in this paper. See *e.g.* [Klo80] for a formal treatment. From now on, whenever we talk about an equational program E, we assume that E is regular.

3.3 Strong Sequentiality and Strong Stability

Definition 3.2 *Given a skeleton α, a node $\mathbf{n} \in N(\alpha)$ such that $LABEL(\mathbf{n}) = ?$, and a symbol σ, we can define the* extended skeleton *$EXT(\alpha, \mathbf{n}, \sigma)$ as a new skeleton β which is α with $\mathbf{n}$ replaced by a node $\mathbf{m}$, such that $LABEL(\mathbf{m}) = \sigma$, and $\mathbf{m}$ has $n = ARITY(\sigma)$ children, all labeled with ?.*

We write $\beta \geq \alpha$ if there exists a sequence of skeletons $\alpha_0, \alpha_1, \ldots, \alpha_n$, where $\alpha = \alpha_0$ and $\beta = \alpha_n$, such that for $i = 0, 1, \ldots, n-1$, $\alpha_{i+1} = EXT(\alpha_i, \mathbf{n}_i, \sigma_i)$ for some node $\mathbf{n}_i$ and symbol σ_i. We write $\beta > \alpha$ if $n > 0$, or equivalently if $\beta \geq \alpha$ and $\beta \neq \alpha$. Sometimes the phrase "β is an *instance* of α" is used for $\beta \geq \alpha$.

Definition 3.3 *Given an equational program E, a term T is said to* omega-reduce *to a term U, written $T \rightarrow_{\{\omega,E\}} U$, if and only if $\exists F, \Sigma(F) = \Sigma(E), \Pi_p(F) = \Pi_p(E), T \rightarrow_F U$ [HL79]. As with ordinary reduction, we write $\rightarrow^*_{\{\omega,E\}}$ for the reflexive and transitive closure of $\rightarrow_{\{\omega,E\}}$. Notice that the reduction system F is still quantified over a single reduction step, so that there may in general be different such systems in each reduction step.*

The concept of omega-reduction represents a total lack of knowledge of the right-hand sides of E. We shall want to talk about reductions not only of terms but also of skeletons.

In that case we treat ? as a new zeroary symbol in Σ that does not otherwise appear in any $\pi \in \Pi_p$. The concepts of normal form (see definition 2.6, Redex, Normal Form) and omega-reduction (see definition 3.3, Omega-Reduction) extend in the obvious way to skeletons.

Definition 3.4 *A skeleton α is in* total normal form *if and only if α is in normal form and $\forall \mathbf{n} \in N(\alpha), LABEL(\mathbf{n}) \neq$? [HL79] [ODo85]. A total normal form is thus a term that contains no uncertainty.*

Definition 3.5 *Let α be a skeleton such that there exists no total normal form T such that $\alpha \rightarrow_{\omega}^{*} T$. A node $\mathbf{n} \in N(\alpha)$ is a* strong index *if and only if $LABEL(\mathbf{n}) = ?$, and $\forall \beta \geq \alpha$, if $\beta \rightarrow_{\omega}^{*} U$ where U is in total normal form, then $\beta \geq EXT(\alpha, \mathbf{n}, \sigma)$ for some σ [HL79] [ODo85].*

Intuitively, if α has a strong index, and α is not extended at that index, then every possible way to extend α at some other locations and subsequently omega-reduce α gives a skeleton with at least one *?*.

Definition 3.6 *An equational program E is said to be* strongly sequential *if and only if $\forall \alpha \in \bar{\Theta}(\Sigma(E) \cup \{?\})$ such that $\alpha \notin \Theta(\Sigma(E))$ and α is in normal form, α contains at least one strong index [HL79].*

The concept of strong sequentiality ensures that every skeleton has an extension that is *safe* to explore, *i.e.* if an attempt to find a normal form for a term by exploring and possibly reducing some strong index ends up in an infinite computation, then *every* attempt ends up in an infinite computation. As an example of a set of patterns that is regular but not strongly sequential, consider (from [HL79]) Π = {f[x;a;b], f[c;x;d], f[e;g;x]}. Obviously no term matches two or more patterns of Π. Still, the skeleton f[*?*;*?*;*?*], does not have any strong index. The example is derived from a simpler example, the *parallel or* with Γ = {(or[true;x],true), (or[x;true],true)}. The *parallel or* program, however, is not regular.

Definition 3.7 *A skeleton α is a* potential redex *if and only if $\exists \beta, \gamma, \pi$ such that $\beta \geq \alpha$, $\beta \rightarrow_{\omega}^{*} \gamma$ and γ matches π [HL79].*

Intuitively, a skeleton is a potential redex if there is a way to extend it, and then omega-reduce it so that it becomes a redex. Notice that a skeleton may reduce without being a redex, since every redex may occur in a proper subskeleton.

Definition 3.8 *A skeleton α is* strongly root stable *if and only if it is not a potential redex [HL79].*

A node $\mathbf{n}$ *is* strongly stable *if and only if the skeleton rooted at* $\mathbf{n}$ *is strongly root stable [HL79].*

Strong stability is essential to the rest of this paper, even more so than the concept of normal forms. A normal form can be thought of as a term where every subterm is strongly stable.

Huet and Lévy [HL79] give an algorithm that may be modified to detect strong stability. Their algorithm uses a structure similar to a finite state automaton. They also present efficient algorithms to construct the finite state automaton for two important subclasses of the strongly sequential programs, called *Constructor Based Programs* and *Simple Programs* respectively [HL79]. They do not give such an algorithm for the general case where the program is strongly sequential, although they show that such a finite state automaton must exist.

The algorithm in [HL79] cannot be used directly for our purpose, since it is designed to find a redex in a term, whereas we are trying to detect strong stability. Thus, we are concerned about what happens when the algorithm has found a redex, reduced it, and continues to try to detect strong stability. The algorithm in [HL79] stops when a redex is found. In order to address these problems, we put restrictions on the skeletons we allow in the tree.

3.4 Automata for Detecting Strong Stability

Definition 3.9 *A skeleton α is said to be a* firm skeleton *if and only if* $\exists \mathbf{n} \in N(\alpha)$ *with* $LABEL(\mathbf{n}) = ?$ *such that* $\forall \mathbf{m} \in N(\alpha), LABEL(\mathbf{m}) \neq ?$ *either* $\mathbf{m}$ *is strongly stable, or* $\mathbf{m}$ *is an ancestor of* $\mathbf{n}$. *We call such a node* $\mathbf{n}$ *a* firm extension node *The set of all nodes that are ancestors of a firm extension node* $\mathbf{n}$ *is called the* spine *of* α induced *by* $\mathbf{n}$. *We simply say* the spine of α *when it is clear which spine is referred to.*

Definition 3.10 *A skeleton α is a* partial redex *if and only if* $\exists \pi \in \Pi_p$ *such that* $\beta > \alpha$, *where* β *is the skeleton obtained by replacing all variables in* π *by ?.*

Definition 3.11 *An* index point *s is a pair (α, p) where α is a firm skeleton and also a partial redex, and p is a path such that* $\mathbf{n} = NODE(p, \alpha)$ *is a firm extension node of α, and* $\mathbf{n}$ *is also a strong index of α. We call* $\mathbf{n}$ *the* query node *of s. The* spine *of s is the spine of α induced by* $\mathbf{n}$.

Definition 3.12 *Given an index point $s = (\alpha, p)$, let* $\mathbf{n}$ *be the query node of s. An index point $t = (\beta, p')$ is a* failure point *of s if and only if β is the skeleton rooted at an ancestor (not necessarily proper) of* $\mathbf{n}$ *other than the root of α, and $q.p' = p$ where q is the path to β in α. A failure point t is the* immediate failure point *of s if and only if every failure point of s other than t is a failure point of t.*

Definition 3.13 *An* index tree *I, for an equational program E, is a* finite state automaton *which, in addition to the usual transfer function, also has a* failure function *as explained below. A state s of I is either an index point or a pattern. The transfer function of I, written $\delta(s, \sigma)$, gives a new state of the automaton, given a state and a symbol. The transfer function is constructed so that $\delta((\alpha, p), \sigma) = (\beta, p')$ (or $\delta((\alpha, p), \sigma) = \beta$ if β is a pattern) only if $\beta = EXT(\alpha, p, \sigma)$. As mentioned above, we define, in addition to the transfer function, a* failure function ϕ, *designed so that $\phi(s) = t$ if and only if t is the immediate failure point of s. For a state $s \in \Pi_p$ the transition function and the failure function are both undefined. We distinguish between such a* matching state *and an ordinary state by a predicate $MATCH(s)$ which is* true *if the state is a matching state and* false *otherwise.*

Our definition of index tree is similar to what is called a *matching DAG* in [HL79]. The major difference is that in [HL79], the automaton may be *nondeterministic*. Despite its name, the index tree is not a tree, but a DAG. In general, an equational program may have several index trees, one, or none.

Definition 3.14 *We call an equational program E a* bounded equational program *if and only if there is an index tree for E.*

Theorem 3.1 *Every strongly sequential program is bounded.*

Proof *See [HL79].*
A valid question at this point is whether an index tree can be constructed simply by starting with $(?, \epsilon)$ and building an index tree according to the definition, arbitrarily choosing one of several possible paths to a strong index for each index point. Unfortunately, the answer is negative. See [Str88] for examples.

3.5 Forward Branching Equational Programs

Definition 3.15 *An index tree is said to be a* forward branching index tree *if and only if every state of the index tree can be reached from the root, without following a failure transition.*

There are equational programs for which we can construct an index tree, but for which no forward branching index tree exists. See [Str88] for examples. In [HL79], the automaton (called the *matching DAG*) is typically nondeterministic. Thus, the concept of forward branching is irrelevant. Every state can be made reachable from the root simply by adding new transitions that cause the automaton to be nondeterministic.

Definition 3.16 *An equational program E is said to be* forward branching *if and only if a forward branching index tree exists for E.*

4 A New Reduction Algorithm

4.1 Introduction

In this section we introduce a new reduction algorithm of our own design. The algorithm works on the entire class of forward branching programs. (See definition 3.16, Forward Branching Equational Program .) Our algorithm uses a somewhat surprising property of the class of forward branching programs. It turns out that for this class of programs the order between recursive stabilization of subterms and transitions in the index tree can be interchanged. This property is treated more formally before we introduce our algorithm.

4.2 Order of Reductions

Definition 4.1 *Given an index tree I for an equational program, let δ be the transition function of I. Let s and t be two index points of I. We say that s is an* ancestor *of t if and only if $\exists s_1, s_2, \ldots, s_n$ and $\sigma_1, \sigma_2, \ldots, \sigma_{n-1}$ $n > 0$ such that $s = s_1$, $t = s_n$ and $s_{i+1} = \delta(s_i, \sigma_i)$ for $1 \leq i < n$. We say that s is a* proper ancestor *of t if and only if s is an ancestor of t and $s \neq t$.*

That is, for s to be ancestor of t, we require that we can reach t without following any failure transition.

Lemma 4.1 *Let $s = (\alpha, p)$ and $t = (\beta, q)$ be index points of some index tree such that s is an ancestor of t. Then $\alpha < \beta$ and $LABEL(NODE(p, \beta)) \neq ?$.*

Proof *By definition 3.13, β is a sequence of extensions of α, the first one at p. It follows immediately that $\alpha < \beta$ and $LABEL(NODE(p, \beta)) \neq ?$.*

Definition 4.2 *Two index points s and t in an index tree are said to be* independent *if and only if neither is an ancestor of the other.*

The following lemma states that the skeletons of two independent index points in a forward branching index tree must differ in that there is a path p where both skeletons have a symbol (rather than a *?*), and the symbol is different in each skeleton.

Lemma 4.2 *Let $s = (\alpha, q)$ and $t = (\beta, r)$ be two independent index points in some forward branching index tree, I. Then $\exists p$ such that $LABEL(NODE(p, \alpha)) \neq ?, LABEL(NODE(p, \beta)) \neq ?$, $LABEL(NODE(p, \alpha)) \neq LABEL(NODE(p, \beta))$.*

Proof *Since I is forward branching, s and t have at least one common ancestor. Consider the youngest such common ancestor, $u = (\alpha, p)$. By lemma 4.1, $LABEL(NODE(p, \alpha)) \neq ?, LABEL(NODE(p, \beta)) \neq ?$. Suppose $LABEL(NODE(p, \alpha)) = LABEL(NODE(p, \beta)) = \sigma$. Then there is a descendant $v = \delta(u, \sigma)$ of u, where v is an ancestor of both s and t, contradicting the assumption that u was a youngest such ancestor.*

Next, we introduce the concepts *intruder*, *weak node*, and *weak subskeleton* Ultimately we show that none of these can exist in a forward branching program. To appreciate the concept of an intruder, consider an index point s. Now consider a proper subskeleton of the skeleton of s. We want to make sure that, if this subskeleton were to be run through the index tree as a term, then no node labeled *?* would be a query point, unless it is also the query point of s.

Definition 4.3 *Let $s = (\alpha, p)$ be an index point of an index tree I. Let $\mathbf{n} = NODE(p, \alpha)$. Let β be a proper subskeleton of α other than ? and let $t = (\gamma, q)$ be an index point of I such that $\gamma \leq \beta$. Then t is called an* intruder *of s if and only if $\mathbf{n} \neq NODE(q, \beta)$ and $LABEL(NODE(q, \beta)) = ?$. We call β a* weak subskeleton *of α with respect to s and t, and we call $NODE(q, \beta)$ a* weak node *of β (and of α).*

Definition 4.4 *An index point s of an index tree I is said to be a* fragile index point *if and only if there exists an intruder of s in I. An index point that is not fragile is said to be a* solid index point.

Intuitively, if an index point is solid, we may apply the stabilization algorithm to any subskeleton of the skeleton known up to that point, and no node labeled *?* in the skeleton will be expanded.

Lemma 4.3 *In a forward branching index tree, two index points $s = (\alpha, p)$ and $t = (\alpha, q)$, where $p \neq q$, cannot exist.*

Proof *Suppose they could. By lemma 4.1, s and t are independent. By lemma 4.2, $\exists p$ such that $LABEL(NODE(p, \alpha)) \neq ?, LABEL(NODE(p, \alpha)) \neq ?$, $LABEL(NODE(p, \alpha)) \neq LABEL(NODE(p, \alpha))$. This, of course, is not possible.*

Theorem 4.1 *A forward branching index tree has no fragile points.*

Proof *Suppose one exists, and let I be such an index tree with fragile index point $s = (\alpha, p)$. Let $t = (\gamma, q)$ be an intruder of s and let β be a weak subskeleton with respect to s and t. Let r be the path to the root of β in α. Since I is forward branching, s and t are both reachable from $s_0 = (?, \epsilon)$. Consider the sequence of index points $A = ((?, \epsilon) = (\alpha_0, p_0), (\alpha_1, p_1), \ldots, (\alpha_n, p_n) = (\alpha, p))$ from the root of I to s. Since β is a proper subskeleton of α and $\beta \neq ?$, there is an i, such that $0 < i \leq n$ for which $p_i = r$. Notice that r is not a prefix of any path that comes before p_i in A. Consider a contiguous subsequence of A called B, where $B = ((\alpha_i, p_i), (\alpha_{i+1}, p_{i+1}), \ldots, (\alpha_{i+j}, p_{i+j}))$ such that r is a prefix of $p_i, p_{i+1}, \ldots, p_{i+j}$ and either $i+j = n$ or r is not a prefix of p_{i+j+1}. Obviously $n \geq i+j$. Intuitively, B consists of the entire first chain of index points in I from the root to s such that the node at path r is on the spine, and the chain ends either because the last element of the chain is s, or next index point does not have the node at r on the spine. Now construct a new sequence B' from B, such that $B' = ((\beta_0, r_0), (\beta_1, r_1), \ldots, (\beta_j, r_j))$, $p_{i+m} = r.r_m$ and β_m is the subskeleton of α_{i+m} rooted at r for $0 \leq m \leq j$. Notice that (β_m, r_m) is a failure point of (α_{i+m}, p_{i+m}) if β_m is a partial redex. Finally consider the sequence $C = ((?, \epsilon) = (\gamma_0, q_0), (\gamma_1, q_1), \ldots, (\gamma_k, q_k) = (\gamma, q))$ from the root if I to t. Notice that A, B, and C consist of index points, whereas B' may not.*

Case 1: *$\exists h \leq min(j, k)$ such that $(\beta_h, r_h) \neq (\gamma_h, q_h)$. Suppose without loss of generality that h is the smallest such number. Obviously $h > 0$.*

Case 1a: *$\beta_h \neq \gamma_h$. Then, since $\beta_{h-1} = \gamma_{h-1}$, β_h and γ_h differ in position $r_{h-1} = q_{h-1}$ in that $LABEL(NODE(\beta_h, r_{h-1})) \neq ?$, $LABEL(NODE(\gamma_h, q_{h-1})) \neq ?$, and $LABEL(NODE(\beta_h, r_{h-1})) \neq LABEL(NODE(\gamma_h, q_{h-1}))$. But $\beta \geq \beta_h$ and $\gamma \geq \gamma_h$, so $LABEL(NODE(\beta, r_{h-1})) \neq ?$, $LABEL(NODE(\gamma, r_{h-1})) \neq ?$, and $LABEL(NODE(\beta, r_{h-1})) \neq LABEL(NODE(\gamma, r_{h-1}))$. But for t to be an intruder, we require that $\gamma \geq \beta$. Contradiction.*

Case 1b: *$\beta_h = \gamma_h$, but $r_h \neq q_h$. By definition 3.12 and definition 3.13 and since γ_h is a partial redex, $(\beta_h, r_h) = (\gamma_h, r_h)$ must be a failure point in I. By lemma 4.3, (γ_h, r_h) and (γ_h, q_h), $r_h \neq q_h$ cannot both be index points in I. Contradiction.*

Case 2: *No h exists such that $h \leq min(j, k)$ and $(\beta_h, r_h) \neq (\gamma_h, q_h)$. In that case, either the list B' or the list C is exhausted before any difference could be found.*

Case 2a: *$k < j$. Then $q_k = r_k = q$. Also, (β_{k+1}, r_{k+1}) exists and $LABEL(NODE(\beta_{k+1}, r_k)) \neq ?$. Therefore $LABEL(NODE(\beta, q)) \neq ?$. But for t to be an intruder, we require that $LABEL(NODE(\beta, q)) = ?$. Contradiction.*

Case 2b: *$k > j$.*

Case 2b1: *$n > i + j$. Consider $s' = (\alpha_{i+j+1}, p_{i+j+1})$, and consider β', the subskeleton rooted at r in α_{i+j+1}. Obviously $\beta' = \gamma_{j+1}$, since $\beta \geq \beta'$, $\gamma \geq \gamma_{j+1}$, and if $\beta' \neq \gamma_{j+1}$, they differ in that $LABEL(NODE(\beta', r_j)) \neq ?$, $LABEL(NODE(\gamma_{j+1}, r_j)) \neq ?$, and $LABEL(NODE(\beta', r_j)) \neq LABEL(NODE(\gamma_{j+1}, r_j))$. Therefore γ and β differ in the same way.*

This is impossible, since it is required that $\gamma \leq \beta$. But β is a proper subskeleton of α_{i+j+1} and not on the spine. By definition 3.9, β' consists of all stable nodes. But $\gamma_{j+1} = \beta'$ is a partial redex. Contradiction.

Case 2b2: *$n = i + j$. Since $(\beta, r) = (\beta_j, r_j) = (\gamma_j, q_j)$, and $k > j$, it follows that $\gamma_k > \beta_j$, or $\gamma > \beta$. But we require that $\gamma \leq \beta$. Contradiction.*

Case 2c: *$k = j$.*

Case 2c1: *$n > i + j$. Then consider again $s' = (\alpha_{i+j+1}, p_{i+j+1})$, and consider β', the subskeleton rooted at r in α_{i+j+1}. Since $s' = (\alpha_{i+j+1}, p_{i+j+1})$ is a descendant of (α_{i+j}, p_{i+j}), and $r_j = q_j = q$, $LABEL(NODE(\alpha_{i+j+1}, pi + j)) \neq ?$. Therefore $LABEL(NODE(\beta, q)) \neq ?$. But then by definition 4.3, t cannot be an intruder of s.*

Case 2c2: *$n = i + j$. Then $(\beta, r) = (\gamma, q)$ and again t cannot be an intruder of s.*

The implication of theorem 4.1 is that in a forward branching system we may recursively call the procedure stabilize on any index point *before* inspecting it. Exactly the same nodes of our term are inspected in both cases.

Corollary 4.1 *In a forward branching program the order of stabilization and extension of an index may be reversed without changing the output behavior.*

We now present a particularly efficient algorithm for forward branching equational programs, using the implications of theorem 4.1. Notice that we may design our algorithm without any provisions for following a failure transition. Since we have reversed the order of stabilization and expansion, a sequence of failure transitions would always end up concluding that the root of the term is stable. Thus, we might as well simply return to the caller immediately. Furthermore, consider what happens when we find a redex. In this case, we may simply replace with the corresponding right-hand side, continue to try to stabilize the new term, and return only when the new term is stable at the root. With this background, we can now present our new algorithm. (See algorithm 4.1, Reduction Algorithm for Forward Branching Programs .)

The advantages of algorithm 4.1 as presented here are not immediately clear. It is easy to come up with cases where performance of this algorithm is worse than that of the algorithm given in [HL79]. In fact, where a single visit may suffice for their algorithm, each node may be visited several times by algorithm 4.1. However, by including a simple test for stability, we may limit the number of visits to every node to two: once to stabilize it and another time to inspect its label. The extra visit is more than made up for by the advantages. The main advantage is that this algorithm enables us to generate extremely efficient code for building right-hand sides [Str88].

5 Summary and Conclusions

We showed how to implement an efficient pattern matching procedure for the class of forward branching equational programs. The algorithm uses a surprising property of the forward branching equational programs, namely that stabilization and extension can be

Algorithm 4.1: Reduction Algorithm for Forward Branching Programs

```
procedure stabilize(n:node);
begin
start:
  T := the term rooted at n;
  σ := LABEL(n);
  if δ(s0, σ) = ϵ
  then
  begin
    mark n as stable;
    return;
  end;
  s := δ(s0, σ);
  do-forever
    if MATCH( s ) then
    begin
      Replace at n with the corresponding right-hand side;
      goto start;
    end;
    let s be (α, p);
    m := NODE(p, T);
    call stabilize(m);
    σ := LABEL(m);
    if δ(s, σ) = ϵ then
    begin
      mark n as stable;
      return;
    end
    s := δ(s, σ);
  end
end;
```

interchanged. This property permits us to use a hardware stack to hold information on the term being processed.

Since the automaton of [HO82a] is just a restricted index tree, the class of strongly left-sequential programs is a subset of the class of forward branching programs. In particular, the reduction algorithm presented in this paper can be used also on the class of strongly left-sequential programs.

References

[AC75] A. V. Aho and M. J. Corasick. Efficient string matching: an aid to bibliographic search. *Communications of the ACM*, 18(6):333–340, 1975.

[HL79] G. Huet and J.-J. Lévy. *Call by Need Computations in Non-ambiguous Linear Term Rewriting Systems.* Technical Report 359, IRIA, 1979.

[HO82a] C. Hoffmann and M. J. O'Donnell. Pattern matching in trees. *Journal of the ACM*, 68–95, 1982.

[HO82b] C. Hoffmann and M. J. O'Donnell. Programming with equations. *ACM, Transactions on Programming Languages and Systems*, 83–112, 1982.

[HOS85] C. Hoffmann, M. J. O'Donnell, and R. Strandh. Programming with equations. *Software, Practice and Experience*, 1985.

[JW76] K. Jensen and N. Wirth. *Pascal, User Manual and Report.* Springer-Verlag, Heidelberg, 1976.

[Klo80] J. W. Klop. *Combinatory Reduction Systems.* PhD thesis, Mathematisch Centrum, Amsterdam, 1980.

[KMP77] D. E. Knuth, J. Morris, and V. Pratt. Fast pattern matching in strings. *SIAM Journal on Computing*, 6(2):323–350, 1977.

[KR78] B. W. Kernighan and D.M. Ritchie. *The C Programming Language.* Prentice-Hall, Englewood Cliffs, New Jersey, 1978.

[McC60] J. McCarthy. Recursive functions of symbolic expressions and their computation by machine. *Communications of the ACM*, 3(4):185–195, 1960.

[ODo85] M. J. O'Donnell. *Equational Logic as a Programming Language.* MIT Press, Cambridge, Mass, 1985.

[Str87] R. I. Strandh. Optimizing equational programs. In *Proceedings of the Second International Conference on Rewriting Techniques and Applications*, pages 13–24, Bordeaux, France, 1987.

[Str88] R. I. Strandh. *Compiling Equational Programs into Efficient Machine Code.* PhD thesis, The Johns Hopkins University, Department of Computer Science, Baltimore, Maryland, 1988.

Fair termination is decidable for ground systems

Sophie Tison *
LIFL-UA 369 CNRS
Université de Lille-Flandres-Artois
59655 VILLENEUVE d'ASCQ Cedex (FRANCE)

Acknowledgement

I sincerly thank Bruno Courcelle for his comments on earlier versions of this paper and for his suggestions.

Introduction

The problem of *fairness* has been raised by the study of concurrency and nondeterminisn. In this paper, we are interested with fairness for *ground term rewriting systems.* More precisely, we prove that *fair termination* is **decidable for ground systems**. The question has been asked by N.Francez and S.Porat who gave a partial answer in [7].

A derivation is said to be *fair* if and only if any rule infinitely often enabled during the derivation is applied infinitely often; a system is said to be *fairly terminating* if and only if there is no fair infinite derivation. E.g. the system $(a \rightarrow f(a))$ does not terminate fairly since the infinite derivation $a \rightarrow f(a) \ldots \rightarrow f^n(a) \ldots$ is fair; but the system $(a \rightarrow f(a), a \rightarrow b)$ fairly terminates. The property is indecidable in the general case; N.Francez and S.Porat have proved that the property is decidable for some subclasses of ground systems [7]. (A ground system is a system where no variable occurs in the rules). Here we prove the property is decidable for an arbitrary ground system. The result is not surprising, since the ground systems are known to have many decidable properties (termination, confluence, reachability ... [2,3,6]). Furthermore, ground systems are very closed to finite state tree automata; indeed, we may construct for every ground system a couple of automata which "simulates" the system -such a pair is called a G.T.T.- [2].

B.Courcelle suggested a nice way to obtain the result in the case of words: let Δ be an alphabet and R be a set of rules on Δ. Roughly speaking, a derivation will be associated with a word of R^∞ and one may define a pushdown automaton -with Δ as pushdown alphabet- which recognizes infinite words corresponding to fair derivations. The case of trees is more complex and here we give a proof which is founded on an extension of G.T.T.'s;

Our proof may be divided in three parts:

. In the first one, we study fair derivations; we note that there are two distinct problems: on one side we have to control the rules enabled, on the other side we have to control the rules effectively applied.

. Secondly, we modify the system in order to encode which rules are applied along the derivation.

---

* *This work is supported in part by the PRC "Mathematiques et Informatique" and the "GRECO de Programmation". Part of the ESPRIT Basic Research Actions, BRA "Algebraic and Syntactic Methods in Computer Science" and BRA "Computing by Graph Transformations"*

. Last, we control which rules are enabled. In fact, we modify the G.T.T. in order to control any ground term which occurrs in a term along the derivation and whose depth is less than some chosen n.

Finally, we reduce the problem of fair termination to the emptiness of an intersection of **recognizable** forests, which proves its decidability. Furthermore, we sketch an algorithm for checking this property.

0. Notations

Let Δ be a finite graduated alphabet. Let $X = \{x_i/i \in N\}$ be a set of variables. $T(\Delta, X)$ denotes the term algebra built up from Δ and X in the usual way. We call its elements terms or trees.

$T(\Delta)_p$ denotes the set of elements of $T(\Delta, X)$ whose arity is p; in particular $T(\Delta)_0$ is the set of ground terms and will be identified with $T(\Delta)$.

Let t be a tree; $|t|$ denotes its depth; $T(\Delta)^{\leq n}$ stands for the set of ground terms whose depth is less than n; $FD(t)$ is the set of its "terminal" subtrees of t, i.e. the trees u such that $t = v(u)$ for some v; So, when t is ground, every term of $FD(t)$ is ground too.

The reader will be supposed familiar with the notions of automata and recognizable forests (cf [4,8] for more developments).

A *ground term rewriting system* R over $T(\Delta)$ is a finite set of rewrite rules, each of the form $l \rightarrow r$, where l and r are ground terms. $\rightarrow_R, \xrightarrow{*}_R$ are defined as usual; $MG(R)$ denotes the set of left-hand side of rules.

Let F be a set of ground terms. $\rightarrow_{R/F}$ denotes the relation defined by:

$$u \rightarrow_{R/F} v \iff_{def} (u \rightarrow v, FD(u) \cap F = FD(v) \cap F = \emptyset);$$

$\xrightarrow{*}_{R/F}$ is the reflexive and transitive closure of $\rightarrow_{R/F}$;

Intuitively, $u \xrightarrow{*}_{R/F} v$ if $u \xrightarrow{*}_R v$ and no term of F occurs as subtree along the derivation. Let $d = u_0 \rightarrow_{R/F} u_1 \rightarrow_{R/F} \dots u_p$ be a derivation in $R_{/F}$ and R' a subset of R; whenever the derivation uses effectively any rule of R', we note $u_0 \xrightarrow{*}_{R/F,eff.R'} u_p$; Last, if $u_0 = t(v_0)$, $u_p = t(v_p)$ and if the derivation does not rewrite t, - i.e. any rewrite applies below t-, we note $\tilde{t}(v_0) \xrightarrow{*}_{S/F,eff.S'} \tilde{t}(v_p)$;

For example, when $R = \{f(a) \rightarrow f(b), b \rightarrow c\}$, $\tilde{f}(f(a)) \xrightarrow{*}_{R_\emptyset,eff.R} f(f(c))$ but $\tilde{f}(a) \not\rightarrow_R f(b)$, $b \not\rightarrow_{eff.R} c$, $f(f(a)) \not\xrightarrow{*}_{R/\{f(b)\}} f(f(c))$.

1. Fair infinite derivations

Let $d = (u_i)_{i \in N}$ be an infinite derivation; $Oc(d)$ (resp. $Infoc(d)$) will denote the set of rules enabled -whose left-hand side occurs- at least once (resp. infinitely often) in the u_i; $Ap(d)$ (resp. $Infap(d)$) will denote the set of rules which are applied at least once (resp. infinitely often) along the derivation. So, by definition, an infinite derivation d is **fair** if and only if $Infap(d) = Infoc(d)$

The two following lemmas are now straightforward:

Lemma 1. *Let $d = (u_i)_{i\in N}$ be an infinite derivation;*

$$d\ is\ fair \iff \exists i_0, Oc((u_i)_{i\geq i_0}) = Infoc((u_i)_{i\geq i_0}) = Infap((u_i)_{i\geq i_0}) = Ap((u_i)_{i\geq i_0})$$

Lemma 2. *There is some fair infinite derivation in R iff for some subset S of R,there exists a sequence $(u_i)_{i\in N}$ such that:*

$$\forall i, \exists j > 0, u_i \xrightarrow{*}_{S/F_0, eff.S} u_{i+j}\ with\ F_0 = MG(R\backslash S) \qquad (a)$$

It just means that there is an infinite fair derivation if and only if there is an infinite derivation where all the rules enabled are applied infinitely often.

So, to decide of the existence of an infinite fair derivation in R, it suffices to decide of (a) for any subset S of R; Now, as for similar problems, we hope to reduce (a) to a "periodicity" property, i.e. to something like: $\exists l, \exists t/l \xrightarrow{*}_{S/F_0,eff.S} t(l)$. But some problems arise:

- First, $l \xrightarrow{*}_{S/F_0,eff.S} t(l)$ does not imply that $t(l)l \xrightarrow{*}_{S/F_0,eff.S} t(l)t^2(l)$; For example, when $R = \{a \to f(a), f^2(a) \to f^2(b)\}$ and $S = \{a \to f(a)\}$, we have $a \xrightarrow{*}_{S/F_0,eff.S} f(a)$ but not $f(a) \xrightarrow{*}_{S/F_0,eff.S} f^2(a)$.
- In the case of trees, the problem is somewhat complicated; indeed, the different branchs may "compensate" each other -e.g. the system $(a \to b, a \to c, g(c) \to g(a), h(b) \to h(a))$ is fairly terminating; but, if you adjoin a symbol of arity 2, there is some infinite derivation- ;
- The third problem is also specific to the trees; indeed, any word whose length is at least n may be "cut" in order to consider only its n last letters; In a term, a ground subterm whose depth is less than n may not necessarily be regarded as occurring in a ground subterm whose depth is exactly n. For example, let $t = f(f(f(f(a,a),a),a),c)$ and $n = 3$; t's depth is superior to n, c is a ground subterm of t but c does not occur in any ground subterm of t whose depth is exactly n;

So in order to study (a), the maximal arity of the alphabet will be considered;

First case:

Lemma 3. *Let n be at least the maximal depth of any side of a rule of R. If the maximal arity of Δ is 1, (a) is equivalent to the following property:*

$$\exists l \in MG(S), \exists (t,v) \in T(\Delta)_1^2 / \tilde{t}(l) \xrightarrow{*}_{S/F_0,eff.S} \tilde{t}(v(t(l))), (|t| = |v| = 0) \text{ or } |t| = n) \quad (b)$$

Proof:

. Suppose (a) verified; Let (u_i) be the infinite corresponding derivation. Let o_i be the occurrence of u_i where the rewriting is applied at step i; We may build a strictly increasing sequence $(j_i)_{i\in N}$, such that $o_k > o_{j_i}$ for any $k > j_i$; Now, we associate with any j_i, the couple (t_i, l_i) in $T(\Delta)^{\leq n} \times MG(S)$, where $u_{j_i} = \alpha(t_i(l_i)), u_{j_{i+1}} = \alpha(t_i(r_i))$ with $|t_i| = n$ *or* $(|t_i| < n$ *and* $|\alpha| = 0)$. As $T(\Delta)^{\leq n} \times MG(S)$ is finite, there is some (t,l) such that $(t_i, l_i) = (t,l)$ infinitely often; now, either the sequence of the corresponding (o_i) is bounded hence quasi-constant, and we can find l with $l \xrightarrow{*}_{S/F_0,eff.S} l$, or the sequence is not bounded and then we can find t, v and l with $\tilde{t}(l) \xrightarrow{*}_{S/F_0,eff.S} \tilde{t}(v(t(l)))$ and $|t| = n$; *q.e.d.*

. Conversely: Let $u_i = (v(t))^i(l)$; Then $\tilde{t}(u_i) \xrightarrow{*}_{S/F_0,eff.S} \tilde{t}(u_{i+1})$; q.e.d.

Remark:

If all symbols are of arity 0, we obtain as expected: there is some l in $MG(S)$, such that $l \xrightarrow{*}_{S/F_0,eff.S} l$!

Second case:

Lemma 4. *Let n be at least the maximal depth of any side of rule of R. Suppose there is some symbol of arity greater (or equal) than 2; (P) denotes the following property: There is some term t whose depth is n such that $FD(t) \cap F_0 = \emptyset$; Then: (1): Either (P) holds; then, the property (a) is satisfied iff the following property (c) holds for some finite recovering $(S_j)_{j \in J}$ of S:*

$$\forall j \in J, \exists l_j \in MG(S), \exists (t_j, v_j) \in T(\Delta)_1^2 / \tilde{t}_j(l_j) \xrightarrow{*}_{S/F_0,eff.S} \tilde{t}_j(v_j(t_j(l_j))) \text{ with :}$$
$$.|v_j| = 0 = |t_j| \text{ or}$$
$$.|t_j| = n \text{ or} \qquad (c)$$
$$.|t_j| < n \text{ and } \mathrm{v_j(x) = w(\alpha(s_1, \ldots, x, \ldots, s_q))}, \text{ with : } \alpha \in \Delta, \max_{\mathrm{k}} |\mathrm{s_k}| \geq \mathrm{n};$$

(2):Either (P) does not hold; Then, (a) is equivalent to the following property:

$$\exists t, |t| < n / t \xrightarrow{*}_{S/F_0,eff.S} t, \qquad (d)$$

Proof:

(1):

.Suppose (a) is verified; Let $((u_i)_{i \in N})$ be the corresponding derivation; Let o_i be the position which is reduced in u_i at the step i;

Let r be a rule of S; as r is applied infinitely often along the derivation, we may build an increasing subsequence $(o_{i_j})_{j \in N}$ such that r is applied at least once below each o_{i_j} at an ulterior step; the corresponding u_{i_j} may be viewed as $h(\alpha(s_1, ..., t_j(l_j), ..., s_q))$ where l_p is the left-hand side of the rule applied at this step, and $(\alpha, t_j, (s_i)_i)$ is "the n-context" of the rewriting, i.e.:

$$\begin{cases} |h| = |\alpha| = 0, & \text{if } |t_i| = 0 \\ \exists k, |s_k| \geq n, & \text{if } |t_i| < n \\ |t_i| = n, & \text{otherwise.} \end{cases}$$

Now, we may associate with each j the corresponding (t_j, l_j); as $T(\Delta)^{\leq n} \times MG(S)$ is finite, there is some (t, l) such that (t_j, l_j) is equal infinitely often to (t, l); So, in the same way as in the case of words, for any $S_r = \{r\}$, we may define the corresponding (t, l) which verifies the required conditions. q.e.d.

.Conversely:

Let f be a symbol whose arity is greater than 2; Let t_0 be a term which satisfies (P); We define a tree g of arity 2 by $g(x, y) = f(x, y, t_0, .., t_0)$. For any j in

J, for any integer i, we define w_j^i by: $w_j^i = t_j((v_j(t_j)^i(l_j))$; Then, we define u_i by $u_i = g(g(....g(t_0, w^i_{card\ J}), ..., w_2^i), w_1^i)$ and the $(u_i)_{i\in n}$ verifies the conditions of (a);

(2):

The sufficiency is obvious; Conversely let us suppose (a) satisfied; necessary the depth of each u_i is inferior to n; as $T(\Delta)^{\leq n}$ is finite, there is some t such that $u_i = t$ infinitely often, *q.e.d.*

So, to decide of (a), it is enough to decide of (b) in the monadic case, of (d),(c) and (P) in the general case;

.(d) is obviously decidable: we just have to build the following graph: the nodes are the terms of depth inferior to n, the arrows are the derivation steps in S/F_0 and the labels are the corresponding rules; now, (d) is satisfied iff there is a cycle in the graph with a label "corresponding" to S, which we can easily decide;

.(b) is a particular case of (c);

.So, the main problem is the decidability of (c).

(c) can be simplified by modifying slightly S; indeed, let $\Delta' = \Delta \cup \bar{\Delta}$, where $\bar{\Delta}$ is a copy of Δ obtained by "overlining" the letters. Define a system S' by:

$$S' =_{def} \{(\bar{t}(g) \rightarrow \bar{t}(d))/(g \rightarrow d) \in S, |\bar{t}| \leq n\}$$

For example, let $X = \{a, b, f(x)\}$, $S = \{a \rightarrow b\}$, $n = 2$; S' will be $\{a \rightarrow b, \bar{f}(a) \rightarrow \bar{f}(b), \bar{f}(\bar{f}(a)) \rightarrow \bar{f}(\bar{f}(b))\}$.

Next, define the application π from S' to S by $\pi((\bar{t}(g) \rightarrow \bar{t}(d))) = (g \rightarrow d)$ and the projection p from $T(\Delta')$ to $T(\Delta)$ by $p(\delta) = \delta = p(\bar{\delta})$. Then, when $F_0' = p^{-1}(F_0)$, (c) is ensured Iff there is some subset S_j' of S which verify $\pi(S_j') = S_j$ and the property (c') defined as follows:

$$\exists l_j, v_j, \bar{t}_j / \bar{t}_j(l_j) \rightarrow_{S'/F_0', eff.S_j'} \bar{t}_j(v_j(t_j(l_j))) \ with:$$
$$.\pi(\bar{t}_j) = t_j \qquad (c')$$
$$.\ v_j, t_j\ verify\ the\ conditions\ of\ (c)$$

So, finally we have reduced the problem to the decision of (c') for any subset T' of S'. In (c') appear two different constraints: on one side, we have to control which rules are effectively applied, on the other side we have to control the ground subterms whose depth is less than n which occurr in terms along the derivation. We show in the next section that we can treat separately the two problems.

Remark:

The properties (c) and (c') look like those obtained by G.Huet in [6] in order to prove the decidability of termination for ground systems; indeed, a similar proof applies to decide the existence of an infinite derivation in $R_{/F_0}$; but when we are interested with the rules effectively applied, we obtain only partial decision by this way.

2. How to control which rules are applied.

Let R be a rewriting system on the alphabet Δ; We define a new alphabet $\Sigma = \Delta \times P(R)$ - i.e. we index the letters of Δ by the rules of R - the idea is to memorize the rules as they are applied-.

Define Π the projection of Σ on Δ, i.e. $\Pi(\delta, S) = \delta$, and for any subset S of R, the canonical injection $i_S : T(\Sigma) \to T(\Delta)$ i.e. $i_S(\delta) = (\delta, S)$; occ:$T(\Delta) \to P(R)$ will denote the application defined by:

$.occ((\delta, s)) = s \ \forall(\delta, s)$

$.occ(f(t_1, t_2, ...t_n)) = occ(f) \cup \bigcup_{i=1}^{n} occ(t_i)$.

Next, construct the system R' by:

$$R' =_{def} \{(l' \to d')/(\Pi(l') \to \Pi(d')) = r \in R; d' = i_S(\Pi(d)) \ with \ S' = occ(l') \cup \{r\}\}$$

Example.

If R is the system $((1) : (a \to b), (2) : (f(b) \to f(a)))$, R' will be defined by:

$$R' = \left\{ \begin{array}{lr} a_\emptyset, a_1 & \to b_1, \\ a_2, a_{1,2} & \to b_{1,2}, \\ f_\emptyset(b_\emptyset), f_\emptyset(b_2), f_2(b_\emptyset), f_2(b_2) & \to f_2(a_2), \\ f_\emptyset(b_1), f_\emptyset(b_{1,2}), & \\ f_1(b_\emptyset), f_1(b_1), f_1(b_2), f_1(b_{1,2}), & \\ f_{1,2}(b_\emptyset), f(_{1,2}(b_1), f_{1,2}(b\emptyset), f_{1,2}(b_{1,2}) & \to f_{1,2}(a_{1,2}) \end{array} \right\}$$

For example, the derivation $g(a, f(a)) \to_R g(a, f(b)) \to_R g(b, f(b)) \to_R g(b, f(a))$ corresponds to $g_\emptyset(a_\emptyset, f_\emptyset(a_\emptyset)) \to_{R'} g_\emptyset(a_\emptyset, f_\emptyset(b_1)) \to_{R'} g_\emptyset(b_1, f_\emptyset(b_1)) \to_{R'} g_\emptyset(b_1, f_{1,2}(a_{1,2}))$.

Now let S be any subset of R, and F_0 any subset of $T(\Delta)_0^{\leq n}$:

Lemma 5. *Let $F_0' = \Pi^{-1}(F_0)$. Then, $F_0' \subset T(\Delta)_0^{\leq n}$ and:*

$$\forall u, v \in T(\Delta), u \xrightarrow{*}_{R/F_0, eff.S} v \iff \exists v'(u' \xrightarrow{*}_{R'/F_0'} v') \Big/ \left\{ \begin{array}{l} \Pi(v') = v, \\ Occ(v') = S, \\ u' = i_\emptyset(u); \end{array} \right.$$

The proof of the lemma is straightforward by induction on the length of the derivation.

So, finally we have reduced the problem to the decision of the following property, for any given (S_j, S, n, F_0):

$$\exists(l, v, t)/t(l) \xrightarrow{*}_{S/F_0} t''(v(t'(l'))) \ with :$$

$$\left\{ \begin{array}{l} \Pi(l) \in MG(S), \Pi(t) = \Pi(t') = \Pi(t''), \ \Pi(l') = l; \\ occ(t) = \emptyset = occ(l), occ(t''(v(t'(l')))) = S_j; \\ (|t| = |v| = 0) \ or \ (|t| = n) \ or \ (|t| < n, v(x) = u(\alpha(s_1, ..x, ..s_q)), \max_i |s_i| > n). \end{array} \right.$$

Now, it is easy to see, that, for any given t and l, the set of $t'(v(t(l')))$ where t', t'', v, l' verify the "good properties" stated above, is *recognizable*; Furthermore it is easy to construct an automaton which recognizes it; Let $S'_{F'_0}(u) = \{v/u \xrightarrow{*}_{S'/F'_0} v\}$; Since $t(l)$ is a left-hand side of a rule of S', we just have to study the intersection of $S'_{F_0}(u)$ with a recognizable forest, for any u in $MG(S')$; if we know how to decide the emptiness of this intersection, we know how to decide the property (e); the following part will indicate us that $S'_{F'_0}(u)$ is recognizable and will give us a way to construct a corresponding automaton, which will achieve the proof.

3. Simulation of $R_{/F_0}$

Let R be a rewriting system on Δ; F_0 denotes a finite forest of $T(\Delta)_0$; n denotes $max(max\{|l|,|r|/(l \rightarrow r) \in R)\}, max\{|t|/t \in F_0\})$; C_0 denotes the forest of ground terms whose no subterm belongs to F_0; We want to simulate the derivations in $R_{/F_0}$.

In the case of the words, the problem is easy: we use an additional symbol $\sharp$ to mark the beginning of the words and we construct a new system as following:

$$R' = \{ul \rightarrow ur/min(|ul|,|ur|) = n, (l \rightarrow r) \in R, ul \in C_0, ur \in C_0\}$$
$$\cup \{\sharp ul \rightarrow \sharp ur/min(|ul|,|ur|) < n, (l \rightarrow r) \in R, ul \in C_0, ur \in C_0\}$$

Then, clearly, $u \xrightarrow{*}_{R/F_0} v$ Iff $\sharp u \xrightarrow{*}_{R'} \sharp v$; So, the existence of an infinite derivation is decidable since terminating of ground term system is decidable [1,2,6]. Furthermore, $R'_{F'_0}(u) = \{v/u \xrightarrow{*}_{R'/F'_0} v\}$ is recognizable for any word u [2,3] q.e.d.

In the case of trees, the situation is more complex; indeed the same problem as in part 1. arises: a ground subterm of t whose depth is less than n may not necessarly be viewed as occurring in a ground subterm of depth n; so, the method used above doesn't apply. Here, the idea is to transform the automaton constructed in [2] for simulation of ground systems by enriching the information contained in states to "control" the ground subterms. Although it is difficult to get overall view of the problem, the following example helps to understand the difficulty:

Let $R = \{a \rightarrow d, a \rightarrow f(b,a), b \rightarrow c,\}$; The corresponding g.t.t. may be represented by:

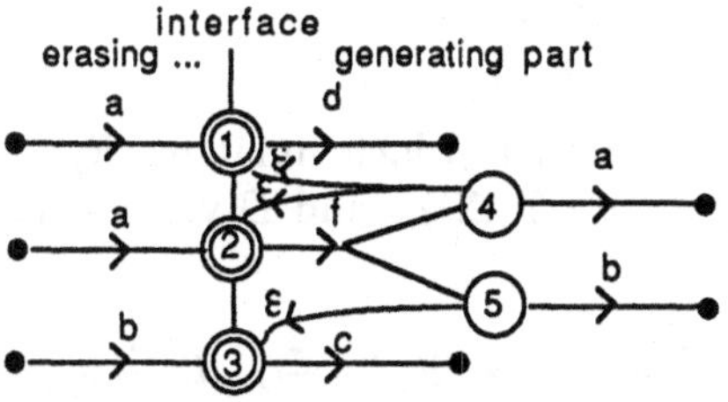

Now, let $F_0 = \{f(b,d), f(c,d)\}$; Roughly speaking, we just have to delete the ε−transition from state 4 to state 1 to simulate $R_{/F_0}$. At last, let $F_0 = \{f(c,a), f(c,d)\}$. We see that the problem is more complex; The following automaton recognizes $\{t/a \xrightarrow{+}_{R/F_0} t\}$:

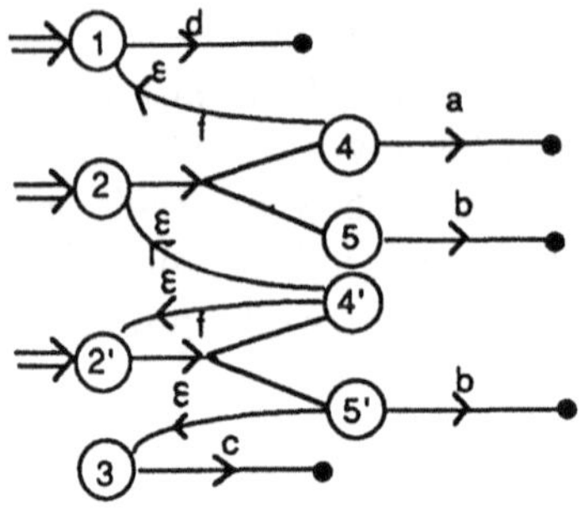

So, we are going to construct a top-down automaton which simulates $R_{/F_0}$ -more precisely which simulates the derivations from any left-hand side of rule-; More precisely let us call D this automaton and Q this set of states; D will verify the following specification:

$$\forall(l \rightarrow r) \in R, \exists Q_0 \subset Q/(\exists q \in Q_O/q \xrightarrow{*}_t \Leftrightarrow l \rightarrow_R r \xrightarrow{*}_{R/F_0} t)$$

Notations:

.Let $\approx$ be the congruence relation defined on ground terms by:

$$t \approx u \iff {}_{def}(t = u \text{ } or \text{ } (|t| > n \text{ } and \text{ } |u| > n))$$

$[t]$ denotes the class of the tree t; ∇ denotes the class of trees whose depth is at least n; when the depth of t is less than n, t will be identified with its class.

.Let $(u_i)_{1 \leq i \leq p}$ be a derivation in $R_{/F_0}$; whenever each u_i's depth for $1 < i < p$ is at least n, we will note $u_1 \xrightarrow{*}_{>n} u_p$;

.Next, define a top-down automaton D_0 as follows: each rule r of R is associated with a top-down automaton D_r, whose initial state is denoted by i_r, and which recognizes d_r, the right hand side of r; D_O is the disjoint union of the D_r's, for r in R; Q_0 denotes the set of states, T_0 the transitions;

Specification:

Let us now define an automaton D which simulates $R_{/F_0}$; Its set of states, Q, is a subset of $Q_0 \times T(\Delta)_{/\approx} \times \{0,1,2\} \times T(\Delta)_{/\approx}$ and D verifies the following specification:

$$S =_{def} \left\{ \begin{array}{l} (q, cl_1, x, cl_2) \xrightarrow{*}_D t \iff \exists t_0/ \ q \xrightarrow{*}_{D_0} t_0 \xrightarrow{*}_{R/F_0} cl_1 \xrightarrow{i}_{>n} t \ with: \\ \quad either: \ .cl_1 \neq \nabla, cl_2 = cl(t), (i = 1 = x \ or \ i > 1, x = 2) \\ \quad either: \ .cl_1 = \nabla, cl_2 \neq \nabla, x = 0, t_0 = t = cl_2 \in C_0; \end{array} \right\}$$

$\{S_\Rightarrow\}$ denotes the direct part of $\{S\}$.

An automaton which verifies this specification will work; indeed $\exists(cl_1, x, cl_2)/$ $(q, cl_1, c, cl_2) \xrightarrow{*}_D t$ If and only if $\exists t_0, q \rightarrow t_0 \xrightarrow{*}_{R/F_0} t$; conversely, this specification may seem too "strong" for our problem: why have we to know (cl_1, x, cl_2)? In fact, the information expressed in (cl_1, x, cl_2) is just useful during the construction of the automaton. Roughly speaking, it enables us to build all the derivations by "combining" (cl_1, x, cl_2) and (cl_2, x, cl_3). So, after the automaton is built, we may reduce the automaton -for example the states (q, cl_1, x, cl_2) and (q, cl_3, y, cl_2) with $cl_2 \neq \nabla$ are equivalent;

Algorithm:

Initialization:

D is obtained from D_0 as follows:

By construction of D_0, for any q in Q_0, there is one and only one term t such that $q \xrightarrow{*}_{D_0} t$; Furthermore the depth of t is less than n; If t belongs to C_0, q is replaced in Q by $(q, \nabla, 0, t)$; in the other case q is forgotten; T_0 is modified in consequence to obtain T;

Iteration step:

While T is modified, we execute the following step:

1: First, we add ε-transitions by the following inference rule:

$\forall d_r \in C_0, \forall r : (l_r \rightarrow d_r) \in R$

$$\frac{(q, cl_1, x, l_r) \xrightarrow{*} l_r}{(q, l_r, 1, d_r) \xrightarrow{\varepsilon} (i_r, \nabla, 0, d_r)}$$

Intuitively, this rule just says: if $t \xrightarrow{*}_{R/F_0} l_r$ and if d_r belongs to C_0, $t \xrightarrow{*}_{R/F_0} l_r \rightarrow_{R/F_0} d_r$.

2: We apply now the folllowing inference rules on the set of transitions:

.Let $cl_2 \neq \nabla, \alpha = (q, cl_1, x, cl_2), \beta = (q, cl_2, 1, cl_3), \alpha_i = (q_i, cl_i^1, x_i, cl_i^2)$

$$\frac{(q_0, cl_1^0, x_0, cl_2^0) \rightarrow f(\alpha_1, .., \alpha, ..\alpha_n),\ t = f(cl_2^1, .., cl_3, .., cl_2^n) \in C_0}{(q_0, cl', x', cl) \rightarrow f(\alpha_1, .., \beta, .., \alpha_n)}$$

with:

$cl = cl(t)$

$$\begin{cases} (cl' = cl_2^0, x' = 1) & \text{if } cl_2^0 \neq \nabla; \\ (cl' = cl_1^0, x' = 2) & \text{if } cl_2^0 = \nabla. \end{cases}$$

$$\frac{(q_0, cl_1^0, x_0, cl_2) \xrightarrow{\varepsilon} (q, cl_1, x, cl_2)}{(q_0, cl_2, 1, cl_3) \xrightarrow{\varepsilon} \beta}$$

Intuitively, it just means you may compose the two derivations $cl_1 \xrightarrow{*}_{R/F_0} cl_2$, and $cl_2 \rightarrow_{R/F_0} cl_3$; Similarly, if $f(\ldots, cl_1, \ldots) \xrightarrow{*}_{R/F_0} f(\ldots, cl_2, \ldots)$, if $cl_2 \rightarrow_{R/F_0} cl_3$ and if $f(\ldots, cl_3, \ldots)$ belongs to C_0, $f(\ldots, cl_1, \ldots) \xrightarrow{*}_{R/F_0} f(\ldots, cl_2, \ldots) \rightarrow_{R/F_0} f(\ldots, cl_3, \ldots)$;

.Let $\alpha = (q, cl_1, x, \nabla), \beta = (q, cl_1, 2, cl_2), \alpha_i = (q_i, cl_i^1, x_i, cl_2^i)$

$$\frac{(q_0, cl_1^0, y, \nabla) \rightarrow f(\alpha_1, .., \alpha, ..\alpha_n), f(cl_2^1, .., cl_2, .., cl_2^n) \in C_0}{(q_0, cl_1^0, 2, cl(t)) \rightarrow f(\alpha_1, .., \beta, .., \alpha_n)}$$

$$\frac{(q_0, cl_1, x, \nabla) \xrightarrow{\varepsilon} \alpha}{(q_0, cl_1, 2, cl_2) \xrightarrow{\varepsilon} \beta}$$

Intuitively, it just enables us to combine the two derivations $cl_1 \xrightarrow{*}_{>n} t$ and $t \rightarrow_{R/F_0} u$ with $|t| > n$.

Example:

Let $\Delta = \{f(x), a\}$, $R = (a \rightarrow f(a))$,$n = 3$, and $F_0 = \emptyset$; Clearly, R is not fairly terminating (R has only one rule and is not terminating!) and the exemple is trivial; but, it illustrates the construction's mechanism. The corresponding automaton is represented in the following figure; let us explain how one obtain it:

. $(i, \nabla, 0, f(a)) \rightarrow f((q, \nabla, 0, a))$ and $(q, \nabla, 0, a) \rightarrow a$ are obtained by initialization. It corresponds to the derivation $a \rightarrow f(a)$.

. $(q, a, 1, f(a)) \xrightarrow{\varepsilon} (i, \nabla, 0, f(a))$ is obtained by the first inference rule;

. We have added the state $(q, a, 1, f(a))$; As $(i, \nabla, 0, f(a)) \rightarrow f((q, \nabla, 0, a))$, we obtain $(i, f(a), 1, f(f(a))) \rightarrow f((q, a, 1, f(a)))$ by the second inference rule; it corresponds to the derivation $a \rightarrow f(a) \rightarrow f^2(a)$.

. We have added the state $(i, f(a), 1, f(f(a)))$; as $(q, a, 1, f(a)) \xrightarrow{\varepsilon} (i, \nabla, 0, f(a))$, we obtain $(q, f(a), 1, f(f(a))) \xrightarrow{\varepsilon} (i, f(a), 1, f(f(a)))$ by the last inference rule.

And, so on.....

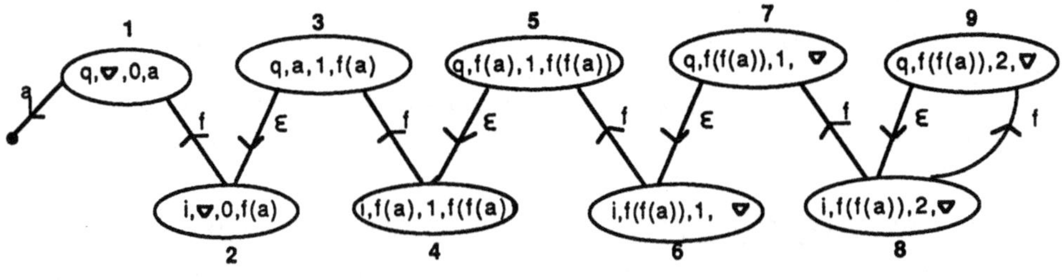

Proof of the correctness of the algorithm:

Clearly, the algorithm terminates: indeed, at each step, T increases (for inclusion); But, T is bounded by the set of possible rules; as Q and Δ are finite, this set is finite and so T becomes stationnary.

Unfortunately, the proof of partial correctness is rather technical and tedious:

.First, we prove that each derivation of $R_{/F_0}$ is simulated by D, more precisely:

$$\exists t_0, q \xrightarrow{*}_{D_0} t_0 \xrightarrow{*}_{R/F_0} t \Rightarrow \exists (cl_1, x, cl_2)/(q, cl_1, x, cl_2) \xrightarrow{*}_D t.$$

Proof:

We will show:

$$(q, cl_1^0, x, cl_2^0) \xrightarrow{*}_D t \rightarrow_{R/F_0} t' \Rightarrow (q, cond) \xrightarrow{*}_D t' \text{ with :}$$

$$\begin{cases} cond = (cl_1^0, 2, cl(t')) & \text{if } cl_2^0 = \nabla \text{ ;} \\ cond = (cl_2^0, 1, cl(t')) & \text{otherwise;} \end{cases}$$

The proof is an induction on the depth of the occurrence of the rewriting in $t \rightarrow t'$:

. *Basis step:* The rule applies at the root hence $t = l_r$, $t' = d_r$ for some rule r; So $(q, cl_1^0, x, cl_2^0) \xrightarrow{*}_D l_r$, $FD(d_r) \in C_0$ and the rule $(q, l_r, 1, d_r) \xrightarrow{\varepsilon} (i_r, \nabla, 0, d_r)$ is added by the first part of the iteration step; *q.e.d.*

. *Induction step:* Suppose the property ensured whenever the rewriting applies at depth less than n; Let $t \rightarrow_{R/F_0} t'$, where the rule applies at depth $n+1$; Hence, there is some f in Δ, some j such that $t = f(t_1, \ldots, t_j, \ldots, t_k), t' = f(t_1, \ldots, t'_j, \ldots, t_k)$ with $t_j \rightarrow_{R/F_0} t'_j$. As $(q, cond_0) \xrightarrow{*}_D t$, we can find q', $(q_i, cond_i)$ such that:

$$(q, cond_0) \xrightarrow{\varepsilon,*}_D (q', cond'_0) \rightarrow_D f((q_1, cond_1), \ldots, (q_k, cond_k)), (q_i, cond_i) \xrightarrow{*}_D t_i;$$

Now, by induction hypothesis, $(q_j, cond'_j) \xrightarrow{*}_D t'_j$ and by the second part of inference rules we obtain $(q, cond) \xrightarrow{\varepsilon,*} (q', cond) \rightarrow f((q_1, cond_1), \ldots, (q_j, cond'_j), \ldots, (q_k, cond_k))$; then, $(q, cond) \xrightarrow{*}_D t$.

So, let $q \xrightarrow{*}_{D_0} t_0 \xrightarrow{k}_{R/F_0} t$; since t_0 belongs to C_0, $(q, \ldots) \xrightarrow{*}_D t_0$; so, by what precedes, $(q, \ldots) \xrightarrow{*}_D t$ and the proof is complete.

. Conversely, we prove each derivation simulated by the automaton D is a derivation in $R_{/F_0}$; more precisely, we prove the automaton D verifies the following property:

$$\exists (cl_1, x, cl_2)/(i_{l \rightarrow d}, cl_1, x, cl_2) \xrightarrow{*}_D t \Rightarrow l \xrightarrow{*}_{R/F_0} t.$$

The interested reader will find the proof in Appendix.

Finally, for any r right hand side of rule which belongs to C_0:

$$(i_{l \rightarrow r}, \ldots) \xrightarrow{*}_D t \iff r \xrightarrow{*}_{R/F_0} t$$

So, let t be any term in $C_0 \cap T(\Delta)_0^{\leq n}$; add to R the rule $(t \rightarrow t)$ and construct the automaton D as above. Define A as the automaton D where the initial states are all the $(i_{t \rightarrow t}, \ldots)$. Then A recognizes exactly $R_{F_0}(t)$. So, for any ground term t, the forest of all terms obtained by rewriting t in $R_{/F_0}$ is **recognizable**; *q.e.d.*

Conclusion.

By summing up, we have reduced the problem of fair termination to the emptiness of the intersection of two constructible and recognizable forests. Since the family of recognizable forests is closed under intersection and since emptiness is decidable in this family, **fair termination is decidable.**

Bibliography:

[1] W.S. BRAINERS : "*Tree-generating regular systems*", Inf. and Control 14 (1969), pp.217-231.

[2] M.DAUCHET & S.TISON :"*Tree automata and decidability in ground term rewriting systems*", Rapport interne, Université de Lille I, et FCT'85, *Lecture Notes in Computer Science* vol.199, pp. 80-84 (1985).

[3] M.DAUCHET,T.HEUILLARD, P.LESCANNE & S.TISON: " *Decidability of the confluence of ground term rewriting systems*", rapport INRIA 675 et LICS'87.

[4] J.ENGELFRIET: "*Bottom-up and top-down tree transformations, a comparison*", Math. Systems Theory 9 (1975), pp. 198-231.

[5] N.FRANCEZ: "*Fairness*, Texts and Monographs in Computer Science, Springer Verlag, 1985.

[6] G.HUET & D.LANKFORD: "*The uniform halting problem for term rewriting systems*", Rapport INRIA 283 (1978).

[7] S.PORAT & N.FRANCEZ: "*Fairness in rewriting term systems*", RTA'85.

[8] J.W. THATCHER: "*Tree automata: an informal survey*", Currents in the Theory of Computing, (A.V. Aho, Ed.) Prentice Hall (1973), pp.14..

Appendix:

First, we need some new definitions:

Definitions:

. Let $q = (q_0, cl, x, cl')$ in Q; $cond(q)$ denotes (cl, x, cl'), $et(q)$ denotes q_0, which is by construction a state of D_0;

. Let $d = (u_i)_0^p$ be a not empty derivation in $R_{/F_0}$:

  . $first(d)$ denotes the first term of the derivation, i.e. u_0;

  . $last(d)$ denotes the last term of the derivation, i.e. u_p;

  . $head(d)$ denotes the derivation d without its last step , i.e. $(u_i)_0^{p-1}$;

  . $body(d)$ denotes the derivation d without its first step, i.e. $(u_i)_1^p$;

  . $cd(d)$ denotes $(cl, x, cl(u_p))$ with:

$$\begin{cases} x = 0 & \text{if p=0} \\ x = 2 & \text{if } (p \geq 2 \text{ et } \mathrm{u_{p-1}} = \nabla), \\ x = 1 & \text{otherwise} \end{cases} \begin{cases} cl = u_{p_0} & \text{if } \exists p_0 = sup\{j/0 \leq j < p, |u_j| \leq n\} \\ cl = \nabla & \text{otherwise} \end{cases}$$

. Let $d_1 = (u_i)_0^{p_1}$ et $d_2 = (v_i)_0^{p_2}$ be two derivations in $R_{/F_0}$; if $u_{p_1} \rightarrow_{R_{/F_0}} v_0$, we define the "product" $d_1.d_2$ as the derivation $(w_i)_i^{p_1+p_2+1}$ with:

$$\begin{cases} w_i & = u_i \quad i \leq p_1 \\ & = v_{p_1+i+1} \quad p_1 < i \leq p_1 + p_2 + 1 \end{cases}$$

Whenever one derivation is empty, the product will be the other derivation. This product is naturally extended to the sets of derivations.

. Finally, define the operator Mel which associate to each letter of arity k and to k not empty derivations, a set of derivations defined by:

. $Mel(f,((u_0^j)_1^k)) = (f(u_0^1,\dots,u_0^k))$

. $Mel(f,(d_j)_1^k) = \cup_{j=1,|d_j|>1}^k f(first(d_1),\dots,first(d_k)).Mel(f(d_j^i)_1^k)$ with:

$$\begin{cases} d_i^i = & body(d_i); \\ d_j^i = & d_j \text{ if } j \neq i; \end{cases}$$

Intuitively, $Mel(f,(d_j)_1^k)$ works as a "shuffle" operator; It construct all the derivations from $f(first(d_1),\dots,first(d_k))$ obtained by "mixing" all the d_j .

We are now able to define the *invariant* of the iteration; first, define the following assertion:

$$\{j\} =_{def} \begin{cases} 1.\forall\theta : (q, cl_1^0, c, cl_2^0) \rightarrow f(\alpha_1,\dots,\alpha_k), \forall t_i : \alpha_i \xrightarrow{*}_D t_i, \\ \exists(d_j)_1^k, \exists d \in Mel(f,(d_j)_1^k)/ \\ \quad .q \rightarrow_{D_0} f(et(\alpha_1),\dots,et(\alpha_k)) \xrightarrow{*}_{D_0} f(d_{1,0},\dots,d_{k,0}); \\ \quad .cond(\alpha_j) = cd(d_j); \\ \quad .cd(d) = (cl_1^0, c, cl_2^0); \\ \quad .last(d) = f(t_1,\dots,t_k); \\ 2.\forall\theta : \alpha = (q, cl_1^0, c, cl_2) \xrightarrow{\varepsilon} (q, cl_1, x, cl_2) = \beta, \forall t : \beta \xrightarrow{*}_D t, \\ \exists(t_j)_0^k, \exists i / \\ \quad .q \rightarrow_{D_0} t_0 \rightarrow_{R/F_0} \cdots \rightarrow_{R/F_0} t_i \dots \rightarrow_{R/F_0} t_k = t \\ \quad .cd((t_j)_0^k) = (cl_1^0, c, cl_2); \\ \quad .cd((t_j)_i^k) = (cl_1, x, cl_2); \end{cases}$$

We prove that the assertion $\{i\} = \{j\}$ *and* $\{S_\Rightarrow\}$ is an **invariant** property of the iteration. The proof needs some preliminary definitions and lemmas;

. For any non empty derivation $d = (u_i)_0^p$ in $R_{/F_0}$, we define:

. d^* as the derivation $(u_i)_0^{q_0}$, where $q_0 = sup\{j/0 \leq j \leq p, |u_j| \leq n\}$, - d^* is empty whenever $q_0$is not defined -;

. d^∇ as the derivation $(u_i)_{q_0+1}^p$ where q_0 is define as above - $d^\nabla = d$ whenever q_0 is not defined -;

. Now,we define an equivalence relation $\sim$ on the derivations in $R_{/F_0}$ by:

$$d_1 \sim d_2 \iff_{def} (d_1^* = d_2^*) and (d_1^\nabla = \epsilon \iff d_2^\nabla = \epsilon)$$

The following lemma is straightforward:

Lemma 6. *Let* $(d_j)_1^k$ *be* k *derivations in* $R_{/F_0}$*; if* $d_i = d_i^\nabla$ *for some* i, $Mel(f,(d_j))$ *contains only derivations of* $R_{/F_0}$.

The following lemma will be very useful :

Lemme 7. *Let* $(d_j)_1^k$, $(\delta_j)_1^k$ *be some derivations in* $R_{/F_0}$ *such that* $d_j \sim \delta_j$, *for any* j. *Then, for any derivation* α *of* $Mel(f,(d_j)_1^k)$ *in* $R_{/F_0}$, *we can find some* β *in* $Mel(f,(\delta_j)_1^k)$ *equivalent to* α.

The proof is an easy induction on $\sum_{j=1}^k |d_j|$:

Now, we are ready to prove the invariance of $\{i\} = \{j\} and \{S_\Rightarrow\}$:

. Clearly, $\{i\}$ is true at the beginning;

. Let $\{j_n\}$ be the assertion $\{j\}$ limited to $\alpha_i \overset{\leq n}{\rightarrow}_D t_i$ and $\beta \overset{\leq n}{\rightarrow}_D t$; similarly $\{S_n\}$ denotes the assertion $\{S_\Rightarrow\}$ limited to $q \overset{\leq n}{\rightarrow}_D t$.

. Clearly, $\{j_n\} \Rightarrow \{S_{n+1}\}$.

. $\{S_1\}$ is always ensured since the rules $\theta : q \rightarrow x, x \in \Sigma$ are rules of D_0.

. Now, prove $\{j\}$ ACTION $\{S_n\} \Rightarrow \{j_n\}$,where ACTION is the addition of any rule θ during a step; we distinguish whether θ is an ϵ-transition or not.

Case 1: θ is not an ϵ-transition. Let $\theta : \alpha_0 = (q_0, cond_0) \rightarrow f(\alpha_1,\ldots,\alpha_k), \alpha_i \overset{\leq n}{\rightarrow} t_i, 1 \leq i \leq k$ with $\alpha_j = (q_j, cl_1^j, x_j, cl_2^j)$.

By $\{S_n\}$, $et(\alpha_i) \overset{*}{\rightarrow}_{D_0} u_i^0$ et $cd(d_i) = cond(\alpha_i)$, $1 \leq i \leq k$ for some $(d_i)_1^k = ((u_i^j))_1^k$.

Again, we distinguish two cases:

First case: θ is an "old" transition; So $\alpha_i \overset{*}{\rightarrow}_D t'_i, 1 \leq i \leq k$ before ACTION, for some $(t'_i)_1^k$. By $\{j\}$, there exists $(\delta_i)_1^p$ and δ_0 in $Mel(f,(\delta_i))$ such that:

. $q \rightarrow_{D_0} f(q_1,\ldots,q_n) \overset{*}{\rightarrow}_{D_0} prem(\delta_0)$;

. $cd(\delta_i) = cond(\alpha_i)$, hence $last(\delta_i^0) = cl_2^i$;

. $cd(\delta_0) = cond_0$, hence $last(\delta_0^0) = cl_2^0$;

Define $\gamma_i = \delta_i^0.d_i^\nabla, 1 \leq i \leq k$: hence $\gamma_i \sim \delta_i$ and $last(\gamma_i) = last(d_i)$. So, by lemma 7, we can find in $Mel(f,(\gamma_i))$ a derivation γ_0 of $R_{/F_0}$ equivalent to δ_0 such that $last(\gamma_0) = f(last(d_1),\ldots,last(d_n))$. *q.e.d.*

Second case: θ is a "new" transition (so the considered step is its creation).

By construction, there is some i, some $\beta_i = (q_i, cl_1'^i, x'_i, cl_2'^i)$, $\theta' : (q_0, cl_1'^0, x'_0, cl_2'^0) \rightarrow f(\alpha_1,\ldots,\beta_i,\ldots,\alpha_p)$ such that θ is obtained from θ' .

Since α_j and β_i are coaccessible, by $\{j\}$ there is (t'_j, δ_j) with:

. $q_j \overset{*}{\rightarrow}_{D_0} prem(\delta_j)$ $1 \leq j \leq k$

. $last(\delta_j) = t'_j$ $1 \leq j \leq k$

. $cd(\delta_j) = cond(\alpha_j) = cd(d_j)$ $j \neq i, 1 \leq j \leq k$

. $cd(\delta_i) = cond(\beta_i)$

Furthermore, $cd(\delta) = (cl_1'^0, x'_0, cl_2'^0)$ for some derivation δ in $R_{/F_0}$ belonging to $Mel(f,(\delta_j))$. Let:

$$\gamma_j = \delta_j^0.d_j^\nabla \text{ if } j \neq i: \text{ then } \gamma_j \sim \delta_j, last(\gamma_j) = last(d_j).$$
$$\gamma_i = \delta_i \text{ if } x_i = 1;$$
$$= \delta_i^0.head(d_i)^\nabla \text{ if } x_i = 2 : then\ \gamma_i \sim \delta_i$$

Now, by the lemma 7, we can find a derivation in $R_{/F_0}$, γ_0, belonging to $Mel(f,(\gamma_j))$ such that $\gamma_0 \sim \delta$; so $cd(\gamma_0) = (cl_1'^0, x'_0, cl_2'^0)$. Let:

$$\gamma_j' = \gamma_j \; if \; j \neq i: \; cd(\gamma_j') = cond(\alpha_j)$$
$$\gamma_i' = \gamma_i.last(d_i): \; cd(\gamma_i') = cond(\alpha_i)$$

The γ_j' are derivations in $R_{/F_0}$.

Now, define $\gamma = \gamma_0.(f(t_1,\ldots,t_n))$:

. $\gamma \in Mel(f,(\gamma_j'))$;

. $last(\gamma) = f(t_1,\ldots,t_n)$;

. $cd(\gamma) = cond_0$;

. Since the γ_i' are derivations in R_{F_0} and $cl(last(\gamma)) = cl_2$, γ is also a derivation in $R_{/F_0}$; q.e.d.

se 2: Let θ: $\alpha \xrightarrow{\epsilon} \beta = (q, cl_1, x, cl_2)$, t: $\beta \xrightarrow{\leq n}_D t$.

First case: θ is "old"; so $\beta \xrightarrow{*} t'$ for some t' before ACTION.

By $\{j\}$, there is some derivation in $R_{/F_0}$, $d = (t_i)$, such that:

. $\alpha \rightarrow_{D_0} t_0' \rightarrow_{R_{/F_0}} \ldots \rightarrow t_n' = t'$;

. $cd(d) = cond(\alpha)$;

. $\exists j, cd((t_i')_{i\geq j}) = cond(\beta)$;

By $\{S_n\}$, there is some derivation in $R_{/F_0}$, $\delta = (u_i)_0^p$, such that:

. $\beta \rightarrow_{D_0} u_0 \ldots \rightarrow_{R_{/F_0}} u_p = t$;

. $cd(\gamma) = cond(\beta)$;

Then, if $cl_2 \neq \nabla$, $t = t'$ and the derivation d' works; otherwise, $cl_2 = \nabla$ and the derivation $d^*.\delta^{\nabla}$ is convenient;

Second case: θ is obtained by the inference rule a); So $\beta = (i_r, \nabla, 0, d_r)$, $\alpha = (q_0, l_r, 1, d_r)$, $t = d_r$ and l_r et d_r belongs to C_0. Moreover, $(q_0, cl_1, x, l_r) \xrightarrow{*}_D l_r$ before ACTION for some (cl_1, x). By $\{j\}$, we can find a derivation in $R_{/F_0}$, $d = (t_i)$, such that:

. $q_0 \rightarrow_{D_0} t_0$.

. $cd(d) = (cl_1, x, l_r)$.

Thus the derivation $\delta = d.(d_r)$ works;

Third case: θ is obtained by the inference rule b):

Then either $x = 1$, either $x = 2$.

. If $x = 1$, θ is obtained from θ' : $(q_0, cl_0', y_0, cl_1) \xrightarrow{\epsilon} (q, cl_0, x_0, cl_1)$ with $\alpha = (q_0, cl_1, 1, cl_2)$. By $\{j\}$, there is some derivation in $R_{/F_0}$, $d = (t_i)$, such that:

. $q_0 \rightarrow t_0$.

. $cd(d) = (cl_0', y_0, cl_1)$.

. $\exists d', d''/d = d'.d'', cd(d'') = (cl_0, x_0, cl_1)$.

By $\{S_n\}$, $cl_1 \rightarrow_{R_{/F_0}} t$. So the derivation $d.(t)$ works;

. If $x = 2$, θ is obtained from θ' : $(q_0, cl_1, x, \nabla) \rightarrow (q, cl_1, x, \nabla)$ with $\alpha = (q_0, cl_1, 2, cl_2)$. By $\{j\}$, there is some derivation in $R_{/F_0}$, $d = (t_i)$, such that:

. $q_0 \rightarrow t_0$.

. $cd(d) = (cl_1, x, \nabla)$.

By $\{S_n\}$, we can find δ a derivation in $R_{>n}$ from cl_1 to t, whose length is at least 1. Then the derivation $d^*.\delta$ works. *q.e.d.*

Termination for the Direct Sum of Left-Linear Term Rewriting Systems
- Preliminary Draft*-

Yoshihito Toyama
NTT Basic Research Laboratories
3-9-11 Midori-cho, Musashino-shi, Tokyo 180, Japan
email: toyama%ntt-20.ntt.jp@relay.cs.net

Jan Willem Klop
Centre for Mathematics and Computer Science
Kruislaan 413, 1098 SJ, Amsterdam, The Netherlands

Hendrik Pieter Barendregt
Informatica, Nijmegen University
Toernooiveld 1, 6525 ED, Nijmegen, The Netherlands

Abstract

The direct sum of two term rewriting systems is the union of systems having disjoint sets of function symbols. It is shown that two term rewriting systems both are left-linear and complete if and only if the direct sum of these systems is so.

1. Introduction

An important concern in building algebraic specifications is their hierarchical or modular structure. The same holds for term rewriting systems [1] which can be viewed as implementations of equational algebraic specifications. Specifically, it is of obvious interest to determine which

*This paper is an abbreviated version of the *IEICE technical report COMP88-30*, July 1988. Now we are preparing a final version for submission based on this draft.

properties of term rewriting systems have a *modular* character, where we call a property *modular* if its validity for a term rewriting system, hierarchically composed of some smaller term rewriting systems, can be inferred from the validity of that property for the constituent term rewriting systems. Naturally, the first step in such an investigation considers the most basic properties of term rewriting systems: confluence, termination, unique normal form property, and similar fundamental properties as well as combinations thereof.

As to the modular structure of term rewriting systems, it is again natural to consider as a start the most simple way that term rewriting systems can be combined to form a larger term rewriting system: namely, as a disjoint sum. This means that the alphabets of the term rewriting systems to be combined are disjoint, and that the rewriting rules of the sum term rewriting system are the rules of the summand term rewriting systems together. (Without the disjointness requirement the situation is even more complicated - see for some results in this direction: Dershowitz [2], Toyama [10].) A disjoint union of two term rewriting systems R_0 and R_1 is called in our paper a direct sum, notation $R_0 \oplus R_1$.

Another simplifying assumption that we will make, is that R_0, R_1 are homogeneous term rewriting systems, i.e. their signature is one-sorted (as opposed to the many sorted or heterogeneous case; for results about direct sums of heterogeneous term rewriting systems, see Ganzinger and Giegerich [3].)

The first result in this setting is due to Toyama [8], where it is proven that confluence is a modular property. (I.e. R_0 and R_1 are confluent iff $R_0 \oplus R_1$ is so. Here $\Leftarrow$ is trivial; $\Rightarrow$ is what we are interested in.) To appreciate the non-triviality of this fact, it may be contrasted with the fact that another fundamental property, termination, is *not* modular, as the following simple counterexample in Toyama [9] shows:

$$R_0 \quad \left\{ F(0,1,x) \triangleright F(x,x,x) \right.$$

$$R_1 \quad \left\{ \begin{array}{l} g(x,y) \triangleright x \\ g(x,y) \triangleright y \end{array} \right.$$

It is trivial that R_0 and R_1 are terminating. However, $R_0 \oplus R_1$ is not terminating, because $R_0 \oplus R_1$ has the infinite reduction sequence:

$$F(g(0,1),g(0,1),g(0,1)) \to F(0,g(0,1),g(0,1)) \to F(0,1,g(0,1))$$

$$\to F(g(0,1),g(0,1),g(0,1)) \to \cdots.$$

The above counterexample uses a non-confluent term rewriting system R_1. A more complicated counterexample to the modularity of *termination*, involving only confluent term rewriting systems, was given by Klop an Barendregt [4] (for ground terms only; for some improved versions, holding for open terms as well, and even using term rewriting systems which are *irreducible*, see Toyama [9]). This means that the important property of *completeness* of term rewriting systems (a term rewriting system is complete iff it is both confluent and terminating) is not modular, i.e. there are complete term rewriting systems R_0, R_1 such that $R_0 \oplus R_1$ is not complete (in fact, not terminating; confluence of $R_0 \oplus R_1$ is ensured by the theorem in Toyama [8]). This counterexample, however, uses non-left-linear term rewriting systems.

The point of the present paper is that left-linearity is essential; if we restrict ourselves to left-linear term rewriting systems, then completeness is modular. Thus we prove: If R_0 and R_1 are left-linear (meaning that the rewriting rules have no repeated variables in their left-hand-sides), then R_0 and R_1 are complete iff $R_0 \oplus R_1$ is so. As left-linearity is a property which is so easily checked, and many equational algebraic specifications can be given by term rewriting systems which are left-linear, we feel that this result is worth while.

The proof, however, is rather intricate and not easily digested. A crucial element in the proof, and in general in the way that the summand term rewriting systems interact, is how terms may *collapse* to a subterm. The problem is that this collapsing behavior may exhibit a *nondeterministic* feature, which is caused by ambiguities among the rewriting rules. We hope that the present paper is of value not only because it establishes a result that in itself is simple enough, but also because of the analysis necessary for the proof which gives a kind of structure theory for disjoint combinations of term rewriting systems and which may be of relevance in other, similar, studies.

Regarding the question of modular properties in the present simple set-up, we mention the recent results by Rusinowitch [7] and Middeldorp [5]; these papers, together, contain a complete analysis of the cases in which termination for $R_0 \oplus R_1$ may be concluded from termination of R_0 R_1, depending on the distribution among R_0, R_1 of so-called collapsing and duplicating rules.

Another useful fact is established in Middeldorp [6], where it is proven that the *unique normal form property* is a modular property.

2. Notations and Definitions

Assuming that the reader is familiar with the basic concepts and notations concerning term rewriting systems in [1,8], we briefly explain notations and definitions for the following discussions.

Let F be a set of function symbols, and let V be a set of variable symbols. By $T(F, V)$, we denote the set of terms constructed from F and V.

A term rewriting system R is a set of rewriting rules $M \triangleright N$, where M and N are terms disjoint function symbols [8].

In this paper, we assume that two disjoint systems R_0 on $T(F_0,V)$ and R_1 on $T(F_1,V)$ both are left-linear and complete. Then we shall prove that the direct sum system $R_0 \oplus R_1$ on $T(F_0 \cup F_1, V)$ is terminating. From here on the notation $\rightarrow$ represents the reduction relation on $R_0 \oplus R_1$.

Lemma 2.1. $R_0 \oplus R_1$ is weakly normalizing, i.e., every term M has a normal form (denoted by $M\downarrow$).

Proof. Since R_0 and R_1 are terminating, M can be reduced into $M\downarrow$ through innermost reduction. □

The identity of terms of $T(F_0 \cup F_1, V)$ (or syntactical equality) is denoted by $\equiv$. $\stackrel{*}{\rightarrow}$ is the transitive reflexive closure of $\rightarrow$, $\stackrel{+}{\rightarrow}$ is the transitive closure of $\rightarrow$, $\stackrel{\equiv}{\rightarrow}$ is the reflexive closure of $\rightarrow$, and $=$ is the equivalence relation generated by $\rightarrow$ (i.e., the transitive reflexive symmetric closure of $\rightarrow$). $\stackrel{m}{\rightarrow}$ denotes a reduction of m $(m \geq 0)$ steps.

Definition. A *root* is a mapping from $T(F_0 \cup F_1, V)$ to $F_0 \cup F_1 \cup V$ as follows: For $M \in T(F_0 \cup F_1, V)$,

$$root(M) = \begin{cases} f & \text{if } M \equiv f(M_1, \ldots, M_n), \\ M & \text{if } M \text{ is a constant or a variable.} \end{cases}$$

Definition. Let $M \equiv C[B_1, \ldots, B_n] \in T(F_0 \cup F_1, V)$ and $C \not\equiv \square$. Then write $M \equiv C[\![B_1, \ldots, B_n]\!]$ if $C[\,,\ldots,\,]$ is a context on F_d and $\forall i, root(B_i) \in F_{\bar{d}}$ ($d \in \{0,1\}$ and $\bar{d} = 1-d$). Then the set $S(M)$ of the special subterms of M is inductively defined as follows:

$$S(M) = \begin{cases} \{M\} & \text{if } M \in T(F_d, V) \ \ (d = 0 \text{ or } 1), \\ \bigcup_i S(B_i) \cup \{M\} & \text{if } M \equiv C[\![B_1, \ldots, B_n]\!] \ (n > 0). \end{cases}$$

The set of the special subterms having the root symbol in F_d is denoted by $S_d(M) = \{N \mid N \in S(M) \text{ and } root(N) \in F_d\}$.

Let $M \equiv C[\![B_1, \ldots, B_n]\!]$ and $M \stackrel{A}{\rightarrow} N$ (i.e., N results from M by contracting the redex occurrence A). If the redex occurrence A occurs in some B_j, then we write $M \underset{i}{\rightarrow} N$; otherwise $M \underset{o}{\rightarrow} N$. Here, $\underset{i}{\rightarrow}$ and $\underset{o}{\rightarrow}$ are called an inner and an outer reduction, respectively.

Definition. For a term $M \in T(F_0 \cup F_1, V)$, the rank of layers of contexts on F_0 and F_1 in M is inductively defined as follows:

$$rank(M) = \begin{cases} 1 & \text{if } M \in T(F_d, V) \ \ (d = 0 \text{ or } 1), \\ max_i\{rank(B_i)\} + 1 & \text{if } M \equiv C[\![B_1, \ldots, B_n]\!] \ (n > 0). \end{cases}$$

Lemma 2.2. If $M \to N$ then $rank(M) \geq rank(N)$.

Proof. It is easily obtained from the definitions of the direct sum. □

Lemma 2.3. Let $M \xrightarrow{*} N$ and $root(M), root(N) \in F_d$. Then there exists a reduction $M \equiv M_0 \to M_1 \to M_2 \to \cdots \to M_n \equiv N$ $(n \geq 0)$ such that $root(M_i) \in F_d$ for any i.

Proof. Let $M \xrightarrow{k} N$ $(k \geq 0)$. We will prove the lemma by induction on k. The case $k = 0$ is trivial. Let $M \to M' \xrightarrow{k-1} N$ $(k > 0)$. If $root(M') \in F_d$ then the lemma holds by the induction hypothesis. If $root(M') \in F_{\bar{d}}$ then there exists a context $C[\,]$ with root $\in F_d$ such that $M \equiv C[M']$ and $C[\,] \to \Box$. Thus, we can obtain a reduction $M \equiv C[M'] \xrightarrow{*} C[N] \to N$ in which all terms have root symbols in F_d. □

The set of terms in the reduction graph of M is denoted by $G(M) = \{N \mid M \xrightarrow{*} N\}$. The set of terms having the root symbol in F_d is denoted by $G_d(M) = \{N \mid N \in G(M) \text{ and } root(N) \in F_d\}$

Definition. A term M is erasable iff $M \xrightarrow{*} x$ for some $x \in V$.

From now on we assume that every term $M \in T(F_0 \cup F_1, V)$ has only x as variable occurrences unless it is stated otherwise. Since $R_0 \oplus R_1$ is left-linear, this variable convention may be assumed in the following discussions without loss of generality. If we need fresh variable symbols not in terms, we use z, z_1, z_2, $\cdots$.

3. Essential Subterms

In this section we introduce the concept of the essential subterms. We first prove the following property:

$$\forall N \in G_d(M)\ \exists P \in S_d(M),\ M \xrightarrow{*} P \xrightarrow{*} N.$$

Lemma 3.1. Let $M \to N$ and $Q \in S_d(N)$. Then, there exists some $P \in S_d(M)$ such that $P \xrightarrow{\equiv} Q$.

Proof. We will prove the lemma by induction on $rank(M)$. The case $rank(M) = 1$ is trivial. Assume the lemma for $rank(M) < k$ $(k > 1)$, then we will show the case $rank(M) = k$. Let $M \equiv C[\![M_1, \ldots, M_n]\!]$ $(n > 0)$ and $M \xrightarrow{A} N$.

Case 1. $M \equiv C[\![M_1, \ldots, M_r, \ldots, M_n]\!] \xrightarrow[o]{A} N \equiv M_r$.
Then $S_d(N) \subseteq S_d(M)$.

Case 2. $M \equiv C[\![M_1, \ldots, M_n]\!] \xrightarrow[o]{\Delta} N \equiv C'[\![M_{i_1}, \ldots, M_{i_p}]\!]$ $(1 \le i_j \le n)$.
If $root(M) \in F_d$ then
$S_d(M) = \{M\} \cup \bigcup_i S_d(M_i)$,
$S_d(N) = \{N\} \cup \bigcup_j S_d(M_{i_j})$.
Thus the lemma holds since $\bigcup_j S_d(M_{i_j}) \subseteq \bigcup_i S_d(M_i)$, and $M \to N$.
If $root(M) \in F_{\bar{d}}$ then $S_d(N) = \bigcup_j S_d(M_{i_j}) \subseteq \bigcup_i S_d(M_i) = S_d(M)$.

Case 3. $M \equiv C[\![M_1, \ldots, M_r, \ldots, M_n]\!] \xrightarrow[i]{\Delta} N \equiv C[M_1, \ldots, M'_r, \ldots, M_n]$ where $M_r \xrightarrow{\Delta} M'_r$.
If $root(M) \in F_d$ then
$S_d(M) = \{M\} \cup S_d(M_r) \cup \bigcup_{i \ne r} S_d(M_i)$,
$S_d(N) \subseteq \{N\} \cup S_d(M'_r) \cup \bigcup_{i \ne r} S_d(M_i)$.
If $root(M) \in F_{\bar{d}}$ then
$S_d(M) = S_d(M_r) \cup \bigcup_{i \ne r} S_d(M_i)$,
$S_d(N) = S_d(M'_r) \cup \bigcup_{i \ne r} S_d(M_i)$.
By the induction hypothesis, $\forall Q \in S_d(M'_r) \exists P \in S_d(M_r), P \xrightarrow{\equiv} Q$ for the both $root(M) \in F_d$ and $root(M) \in F_{\bar{d}}$. Thus the lemma holds. □

R_e consists of the single rule $e(x) \triangleright x$. $\xrightarrow[e]{}$ denotes the reduction relation of R_e, and $\xrightarrow[e']{}$ denotes the reduction relation of $R_e \oplus (R_0 \oplus R_1)$ such that if $C[e(P)] \xrightarrow[e']{\Delta} N$ then the redex occurrence Δ does not occur in P. It is easy to show the confluence property of $\xrightarrow[e']{}$.

From here on, $C[e(P_1), \cdots, e(P_p)]$ denotes a term such that $C[P_1, \cdots, P_p] \in T(F_0 \cup F_1, V)$, i.e., C and P_i contain no e.

Lemma 3.2. Let $C[e(P_1), \cdots, e(P_{i-1}), e(P_i), e(P_{i+1}), \cdots, e(P_p)] \xrightarrow[e']{k} e(P_i)$. Then $C[P_1, \cdots, P_{i-1}, e(P_i), P_{i+1}, \cdots, P_p] \xrightarrow[e']{k'} e(P_i)$ $(k' \le k)$.

Proof. It is easily obtained from the definition and the left-linearity of the reduction $\xrightarrow[e']{}$. □

Let $M \equiv C[P] \in T(F_0 \cup F_1, V)$ be a term containing no function symbol e. Now, consider $C[e(P)]$ by replacing the occurrence P in M with $e(P)$. Assume $C[e(P)] \xrightarrow[e']{*} e(P)$. Then, by tracing the reduction path, we can also obtain the reduction $M \equiv C[P] \xrightarrow{*} P$ (denoted by $M \xrightarrow[pull]{*} P$) under $R_0 \oplus R_1$. We say that the reduction $M \xrightarrow[pull]{*} P$ pulls up the occurrence P from M.

Example 3.1. Consider the two systems R_0 and R_1:

$$R_0 \quad \begin{cases} F(x) \to G(x,x) \\ G(C,x) \to x \end{cases}$$

$$R_1 \quad \left\{ h(x) \to x \right.$$

Then we have the reduction:

$F(e(h(C))) \underset{e'}{\to} G(e(h(C)), e(h(C))) \underset{e'}{\to} G(h(C), e(h(C))) \underset{e'}{\to} G(C, e(h(C))) \underset{e'}{\to} e(h(C))$.

Hence $F(h(C)) \underset{pull}{\overset{*}{\longrightarrow}} h(C)$. However, we cannot obtain $F(z) \underset{pull}{\overset{*}{\longrightarrow}} z$. Thus, in general, we cannot obtain $C[z] \underset{pull}{\overset{*}{\longrightarrow}} z$ from $C[P] \underset{pull}{\overset{*}{\longrightarrow}} P$. □

Lemma 3.3. Let $P \overset{*}{\to} Q$ and let $C[Q] \underset{pull}{\overset{*}{\longrightarrow}} Q$. Then $C[P] \underset{pull}{\overset{*}{\longrightarrow}} P$.

Proof. Let $M \equiv C[e(Q)] \underset{e'}{\overset{k}{\longrightarrow}} e(Q)$. We will prove the lemma by induction on k. The case $k = 0$ is trivial. Let $M \equiv C[e(Q)] \underset{e'}{\longrightarrow} C'[e(Q), \cdots, e(Q), \cdots, e(Q)] \underset{e'}{\overset{k-1}{\longrightarrow}} e(Q)$. Then, from Lemma 3.2 we can obtain the following reduction:

$C'[Q, \cdots, e(Q), \cdots, Q] \underset{e'}{\overset{k'}{\longrightarrow}} e(Q)$ $(k' \leq k-1)$.

By using the induction hypothesis, $C'[Q, \cdots, e(P), \cdots, Q] \underset{e'}{\overset{*}{\longrightarrow}} e(P)$. Therefore, we can obtain

$C[e(P)] \underset{e'}{\longrightarrow} C'[e(P), \cdots, e(P), \cdots, e(P)] \underset{e'}{\overset{*}{\longrightarrow}} C'[Q, \cdots, e(P), \cdots, Q] \underset{e'}{\overset{*}{\longrightarrow}} e(P)$

from $P \overset{*}{\to} Q$. □

Lemma 3.4. $\forall N \in G_d(M)\ \exists P \in S_d(M),\ M \underset{pull}{\overset{*}{\longrightarrow}} P \overset{*}{\to} N$.

Proof. If $root(M) \in F_d$ then the above property is trivial by taking M as P. Thus we consider only the non trivial case of $root(M) \in F_{\bar{d}}$. Let $M \overset{k}{\longrightarrow} N$. We will prove the lemma by induction on k. The case $k = 1$ is trivial since $M \equiv C[\![M_1, \ldots, M_r, \ldots, M_n]\!] \to N \equiv M_r$ for some r (i.e., take $P \equiv M_r$). Assume the lemma for $k-1$. We will prove the case k. Let $M \to M' \overset{k-1}{\longrightarrow} N$.

Case 1. $root(M') \in F_d$.

Then $M \equiv C[\![M_1, \ldots, M_r, \ldots, M_n]\!] \to M' \equiv M_r$ for some r. Take $P \equiv M_r$.

Case 2. $root(M') \in F_{\bar{d}}$.

By using the induction hypothesis, $\exists P' \in S_d(M')$, $M' \xrightarrow[pull]{*} P' \xrightarrow{*} N$. Here, from Lemma 3.1, here exists some $P \in S_d(M)$ such that $P \xrightarrow{\equiv} P'$. We will consider the following two subcases:

Case 2.1. $P \to P'$. Then $M \equiv C[P] \to M' \equiv C[P']$. Thus, by using Lemma 3.3, $M \equiv$ $C[P] \xrightarrow[pull]{*} P \to P' \xrightarrow{*} N$.

Case 2.2. $P \equiv P'$. Then, for some context $C'[\,,\cdots,\,]$, $M \equiv C[P] \to M' \equiv C'[P,\cdots,P,\cdots,P]$ nd $C'[P,\cdots,e(P),\cdots,P] \xrightarrow[e']{*} e(P)$. Therefore

$C[e(P)] \to C'[e(P),\cdots,e(P),\cdots,e(P)] \xrightarrow[e]{*} C'[P,\cdots,e(P),\cdots,P] \xrightarrow[e']{*} e(P)$. Thus $M \equiv C[P]$ $\xrightarrow[pull]{*} P \xrightarrow{*} N$. □

Now, we introduce the concept of the essential subterms. The set $E_d(M)$ of the essential ubterms of the term $M \in T(F_0 \cup F_1, V)$ is defined as follows:

$E_d(M) = \{P \mid M \xrightarrow[pull]{*} P \in S_d(M) \text{ and } \neg\exists Q \in S_d(M)\ [M \xrightarrow[pull]{*} Q \xrightarrow{+} P]\}$.

The following lemmas are easily obtained from the definition of the essential subterms and Lemma 3.4.

Lemma 3.5. $\forall N \in G_d(M)\ \exists P \in E_d(M),\ P \xrightarrow{*} N$.

Lemma 3.6. $E_d(M) = \phi$ iff $G_d(M) = \phi$.

We say M is deterministic for d if $|E_d(M)| = 1$; M is nondeterministic for d if $|E_d(M)| \geq 2$. The following lemma plays an important role in the next section.

Lemma 3.7 If $root(M\downarrow) \in F_d$ then $|E_d(M)| = 1$, i.e., M is deterministic for d.

Proof. See Appendix in [11]. □

4. Termination for the Direct Sum

In this section we will show that $R_0 \oplus R_1$ is terminating. Roughly speaking, termination is proven by showing that any infinite reduction $M_0 \to M_1 \to M_2 \to \cdots$ of $R_0 \oplus R_1$ can be translated into an infinite reduction $M'_0 \to M'_1 \to M'_2 \to \cdots$ of R_d.

We first define the term $M^d \in T(F_d, V)$ for any term M and any d.

Definition. For any M and any d, $M^d \in T(F_d, V)$ is defined by induction on $rank(M)$:

(1) $M^d \equiv M$ if $M \in T(F_d, V)$.

(2) $M^d \equiv x$ if $E_d(M) = \phi$.

(3) $M^d \equiv C[M_1^d, \cdots, M_m^d]$ if $root(M) \in F_d$ and $M \equiv C[\![M_1, \cdots, M_m]\!]$ $(m > 0)$.

(4) $M^d \equiv P^d$ if $root(M) \in F_{\bar{d}}$ and $E_d(M) = \{P\}$. Note that $rank(P) < rank(M)$.

(5) $M^d \equiv C_1[C_2[\cdots C_{p-1}[C_p[x]]\cdots]]$ if $root(M) \in F_{\bar{d}}$, $E_d(M) = \{P_1, \cdots, P_p\}$ $(p > 1)$, and every P_i^d is erasable. Here $P_i^d \equiv C_i[x] \xrightarrow[pull]{*} x$ $(i = 1, \cdots, p)$. Note that, for any i, $rank(P_i) < rank(M)$ and $M^d \xrightarrow{*} P_i^d$.

(6) $M^d \equiv x$ if $root(M) \in F_{\bar{d}}$, $|E_d(M)| \geq 2$, and not (5).

Note that M^d is not unique if a subterm of M^d is constructed with (5) in the above definition.

Lemma 4.1. $root(M\downarrow) \notin F_d$ iff $M^d \downarrow \equiv x$.

Proof. Instead of the lemma, we will prove the following claim:

Claim. If $root(M\downarrow) \notin F_d$ then $M^d \downarrow \equiv x$. If $root(M\downarrow) \in F_d$ and $M\downarrow \equiv \hat{C}[\![M_1, \cdots, M_m]\!]$ then $M^d \downarrow \equiv \hat{C}[x, \cdots, x]$.

Proof of the Claim. We will prove the lemma by induction on $rank(M)$. The case $rank(M) = 1$ is trivial by the definition of M^d. Assume the lemma for $rank(M) < k$ $(k \geq 2)$. Then we will prove the case $rank(M) = k$.

Case 1. $root(M) \in F_d$.

Let $M \equiv C[\![M_1, \cdots, M_m]\!]$. Then $M^d \equiv C[M_1^d, \cdots, M_m^d]$. We may assume that $root(M_i\downarrow) \notin F_d$ $(1 \leq i < p)$ and $root(M_j\downarrow) \in F_d$ $(p \leq j \leq m)$ without loss of generality. Let $M_j \downarrow \equiv \hat{C}_j[\![N_{j,1}, \cdots, N_{j,n_j}]\!]$ $(p \leq j \leq m)$. Then, by using the induction hypothesis, $M_i^d \downarrow \equiv x$ $(1 \leq i < p)$ and $M_j^d \downarrow \equiv \hat{C}_j[x, \cdots, x]$ $(p \leq j \leq m)$. Thus $M\downarrow \equiv C[M_1\downarrow, \cdots, M_m\downarrow]\downarrow$
$\equiv C[M_1\downarrow, \cdots, M_{p-1}\downarrow, \hat{C}_p[\![N_{p,1}, \cdots, N_{p,n_p}]\!], \cdots, \hat{C}_m[\![N_{m,1}, \cdots, N_{m,n_m}]\!]]\downarrow$
and $M^d \downarrow \equiv C[M_1^d\downarrow, \cdots, M_m^d\downarrow]\downarrow \equiv C[x, \cdots, x, \hat{C}_p[x, \cdots, x], \cdots, \hat{C}_m[x, \cdots, x]]\downarrow$. Note that $M_i\downarrow$ $(1 \leq i < p)$, $N_{p,1}, \cdots, N_{m,n_m}$ are normal forms having root symbols not in F_d. Therefore, if $root(M\downarrow) \notin F_d$ then $C[x, \cdots, x, \hat{C}_p[x, \cdots, x], \cdots, \hat{C}_m[x, \cdots, x]]\downarrow \equiv x$; if $root(M\downarrow) \in F_d$ then we have a context
$\hat{C}[\ ,\cdots,\ ] \equiv C[\ ,\cdots,\ ,\hat{C}_p[\ ,\cdots,\ ],\cdots,\hat{C}_m[\ ,\cdots,\ ]]\downarrow$ such that $M\downarrow \equiv \hat{C}[\![N_1, \cdots, N_n]\!]$ where $N_i \in \{M_1\downarrow, \cdots, M_{p-1}\downarrow, N_{p,1}, \cdots, N_{m,n_m}\}$ and $M^d \downarrow \equiv \hat{C}[x, \cdots, x] \not\equiv x$.

Case 2. $root(M) \notin F_d$.

Consider three subcases:

Case 2.1. $E_d(M) = \phi$.

From Lemma 3.6, $root(M\downarrow) \notin F_d$. Since $M^d \equiv x$, $M^d \downarrow\equiv x$.

Case 2.2. $E_d(M) = \{P\}$.

Then $M^d \equiv P^d$. Note that $rank(P) < k$. Since $M \downarrow\equiv P\downarrow$ and $M^d \downarrow\equiv P^d\downarrow$, the claim follows by using the induction hypothesis.

Case 2.3. $E_d(M) = \{P_1, \cdots, P_p\}$ $(p > 1)$.

Note that $rank(P_i) < k$ for any i. From Lemma 3.7, $root(M\downarrow) \notin F_d$. Since $M\downarrow\equiv P_i\downarrow$, it is clear that $root(P_i\downarrow) \notin F_d$ for all i. Thus, we have $P_i^d \downarrow\equiv x$ by the induction hypothesis. From case (5) in the definition of M^d, it follows that $M^d \downarrow\equiv x$. □

Note. Let $E_d(M) = \{P_1, \cdots, P_p\}$ $(p > 1)$. Then, from Lemma 3.7 and Lemma 4.1, it follows that every P_i^d is erasable. Hence case (6) in the definition of M^d can be removed.

Lemma 4.2. If $P \in E_d(M)$ then $M^d \xrightarrow{*} P^d$.

Proof. Obvious from the definition of M^d and the above note. □

We wish to translate directly an infinite reduction $M_0 \to M_1 \to M_2 \to \cdots$ into an infinite reduction $M_0^d \xrightarrow{*} M_1^d \xrightarrow{*} M_2^d \xrightarrow{*} \cdots$. However, the following example shows that $M_i \to M_{i+1}$ cannot be translated into $M_i^d \xrightarrow{*} M_{i+1}^d$ in general.

Example 4.1. Consider the two systems R_0 and R_1:

$$R_0 \quad \begin{cases} F(C,x) \to x \\ F(x,C) \to x \end{cases}$$

$$R_1 \quad \begin{cases} f(x) \to g(x) \\ f(x) \to h(x) \\ g(x) \to x \\ h(x) \to x \end{cases}$$

Let $M \equiv F(f(C),h(C)) \to N \equiv F(g(C),h(C))$. Then $E_1(M) = \{f(C)\}$ and $E_1(N) = \{g(C),h(C)\}$. Thus $M^1 \equiv f(x)$, $N^1 \equiv g(h(x))$. It is obvious that $M^1 \xrightarrow{*} N^1$ does not hold. □

Now we will consider to translate indirectly an infinite reduction of $R_0 \oplus R_1$ into an infinite reduction of R_d.

We write $M \underset{o}{\equiv} N$ when M and N have the same outermost-layer context, i.e., $M \equiv C[\![M_1, \cdots, M_m]\!]$ and $N \equiv C[\![N_1, \cdots, N_m]\!]$ for some M_i, N_i.

Lemma 4.3. Let $A \xrightarrow[i]{*} M$, $M \xrightarrow[o]{} N$, $A \underset{o}{\equiv} M$, and $root(M), root(N) \in F_d$. Then, for any A^d there exist B and B^d such that

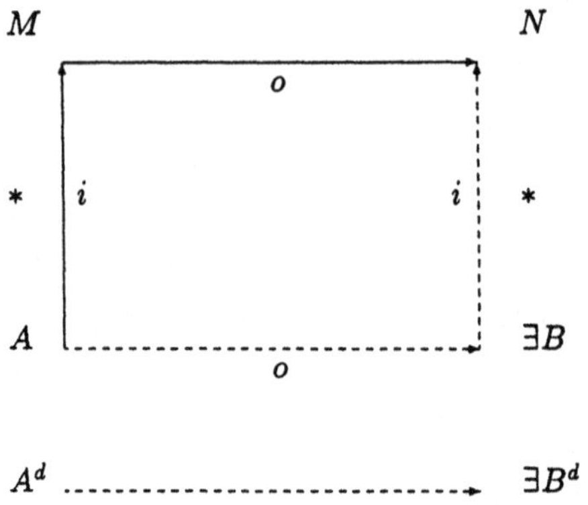

Proof. Let $A \equiv C[\![A_1, \cdots, A_m]\!]$, $M \equiv C[\![M_1, \cdots, M_m]\!]$, $N \equiv C'[\![M_{i_1}, \cdots, M_{i_n}]\!]$ ($i_j \in \{1, \cdots, m\}$). Take $B \equiv C'[\![A_{i_1}, \cdots, A_{i_n}]\!]$. Then, we can obtain $A \xrightarrow[o]{} B$ and $B \xrightarrow[i]{*} N$. From $A^d \equiv C[A_1^d, \cdots, A_m^d]$ and $B^d \equiv C'[A_{i_1}^d, \cdots, A_{i_n}^d]$, it follows that $A^d \to B^d$. □

Lemma 4.4. Let $M \xrightarrow{*} N$, $root(N) \in F_d$. Then, for any M^d there exist A ($A \underset{o}{\equiv} N$) and A^d such that

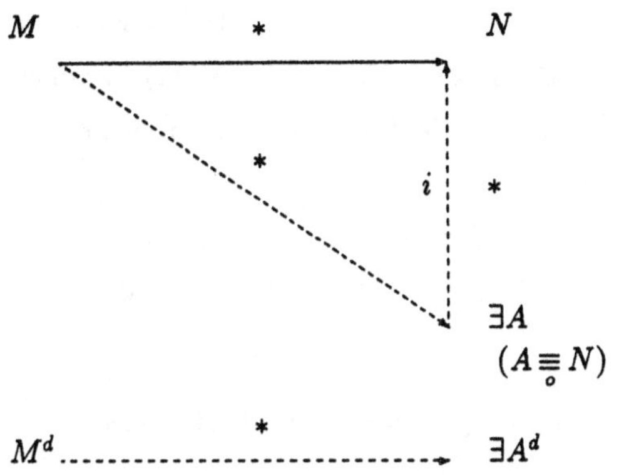

Proof. We will prove the lemma by induction on $rank(M)$. The case $rank(M) = 1$ is rivial by taking $A \equiv N$. Assume the lemma for $rank(M) < k$. Then we will prove the case $rank(M) = k$. We start from the following claim.

Claim. The lemma holds if $M \xrightarrow[i]{*} N$.

Proof of the Claim. Let $M \equiv C[\![M_1, \cdots, M_m]\!] \xrightarrow[i]{*} N \equiv C[N_1, \cdots, N_m]$ where $M_i \xrightarrow{*} N_i$ for every . We may assume that $N_1 \equiv x, \cdots, N_{p-1} \equiv x$, $root(N_i) \in F_d$ $(p \leq i \leq q-1)$, and $root(N_j) \in F_{\bar{d}}$ $(q \leq j \leq m)$ without loss of generality. Thus $N \equiv C[x, \cdots, x, N_p, \cdots, N_{q-1}, N_q, \cdots, N_m]$. Then, by using the induction hypothesis, every M_i $(p \leq i \leq q-1)$ has A_i $(A_i \underset{o}{\equiv} N_i)$ and A_i^d uch that

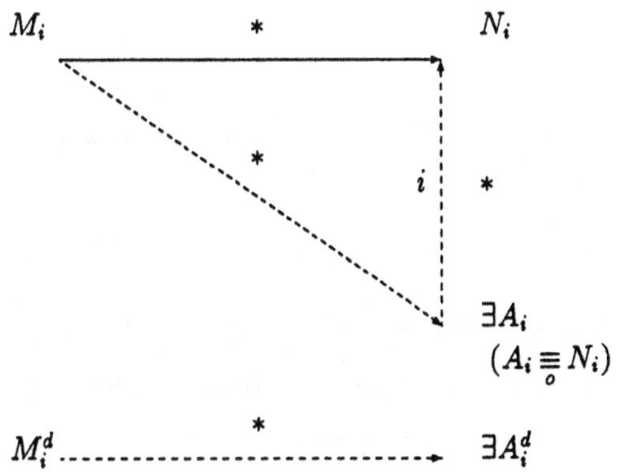

Now, take $A \equiv C[x, \cdots, x, A_p, \cdots, A_{q-1}, M_q, \cdots, M_m]$. It is obvious that $M \overset{*}{\to} A$. From Lemma 2.3, we can have the reductions $M_j \overset{*}{\to} N_j$ $(q \leq j \leq m)$ in which every term has a root symbol in $F_{\bar{d}}$. Thus it follows that $A \overset{*}{\underset{i}{\to}} N$ and $A \underset{o}{\equiv} N$. From Lemma 4.1 and $M_i \downarrow \equiv x$ $(1 \leq i < p)$, $M_i^d \downarrow \equiv x$. Therefore, since

$M^d \equiv C[M_1^d, \cdots, M_{p-1}^d, M_p^d, \cdots, M_{q-1}^d, M_q^d, \cdots, M_m^d]$

and $A^d \equiv C[x, \cdots, x, A_p^d, \cdots, A_{q-1}^d, M_q^d, \cdots, M_m^d]$, it follows that $M^d \overset{*}{\to} A^d$. *(end of the claim)*

Now we will prove the lemma for $rank(M) = k$. Consider two cases.

Case 1. $root(M) \in F_d$.

From Lemma 2.3, we may assume that every term in the reduction $M \overset{*}{\to} N$ has a root symbol in F_d. By splitting $M \overset{*}{\to} N$ into $M \overset{*}{\underset{i}{\to}} \underset{o}{\to} \overset{*}{\underset{i}{\to}} \underset{o}{\to} \cdots \overset{*}{\underset{i}{\to}} N$ and using the claim for diagram (1) and Lemma 5.1 for diagram (2), we can draw the following diagram:

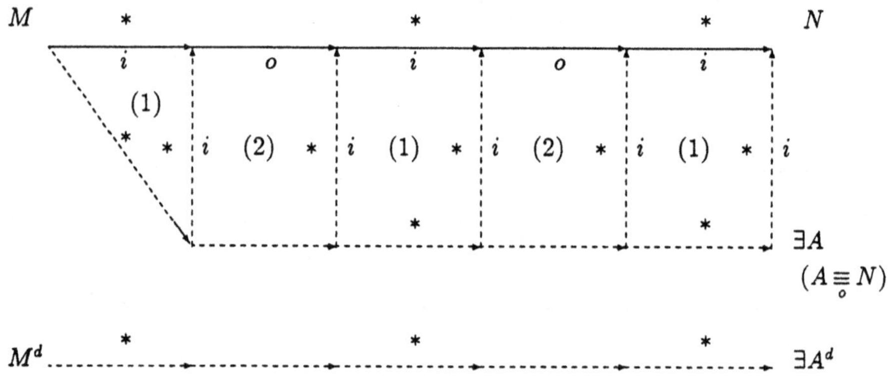

Note that if $M' \overset{*}{\underset{i}{\to}} M'' \overset{*}{\underset{i}{\to}} M'''$ then $M' \overset{*}{\underset{i}{\to}} M'''$; thus, the claim can be applied to diagram (1) in the above diagram.

Case 2. $root(M) \in F_{\bar{d}}$.

Then we have some essential subterm $Q \in E_d(M)$ such that $M \overset{*}{\to} Q \overset{*}{\to} N$. From Lemma 4.2 it follows that $M^d \overset{*}{\to} Q^d$. It is obvious that $rank(Q) < k$. Hence, we have the following diagram where diagram (1) is obtained by the induction hypothesis:

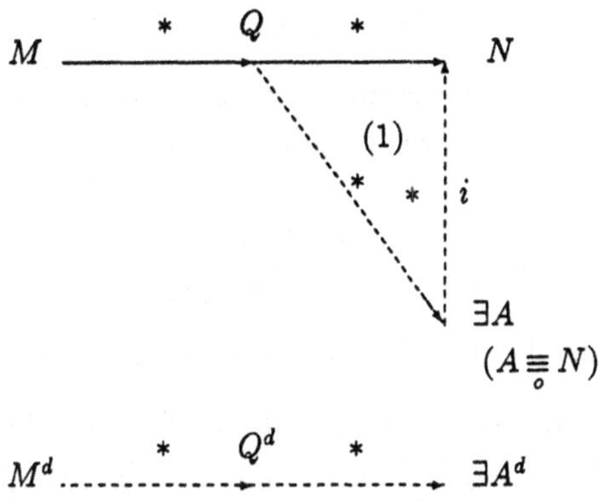

□

Now we can prove the following theorem:

Theorem 4.1. No term M has an infinite reduction.

Proof. We will prove the theorem by induction on $rank(M)$. The case $rank(M) = 1$ is trivial. Assume the theorem for $rank(M) < k$. Then, we will show the case $rank(M) = k$. Suppose M has an infinite reduction $M \to\to\to \cdots$. From the induction hypothesis, we can have no infinite inner reduction $\underset{i}{\to}\underset{i}{\to}\underset{i}{\to}\cdots$ in this reduction. Thus, $\underset{o}{\to}$ must infinitely appear in the infinite reduction. From the induction hypothesis, all of the terms appearing in this reduction have the same rank; hence, their root symbols are in F_d if $root(M) \in F_d$. Hence, by a similar construction of diagrams as for Case 1 in the proof of Lemma 4.4, it follows that M^d has an infinite reduction. This contradicts that R_d is terminating. □

Corollary 4.1. Two term rewriting systems R_0 and R_1 are left-linear and complete iff the direct sum $R_0 \oplus R_1$ is so.

Proof. $\Leftarrow$ is trivial. $\Rightarrow$ follows from Theorem 4.1 and the theorem in Toyama [8] stating that two term rewriting systems R_0 and R_1 are confluent iff the direct sum $R_0 \oplus R_1$ is so. □

References

[1] G. Huet and D. C. Oppen, Equations and rewrite rules: a survey, in: R. Book, ed., Formal

languages: perspectives and open problems (Academic Press, 1980) 349-393.

[2] N. Dershowitz. Termination of linear rewriting systems: Preliminary version, *Lecture Notes in Comput. Sci. 115* (Springer-Verlag, 1981) 448-458.

[3] H. Ganzinger and R. Giegerich, A note on termination in combinations of heterogeneous term rewriting systems, *EATCS Bulletin 31* (1987) 22-28.

[4] J. W. Klop and H. P. Barendregt, Private communication (January 19, 1986).

[5] A. Middeldorp, A sufficient condition for the termination of the direct sum of term rewriting systems, Preliminary Draft, *Report IR-150*, Free University, Amsterdam (1988).

[6] A. Middeldorp, Modular aspects of properties of term rewriting systems related to normal forms, *Preliminary Draft*, Free University, Amsterdam (September 1988).

[7] M. Rusinowitch, On termination of the direct sum of term rewriting systems, *Inform. Process. Lett. 26* (1987) 65-70.

[8] Y. Toyama, On the Church-Rosser property for the direct sum of term rewriting systems, *J. ACM 34* (1987) 128-143.

[9] Y. Toyama, Counterexamples to termination for the direct sum of term rewriting systems, *Inform. Process. Lett. 25* (1987) 141-143.

[10] Y. Toyama, Commutativity of term rewriting systems, in: K. Fuchi and L. Kott, eds., Programming of Future Generation Computer II (Norht Holland 1988) 393-407.

[11] Y. Toyama, J.W. Klop, H.P. Barendregt, Termination for the direct sum of left-linear term rewriting systems: Preliminary draft, *IEICE technical report COM88-30* (July 1988) 47-55.

CONDITIONAL REWRITE RULE SYSTEMS WITH BUILT-IN ARITHMETIC AND INDUCTION

Sergey G. Vorobyov

Program Systems Institute
USSR Academy of Sciences
152140 Pereslavl-Zalessky
SOVIET UNION

ABSTRACT. Conditional rewriting systems, conditions being the formulae of decidable theories, are investigated. A practical search space-free decision procedure for the related class of unquantified logical theories is described. The procedure is based on cooperating conditional reductions, case splittings and decision algorithms, and is able to perform certain forms of inductive inferences. Completeness and termination of the procedure are proved.

1. INTRODUCTION

Rewrite rule systems and related techniques appeared to be extremely suitable for applications in data type specification, program verification, automated theorem proving, etc. Owing to the search space-free character inherent to rewrite rule-based decision procedures, it is possible to maintain theories with large number of axioms and conduct tedious proofs of long formulae arising often in practice, e.g. in verification. However, to obtain the above advantages, preliminary compilation of axioms into uniquely terminating (or canonical) rewrite rules must be accomplished. As it is well known, such a compilation may fail without any result. This is especially probable for rich data types, containing arithmetic or induction principles. There exist at least two acknowledged approaches to overcome the low expressiveness barrier of canonical term rewriting: the "inductionless" inductive method [HuHu, JoKo] and rewrite rule-based refutational strategy for first-order logic [Hs]. It must be emphasized that both methods exploit the Knuth-Bendix completion algorithm [KB, HuOp] in an interpretive mode, running it every time proving a new formula. Although successful, these methods loose the search space freedom, the main value of canonical term rewriting.

In this paper we develop a new approach to increase the expressive power of canonical term rewriting, based on building-in decision algorithms for fragments of arithmetic and special inductive inference procedures. It turns out that building-in decision algorithms into conditional rewrite rule systems, as a tool to check conditions and to decide formulae reduced to normal forms, considerably increases their expressive power. We proved in [Vo1], that the quantifier-free first-order theory of $\mathfrak{N} = \langle N; 0, Succ\rangle$ as well as any its enrichment can not be axiomatized by a canonical context-free (i.e. containing no boolean connectives in left-hand sides of non-logical rules) term rewriting system. Also, the unquantifier-free theory of partially ordered structures can not be axiomatized by a canonical system admitting only conjunctions in left-hand sides of non-logical rules [Vo6]. Therefore, building-in decision algorithms into rewrite rule systems is worth investigating. We introduce a class of quantifier-free logical theories, which we call E-theories, axiomatized by formulae of the form **Cond** $\supset$ (**Atom** $\Leftrightarrow$ **Formula**), where **Cond** is a formula of some basic theory T_0 assumed to be decidable (say, Presburger arithmetic **PAR**), **Atom** is an atom of the extended language, and **Formula** is an arbitrary unquantified formula. An $E(T_0)$-theory may be thought of as an extension of T_0, possessing a clear operational semantics. Converting E-axioms into rewrite rules yields conditional rules of the form **Cond** :: **Atom** ---> **Formula**, and it is valid to ask, whether the universal fragment of an $E(T_0)$-theory can be decided by rewriting using the rules generated by its axioms. We show that the answer is positive, but it is necessary to generalize the main notions of term rewriting appropriately, because proper interconnections between a rewriting and a decision parts of a conditional rewrite rule system must be provided to achieve completeness. For example, the case splitting inference rule must be introduced: suppose we reduce a formula G in a global context C, and a condition **Cond** of a rule to be applied next to G does not follow from C; then the proof of G must be splitted into two proofs of G, one in the context C & **Cond**, and the second in the context C & $\neg$**Cond**. Also, formulae in complete normal forms, i.e. neither reducible, nor case splittable must be decided using built-in decision algorithms. The definitions of confluency and finite termination are modified substantially to capture case splittings and decision algorithms invocations. We describe here the decision procedure for $E(T_0)$-theories and prove corresponding completeness and termination results.

The key idea validating our decision method is axiom omitting which can be roughly described as follows. Suppose a theory T is partitioned into two parts: a reductive part T_1 generating a conditional rewriting system R_1, and a non-reductive part T_2, decidable by some algorithm. We explore conditions guaranteeing that any valid in $T_1 \cup T_2$ unquantified formula H, which is neither R_1-reducible, nor case splittable, does not depend on T_1, i.e. H is valid in T_2 only. Fortunately, these conditions, viz. confluency and termination of R_1, are not very strong or restrictive in the context of term rewriting. Generalizing the idea of axiom omitting, we obtain a possibility to incorporate special kinds of proofs by induction using appropriate tools of inductive reasoning, e.g. decision procedures for fragments of second-order successor arithmetic S1S [Ra], or explicit definability. This differs radically from traditional "inductionless" inductive proof methods by means of the Knuth-Bendix completion algorithm [HuHu, JoKo]. We present herein the results on completeness, decidability and termination of our method.

The scenario of the presentation is as follows. Each Section has a corresponding prelude. In Section 2 we define the class of $E(T_0)$-theories, the related decision problems, the class of conditional rewrite rule systems with built-in decision algorithms and present rewrite rule-based theorem proving techniques for $E(T_0)$-theories. Completeness is investigated in Section 3, it is there we state our main completeness result, the axioms omitting theorem. Section 4 is devoted to the problem of inductive definitions, proofs by induction and related termination issues, there we generalize the main theorem. Section 5 is devoted to inductive proofs and decidability of inductive theories, axioms omitting applies again. Appendix contains sketches of the proofs of all theorems presented.

2. E-THEORIES AND CONDITIONAL REWRITE RULE SYSTEMS

In this Section we introduce denotational conventions and define a class of quantifier-free logical theories together with the associated conditional rewrite rule systems. Also, we formulate the decision problem for these theories, describe the rewrite rule-based decision procedure and give examples both of $E(T_0)$-axiomatizations and decision sessions.

Conditional equivalences of the form **Cond** $\supset$ (**Atom** $\Leftrightarrow$ **Formula**) possess the following obvious operational semantics: while proving a theorem, replace instances of the **Atom** by corresponding instances of the **Formula** whenever the condition **Cond** is satisfied in the current set of assumptions. The conditions **Cond** can be represented by formulae of some decidable theory T_0, which we call a basic theory. We suppose that there exists an algorithm $\mathfrak{A}$ deciding T_0 which we use to decide the validity of conditions to perform the replacements. In §2.2 we will rigorously define a class of theories, which we will call E-theories, specified by axioms of the above form. We give examples of E-theories in §2.3, and in §2.4 define a class of conditional rewrite rule systems closely related to E-theories. The decision procedure based on cooperating conditional reductions and decision algorithms is described in §2.5, and in §2.6 we give an example of its work. First of all we introduce

2.1. Denotational conventions. We assume familiarity with basic notions of many-sorted logic and term rewriting systems, see [KB, HuOp]. For clarity we will use the following unified notation till the end of the paper. All symbols introduced in definitions below will also have fixed meaning throughout the text.

Let S_0 and S be two non-empty sets of sorts such that $S_0 \subseteq S$. Let FP_0 be an S_0-sorted similarity type (signature) and FP be an S-sorted similarity type such that the difference $FP \setminus FP_0$ does not contain function symbols with the S_0-sorted codomain. By T_0 we denote an arbitrary first-order theory with the decidable set of quantifier-free logical consequences:

$$Th^{\forall}(T_0) \equiv_{df} \{ B \mid T_0 \models B, \text{ B is quantifier-free } \} .$$

Hereafter, T_0 is referred to as the basic theory. A good heuristic is to think of T_0 as of Presburger arithmetic PAR or successor arithmetic SA (see below). The letters B and C denote T_0-consistent quantifier-free formulae of the FP_0 similarity type, further referred to as contexts, whereas f and g stand for atomic formulae of FP with leading predicate symbols from $FP \setminus FP_0$. Quantifier-free formulae of FP are denoted by F, G. By $G[g]$ we express the fact that a formula G contains a distinguished occurence of an atomic formula g, and by $G[F / g]$ we denote the result of replacing the distinguished occurence of g in G by F.

Let Var(F) be the set of variables in F, σ be a substitution of terms for variables, and $\sigma(F)$ be the result of substituting to a formula. By &, $\vee$, $\supset$, $\neg$, and $\Leftrightarrow$ we denote the usual boolean connectives. FP_{PAR} = < 0, Succ, +, −, $\leq$ > is the similarity type of Presburger arithmetic PAR, and Stm(PAR) is the standard model < $\mathbb{N}$, FP_{PAR} > of PAR, $\mathbb{N}$ being the set of natural numbers. FP_{SA} = < 0, Succ, $\leq$ > is the similarity type of successor arithmetic SA, and Stm(SA) is the standard model < $\mathbb{N}$, FP_{SA} > of SA. Let $T_0^=$ denote the basic theory T_0 enriched with the uninterpreted symbols of $FP \setminus FP_0$, augmented with equality axioms of the form $x = y$ & $f(x) \supset f(y/x)$. By STM($PAR^= \cup$ T) we denote the class of standard models of $PAR^= \cup$ T, with the carriers of the arithmetic sort isomorphic to Stm(PAR). Similarly, by STM($SA^= \cup$ T) we denote the class of standard models of $SA^= \cup$ T.

2.2. E-theories. In this paragraph we define syntactically a class of "operationally tractable" theories with axioms easily transformable into conditional rewrite rules, and to pose decision problems for these theories.

Definition. A theory T of the similarity type FP is called a $E(T_0)$-theory, iff each axiom of T is of the form

$$B \supset (f \Leftrightarrow F) , \qquad (\text{E-axiom})$$

and the variable restriction $Var(B) \cup Var(F) \subseteq Var(f)$ holds for each. ■

<u>Definition</u>. A <u>quantifier-free</u> <u>decision</u> <u>problem</u> for an $E(T_0)$-theory T consists in determining, whether or not an unquantified formula belongs to the set

$$\mathrm{Th}^{\forall}(T_0^{=} \cup T) =_{df} \{ H \mid T_0^{=} \cup T \models H \text{ , } H \text{ is unquantified } \} .$$

Similarly, an <u>inductive</u> <u>quantifier-free</u> <u>decision</u> <u>problem</u> for an E(PAR)-theory T consists in determining, whether or not an unquantified formula belongs to the set

$$\mathrm{Th}^{\forall}(\mathrm{STM}(\mathrm{PAR}^{=} \cup T)) =_{df} \{ H \mid \mathrm{STM}(\mathrm{PAR}^{=} \cup T) \models H, H \text{ is unquantified } \}. ■$$

The decision problem will be treated in Section 3, and the inductive decision problem will be considered in Sections 4 and 5.

2.3. <u>Examples</u> <u>of</u> <u>E-theories</u>. A diversity of data types and problems in artificial intelligence, program specification and verification, etc. can be presented in the form of $E(T_0)$-theories, as it is seen from the following examples, see also [Vo2, Vo3].

<u>Example</u> 1 (From linear algebra). Let A denote an $n \times (n+1)$-matrix representing a system of linear algebraic equations. Let the predicates zero(A, i, j) and one(A, i, j) express $A_{ij} = 0$ and $A_{ij} = 1$ respectively, let P(k, l) be the identity matrix E with k-th and l-th rows interchanged, E(A, l) coincide with the identity matrix E, but $E(A, l)_{ll} = {}^{1}/_{A_{ll}}$, and G(A, l) be the Gauss' extraction matrix (coincides with E, but $G(A, l)_{il} = -A_{il}$ for $i > l$). Let triang(A, l) express the upper triangularity of $n \times l$-submatrix of A, and * denote the usual matrix multiplication. Below we give some typical axioms:

$1 \leq k, l \leq n \supset (\mathrm{zero}(P(k, l) * A, 1, 1) \Leftrightarrow \mathrm{zero}(A, k, 1));$

$1 \leq l \leq n \supset (\mathrm{one}(E(A, l) * A, l, l) \Leftrightarrow \neg\, \mathrm{zero}(A, l, l));$

$l = n \supset (\mathrm{triang}(A, l) \Leftrightarrow \mathrm{one}(A, l, l) \,\&\, \mathrm{triang}(A, l-1));$

$1 \leq l \leq n \supset (\mathrm{triang}(G(A, l) * A, l) \Leftrightarrow \mathrm{one}(A, l, l) \,\&\, \mathrm{triang}(A, l-1)).$

<u>Example</u> 2 (Sorting). Let M denote an array, M[i..j] be the segment of M from i-th to j-th element, exch(M, i, j) be the result of swapping i-th and j-th elements of M, $M_1 < M_2$ express the fact that the maximal element in M_1 does not exceed the minimal in M_2, $i \in [p..q]$ abbreviate $p \leq i \leq q$:

$i,j \in [p..q] \vee i,j \notin [p..q] \supset$
$\quad \supset (\mathrm{exch}(M_1, i, j)[p..q] < M_2 \Leftrightarrow M_1[p..q] < M_2);$

$i \in [p..q] \,\&\, j \notin [p..q] \supset$
$\quad \supset (\mathrm{exch}(M_1, i, j)[p..q] < M_2 \Leftrightarrow$
$\quad \Leftrightarrow M_1[p..i-1] < M_2 \,\&\, M_1[i+1..q] < M_2 \,\&\, M_1[j..j] < M_2).$

Example 3 (Theory of arrays). Although we do not preassume equality axioms for non-basic sorts, conditional axioms of the form $B \supset l = r$ can be perceived as schemata of all axioms $B \supset (f(l) \Leftrightarrow f(r))$. Let A be an array, <A, i, e> be the result of replacing i-th element of A by e, and A[i] denote the i-th element of A:

$i = j \supset \langle A, i, e\rangle [j] = e$,

$\neg\ i = j \supset \langle A, i, e\rangle [j] = A[j]$,

$i = j \supset \langle\langle A, i, e\rangle, j, f\rangle = \langle A, j, f\rangle$,

$i > j \supset \langle\langle A, i, e\rangle, j, f\rangle = \langle\langle A, j, f\rangle, i, e\rangle$,

$\neg\ i = j \supset \langle\langle A, i, e\rangle, k, f\rangle [j] = \langle A, k, f\rangle [j]$.

2.4. *Conditional rewrite rules*. Now we will introduce a class of conditional rewrite rule systems closely related to $E(T_0)$-theories, and define the notions concerning rewriting techniques, such as *redex*, *convolution*, *conditional reducibility*, *normal form*.

Definition. An expression r of the form

$$B :: f \dashrightarrow F$$

such that $Var(B) \cup Var(F) \subseteq Var(f)$ is called a *conditional rewrite rule*; the formulae B, f, F are called respectively the *condition*, the *left-hand side* and the *right-hand side* of the rule. A set of rewrite rules (usually finite) is called a *conditional rewrite rule system*. Subsequently the letter R denotes an arbitrary rewrite rule system, and R_T stands for the system obtained by the obvious transformation of the axioms of an $E(T_0)$-theory T. ■

To formulate a notion of conditional reducibility we need an auxiliary definition of *equality modulo a basic theory*.

Definition. Let t, p be terms or formulae of the similarity type FP. The condition $Eq(t, p)$ is defined inductively:

- if $t \equiv f(t_1, \ldots, t_n)$ and $p \equiv f(p_1, \ldots, p_n)$ for $f \in FP \setminus FP_0$, then $Eq(t, p) \equiv_{df} \&_{i=1}^{n} Eq(t_i, p_i)$;

- if t, p are terms (formulae) of the similarity type FP_0, then $Eq(t, p) \equiv_{df} t = p$ ($Eq(t, p) \equiv_{df} t \Leftrightarrow p$ respectively);

- otherwise $Eq(t, p)$ is defined as the false condition. ■

Definition. Let g be an occurence of an atomic formula in a formula G and δ be a substitution. We say that g is an $R(r, \delta)$-*redex* in G, iff

1) the rule $r \equiv B :: f \dashrightarrow F$ belongs to R,

2) the formula $Cond(g, r, \delta) \equiv_{df} \delta(B) \ \&\ Eq(g, \delta(f))$ is T_0-consistent.

If g is an $R(r, \delta)$-redex in a formula G, then the formula G' obtained by the replacement of g with $\delta(F)$ is called an $R(g, r, \delta)$-*convolution* of G. ■

Note that the definition of a redex is not purely syntactical, as opposed to the unconditional case. Now we can define conditional reducibility.

Definition. Suppose that G' is an $R(\mathcal{G}, r, \delta)$-convolution of G and the condition $T_0 \models C \supset Cond(\mathcal{G}, r, \delta)$ is valid. We then say that the formula G $R(C)$-reduces to the formula G', and write $G \xrightarrow{R(C)} G'$. By $\xrightarrow{R(C)}{}^{+}$ and $\xrightarrow{R(C)}{}^{*}$ we denote the transitive and the reflexive-transitive closures of the $\xrightarrow{R(C)}$ relation. ■

2.5. The decision procedure. We are going to describe the decision procedure for $Th^{\forall}(T_0^{=} \cup T) =_{df} \{ H \mid T_0^{=} \cup T \models H, H \text{ is unquantified} \}$, where T is an $E(T_0)$-theory. The procedure will combine conditional reductions and invocations of a decision algorithm (as an oracle) for the basic theory T_0. The latter is needed to search redexes and check conditions. Also, an additional rule, called case splitting, will be introduced to achieve completeness. Moreover, since such formulae as $0 \leq x \leq 1$ & $P(x) \supset P(0) \vee P(1)$ may arise, we need incorporating a decision algorithm for $T_0^{=}$ (i.e. for the basic theory enriched with uninterpreted symbols) into our decision procedure.

Definition (reduction). Let a formula H has the form $C \supset G$ and $G \xrightarrow{R(C)} G'$. We then say that the formula H R-reduces to the formula H' of the form $C \supset H'$, symbolically $H \xrightarrow{R} H'$. ■

Definition (case splitting). Let a formula H has the form $C \supset G$ and for some $R(r, \delta)$-redex $\mathcal{G}$ in G the both formulae $C_1 \equiv C$ & $Cond(\mathcal{G}, r, \delta)$ and $C_2 \equiv C$ & $\neg Cond(\mathcal{G}, r, \delta)$ are T_0-consistent. We then say that the formulae $H_1 \equiv C_1 \supset G$ and $H_2 \equiv C_2 \supset G$ are obtainable as the result of a R_T-case splitting applied to the formula H. ■

Definition. We say that a formula H is in the R-normal form, iff it cannot be R-reduced. We say that a formula H is in the R-complete normal form, iff it is neither R-reducible, nor R-case splittable. ■

Definition. An R-reduction tree $\mathfrak{T}$ for a formula H_0 is an oriented tree with nodes labelled by formulae such that:

a) H_0 labelles the root of $\mathfrak{T}$;

b) each node of $\mathfrak{T}$ has at most two successors;

c) if H' is the unique successor of H then $H \xrightarrow{R} H'$;

d) if H_i, H_j are the successors of H_k then H_i and H_j are obtainable by the R-case splitting applied to the formula H.

e) the leaves of $\mathfrak{T}$ are labelled by formulae being in R-complete normal forms. ■

We are now ready to describe

<u>The decision procedure</u> $DP_{T_0}(T)$ for $Th^{\forall}(T_0^{=} \cup T)$:

<u>Entry</u>: a quantifier-free formula H.

<u>Result</u>: true or false according to the validity of

$$Th^{\forall}(T_0^{=} \cup T) \models H .$$

<u>Algorithm</u>. Construct any R_T-reduction tree $\mathfrak{T}$ for H, then check the validity of all the nodes of $\mathfrak{T}$ using the decision procedure for $Th^{\forall}(T_0^{=})$. Return <u>false</u> if some node is not true in $Th^{\forall}(T_0^{=})$. Otherwise, return <u>true</u>. ■

<u>Remark</u>. There is the method due to Shostak (see [Sh2]) which can be applied to construct the decision procedure for $Th^{\forall}(T_0^{=})$ starting with the decision algorithm for $Th^{\forall}(T_0)$.

We will show in the next Section that the procedure $DP_{T_0}(T)$ is strongly complete for $Th^{\forall}(T_0^{=} \cup T)$ provided that T generates the uniquely and finitely terminating conditional rewrite system R_T. <u>Strong completeness</u> means that a result of $DP_{T_0}(T)$ is independent of a reduction tree constructing strategy chosen, in other words, $DP_{T_0}(T)$ is <u>search space-free</u>.

2.6. <u>Proof example</u>. To illustrate the notions introduced above and the capabilities of the decision procedure, let us consider the following example. The formula below is the verification condition for the Hoare's program FIND [Ho], see also the example of its proof using rewrite rule based strategy by Hsiang [Hs, Example 3.6]:

$$\forall xy(x \le y \lor y < x) \;\&\; i < j \;\&\; m \le p \le q \le n \;\&$$
$$\forall xy(m \le x < i \;\&\; j < y \le n \supset A[x] \le A[y]) \;\&$$
$$\forall xy(m \le x \le y \le j \supset A[x] \le A[y]) \&$$
$$\forall xy(i \le x \le y \le n \supset A[x] \le A[y]) \supset A[p] \le A[q] .$$

Here T_0 is SA and the role of ET(SA)-theory T play three implications in the premise of the formula. The rewrite rule system R_T generated by T consists of the rules $r_i \equiv B_i :: A[x] \le A[y] \longrightarrow true$, where B_i is the premise of the corresponding implication (i = 1, 2, 3). Let us denote by C_0 the formula $j < i \;\&\; m \le p \le q \le n$ and by H the formula $C_0 \supset A[p] \le A[q]$. We must prove (or refute) that $SA^{=} \cup T \models H$.

The conclusion of H is the $R_T(A[p] \le A[p], r_i, [p/x, q/y])$-redex for i=1, 2, 3, since all three formulae $Cond_i \equiv Cond(A[p] \le A[p], r_i, [p/x, q/y]) \equiv B_i[p/x, q/y]$ are SA-consistent. At the same time H is not R_T-reducible, because $SA \not\models C_0 \supset B[p/x, q/y]$ (for i = 1, 2, 3). But every pair of the formulae $C_0 \;\&\; Cond_i$ and $C_0 \;\&\; \neg Cond_i$ is SA-consistent. So H is R_T-splittable using any of the three rules r_i. Let us choose the rule r_1 (in fact, there is no difference which rule to choose). The first case in splitting reduces directly to $C_0 \;\&\; B_1[p/x,q/y] \supset true$ (and this finishes its proof). The second case will be $H_2 \equiv C_1 \supset A[p] \leq A[q]$, where C_1 is the formula $C_0 \,\&\, \neg B_1[p/x,q/y]$. It is easy to check that H_2 is R_T-irreducible, but for i = 2, 3 both formulae $C_0 \,\&\, Cond_i$ and $C_0 \,\&\, \neg Cond_i$ are SA-consistent. Hence, H_2 is R_T-splittable by r_2 or r_3. Using r_2, as before, we reduce the first case to the tautology, the remaining case is $H_{22} \equiv C_1 \,\&\, \neg B_2 \supset A[p] \leq A[q]$. Using the decision algorithm for $Th^{\forall}(SA)$ we check that $SA \models C_1 \,\&\, \neg B_2 \supset B_3[p/x,q/y]$ and therefore H_{22} R_T-reduces (by r_3) to the tautology. This completes the exhaustive and search space free proof of Hoare's formula, which consists of two case splits and three reductions.

3. COMPLETENESS, AXIOM OMITTING

In this Section we will state the main theorem which guarantees the completeness of the $DP_{T_0}(T)$ decision procedure. Since conditional reducibility and case splitting obviously preserve validity of formulae in $T_0^{=} \cup T$, we only need proving that $T_0^{=} \cup T \models H$ implies $T_0^{=} \models H$ whenever H is in the R_T-complete normal form. The above can be seen as a result on seperating the reductive part T and non-reductive part T_0 of $T_0^{=} \cup T$. Of course, canonicity of the system R_T generated by T is needed, see Theorem 1 below. This is usual condition for the search space-free compleleness of rewrite rule based decision procedures. However the notion of canonicity must be transformed substantially for the class of conditional systems we consider.

3.1. Termination. We will define two slightly different kinds of the finite termination property. As we will see in Section 4 this difference becomes crucial when we turn to inductive decision problems.

Definition. A conditional rewrite rule system R is said to be

a) v-noetherian, iff there does not exist an infinite reduction chain of the form

$$G_0 \xrightarrow{R(C)} G_1 \xrightarrow{R(C)} G_2 \xrightarrow{R(C)} \dots ;$$

b) c-noetherian, iff there are no such infinite sequences of contexts $C_0, C_1, C_2, \dots$ and formulae $G_0, G_1, G_2, \dots$ that for all $i \in \mathbb{N}$

- the context C_i is T_0-consistent;

- $T \models C_{i+1} \supset C_i$;

- $G_i \xrightarrow{R(C_i)} G_{i+1}$. ■

Evidently, the c-noetherianity implies the v-noetherianity. The converse is, however, not true, as shows

Proposition 1. There exist a conditional rewrite rule system which is simultaneously v-noetherian but not c-noetherian. ■

Proof. Consider the system $R_i = \{ x > 0 :: P(x) \dashrightarrow P(x-1) \}$ and the sequence of contexts $C_0, C_1, C_2, \ldots$, where $C_i \equiv x > i$ for all $i \in \mathbb{N}$. It is easy to see that for every $i \in \mathbb{N}$ $P(x-i) \xrightarrow{R(C_i)} P(x-i-1)$. Hence, R_i is not c-noetherian.

To prove that R_i is v-noetherian, let us suppose, on the contrary, that R_i is not v-noetherian. Then there must exist a context C such that $PAR \models C \supset x > i$ for every $i \in \mathbb{N}$. Since C being the PAR-consistent formula (by the definition), and PAR is known to be a complete theory, the existential closure $\exists x C$ of C must be valid in every model of PAR, in particular, in the standard model Stm(PAR). Hence, there exists the variable interpretation validating C in Stm(PAR). But the same interpretation must validate all the formulae $x > i$ $(i \in \mathbb{N})$ in Stm(PAR). This means, however, that x is interpreted as a non-standard natural number in Stm(PAR). This contradiction shows that R_i is v-noetherian and concludes the proof.■

3.2. Confluency. Another condition necessary for strong completeness is well known in rewrite rule theory. This is the confluence property generalized for the class of rewrite rule systems we consider, cf. [Ka,ZRe]. Our formulation of confluency, however, attracts the notion of case splitting, see [Vo4].

Definition. A conditional rewrite rule system R is confluent, iff whenever $G \xrightarrow{R(C)}{}^* G_1$ and $G \xrightarrow{R(C)}{}^* G_2$, there exist such sets I, $\{ C_i \}_{i\in I}$, $\{ G_1^i \}_{i\in I}$ and $\{ G_2^i \}_{i\in I}$ that

a) for all $i \in I$ $Var(C_i) \subseteq Var(C)$;

b) $T_0 \models C \Leftrightarrow (\bigvee_{i\in I} C_i)$;

c) for all $i \in I$ $G_j \xrightarrow{R(C_i)}{}^* G_j^i$ $(j = 1, 2)$;

d) for all $i \in I$, $T_0^= \models C_i \supset (G_1^i \Leftrightarrow G_2^i)$.

We say that an $E(T_0)$-theory is confluent iff it generates a confluent conditional rewrite rule system. An $E(T_0)$-theory is said to be canonical iff it generates both confluent and c-noetherian conditional rewrite rule system.■

Note that the above definition of confluency attracts case splittings (the formulae C_i are, in fact, cases which occur in splittings). This is essential, as shows the following

Example. Let us consider the system R generated by the E-theory from Example 3 of §2.3. Let $C \equiv \neg\, 1 = \blacksquare \;\&\; k > 1$. Then $G \equiv f(\langle\langle A, k, e\rangle, 1, f\rangle[\blacksquare])$

R(C)-reduces to $G_1 \equiv f(\langle A, k, e\rangle[\blacksquare])$ by the second rule and to $G_2 \equiv f(\langle\langle A, 1, f\rangle, k, e\rangle[\blacksquare])$ by the fourth. But G_1 and G_2 are neither R(C)-reducible, nor equivalent (i.e. $T_0^= \nvDash C \supset (G_1 \Leftrightarrow G_2)$). However, if we take $C_1 \equiv C \mathbin{\&} \neg\, k = \blacksquare$ and $C_2 \equiv C \mathbin{\&} k = \blacksquare$ (these formulae arise when we split using the second rule), then G_1 and G_2 can be $R(C_1)$-reduced to $f(A[\blacksquare])$ and $R(C_2)$-reduced to $f(e)$. So, all the conditions of confluency are true for G_1, G_2. Similarly, it can be shown that R is confluent. ■

Suppose, we have some method to complete (by adding axioms) non-confluent $E(T_0)$-theories, transforming them into equivalent confluent ones (this is, in fact, the well-known Knuth-Bendix completion method, not discussed here, see [Vo4]). A canonical theory was demonstrated in Example 2.6 above.

Here is the proper place to formulate the main

<u>Theorem</u> 1 [Vo1, Vo2] (<u>Axioms omitting theorem</u>). Let T be a canonical $E(T_0)$-theory and H be a formula in the R_T-complete normal form. Then

$$T_0^= \cup T \vDash H \quad \text{iff} \quad T_0^= \vDash H\ . \ \blacksquare$$

<u>Proof</u>. See Appendix. ■

Roughly speaking, this theorem states that the validity of a completely reduced formula is independent of the axioms. It must be emphasised that the condition of canonicity is essential, as shows a rather simple

<u>Example</u>. Let T be axiomatized by two non-confluent unconditional axioms $f \Leftrightarrow F$ and $f \Leftrightarrow G$. Then $f \supset f$ may be reduced to $F \supset G$ being in the complete normal form, but axioms cannot be omitted, since $F \supset G$ is not a tautology. ■

<u>Corollary</u>. The $DP_{T_0}(T)$ decision procedure is <u>strongly complete</u> for the class of canonical $E(T_0)$-theories, i.e. able to prove any theorem whatever reduction tree construction strategy is used. ■

<u>Proof</u>. Obvious consequence of Theorem 1. ■

4. <u>INDUCTION</u>, <u>GENERALIZED AXIOM OMITTING</u>

In this Section we consider the inductive decision problem for E(PAR)-theories, generalize correspondingly the axiom omitting techniques and the decision procedure. It turns out that inductive definitions present in E(PAR)-theories can lead to non-termination of $DP_{T_0}(T)$. In §4.1 we analyze reasons for such a behaviour, and in §4.2 formulate restrictive conditions and state the Generalized axiom omitting theorem.

4.1. Non-termination. The main problem with the decision procedure $DP_{T_0}(T)$ described above is to create problem domains specifications in the form of canonical $E(T_0)$-theories. In majority of applications the role of a basic theory T_0 plays the unquantified Presburger arithmetic PAR or even the successor arithmetic SA. It turns out that in this case the difference between the two kinds of finite termination properties (v- and c-noetherianity) becomes crucial. Some natural forms of E-axioms, viz. inductive definitions, give rise to rewrite rules which violate the c-noetherian property, generating only v-noetherian systems. The last property is, however, insufficient for the finite termination of the $DP_{T_0}(T)$ decision procedure, as shows Proposition 2 below. First of all we formalize the notion of the inductive E-theory and inductive rewrite rule system.

Definition. A conditional rewrite rule system is called inductive iff it is simultaneously v-noetherian and not c-noetherian. An E(PAR)-theory is inductive, iff it generates the inductive conditional rewrite system. ■

Proposition 1 from §3.1 can now be reformulated as follows.

Proposition 1. There exist inductive rewrite systems. ■

Inductive definitions (and rewrite systems resp.) lead to the non-termination of the decision procedure $DP_{T_0}(T)$, as shows the following

Proposition 2. The v-noetherian property is insufficient for the finite termination of $DP_{T_0}(T)$. ■

Proof. Consider the one-axiom system R_i described in the proof of Proposition 1 in §3.1. Reducing the formula $H_0 \equiv z > 0 \supset P(z)$ we obtain $H_1 \equiv z > 0 \supset P(z-1)$. Case splitting for H_1 produces the formula $H_2 \equiv z > 0 \;\&\; z-1 > 0 \supset P(z-1)$. Reducing again, we obtain the formula $H_3 \equiv z > 0 \;\&\; z-1 > 0 \supset P(z-1-1)$. Obviously, this alternation of reduction steps and case splits may run ad infinitum. ■

By the same argument each of the following axioms can generate infinite proof trees when allowed to be used in case splitting:

$x > 0 \supset (\text{Even}(x) \Leftrightarrow \neg\, \text{Even}(x-1))$,

$x > 0 \;\&\; y > z \supset (\text{Mult}(x, y, z) \Leftrightarrow \text{Mult}(x-1, y, z-y))$,

$i > p \supset (\text{Ord}(A[p..i] \Leftrightarrow \text{Ord}(A[p..i-1]) \;\&\; A[p..i-1] \leq A[i])$.

Note that all axioms above are in fact inductive definitions. Usually inductive definitions have the form

$x > 0 \supset (f(x) \Leftrightarrow F [f(x-1)])$,

or even the more complex one (see inductive definition for Mult).

As it is clearly seen in the example above the reason of the non-termination of $DP_{T_0}(T)$ - is the unlimited use of the case splittings with the inductive axioms (rewrite rules).

4.2. Partial axioms omitting. One possible approach to solve the problem of inductive definitions consists in restricting unbounded case splitting. Since inductive definitions used in case splitting make possible infinite alternations of case splits and reduction steps, we must limit the case split applications appropriately, e.g. use it only with non-inductive axioms (rules). Instead of the complete formulae normalization we may carry out only the semi-complete normalization, allowing reduction steps by all the rules and case splitting by only the non-inductive ones. Of course, this may prevent axiom omitting when deciding formulae in semicomplete normal forms, because inductive definitions of the theory must be taken into account. We will prove, however, that under certain conditions all non-inductive axioms can be omitted on semicomplete normal forms, see Theorem 2. In fact, this reduces the problem to the decision of the purely inductive subtheory of the whole theory. To decide such a subtheory we use the induction principle implicitely present in the built-in decision algorithm for PAR, see Section 5. Of course, such an approach is suitable iff only a small part of the theory axioms being inductive definitions. Let us make precise the above arguments.

Assumption. From now on we will assume that any E(PAR)-theory T (and any conditional rewrite rule system R_T respectively) in consideration is partitioned into two parts: the non-inductive subtheory $T_{non-ind}$ ($R_{non-ind}$ subsystem resp.) and inductive subtheory T_{ind} (R_{ind} subsystem resp.) such that

$$T = T_{non-ind} \cup T_{ind}$$

$$(R = R_{non-ind} \cup R_{ind} \quad \text{resp.})$$

and any inductive R_T-reduction tree defined below is finite. We will call such a system inductively noetherian.

This partition will be fixed in the further definitions. ■

Definition. Let R_T be the system corresponding to a E(PAR)-theory T, and $R_T = R_{non-ind} \cup R_{ind}$. We say that a formula H is in the R_T-semicomplete normal form, iff it is both in the $R_{non-ind}$-complete normal form and in the R_{ind}-normal form. ■

Definition. Let T be a E(PAR)-theory. An inductive R_T-reduction tree $\mathfrak{T}$ for a formula H is an R_T-reduction tree for H with the following properties:

the leaves of $\mathfrak{T}$ are formulae being in R_T-semicomplete normal forms;

if H_1, H_2, H_3 are any three successive nodes of $\mathfrak{T}$ and $H_2 \xrightarrow{R_{ind}} H_3$, then $H_1 \xrightarrow{R_T} H_2$. ■

In other words, inductive trees do not contain case splits which prepare (immediately precede) inductive reduction steps, i.e. inductive rules are never used to find redexes for the case splitting. The following condition guarantees the finiteness of inductive reduction trees.

Definition. A rewrite system R is inductively noetherian iff there are no infinite sequences of contexts $C_0, C_1, C_2, \ldots$, and formulae $G_0, G_1, G_2, \ldots$ such that for each $i \in \mathbb{N}$

a) the context C_i is T_0-consistent;

b) $PAR \models C_{i+1} \supset C_i$;

c) $G_i \xrightarrow{R_{non-ind}(C_i)}{}^{+} G_i' \xrightarrow{R(C_i)}{}^{*} G_{i+1}$ ■

Intuitively, this means that we cannot iterate case splits with non-inductive reductions ad infinitum.

4.3. Separability, Completeness theorem. We now formulate sufficient conditions to omit all non-inductive axioms $T_{non-ind}$ of the theory T deciding formulae in semicomplete normal forms.

Definition. We say that the T_{ind} subtheory of a E(PAR)-theory T is T-separable, iff for every formula $H \equiv C \supset G$, being in R_T-semicomplete normal form, there exist sets I, $\{ C_i \}_{i \in I}$ and $\{ G_i \}_{i \in I}$ such that:

a) for every $i \in \mathbb{N}$ $Var(C_i) \subseteq Var(C)$;

b) $Stm(PAR) \models C \Leftrightarrow (\bigvee_{i \in I} C_i)$;

c) for each $i \in \mathbb{N}$ the formula $C_i \supset G_i$ is in the R_T-complete normal form and $G \xrightarrow{R_T(C_i)}{}^{*} G_i$;

d) for any $i \in \mathbb{N}$ $Stm(PAR^{=} \cup T_{ind}) \models C_i \supset (G \Leftrightarrow G_i)$.■

The separability conditions may be intuitively perceived as follows. Although case splits using inductive rewrite rules are prohibited, we can, at least in principle, capture all the cases which may occur if will construct the complete (may be infinite) reduction tree by the fair case splitting, using all the rules of the system. The condition d) means that the fair (inabridged) reduction tree construction from the inductive one attracts nothing but inductive subtheory. This clarifies the term "separability". Now we can formulate the inductive completeness theorem.

Theorem 2 (The generalized axiom omitting theorem). Let an E(PAR)-theory $T = T_{ind} \cup T_{non-ind}$ generate the inductively noetherian and confluent conditional rewrite rule system R_T, and T_{ind} be T-separable. Then for every formula H being in an R_T-semicomplete normal form

$$STM(PAR^{=} \cup T_{ind} \cup T_{non-ind}) \models H \quad \text{iff} \quad STM(PAR^{=} \cup T_{ind}) \models H. ■$$

Proof. See Appendix. ■

By means of this theorem, decision problems for some E(PAR)-theories with inductive axioms can be reduced to decision problems for purely inductive theories. This is, in fact, the result on relative decidability.

5. DECIDING PURELY INDUCTIVE SUBTHEORIES

This Section is devoted to inductive decision problems. We remind that the inductive decision problem for a theory Tind consists in deciding the set of formulae $Th^{\nabla}(\ STM\ (\ PAR^{=} \cup T_{ind}\ )\ )$, where an E(PAR)-theory T_{ind} generates the inductive rewrite rule system. In §5.1 we will show the undecidability of the problem in general, and in §§5.2 and 5.3 try to single out decidable cases.

5.1. Undecidability. The generalized axiom omitting theorem can be seen as a result on separating inductive and non-inductive parts of proofs. It is an interesting problem to develop decision algorithms for some classes of inductive definitions. It must be emphasized that generally inductive E(PAR)-theories are undecidable. This is not very strange, since they are expressive enough to axiomatize, say, integer multiplication.

Let Mult(x, y, z) be the ternary predicate symbol with the assumed interpretation "z is the product of x and y". Here are inductive definitions for Mult (note, that $B \supset \neg f$ abbreviates the formula $B \supset (\ f \Leftrightarrow g\ \&\ \neg g\ )$):

$$(x = 0 \vee y = 0)\ \&\ z = 0 \supset Mult(x,\ y,\ z),$$
$$(x = 0 \vee y = 0)\ \&\ \neg\ z = 0 \supset \neg\ Mult(x,\ y,\ z),$$
$$0 < y\ \&\ 0 < x \leq z \supset (Mult(x,\ y,\ z) \Leftrightarrow Mult(x,\ y\text{-}1,\ z\text{-}x)),$$
$$0 < y\ \&\ 0 < z < x \supset \neg\ Mult(x,\ y,\ z).$$

Then, for example, the "undecidability" of the diophantine equation $x^2 + y^2 = z^2$ can be expressed by the formula

$$x > 0\ \&\ y > 0\ \&\ z > 0\ \&\ Mult(x,\ x,\ x_2)\ \&$$
$$\&\ Mult(y,\ y,\ y_2) \supset \neg\ Mult(z,\ z,\ x_2 + y_2).$$

Together with the undecidability of Hilbert's tenth problem this leads to undecidability of the inductive decision problem in general.

5.2. Decidability. The general undecidability result stimulates investigations of decidable classes of inductive E(PAR)-theories. We can, for example, impose some restrictions on the form of inductive definitions, e.g. prohibit negations in the premises of axioms (note, that the axioms for Mult contain such negations).

One decidable class of inductive theories can be characterized as follows. Suppose, the role of the basic theory T_0 plays the successor arithmetic SA, and all predicate symbols in inductive definitions are monadic. In this case the decision procedure for the monadic second-order successor arithmetic S1S applies, see [Ra], and we can use it to decide formulae of reasonable length. However, to achieve practical effectiveness the general decision procedure for S1S must be specialized to the particular form of quantifier-free inductive definitions. We will report on this topic elsewhere.

Another decidable class of inductive definitions can be characterized in terms of rewrite rules.

Definition. Let us call an inductive axiom simple iff it has the form

$$f(Sx) \Leftrightarrow F\,[\,f(x)\,]\ .\ \blacksquare$$

Simple inductive axioms arise often in practice. Consider, for example, the following schemata of inductive axioms:

- $P(Sx) \Leftrightarrow P(x)$, (1)
- $P(Sx) \Leftrightarrow \neg P(x)$, (2)
- $P(Sx) \Leftrightarrow P(x)\ \&\ Q(x)$, (3)
- $P(Sx) \Leftrightarrow P(x) \vee Q(x)$. (4)

All of them give rise to inductive theories.

Definition. Let T_{ind} be an inductive E(PAR)-theory generating the conditinal rewrite rule system R_{ind}. An atom f is called T_{ind}-inductive iff for some ground substitution δ the variable-free atom $\delta(f)$ is R_{ind}-reducible.

Definition. Let T_{ind} be an inductive theory and f be an inductive atom. A formula Ξ (possibly containing quantifiers on variables of the basic sort) is called the explicit definition of f w.r.t. T_{ind} iff:

1) $STM(PAR^{=} \cup T_{ind}) \models f \Leftrightarrow \Xi$;

2) Ξ does not contain inductive atoms. ■

Definition. Let H be an unquantified formula. A formula H^* obtained from H by replacing all occurences of T_{ind}-inductive atoms in H by their explicit definitions w.r.t. T_{ind} is called a T_{ind}-explication of H. ■

Explications of a formula can be seen as solutions of corresponding inductive equations.

Example. Let T_{ind} consist of two axioms: $\supset$ even(0) and even(SSx) $\Leftrightarrow$ even(x). Then $f(z)$ is the inductive atom, $\Xi_1(z) \equiv \exists u(2u = z)$ and $\Xi_2(z) \equiv \forall v\, \neg(2v + 1 = z)$ are two explicit definitions of f.

Of course, explicit definitions may not contain quantifiers, e.g. simple inductive definitions (1) (3) and (4) above give explicit definitions of a transparent form:

$$P(x) \Leftrightarrow P(0), \qquad (1)$$

$$P(x) \Leftrightarrow P(0) \;\&\; \mathop{\&}_{i=0}^{x-1} Q(i) \qquad (3)$$

$$P(x) \Leftrightarrow P(0) \vee \bigvee_{i=0}^{x-1} Q(i) \qquad (4)$$

Our aim can be stated as follows. We need conditions which will guarantee that for any unquantified formula H

$$STM(PAR^{=} \cup T_{ind}) \models H \quad \text{iff} \quad STM(PAR^{=}) \models H^{*},$$

where H^{*} is an arbitrary T_{ind}-explication of H. In other words, we need conditions to omit inductive theory T_{ind}. Intuitively, T_{ind} must provide uniqueness of explications. It turns out that the following well-known property suffices for inductive axioms omitting.

Definition. We say that a conditional rewrite rule system R is confluent on ground atoms iff for every variable-free atom f, whenever f is R-reducible to different formulae F_1 and F_2, the latter are equivalent modulo $STM(PAR^{=})$, i.e. $STM(PAR^{=}) \models F_1 \Leftrightarrow F_2$. ■

Now we are ready to state

Theorem 3 (inductive axioms omitting theorem). Let T_{ind} be an inductive theory generating a conditional rewrite rule system which is confluent on ground atoms. Then for every unquantified formula H and any its T_{ind}-explication H^{*} the following is true:

$$STM(PAR^{=} \cup T_{ind}) \models H \quad \text{iff} \quad STM(PAR^{=}) \models H^{*} \;■$$

Remarks. 1) Of course, there may exist formulae which do not possess T_{ind}-explications at all. Also there may exist formulae with uniquely determined explications, as well as with multiple explications. In any case, it is the responsibility of a problem domain specifier to provide explicit definitions and to prove that they indeed satisfy the corresponding conditions. This is similar to the compilation of a rewrite rule system into canonical one via completion and may be achieved either manually or automatically. When explicit definitions are given, the following inductive proofs are conducted completely automatically (modulo the decision of explications in $STM(PAR^{=})$).

2) Often it is useful to have multiple explicit definitions of the same atom. Suppose, $f(x)$ has two different explicit definitions, one the existential $\exists y\, H_1(x, y)$ and the other universal $\forall z\, H_2(x, z)$. In order to prove the formula $f(u) \supset f(u)$, we can use the existential definition for the first

occurence of $f(u)$ and the universal one for the second, obtaining the universally quantified formula $\forall y \forall z(H_1(x, y) \supset H_2(x, z))$ as a result. This remark can be easily transformed into the rule guiding the choice of $\forall$- or $\exists$-type definitions for positive and negative occurences of inductive atoms.

Example. Let the predicate Three(x) be defined by inductive axioms: $\supset$ Three(0), $\supset \neg$Three(s0), $\supset \neg$Three(ss0), Three(sssx) $\Leftrightarrow$ Three(x) (i.e. "x is a multiple of 3"). There are at least two possible explicit definitions of Three(x), $\exists u(3u = x)$ and $\forall v(\neg 3v+1=x \ \& \ \neg 3v+2=x)$. Using them we can easily decide the corresponding inductive problem, e.g. to prove such formulae as $\neg$Three(x) $\supset$ Three(sx) $\vee$ Three(ssx), Three(x) $\supset \neg$Three(ssx) etc. reducing them to unquantified Presburger formulae.

For explications containing non-basic predicate and function symbols, the decision process is more complicated but still possible, as we will show in the next paragraph.

5.3. Deciding explications with uninterpreted symbols. Suppose we deal with simple inductive axioms of the form (1) - (4) described in the previous subsection. Then explications arising will necessarily contain subformulae of the form $\underset{i=0}{\overset{x}{\&}} Q(i)$ and $\bigvee_{i=0}^{x} Q(i)$. Some decision mechanism is needed in this case. The direct method is to use bounded quantifiers, but it will produce formulae with alternating quantifiers which are undesirable. Now we will describe less direct algorithm, producing only universal formulae, decidable using Shostak's procedure [Sh2]. The key idea of the method is similar to that of Shostak [Sh2].

Let a formula H contain occurences of $\underset{i=0}{\overset{x}{\&}} Q(i)$ and $\bigvee_{i=0}^{x} Q(i)$. We can assume without the loss of generality, that the upper limits in these formulae are always variables, since $F(t)$ is equivalent to $x = t \supset F(x)$. We may replace $\underset{i=0}{\overset{x}{\&}} Q(i)$ and $\bigvee_{i=0}^{x} Q(i)$ everywhere in H by atomic formulae K(x) and D(x) with new predicate symbols K and D, thus obtaining the formula H_1. Then H_1 is decided using the algorithm for $Th^{\forall}(PAR^{=})$. If H_1 is valid, then H is also valid. Suppose, a counterexample for H_1 is found. We must check whether this counterexample refutes H. E.g. K(2) = true and K(1) = false may disprove H_1 but cannot refute H, owing to the semantics of conjunction. So we must compute the violated conjunction property for the counterexample found. For example, if the variables x and y receive values a and b, $a < b$, but K(x) and K(y) receive values false and true respectively, then the violated property will be $F \equiv x < y \supset (K(y) \supset K(x))$. We then transform H_1 into $F \supset H_1$ and repeat the process of searching for a counterexample. Note that adding F as a premise to H_1 excludes the counterexample found at the previous step. The above loop must eventually terminate either with proof of the source formula H, or with a valid counterexample. The arguments behind the termination proof are as follows. We can suppose that H_1 is of the form $B \supset (F_1 \supset F_2)$, F_1 is a conjunction and F_2 is a disjunction of atomic formulae. The transformation of H_1 into $F \supset H_1$ yields three formulae:

$B \,\&\, \neg\, x < y \supset (F_1 \supset F_2)$, and $\neg\, x < y$ is not a consequence of B in PAR,
$B \supset (F_1 \supset F_2 \vee K(y))$,
$B \supset (F_1 \,\&\, K(x) \supset F_2)$.

Since starting from any formula H_1 such transformations can produce only finite number of inequivalent formulae, the above decision process will always terminate. Of course, all the above must be repeated mutatis mutandis for formulae containing D.

Acknowledgements. This paper could not appear without friendly support of Nachum Dershowitz and Hans Woessner.

6. REFERENCES

[Hs] Hsiang J. Refutational Theorem proving using term rewriting systems. - Artificial Intelligence, 1985, vol. 25, no.2, pp. 255-300.

[HuHu] Huet G., Hullot J.M. Proofs by Induction in Equational Theories with Constructors. - Proc. 21st Symp. on Foundations of Computer Science, 1980, pp. 96-107.

[HuOp] Huet G., Oppen D.C. Equations and Rewrite Rules: A Survey. - In: Formal Language Theory: Perspectives and Open Problems. - New-York, Academic Press, 1980, pp. 349-406.

[Jo] Jouannaud J.-P. Confluent and Coherent Equational Term Rewriting Systems: Applications to Proofs in Abstract Data Types.- Lecture Notes in Computer Science, 1983, vol. 159, pp. 256-283.

[JoKo] Jouannaud J.P., Kounalis E. Automatic Proofs by Induction in Equational Theories Without Constructors. - Proc. Symp. "Logic in Computer Science", 1986, pp. 358-366.

[Ka] Kaplan S. Conditional Rewrite Rules. - Theoretical Computer Science, 1984, vol. 33, no. 2-3, pp. 175-193.

[KB] Knuth D.E., Bendix P.B. Simple Word Problems in Universal Algebras. - In: Computational Problems in Universal Algebras. - Pergamon Press, 1970, pp. 263-297.

[Ra] Rabin M.O. Decidability of second-order theories and automata on infinite trees. - Transactions of the American Mathematical Society, 1969, vol.141, no.7, pp.1-35.

[Sh1] Shostak R.E. On the SUP-INF Method for Proving Presburger Formulas. - Journal of the ACM, 1977, vol. 24, no. 4, pp. 529-543.

[Sh2] Shostak R.E. A Practical Decision Procedure for Arithmetic with Function Symbols. - Journal of the ACM, 1979, vol.26, no. 2, pp. 351-360.

[Vo1] Vorobyov S.G. On the Arithmetic Inexpressiveness of Term Rewriting Systems. - Proc. 3rd Symp. "Logic in Computer Science", 1988, pp. 212-217.

[Vo2] Vorobyov S.G. On the Use of Conditional Term Rewriting Systems in Program Verification. - **Programmirovanie**, 1986, no. 4, pp. 3-14 (in Russian).

[Vo3] Vorobyov S.G. Applying Conditional Term Rewriting Systems in: Automated Theorem Proving. - **Izvestija Akademii Nauk SSSR, Technicheskaya Kibernetika**, 1988, N 5, pp. 25-39 (in Russian).

[Vo4] Vorobyov S.G. On the Completion of Conditional Rewrite Rule Systems with Built-in Decision Algorithms. - In: Theory and Methods of Parallel Information Processing. - Novosibirsk, Computer Center, 1988, pp.131-146 (in Russian).

[Vo5] Vorobyov S.G. A Structural Completeness Theorem for a Class of Conditional Rewrite Rule Systems. - Proc. Conf. on Computer Logic, COLOG'88, Vol. 2, pp. 194-208, Tallinn, Estonia, 1988 (to appear in Journal of Symbolic Computation).

[Vo6] Vorobyov S.G. The quantifier-free theory of partial order cannot be axiomatized by a canonical rewrite rule system. - Submitted to "Foundations of Computational Theory'89"

[ZRe] Zhang H., Remy J.-L. Contextual Rewriting. - Lecture Notes in Computer Science, 1985, vol. 202, pp. 46-62.

7. APPENDIX: PROOFS OF COMPLETENESS THEOREMS

Proof of Theorem 1 can be obtained as a direct simplification of the proof of Theorem 2 below. The only difference consists in classes of models which must be considered. Proving Theorem 1 we must work with all models of a E-theory, whereas proving Theorem 2 only with standard ones. ■

Proof of Theorem 2. We will prove only the nontrivial part of Theorem 2 - the implication

$$STM(PAR^{=} \cup T_{ind} \cup T_{non\text{-}ind}) \models H \text{ implies } STM(PAR^{=} \cup T_{ind}) \models H ,$$

where H is in the R_T-semicomplete normal form (the inverse implication is trivial).

Let us assume, on the contrary, that for some H being in the R_T-semicomplete normal form:

(1) there exists a model $\mathfrak{M}_0 \in STM(PAR^{=} \cup T_{ind})$ such that H is not valid in $\mathfrak{M}_0$, i.e. for some ground instance H^c of H we have $\mathfrak{M}_0 \models \neg H^c$;

(2) for all $\mathfrak{M} \in STM(PAR^{=} \cup T_{ind} \cup T_{non\text{-}ind})$ we have $\mathfrak{M} \models H$.

To obtain the contradiction we will describe a transformation of an arbitrary model $\mathfrak{M}_0 \in STM(PAR^{=} \cup T_{ind})$ into a model $\mathfrak{M}_T(\mathfrak{M}_0) \in STM(PAR^{=} \cup T_{ind} \cup T_{non\text{-}ind})$ preserving the invalidity (and validity) of all ground instances of the formulae in R_T-semicomplete normal forms. This will conclude the proof.

Without the loss of generality we can suppose that a model $\mathfrak{M}_0$ giving a counterexample to H (i.e. satisfying (1)) possesses the following properties:

a) the signature of $\mathfrak{M}_0$ is $FP' = FP \cup FP_c$, where FP_c consists of constant symbols only;

b) every element of $\mathfrak{M}_0$ is denoted by (i.e. is a value of) some constant (ground) term of FP;

c) for any sort s different from the sort of natural numbers and for any pair t, r of distinct ground terms of the sort s the values of t and s are distinct (this can be achieved by dublicating the elements denoted by different terms, preserving the values of predicates and functions).

The construction of $\mathfrak{M}_T(\mathfrak{M}_0)$ is as follows:

1) the signatures and the carriers of $\mathfrak{M}_0$ and $\mathfrak{M}_T(\mathfrak{M}_0)$ coincide;

2) a ground (variable-free) atom f^c is true in $\mathfrak{M}_T(\mathfrak{M}_0)$ iff its $R_T(\mathfrak{M}_0)$-normal form (see the definition below) is true in $\mathfrak{M}_0$.

The $R_T(\mathfrak{M}_0)$-reduction relation and the $R_T(\mathfrak{M}_0)$-normal form are defined directly: if g^c is an $R_T(r, \delta)$-redex in G^c such that $M_0 \models Cond(g^c, r, \delta^c)$, then G^c $R_T(\mathfrak{M}_0)$-reduces to $G^{c'}$, the last formula being the corresponding $R_T(g^c, r, \delta^c)$-convolution of G.

It can be proved straightforwardly that the imposed confluence and finite termination properties guarantee the correctness of the $\mathfrak{M}_T(\mathfrak{M}_0)$ construction, and $\mathfrak{M}_T(\mathfrak{M}_0) \models PAR^= \cup T_{ind} \cup T_{non\text{-}ind}$. It remains to show that the invalidity of H^c (H^c is the counterexample, $\mathfrak{M}_0 \models \neg H^c$) is preserved when we transform $\mathfrak{M}_0$ to $\mathfrak{M}_T(\mathfrak{M}_0)$. Let $H^c \equiv c^c \supset G^c$. It is easy to see that c^c remains true in $\mathfrak{M}_T(\mathfrak{M}_0)$, since it is in the $R_T(\mathfrak{M}_0)$-normal form. By the first separability condition there must exist $i_0 \in I$ such that $\mathfrak{M}_0 \models c^c \Leftrightarrow c^c_{i_0}$. It follows from the definitions that G^c $R_T(\mathfrak{M}_0)$-reduces to $G^c_{i_0}$, and, moreover, $G^c_{i_0}$ is in the $R_T(\mathfrak{M}_0)$-normal form (because $H^c_{i_0} \equiv c^c_{i_0} \supset G^c_{i_0}$ is the R_T-complete normal form). Since G^c is equivalent to $G^c_{i_0}$ modulo $STM(PAR^= \cup T_{ind})$ (by the third condition) and $\mathfrak{M}_0 \in Stm(PAR^= \cup T_{ind})$, G^c must remain false in $\mathfrak{M}_T(\mathfrak{M}_0)$, hence H^c remains false in $\mathfrak{M}_T(\mathfrak{M}_0)$. This contradiction completes the proof. ■

<u>Proof of Theorem</u> 3. Let a formula H be valid in all models of $STM(PAR^= \cup T_{ind})$, but some its explication H^* is not true in some model $\mathfrak{M}_0 \in STM(PAR^=) \models H^*$. However H^* is still valid in $STM(PAR^= \cup T_{ind})$. We can transform $\mathfrak{M}_0$ into a model $\mathfrak{M}_{ind}(\mathfrak{M}_0)$ using the construction from the proof of Theorem 2 above. It can be proved directly that $\mathfrak{M}_{ind}(\mathfrak{M}_0)$ is a model of T_{ind}, since R_{ind} is conluent on ground atoms, and H^* is false in $\mathfrak{M}_{ind}(\mathfrak{M}_0)$, because it does not contain inductive atoms. This gives the contradiction which concludes the proof. ■

Consider Only General Superpositions in Completion Procedures*

Hantao Zhang
Department of Computer Science
The University of Iowa
Iowa City, IA 52242
hzhang@herky.cs.uiowa.edu

Deepak Kapur
Department of Computer Science
State University of New York at Albany
Albany, NY 12222
kapur@albanycs.albany.edu

Abstract

Superposition or critical pair computation is one of the key operations in the Knuth-Bendix completion procedure and its extensions. We propose a practical technique which can save computation of some critical pairs where the most general unifiers used to generate these critical pairs are *less general* than the most general unifiers used to generate other joinable critical pairs. Consequently, there is no need to superpose identical subterms at different positions in a rule more than once and there is also no need to superpose *symmetric* subterms in a rule more than once. The combination of this technique with other critical pair criteria proposed in the literature is also discussed. The technique has been integrated in the completion procedures for ordinary term rewriting systems as well as term rewriting systems with associative-commutative operators implemented in *RRL*, *Rewrite Rule Laboratory*. Performance of the completion procedures with and without this technique is compared on a number of examples.

1 Introduction

The Knuth-Bendix completion procedure [15] was originally introduced as a means of deriving canonical term rewriting systems from a set of equations. A canonical rewriting system can serve as a decision procedure for the equational theory specified by a finite set of equations. Extensions of the Knuth-Bendix procedure have been developed to handle associative and commutative operators by Lankford and Ballantyne [17] as well as Peterson and Stickel [20]. For a survey on some applications of completion procedures, see [7] and [8].

Inspired by Stickel's impressive results [23], we began investigations in the fall of 1987 on the use of the extended Knuth-Bendix completion procedure implemented in *Rewrite Rule Laboratory* (*RRL*) ([12], [13] and [14]) for proving that a free associative ring is commutative if every element x in the ring satisfies $x^3 = x$. This theorem has been considered as one of the most challenging problems for automated reasoning programs (see for instance, [4], [19] and [29]).

During the generation of a proof of this theorem, *RRL* produces the following rewrite rule:

$$\mathbf{r}_1 : (x * y * z) + (x * z * y) + (y * x * z) + (y * z * x) + (z * x * y) + (z * y * x) \rightarrow 0$$

*Part of the work was done while the first author was at Rensselaer Polytechnic Institute, New York and was partially supported by the National Science Foundation Grant no. CCR-8408461.

where $+$ satisfies the commutativity and associativity laws (in that case $+$ is called an *AC function*) and $*$ is assumed to be right associative; this rule is also generated in Stickel's proof [23]. According to the standard definition of the completion procedure, we need to superpose other rules like $u * u * u \rightarrow u$ and $(x + y) * z \rightarrow (x * z) + (y * z)$ at the six non-variable proper subterms of the left side of $\mathbf{r}_1$:

$$(x * y * z), (x * z * y), (y * x * z), (y * z * x), (z * x * y) \text{ and } (z * y * x)$$

For example, the critical pair obtained by superposing $u * u * u \rightarrow u$ at $(x * y * z)$ is

$$u + (u * u * u) + (u * u * u) + (u * u * u) + (u * u * u) + (u * u * u) = 0.$$

It is easy to see that every other critical pair generated by superposing $u * u * u \rightarrow u$ at any other subterm will be the same (under the associative and commutative laws for "+").

Let us also consider the critical pair obtained by superposing $u * u * u \rightarrow u$ at the subterm $y * z$ in $(x * y * z)$:

$$(x * u) + (x * (u * u) * u) + (u * x * (u * u)) + (u * (u * u) * x) + \\ ((u * u) * x * u) + ((u * u) * u * x) = 0.$$

Again, the critical pair obtained by superposing $u * u * u \rightarrow u$ at any proper subterm of any of the other five subterms of $\mathbf{r}_1$ is equivalent to the above critical pair (after renaming variables).

Except for the first of the above six subterms, the superpositions at the rest of five subterms as well as the superpositions at their subterms are fruitless in the sense that no new rules could be produced from these superpositions. In fact, any of the six subterms can be picked for superpositions without having to consider superpositions at other subterms. This is so because the variables x, y and z are *symmetric* in $\mathbf{r}_1$, in the sense that renaming x by y and y by x simultaneously in $\mathbf{r}_1$, the result remains equivalent to the original. The above six subterms are also *symmetric* in $\mathbf{r}_1$, in the sense that one can derive the other by switching properly the names of the variables appearing in these terms. And, this phenomenon recurs considerably for such problems. Hence we conjectured that it is enough to superpose at one of the symmetric terms of a rule in the completion procedure. In an attempt to prove this conjecture, we found that the conjecture is just a consequence of the main technique described in the paper.

Our main technique can be roughly described as follows: Given three rewrite rules $\mathbf{r}_1 = l_1 \rightarrow r_1$, $\mathbf{r}_2 = l_2 \rightarrow r_2$ and $\mathbf{r}_3 = l_3 \rightarrow r_3$. Suppose the most general unifier of l_1/p_1 and l_2 is σ_1 and the most general unifier of l_1/p_2 and l_3 is σ_2. If p_1 and p_2 are disjoint positions and σ_1 is more general than σ_2 in the sense that there exists a substitution θ such that $\sigma_2 = \theta\sigma_1$, then there is no need to test the joinability of the critical pair obtained by superposing $\mathbf{r}_3$ at p_2 of l_1 if we have superposed $\mathbf{r}_2$ at p_1 of l_1 in a completion procedure. This situation is depicted in Figure 1 where $(\sigma_2(r_1), t[p_2 \leftarrow \sigma_2(r_3)])$ is the critical pair of the superposition between $\mathbf{r}_1$ and $\mathbf{r}_3$ and $(\theta\sigma_1(r_1), t[p_1 \leftarrow \theta\sigma_1(r_2)])$ is an instance of the critical pair of the superposition between $\mathbf{r}_1$ and $\mathbf{r}_2$.

Further, this technique can be combined with the result about prime superpositions reported in [10] or the one proposed in [28]. However, we show in Subsection 3.4 that Winkler and Buchberger's method cannot be combined with that of Kapur et al's method, and some critical pair criterion cannot be combined with any technique mentioned above, including the one proposed in this paper.

Let us illustrate the technique with a simple example.

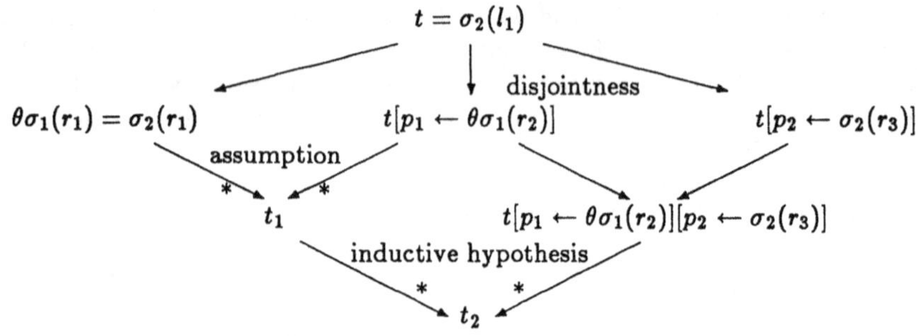

Figure 1.

Example 1.1 Suppose R contains the rules

$$\mathbf{r}_4 : f(a,a) \rightarrow e \qquad \text{and} \qquad \mathbf{r}_5 : g(f(x,y), f(y,x)) \rightarrow e.$$

Superposing $f(a,a) \rightarrow e$ at the positions 1 and 2 of $g(f(x,y), f(y,x))$, we get two superpositions with the same unifier $\beta_1 = \beta_2 = \{x \leftarrow a, y \leftarrow a\}$:

$$(\beta_1, \mathbf{r}_4, 1, \mathbf{r}_5) \qquad \text{and} \qquad (\beta_2, \mathbf{r}_4, 2, \mathbf{r}_5).$$

Using the technique, at most one of these two superposition need to be considered in the completion procedure because 1 and 2 are disjoint positions in $g(f(x,y), f(y,x))$ and β_1 is more general than β_2 as well as β_2 is more general than β_1. There is no need to test the joinability of both superpositions.

As an application of the main technique, we will define a *symmetry* relation on terms and show how this relation can help in avoiding many unnecessary superpositions. For instance, for the rewrite rule $\mathbf{r}_1$ given in the associative ring example above, we may gain by a factor of 6 by recognizing that the six subterms in the left side of $\mathbf{r}_1$ are symmetric when superposing rules like $u * u * u \rightarrow u$ and $(x + y) * z \rightarrow (x * z) + (y * z)$ into $\mathbf{r}_1$. The symmetry relation can be tested automatically from the input with little expense and can be easily implemented.

Using the checking for blocked superpositions discussed in [10], the technique described in this paper, we were able to achieve substantial gains in the performance of the completion procedure for associative-commutative function symbols. As a result, we have been successful in solving the associative ring problem mentioned earlier and related problems using relatively less computer time. *RRL* produced the proof of the above associative ring theorem in 2 minutes on a Symbolics Lisp machine. With some additional heuristics and techniques, *RRL* produced a proof of the theorem in 5 seconds. More interestingly, *RRL* also proved in 4 minutes the commutativity of a free associative ring in which every element satisfies $x^6 = x$. To our knowledge, this is the first computer proof of this theorem. This paper is the first in a series of papers describing our results in attacking this family of problems about associative rings. Subsequent papers will discuss additional results employed and heuristics used to obtain computer proofs of such theorems. For other computer proofs on the commutativity problem of associative rings, the reader please refer to [24], [23] and [25].

1.1 Related Work

Different criteria have been proposed for eliminating unnecessary critical pairs in the Knuth-Bendix completion procedures. The following material is based on a subsection on related work in [10].

Slagle [21] first introduced the concept of *blocked substitutions*, whose unifying substitution contains terms that are reducible. Lankford and Ballantyne [18] generalized it and reported favorable experiments by throwing away unblocked critical pairs in the Knuth-Bendix completion procedure; see also [4]. A generalization of blocking methods to theories with permutative axioms (including associative-commutative axioms) with favorable experiments was reported in [18]. In [10], a correctness proof was given, showing that certain kinds of superpositions, called *composite* superpositions, which include all unblocked superpositions, do not have to be considered at all in the completion procedure.

Independently Buchberger [5] developed a criterion for detecting unnecessary reductions in the computation of a Gröbner basis of a polynomial ideal over a field; a Gröbner basis of a polynomial ideal is a complete system for the polynomial ideal when polynomials are viewed as rewrite rules. Subsequently, Winkler and Buchberger [28] generalized Buchberger's criterion to include term rewriting systems; see also [26] and [27]. Küchlin [16] extended their results and further developed their criterion based on the generalized Newman's lemma proposed by Buchberger as well as the notions of connectedness and subconnectedness. In [27], the combination of several criteria is discussed.

More recently, Bachmair and Dershowitz used proof ordering to provide a general framework in which various criteria for unnecessary critical pairs can be studied and the correctness of completion procedures using such criteria can be established ([1], [2] and [3]).

Our technique can be implemented, just like those proposed in [28], [16] and [10]. Moreover, our approach is different from the above methods in that we do not first compute unifiers in a superposition, then throw away some of them according to some criteria. Our technique simply allows us to avoid generating unnecessary superpositions; no unification need be computed in these saved superpositions. Moreover, the critical pairs created from these saved superpositions are not covered by any technique reported in the literature.

2 Definitions

We will assume the standard definitions of *term*, *size* of a term, *equation*, *rewrite rule*, *substitution*, the *most general unifier (mgu)* and etc. A substitution can be naturally extended to a mapping from terms to terms, from equations to equations, etc. We will use, with possible subscripts, the bold case $\mathbf{r}$ to denote a rewrite rule; the bold case $\mathbf{e}$ for an equation; x, y and z for variables and greek letters for substitutions.

Some function symbols can have the associative and commutative properties; such functions are said to be *AC* (associative-commutative) functions. Without any loss of generality, an AC function f can be assumed to be of an arbitrary arity; this allows us to consider terms constructed using AC functions in *flattened* form, that is, no argument to an AC function f is a term whose outermost symbol is f itself.

A *subterm position* within a flattened term is similar to a *subterm position* within an ordinary term except that we must take care of the case where any non-empty subset of the arguments of an AC function f is a subterm. A subterm position used in this paper is a sequence of integers possiblely ending with a set of integers. For example, given a flatten term $t = h(+(a,b,c))$, where $+$ is an AC function, the subterm at the position ϵ, the empty sequence, is t itself, at 1 of t is $+(a,b,c)$, at 1.1 is a, at 1.{1,3} is $+(a,c)$, at 1.{2,3} is $+(b,c)$. There are four cases regarding the relation between two positions p_1 and p_2 in a flatten term:

(1) p_1 is a prefix of p_2; (2) p_2 is a prefix of p_1; (3) p_1 and p_2 are properly intersecting; (4) p_1 and p_2 are disjoint. For the term t given above, 1 is a prefix of 1.{1, 3}. Positions 1.{1, 3} and 1.{2, 3} are properly intersecting, whereas positions 1.1 and 1.2 are disjoint. For a precise definition of position and its relation, the reader may refer to [10].

We write t/p for the subterm of t at position p. Note that $t/p.q = (t/p)/q$. We also write $t[p \leftarrow s]$ for the term obtained by replacing the subterm t/p at the position p by another term s. Given two terms t and s, $t =_{AC} s$ if and only if they are identical after flattening Henceforth, by equality on terms, we mean $=_{AC}$.

2.1 Symmetry Relations

A substitution σ is said to be a *variable renaming* iff for any variable x, y, (i) $\sigma(x)$ is a variable and (ii) $x \neq y$ implies $\sigma(x) \neq \sigma(y)$.

Definition 2.1 *Let A be a set of axioms and t a term (or equation, rewrite rule, etc.). Variables x_1, x_2, ..., x_n in t are symmetric with respect to A if $t =_A \theta(t)$, where θ is a permutation on $\{x_1, x_2, ..., x_n\}$.*

Often, we are interested in the case of two distinct variables. The above θ will be called a *symmetric renaming* of **f**, because θ simultaneously renames one by another in **f** and preserves equivalence relation under $=_A$. Note that there is a number k such that θ^k is a trivial (identical) substitution; further, θ^{-1} is also a symmetric renaming. Note also that A may contain any kind of equations. However in this paper, we consider the case where A contains only AC axioms. In the following, $=_A$ is assumed to be identical to $=_{AC}$.

Example 2.2 Suppose $+$ is an AC function. Then x and y are symmetric in the following equation

$$(x + y) * z = (x * z) + (y * z)$$

and x, y and z are symmetric to each other in the following rewrite rule:

$$(x * y * z) + (x * z * y) + (y * x * z) + (y * z * x) + (z * x * y) + (z * y * x) \rightarrow 0.$$

It is easy to see that the symmetry relation on variables is an equivalence relation on the variables in a formula.

Definition 2.3 *Two terms t_1 and t_2 in a formula **f** are said to be symmetric if there exists a symmetric renaming θ of **f** (i.e. θ^n is trivial for some n and $\theta(\mathbf{f}) =_{AC} \mathbf{f}$) such that $\theta(t_1) = t_2$.*

Example 2.4 It is easy to see that x and y are symmetric in the rewrite rule

$$(x * x * y) + (x * y * x) + (x * y * y) + (y * x * x) + (y * x * y) + (y * y * x) \rightarrow 0$$

when $+$ is an AC function. Subterms $(x * x * y)$ and $(y * y * x)$ are symmetric. So are $(x * y * x)$ and $(y * x * y)$. However, $(x * x * y)$ and $(x * y * x)$ are not symmetric.

Lemma 2.5 *The symmetry relation on terms is an equivalence relation.*

2.2 Canonical Relation

Let $\rightarrow$ be a binary relation on a set T of objects, referred to as a *rewriting relation*. The reflexive, transitive closure of a rewriting relation $\rightarrow$ is denoted by $\rightarrow^*$. We write $t \leftarrow s$ iff $s \rightarrow t$, and $t \leftarrow^* s$ iff $s \rightarrow^* t$. An object t in T is said to be *in normal form* if and only if there is no t' such that $t \rightarrow t'$. Objects t_1 and t_2 are *joinable* if and only if there exists an object t such that $t_1 \rightarrow^* t \leftarrow^* t_2$.

A rewriting relation $\rightarrow$ is said to be *terminating* (or *Noetherian*, *well-founded*) if and only if there is no infinite sequence of the form $a_1 \rightarrow a_2 \rightarrow a_3 \rightarrow \cdots$. The relation $\rightarrow$ is said to be *confluent* at a term t if and only if for all t_1, t_2, if $t_1 \leftarrow^* t \rightarrow^* t_2$ then t_1 and t_2 are joinable. The relation $\rightarrow$ is *confluent* iff it is confluent at any term. $\rightarrow$ is said to be *canonical* (or *complete*) if it is both terminating and confluent. If $\rightarrow$ is canonical, each object of T has a unique normal form.

The reflexive, symmetric and transitive closure of $\rightarrow$, denoted by $\leftrightarrow^*$, is a congruence relation. A rewriting relation $\rightarrow$ is said to have the *Church-Rosser property* if and only if for all t_1, t_2 such that $t_1 \leftrightarrow^* t_2$, t_1 and t_2 are joinable. It can be easily shown that $\rightarrow$ has the Church-Rosser property if and only if $\rightarrow$ is confluent [9].

2.3 Completion Procedures

Let $R = \{l_i \rightarrow r_i\}$ be a term rewriting system where l_i and r_i are terms and $l_i > r_i$ for a well-founded ordering $>$. The term rewriting system R induces a rewriting relation, denoted as $\rightarrow$, over the set of terms in the following way: $t \rightarrow t'$ if and only if there is a position p in t, a rule $l_i \rightarrow r_i$ in R, and a substitution σ such that $t/p = \sigma(l_i)$, where t/p stands for the subterm position p in t, and $t' = t[p \leftarrow \sigma(r_i)]$, the term obtained by replacing the subterm at position p in t by the term $\sigma(r_i)$. In case we need to indicate that t is rewritten to t' by a rule $\mathbf{r}$, we write $t \xrightarrow{\mathbf{r}} t'$.

In order to handle certain cases that arise with AC functions, it is useful to define, corresponding to R, an *extended* term rewriting system R^e, obtained by adding certain rules to R: for every rule $f(t_1, ..., t_n) \rightarrow t$ in R where f is an AC function, the rule $f(t_1, ..., t_n, z) \rightarrow t'$ is in R^e, where z is a variable not occurring in $f(t_1, ..., t_n)$ and t' is the flattened term corresponding to $f(t, z)$. Clearly (i) if R has no rules having AC function symbols as their outermost function symbols, then $R^e = R$, (ii) R^e has the same rewriting relation as R [10].

Let us define a *superposition* of two rewrite rules $l_1 \rightarrow r_1$ and $l_2 \rightarrow r_2$ in the usual way as a 4-tuple $(\sigma, l_1 \rightarrow r_1, p, l_2 \rightarrow r_2)$, where p is a non-variable position of l_1 and σ is a most general unifier of l_1/p and l_2. As usual, we assume that the two rules in a superposition have no variables in common. Associated with each superposition $(\sigma, l_1 \rightarrow r_1, p, l_2 \rightarrow r_2)$, there exists a pair of terms $(\sigma(l_1[p \leftarrow r_2]), \sigma(r_1))$, called a *critical pair* of $l_1 \rightarrow r_1$ and $l_2 \rightarrow r_2$. Given a critical pair (s, t), by $top(s, t)$ we denote the term $\sigma(l_1)$ if $s = \sigma(l_1[p \leftarrow r_2])$ and $t = \sigma(r_1)$. A superposition is said to be *joinable* iff the associated critical pair is joinable.

Lemma 2.6 ([15], [9], and [10]) *For any terms t, t_1 and t_2 such that $t_1 \xleftarrow{\mathbf{r}_1} t \xrightarrow{\mathbf{r}_2} t_2$, then either (i) there exists a term t' such that $t_1 \rightarrow^* t' \leftarrow^* t_2$, or (ii)$(t_1, t_2)$ is an instance of a critical pair of $\mathbf{r}_1$ and (the extension of) $\mathbf{r}_2$.*

As in [10], we say $t_1 \xleftarrow{\mathbf{r}_1} t \xrightarrow{\mathbf{r}_2} t_2$ is a *peak*, and $t_1 \rightarrow^* t' \leftarrow^* t_2$ is a *valley*. The extension of $\mathbf{r}_2$ is needed because $\mathbf{r}_1$ and $\mathbf{r}_2$ may rewrite t at properly intersecting positions in the presence of AC functions (see Lemma 1 in [10]). In [10], the notion of *generalized superposition* (a formal description of a peak) is introduced to handle this case. Because of this, the Church-Rosser property of R^e (hence R) can be proved in the same way as in the case when there are no AC function symbols involved.

Theorem 2.7 ([15], [17], [20], and [10]) *Let R be a terminating term rewriting system and R^e, an extension of R. If every critical pair of R^e is joinable, then R has the Church-Rosser property.*

This theorem was first proved in [15] where AC functions are empty. Lankford and Ballantyne [17] and Peterson and Stickel [20] gave proofs of the theorem when AC functions are considered.

Based on Theorem 2.7, a *completion procedure* is proposed, which in many cases generates a canonical (complete) term rewriting system from a given set of equations. The performance of a completion procedure depends mainly upon the number of critical pairs generated from all the rules and the time involved in checking whether or not they are joinable. Let S be the set of all the superposition of a rewriting system R. That means for any pair of rules $l_1 \rightarrow r_1$ and $l_2 \rightarrow r_2$ in $R \times R$, for any non-variable position p of l_1, for any most general unifier σ of l_1/p and l_2, the superposition

$$(\sigma, l_1 \rightarrow r_1, p, l_2 \rightarrow r_2)$$

is in S. Let $|R|$ denote the number of the rules in R, $\overline{|lhs|}$, the average size of the left sides of R and $\overline{|Unifiers|}$, the average number of the complete set of unifiers of two terms in a unification, then the total number of superpositions will approximately be

$$|R|^2 \times \overline{|lhs|} \times \overline{|Unifiers|}.$$

In the case where there are no AC functions, $\overline{|Unifiers|}$ is a real number falling between 0 and 1 (some terms are not unifiable) and the number of critical pairs to be computed is approximatedly equal to the square of the number of rules. When AC functions are involved, this number increases dramatically because of many most general unifiers of two terms (see [6] for examples with thousands of unifiers).

3 Detecting Unnecessary Superpositions

3.1 General Superpositions

We say a substitution θ is *more general* than another substitution σ if there exists a substitution β such that $\sigma = \beta\theta$. There exists the case where θ is more general than σ and σ is more general than θ. In this case, β must be a variable-renaming.

Definition 3.1 *Let $s_1 = (\sigma, l_1 \rightarrow r_1, p_1, l_2 \rightarrow r_2)$ and $s_2 = (\theta, l_1 \rightarrow r_1, p_2, l_3 \rightarrow r_3)$ be two superpositions, where the first rule is the same, we say s_1 is more general than s_2 if σ is more general than θ, or equivalently, s_2 is less general than s_1.*

Lemma 3.2 *Let $s_1 = (\sigma, l_1 \rightarrow r_1, p, l_2 \rightarrow r_2)$ and $s_2 = (\theta, l_1 \rightarrow r_1, p, l_3 \rightarrow r_3)$, be two superpositions. If s_2 is more general than s_1, $\sigma(r_2) = \sigma(r_3)$ and s_2 is joinable, then s_1 is joinable.*

Proof. Suppose $\sigma = \beta\theta$ for some β. Because $\sigma(r_2) = \sigma(r_3)$, $(\sigma(l_1)[p \leftarrow \sigma(r_2)], \sigma(r_1))$, the critical pair of s_1, can be written as $(\beta\theta(l_1)[p \leftarrow \beta\theta(r_3)], \beta\theta(r_1))$, which is an instance of the critical pair of s_2. □

Theorem 3.3 *R is Church-Rosser iff R is terminating and for any superposition $s_1 = (\sigma, l_1 \rightarrow r_1, p_1, l_2 \rightarrow r_2)$ of R^e, either (1) s_1 is joinable, or (2) there exists another joinable superposition $s_2 = (\theta, l_1 \rightarrow r_1, p_2, l_3 \rightarrow r_3)$, where p_1 and p_2 are disjoint positions in l_1 and s_2 is more general than s_1.*

Though its proof is quite simple using the connectedness result of [28] and [16] (to be discussed in Subsection 3.3), it provides us a practical technique to detect unnecessary critical pairs in completion procedures. That is, we may skip the joinability test of superpositions which are less general than other joinable superpositions at disjoint positions. The above result is shown for AC-completion procedure, it can be extended to E−completion procedures, too.

An immediate application of Theorem 3.3 is that if the left side of a rewrite rule has several identical subterms, then it is sufficient to superpose other rules at one of these subterm positions in a completion procedure.

Corollary 3.4 *Suppose that $l_1/p_1 = l_1/p_2$ in a rewrite rule $l_1 \rightarrow r_1$ where p_1 and p_2 are disjoint positions, then there is no need to consider the superpositions of the form $(-, l_1 \rightarrow r_1, p_2, -)$ if all the superpositions of the form $(-, l_1 \rightarrow r_1, p_1, -)$ are under consideration in a completion procedure.*

It is also straightforward to replace the condition $l_1/p_1 = l_1/p_2$ in the above corollary by $l_1/p_1 =_E l_1/p_2$ if we like to use this result in E−completion procedures. In the next subsection, we will show that the above corollary still holds if we replace $l_1/p_1 = l_1/p_2$ by the condition that l_1/p_1 and l_1/p_2 are symmetric in $l_1 \rightarrow r_1$.

3.2 Avoiding Superpositions on Symmetric Terms

As an application of Theorem 3.3, we show that if several subterms in the left side of a rewrite rule are symmetric, then only one of them is needed to be superposed in a completion procedure. To establish this result, we need first a technical lemma.

Lemma 3.5 *Given three terms t_1, t_2 and t_3, if (i) t_1 is a renaming of t_2, i.e., there exists a variable renaming θ such that $t_1 = \theta(t_2)$; (ii) none of the variables in t_3 appears in t_1 or t_2; (iii) σ is the mgu of t_1 and t_3, then $\sigma\theta$ is the mgu of t_2 and t_3.*

Proof. (a) At first, $\sigma\theta$ is a unifier of t_2 and t_3 because $\sigma\theta t_2 = \sigma t_1 = \sigma t_3$ by (iii) and $\sigma\theta t_3 = \sigma t_3$ by (ii).

(b) Now let us show $\sigma\theta$ is the most general unifier (unique upto renaming) of t_2 and t_3. For any unifier σ' of t_2 and t_3, using the same proof given in (a) above, we know that $\sigma'\theta^{-1}$ is a unifier of t_1 and t_3. So $\sigma'\theta^{-1} = \beta\sigma$ for some β. Hence $\sigma' = (\sigma'\theta^{-1})\theta = (\beta\sigma)\theta = \beta(\sigma\theta)$. □

Theorem 3.6 *Suppose that l_1/p_1 and l_1/p_2 are symmetric in a rewrite rule $l_1 \rightarrow r_1$, then there is no need to consider the superpositions of the form $(-, l_1 \rightarrow r_1, p_1, -)$ if all the superpositions of the form $(-, l_1 \rightarrow r_1, p_2, -)$ are under consideration in a completion procedure.*

Proof. Obviously, p_1 and p_2 are disjoint positions. Let $\mathbf{s}_1 = (\sigma_1, l_1 \to r_1, p_1, l_2 \to r_2)$, $t_1 = l_1/p_1$ and $t_2 = l_1/p_2$. Since t_1 and t_2 are symmetric, there exists a symmetric renaming θ such that $\theta(t_1) = t_2$, $\theta(l_1) = l_1$ and $\theta(r_1) = r_1$.

Suppose t_1 and l_2 are unifiable with a mgu σ, by Lemma 3.5, $\sigma\theta$ is a mgu of t_2 and l_2. Hence there exists the following superposition $s' = (\sigma\theta, l_1 \to r_1, p_2, l_2 \to r_2)$. Because θ is a symmetric renaming of $l_1 \to r_1$, s' is equivalent to $\mathbf{s}_2 = (\sigma, \theta(l_1) \to \theta(r_1), p_2, l_2 \to r_2)$. In other words, s' is joinable iff $\mathbf{s}_2$ is joinable. Because $\mathbf{s}_2$ is more general than $\mathbf{s}_1$, by Theorem 3.3, there is no need to test the joinability of $\mathbf{s}_1$. □

Now let us look back the rewrite rule given in the beginning of the paper:

$$\mathbf{r}_1 : (x*y*z) + (x*z*y) + (y*x*z) + (y*z*x) + (z*x*y) + (z*y*x) \to 0$$

where $+$ is an AC function. Because $(x*y*z)$, $(x*z*y)$, $(y*x*z)$, $(y*z*x)$, $(z*x*y)$ and $(z*y*x)$ are symmetric, it is indeed sufficient to superpose other rules at one of these six subterms.

3.3 Correctness Proof via Connectedness

Theorem 3.3 can be proved using connectedness or using Bachmair and Dershowitz's critical pair criterion. We choose the former in this subsection and leave the later to the interested reader.

Definition 3.7 ([5], [28], [16]) *Given an ordering $>$ and a term u, two terms s and t are connected below u in R if there exist $n+1$ terms $t_0, t_1, ..., t_n$ such that $s = t_0$, $t = t_n$, $u > t_i$ and $t_{i-1} \leftrightarrow t_i$ for $0 < i \le n$, where $\leftrightarrow$ is $\to \cup \leftarrow$.*

Theorem 3.8 ([5], [28], [16], [2]) *Suppose $>$ is well-founded and $\to \subseteq <$. R is Church-Rosser iff R is terminating and for every critical pair (s,t) of R^e, s and t are connected below $top(s,t)$ in R.*

Proof. We like to prove that $\to$ is confluent at every term by induction on $>$. The basis case is trivial since $\to$ is confluent at every normal form. As the induction hypothesis, assume that $\to$ is confluent at t for any $t < u$.

As a special case, suppose $u = top(s,t)$ for a critical pair (s,t). Because s and t are connected below u in R, there exist $n+1$ terms $t_0, t_1, ..., t_n$ such that $s = t_0$, $t = t_n$, $u > t_i$ and $t_{i-1} \leftrightarrow t_i$ for $0 < i \le n$. We prove that t_0 and t_n are joinable by induction on n. Suppose as induction hypothesis that $t_0 \to^* s_{n-1} \leftarrow^* t_{n-1}$ for some term s_{n-1}. If $t_{n-1} \leftarrow t_n$, then $t_0 \to^* s_{n-1} \leftarrow^* t_{n-1} \leftarrow t_n$. Otherwise, $t_{n-1} \to t_n$. Because $u > t_{n-1}$, $\to$ is confluent at t_{n-1}. Hence, from $s_{n-1} \leftarrow^* t_{n-1} \to t_n$, we have $t_0 \to^* s_{n-1} \to^* s_n \leftarrow^* t_n$ for some s_n.

For a general peak like $s \leftarrow u \to t$, Lemma 2.6 and the above special case apply. □

There are no general methods known to test the connectedness of two terms below another term. It is trivial to see that two joinable terms are always connected; but this test does not reduce any complexity of a completion procedure. Even though the above result does not tell us which superpositions are unnecessary, it provides a powerful tool for both designing and proving different critical pair criteria which are oriented towards implementation, like those proposed in [28], [16], [10], and this paper. The above discussion on the connectedness applies to Bachmair and Dershowitz's critical pair criterion, too [2].

Proof of Theorem 3.3. Suppose for the superposition $\mathbf{s}_1 = (\sigma, l_1 \to r_1, p_1, l_2 \to r_2)$ of R^e, there exists another joinable superposition $\mathbf{s}_2 = (\theta, l_1 \to r_1, p_2, l_3 \to r_3)$, where p_1 and p_2 are disjoint positions in l_1 and $\mathbf{s}_2$ is more general than $\mathbf{s}_1$. Let (s,t) be the critical pair of $\mathbf{s}_1$ where $s = \sigma(l_1[p_1 \leftarrow r_2])$ and $t = \sigma(r_1)$. By Theorem 3.8, it is sufficient to prove that s and t are connected below $top(s,t) = \sigma(l_1)$ in R. Because θ is more general than σ, $\sigma(l_1)$ is reducible by $l_3 \to r_3$ at the position p_2. Let $s' = \sigma(l_1)[p_2 \leftarrow \beta(r_3)]$ for some β, then (s',t) is an instance of the critical pair of $\mathbf{s}_2$. By the assumption, $s' \to^* t' \leftarrow^* t$ for some term t'. Hence we have a proof that s and t are connected below $\sigma(l_1)$:

$$s \overset{l_3 \to r_3}{\longrightarrow} \sigma(l_1)[p_1 \leftarrow \sigma(r_2)][p_2 \leftarrow \beta(r_3)] \overset{l_2 \to r_2}{\longleftarrow} s' \to^* t' \leftarrow^* t$$

(A direct proof is also possible, which has the same proof structure as that of Theorem 3.8 and was illustrated by Figure 1 in the introduction.) □

In [10], Kapur, Musser and Narendran suggested that only prime superpositions need to be considered in completion procedures. Their criterion can be established by connectedness as well.

Definition 3.9 ([10]) *A superposition $\mathbf{s} = (\sigma, l_1 \to r_1, p, l_2 \to r_2)$ of R^e is said composite if there exists a non-empty position q such that such that $\sigma(l_2)/q$ is reducible. $\mathbf{s}$ is said prime iff $\mathbf{s}$ is not composite.*

They proved that any composite generalized superposition (a peak) can be factored into two prime generalized superpositions. Let us say that a critical pair of a composite superposition is *composite*, using Lemma 2.6, their result can be stated as follows:

Lemma 3.10 ([10]) *For any composite critical pair (t_1, t_2), there exists a term t such that $top(t_1, t_2) \to t$, and both (t_1, t) and (t, t_2) are either an instance of a prime critical pair or a joinable pair of terms.*

Theorem 3.11 ([10]) *R is Church-Rosser iff R is terminating and every prime superposition of R^e is joinable.*

Proof. It is sufficient to prove that for any composite critical pair (t_1, t_2), t_1 and t_2 are connected below $top(t_1, t_2)$. By assumption and lemma 3.10, there exists a term t such that both (t_1, t) and (t, t_2) are joinable pairs of terms. Hence

$$t_1 \to^* s_1 \leftarrow^* t \to^* s_2 \leftarrow^* t_2$$

for some terms s_1 and s_2. □

The above proof is much simpler than that of [10] and that of [2]. In [2], Bachmair and Dershowitz used proof orderings to prove Theorem 3.8 and the above theorem. They considered connectedness and compositeness as different criteria. As revealed by the proof of the above theorem, connectedness is more general than compositeness. Our opinion is that connectedness and Bachmair and Dershowitz's result are proof-oriented in the sense that they are useful to establish the correctness of implementation-oriented techniques like those proposed by Winkler and Buchberger, Küchlin, Kapur et al, and the one suggested by Theorem 3.3 of this paper.

3.4 Combination of Different Critical Pair Criteria

Because Theorem 3.3 and Theorem 3.11 handle disjoint cases, the techniques suggested by these two theorems can be combined safely.

Theorem 3.12 *R is Church-Rosser iff R is terminating and for any superposition $\mathbf{s}_1 = (\sigma, l_1 \to r_1, p_1, l_2 \to r_2)$ of R^e, either (1) $\mathbf{s}_1$ is joinable, or (2) there exists another joinable superposition $\mathbf{s}_2 = (\theta, l_1 \to r_1, p_2, l_3 \to r_3)$ which is more general than $\mathbf{s}_1$ and one of the following is true: (a) $p_1 = p_2$ and $\sigma(r_2) = \sigma(r_3)$; (b) p_1 and p_2 are disjoint positions in l_1; (c) p_1 is a proper prefix of p_2.*

Proof. It is sufficient to prove that every superposition is joinable. Case 1 is trivial. Cases 2.a and 2.b are covered by Lemma 3.2 and Theorem 3.3, respectively. In Case 2.c, because $\sigma(l_1)$ is reducible at p_2 by $l_3 \to r_3$, $\mathbf{s}_1$ is thus composite. $\mathbf{s}_1$ can be proved joinable using the same proof of Theorem 3.11. □

Note also that the condition in Case 2.c can be further generalized to that $\mathbf{s}_1$ *is a composite superposition* (or equivalently, $\sigma(l_2)$ *is reducible at a non-empty position*), which does not require the existence of $\mathbf{s}_2$.

As mentioned earlier, Winkler and Buchberger [28] proposed a technique for detecting unnecessary critical pairs based on connectedness. A modification of their result can be summarized as follows ([28], [10]):

Definition 3.13 *A superposition $\mathbf{s} = (\sigma, l_1 \to r_1, p, l_2 \to r_2)$ of R^e is said up-prime if for any non-empty proper prefix q of p, $\sigma(l_1)/q$ is in normal form. $\mathbf{s}$ is said up-composite iff $\mathbf{s}$ is not up-prime.*

Let us call a prime superposition defined in Definition 3.10 as being *down-prime* for the reason that a superposition $\mathbf{s} = (\sigma, l_1 \to r_1, p, l_2 \to r_2)$ of R^e is prime by Definition 3.10, iff any subterm of $\sigma(l_1)$ at a position below p is in normal form; while $\mathbf{s}$ is up-prime iff any proper subterm of $\sigma(l_1)$ at a position above p is in normal form. Accordingly, $\mathbf{s}$ is down-composite iff $\mathbf{s}$ is not down-prime.

Example 3.14 Suppose R consists of the following three rules:

$$\begin{array}{lrcl} \mathbf{r}_1: & h(a) & \to & a \\ \mathbf{r}_2: & g(a,y) & \to & a \\ \mathbf{r}_3: & f(g(x,h(x))) & \to & a \end{array}$$

The superposition $\mathbf{s}_1 = (\{x/a, y/h(a)\}, \mathbf{r}_3, 1, \mathbf{r}_2)$ is down-composite and up-prime because $f(g(a,h(a)))/1.2 = g(a,h(a))/2 = h(a)$ is reducible by $\mathbf{r}_1$ and no non-empty position can be a proper prefix of 1. Naturally, $\mathbf{s}_1$ is not down-prime and not up-composite.

Theorem 3.15 ([28]) *R is Church-Rosser iff R is terminating and every up-prime superposition of R^e is joinable.*

This result can be easily proved using connectedness and is similar to that of Theorem 3.11. The following proposition states that Kapur, Musser and Narendran's technique and that of Winkler and Buchberger cannot be combined together.

Proposition 3.16 *If both up-composite and down-composite superpositions of R^e are discarded in a completion procedure, the Church-Rosser property of R is not guaranteed.*

Proof. The other superposition of R in Example 3.14 is $\mathbf{s}_2 = (\{x/a\}, \mathbf{r}_3, 1.2, \mathbf{r}_1)$, which is up-composite and down-prime. If $\mathbf{s}_1$ is discarded because of being down-composite and $\mathbf{s}_2$ is discarded because of being up-composite, it is wrong to say R is Church-Rosser because $f(a) \leftrightarrow^* a$ but $f(a)$ and a are not joinable in R. □

Despite of the above negative result, Winkler and Buchberger's technique can be combined with our technique:

Theorem 3.17 *Theorem 3.12 still holds if the condition* 2.c *is replaced by that p_2 is not empty and is a proper prefix of p_1.*

A simple and useful technique based on an order of rules instead of the restrictions on positions was proposed, we think, by Küchlin . The technique can be summarized as follows: Let $\mathbf{r}_1, \mathbf{r}_2, ..., \mathbf{r}_n$ be a listing of rewrite rules of R. Suppose that $s \xleftarrow{\mathbf{r}_i} u \xrightarrow{\mathbf{r}_j} t$, where (s,t) is a critical pair and $u = top(s,t)$. If $u \xrightarrow{\mathbf{r}_k} v$ and $k < min(i,j)$, then the joinability test of (s,t) can be omitted in a completion procedure. Its correctness can be established by connectedness and an induction on $min(i,j)$ because critical pairs between $\mathbf{r}_i$ and rr_k as well as that of $\mathbf{r}_j$ and $\mathbf{r}_k$ are joinable by induction hypothesis. This technique cannot, however, be combined with any technique mentioned above. For instance, $\mathbf{s}_1$ in Example 3.14 can be discarded by the above criterion and $\mathbf{s}_2$ can be discarded by Winkler-Buchberger's criterion. The result is incorrect as indicated in the proof of Proposition 3.16. Counter examples are easy to construct to show that above criterion cannot be combined with that of Kapur, Musser and Narendran and ours.

The problem of combining different criteria is also discussed in [27] and [2]. In [2], Bachmair and Dershowitz stated that connectedness and compositeness can be combined freely. We have seen that compositeness can be established by connectedness and some techniques, which can be established by connectedness, cannot be combined correctly.

4 Experimental Results

To test whether a superposition is more general than another in different ways requires us to keep a record of all superpositions computed so far in a completion procedure. We have not yet been able to work out an inexpensive way of performing this test. Hence we have only incorporated in *RRL* the check for symmetric terms to avoid unnecessary superpositions in the completion procedure as well as of the AC completion procedure.

For the case there are no AC functions, the symmetric relation is reduced to the identical relation. We observed the performance of the completion procedure with the check for identical subterms has almost no change on a number of examples including various axiomatizations of free groups discussed in [11], this is due to the facts that (i) checking for identical subterms is not a big overhead and (ii) there are not many identical subterms in the rewrite rules involved in these examples.

We found that when used in the presence of AC functions, the check for symmetric terms gains time varying on different examples. At first, to test whether two variables x and y are symmetric in a rewrite rule $\mathbf{r}$, we may construct first the following two substitutions:

$$\sigma_1 = \{x \leftarrow z_1, y \leftarrow z_2\} \qquad \text{and} \qquad \sigma_2 = \{y \leftarrow z_1, x \leftarrow z_2\}$$

where z_1 and z_2 are new variable names, then check if $\sigma_1(\mathbf{r}) =_{AC} \sigma_2(\mathbf{r})$. Noticing that σ_1 and σ_2 have fix-point property ($\sigma^n = \sigma$ for any n). This test has to be performed only once, when each new rule is produced in the completion procedure. The symmetry relation among variables and terms is stored as a part of the data of a rewrite rule. During the superposition process, this information is retrieved and only one of the symmetric terms is needed to be superposed. This technique is easy to implement and contrasts with other known techniques in that superpositions are not first computed and then later discarded.

Secondly, as indicated by Lemma 3.2, there is no need to compute unifiers which are equal after renaming in the AC unification algorithm. There is a simple technique available, also implemented in *RRL*, to avoid computing such kind of unifiers in Stickel's "variable-abstraction" AC unification algorithm; because of space limit, this technique will be presented in another paper. Moreover, as discussed in Section 3.4, our technique can be combined with some other techniques.

For small examples such as Abelian group, the saved superpositions are 16 and the runtime to generate a canonical system drops from 5.12 seconds to 3.89 seconds; for a free commutative ring with unity, the saved superpositions are also 16 and the runtime drops from 19 seconds to 18 seconds. Note that the canonical system for a free commutative ring mentioned above contains the rule $i(x+y) \rightarrow i(x) + i(y)$. If the rule is made as $i(x) + i(y) \rightarrow i(x+y)$, we can still get a canonical system. In this case, the runtime is reduced from 40 seconds to 32 seconds and the saved superpositions are 28. Similarly, for the boolean ring example, the runtime is reduced from 22 seconds to 20 seconds. For the free distributive lattice example (FDL) in [20], the check for symmetric superpositions reduces the total time only marginally. However, for the associative ring problem, the runtime to generate a canonical system for an associative ring plus the axiom $x * x * x = x$ is reduced from 312 seconds to 124 seconds, approximately by a factor of 2.5. The saved superpositions are 406. In the proof of the theorem that $x^4 = x$ implies the commutativity of an associative ring, the run time is reduced by a factor of 2.2. For the case where $x^6 = x$, the run time is saved halfly. The above timings are taken from a Symbolics Lisp Machine 3670.

References

[1] Bachmair, L., Dershowitz, N. and Hsiang, J. (1986) Orderings for equational proofs. *Proc. IEEE Symp. Logic in Computer Science*, Cambridge, Massachusetts, pp. 346-357.

[2] Bachmair, L., and Dershowitz, N. (1986) Critical pair criteria for the Knuth-Bendix completion procedure. Proc. of *SYMSAC*. See also *J. of Symbolic Computation*, (1988) 6, 1-18.

[3] Bachmair, L., and Dershowitz, N. (1987) Critical pair criteria for rewriting modulo a congruence, In Proc. Eurocal 87, Leipzig, GDR.

[4] Bledsoe, W., (1977) "Non-resolution theorem proving", Artificial Intelligence 9, 1, pp.1-35.

[5] Buchberger, B. (1979) A criterion for detecting unnecessary reductions in the construction of Gröbner Bases. Proc, of *EUROSAM 79*, Springer Verlag, LICS 72, 3-21.

[6] Bürckert, H.-J., Herold, A., Kapur, D., Siekmann, J.H., Stickel, M., Tepp, M., and Zhang, H., (1988) Opening the AC-Unification Race, *J. of Automated Reasoning*, 4, 465-474.

[7] Dershowitz, N., (1983) *Applications of the Knuth-Bendix Completion Procedure*. Laboratory Operation, Aerosapce Corporation, Aerospace Report No. ATR-83(8478)-2, 15.

[8] Dershowitz, N. and Jouannaud J.P. (1988). Rewriting Systems, *Handbook of Theoretical Computer Science.*

[9] Huet, G. (1980) Confluent reductions: abstract properties and applications to term rewriting systems. *J. ACM*, 27/4, 797-821.

[10] Kapur, D., Musser, D.R., and Narendran, P. (1984). Only prime superpositions need be considered for the Knuth-Bendix completion procedure. GE Corporate Research and Development Report, Schenectady. Also in *J. of Symbolic Computation* (1988) 4, 19-36.

[11] Kapur, D. and Sivakumar, G. (1984). Architecture of and experiments with RRL, a Rewrite Rule Laboratory. Proceedings of an *NSF Workshop on the Rewrite Rule Laboratory* Sept. 6-9 Sept. 1983, General Electric Research and Development Center Report 84GEN008.

[12] Kapur, D., Sivakumar, G., and Zhang, H. (1986). RRL: A Rewrite Rule Laboratory. *Eighth International Conference on Automated Deduction (CADE-8)*, Oxford, England, July 1986, LNCS 230, Springer Verlag, 692-693.

[13] Kapur, D., and Zhang, H. (1987). *RRL: A Rewrite Rule Laboratory - User's Manual.* GE Corporate Research and Development Report, Schenectady, NY.

[14] Kapur, D., and Zhang, H. (1988). *RRL: A Rewrite Rule Laboratory.* Proc. of *Ninth International Conference on Automated Deduction* (CADE-9), Argonne, IL.

[15] Knuth, D.E. and Bendix, P.B. (1970). Simple word problems in universal algebras. In: *Computational Problems in Abstract Algebras.* (ed. J. Leech), Pergamon Press, 263-297.

[16] Küchlin, W. (1985). A confluence criterion based on the generalized Newman Lemma. In: *EUROCAL'85* (ed. Caviness) 2, Lecture Notes in Computer Science, 204, Springer Verlag, 390-399.

[17] Lankford, D.S., and Ballantyne, A.M. (1977). Decision procedures for simple equational theories with commutative-associative axioms: complete sets of commutative-associative reductions. Automatic Theorem Proving Project, Dept. of Math. and Computer Science, University of Texas, Austin, Texas, Report ATP-39.

[18] Lankford, D.S., and Ballantyne, A.M. (1979). The refutational completeness of blocked permutative narrowing and resolution. Proc. *4th Conf. on Automated Deduction*, Austin, Texas.

[19] Lusk E., and Overbeek, R., (1985) "Reasoning about equality", Journal of Automated Reasoning 6, pp.209-228.

[20] Peterson, G.L., and Stickel, M.E. (1981). Complete sets of reductions for some equational theories. *J. ACM*, 28/2, 233-264.

[21] Slagle, J. (1974) Automated theorem proving for theories with simplifiers, commutativity and associativity. *J. ACM*, 4, 622-642.

[22] Stickel, M. (1981). A complete unification algorithm for associative-commutative functions. *J. ACM*, 28, 3, 423-434.

[23] Stickel, M. (1984). A case study of theorem proving by the Knuth-Bendix method: Discovering that $x^3 = x$ implies ring commutativity. Proc. of *7th Conf. on Automated Deduction,* NAPA, Calif., LNCS 170, Springer Verlag, 248-258.

[24] Veroff, R.L. (1981). Canonicalization and demodulation. Report ANL-81-6, Argonne National Lab., Argonne, IL.

[25] Wang, T.C. (1988). Case studies of Z-module reasoning: Proving benchmark theorems for ring theory. *J. of Automated Reasoning,* 3.

[26] Winkler, F. (1984). The Church-Rosser property in computer algebra and special theorem proving: an investigation of critical pair, completion algorithms". Dissertation der J. Kepler-Universitaet, Linz, Austria.

[27] Winkler, F., (1985) Reducing the complexity of the Knuth-Bendix completion algorithm: a 'unification' of different approaches. In: *EUROCAL'85* (ed. Caviness) 2, Lecture Notes in Computer Science, 204, Springer Verlag, 378-389.

[28] Winkler, F., and Buchberger, B. (1983). A criterion for eliminating unnecessary reductions in the Knuth-Bendix algorithm. In: Proc. of the *Coll. on Algebra, Combinatorics and Logic in Computer Science.* Györ, Hungry.

[29] Wos, L. (1988). *Automated Reasoning: 33 basic research problems.* Prentice Hall, NJ.

SYSTEM DESCRIPTIONS

Solving Systems of Linear Diophantine Equations and Word Equations

Habib Abdulrab* Jean-Pierre Pécuchet**
Faculté des Sciences, B.P. 118, 76134 Mont-Saint-Aignan Cedex †

Solving systems of linear diophantine equations (SLDE) consists in finding a non-negative integer vector $X = (x_1, \ldots x_n)$ of $AX = B$; A and B are $m \times n$ and $m \times 1$ matrices respectively with integer entries.

Solving SLDE arises in many formal calculus systems, and particularly in Makanin's algorithm solving word equations. This algorithm is one of the most important and major results in theoretical computer science.

Several results give bounds on minimal solutions of $AX = B$ and describe the rational set of its solutions.

Consequently, an algorithm solving SLDE is simply obtained by enumerating all n-tuples $V = (x_1, \ldots, x_n)$ such that $x_i \leq \alpha$, where α is a bound on minimal solutions of $AX = B$, and checking if V is a solution of $AX = B$. This immediate enumerating algorithm cannot provide any efficient computer implementation for solving SLDE, because of the value of α and the potentially $(\alpha + 1)^n$ vectors to check.

In order to realize an efficient implementation of the resolution of SLDE to be used as a module in our implementation of Makanin's algorithm [Abd1] we have experimented two algorithms solving SLDE. We describe here the results of this experimentation.

The first one is the "Cutting Planes" method [Gom1]. It consists in a loop of two steps.

1) Finding a non-negative real vector X, solution of $AX = B$, by using the simplex method.

2) If X is a non-negative integer vector, then $AX = B$ has a non-negative integer solution, otherwise a new constraint is imposed to the resolution. This new constraint is a new equation computed from X, and the entries of the system to be solved. More precisely, it is the equation of an hyperplan whose intersection with the solution space S of $AX = B$ gives a new solution space S', such that the following three conditions are realized :

a) All the integer solutions in S remain in S'.

b) A portion of S, containing the optimal solution of 1) falls out S'.

c) the number of constraints is only finite. That is, after a finite number of steps, either there will be no positive real solution in the first step, or its optimal solution will be integer valued. (this condition ensures the convergence and the finiteness of the algorithm).

The second algorithm which we have experimented is the "all integer" algorithm, which differs from the first one by the fact that the simplex method is not used. The basic idea of this algorithm consists in making a series of changes of variables :

---

* Laboratoire d'Informatique de Rouen et LITP. E.m.: mcvax!inria!geocub!abdulrab.

** Laboratoire d'Informatique de Rouen, LITP et CNRS. E.m.: mcvax!inria!geocub!pecuchet.

† This work was also supported by the Greco de Programmation du CNRS and the PRC Programmation Avancee et Outils pour l'Intelligence Artificielle.

$$(*) : x_j = d_j + \sum_{k=1}^{n} d_{jk} y_k, \ j = 1,2,\ldots n. \ \ y_k \geq 0, d_j \in Z.$$

It is clear that if $d_j \geq 0$, then the minimal solution is given by the conditio[n] $y_j = 0, j = 1,2\ldots n$.

The "all-integer" algorithm finds the transformations (*) that leads in a finite numbe[r] of steps to the condition $d_j \geq 0$.

Our experimentation of both algorithms consists in comparing the resolution of system[s] of linear diophantine equations given in literature, and especially by random generating, durin[g] many days, a lot of systems to be solved by both algorithms.

We have observed that the first algorithm provides faster results. This may be explaine[d] by :

1) It is well-known that the simplex method is an efficient algorithm whose compu[-] tational results are particularly promising. Thus, it can be used to advantage for the intege[r] problem.

2) If the system to be solved has no non-negative integer solution, because simply ther[e] is no positive real solution in the first part of the loop, (this is rather frequent in this case), th[e] halt of the resolution is obtained fast at the first iteration of the "cutting plane" method.

3) Especially for grand systems, the convergence is, in general, faster in the firs[t] algorithm, because of both the number of iterations and the computational time of eac[h] iteration.

On the other hand, the second algorithm can be implemented, with less difficulty, an[d] gives interesting results in the case of small systems.

The role of solving SLDE within Makanin's algorithm is studied in [Abd1][Mak,][Pec The construction of efficient systems of linear diophantine equations in all the steps of Makanin' algorithm is presented in [Abd1]. in order to reduce, as many as possible, the resolution of thes[e] systems in our implementation, we gave [Abd1] some conditions characterizing when there i[s] no need to solve such systems, i.e. when one can deduce the result of the resolution from th[e] previously solved systems. In practice, these conditions permit to avoid the resolution in abou[t] 60 per cent of cases.

References :

[Abd1] H. Abdulrab : Résolution d'équations sur les mots: étude et implémentation LISP d[e] l'algorithme de Makanin. (Thèse), University of Rouen (1987). And Rapport LITP 87-2[5] University of Paris-7 (1987).

[Abd2] H. Abdulrab : Implementation of Makanin's algorithm. Rapport LITP 87-72, Universit[y] of Paris-7 (1987).

[Abd3] H. Abdulrab : Equations in words. Proceedings of ISSAC-88, to appear in LNCS.

[Gom] R.E. Gomory : An algorithm for integer solutions to linear programs. Recent advanc[es] in mathematical programming, Eds R.L Graves et p. Wolfe. p. 269-302, (1963).

[Mak] G.S. Makanin : The problem of solvability of equations in a free semigroup. Mat. Sb. 103(145) (1977) p. 147-236 English transl. in Math. USSR Sb. 32, (1977).

[Pec1] J.P. Pécuchet : Équations avec constantes et algorithme de Makanin. (Thèse), University of Rouen, (1981).

SbReve2: A Term Rewriting Laboratory with (AC-)Unfailing Completion

Siva Anantharaman
LIFO, Dépt. Math-Info.
Université d'Orléans
45067 ORLEANS Cedex 02
FRANCE
E-mail: siva@univ-orleans.fr

Jieh Hsiang* **Jalel Mzali**
L. R. I. Bât 490
Université Paris-Sud Orsay
91405 ORSAY Cedex
FRANCE
hsiang@sbcs.sunysb.edu, mzali@lri.lri.fr

January 21, 1989

1 Introduction

In this paper we describe the functionalities and features of *SbReve2*, a system for automated deduction in equational theories.

1.1 Background

In the last fifteen years the family of Knuth-Bendix completion procedures has been studied extensively and its applicability has been extended considerably, and to different domains. One of the important theoretical developments recently is the so-called *unfailing Knuth-Bendix completion* (UKB for short) [BDP-87,HR-87]. Intuitively, UKB imposes a well-founded ordering on the (ground) term structure rather than on the orientation of the equations. Consequently, UKB does not have the abnormal failure frequently caused by un-orientable equations in Knuth-Bendix completion. Thus, UKB provides a complete semi-decision procedure for word problems while retaining to a large extent the efficiency of the original Knuth-Bendix procedure.

In addition to providing a solution to the failure problem of Knuth-Bendix completion, UKB also opens a new perspective of the term rewriting enterprise. That is, Knuth-Bendix type completion procedures have usually been considered as a tool for generating *decision procedures* – canonical systems. UKB, on the other hand, is a *refutational* method targeted on proving a given (equational) theorem. If the procedure does indeed terminate and returns a canonical system, then so much the better. However, even if a canonical system for the given theory does not exist, UKB can still provide a proof for any valid equational theorem (at least theoretically).

Although the commutative/associative axioms can be treated *as is* in UKB, they frequently permute existing equations/rules and cause inefficiency. Thus, it is desirable to treat commutative-associative axioms in the unification process, as in the case of the classical extension of Knuth-Bendix completion to AC-completion. Such an extension has been developed ([AM-88]) which we term *AC-UKB*.

*On leave from SUNY Stony Brook. Research supported in part by grants CCR-8805734 and INT-8715231, both funded by the National Science Foundation.

Another extension for improving the efficiency of refutational theorem proving is to incorporate the *cancellation axioms* into the inference process ([HRS-87]). Cancellation reduces the excess of equations of un-manageable sizes. The potential of such a technique was first demonstrated by Stickel's proof of the Jacobson's Theorem for $n = 3$ ([St-84]).

2 Overview of *SbReve2*

SbReve2 is an extension of *SbReve1*, which is in itself an extension of *Reve2.4* ([FG-84]). All of them are written in CLU and run on the Vaxen and Sun3. *SbReve1* was first developed by Mzali and Hsiang at Stony Brook in 1986 ([HM-88]). *SbReve2* is developed by Anantharaman at the Université d'Orléans. The *SbReve* family maintains the rigorous modularity and user-friendliness requirements which are the trademarks of *Reve2.n* ($1 \leq n \leq 5$). However, the basic completion algorithm employed by the *SbReve* family is quite different from *Reve2*. Roughly speaking, in *Reve2* critical pairs are generated in batches, then they are oriented and processed one after another. Even if an extremely simple critical pair which can be used to simplify the entire set of rules to next-to-nothing it still has to wait until its turn to be processed. Thus in many cases *Reve2* generates much more critical pairs than necessary. In the *SbReve* family we use a different algorithm which makes a much better utilization of simplification. Critical pairs are generated one at a time, and they are processed and used for simplification immediately. Through a careful design of inference rules ([Mz-86]), we did not compromise the modularity requirement of *Reve2*.

Such a simplification-first approach enables us to gain considerably in efficiency over *Reve2.4*. It is even more significant for UKB type procedures since in such procedures the only desirable critical pairs are those that lead more directly to the proof of the intended theorem.

The UKB procedure and its extension to the theory of equalities and inequalities are implemented in *SbReve1*, based on the "simplification" algorithm discussed above. All the other features in the original *Reve2.4* are essentially intact. And *SbReve2* completely subsumes *SbReve1*. In addition, AC-UKB and two ways of treating cancellation axioms are incorporated in *SbReve2*. Other measures for further improving efficiency (such as some critical pair criteria) are also implemented in *SbReve2*. In the following we describe some of the features of *SbReve2*.

3 A Brief Description of *SbReve2*

Since the main new features of *SbReve2* are related to UKB and its extensions, we do not discuss the facilities which already exist in *Reve2.4*, such as standard Knuth-Bendix completion, induction-less induction, and the extensive user-interface as well as the termination ordering library. We describe only the features not present in *Reve2.4*.

(i) Refutational Theorem Proving via (AC-)UKB

Since we intend (AC-)UKB as mainly a semi-decision procedure, the user is encouraged to provide both an equational theory and an equational theorem to be proved. The intended theorem is presented as an inequality which is construed as the skolemization of its negation. For completeness we assume (in theory) that a *complete simplification ordering* (a simplification ordering which is total on ground terms) is given by the user. To allow flexibility the system does not check the totality of this user given ordering. Usually one provides an order on the operators as well as their status (lexicographic or multiset, etc), and the system automatically uses either recursive path ordering or lexicographic path ordering. One can also choose one-component polynomial ordering if one feels like.

The user can either declare certain operators to be AC, or treat the AC axioms as part of the user input equations. In the former case AC-unification and AC-UKB are invoked, while in the latter, the axioms are treated as any other equations.

If UKB can no longer find critical pairs, it returns a *finite* set of equations/rules R. R is canonical on all terms if all equations in R are oriented. Otherwise it is only canonical on *ground* terms ([HR-87]). The same behaviour is true for AC-UKB, provided that in the latter case the ordering is total modulo AC on ground terms. In the current implementation, however, the orderings used do not have this property.

We should also mention that in addition to superposition and simplification, UKB and AC-UKB also need the inference rule of *functional subsumption*, which is also implicitly incorporated in our system.

(ii) Structural Cancellation Laws

An operator f is *right cancellable* if for every x, y, z, $f(x,z) = f(y,z)$ implies $x = y$. *Left cancellation* and *cancellation with an AC-operator* can be defined similarly. Such cancellation properties are enjoyed by many algebraic structures and they can lead to the generation of shorter rules/equations quickly. In *SbReve*2 we allow the user to declare an AC-operator to be cancellable. We incorporated inference rules which automatically check whether a critical pair generated satisfy the properties and, if so, perform cancellation and retain the shorter equation. For the theory behind cancellation inference rules, see [HRS-87].

(iii) User-defined Cancellation Laws for Inequalities

*SbReve*2 also provide the user the ability to do simple cancellation with respect to inequality in a direct way. For instance, one can input, for a binary operator g and a constant c, the "second order" equation $(g(c,x) = / = g(c,z)) == (x = / = z)$, which simplifies the goal $(g(c,x) = / = g(c,z))$ to $(x = / = z)$. In *SbReve*2 such a user provided equation is *never* used for superpositions with the other equations.

This facility provides a limited way of doing *backward reasoning* in *SbReve*2. However, the user should be careful in providing such quations since they may lead to incompleteness. As an example, suppose one gives a theory with only one equation $f(a,b) == f(a,c)$, with the intended theorem also $f(a,b) = / = f(a,c)$ (in its negated form, of course). Then declaring the equation $f(a,b) = / = f(a,c) == (b = / = c)$ will reduce the target theorem to $b = / = c$ which does not yield a refutation with the given equation.

(iv) Elimination of Redundant Critical Pairs

One difficulty with AC-completion (as well as AC-UKB) is the large number of redundant critical pairs generated from the extended rules ([PS-81]) incurred from the use of the AC-axioms. This problem is particularly acute in AC-UKB since some of the equations may not be oriented. We have implemented some critical pair criteria which eliminate many redundant critical pairs implicitly. For the theory behind the method, see [AM-88]. Now, *something very specific to* (some versions of) *Sbreve*2: Critical pairs which have "no chance" of proving the given theorem, are not kept. This is done via a quantitative study of the operators in the critical pair, with respect to the theorem. Of course, this approach is not (theoretically) complete.

(v) Strategies for Ordering Rules/Equations

When a new equation is generated and kept, it is inserted in the existing set of equations waiting to be chosen for generating the next critical pair. The choice of the *next* equation can be crucial to the efficiency of the final proof. We have chosen a mixed strategy for the version presented. Future versions will probably offer the user some choice.

4 Examples

As is perhaps clear now, *Sbreve*2 is not oriented towards a special domain of applications. So its efficiency is surely not optimum for *every* envisageable problem. (On the other hand the orderings available at the moment, and the AC-unification algorithm, all borrowed from *Reve*2, will surely have to be improved). All the same we give below a few examples coming from various fields, executed on a SUN 3-50, at Orléans(Fr). More will be avilable on line.

(i) An algebraic calculation under UKB.
User Equations:
(x*x)*y == y and (x*y)*z == (y*z)*x

Theorems Proved: Associativity of *, (6 secs); its Commutativity, (10 secs).

(ii) A theorem in propositional calculus, under AC-UKB.
(It takes much longer to proceed via AC-Completion, although the proofs obtained here are just by AC-rewriting).
User Equations:

```
or(x, y) == (x * y) + x + y
impl(x, y) == (x*y)+x+1
iff(x, y) == x+y+1
not (x) == x + 1
x+0 == x
x+x == 0
x*1 == x
x*x == x
x*0 == 0
x*(y+z)==(x*y)+(x*z)

Declared AC-operators: ''+'', ''*''.
Theorems Proved:
  or(p,(impl(p,p*q))) == or(not(p),(or(not(q),p*q))),    (2 minutes).

  impl((impl(p,q)),(p*q))==((impl(not(p),q))*(impl(q,p))), (14 secs).
```

(iii) An algebraic calculation using AC-UKB and Cancellation.
User Equations:

```
0+x==x
0*x==0
x*0==0
g(x)+x==0
g(x+y)==g(x)+g(y)
g(g(x))==x
x*(y+z)==(x*y)+(x*z)
(y+z)*x==(y*x)+(z*x)
```

```
(x*y)*y==x*(y*y)
(x*x)*y==x*(x*y)
g(x)*y==g(x*y)
x*g(y)==g(x*y)
```

Declared AC-operator *with cancellation*: "+".

```
Theorem Proved: (a*b)*a == a*(b*a), (55 secs).
```

References

[AM-88] S. ANANTHARAMAN, J. MZALI, Unfaling Completion Modulo a set of Equations, *Technical report, LRI, 1989*

[BDP-87] L. BACHMAIR, N. DERSHOWITZ, D. PLAISTED, Completion without failure, *Proc. Coll. on Resolution of Equations in Algebraic Structures, Lakeway, Texas, 1987*

[FG-84] R. FORGAARD, J. GUTTAG, REVE: A term rewriting system generator with failure-resistant Knuth-Bendix, *MIT-LCS technical report, 1984*

[HM-88] J. HSIANG, J. MZALI, SbReve users guide, *Technical report, LRI, 1988*

[HR-87] J. HSIANG, M. RUSINOWITCH, On word problems in equational theories, *14th ICALP, Springer-Verlag LNCS Vol 267, pp54–71, 1987*

[HRS-87] J. HSIANG, M. RUSINOWITCH, K. SAKAI, Complete set of inference rules for the cancellation laws, *IJCAI 87, Milan, Italy, 1987*

[St-84] M.E. STICKEL, A case study of theorem proving by the Knuth-Bendix method: Discovering that $x^3 = x$ implies ring commutativity, *7th CADE, Springer-Verlag, LNCS Vol 170, pp248-258, 1984*

[Mz-86] J. MZALI, Methodes de filtrage equationnel et de preuve automatique de theoremes, *Thesis, Université de Nancy, 1986*

[PS-81] G. PETERSON, M.E. STICKEL, Complete sets of reductions for some equational theories, *JACM Vol 28, pp 233-264, 1981*

THEOPOGLES - An efficient Theorem Prover based on Rewrite-Techniques

J. Avenhaus, J. Denzinger, J. Müller

Department of Computer Science
University of Kaiserslautern
6750 Kaiserslautern (FRG)

Introduction

THEOPOGLES (german acronym for Theorem Prover for First Order Polynomial Equalitions) is an automated theorem prover for first order predicate logic with equality. It´s main feature is the use of rewriting techniques, e.g. reduction and critical pair completion, as inference rules on a two level Knuth-Bendix Completion Procedure. Using THEOPOGLES without equality means working with a variation of the methods of Hsiang [H85], Kapur and Narendran [KN85] and Bachmair, Dershowitz [BD87]. That is, a first order formula F to be proved valid is transformed into a system E of polynomial equations, s.th. F is valid iff E is unsolvable. The ideas to handle equality with THEOPOGLES stems from approaches of Hsiang [H87] and Rusinowitch [R87]. But as in the definition for pure first order predicate calculus THEOPOGLES avoids the use of unnecessary inference steps. In general E is then divided into a system $(\mathcal{R},\varepsilon)$ of polynomial rewrite rules $\mathcal{R}$ and polynomial equations ε, and into a set R of term rewrite rules. Completion is performed separately on $(\mathcal{R},\varepsilon)$ and on R, and special inference rules are applied on elements of $(\mathcal{R},\varepsilon)$ and R to get the connection of the whole system.

In the following we will give a sketch of the theoretical technicalities and then we will describe the general structure of the system.

Details can be found in [M87], [M88], [D88] for the theoretical aspects especially the completeness proofs, in [M88], [MS88a], [MS89] for the correspondence with resolution ATPs, and in [D87], [DM87], [MS88b] for technical descriptions of the system.

Theoretics

Polynomial equations are of the form $m_1+m_2 + ... + m_k = 0$, where + is the operation XOR and the monomials m_i are conjunctions $L_1*L_2* ... *L_n$ of atoms L_j. Given a formula F, it can be transformed into a set $E = \{p_i = 0\}_{i=1..n}$ of polynomial equations, s.th. F is valid iff E has no

solution, i.e. there is no interpretation $\mathcal{J}$, s.th. $\mathcal{J}(p)$ = false for all $p = 0 \in E$.
THEOPOGLES divides the equational system E into three sets of rewrite rules and equations.

E

$\mathcal{R}$:	ε:	R:
$m \rightarrow 0$	$m_1+m_2+...+m_k=0$	$t_1 \rightarrow t_2$
$EQ*m \rightarrow m$		
$L \rightarrow 1$		
$L*m \rightarrow m$		

If an equation in E has the form $m = 0$, $L+1 = 0$, $L*m+m = 0$ or $EQ*m+m = 0$, then it is transformed into a polynomial rewrite rule in $\mathcal{R}$ (m is a monomial, L an (non-EQ) atom and EQ an equality atom). If an equation is of the form $EQ(t_1,t_2)+1 = 0$ and $t_1 > t_2$ for a simplification ordering > on terms, then the equation is turned into $t_1 \rightarrow t_2$ in R. All equations that do not correspond to a rewrite rule belong to ε.
As usual the rewrite rules in $\mathcal{R}$ are used for mutual normalization and to simplify the polynomials in ε. The rewrite rules in R reduce the terms of the atom arguments and they are also used for the interreduction of R itself. This constitutes the first class of inference rules. Note that the rewrite rules are very simple and can be efficiently implemented. On the other hand they are strong enough to reduce the search space for the (more expersive) critical pair generation drastically.
The second class of inference rules for $(\mathcal{R},\varepsilon,R)$ is the critical pair generation. Let p_{EQ} denote polynomials consisting of EQ-atoms and/or the atom 1. Then we have:

- Superposition of **one** atom in m from an equation (rule) of the form $m*p_{EQ} = 0$ with an atom of a rule in $\mathcal{R}$ or an equation in ε.
- Paramod-Superposition with EQ from an equation (rule) $EQ*p_1+p_2 = 0$ with a term in an atom of a rule in $\mathcal{R}$ or an equation in ε.
- Factorization of equations (rules) of the form $m*p_{EQ} = 0$.
- Standard superposition of rules in R .
- Superposition of a rule in R and a term of an atom of a rule in $\mathcal{R}$ or an equation in ε.

These superpositions for critical pair generation together with the simplification with $\mathcal{R}$ and R present the complete theorem prover THEOPOGLES. Note here, that the application of the critical pair generation is very restricted compared with other methods. For example, the first superposition rule might save an exponential factor relative to Hsiangs approach.

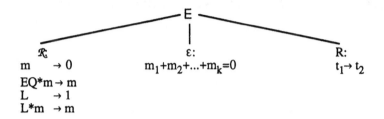

The System

THEOPOGLES is implemented on an APOLLO-Workstation in Common-LISP. The system's modular structure is shown below.

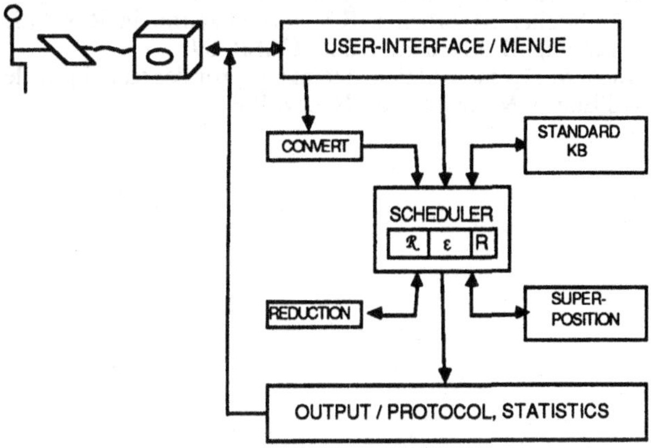

CONVERT transforms a first order formula to be proved into the corresponding equational system E and divides E into the working triple ($\mathcal{R}$,ε,R).

The SCHEDULER administers ($\mathcal{R}$,ε,R) by using a link structure to represent potential reduction and superposition occurrences. It controls the order of the reduction steps and of the critical pair generation and the standard completion. The I/O-devices of the system are a comfortable user interface which is menue driven and which includes a help system. For an output the user may choose between several modes with different information depth. Additionally there is a statistical preparation of the proofs.

References

[BD87] Bachmair, L., Dershowitz, N.: Inference Rules for Rewrite-Based First-Order Theorem Proving, 2nd LICS, May 1987.

[D87] Denzinger, J.: Implementation of a Theorem Prover Based on Rewriting-Techniques (in German), project-report, University of Kaiserslautern.

[D88] Denzinger, J.: EQTHEOPOGLES (in German), M.S. thesis, University of Kaiserslautern.

[DM87] Denzinger, J., Müller, J.: THEOPOGLES user manual, University of Kaiserslautern.

[H85] Hsiang, J.: Refutational Theorem Proving using Term Rewriting Systems, AI 25, 1985, 255-300.

[H87] Hsiang, J.: Rewrite Method for Theorem Proving in First Order Theory with Equality, J. of Symb. Comp., 1987.

[KN85] Kapur, D., Narendran, P.: An Equational Approach to Theorem Proving in First-Order Predicat Calculus, 84CRD322, GEC Research and Dev. Report. Schenactady, N.Y., Sept. 85.

[M87] Müller, J.: THEOPOGLES - A Theorem Prover Based on First-Order Polynomials and a Special Knuth-Bendix Procedure, Proc. 11th GWAI, Sept. 87, Springer IFB 152.

[M88] Müller, J.: Theorem Proving with Rewrite-Techniques, - Methods, Strategies and Comparisons - (in German), Ph.D.Thesis, University of Kaiserslautern, Nov. 1988.

[MS88a] Müller, J., Socher-Ambrosius, R.: On the Unnecessity of Multiple-Overlaps in Completion Theorem Proving, Proc. 12th GWAI, Sept. 88, Springer IFB 181.

[MS88b] Müller, J., Socher-Ambrosius, R.: Topics in Completion Theorem Proving, SEKI-Report SR-88-13, University of Kaiserslautern.

[MS89] Müller, J., Socher-Ambrosius, R.: Normalization in Resolution and Completion Theorem Proving, Draft, submitted to ICALP ´89.

[R87] Rusinowitch, M.: Demonstration automatique par des techniques de reecriture, These de Doctorat d'Etat en Mathematique, Nancy, 1987.

COMTES – An experimental environment for the completion of term rewriting systems

Jürgen Avenhaus
avenhaus@uklirb.uucp

Klaus Madlener
madlener@uklirb.uucp

Joachim Steinbach
steinba@uklirb.uucp

Universität Kaiserslautern
Fachbereich Informatik
Postfach 30 49
D - 6750 Kaiserslautern
F.R.G.

1. Overview

COMTES can be viewed as a parametrized completion algorithm that is particularly suited for efficiency experiments. The kernel of the system is a slightly modified version of Huet's method ([Hu81]). Various strategies for generating critical pairs, reducing terms and for proving the termination of rules are integrated.

COMTES is implemented in Common-Lisp and currently running on Apollo Domain systems under Aegis. It consists of 6500 lines of source code without documentation corresponding to 270 KB of compiled code. The architecture of COMTES is described in [St89].

2. Completion procedure

The Knuth-Bendix completion algorithm tries to transform a system of equations into a confluent and terminating system of rules by orienting equations and by adding new rules. During this completion process, some newly introduced rule may simplify some old rule, either on its left or on its right-hand side. When a set of equations can be oriented in such a way that the completion process terminates, the resulting system defines a decision procedure for the word problem in the corresponding equational theory.

The implemented method chooses the shortest (according to the number of symbols) unmarked rule $l \rightarrow r$ of the actual rule system R. After marking $l \rightarrow r$ the critical pairs between this and all other rules that are already marked are generated. The non-trivial oriented normal forms (relative to the selected ordering) of each critical pair represent a new rule $l' \rightarrow r'$ (incomparable terms l' and r' cause a break of the algorithm). The optional process of interreduction exchanges the right-hand sides of R for their normal forms relative to $l' \rightarrow r'$. The iteration of the whole process will stop with success, if the set of critical pairs of R is empty.

3. Special features

COMTES provides various tools for generating critical pairs, reducing terms and orienting rules. A brief survey of these features indicates the performance of the system.

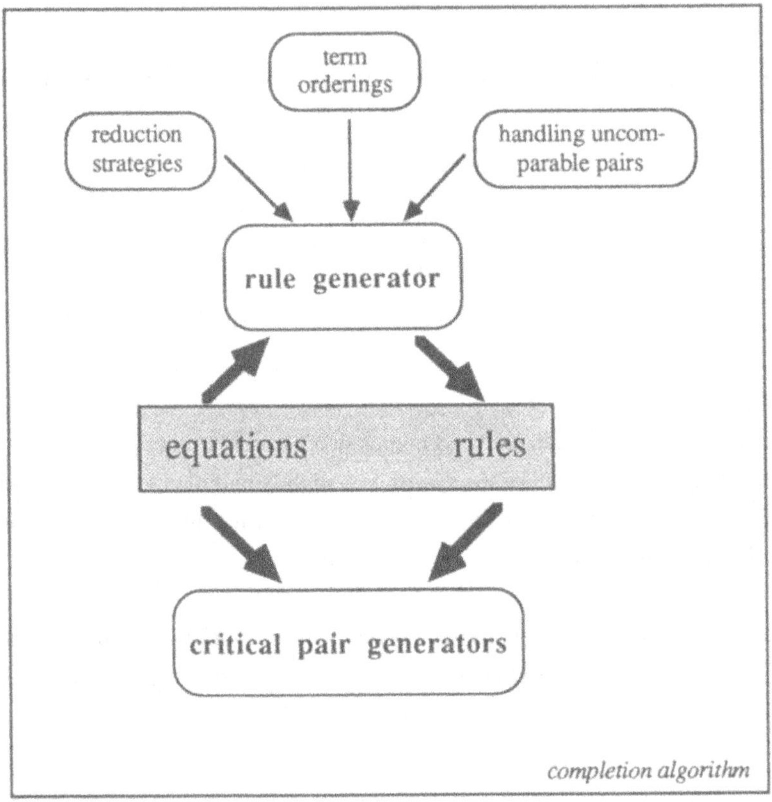

Critical pair algorithms

The efficiency of the completion process depends on the number of critical pairs that are generated. In general, a lot of them are redundant. Thus, they might be eliminated if they could be characterized. Such characterizations of special redundant critical pairs are called critical pair criteria (see [BD88], [MSS89]). Critical pair criteria that are based on the generalized Newman lemma have been proposed by Winkler and Buchberger ([WB83], [Wi84]), by Küchlin ([Ku85]) and by Kapur, Musser, Narendran ([KMN84]). These criteria have in common that they check whether a special rule is applicable. The criterion of Kapur restricts the positions at which this rule may be applied. The other criteria test whether certain critical pairs have already been computed. The strategy of Küchlin is an extension of the method developed by Winkler and Buchberger. All these methods are implemented together with extensions of Küchlin's and Winkler's strategy (cf. [So88], [Sa89]).

Reduction strategies

The algorithms presented here do not impose any restrictions on the rewriting systems. We have implemented a nondeterministic reduction algorithm and the four deterministic strategies leftmost-innermost, bottom-up, leftmost-outermost and top-down (see [Ku82], [St83], [GB85], [Wa86]). The last two ones visit the nodes of a term from the top to the bottom whereas leftmost-innermost and bottom-up take the opposite direction. The difference between leftmost-innermost (leftmost-outermost) and bottom-up (top-down) is that the latter works parallel and the former depth-first. These four strategies are implemented in two ways, working with or without explicitly marking irreducible subterms (the order of applying rules to subterms gives some information about which subterms may be reducible and which are not).

Term orderings

The effective computation with term rewriting systems presumes termination. Orderings on terms are able to guarantee termination. COMTES provides seven different term orderings: Several path and decomposition orderings, the Knuth-Bendix ordering and the ordering on polynomial interpretations. The recursive path ordering of Dershowitz ([De82], [St88a]) decomposes terms into subterms which are compared relative to an underlying multiset ordering. The path orderings of Plaisted ([Pl78], [St88a]) and Kapur, Narendran, Sivakumar ([KNS85], [St88a]) decompose terms into paths which are compared relative to an underlying multiset ordering. Two new decomposition orderings ([St88a]) based on the concept of the decomposition ordering of Jouannaud, Lescanne, Reinig ([JLR82]) decompose terms into three (or even four) parts which are compared lexicographically. The ordering on polynomials ([La79], [BL87], [Pa89]) transforms terms into polynomials over natural numbers which are compared relative to the greater relation on the natural numbers. Finally, the Knuth-Bendix ordering is a special version of the latter considering only linear polynomials of the form $x_1+...+ x_n+ a$.

Handling incomparable pairs

COMTES possesses some special operations for handling incomparable pairs. Extending the precedence of the ordering (the system suggests some suitable possibilities), postponing a pair, introducing new operators or orienting a pair by hand may prevent a break of the algorithm.

4. Input / Output

The input for COMTES consists of the initial system of equations, a detailed specification of the used function symbols (a signature) and a term ordering. A special tool of COMTES - the parser - transforms this external specification into an internal representation checking the syntax of the specification.

COMTES creates several output files. A statistical file contains term measurements for reduction, orientation and interreduction. For each specification file an output file is created which contains the initial and canonical rule system. Furthermore, there is the possibility to make a snapshot during the completion process, i.e. the actual rule system, the set of incomparable pairs and detailed information about the current ordering will be printed out.

5. References

[AGGMS87] Jürgen Avenhaus, Richard Göbel, Bernhard Gramlich, Klaus Madlener, Joachim Steinbach: *TRSPEC - a term rewriting based system for algebraic specifications,* Proc. 1st int. workshop on conditional term rewriting systems, LNCS 308, Orsay, France, July 1987, pp. 245-248

[AGM88] Jürgen Avenhaus, Bernhard Gramlich, Klaus Madlener: *Experimental tools for rewriting techniques and applications at Kaiserslautern - An overview,* Draft, Dept. of computer science, Univ. of Kaiserslautern, W. Germany, November 1988

[Av83] Jürgen Avenhaus: *Reduction systems,* Manuscript of a lecture, Dept. of computer science, Univ. of Kaiserslautern, W. Germany, Summer 1983 (in German)

[BD88] Leo Bachmair, Nachum Dershowitz: *Critical pair criteria for completion,* J. symbolic computation 6, 1988, pp. 1-18

[BL87] Ahlem Ben Cherifa, Pierre Lescanne: *Termination of rewriting systems by polynomial interpretations and its implementation,* Science of computer programming 9(2), October 1987, pp. 137-160

[Bu85] Bruno Buchberger: *Basic features and development of the critical-pair/completion procedure,* Proc. 1st int. conf. on rewriting techniques and applications, LNCS 202, Dijon, France, May 1985, pp. 1-45

[De85] Nachum Dershowitz: *Termination,* Proc. 1st int. conf. on rewriting techniques and applications, LNCS 202, Dijon, France, May 1985, pp. 180-224

[De82] Nachum Dershowitz: *Orderings for term rewriting systems,* J. theoretical computer science, Vol. 17, No. 3, March 1982, pp. 279-301

[Fa89] Christian Falter: *User interface for softwaretools,* Internal report, Dept. of computer science, Univ. of Kaiserslautern, W. Germany, 1989, to appear (in German)

[Fe88] Roland Fettig: *Dynamic multiset orderings for ground terms,* Internal report, Dept. of computer science, Univ. of Kaiserslautern, W. Germany, May 1988 (in German)

[GB85] Jean H. Gallier, Ronald V. Book: *Reductions in tree replacement systems,* Theoretical computer science 37, 1985, pp. 1-55

[Gr88] Bernhard Gramlich: *Unification of term schemes - Theory and Applications,* SEKI-Report SR-88-12, Artificial intelligence laboratories, Dept. of computer science, Univ. of Kaiserslautern, W. Germany, December 1988

[Hu81] Gerard Huet: *A complete proof of correctness of the Knuth-Bendix completion algorithm,* J. computer and system science 23, 1981, pp. 11-21

[JLR82] Jean-Pierre Jouannaud, Pierre Lescanne, Fernand Reinig: *Recursive decomposition ordering,* I.F.I.P. Working conf. on formal description of programming concepts II (D. Bjørner, ed.), North Holland, Garmisch Partenkirchen, W. Germany, 1982, pp. 331-348

[KB70] Donald E. Knuth, Peter B. Bendix: *Simple word problems in universal algebras,* Computational problems in abstract algebra (J. Leech, ed.), Pergamon Press, 1970, pp. 263-297

[KMN84] Deepak Kapur, David R. Musser, Paliath Narendran: *Only prime superpositions need be considered in the Knuth-Bendix completion procedure,* J. symbolic computation 6, 1988, pp. 19-36

[KNS85] Deepak Kapur, Paliath Narendran, G. Sivakumar: *A path ordering for proving termination of term rewriting systems,* Proc. 10th coll. on trees in algebra and programming, LNCS 185, 1985, pp. 173-187

[Ku85] Wolfgang Küchlin: *A confluence criterion based on the generalized Newman lemma,* Proc. european conf. of computer algebra, Linz, 1985, pp. 390-399

[Ku82] Wolfgang Küchlin: *Some reduction strategies for algebraic term rewriting,* SIGSAM Bulletin 16 (4), November 1982, pp. 13-23

[La79] Dallas Lankford: *On proving term rewriting systems are noetherian,* Memo MTP-3, Mathematics dept., Louisiana tech. univ., Ruston, LA, May 1979

[MS86] Jürgen Müller, Joachim Steinbach: *Topographical multiset orderings,* Proc. 10th German workshop and 2nd Austrian conf. on artificial intelligence, Ottenstein, Austria, September 1986, pp. 254-264 (in German)

[MSS89] Jürgen Müller, Andrea Sattler-Klein, Inger Sonntag: *Comparison of subconnectedness criteria for term rewriting systems,* Internal report, Dept. of computer science, Univ. of Kaiserslautern, W. Germany, 1989, to appear

[MW86] Jürgen Müller, Elvira Wagner: *Efficient reduction strategies for term rewriting systems,* Proc. 10th German workshop and 2nd Austrian conf. on artificial intelligence, Ottenstein, Austria, September 1986, pp. 242-253 (in German)

[Pa89] Doerthe Paul: *Implementing the ordering on polynomials,* Internal report, Dept. of computer science, Univ. of Kaiserslautern, W. Germany, 1989, to appear (in German)

[Pl78] David A. Plaisted: *A recursively defined ordering for proving termination of term rewriting systems,* Report UIUCDCS-R-78-943, Dept. of computer science, Univ. of Illinois, Urbana, IL, September 1978

[Sa89] Andrea Sattler-Klein: *Critical pair criteria - comparison and improvement,* Internal report, Dept. of computer science, Univ. of Kaiserslautern, W. Germany, 1989, to appear

[Sa88] Andrea Sattler-Klein: *The SMOV-strategy - A strategy that can be used to complete a string rewriting system by the Knuth-Bendix algorithm,* Internal report, Dept. of computer science, Univ. of Kaiserslautern, W. Germany, 1988

[Sa87] Andrea Sattler-Klein: *Minimization of critical pairs in the completion system COSY,* Internal report, Dept. of computer science, Univ. of Kaiserslautern, November 1987 (in German)

[So88] Inger Sonntag: *Subconnectedness criteria - Implementation and comparison,* Internal report, Dept. of computer science, Univ. of Kaiserslautern, W. Germany, Summer 1988 (in German)

[St89] Joachim Steinbach: *COMTES - completion of term rewriting systems,* to appear as SEKI-Workingpaper, Univ. of Kaiserslautern, W. Germany, 1989 (in German)

[St88] Joachim Steinbach: *Comparison of simplification orderings,* Proc. of the term rewriting workshop, Bristol, U.K., September 1988, to appear

[St88a] Joachim Steinbach: *Term orderings with status,* SEKI-Report SR-88-12, Artificial intelligence laboratories, Dept. of computer science, Univ. of Kaiserslautern, W. Germany, September 1988

[St88b] Joachim Steinbach: *Comparison of simplification orderings,* SEKI-Report SR-88-02, Artificial intelligence laboratories, Dept. of computer science, Univ. of Kaiserslautern, W. Germany, Feb. 1988

[St86] Joachim Steinbach: *Orderings for term rewriting systems,* Internal report, Dept. of computer science, Univ. of Kaiserslautern, W. Germany, June 1986 (in German)

[St83] Mark E. Stickel: *A note on leftmost-innermost reduction,* SIGSAM Bulletin 17, 1983, pp. 19-20

[Wa86] Elvira Wagner: *An extended Knuth-Bendix algorithm,* Internal report, Dept. of computer science, Univ. of Kaiserslautern, W. Germany, June 1986 (in German)

[WB83] Franz Winkler, Bruno Buchberger: *A criterion for eliminating unnecessary reductions in the Knuth-Bendix algorithm,* Proc. coll. on algebra, combinatorics and logic in computer science, Györ, Hungary, 1983

[Wi84] Franz Winkler: *The Church-Rosser property in computer algebra and special theorem proving: An investigation of critical pair completion algorithms,* Thesis, Univ. of Linz, Austria, 1984

[WS84] Elvira Wagner, Joachim Steinbach: *Implementation of a basic version of the Knuth-Bendix completion algorithm,* Internal report, Dept. of computer science, Univ. of Kaiserslautern, W. Germany, Summer 1984 (in German)

ASSPEGIQUE : an integrated specification environment

Michel BIDOIT, Francis CAPY, Christine CHOPPY
Stephane KAPLAN, Francoise SCHLIENGER, Frederic VOISIN
Laboratoire de Recherche en Informatique
U.A. C.N.R.S. 410
Universite Paris-Sud, Bat 490
F-91405 ORSAY Cedex, FRANCE

Asspegique is an integrated environment for the development of large algebraic specifications and the management of a specification data base. The aim of the Asspegique specification environment is to provide the user with a “specification laboratory” where a large range of various tools supporting the use of algebraic specifications are closely integrated. The main aspects adressed in the design o Asspegique are the following: dealing with modularity and reusability, providing ease of use, flexibility of the environment and user-friendly interfaces. The Asspegique specification environment supports (a subset of) the specificatio langage Pluss.

The tools available in the Asspegique specification environment include a specia purpose syntax directed editor, a compiler, symbolic evaluation tools, theorem proving tools, an assistant for deriving Ada implementations from specifications and interfaces with the Reve and Slog systems. All these tools are available to the user through a full-screen, multi-window user interface and access the specification data base through the hierarchical management tool. Most of these use Cigale, a system for incremental grammar construction and parsing, which has especially designed in order to cope with a flexible, user-oriented way of defining operators, coercion and overloading of operators.

The symbolic evaluation tools take conditional rewrite rules into account. There are three different possibilities for evaluation:

1. “standard” rewriting, with textual and graphic trace options, and strategy choice
2. evaluation with compiled rules (much faster)
3. mixed evaluation (rewriting + user code).

KBlab: an equational theorem prover for the Macintosh

Maria Paola Bonacina, Giancarlo Sanna
Dipartimento di Scienze dell'Informazione, Università degli Studi di Milano
Via Moretto da Brescia 9, 20133 Milano
Mail: gdantoni@imisiam.bitnet

KBlab is a Completion based theorem prover for equational logic, written in the language C and developed on the Macintosh in the MPW (Macintosh Programmer Workshop) programming environment.

The core of KBlab is the Knuth–Bendix Completion Procedure (*KB*) [9,7,1], extended to Unfailing Knuth–Bendix (*UKB*) [5], *S–strategy* [5] and inductive theorem proving (*IKB*). *IKB* implements the Huet–Hullot method for inductionless induction [8] and the Fribourg linear strategy [3]. The Knuth–Bendix ordering [9] and both the multiset extension and the lexicographic extension of the recursive path ordering are available.

KBlab couples ease of use, portability and low cost of a small Macintosh application with advanced features for experimenting in automated reasoning. It is possible to edit theories, execute the Completion procedures, store the proof traces and even modify the search strategy, all within the same environment.

Experimentation with search strategies is one of the new features of KBlab. The user chooses the search strategy for selecting axioms during the Completion process. Strategy selection is a key feature for a Completion based theorem prover, since it affects termination and the number of critical pairs generated. KBlab also allows the user to modify the strategy during a proof session, to direct it towards a positive result. Nine different strategies are available: *FIFO*, *LIFO*, *smallest components + FIFO or LIFO*, *by size + FIFO or LIFO*, *ordered axioms + FIFO or LIFO*, *linear*. The *smallest components* and *by size* strategies are based on counting the number of symbols in the axiom terms. The *ordered axioms* strategy extends to axioms the selected simplification ordering on terms. The FIFO or LIFO strategy solves possible conflicts.

The results of the experimentation performed with KBlab are collected in a small data base of solved problems. We have extensively tested KBlab on problems in combinatorial logic taken from [12]. So far we have succeded in proving about 23% of them, including some rather large examples. The availability of several strategies turned out to be a key feature in solving these problems, since many of them could be solved, or solved in shorter time, by modifying the selected strategy during the run (e.g. es.6 p.97, es.1 p.118, es.13 p.132, es.1 p.151, es.2 p.182). Search strategies interaction allowed also to prove that Grau's three axioms are sufficient to define a ternary Boolean algebra (example 6, page 158 in [2]).

The following table shows results obtained from running KBlab on some problems with different search strategies.

Problem A: completion of the abstract loop axioms [9] by KB.

Problem B: completion of the central groupoid axioms ($(X*Y)*(Y*Z) = Y, (X*(X*X))*Y = X*Y$) by UKB.

Problem C: proving by UKB that in group theory $x^2 = e$ implies commutativity.

Problem D: proving by S–strategy that in group theory $x^3 = e$ implies $h(h(X,Y),Y) = e$ where the commutator h is defined by $h(X,Y) = XYX^{-1}Y^{-1}$.

Problem E: proving by S–strategy that given combinators t, $tXY = YX$, and q, $qXYZ = Y(XZ)$, there exists a combinator b such that $bXYZ = X(YZ)$ [12].

Problem F: proving by S–strategy that given combinators b, $bXYZ = X(YZ)$ and m, $mX = XX$, for all X exists Y such that $XY = Y$, i.e. a fixed point [12,11].

Problem G: proving by S–strategy that given combinators b, $bXYZ = X(YZ)$ and s_2, $s_2XYZ = XZ(YY)$, for all X exists Y such that $XY = Y$, i.e. a fixed point [12,10].

Experiments with search strategies in KBlab

	FIFO	*LIFO*	*smallest compo- nents FIFO*	*smallest compo- nents LIFO*	*by size FIFO*	*by size LIFO*	*ordered axioms FIFO*	*ordered axioms LIFO*
A *KB*	*6* *40* *9* *14* *1.20*	*6* *40* *10* *14* *1.20*	*6* *41* *10* *14* *1.20*	*6* *41* *10* *14* *1.20*	*6* *41* *10* *14* *1.20*	*6* *41* *10* *14* *1.20*	*6* *41* *10* *14* *1.35*	*6* *41* *10* *14* *1.32*
B *UKB*	*2* *891* *43* *6* *101.92*	*2* ↑	*2* *320* *29* *6* *22.13*	*2* *342* *37* *6* *24.25*	*2* *320* *29* *6* *23.28*	*2* *342* *37* *6* *26.98*	*2* *709* *40* *6* *78.67*	*2* *318* *25* *6* *26.93*
C *UKB*	*5* *89* *15* *13* *4.80*	*5* *24* *10* *11* *1.12*	*5* *33* *11* *11* *1.42*	*5* *30* *10* *11* *1.25*	*5* *33* *11* *11* *1.42*	*5* *30* *10* *11* *1.22*	*5* *30* *10* *11* *1.03*	*5* *30* *10* *11* *1.02*
D *S*	*5* *405* *65* *36* *102.53*	*5* ↑	*5* *401* *76* *55* *136.05*	*5* *248* *51* *37* *41.28*	*5* *401* *76* *55* *135.93*	*5* *248* *51* *37* *41.35*	*5* ↑	*5* ↑
E *S*	*3* *119* *101* *105* *352.17*	*3* ↑	*3* *55* *49* *53* *51.58*	*3* *35* *35* *39* *25.28*	*3* ↑	*3* ↑	*3* ↑	*3* ↑
F *S*	*3* *9* *9* *12* *1.33*	*3* ↑	*3* *22* *20* *20* *6.78*	*3* ↑	*3* *21* *19* *19* *6.65*	*3* ↑	*3* *22* *20* *20* *6.87*	*3* *22* *20* *20* *6.90*
G *S*	*3* *18* *16* *19* *7.38*	*3* *2* *2* *5* *0.13*	*3* *2* *2* *5* *0.20*	*3* *2* *2* *5* *0.15*	*3* ↑	*3* *2* *2* *5* *0.13*	*3* ↑	*3* ↑

Each entry gives, in order, the number of axioms in the input, the number of critical pairs generated, the number of non trivial critical pairs generated, the number of axioms in the output and the running time in seconds on the Macintosh II. The ↑ entry means that the prover was interrupted after running without yielding an answer for a much longer time than that required by the same problem with other strategies. The KBO ordering was selected on problems A, C, D, E, F and the RPO ordering on problems B and G.

All but the first problem in the table above turned out to be strongly sensitive to the different strategies. Problem E was the hardest one: only three search strategies lead KBlab to find the solution in a reasonable time. The smallest components with LIFO strategy gave a very good result, whereas the FIFO strategy required a much longer time. The running time of the FIFO strategy on this problem was higher than that on problems B and D although in these two examples KBlab generated many more critical pairs, because equations generated in solving problem E involved very long terms. The FIFO strategy yielded a result on all the listed problems, but it was always slower than the others strategies with the exception of problem F. The LIFO strategy halted only on problems A, C and G. Both strategies are very useful for experimentation: the former is a safe, exhaustive strategy; the latter is certainly not fair, but when it works it can yield very good results, as shown by examples C and G, where it yielded the fastest running time. The ordered axioms strategy worked very well on example C but not in cases D, E, F. Moreover its behaviour on problem B was affected by the choice between FIFO and LIFO as strategy to solve conflicts among axioms having the same position in the ordering. The four smallest components and by size strategies behaved very similarly on examples A, B and C. The choice of FIFO or LIFO made a significant difference in problems D and F. Strategies in the LIFO family were faster on problem D, but they did not halt on problem F, because they are not fair. It is interesting to note that, the ordered axioms strategy, the only strategy which orders the terms and the axioms coherently, did not fare as well as we had expected.

Acknowledgements

KBlab has been developed at Dipartimento di Scienze dell'Informazione, Università degli Studi di Milano. Giovanni Degli Antoni followed, encouraged and supported the development of KBlab since its very beginning.

Jieh Hsiang suggested several improvements to KBlab and its presentation and gave many interesting problems for testing.

The experimentation done with KBlab while the first author was visiting Laboratoire de Recherche en Informatique, Université de Paris Sud at Orsay, contributed to improve it significantly: we thank Jean Pierre Jouannaud and Emmanuel Kounalis.

References

[1] L.Bachmair, N.Dershowitz, J.Hsiang – Orderings for Equational Proofs
In Proceedings 1st Annual IEEE Symp. on Logic in Computer Science, 346–357, Cambridge, MA, June 1986

[2] L. Fribourg – A superposition oriented theorem prover
J. of Theoretical Computer Science, Vol. 35, 129–164, 1985

[3] L.Fribourg – A Strong Restriction to The Inductive Completion Procedure
In Proceedings 13th Int. Conf. on Automata, Languages and Programming, Rennes, France, July 1986, Lecture Notes in Computer Science 226, 1986

[4] Glickfield, R.Overbeek – A Foray Into Combinatory Logic
J. of Automated Reasoning, Vol. 2, No. 4, Dec. 1986

[5] J.Hsiang, M.Rusinowitch – On Word Problems in Equational Theories
In Th.Ottman ed., Proceedings 14th Int. Conf. on Automata, Languages and Programming, Karlsrhue, W.Germany, July 1987, Lecture Notes in Computer Science 267, 1987

[6] J.Hsiang, J.Mzali – SbREVE User's Guide
To appear as Technical Report L.R.I. Universitè de Paris Sud, Orsay, France

[7] G.Huet – A Complete Proof of Correctness of Knuth–Bendix Completion Algorithm
J. of Computer and System Sciences, Vol. 23, 11–21, 1981

[8] G.Huet, J.M.Hullot – Proofs by Induction in Equational Theories with Constructors
J. of Computer and System Sciences, Vol. 25, 239–266, 1982

[9] D.E.Knuth, P.Bendix – Simple Word Problems in Universal Algebras
In J.Leech ed., Proceedings of the Conf. on Computational Problems in Abstract Algebras, Oxford, 1967, Pergamon Press, Oxford, 263–298, 1970

[10] B.McCune, L.Wos – Some Fixed Points Problems in Combinatory Logic
AAR Newsletter

[11] A.Ohsuga, K.Sakai – Refutational Theorem Proving for Equational Systems Based on Term Rewriting
Technical Report COMP86–40, ICOT, 1986

[12] R.Smullyan – How to mock a mocking bird
Alfred A. Knopf, New York 1985

Fast Knuth-Bendix Completion: Summary

Jim Christian [1]
jimc@rascal.ics.utexas.edu

1 Introduction

By the use of a linear term representation, and with discrimination nets for subterm indexing, the efficiency of Knuth-Bendix completion [3] can be dramatically improved. The speedup factor is typically ten to twenty on small problems, and increases indefinitely as the problem becomes harder. We show one problem for which a speedup factor of more than 60 was achieved; the run time for the problem was reduced from 80 minutes to 80 seconds. Of course, the speedup comes at the price of extra memory consumption — typically five to six times that required for the slow version.

To make the needed comparisons, I implemented a "control" version of completion in C, using conventional techniques. Terms are represented in the system as tagged 32-bit objects, where the upper few bits determine the type of the term (variable, constant, or functional term of arity 1 through 4), and the lower bits contain data or the address of a functional term record, depending upon the type of the term. Memory management is explicit; there is a separate freelist for each data structure used by the system, and a central dynamically grown heap from which new structures are allocated when a freelist becomes empty. Terms don't share structure, so garbage collection is unnecessary — a structure is simply returned to the appropriate freelist when it is no longer needed. As in Prolog systems [8], a trail stack is used to record and undo variable bindings. The standard Knuth-Bendix ordering is used to orient equations. Critical pairs are generated by choosing an unmarked rule with the smallest total number of symbols in its left and right sides, and superposing it with other marked rules in the order in which they were marked.

2 A Linear Term Representation

Terms in the fast implementation are represented as doubly-linked lists with skip pointers, instead of as trees. (See Figure 1a.) The basic term structure, called a *flatterm*, comprises four fields. The *offset* field (so called because of its use in discrimination nets) is a short integer. Values from 0 to 255 are understood to represent function symbols (including atomic constants). The are an index into the system symbol table, where the symbol's arity and weight can be found if needed. Hence, two flatterms representing the same function symbol have identical offset fields. Variables are represented by values greater than 255. In this case, a value is an index into an array of registers which hold the variable bindings. A register associated with an unbound variable is represented by the constant `NULL`. Normally, variable offsets range from 256 to 319. To rename the variables of a term for unification, then, the term is simply scanned and 64 is added to the offset of each variable. (Since unification occurs infrequently compared to matching and equality, renaming is seldom necessary.)

The *next* and *prev* fields simply point to the next and previous flatterm records in the term. This allows subterms to be rewritten, copied, and spliced into other terms easily; also, terms can be conveniently traversed forward or backward as necessary. The *end* field of a flatterm points to the subterm following the subterm headed by that flatterm. The *end* fields are useful during matching, unification, and lookup operations, where a subterm must be bound to a variable, and then a pointer must be advanced to the next subterm.

With the linear representation, subroutines for equality, matching, unification, variable renaming, and several others run more than twice as fast as they do with a tree representation. The space penalty is in the neighborhood of 20 percent. Overall, the fast version of completion runs about 25 percent faster using flatterms than it does with the tree representation.

---

[1] The University of Texas at Austin, Dept. of Computer Sciences, Austin, TX 78712

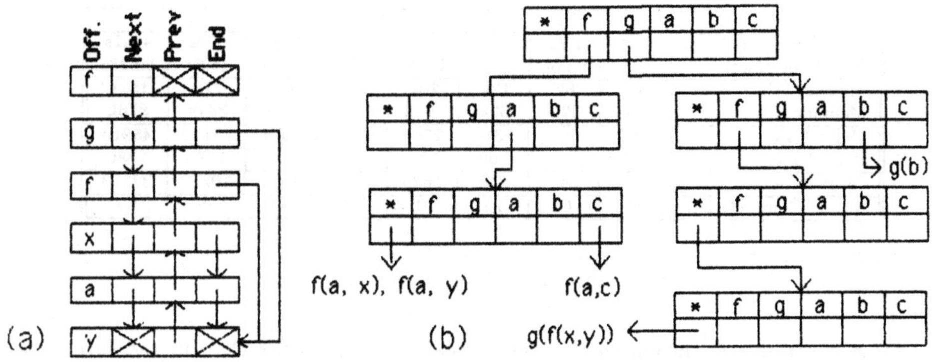

Figure 1: (a) A Linear Term. (b) A Discrimination Net

3 Discrimination Nets

The discrimination net is a variant of the trie data structure used to index dictionaries. I'm not certain of their actual origin, though I know that both Mike Ballantyne and the Argonne Theorem Proving Group [6] have studied them for various purposes. The basic idea is to partition terms based upon their structure. This is most easily explained pictorially. In Figure 1b, we see a discrimination net for the terms $f(a,x)$, $f(a,y)$, $f(a,c)$, $g(f(x,y))$, and $g(b)$. At the end of each path in the tree, there is a linked list of terms sharing the same structure. Notice that no distinction is made among different variables; they are treated as a single wildcard symbol. I decided to treat variables in this manner after some experiments while tuning the slow implementation indicated that almost all matches fail because of structural differences and not because of variable binding conflicts. (Because of this, the match algorithm in the slow version is a two-phase one which checks only function symbol compatibility in the first phase, and then checks variable binding consistency in the second phase.) Also, the discrimination net routines are considerably simplified when variables are considered in this way, and memory consumption is substantially reduced.

In each node of a net, there is a slot for each function symbol (the number of function symbols is known after the initial equations are read in). There is an additional slot for variables. A non-variable slot is indexed by the offset field of a flatterm. There are also fields pointing to the parent node and parent slot, and a counter indicating how many slots are currently in use.

The linked list at the end of a path is composed of structures of type *leaf*. A leaf contains a pointer to a subterm, a pointer to the equation or rule containing that subterm, and pointers to the node and node slot containing the list. To speed deletion from nets, each rule contains an array of pointers to leaves containing its subterms.

Besides insertion and deletion, there are two basic operations on discrimination nets. One is to find the left hand side of a rule which can reduce a given subterm. A separate net, `LHSNet`, indexing only the left hand sides of rules, is used for this purpose. A backtracking algorithm is used to walk the net. Whenever a node is encountered with a non-null variable slot, the current subterm being matched is pushed onto a stack of bindings for later checking, and a choicepoint is added to the backtrack stack. Otherwise, or during backtracking, the slot indexed by the current subterm's offset field is followed. At the end of a path, a quick check is made for variable binding consistency. The entire lookup operation is quite fast. For each of Problems 3 through 6, the average time for a lookup is on the order of 17 microseconds, even though the number of terms indexed in the net varies from as many as 359 in Problem 3 to as many as 1797 in Problem 6. Overall, rule lookup takes about 20 percent of the total runtime.

The other use of the nets is to locate subterms which can be reduced, given a rule. For this purpose, an additional net, `TermNet`, is used which contains all subterms not indexed by `LHSNet`. Another backtracking algorithm is used, similar to the other one, except that choicepoints are

	P1	P2	P3	P4	P5	P6	P7	P8
Run Time (slow)	0.2s	1.7s	7.9s	26.4s	72.9s	172s	670s	4842s
Run Time (fast)	0.08s	0.26s	0.68s	1.43s	2.84s	5.11s	15.0s	80.38s
Memory (slow)	4k	8k	12k	20k	32k	40k	72k	152k
Memory (fast)	12k	32k	60k	100k	152k	216k	384k	876k
Num. Pairs	189	566	1315	2564	4441	7074	15120	45916
Max. Rules	26	80	166	384	434	616	1076	2380
Useless Pairs	115	343	819	1639	2899	4694	10279	32103
Size of Complete Set	13	19	25	31	37	43	55	79

Table 1: Run time statistics

pushed when a variable is encountered in the current subterm rather than in the current node. This operation is slower than looking for left-hand sides of rules, since a variable can match a large number of sub-paths in the net. Moreover, as the size of the net grows, the time required for this type of lookup seems to increase roughly linearly. Term lookup accounts for 10 to 20 percent of the time required for the problems in the appendix.

The indexing nets aren't used during the generation of critical pairs. This is because the order in which critical pairs are generated would be randomized. Completion is very sensitive to the order in which pairs are generated; a slight perturbation might result in good behavior for one problem, and very poor behavior for another.

Discrimination nets tend to be memory hogs. The fast implementation typically consumes five to six times as much memory as does the slow version. However, discrimination nets also tend to be very sparse — most nodes usually have only one or two active slots. Overall memory consumption could be cut by half or more if nodes weren't allocated with empty slots. But the insertion, deletion, and lookup routines would become more complicated, and the runtime would degrade.

4 Where Does all the Time Go?

Table 1 is a summary of statistics for the two versions of completion running on problems P1 through P8. A description of the problems can be found in the appendix. The problems are identical except for the number of identities and inverses in each, and in the particular weights assigned to symbols. There are 2, 4, 6, 8, 10, 12, 16, and 24 identities/inverses in P1, P2, P3, P4, P5, P6, P7, and P8, respectively. These problems were intentionally studied so that the behavior of the systems can be observed as the input changes in a regular fashion. Naturally, other problems will yield different results, but in my experience the results presented here are typical.

Run times are given for a Sun 4/280, and they include the time necessary to parse the input file, and to pretty-print the initial equations and the complete sets. For comparison with a Sun 3/280, 3/160, or 3/50, multiply run times by 3.0, 5.6, or 7.4, respectively. For both versions, the standard Sun C compiler was used with optimization enabled. For each problem, the number of equations introduced during the run includes the initial input equations. The maximum number of rules at any point during the run is listed, along with the number of equations which were reduced to trivial equalities and deleted without ever being used as rules. I would be interested in hearing from anyone with a substantially faster system. (If you think you have a fast system, try running it on P8!)

Figure 2 illustrates the throughput of the system, in terms of the number of equations processed per second. This is simply the number of equations generated divided by the run time. Notice that the throughput of the slow system rapidly approaches zero, while the fast system maintains a much higher throughput. For P8, the fast version of completion runs 60 times faster than the slow one. Even on small problems, the speedup is substantial.

In Figure 3, typical runtime profiles are displayed. These are based upon information provided by the **gprof** profiling utility. While in the slow system the match operation is the bottleneck, in the

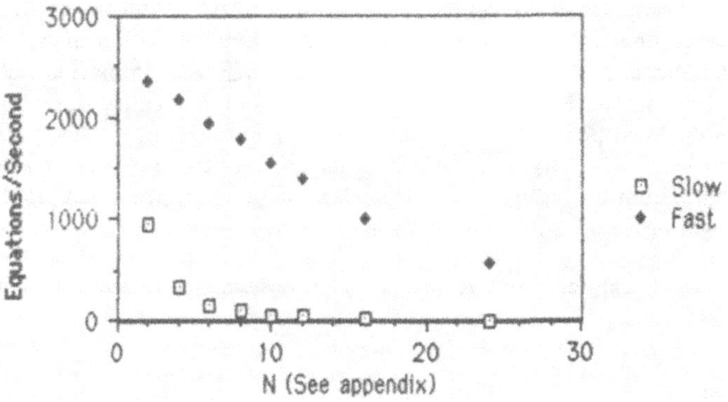

Figure 2: System throughput

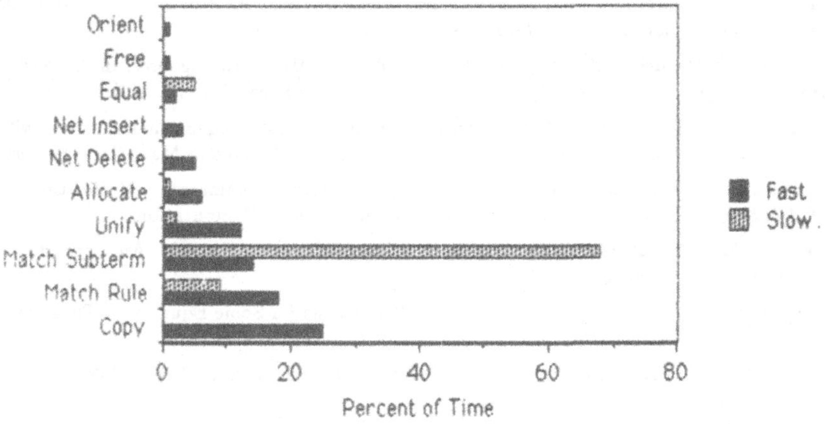

Figure 3: Runtime profiles

fast system, several operations consume significant percentages of time. The single most expensive operation in the fast system is copying of terms during rewriting and pair generation; this indicates that time is being well spent in finding successful matches. In the slow system, roughly 99 percent of match attempts fail.

5 E-Completion

While the methods used here are clearly a win for ordinary Knuth-Bendix completion, one naturally wonders whether this is so for more sophisticated E-Completion procedures [4], [7], [1]. The answer is that it depends upon whether the match operation for the procedure is compatible with the index lookup operation. In the case where the match operation is associative-commutative (AC), it seems that indexing is not very useful — AC matching apparently requires both terms to be known before even a partial match can be made. When there is a mixture of AC and non-AC function symbols, indexing is probably of some value; but in the case that all function symbols are AC, the best hope seems to be to find a fast AC matching algorithm. (Because completion requires only a single matching substitution, the AC matching algorithm by myself and Pat Lincoln [5] seems particularly well suited, since it can generate a single matching substitution efficiently; other AC matching algorithms are designed to generate entire sets of unifiers.)

On the other hand, for the class of equations variously called *simple* or *decomposable*, the match operation is indeed compatible; the key term can be permuted — "mutated", in Kirchner's terminology [2] — and another lookup operation tried. (Actually, there is a more efficient way which interleaves the lookup and mutation operations). Moreover, discrimination nets can be used to index equations used in the E-matching, E-unification, and E-equality operations, making them faster. Examples of decomposable theories include commutativity and other permutative theories, distributivity and commutativity, and certain other collapse-free theories. I am currently implementing an E-completion procedure to use this class of theories, and I anticipate that the results from this paper will prove useful in that system.

Many thanks to Dallas Lankford, Mike Ballantyne, Ross Overbeek, and Hantao Zhang for useful comments and other information.

References

[1] J-P. Jouannaud, H. Kirchner. (1984) Completion of a Set of Rules Modulo a Set of Equations. *Proceedings 11th POPL*, Salt Lake City.

[2] C. Kirchner. (1987) Methods and Tools for Equational Unification. *Proceedings of the Colloquium on the Resolution of Equations in Algebraic Structures*, Lakeway, Texas.

[3] D. Knuth and P. Bendix. (1970) Simple Word Problems in Universal Algebras. in J. Leech, ed., *Computational Problems in Abstract Algebra*, 263-297, Pergamon Press, Oxford, U.K.

[4] D. Lankford, M. Ballantyne. (1977) Decision Procedures for Simple Equational Theories with Commutative Axioms: Complete Sets of Commutative Reductions. U.T. Austin Math Tech Report ATP-35.

[5] P. Lincoln and J. Christian. (1988) Adventures in Associative-Commutative Unification. *Proceedings CADE-9* 358-367. Full version to appear in Journal of Symbolic Computation.

[6] R. Lusk, W. McCune, R. Overbeek. Implementation of Theorem Provers: An Argonne Perspective. Lecture notes from a seminar at CADE-9.

[7] G. Peterson, M. Stickel. (1981) Complete Sets of Reductions for Some Equational Theories. *JACM* **28**, 233-264.

[8] D. H. Warren. (1983) An Abstract Prolog Instruction Set. SRI Technical Note 309.

6 Appendix

Listed here is the schema for problems P1 through P8. The parameter n equals 2, 4, 6, 8, 10, 12, 16, and 24 for P1, P2, P3, P4, P5, P6, P7, and P8, respectively.

$f(f(x,y),z) = f(x,f(y,z))$.
$f(e_j,x) = x$ for $1 \leq j \leq n$.
$f(x,i_j(x)) = e_j)$ for $1 \leq j \leq n$.

$weight(e_j) = 1$ for $1 \leq j < n$.
$weight(e_n) = 2$.
$weight(f) = 0$
$weight(i_j) = 2(n-j)$.

Lexical ordering: $e_1 < \ldots < e_n < f < i_1 < \ldots < i_n$.

Compilation of ground term rewriting systems and applications#
(DEMO)

M. DAUCHET and A. DERUYVER
LIFL (UA 369 CNRS)
Universite des sciences et techniques de LILLE FLANDRES ARTOIS
U.F.R. d' I.E.E.A. Bat. M3 59655 VILLENEUVE D'ASCQ CEDEX FRANCE
dauchet@frcitl71.bitnet

ABSTRACT OF THE DEMO:
We get an algorithm (based on tree automata technics) which "compiles" ground term rewriting systems to solve reachability problem in linear time relatively to the terms. The name of the software is VALERIAAN: V=Verification, A= Algebraic, L=Logic, E=Equation, R=Rewrite, I = Inference, A = (tree) Automata , N= Normal (form). VALERIAN is a comics character who travels across time and space.

The problem:
The reachability problem for term rewriting systems (TRS) is deciding, for given TRS S and two terms M and N, whether M can reduce to N by applying the rules of S. It is well-known that this problem is undecidable for general TRS's. A TRS is *ground* if its set of rewriting rules R={ li->ri | iŒI} (where I is finite) is such that li and ri are ground terms(no variable occurs in these terms). The decidability of the reachability problem for ground TRS was studied by Dauchet and all [4],[5] as a consequence of decidability of confluence for ground TRS. Oyamaguchi [15] and Togushi-Noguchi have shown this result too for ground TRS and in the same way for quasi-ground TRS. We take again this study with two innovator aspects:
- the modulary aspect of the decision algorithm which use all algebraical tools of tree automata, that permits to clearly describe it.
- the exchange between time and space aspect which have permitted to obtain some time complexities more and more reduced.
Therefore we have proceeded in three steps:
1- We begin with the TRS S not modified which gives the answer to the problem with a time complexity non bounded time complexity .
2- We transform the system S in a GTT (ground tree transducer, [4]) which simulates it, we call S' this system. Then the decision algorithm time and space complexity are quadratic.
3- Then we obtain, after a compilation of S' which could be realised in an exponential time (reduction of nondeterminism), a real time decision

---

supported by "Greco Programmation" and "PRC Mathematique et Informatique". Part of ESPRIT Basic Reasearch Action n° 3166 - "Algebraic and Syntactic Methods in Computer Science" and of ESPRIT BRA "Computing by Graph Transformations".

algorithm (linear complexity). But the necessary memory space, after the compilation of S', is in O(exp(number of rules of S)).

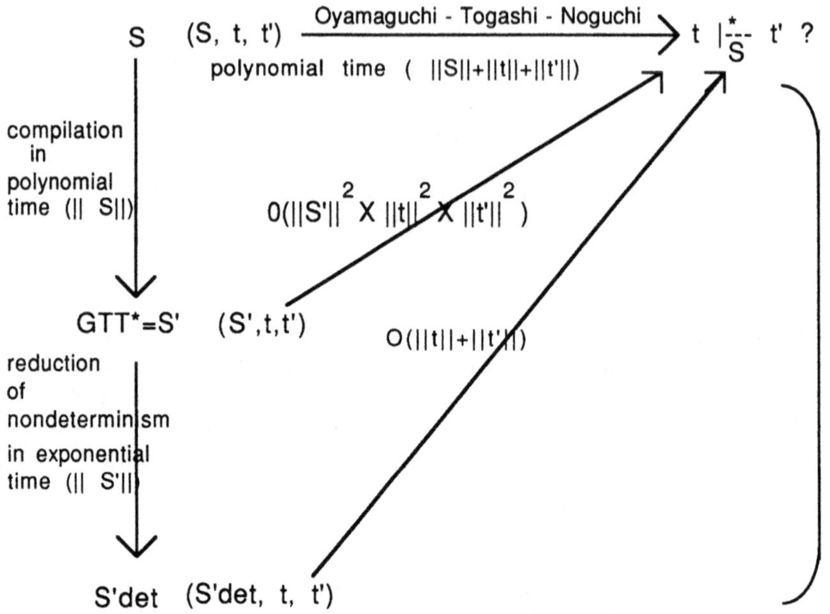

S = a rewriting system
S'= our system S after a first compilation
S'det = our system S' after the reduction of nondeterminism
t,t' = the given trees
||S|| = size of the rewriting system S
||t|| = size of the tree t (number of nodes).

figure 1: A comparison with the result of Oyamaguchi's results

A program, which is called VALERIANN, written in PROLOG realizes at the present time this algorithm. Some results on a SUN 3/50:
First system S1: 9 rules, 50 nodes. With S'1, reachability problem is solved in about 40s for ||t|| = ||t'|| = 5, about 4mn for ||t|| = 10 and ||t'|| = 5, about 12mn for ||t|| = 10 and ||t'|| = 10. With S'1det, each solution seems instantaneous. Second system S2: 12 rules, 70 nodes. With S'2, reachability problem is solved in about 4mn 40s for ||t|| = ||t'|| = 7, about 42 mn for ||t|| = 10 and ||t'|| = 10. With S'2det, each solution seems also instantaneous.

The sofware will be extended to explain the results, i.e. to give sequences of rewriting solving the reachability problem; the methods are also to extend to other TRS, as ground + Commutativity, or ground + Commutativity + Associativity [16].

BIBLIOGRAPHY

[1] BRAINERS : Tree-generating regular systems, Info and control (1969)
[2].G.W.BRAMS : Reseaux de Petri:theorie et pratique,tomes 1&2, Masson,Paris (1983)

[3] CHEW : An improved algorithm for computing with equations,21st FOCS (1980)

[4] DAUCHET, HEUILLARD, LESCANNE, TISON : The confluence of ground term tewriting systems is decidable,2nd LICS (1987)

[5] DAUCHET & TISON : Tree automata and decidability in ground term rewriting systems FCT' 85 (LNCS n° 199)

[6] N.DERSHOWITZ,J.HSIANG,N.JOSEPHSON and D.PLAISTED : Associative-commutative rewriting,Proc.10th IJCAI, LNCS 202 (1983)

[7] GECSEG F. & STEINBY M : tree automata,Akadémiai Kiado, Budapest (1984)

[8] HUET.G & OPPEN.D.: Equations and rewrite rules:a survey,in formal languages :perspective and open problems,Ed.Book R.,Academic Press(1980)

[9] J.P.JOUANNAUD : Church-Rosser computations with equational term rewriting systems,Proc.4th Conf on Automata,Algebra and programming,LNCS 159 (1983)

[10] C.KIRCHNER : Methodes et outils de conception systematique d'algorithmes d'unification dans les théories équationnelles,These d'etat de l'universite de Nancy I (1985)

[11] S.R.KOSARAJU : Decidability of reachability in vector addition systems, Proc.14th Ann.Symp.on Theory of Computing,267-281.(1982)

[12] KOZEN : Complexity of finitely presented algebra, 9th ACM th. comp. (1977)

[13] E.W.MAYR : An algorithm for the general Petri net reachability problem, Siam J Comput.13 441.- 460

[14] NELSON & OPPEN : Fast decision algorithms based on congruence closure, JACM 27 (1980)

[15] OYAMAGUCHI M.: The reachability problem for quasi-ground Term Rewriting Systems, Journal of Information Processing , vol 9 , n°4 (1986)

[16] DERUYVER, A. and R. GILLERON, The reachability problem for ground TRS and some extensions, to appear in CAAP' 89 proceedings, Diaz editor, LNCS, Barcelona, march 1989.

An Overview of Rewrite Rule Laboratory (RRL)*

Deepak Kapur
Department of Computer Science
State University of New York at Albany
Albany, NY 12222
kapur@albanycs.albany.edu

Hantao Zhang
Department of Computer Science
The University of Iowa
Iowa City, IA 52242
hzhang@herky.cs.uiowa.edu

1 What is *RRL*?

Rewrite Rule Laboratory (*RRL*) is a theorem proving environment for experimenting with existing automated reasoning algorithms for equational logic based on rewriting techniques as well as for developing new reasoning algorithms [12, 13, 14]. Reasoning algorithms are used in many application areas in computer science and artificial intelligence, in particular, analysis of specifications of abstract data types, verification of hardware and software, synthesis of programs from specifications, automated theorem proving, expert systems, deductive data basis, planning and recognition, logic programming, constraint satisfaction, natural language processing, knowledge-based systems, robotics, vision.

The project for developing rewrite rule laboratories was jointly conceived in 1983 by the first author and David Musser at Rensselaer Polytechnic Institute (RPI) and John Guttag of MIT when rewrite methods were gaining attention because of their use in the analysis of algebraic specifications of user-defined data types. The work on the *RRL* project was started at General Electric Corporate Research and Development (GECRD) and RPI in Fall 1983, following a workshop on rewrite rule laboratories in Schenectady, New York [2]. Since then, *RRL* has continued to evolve. *RRL* is written in a subset of Zeta Lisp that is (almost) compatible with Franz Lisp and is more than 15,000 lines of code. In order to make *RRL* portable and available for wider distribution, conscious attempt was made not to use special features of Zeta Lisp and the Lisp machine environment.

RRL was initially used to study existing algorithms, including the Knuth-Bendix completion procedure [16], and Dershowitz's algorithm for recursive path ordering [1] to orient equations into terminating rewrite rules [12]. However, this environment soon began serving as a vehicle for generating new ideas, concepts and algorithms which could be implemented and tested for their effectiveness. Experimentation using *RRL* has led to development of new heuristics for decision procedures and semi-decision procedures in theorem proving [7, 24], complexity studies of primitive operations used in *RRL* [9], and new approaches for first-order theorem proving [8, 23], proofs by induction [11, 25].

RRL has been distributed to over 20 universities and research laboratories. Any one interested in obtaining a copy of *RRL* can write to the authors.

*Partially supported by the National Science Foundation Grant no. CCR-8408461.

2 What can *RRL* do?

RRL currently provides facilities for

- automatically proving theorems in first-order predicate calculus with equality,
- generating decision procedures for first-order (equational) theories using completion procedures,
- proving theorems by induction using different approaches - methods based on the *proof by consistency* approach [18, 6] (also called the *inductionless-induction* approach) as well as the explicit induction approach, and
- checking the consistency and completeness of equational specifications.

The input to *RRL* is a first-order theory specified by a finite set of axioms. Every axiom is first transformed into an equation (or a conditional equation). The kernel of *RRL* is the *extended Knuth-Bendix completion procedure* [16, 17, 21] which first produces terminating rewrite rules from equations and then attempts to generate a canonical rewriting system for the theory specified by a finite set of formulas. If *RRL* is successful in generating a canonical set of rules, these rules associate a unique normal form for each congruence class in the congruence relation induced by the theory. These rules thus serve as a decision procedure for the theory.

For proving a single first-order formula, *RRL* also supports the proof-by-contradiction (refutational) method. The set of hypotheses and the negation of the conclusion are given as the input, and *RRL* attempts to generate a contradiction, which is a system including the rule $true \rightarrow false$.

New algorithms and methods for automated reasoning have been developed and implemented in *RRL*. *RRL* has two different methods implementing the proof by consistency approach for proving properties by induction: the *test set* method developed in [11] as well as the method using *quasi-reducibility* developed in [5]. These methods can be used for checking structural properties of equational specifications such as the *sufficient-completeness* [10] and consistency properties [18, 6].

Recently, a method based on the concept of a *cover set* of a function definition has been developed for mechanizing proofs by explicit induction for equational specifications [25]. The method is closely related to Boyer and Moore's approach. It has been successfully used to prove over a hundred theorems about lists, sequences, and number theory including the unique prime factorization theorem for numbers; see [25].

RRL supports an implementation of a first-order theorem proving method based on polynomial representation of formulas using the exclusive-or connective as '+' and the and connective as '.' [3], and using Gröbner basis like completion procedure [8]. Its performance compares well with theorem provers based on resolution developed at Argonne National Lab as well as theorem provers based on the connection graph approach developed at SRI International and University of Kaiserslautern. Most of the hardware verification work mentioned later in the section on applications of *RRL* was done using this method of theorem proving.

Recently, another method for automatically proving theorems in first-order predicate calculus with equality using conditional rewriting has also been implemented and successfully tried on a number of problems [23]. This method combines ordered resolution and oriented paramodulation of Hsiang and Rusinowitch [4] into a single inference rule, called *clausal superposition*, on conditional rewrite rules. These rewrite rules can be used to simplify clauses using conditional rewriting.

RRL supports different strategies for normalization and generation of critical pairs and other features. These strategies play a role similar to the various strategies and heuristics available in resolution-based theorem provers. *RRL* provides many options to the user to adapt it to a particular application domain. Because of numerous options and strategies supported by *RRL*, it can be used as a laboratory for testing algorithms and strategies.

3 Some Applications of *RRL*

RRL has been successfully used to attack difficult problems considered a challenge for theorem provers. An automatic proof of a theorem that an associative ring in which every element x satisfies $x^3 = x$, is commutative, was obtained using *RRL* in nearly 2 minutes on a Symbolics machine. Previous attempts to solve this problem using the completion procedure based approach required over 10 hours of computing time [22]. With some special heuristics developed for the family of problems that for $n > 1$, for every x, $x^n = x$ implies that an associative ring is commutative, *RRL* takes 5 seconds for the case when $n = 3$, 70 seconds for $n = 4$, and 270 seconds for $n = 6$ [7, 24]. To our knowledge, this is the first computer proof of this theorem for $n = 6$. Using algebraic techniques, we have been able to prove this theorem for many even values of n. *RRL* has also been used to prove the equivalence of different non-classical axiomatizations of free groups developed by Higman and Neumann to the classical three-law characterization [15].

RRL can be also used as a tool for demonstrating approaches towards specification and verification of hardware and software. *RRL* is a full-fledged theorem prover supporting different methods for first-order and inductive theorem proving in a single reasoning system. It has been successfully used at GECRD for hardware verification. Small combinational and sequential circuits, leaf cells of a bit-serial compiler have been verified [19]. The most noteworthy achievement has been the detection of 2 bugs in a VHDL description of a 10,000 transistor chip implementing Sobel's edge detection algorithm for image processing and after fixing those bugs, verifying the chip description to be consistent with its behavioral description [20].

RRL has also been used in teaching a course on theorem proving methods based on rewrite techniques, providing students with an opportunity to try methods and approaches discussed in the course. It has served as a basis for class projects and thesis research.

Additional details about the capabilities of *RRL* and the associated theory as well as possible applications of *RRL* can be found in the papers listed in the selected bibliography at the end of this write-up.

4 Future Plans

We plan to extend *RRL* to make it more useful and convenient for reasoning about software and hardware. The test set and cover set methods for supporting automatic proofs by induction in *RRL* will be further developed and integrated, with a particular emphasis on the cover set method. Approaches for reasoning with incomplete specifications will also be investigated.

In addition, theoretical and experimental research in developing heuristics, including identifying unnecessary computations for completion procedures as well as for first-order theorem proving will be undertaken to improve their performance. Complexity studies and efficient implementations of primitive operations will also be continued and results will be incorporated into *RRL*.

References

[1] Dershowitz, N. (1987). Termination of rewriting. *J. of Symbolic Computation.*

[2] Guttag, J.V., Kapur, D., Musser, D.R. (eds.) (1984). Proceedings of an *NSF Workshop on the Rewrite Rule Laboratory* Sept. 6-9 Sept. 1983, General Electric Research and Development Center Report 84GEN008.

[3] Hsiang, J. (1985). Refutational theorem proving using term-rewriting systems. *Artificial Intelligence* Journal, 25, 255-300.

[4] Hsiang, J., and Rusinowitch, M. (1986). "A new method for establishing refutational completeness in theorem proving," *Proc. 8th Conf. on Automated Deduction*, LNCS No. 230, Springer Verlag, 141-152.

[5] Jouannaud, J., and Kounalis, E. (1986). Automatic proofs by induction in equational theories without constructors. Proc. of *Symposium on Logic in Computer Science*, 358-366.

[6] Kapur, D., and Musser, D.R. (1984). Proof by consistency. In Reference [2], 245-267. Also in the *Artificial Intelligence* Journal, Vol. 31, Feb. 1987, 125-157.

[7] Kapur, D., Musser, D.R., and Narendran, P. (1984). Only prime superpositions need be considered for the Knuth-Bendix completion procedure. GE Corporate Research and Development Report, Schenectady. Also in *Journal of Symbolic Computation* Vol. 4, August 1988.

[8] Kapur, D., and Narendran, P. (1985). An equational approach to theorem proving in first-order predicate calculus. Proc. of *8th IJCAI*, Los Angeles, Calif.

[9] Kapur, D., and Narendran, P. (1987). Matching, Unification and Complexity. *SIGSAM Bulletin.*

[10] Kapur, D., Narendran, P., and Zhang, H (1985). On sufficient completeness and related properties of term rewriting systems. GE Corporate Research and Development Report, Schenectady, NY. *Acta Informatica,* Vol. 24, Fasc. 4, August 1987, 395-416.

[11] Kapur, D., Narendran, P., and Zhang, H. (1986). Proof by induction using test sets. *Eighth International Conference on Automated Deduction* (CADE-8), Oxford, England, July 1986, Lecture Notes in Computer Science, 230, Springer Verlag, New York, 99-117.

[12] Kapur, D. and Sivakumar, G. (1984) Architecture of and experiments with RRL, a Rewrite Rule Laboratory. In: Reference [2], 33-56.

[13] Kapur, D., and Zhang, H. (1987). *RRL: A Rewrite Rule Laboratory - User's Manual.* GE Corporate Research and Development Report, Schenectady, NY, April 1987.

[14] Kapur, D., and Zhang, H. (1988). *RRL: A Rewrite Rule Laboratory.* Proc. of *Ninth International Conference on Automated Deduction* (CADE-9), Argonne, IL, May 1988.

[15] Kapur, D., and Zhang, H. (1988). Proving equivalence of different axiomatizations of free groups. *J. of Automated Reasoning* 4, 3, 331-352.

[16] Knuth, D.E. and Bendix, P.B. (1970). Simple word problems in universal algebras. In: *Computational Problems in Abstract Algebras.* (ed. J. Leech), Pergamon Press, 263-297.

[17] Lankford, D.S., and Ballantyne, A.M. (1977). Decision procedures for simple equational theories with commutative-associative axioms: complete sets of commutative-associative reductions. Dept. of Math. and Computer Science, University of Texas, Austin, Texas, Report ATP-39.

[18] Musser, D.R. (1980). On proving inductive properties of abstract data types. Proc. *7th Principles of Programming Languages (POPL).*

[19] Narendran, P., and Stillman, J. (1988). Hardware verification in the Interactive VHDL Workstation. In: *VLSI Specification, Verification and Synthesis* (eds. G. Birtwistle and P.A. Subrahmanyam), Kluwer Academic Publishers, 217-235.

[20] Narendran, P., and Stillman, J. (1988). Formal verification of the Sobel image processing chip. GE Corporate Research and Development Report, Schenectady, NY, November 1987. Proc. *Design Automation Conference.*

[21] Peterson, G.L., and Stickel, M.E. (1981). Complete sets of reductions for some equational theories. *J. ACM*, 28/2, 233-264.

[22] Stickel, M.E. (1984). "A case study of theorem proving by the Knuth-Bendix method: discovering that $x^3 = x$ implies ring commutativity", Proc. of 7th Conf. on Automated Deduction, Springer-Verlag LNCS 170, pp. 248-258.

[23] Zhang, H., and Kapur, D. (1987). First-order theorem proving using conditional rewriting. Proc. of *Ninth International Conference on Automated Deduction* (CADE-9), Argonne, IL, May 1988.

[24] Zhang, H., and Kapur, D. (1989). Consider only general superpositions in completion procedures. *This Proceedings.*

[25] Zhang, H., Kapur, D., and Krishnamoorthy, M.S. (1988). A mechanizable induction principle for equational specifications. Proc. of *Ninth International Conference on Automated Deduction* (CADE-9), Argonne, IL, May 1988.

InvX: An Automatic Function Inverter

H. Khoshnevisan and K. M. Sephton
Department of Computing, Imperial College, London, SW7 2BZ, England
Tel: 01 589 5111, Telex: 261503, Email: {hk,kms}@doc.ic.ac.uk

1. Overview

InvX is a mechanised system for synthesising recursive inverse function definitions from first-order recursive functions defined in terms of the construction, composition and conditional combining forms. Since inverse functions are not in general single-valued, we have used an already established powerdomain [HK88], in terms of which we have expressed their transformation into recursive form. Inverses that require unification at run-time are synthesised for a large class of functional expressions, and are then optimised by term rewriting, using a set of axioms which include a form of compile-time unification. The optimisations reduce the dependency on run-time unification, in many instances removing it entirely to give a recursive form.

Our work fits in with work done on the integration of logic programming features into functional languages in that it can be used as the basis of a transformation tool which minimises the use of resolution. Moreover, the synthesised recursive inverses can often be further optimised by transforming them into tail recursive form or loops, as described in [HK86]. As well as providing double-mode use of functions, inverse functions have an important role in the transformation of abstract data types [KS89].

In this paper we briefly describe a version of InvX and its incorporation into the Flagship programming environment [Da88].

1.1. The Flagship Environment

The Flagship programming environment supports the development, implementation and modification of programs written in the functional language Hope$^+$ [Pe87], which is a lambda-calculus based functional language allowing user-defined data types and pattern matching.

The programming environment currently consists of an interpreter for the language Hope$^+$ with extensions to support absolute set abstraction; a compiler; a script driven partial evaluator and fold/unfold transformation system; a method of enforcing temporal synchronisation on the execution of programs; and a number of algebraic transformation techniques including memoisation and the InvX function inversion system. Transformations are performed by using a set of meta-functions which can be applied to modules within the programming environment.

The transformation environment is written in Hope+ and is compiled to run on a SUN3 machine using a compiler also developed at Imperial College [PS87]. It utilises an interface to the X10 windowing system in order to provide a convenient user interface.

2. The intermediate language

We use FP [Ba78] as our intermediate language because of its amenability to formal reasoning, [Ba81], but make a number of extensions to it in order to cater for inverses. In general the inverse, f^{-1}, of a function f will return several values when applied to an argument, x, in the range of f; when f is applied to *any* of these values, the result will be x. We therefore consider functions defined on *powerdomains*, as opposed to domains of objects [KS89]. By treating all functions as mapping a set of values onto a new set of values, the distinction between "ordinary" and inverse functions is removed. However, these extensions have considerable repercussions on the *program forming operations* (PFOs) composition, construction and conditional of FP. We have lifted these PFOs to the powerdomain level, i.e. suitably generalised to operate on functions mapping sets to sets, and these become primitive functionals of our formalism. We also define a new PFO, «...», called "unify at function level" used for the inversion of *constructions* of functions and another new PFO called "set union", which we denote by {}. {} takes two or more functions as arguments and returns a new function which has the effect of applying all the argument functions to a set of objects and returning the union of the resultant objects as its result. Hope+ data types are represented by constructor, destructor and predicate functions which are provided for each data type. Details of these extensions and the repercussions on Backus' axioms can be found in [HK88,KS89].

3. The mechanised system

The various processes of InvX when called by the Flagship Environment are illustrated below:

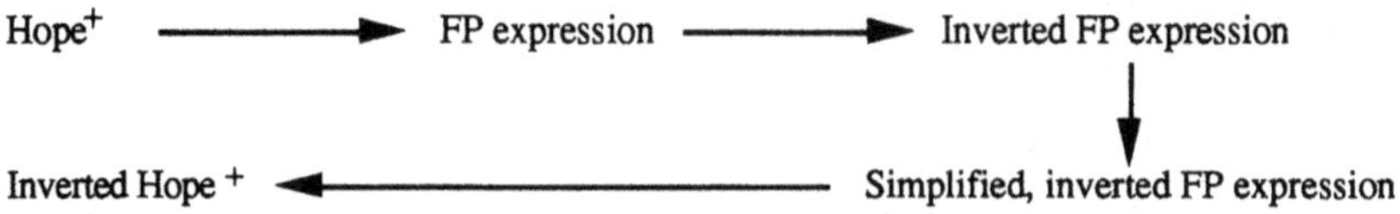

3.1. The translation of Hope+ functions into a variable-free form

This translation is carried out simply by replacing pattern matching by a conditional tree and then performing object abstraction over the entire definition of every function. At present all first-order Hope+ functions except for recursive lets, relative set abstraction and absolute set abstraction can be converted into FP.

3.2. Inversion

Our results specify how inverses can be constructed from various expressions built from primitives and the application of the three elementary PFOs. For example, $(f \circ g)^{-1}$

$= g^{-1} \circ f^{-1}$ and $cons^{-1} = [\ hd,\ tl\ ]$. In addition to the inverses of FP's primitive functions and PFOs, inverses are predefined for a number of standard Hope+ functions and rules are provided to invert the constructors, selectors and predicate functions which form data types of the Hope+ language. The user can specify the inverses for any functions which are not built in to the inversion system.

The class of functions for which we can automatically generate inverses has functions f defined recursively by the equation f = E which has the following syntax

$$E ::= f \mid g \mid E \circ E \mid [E,E,\ldots,E] \mid E \rightarrow E;\ E$$

where the syntactic type g denotes fixed functions and f is the function identifier. In other words, the simple use of our inversion rules enables our system to generate an inverse for any first-order function. This is because any such function can always be expressed in terms of the three elementary PFOs.

3.3. The simplification of inverses

In general the synthesised inverse function will involve logical variables and therefore require unification at run-time for its execution. However, the inverse function can often be simplified using our axiom-based rewrite system. Since we have introduced set-valued functions and correspondingly generalised some of the definitions of the PFOs, some of the FP axioms of [Ba78] no longer hold as they stand. We also introduce additional axioms which are established by deriving corresponding semantic equalities on the powerdomain. Our collection of axioms is used to simplify functional expressions *syntactically*, and in particular reduce the dependence on run-time unification by removing occurrences of the *unification* PFO through transformation-time, 'function-level' first-order unification. A full list of the axioms used by InvX is given in [KS89].

The rewrite strategy we adopted was to perform an innermost first search over the structure of the FP program. Each time an axiom is applied the lower part of the tree would be rescanned to check if any more axioms are applicable. This rewrite strategy was chosen mainly for reasons of efficiency and it is analogous to a call by value mechanism in a functional language. One advantage of this approach is that the axioms can assume that the inner parts of the equations have been simplified. The rewriting terminates when no more axioms can be applied. Different rewrite sequences may produce different normal forms for the same expressions, however all the normal forms will be equivalent.

The axioms are currently specified in Hope+ and compiled into the inversion system. They are, however, totally separate from the rewrite engine itself. This has the advantage of being a very powerful specification mechanism but means that the set of available axioms is static; therefore application-specific axioms can not be specified by the user. Care has been taken to ensure that the axioms will terminate, though no proof has been developed.

3.3.1. Function Level Unification

The function level unification is carried out by the application of an axiom. The routine for performing function level unification works purely on the structure of the expressions it is asked to unify. Two expressions are unified if one of the following rules can be applied:

1. The outermost PFOs are identical and the sub-expressions of the PFOs can be unified.
2. One of the expressions to be unified is a logical variable, in which case the unified expression is the other expression.
3. One of the expressions to be unified is $\perp$ in which case the unified expression is also $\perp$.

For simplicity, all logical variables are assumed to be unique, and the axioms are not allowed to duplicate logical variables.

3.4. The back-translation of inverse functions into Hope+

This is carried out as a two stage process: (1) Pattern generation, where a set of patterns is produced corresponding to the conditional tests in the FP expression. (2) Symbolic evaluation, where the FP expression is symbolically evaluated given each of the patterns in turn. This builds up a Hope+ expression.

The translation back to Hope+ is complicated by the fact that unlike the FP derivative, Hope+ functions do not work on powerdomains of objects, therefore the system has to keep track of whether it is dealing with a set of objects or plain objects at each stage. A side effect of this is that it is often possible to detect when an inverse function is single-valued.

Where the term-rewriting system has failed to remove all of the logical variables from an inverted function, a Hope+ function can still be generated by using absolute set abstraction [DFP86]. This will occur when inverting functions such as length whose inverse returns non-ground objects.

4. An example

In this section we illustrate the operations of InvX by working through an example in which the inverse of the function `append` is synthesised automatically.

Original Hope+ expression:

```
dec append : list alpha # list alpha -> list alpha;
--- append( x :: l1, l2 )
      <= x :: append( l1, l2 );
--- append( nil, l2 )
      <= l2;
```

Original FP expression:

$$\texttt{isnil} \circ 1_2 \rightarrow 2_2 \; ; \; \texttt{cons} \circ [\, \texttt{hd} \circ 1_2,\ \texttt{append} \circ [\, \texttt{tl} \circ 1_2,\ 2_2 \,]\,]$$

Inverted FP expression:

```
{ ( isnil ∘ 1₂ → id ; ⊥ ) ∘ [ ?, id ],
  ( isnil ∘ 1₂ → ⊥ ; id )
     ∘ « [ id, ? ] ∘ cons ∘ [ id, ? ] ∘ hd,
```

$$\langle\langle\ [\ id,\ ?\ ]\circ cons \circ [\ ?,\ id\ ]\circ 1_2,\ [?,\ id\ ]\circ 2_2\ \rangle\rangle \circ append^{-1}\circ tl\ \rangle\rangle\ \}$$

Note that each ? denotes a unique logical variable. The simplification steps for this example can be found in [KS89].

Simplified, inverted FP expression:

$$\{\ [\ nil,\ id\ ],\ [\ cons\circ [\ 1_2,\ 1_2\circ 2_2\ ],\ 2_2\circ 2_2\ ]\circ [\ hd,\ append^{-1}\circ tl\ ]\ \}$$

Resulting Hope$^+$ expression (naming append^{-1} as split):

```
dec split : list alpha -> set( list alpha # list alpha );
--- split( nil )
      <= { ( nil, nil ) };
--- split( x1 :: x2 )
      <= set_add( ( nil, x1 :: x2 ),
                  { ( x1 :: y1, y2 ) | ( y1, y2 ) in split x2 });
```

Note that the second equation for split makes use of relative set abstraction, which is a totally functional feature of Hope$^+$ and could be replaced by a call to the function map_set.

5. Conclusions

We have implemented InvX, a system that will mechanically generate inverses for a substantial class of functions. By applying an extended set of function-level axioms at compile-time, expressions for the inverses are transformed so that no unification is required at run-time in many cases. This makes the cost of their execution comparable with that of reduction-based semantics. We have also described the incorporation of InvX in the FLAGSHIP programming environment. InvX has been used successfully on a wide range of examples.

References

[Ba78] J W Backus; *Can Programming be Liberated from the von Neumann Style? A Functional Style and its Algebra of Programs*, CACM 21,8, pp. 613-641, (1978)

[Ba81] J W Backus; *The Algebra of functional programs: Function level reasoning, linear equations and extended definitions*, In Lecture Notes in Computer Science, Vol 107: Formalization of Programming Concepts, Springer-Verlag, pp. 1-43, (1981)

[Da88] J Darlington et al; *An Introduction to the Flagship Programming Environment*, In Proc. CONPAR 88, International Conference Drawing the Threads of Parallelism in Research and Practice, Manchester, pp (A)75-(A)88, (September 1988)

[DFP86] J Darlington, A J Field, H Pull; *The Unification of Functional and Logic Languages*, D DeGroot, G Lindstrom (editors), "Functional and Logic Programming", Prentice-Hall, (1986)

[HK86] P G Harrison, H Khoshnevisan; *Efficient Compilation of Linear Recursive Functions into Object-level Loops*, SIGPLAN 86 Symposium on Compiler Construction, Palo Alto, (June 1986)

[HK88] P G Harrison, H Khoshnevisan; *On the Synthesis of Function Inverses*, Research Report, Department of Computing, Imperial College, London, (1988)

[KS89] H Khoshnevisan, K M Sephton; *An Automated System for Inverting Functions*, Research Report, Department of Computing, Imperial College, London, (1989)

[Pe87] N Perry; *Hope$^+$*, Functional Programming Research, Internal Document (IC/FPR/LANG/2.5.1/7), Imperial College, London, (1987)

[PS87] N Perry, K M Sephton; *Hope+ Compiler*, Functional Programming Research, Internal Document (IC/FPR/LANG/2.5.1/14), Imperial College, London, (1987)

A Parallel Implementation of Rewriting and Narrowing

Naomi Lindenstrauss
The Hebrew University
Jerusalem, Israel

Abstract

A parallel implementation of rewriting and narrowing is described. This implementation is written in Flat Concurrent Prolog, but the ideas involved are applicable to any system where processes are capable of creating other processes and communicating with each other. Using FCP enables one to write very short programs, virtually no longer than the verbal description of the algorithms. Running programs under the FCP interpreter and using facilities provided by it, one can compare the efficiencies of various strategies. Theoretical results about the efficiency of strategies in certain cases are also mentioned.

1 Introduction

Our aim was to get a working parallel implementation of rewriting and narrowing, so that one could get empirical information on the efficiency of various strategies. In some cases we could also get theoretical estimates. The main idea of the implementation is to rewrite different subterms in parallel in a coordinated way, while using or-parallelism to simultaneously explore alternative narrowing substitutions. The implementation was written in Flat Concurrent Prolog (cf. [Shap 86, Logix 87]).

We assume a denumerable set of variables is given and also a set of constants, each with its arity, and consider terms in them. We further assume that a (ground) canonical term rewriting system $\mathcal{R}$ is given by a finite number of rules $Left \longrightarrow Right$. By saying that the system is (ground) canonical we mean that for ground terms the system is terminating and satisfies the Church-Rosser property. In other words—as long as we limit ourselves to rewriting steps that do not involve proper narrowing (i.e. that do not instantiate term variables), the order in which steps are applied to a term, and even which steps are chosen, is immaterial, and we can be sure that the number of steps will be finite. A term which cannot be rewritten (without proper narrowing) is said to be in *normal form.* In a (ground) canonical rewrite system each ground term has a unique normal form.

Given two terms T_1 and T_2 we want to 'solve' the equation

$$T_1 = T_2 \text{ modulo } \mathcal{R}$$

i.e. find a substitution σ for the variables in T_1 and T_2 such that $T_1\sigma$ and $T_2\sigma$ both have the same normal form. Without loss of generality we may assume that T_2 is the constant *true.* There are several approaches to the problem (cf. [Hul 80, Rety 87, DerPl 88, JoDer 89]). In our approach we rewrite the given term as much as possible. Then we find the possible proper narrowing substitutions and choose some of them (using various meta-considerations) for further investigation in parallel, while others are kept for future reference (for a similar approach, which replaces backtracking by or-parallelism, cf. [Shap 87]).

2 Rewriting

2.1 The basic algorithm

The main idea of our rewriting algorithm, suggested by Ehud Shapiro, is to consider terms as processes that may spawn other processes associated with their arguments. Thus the division of the rewriting task into subtasks is automatically done in correspondence with the intrinsic structure of the problem.

Given a term T which we want to rewrite, we create a process associated with it. This process may create, according to the rewriting strategy, processes associated with the arguments of T (e.g. if $T = f(T_1, ..., T_n)$ these will be processes associated with T_1 , ... , T_n), and so on. Thus a dynamic system of processes will develop, with processes being created and terminating, until finally only the root process will remain, now labelled by a term which is the normal form of T.

The processes we use can do several things, depending on the state they are in. They can be in one of four states, which we call 'self', 'sons', 'wait', and 'stop'. The initial state is determined when the process is created and will differ with different rewriting strategies, such as innermost-first, outermost-first etc. How the state will change also depends on the rewriting strategy.

A process in the state 'self' tries to rewrite its term according to some rewrite rule. Coordination between processes is achieved by having 'annotated' rules. Instead of the usual rewrite rules of the form $rule(Left, Right)$, meaning that $Left$ can be rewritten to $Right$, we use rules of the form

$$rule(Left, Right, State)$$

where 'State' specifies into what state the process will go after applying the rule. More will be said on this subject later.

In the state 'sons', the process creates sons corresponding to its arguments and then goes into state 'wait' to wait for their answers. When the answers have arrived it applies a test, which is supplied with the rules, to decide into which state to go.

There is also a state 'stop', in which a process sends the value of its term to its father and terminates. This usually happens when it is clear that the term associated with the process has reached its normal form.

The following diagram describes the transitions:

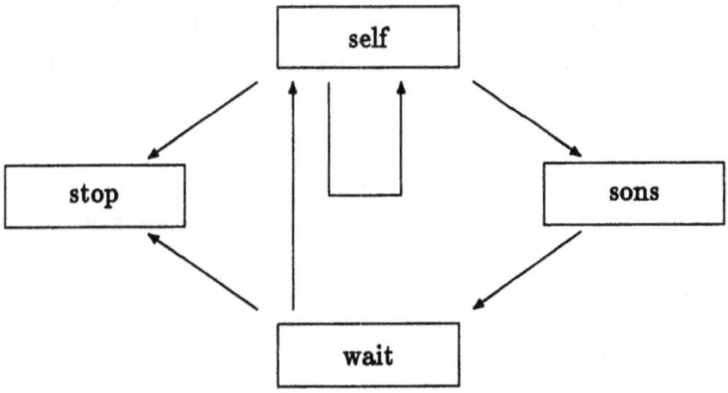

Included with the rules there also is a predicate 'initialstate' which says what the initial state of a process is (this will be different for different rewriting methods and may also depend on the term). In the case where we want to rewrite terms outermost-first, the initial state for all terms is 'self' and the test is: if the answer from the sons has changed the term go to 'self', otherwise go to 'stop'. In the case where we want to rewrite terms innermost-first, the initial state is 'sons' and the decision is "always go after 'wait' to 'self' ".

Innermost-first and outermost-first are valid strategies in the sense that when we follow them we end up with the normal form of the term we started with. When using other strategies we have to make sure they are valid.

2.2 More efficient variations of the algorithm

There are many ways in which the rewrite program can be made more efficient —

- There is no need to create a process for a term whose main functor is a constructor, that is, a functor for which there are no rewrite rules.
- It is possible to mark terms which have already reached their normal form and thus avoid reprocessing them.
- Sometimes we do not need answers from all sons. For example if we want to rewrite $if\_then\_else(C, A, B)$ we may want to work only on the son associated with C till we get the answer from it (cf. the concurrent E-strategies of [GoKiMe 87]).
- There is also a possibility that a process will launch all its sons but will not wait for answers from all of them, but proceed when it has enough information. For example if one rewrites $if\_then\_else(C, A, B)$ and C has been rewritten to *true* one can stop working on B ; or if one computes $mult(A, B)$ and has discovered that A rewrites to 0 there is no more need to rewrite B ; or if one rewrites $and(A, B)$ and A has been rewritten to *false* there is no more need to work on B.

All these possibilities can be investigated by making only minor changes in the original formulation of the program.

We can also treat conditional rewrite systems, where the rules are of the form

$$Left \longrightarrow Right \text{ if } Condition$$

meaning that if *Condition* is satisfied (that is, can be rewritten to *true*) then *Left* is equivalent to *Right* and can be replaced by it. This is achieved by enforcing a certain way of rewriting.

For the case of 'flat' conditional rewrite systems, that is systems where the conditions satisfy the requirements for a guard in FCP, we can use our basic algorithm as is, by just writing rules in the form

$$rule(Left, Right, State) :- Flatcondition \mid true.$$

It is also possible to use the formalism of annotated rewrite rules for enforcing certain ways of rewriting in systems which would not be terminating if all ways of rewriting would be allowed.

2.3 Examples

2.3.1 Towers of Hanoi

The Towers of Hanoi problem can be formulated in terms of a rewrite system:

$$hanoi(0, A, B, C) \longrightarrow move(A, B)$$

$$hanoi(s(N), A, B, C) \longrightarrow t(hanoi(N, A, C, B), move(A, B), hanoi(N, C, B, A))$$

Given a term $hanoi(N, A, B, C)$ where N is instantiated to an integer written in successor notation (e.g. $N = s(s(s(0)))$), and rewriting it to its normal form, we'll get a 'list' of the moves to perform in order to move $s(N)$ disks from pole A to pole B using auxiliary pole C, subject to the well known requirements. Note that we use here a notation for lists which is different from the usual notation. Here we represent the 'list' with the help of a constructor t with arity 3. The representation of the 'list' as a tree in this case is more balanced than in the case of the usual list representation, which is completely lopsided (cf. [GoKiMe 87, p. 3] for the importance of replacing list structures by more balanced structures in a concurrent environment). In this case the parallel time is $\Theta(N)$, while the sequential time is $\Theta(2^N)$.

2.3.2 Evaluating algebraic expressions

Evaluating algebraic expressions can be seen as rewriting N *op* M to its value (where N and M are integers and *op* is one of the algebraic operators). The best strategy in this case is innermost-first, and the time for parallel execution is big-O of the depth of the expression if it is considered as a tree. If we consider all binary trees with n inner nodes all of which are of degree 2 to have equal probability, then the average height $\bar{H}_n$ of these trees satisfies (cf. [FlaOd 82] for the proof)

$$\bar{H}_n \sim 2\sqrt{\pi n}$$

So we may say that the average time for the parallel evaluation of algebraic expressions is $O(\sqrt{n})$.

2.3.3 Evaluation of Boolean expressions

Consider all Boolean expressions made up of *or*, *and*, *true*, and *false* (*not* may be included or excluded). The length of such an expression is the number of the above symbols appearing in it. Consider all such expressions of length n as having the same probability, and use the 'clever' parallel method for evaluation, that is, if either X or Y rewrites to *false* then we stop rewriting $and(X, Y)$, and analogously for *or*. It turns out that in that case the average cost for evaluating an expression of length n tends to a constant as $n \longrightarrow \infty$. A similar result holds for 'clever' sequential rewriting, only with a larger constant. These results will be reported elsewhere.

3 Narrowing

The crux of our narrowing algorithm is the alternate freezing and melting of terms, where by freezing of a term we mean replacing its variables by suitable constants so that it will be possible to reconstruct the original term up to renaming of variables.

Given a term T, which was obtained by alternate rewritings and narrowings from the original goal, we first 'freeze' it, and get the frozen term FT. When the rewrite rules are applied to a frozen term, no proper narrowing can occur because it does not contain variables. So we rewrite FT to its normal form NFT. If NFT is *true* or is a frozen variable we have finished. If it is *false* then we know that it is impossible to find a substitution σ such that $T\sigma$ can be rewritten to *true*. If none of the above has happened we start a phase of investigation. We collect meta-information about NFT and find all possible narrowing substitutions of NFT, say $\sigma_1, ..., \sigma_n$. If this list is empty we know T cannot be 'solved'. Otherwise we choose, according to the meta-information we collected, a subset of $\{\sigma_1, ..., \sigma_n\}$, say $\{\sigma_{i_j}\}$, and apply the above algorithm to the $T\sigma_{i_j}$, while the remaining $T\sigma_i$ are kept for later consideration.

In our program each subterm of a frozen term carries with it a copy of the whole frozen term and an instance of the original goal. When we want to find narrowings we melt the three together (so that their variables correspond to each other). We then try to unify the melted subterm with the left side of some rule. If this unification succeeds it also affects the melted whole term and possibly the instance of the original goal. Now we freeze together the possibly changed instance of the original goal and the new instance of the whole term. By comparing the original frozen goal and the new frozen instance of the goal the substitution can be deduced.

Acknowledgements

This work was done under the supervision of Prof. Nachum Dershowitz, to whom I owe the deepest gratitude for his inspiration, guidance and encouragement. Many thanks are also due to Dr. Ehud Shapiro for introducing me to FCP and for very helpful discussions.

References

[DerPl 88] Dershowitz, N. and Plaisted, D.A., *Equational Programming*, Machine Intelligence 11, eds J.E. Hayes, D. Michie, and J. Richards, pp. 21-56, 1988.

[FlaOd 82] Flajolet, P. and Odlyzko, A., *The Average Height of Binary Trees and Other Simple Trees*, Journal of Computer and System Sciences 25, pp. 171-213, 1982.

[GoKiMe 87] Goguen, J., Kirchner C., and Meseguer, J., *Concurrent Term Rewriting as a Model of Computation*, SRI-CSL-87-2, 1987.

[Hul 80] Hullot, J.-M., *Canonical Forms and Unification*, 5th Conference on Automated Deduction, Lecture Notes in Computer Science 87, 1980.

[JoDer 89] Josephson, A. and Dershowitz, N., *An Implementation of Narrowing*, to appear in the Journal of Logic Programming, 1989.

[Rety 87] Réty, P., *Improving basic narrowing techniques and commutation properties*, Rapports de Recherche No. 681, INRIA, 1987.

[Shap 86] Shapiro, E., *Concurrent Prolog: A progress report*, IEEE Computer, August 1986.

[Shap 87] Shapiro, E., *Or-Parallel Prolog in Flat Concurrent Prolog*, Concurrent Prolog Collected Papers, ed. E. Shapiro, Vol. 2, pp. 415-441, 1987.

[Logix 87] Silverman, W., Hirsch, M., Houri A., and Shapiro, E., *The Logix System User Manual Version 1.22*, Weizmann Institute Technical Report CS87-21, 1987.

Morphocompletion for One–Relation Monoids

John Pedersen

Department of Mathematics
University of South Florida
Tampa, Florida 33620–5700
jfp@usf.edu

This reports on a relatively short C program designed specifically to explore "morphocompletion" strategies for one–relation monoids. The aim was to solve the word problem for some families of one–relation monoids by finding finite canonical rewriting systems for them. Morphocompletion refers to systematically introducing new generators to stand for certain subwords of existing relations during the completion process. It is conjectured that such an approach is always sufficient to find a finite canonical presentation isomorphic to a given one, if such exists (that is, compared with Tietze's theorem, we need only consider critical pair consequences and new generators for subwords of existing relations, not for arbitrary words). The program finds enough examples to lend support to this conjecture.

There are too many ways of introducing new generators for subwords for the program to try them all. Instead it focuses on three strategies which have shown some success. It automatically attempts each strategy in order from the fastest to the most time–consuming. The user can also interrupt at any point to indicate a subword or move on to the next trial. The automatic strategies are:

(S1) The presentation $\langle a,b;\ r = s\rangle$ is changed to $\langle a,b,c;\ r = c,\ s = c\rangle$ (single new generator), and ordinary completion is then applied.

(S2) The presentation is converted to tabular form (see [Ev51,Pe85]), then ordinary completion is applied.

(S3) The original system was converted to tabular form. Ordinary completion was applied for two steps. Then, for each (unresolved) critical pair $u = v$ in turn, it was replaced by $u \longrightarrow x$, $v \longrightarrow x$ for a new generator, x. Ordinary completion was then applied.

Adyan and Oganesyan have shown that many one–relation semigroups have solvable word problems [Ad66, Ad76, AO78, Og79, Og82, AO87]. The remaining unsolved cases are of the form $<a,b;\ aAa = bBa>$ where b occurs in A or B, bBa is not "hypersimple" (i.e. bBa has the form TCT for nonempty T), or bBa does not occur in Aa (and A and B have different lengths). A table of these cases appears below, indicating moderate success for the strategies (S1)–(S3) in previously unknown cases. New infinite families of one–relation semigroups are obtained as a result, as given in the theorems below. all completion was with respect to length/dictionary order ($u > v$ iff u is longer than v or u succeeds v in the dictionary). Other orderings and interactive user ordering may be added to the program, but care must then be taken that it is terminating.

It is perhaps surprising that such a simple strategy as (S1) should give relatively many new cases. Cases in which it works can be described in terms of these definitions.

__Definition.__ Let $s,\ t \in \{a,b\}^*$. Say $s = s_1 s_2 \cdots s_n$ and $t = t_1 t_2 \cdots t_m$ with $s_i, t_i \in \{a,b\}$. The *correlation polynomial* $C(s,t)$ is $c_1 + c_2 x + c_3 x^2 + \cdots + c_n x^{n-1}$ where the coefficients c_i are given by

$$c_i = \begin{cases} 1, \text{ if } t_1 t_2 \cdots t_{m-i+1} \equiv s_i s_{i+1} \cdots s_k \\ 0, \text{ otherwise,} \end{cases}$$

where $k = min(n,\ i + m - 1)$ (cf. [GO81]). The *correlation matrix* of s and t is

$$Corr(s,t) = \begin{bmatrix} C(s,s) & C(s,t) \\ C(t,s) & C(t,t) \end{bmatrix}.$$

__Theorem 1.__ *Let $s = aAa$, $t = bBa$, with $A,\ B \in \{a,b\}^*$. Let the lengths of s and t be m and n respectively and suppose $m > n > \frac{m-1}{2}$. Suppose further that*

$$Corr(s,t) = \begin{bmatrix} 1+x^{m-1} & 0 \\ x^{n-1} & 1 \end{bmatrix}.$$

Then the system $<a,b,c;\ s = c,\ t = c>$ completes finitely under length/dictionary ordering, and thus $<a,b;\ s = t>$ has solvable word problem..

__Proof.__ The finite canonical system generated is $aAa = c$, $bBa = c$, $cAa = bBc$,

$aAc = bBc$, $bBbBc = cAc$, $aAbBc = cAc$, $cAbBc = bBcAc$. It is routine to verify that indeed all of the critical pairs generated from obvious overlaps are resolved. The condition on the correlation matrix prevents unexpected overlaps from occuring. The condition $n > \frac{m-1}{2}$ ensures that the fifth rule is length–reducing. ▌

One family of examples covered is $a^n ba = b^n a$, $n \geq 2$. Another case in which (S1) works is as follows.

THEOREM 2. *The system* $<a,b,c;\ a^p = c,\ bBa^q = c>$, $p \geq q+3$, *completes finitely if* B *ends in* b *and* bBa^q *is hypersimple, and thus* $<a,b;\ a^p = bBa^q>$ *has solvable word problem in this case.*

PROOF. In this case the equivalent finite canonical system produced is

$$ca \longrightarrow ac,\ a^p \longrightarrow c,\ bBa^q \longrightarrow c,$$

$$a^{p-qi}c^i \longrightarrow (bB)^i c, i = 1,2,\cdots,\ k = \left\lfloor \frac{p-2}{q+1} \right\rfloor$$

$$c(bB)^i c \longrightarrow (bB)^i c^2, i = 1,2,\cdots,\ k$$

$$a^q (bB)^i c \longrightarrow (bB)^{i-1} c^2,\ i = 1,2,\cdots,\ k$$

$$bBa^j (bB)^i c \longrightarrow a^j (bB)^{i+1} c,\ i = 1,2,\cdots,\ k-1,\ j = 1,2,\cdots,\ q-1,\ i = 0 \text{ if } k > 2.$$

$$bBa^i (bB)^k\, c \longrightarrow a^{p-qk-q+i} c^{k+1},\ i = 0,1,\cdots,\ q-1.$$

Again the assumption that bBa^q is hypersimple prevents any unexpected overlaps from occuring. ▌

The next theorem gives what appears be all the cases in which (S2) works. Let $|p(x)|$ denote the number of (non–zero) terms in a polynomial $p(x)$.

THEOREM 3. *Let* $s = aAa$, $t = bBa$, *with lengths* m *and* n *respectively. Suppose*

$$Corr(s,t) = \begin{bmatrix} p(x) + x^{m-1} & q(x) \\ r(x) + x^{n-1} & 1 \end{bmatrix}$$

where $|r(x)| \leq |p(x)|$ *and* $|q(x)| \leq |p(x)| + 1$. *Let the highest coefficients in* $p(x)$ *and* $r(x)$ *be* x^{m-d} *and* x^{n-e} *respectively. If* $r(x) = 0$ *or* $e \lesssim d$ *then* (S2) *applied to* $<a,b;\ s = t>$ *produces a finite canonical system, and thus* $<a,b;\ s = t>$ *has solvable word problem.*

Proof The finite canonical systems produced typically have 30–50 rules for the cases shown in Table 1. Displaying their form here is not very useful since the checking of confluence is best achieved with a computer. ∎

Finally, (S3) had more limited success, and often produced sets of over 100 rules for the range displayed in the table, but it did break into the cases where bBa was not hypersimple. In fact it appears only to work in a subset of that case.

TABLE 1

Cases of $aAa = bBa$

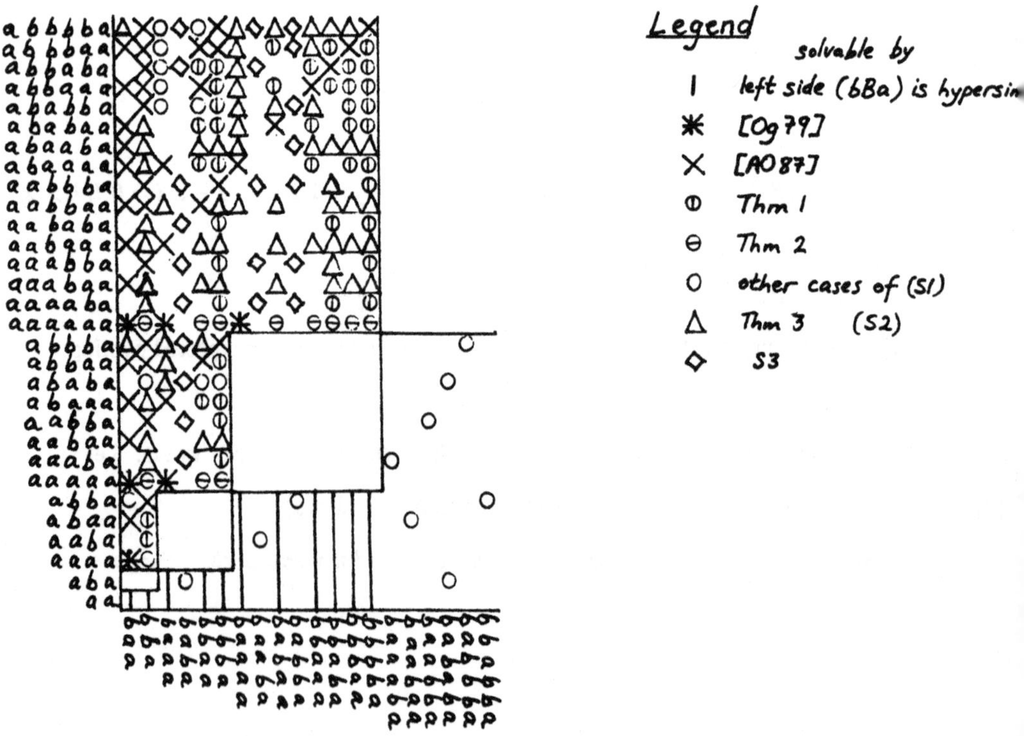

REFERENCES

[Ad66] ADYAN, S.I., *Defining relations and algorithmic problems for groups and seimtroups*, Tr. Mat. Inst. Akad. Nauk SSSR **85** (1966), 1–124.

[Ad76] ADYAN, S.I., *On transformations of words in semigroups presented by a system of defining relations*, Alg. i Log. **15**, 6 (1976), 611–621.

[AO78] ADYAN S.I. and OGANESYAN, G.U., *On the problems of equality and divisibility in semigroups with a single defining relation*, Izv. Akad. Nauk SSSR, Ser. Mat. **42** , 2 (1978), 219–225.

[AO87] ADYAN, S.I. and OGANESYAN, G.U., *Problems of equality and divisibility in semigroups with a single defining relation*, Mat. Zam. **41**, 3 (1987), 412–421.

[Ev51] EVANS, T., *On multiplicative systems defined by generators andrelations, I. Normal form theorems*, Proc. Camb. Phil. Soc. **47** (1951), 637–649.

[Og79] OGANESYAN, G.U., *The solvability of the word problem for semigroups with a defining relation of the form* $A = BtC$, Izv. Akad. Nauk Armjan SSSR, Ser. Mat. **14**, 4 (1979), 288–291, 315 MR #81g: 20106.

[Og82] OGANESYAN, G.U., *On semigroups with a single defining relation and semigroups without cycles*, Izv. Akad. Nauk SSSR, Ser. Mat. **46**, 1 (1982), 84–94.

[Pe85] PEDERSEN, J., *The word problem in absorbing varieties*, Houston J. Math. **11**, 4 (1985), 575–590.

AUTHOR INDEX

Vol. 324: M.P. Chytil, L. Janiga, V. Koubek (Eds.), Mathematical Foundations of Computer Science 1988. Proceedings. IX, 562 pages. 1988.

Vol. 325: G. Brassard, Modern Cryptology. VI, 107 pages. 1988.

Vol. 326: M. Gyssens, J. Paredaens, D. Van Gucht (Eds.), ICDT '88. 2nd International Conference on Database Theory. Proceedings, 1988. VI, 409 pages. 1988.

Vol. 327: G.A. Ford (Ed.), Software Engineering Education. Proceedings, 1988. V, 207 pages. 1988.

Vol. 328: R. Bloomfield, L. Marshall, R. Jones (Eds.), VDM '88. VDM – The Way Ahead. Proceedings, 1988. IX, 499 pages. 1988.

Vol. 329: E. Börger, H. Kleine Büning, M.M. Richter (Eds.), CSL '87. 1st Workshop on Computer Science Logic. Proceedings, 1987. VI, 346 pages. 1988.

Vol. 330: C.G. Günther (Ed.), Advances in Cryptology – EUROCRYPT '88. Proceedings, 1988. XI, 473 pages. 1988.

Vol. 331: M. Joseph (Ed.), Formal Techniques in Real-Time and Fault-Tolerant Systems. Proceedings, 1988. VI, 229 pages. 1988.

Vol. 332: D. Sannella, A. Tarlecki (Eds.), Recent Trends in Data Type Specification. V, 259 pages. 1988.

Vol. 333: H. Noltemeier (Ed.), Computational Geometry and its Applications. Proceedings, 1988. VI, 252 pages. 1988.

Vol. 334: K.R. Dittrich (Ed.), Advances in Object-Oriented Database Systems. Proceedings, 1988. VII, 373 pages. 1988.

Vol. 335: F.A. Vogt (Ed.), CONCURRENCY 88. Proceedings, 1988. VI, 401 pages. 1988.

Vol. 336: B.R. Donald, Error Detection and Recovery in Robotics. XXIV, 314 pages. 1989.

Vol. 337: O. Günther, Efficient Structures for Geometric Data Management. XI, 135 pages. 1988.

Vol. 338: K.V. Nori, S. Kumar (Eds.), Foundations of Software Technology and Theoretical Computer Science. Proceedings, 1988. IX, 520 pages. 1988.

Vol. 339: M. Rafanelli, J.C. Klensin, P. Svensson (Eds.), Statistical and Scientific Database Management. Proceedings, 1988. IX, 454 pages. 1989.

Vol. 340: G. Rozenberg (Ed.), Advances in Petri Nets 1988. VI, 439 pages. 1988.

Vol. 341: S. Bittanti (Ed.), Software Reliability Modelling and Identification. VII, 209 pages. 1988.

Vol. 342: G. Wolf, T. Legendi, U. Schendel (Eds.), Parcella '88. Proceedings, 1988. 380 pages. 1989.

Vol. 343: J. Grabowski, P. Lescanne, W. Wechler (Eds.), Algebraic and Logic Programming. Proceedings, 1988. 278 pages. 1988.

Vol. 344: J. van Leeuwen, Graph-Theoretic Concepts in Computer Science. Proceedings, 1988. VII, 459 pages. 1989.

Vol. 345: R.T. Nossum (Ed.), Advanced Topics in Artificial Intelligence. VII, 233 pages. 1988 (Subseries LNAI).

Vol. 346: M. Reinfrank, J. de Kleer, M.L. Ginsberg, E. Sandewall (Eds.), Non-Monotonic Reasoning. Proceedings, 1988. XIV, 237 pages. 1989 (Subseries LNAI).

Vol. 347: K. Morik (Ed.), Knowledge Representation and Organization in Machine Learning. XV, 319 pages. 1989 (Subseries LNAI).

Vol. 348: P. Deransart, B. Lorho, J. Maluszyński (Eds.), Programming Languages Implementation and Logic Programming. Proceedings, 1988. VI, 299 pages. 1989.

Vol. 349: B. Monien, R. Cori (Eds.), STACS 89. Proceedings, 1989. VIII, 544 pages. 1989.

Vol. 350: A. Törn, A. Žilinskas, Global Optimization. X, 255 pages. 1989.

Vol. 351: J. Díaz, F. Orejas (Eds.), TAPSOFT '89. Volume 1. Proceedings, 1989. X, 383 pages. 1989.

Vol. 352: J. Díaz, F. Orejas (Eds.), TAPSOFT '89. Volume 2. Proceedings, 1989. X, 389 pages. 1989.

Vol. 354: J.W. de Bakker, W.-P. de Roever, G. Rozenberg (Eds.) Linear Time, Branching Time and Partial Order in Logics and Models for Concurrency. VIII, 713 pages. 1989.

Vol. 355: N. Dershowitz (Ed.), Rewriting Techniques and Applications. Proceedings, 1989. VII, 579 pages. 1989.